AF595090

Strategies to Improve Antineoplastic Activity of Drugs in Cancer Progression

Strategies to Improve Antineoplastic Activity of Drugs in Cancer Progression

Editor

Angela Stefanachi

MDPI • Basel • Beijing • Wuhan • Barcelona • Belgrade • Manchester • Tokyo • Cluj • Tianjin

Editor
Angela Stefanachi
Farmacia-Scienze del
Farmaco
Università degli studi di Bari
"Aldo Moro"
Bari
Italy

Editorial Office
MDPI
St. Alban-Anlage 66
4052 Basel, Switzerland

This is a reprint of articles from the Special Issue published online in the open access journal *International Journal of Molecular Sciences* (ISSN 1422-0067) (available at: www.mdpi.com/journal/ijms/special_issues/drugs_cancer).

For citation purposes, cite each article independently as indicated on the article page online and as indicated below:

LastName, A.A.; LastName, B.B.; LastName, C.C. Article Title. *Journal Name* **Year**, *Volume Number*, Page Range.

ISBN 978-3-0365-5928-5 (Hbk)
ISBN 978-3-0365-5927-8 (PDF)

Contents

About the Editor

Angela Stefanachi

Angela Stefanachi graduated in 1998 summa cum laude in Chemistry and Pharmaceutical Technology at the University of Bari and, after two years as postgraduate fellow, in 2000 she started her Ph.D. studies and received the Ph.D. in Medicinal Chemistry in 2004. She worked as Post-doc in the Department of Pharmaceutical Chemistry at the University of Bari from March 2004 to November 2008. In February 2009, she became Assistant Professor at the University of Bari in the Faculty of Pharmacy. She had two wide experiences as visiting scientist abroad: the first in the group of prof Enrique Ravina (Universidad de Santiago de Compostela, Spain) and the second in the group of prof. Steven Ley (University of Cambridge, UK). From March 2022, she has been associate professor at the department of pharmacy- druf sciences. Her research on the design and synthesis of small molecule through parallel organic synthesis, both on solid phase and in solution, with potential pharmacological activity toward pathologies such as cancer (CB2 Ligands, Aromatase, Protein Kinases, MDR and HDAC inhibitors), neurodegenerative diseases (AChE-BChE and Aβ1-40 aggregation inhibitors), and asthma (selective Adenosine2B receptor antagonists).

International Journal of
Molecular Sciences

Article

Hydroxy-Propil-β-Cyclodextrin Inclusion Complexes of two Biphenylnicotinamide Derivatives: Formulation and Anti-Proliferative Activity Evaluation in Pancreatic Cancer Cell Models

Rosa Maria Iacobazzi [1,†], Annalisa Cutrignelli [2,†], Angela Stefanachi [2], Letizia Porcelli [1], Angela Assunta Lopedota [2], Roberta Di Fonte [1], Antonio Lopalco [2], Simona Serratì [3], Valentino Laquintana [2], Nicola Silvestris [1,4], Massimo Franco [2], Saverio Cellamare [2], Francesco Leonetti [2], Amalia Azzariti [1,‡] and Nunzio Denora [2,*,‡]

1 Laboratory of Experimental Pharmacology, IRCCS Istituto Tumori Giovanni Paolo II, 70124 Bari, Italy; rosamaria.iacobazzi@gmail.com (R.M.I.); porcelli.letizia@gmail.com (L.P.); difonte.roberta@gmail.com (R.D.F.); n.silvestris@oncologico.bari.it (N.S.); a.azzariti@oncologico.bari.it (A.A.)
2 Department of Pharmacy-Drug Sciences, University of Bari "Aldo Moro", 70125 Bari, Italy; annalisa.cutrignelli@uniba.it (A.C.); angela.stefanachi@uniba.it (A.S.); angelaassunta.lopedota@uniba.it (A.A.L.); antonio.lopalco@uniba.it (A.L.); valentino.laquintana@uniba.it (V.L.); massimo.franco@uniba.it (M.F.); saverio.cellamare@uniba.it (S.C.); francesco.leonetti@uniba.it (F.L.)
3 Laboratory of Nanotechnology, IRCCS Istituto Tumori Giovanni Paolo II, 70124 Bari, Italy; simonaserrati@hotmail.com
4 Department of Biomedical Sciences and Human Oncology, University of Bari "Aldo Moro", 70124 Bari, Italy
* Correspondence: nunzio.denora@uniba.it; Tel.: +39-080-544-2767
† These authors contributed equally to the work.
‡ These authors contributed equally to the work.

Received: 20 July 2020; Accepted: 3 September 2020; Published: 7 September 2020

Abstract: Pancreatic ductal adenocarcinoma (PDAC) is one of the most aggressive malignancies, with poor outcomes largely due to its unique microenvironment, which is responsible for the low response to drugs and drug-resistance phenomena. This clinical need led us to explore new therapeutic approaches for systemic PDAC treatment by the utilization of two newly synthesized biphenylnicotinamide derivatives, PTA73 and PTA34, with remarkable antitumor activity in an in vitro PDAC model. Given their poor water solubility, inclusion complexes of PTA34 and PTA73 in Hydroxy-Propil-β-Cyclodextrin (HP-β-CD) were prepared in solution and at the solid state. Complexation studies demonstrated that HP-β-CD is able to form stable host–guest inclusion complexes with PTA34 and PTA73, characterized by a 1:1 apparent formation constant of 503.9 M^{-1} and 369.2 M^{-1}, respectively (also demonstrated by the Job plot), and by an increase in aqueous solubility of about 150 times (from 1.95 μg/mL to 292.5 μg/mL) and 106 times (from 7.16 μg/mL to 762.5 μg/mL), in the presence of 45% *w/v* of HP-β-CD, respectively. In vitro studies confirmed the high antitumor activity of the complexed PTA34 and PTA73 towards PDAC cells, the strong G2/M phase arrest followed by induction of apoptosis, and thus their eligibility for PDAC therapy.

Keywords: pancreatic ductal adenocarcinoma; cyclodextrin inclusion complex; phase solubility studies; preformulation studies; biphenylnicotinamide derivatives

---

1. Introduction

Pancreatic ductal adenocarcinoma (PDAC) is the most common type of pancreatic cancer, which kills more patients every year than any other type of cancer excluding lung and colorectal. Although accounting for only 3% of new cancer cases in the United States, it is responsible for over 7% of all cancer deaths, with an overall five-year survival of less than 5% [1]. In 2019, in Italy, 13,500 new cases were expected (6800 in men and 6700 in women), about 3% of all male and female cancers [2]. The American Cancer Society estimates that in 2020 about 57,600 people (30,400 men and 27,200 women) will be diagnosed with pancreatic cancer, and that about 47,050 people (24,640 men and 22,410 women) will die of pancreatic cancer [1].

Since PDAC is generally diagnosed at an advanced stage, systemic therapy is the main strategy of treatment. Currently, the most successful chemotherapy regimens for this type of tumor are gemcitabine, FOLFIRINOX, and the combination gemcitabine/nabpaclitaxel. However, the clinical management of patients still remains an open challenge, because in most cases patients have inherent resistance to therapies. The poor outcome for PDAC patients is mainly due to the peculiarity of the desmoplastic stroma that represents up to 90% of the tumor mass, and is characterized by fibrosis, poor vascularization, high intratumoral pressure, immune infiltrates, and hypoxia, with consequent reduction of the bioavailability of the drugs also hindered by rapid elimination, metabolic inactivation, and not specific systemic toxicity [3,4].

The interaction between pancreatic cancer cells and the tumor microenvironment, including immune cells, endothelial cells, and fibroblasts, plays a crucial role in PDAC development and progression and in drug-resistance phenomena [5].

In this scenario, in order to identify new therapeutic strategies for PDAC, we planned to investigate, in a PDAC cells panel, the pharmacological efficacy of two newly synthesized N-biphenylnicotinamides, namely PTA34 and PTA73 [6,7], formulated as hydroxy-propil-β-cyclodextrin (HP-β-CD) inclusion complexes (Figure 1).

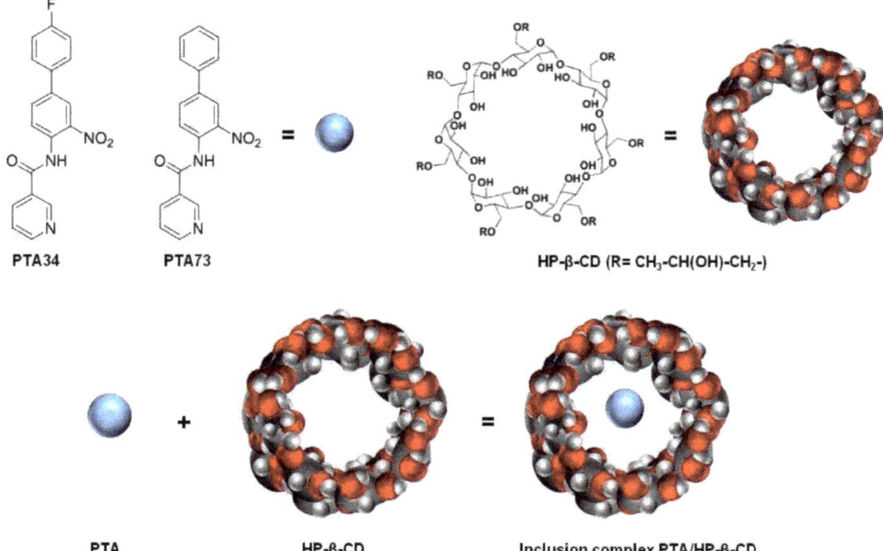

Figure 1. Chemical structures of PTA34, PTA73, hydroxy-propil-β-cyclodextrin (HP-β-CD) and graphical representation of the inclusion complex.

PTA34 and PTA73 molecules have been already classified as a novel, highly potent, and selective class of microtubule targeting agents (MTAs) and potential anti-angiogenic and vascular-disrupting agents in the Hodgkin lymphoma model [7]. Moreover, a remarkable antitumor activity of these molecules at low doses was assessed also in a PDAC model, MIA PaCa-2 cells [7], confirming that targeting microtubule dynamics could be effective against the abnormal proliferation of PDAC cancer cells [8–10].

However, preformulation studies conducted on PTA34 and PTA73 showed very low water solubility, which strongly limits the potential pharmaceutical development for these compounds due to the poor bioavailability of the drug. Therefore, the improvement of the aqueous solubility for the new PTA's formulations was an urgent need.

Different formulation strategies allow overcoming the limits of poorly soluble drugs, such as solid dispersions [11,12], addition of cosolvents [13], complexation, and size reduction [14,15], however the most studied and applied approach to improve the solubility and bioavailability of drugs is the complexation in cyclodextrins [16–18].

Cyclodextrins (CDs), are cyclic oligosaccharides containing at least 6 D - (+) glucopyranose units attached by α-(1,4) glucosidic bonds, with lipophilic inner cavities and hydrophilic outer surfaces. They are able to entrap hydrophobic drugs in their cavities, forming non-covalent inclusion complexes, thus allowing the dissolution in the aqueous phase of the drug included, making it suitable to diffuse in an aqueous medium, to come in contact with the membrane surface, and to permeate through the membrane. CDs are also able to interact with membrane components and to solubilize cholesterol, inducing perturbation in the lipid bilayer, and affect the membrane properties, such as fluidity and permeability. Moreover, the encapsulation in CDs protects the drug from chemical and enzymatic degradation [19–22].

Here, inclusion complexes of both PTA34 and PTA73 in hydroxy-propil-β-cyclodextrin (HP-β-CD), a semisynthetic cyclodextrin approved by the Food and Drug Administration (FDA) as an excipient for parenteral formulations, were developed (Figure 1) [23]. In detail, the inclusion complexes among the HP-β-CD and these biphenylnicotinamide derivatives were studied first in solution, by the analysis of the phase solubility diagram, according to Higuchi–Connors [24], and the construction of the Job plot [18] for the identification of the host–guest stoichiometric ratio. Next, the inclusion complexes were prepared at the solid state by freeze-drying and characterized in terms of incorporation degree and dissolution profiles.

Finally, in order to evaluate the effectiveness of the PTA's complexation in HP-β-CD, in terms of antitumor activity improvement, cytotoxicity studies, cell cycle analysis, and apoptosis determination were conducted in a panel of PDAC cell lines, AsPC-1, PANC-1, and MIA PaCa-2. The activities of the complexes PTA34/HP-β-CD and PTA73/HP-β-CD were compared to those of the corresponding pure molecules, showing a higher antiproliferative efficacy and an unaltered activity in terms of modulation of the cell cycle.

In conclusion, the two new cyclodextrin inclusion complexes have proven in vitro to be promising candidates for PDAC therapy, even if in vivo studies are needed in order to validate an actual clinical use of these formulations.

2. Results and Discussion

2.1. Solubility and Phase-Solubility Studies of PTA73 and PTA34

First of all, the solubility of PTA73 and PTA34 was determined at 37 ± 0.5 °C in ultra-pure water, and the results showed that the solubility is critical for the bioavailability of these compounds. In detail, the intrinsic solubility (S_0), was equal to 1.95 μg/mL (5.78×10^{-6} M) and 7.16 μg/mL (2.24×10^{-5} M) for PTA 34 and PTA 73, respectively, thus complexation with a cyclodextrin could represent a valid solubilization strategy for these compounds. In Figure 2a,b the phase solubility diagrams of both the compounds are reported. It is evident that there is a linear correlation between solubility and HP-β-CD concentration, therefore both diagrams are of the A_L type, according to the classification proposed by Higuchi–Connors [24]. The solubility values obtained at 37 ± 0.5 °C in the presence of different HP-β-CD concentrations are shown in Table 1. The linear trend of the Higuchi–Connors diagram certifies the formation of an inclusion complex, with 1:1 host:guest stoichiometry, so that using the relative equation, as reported in the materials and methods section, it was possible to calculate the relative equilibrium constant (Ks). In particular, for PTA34 the presence of HP-β-CD at a maximum concentration of 45% *w/v* led to an increase in terms of aqueous solubility of about 150 times, bringing it from an initial solubility value of 1.95 μg/mL to a final solubility value of 292.5 μg/mL, and the calculated equilibrium constant was 503.9 M^{-1}. For PTA73, instead, the presence of HP-β-CD at a maximum concentration of 45% *w/v* led to an increase in terms of aqueous solubility of about 106 times, bringing it from an initial solubility value of 7.16 μg/mL to a final solubility value of 762.5 μg/mL, and the calculated equilibrium constant was 369.2 M^{-1}.

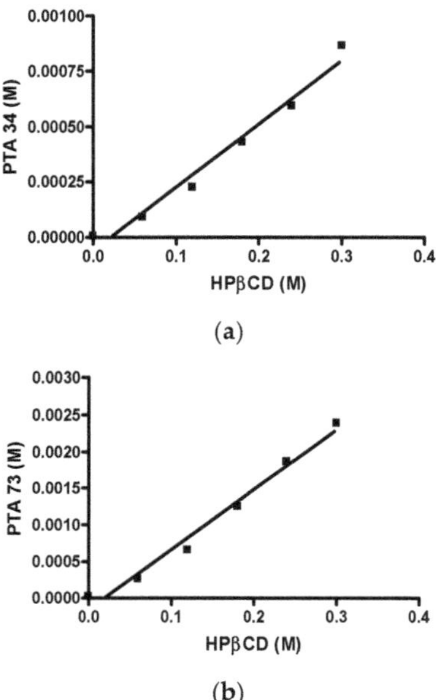

Figure 2. Phase solubility diagrams. (**a**) PTA34/HP-β-CD; (**b**) PTA73/HP-β-CD.

Table 1. PTA34 and PTA73 Water Solubility Values in Presence of HP-β-CD.

HP-β-CD (% *w/v*)	PTA34 (μg/mL)	PTA73 (μg/mL)
0	1.95 ± 0.35	7.16 ± 0.34
9	30.4 ± 4.3	84.1 ± 4.1
18	76.1 ± 3.9	209.5 ± 9.2
27	145.0 ± 7.1	398.4 ± 19.1
36	200.2 ± 28.3	596.0 ± 48.1
45	292.5 ± 24.7	762.5 ± 137.9

Data are the means of at least three determinations.

The calculated Ks values indicated a good complexing capacity of the HP-β-CD towards these two new N-biphenylnicotinamide derivatives. Furthermore, PTA34 presented a higher complexation constant value, and this result is explained by the presence on aromatic ring of the para fluorine atom. Since the fluorine atom is a lipophilic substituent, PTA34 also showed a lower water solubility.

2.2. Job's Plot Method

In order to investigate the host-guest stoichiometric ratio and to confirm the linear behavior of the Higuchi–Connors diagrams, the construction of the Job plot was carried out. Since both compounds under analysis exhibited absorption in the visible spectrum, this determination was conducted via UV-VIS spectrophotometry and the graphs obtained are shown in Figure 3a,b. In both cases, a highly symmetrical trend, with a maximum value recorded at $r = 0.5$, is observed, highlighting the formation of a 1:1 inclusion complex. This result is fully in agreement with the A_L trend of the solubility diagrams.

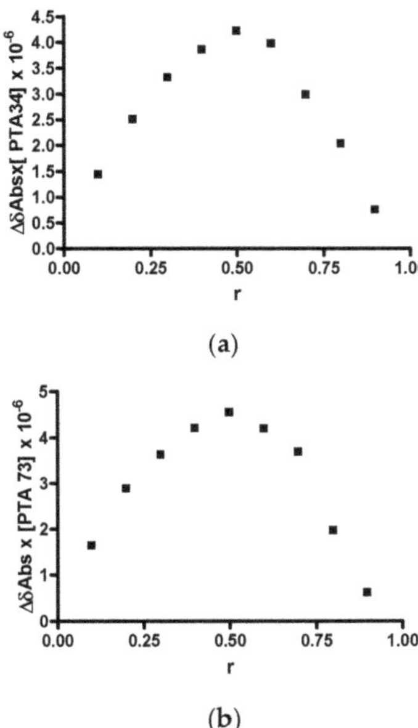

Figure 3. Job plots. (**a**) PTA 34/HP-β-CD; (**b**) PTA 73/HP-β-CD.

2.3. Preparation of PTA34 or PTA73/HP-β-CD Inclusion Complexes at the Solid State and Determination of Their Incorporation Degree

The inclusion complexes PTA34/HP-β-CD and PTA73/HP-β-CD were also prepared in the solid state by freeze-drying. A solid-state drug-cyclodextrin inclusion complex is certainly the most suitable tool to allow the administration of the drug, both orally and parenterally, overcoming the limit represented by its low solubility in water [17]. The lyophilized complexes were characterized through the determination of the incorporation degree, expressed as mg of PTA34 or PTA73 per 1 g of product, and were found to be 1.23 ± 0.42 and 2.91 ± 1.0 mg of drug per 1 g of lyophilized powder for PTA34 or PTA73, respectively.

2.4. In Vitro Dissolution Studies

In Figure 4 in vitro dissolution profiles, in 0.05 M phosphate buffer at pH 7.4, of PTA34 and PTA73, from their respective solid-state inclusion complexes, are shown. In the same graph the dissolution profiles of the two uncomplexed compounds are not reported because, due to their very low solubility, the dissolved quantity was well below the detection limit, and this prevented the quantitative determination via UV-VIS spectrophotometry in the dissolution medium. The lipophilic nature of both drugs limits their contact with the dissolution medium, causing them to float on the surface and hindering their dissolution. On the contrary, the freeze-dried complexes dissolve very quickly once they are placed in the dissolution medium and, in both cases, 100% of the complexed drug was solubilized within the first 20 min of the dissolution process.

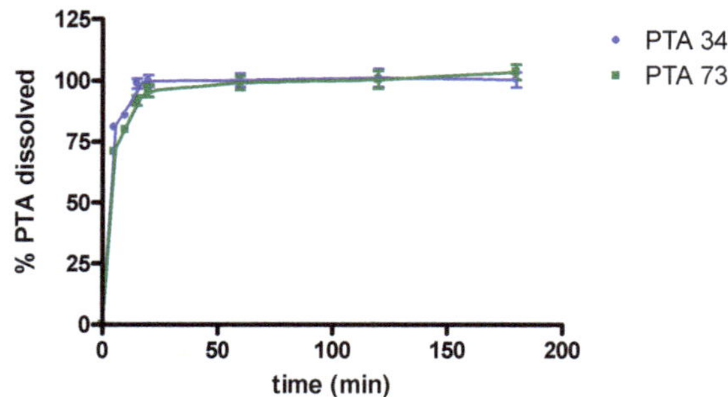

Figure 4. Dissolution profiles of PTA/HP-β-CD solid complexes.

Consequently, the complexation with HP-β-CD certainly represents a valid strategy for improving the solubility characteristics and the dissolution profile of these two biphenylnicotinamide derivatives.

2.5. In Vitro Studies

2.5.1. Cytotoxicity

The effectiveness of PTA34 and PTA73, both complexed in HP-β-CD and as pure compounds, was evaluated in three human pancreatic cancer cell lines AsPC-1, Panc-1, and MIA PaCa-2 cells, by MTT assay after 72 h of treatment. In order to demonstrate that the complexed drugs (PTA 34/HP-β-CD and PTA 73/HP-β-CD) did not lose their antiproliferative activity against tumor cells, in respect to uncomplexed ones (PTA34 and PTA73, respectively), the proliferation of all cells was determined after each drug exposure, and IC_{50} was calculated. The dose/effect curves of PTA34/HP-β-CD vs. PTA34 (Figure 5, Panel (a)) and of PTA73/HP-β-CD vs. PTA73 (Figure 5, Panel (b)), as well as the IC_{50} values reported in Table 2, show that both the complexes were even more active than the non-complexed ones, in AsPC- 1 and in PANC-1 cells, while in MIA PaCa-2 cells, where the lowest IC_{50} values for PTA34 and PTA73 alone were recorded, the activity of the complexed compounds and pure molecules was comparable.

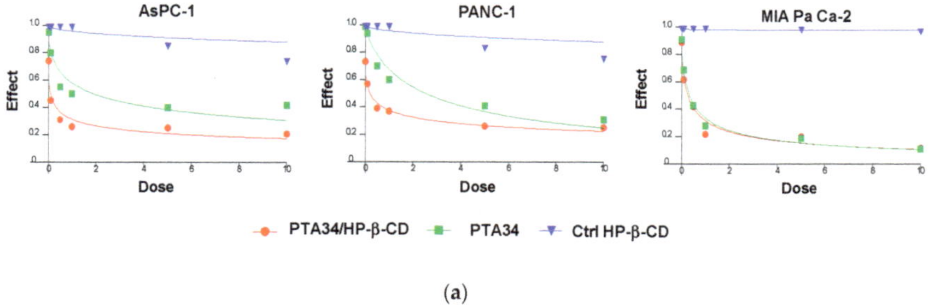

(a)

Figure 5. *Cont.*

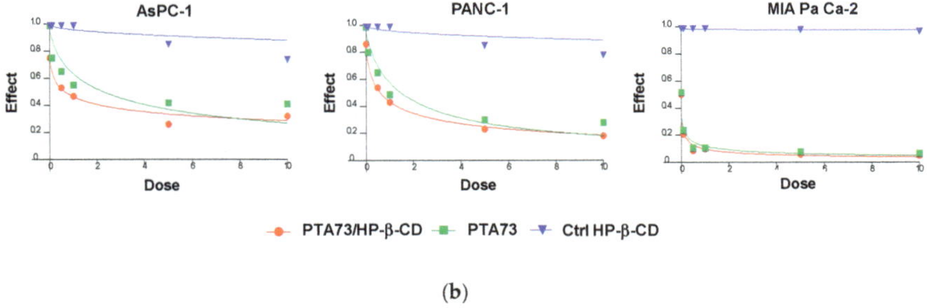

(b)

Figure 5. Dose/effect plots of the mean of three different cell proliferation experiments, conducted in PDAC cell lines incubated for 72 h with HP-β-CD (Ctrl HP-β-CD), PTA34, and PTA34/HP-β-CD (panel (**a**)) or with HP-β-CD (Ctrl HP-β-CD, PTA73, and PTA73/HP-β-CD (Panel (**b**)). Dose was expressed in each graph as 0.01–10 μM concentration range, in terms of PTA34 or PTA73, corresponding to 2.9–2900 μg/mL concentration range for HP-β-CD alone.

Table 2. Inhibitory Effect of PTA Compounds and PTA/HP-β-CD Complexes on Pancreatic Ductal Adenocarcinoma (PDAC) Cancer Cells.

	IC_{50} (μM ± SD)			
Cell Lines	**PTA34/HP-β-CD**	**PTA34**	**PTA73/HP-β-CD**	**PTA73**
AsPC-1	0.09 ± 0.02	1.9 ± 0.3	0.66 ± 0.20	2.3 ± 0.6
Panc-1	0.20 ± 0.02	2.5 ± 0.5	0.62 ± 0.02	1.5 ± 0.2
MIA PaCa-2	0.27 ± 0.05	0.31 ± 0.04	0.005 ± 0.001	0.006 ± 0.002

Furthermore, to exclude that the higher cytotoxic activity of the complexed compounds was to some extent attributable to the HP-β-CD, this was tested in the same concentration ranges utilized in the drug complexes. The dose/effect curves in Figure 5 for cyclodextrin (Ctrl HP-β-CD) clearly show that it was not toxic at the analyzed concentration range. These results evidenced a general improvement in the cytotoxic activity of these drugs following complexation, due to the increase in solubility in aqueous medium and to the enhancement of the plasmatic membrane permeability by HP-β-CD. As reported in the literature, this FDA approved excipient is characterized by a hydrophobic pocket capable of binding and solubilizing cholesterol, a critical component of the plasma membrane, thus having a role in the efflux and redistribution of cholesterol in mammalian cells [23,25–27]. Thus, the interaction of HP-β-CD with cholesterol in the plasma membrane can induce perturbation in the lipid bilayer, affecting the membrane properties such as fluidity and permeability. For these reasons, it is plausible to speculate that the cyclodextrin in the formulation of PTA compounds triggered the increase of intracellular uptake of the complexed drugs by enhancing the plasma membrane permeability. Ultimately, HP-β-CD could be considered a drug delivery enhancer for PTA 34 and PTA73.

2.5.2. Cell Cycle Modulation

In order to investigate whether the PTA's complexation in HP-β-CD altered their mechanism of action, in terms of cell cycle modulation compared to uncomplexed drugs, we conducted the flow cytometry (FCM) analysis, after staining with propidium iodide, of cells previously treated for 24 h with compounds at 1 μM in terms of PTA's concentration. In Figure 6, representative FCM histograms of cell cycle modulation by complexed drugs, or pure compounds in the three cell lines, are reported, as well as the quantification of three independent experiments. Results showed a strong arrest in the G2/M phase of AsPC-1 cells, induced by complexed and uncomplexed PTAs, with 93% vs. 95.50% and 94.1% vs. 95.58% for PTA34 vs. PTA34/HP-β-CD and PTA73 vs. PTA73/HP-β-CD, respectively. PANC-1 and MIA PaCa-2 cells were found to be less responsive to drugs with 57.9% vs. 66.9% and 64.9% vs. 67.3% for PTA34 vs. PTA34/HP-β-CD and PTA73 vs. PTA73/HP-β-CD, respectively in

PANC-1 cells, and 36.7% vs..75.74% and 45.2% vs. 65.7% for PTA34 vs. PTA34/HP-β-CD and PTA73 vs. PTA73/HP-β-CD, respectively, in MIA PaCa-2 cells. However, in PANC-1 and MIA PaCa-2 cells, the complexed drugs induced an increase in the percentage of cells arrested in G2/M phase, compared to uncomplexed drugs, confirming an improvement of the drugs activity. In conclusion, PTA34/HP-β-CD resulted more active than PTA34 in inducing the arrest in G2/M phase in all PDAC cell lines while PTA73/HP-β-CD resulted more active than its pure counterpart only in MIA PaCa2 cells. Moreover, HP-β-CD alone (Ctrl HP-β-CD), tested at a concentration of 290 μg/mL, did not alter the cell cycle in all three investigated cell lines. These results allowed us to confirm that the complexation of these two biphenylnicotinamide derivatives with HP-β-CD, per se, did not modify the blocking activity in the G2/M phase of the cell cycle but may in some cases increase its effectiveness.

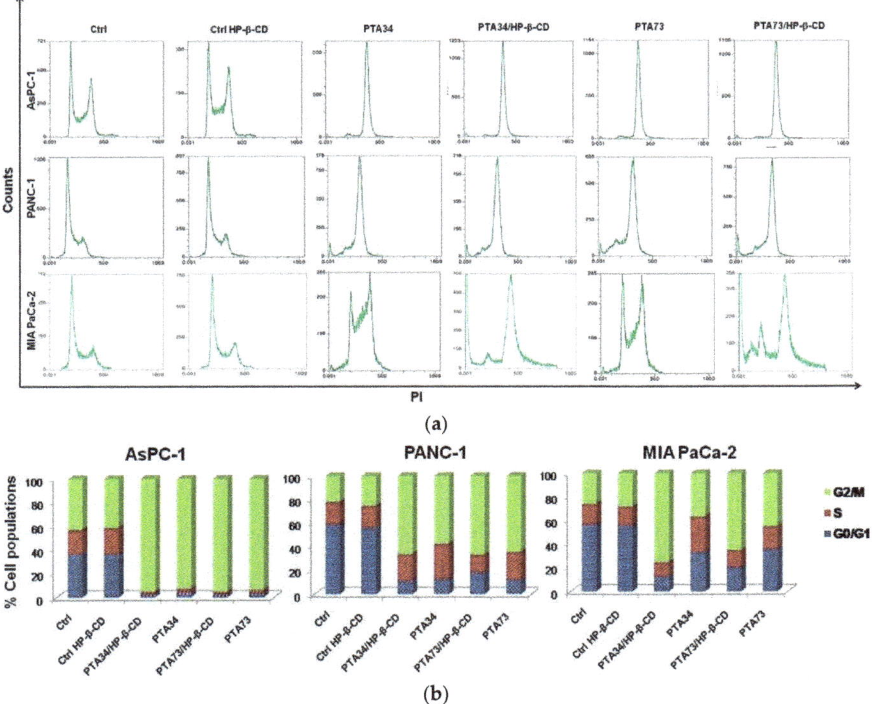

Figure 6. Effects on cell cycle of PDAC cells by CD (Ctrl-CD), complexed PTA34 and PTA73 (PTA34/HP-β-CD and PTA73/HP-β-CD, respectively) and PTA34 and PTA73, alone. The modulation of cell cycle phases was evaluated by flow cytometry (FCM) analysis after staining cells with propidium iodide. In panel (**a**), a representative analysis of three independent experiments is reported for all cell lines investigated. In panel (**b**) the bar graph shows cell population percentage in each phase of the cell cycle.

2.5.3. Apoptosis Assay

The results of cell cycle analysis prompted us to investigate if the observed phase arrest in G2/M was followed by induction of apoptosis. Thus, the FITC Annexin V staining was carried out after 24 h of incubation of PDAC cells with PTA34 and PTA73 free and complexed in HP-β-CD at the same concentration used for cell cycle analysis (1 μM in terms of PTA's). Untreated cells and HP-β-CD treated ones, were used as references for PTAs and PTA's/HP-β-CD complexes, respectively. As representative of apoptosis analysis, the FCM dot plots of Annexin V/PI positive PDAC cells are reported in Figure 7a, whereas fold changes of Annexin V positive cells (early apoptosis plus late apoptosis) in treated samples, versus their specific reference compound, are summarized in Figure 7b. The induction of apoptosis was more evident in AsPC-1 cells when treated with PTA34 complexed and not in respect to PTA73 and PTA73/HP-β-CD, conversely, in MIA PaCa-2 cells both the complexes showed a higher effectiveness than the uncomplexed counterpart. PANC-1 cells showed a completely different behavior, the arrest in G2/M phase after 24 h of treatment did not trigger the marked induction of apoptosis, rather, only a slowdown in cell growth, as observed in cell viability assay. Finally, the HP-β-cyclodextrin alone did not affect the basal condition of cells in terms of apoptotic events (Figure 7a).

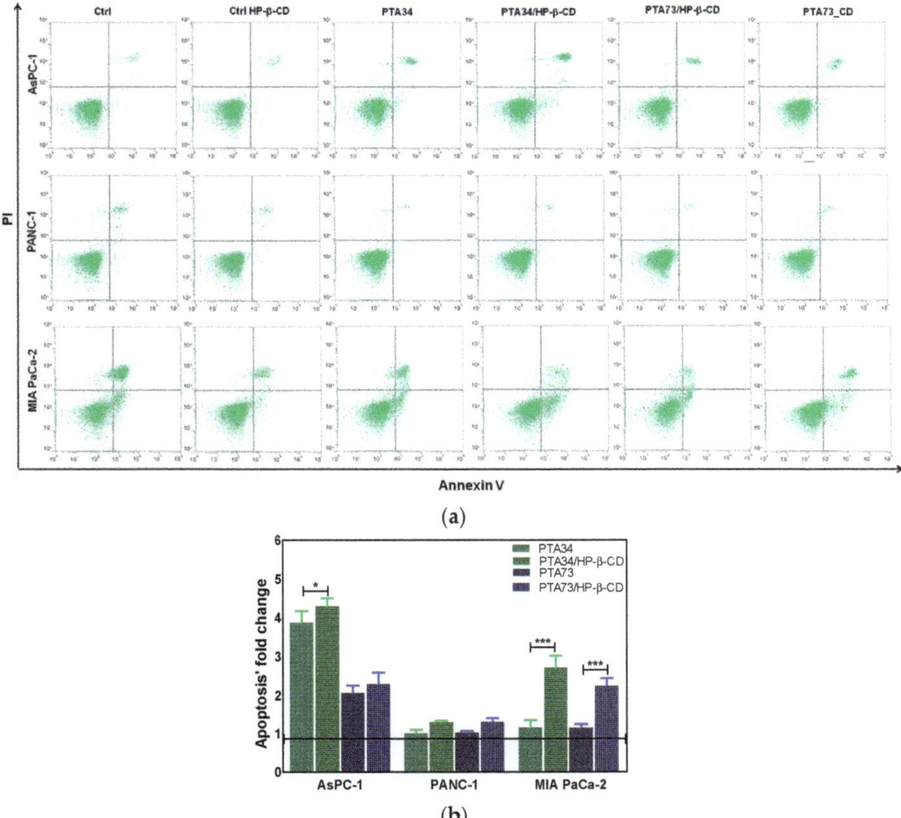

Figure 7. (**a**) Representative apoptosis analysis of PDAC cells after 24 h treatment with HP-β-CD alone (Ctrl HP-β-CD), PTA34, PTA34/HP-β-CD, PTA73, and PTA73/HP-β-CD. Apoptosis was evaluated by using Annexin V/PI staining followed by FCM analysis. (**b**) Fold change of Annexin V positive cells (early plus late apoptosis) in treated samples versus their corresponding reference compounds (Ctrl cells and Ctrl HP-β-CD for PTAs and complexed-PTAs treated cells, respectively). The results are the mean ± SD of three independent experiments (* $p < 0.05$; *** $p < 0.001$).

A possible explanation for the different sensitivity to these drugs, pure or complexed, of PDAC cell lines may lie in their cellular cholesterol levels. Li et al. documented an aberrant accumulation of cholesteryl ester (CEs) in human pancreatic cancer tissues and cell lines, but not in normal counterparts and, specifically, MIA PaCa-2 and PANC-1 cells had much higher levels of CEs than AsPC-1 [28]. Briefly, mammalian cells obtain cholesterol either from de novo synthesis or from the uptake of low-density lipoprotein (LDL). Overaccumulation of free (unesterified) cholesterol can be toxic to cells, thus to prevent accumulation, cholesterol is converted primarily by the enzyme ACAT1, to CEs, which mainly exists as cytosolic lipid droplets. A different enzyme, the cholesteryl ester hydrolase (CEH), acts in the opposite direction by liberating cholesterol from CEs, which now can reach the plasma membrane. Biosynthesis and hydrolysis of CEs occur continuously, forming the cholesterol/CE cycle [29]. Thus, given the existence of the free cholesterol/esterified cholesterol cycle, it is plausible to hypothesize that low levels of esterified intracellular cholesterol correspond to low levels of cholesterol in the plasma membrane and vice versa. This might explain why AsPC-1 cells, having the lowest esterified-cholesterol levels compared to the other two PDAC cell lines, are more sensitive to the cyclodextrin-dependent solubilizing action of cholesterol, which in turn results in both increased intracellular uptake and antitumor activity of the delivered PTAs.

3. Materials and Methods

3.1. Materials

The synthesis of PTA34 (MW = 337.30) and PTA73 (MW = 319.30) was performed according to the already reported literature procedure [6,7]. HP-β-CD (hydroxypropil-β-cyclodextrin, MW = 1396 Dalton, substitution degree 0.65) was purchased from Farmalabor (Canosa, Italy). HCl and phosphate salts for the preparation of buffers were purchased from Fluka (Sigma Aldrich, Milan, Italy). Bidistilled water was bought from Carlo Erba (Milan, Italy). All other products and reagents used in this work were of analytical grade. Pancreatic cancer cell lines PANC-1, AsPC-1 and MIAPaCa-2 cells were purchased from ATCC. The MTT used for cytotoxicity studies and PI for cell cycle analysis were purchased from Sigma Aldrich (Milan, Italy). FITC Annexin V Apoptosis Detection Kit I was from BD PharmingenTM (San Jose, CA, USA, 556547).

3.2. Quantitative Analysis of PTA34 and PTA73

Quantitative analysis of PTA34 and PTA73 was carried out by spectrophotometry using a dual-beam UV-VIS Lambda 20 Bio spectrophotometer (Perkin Elmer, Milan, Italy) and quartz cuvettes, with a capacity of 1 mL, using methanol as solvent. In the case of PTA34 the reading was carried out at a wavelength of 320 nm, and a calibration straight line (r^2 = 0.9979), in the concentration range between 0.15 mg/mL (4.45 × 10^{-4} mM) and 0.015 mg/mL (4.45 × 10^{-5} mM), was obtained. For PTA73 the reading was carried out at a wavelength of 340 nm, and a calibration straight line (r^2 = 0.9999) in the concentration range between 0.16 mg/mL (5.01 × 10^{-4} mM) and 0.016 mg/mL (5.01 × 10^{-5} mM), was obtained.

3.3. Solubility and Phase-Solubility Studies of PTA34 and PTA73

The phase solubility study was conducted according with Higuchi and Connors. In detail, an excess of PTA34 or PTA73 was added to 1 mL of solutions containing HP-β-CD in the appropriate concentration (0–45% *w/v*) in 4 mL vials, with screw caps to avoid evaporation. The obtained mixtures were vigorously vortexed for about 5 min, and the suspensions thus obtained were placed under constant stirring in a thermostat bath at 37 ± 0.5 °C for about 48 h, until the saturation of the system was achieved. Then, each suspension was centrifuged at 13,200 rpm (MIKRO 22 R, Hettich Zentrifugen, Germany) in order to remove the uncomplexed drug and the supernatant was analyzed by spectrophotometry for the quantification of the drug, after appropriate dilution. All determinations were conducted at least in triplicate. The obtained data were used to determine the apparent stability 1:1 constant ($K_{1:1}$) of the HP-β-CD inclusion complexes with the biphenylnicotinamide derivatives, using the Higuchi and Connors equation:

$$K_{1:1} = \frac{slope}{S_0(1 - slope)}$$

where *slope* is the slope of the obtained phase solubility diagrams straight line, and S_0 represents PTA34 or PTA73 intrinsic solubility determined in the same way.

3.4. Job's Plot Method

The stoichiometric ratio between PTA34 or PTA73 and HP-β-CD in the inclusion complexes was determined by the continuous variation method or Job's plot method [18]. In detail, starting from CH_3OH/H_2O (55/45 *v/v*) equimolar solutions (1.0×10^{-4} M) of PTA34/PTA73 and HP-β-CD, and keeping the total molar concentration of the species constant, a set of intermediate solutions with fixed volume was obtained by mixing them in the molar ratio ranged from 0 to 1. After stirring for 1 h, for each solution the absorbance (abs) was registered by UV–VIS spectroscopy at 320 nm for PTA34, and at 340 nm for PTA73. Then, ΔAbs × [PTA34 or PTA73] was plotted versus r, where

$$r = \frac{[PTA34orPTA73]}{[PTA34orPTA73] + [HP - \beta - CD]}$$

3.5. Preparation of PTA34 or PTA73/HP-β-CD Inclusion Complexes at the Solid State and Determination of Their Incorporation Degree

The PTA34 or PTA73 and HP-β-CD (PTA34/HP-β-CD or PTA73/HP-β-CD) inclusion complex was prepared in the solid state by freeze drying. Briefly, PTA34 or PTA73 and HP-β-CD were placed in water in a 1:1 molar ratio and the suspension was at first vigorously vortexed for about 5 min and then left under stirring for two days. At the end of the two days the samples were filtered through 0.22 μM cellulose acetate filters (Millipore), frozen, and freeze-dried (Lio 5P, Milan, Italy). The amount of PTA34 or PTA73 present in the PTA34 or PTA73/HP-β-CD solid complex was determined by solubilizing about 5 mg of each sample in 5 mL of deionized water. The obtained solution was filtered with 0.22 μM cellulose acetate filters (Millipore®), and the quantification of the drug was carried out via UV-VIS spectrophotometry, as described in Section 3.2.

The incorporation degree of PTA34 or PTA73 into the inclusion complexes was determined from the obtained absorbance, using the calibration straight line, and expressed as mg of PTA34 or PTA73 per 1 g of complex.

3.6. In Vitro Dissolution Studies

Dissolution experiments were performed at 37 °C according to the paddle method [30] and maintaining a rotational speed of 100 rpm during the test. Suitable quantities of PTAs/HP-β-CD solid complexes were suspended in the dissolution medium (0.05 M phosphate buffer at pH 7.4). At predetermined time intervals, 600 μL of suspension was collected and, in order to keep constant the initial volume, 600 μL of the same dissolution medium previously thermostated at the same temperature was added. Samples were subsequently filtered using a 0.22 μm membrane filter (Millipore® cellulose acetate), and the filtrates thus obtained were subjected to UV-VIS analysis after appropriate dilution. For quantitative analysis the calibration curve previously constructed was used, and the dissolution profiles shown correspond to the average of three determinations.

3.7. Cell Culture

Human pancreatic ductal adenocarcinoma (PDAC) PANC-1 cells, human pancreas adenocarcinoma ascites metastasis cells AsPC-1, and undifferentiated human pancreatic carcinoma MIAPaCa-2 cells, were purchased from ATCC®. PANC-1 and AsPC-1 cells were cultured in Dulbecco's modified Eagle's medium (DMEM), supplemented with fetal bovine serum (FBS), to a final concentration of 10%, L-glutamine 1% (*v/v*), and penicillin/streptomycin 1% (*v/v*). MIA PaCa-2 cells were grown in Roswell Park Memorial Institute (RPMI) medium, supplemented as above, and in addition with horse serum, to a final concentration of 2.5%. Cells were maintained in a humid atmosphere of 95% air and 5% CO_2 at 37 °C. All materials for cell culturing were purchased from EuroClone (Pero, Milan, Italy).

3.8. Cell Proliferation Assay

The 3-(4,5-dimethylthiazol-2-yl)-2,5- diphenyltetrazoliumbromide (MTT) assay was performed as previously described [5] to investigate the effect of hydroxy-propil-β-cyclodextrin-complexed PTA34 and PTA73 (PTA34/HP-β-CD and PTA73/HP-β-CD, respectively) on cell viability of PANC-1, AsPC-1, and MIA PaCa-2 cell lines. The same uncomplexed compounds and HP-β-CD alone were also tested as reference compounds. Untreated cells were used as a positive control. In particular, cells were seeded in 96 well plates at a density of 5000 cells/well, incubated for 24 h in culturing medium, and subsequently treated for 72 h, with compounds at 0.01–10 μM concentrations range, in terms of PTA34 or PTA73, included. HP-β-CD–cyclodextrin alone was tested in the concentrations range corresponding to that present in the complex, namely 2.9–2900 μg/mL. After the incubation period 10 μL of a 0.5% (*w/v*) MTT/PBS solution were added to each well, and the incubation was prolonged further for 2 h. After this period, medium was removed and replaced with 100 μL of DMSO. The absorbance in each well was measured by a microplate reader (MULTISKAN Sky, ThermoScientific). The results are shown as dose/effect plots of the mean of three different experiments. The IC_{50} was defined as the drug concentration yielding a fraction of affected (no surviving) cells = 0.5, normalized with untreated cells, and was calculated utilizing CalcuSyn v.1.1.1 software (Biosoft, UK).

3.9. Cell Cycle Analysis

For the cell cycle analysis, human pancreatic cancer cells were seeded in 60 mm dishes at a density of 500,000/well and incubated for 24 h at 37 °C, with PTA34, PTA73 both complexed and not at concentration 1 μM in terms of PTA's, and with cyclodextrin alone at concentration equal to 290 μg/mL. Cells were then harvested, washed twice in ice-cold PBS pH 7.4, fixed in 4 mL of 70% ethanol, and stored at -20 °C until analysis. The cell cycle modulation induced by treatments was studied as previously described [7] by propidium iodide (PI) staining; the pellet was resuspended in PBS without Ca^{2+} and Mg^{2+}, containing 1 mg/mL RNase, 0.01% NP40, and 50 ĉg/mL PI (Sigma), and then flow cytometry analysis was performed on Attune NxT acoustic focusing cytometer (Life Technologies). Data were interpreted using the CytExpert software v.1.2, provided by the manufacturer.

3.10. Annexin V/PI Apoptosis Assay

Annexin V/PI assay was conducted as previously described [31] to understand the mechanism of cell death, occurring after cell cycle arrest in phase G2/M. In particular after 24 h of treatment with compounds at 1 μM in terms of PTA's concentration, cells were subjected to Annexin-V FITC/propidium iodide staining, which allowed detection of early and late apoptosis as Annexin V positive cells, and late apoptosis as Annexin V/PI positive cells, and summarized in a graph as fold change of Annexin V positive cells (Annexin V plus Annexin V/PI) in treated samples versus their specific reference compound. FITC Annexin V Apoptosis Detection Kit I was from BD PharmingenTM (San Jose, CA, USA, 556547) and the analysis was performed on Attune NxT acoustic focusing cytometer (Life Technologies). Data were interpreted using the CytExpert software v. 1.2 provided by the manufacturer.

4. Conclusions

In conclusion, the complexation of PTA34 and PTA73 with hydroxy-propil-β-cyclodextrin was successful both in solution and in solid state, allowing an increase of the PTA's water solubility and a favorable dissolution profile with respect to the uncomplexed drug. In addition, the property of hydroxy-propil-β-cyclodextrin, which allows the enhancement of the plasma membrane permeability, was responsible for the increase of intracellular uptake of the complexed drugs, and consequently of their antitumor efficacy, as evidenced in studies conducted on PDAC cells model. Considering the promising results obtained, and that hydroxy-propil-β-cyclodextrin is an excipient already approved by the FDA for parenteral formulations, the inclusion complexes of PTA34 and PTA73, after an in vivo validation step, could be considered for parenteral administration in PDAC therapy.

Author Contributions: Conceptualization, N.D. and A.A.; methodology, A.C., R.M.I. and A.S.; software, R.D.F. and V.L.; validation, N.D., A.A., and F.L.; formal analysis, M.F., A.A.L., L.P. and S.S.; investigation, R.M.I., A.C., A.S., S.C., and A.L.; data curation, A.C., R.M.I.; writing—original draft preparation, R.M.I. and A.C.; writing—review and editing, A.C., R.M.I., A.S., A.A., N.S.; visualization, A.C., R.M.I., A.L., L.P.; supervision, A.A., F.L. and N.D.; funding acquisition, N.S. All authors have read and agreed to the published version of the manuscript.

Funding: This research was partially funded by Italian Ministry of Health, grant number RC 2018-2020: "Marcatori predittivo/prognostici tissutali e circolanti in pazienti con carcinoma del pancreas e delle vie biliari: dal nanostring alla biopsia liquida".

Acknowledgments: Authors acknowledge the University "Aldo Moro" of Bari for its finantial support.

Conflicts of Interest: The authors declare no conflict of interest.

Abbreviations

PDAC	Pancreatic ductal adenocarcinoma
GEM	Gemcitabine
NAB	Nabpaclitaxel
HP-β-CD	Hydroxy-Propil-β-Cyclodextrin
CDs	Cyclodextrins
MTAs	Microtubule targeting agents
FDA	Food and Drug Administration
FACS	Fluorescence-Activated Cell Sorter

References

1. American Cancer Society Journal. Cancer Facts & Figures 2020. *CA Cancer J. Clin.* **2020**, 1–76.
2. Associazione Italiana di Oncologia Medica (AIOM). Linee Guida: Carcinoma del Pancreas Esocrino. *Aiom* **2019**, 1–70.
3. Chen, X.; Zhou, W.; Liang, C.; Shi, S.; Yu, X.; Chen, Q.; Sun, T.; Lu, Y.; Zhang, Y.; Guo, Q.; et al. Codelivery Nanosystem Targeting the Deep Microenvironment of Pancreatic Cancer. *Nano Lett.* **2019**, *19*, 3527–3534. [CrossRef] [PubMed]
4. Veenstra, V.; Garcia-Garijo, A.; van Laarhoven, H.; Bijlsma, M. Extracellular Influences: Molecular Subclasses and the Microenvironment in Pancreatic Cancer. *Cancers (Basel)* **2018**, *10*, 34. [CrossRef]
5. Porcelli, L.; Iacobazzi, R.; Di Fonte, R.; Serratì, S.; Intini, A.; Solimando, A.; Brunetti, O.; Calabrese, A.; Leonetti, F.; Azzariti, A.; et al. CAFs and TGF-β Signaling Activation by Mast Cells Contribute to Resistance to Gemcitabine/Nabpaclitaxel in Pancreatic Cancer. *Cancers (Basel)* **2019**, *11*, 330. [CrossRef]
6. Majellaro, M.; Stefanachi, A.; Tardia, P.; Vicenti, C.; Boccarelli, A.; Pannunzio, A.; Campanella, F.; Coluccia, M.; Denora, N.; Leonetti, F.; et al. Investigating Structural Requirements for the Antiproliferative Activity of Biphenyl Nicotinamides. *ChemMedChem* **2017**, *12*, 1380–1389. [CrossRef]
7. Porcelli, L.; Stolfa, D.; Stefanachi, A.; Di Fonte, R.; Garofoli, M.; Iacobazzi, R.M.; Silvestris, N.; Guarini, A.; Cellamare, S.; Azzariti, A. Synthesis and biological evaluation of N-biphenyl-nicotinic based moiety compounds: A new class of antimitotic agents for the treatment of Hodgkin Lymphoma. *Cancer Lett.* **2019**, *445*, 1–10. [CrossRef]
8. Gagliano, N.; Volpari, T.; Clerici, M.; Pettinari, L.; Barajon, I.; Portinaro, N.; Colombo, G.; Milzani, A.; Dalle-Donne, I.; Martinelli, C. Pancreatic cancer cells retain the epithelial-related phenotype and modify mitotic spindle microtubules after the administration of ukrain in vitro. *Anticancer Drugs* **2012**, *9*, 935–946. [CrossRef]
9. Chan, K.-S.; Koh, C.-G.; Li, H.-Y. Mitosis-targeted anti-cancer therapies: Where they stand. *Cell Death Dis.* **2012**, *3*, e411. [CrossRef]
10. Vila-Navarro, E.; Fernandez-Castañer, E.; Rovira-Rigau, M.; Raimondi, G.; Vila-Casadesus, M.; Lozano, J.J.; Soubeyran, P.; Iovanna, J.; Castells, A.; Fillat, C.; et al. MiR-93 is related to poor prognosis in pancreatic cancer and promotes tumor progression by targeting microtubule dynamics. *Oncogenesis* **2020**, *9*, 43. [CrossRef]
11. Kumar, S.; Parkash, C.; Kumar, P.; Singh, S.K. Application of some novel techniques for solubility enhancement of mefenamic acid, a poorly water soluble drug. *Int. J. Pharm. Sci. Drug Res.* **2009**, *1*, 164–171.
12. Dora, C.P.; Singh, S.K.; Kumar, S.; Datusalia, A.K.; Deep, A. Development and characterization of nanoparticles of glibenclamide by solvent displacement method. *Acta Pol. Pharm.—Drug Res.* **2010**, *67*, 283–290.

13. Tang, C.; Guan, Y.-X.; Yao, S.-J.; Zhu, Z.-Q. Solubility of Dexamethasone in Supercritical Carbon Dioxide with and without a Cosolvent. *J. Chem. Eng. Data* **2014**, *59*, 3359–3364. [CrossRef]
14. Iacobazzi, R.M.; Letizia, P.; Assunta, L.A.; Valentino, L.; Antonio, L.; Annalisa, C.; Emiliano, A.; Roberta, D.F.; Amalia, A.; Massimo, F.; et al. Targeting human liver cancer cells with lactobionic acid-G(4)-PAMAM-FITC sorafenib loaded dendrimers. *Int. J. Pharm.* **2017**, 485–497. [CrossRef] [PubMed]
15. Lopedota, A.; Cutrignelli, A.; Laquintana, V.; Denora, N.; Iacobazzi, R.M.; Perrone, M.; Fanizza, E.; Mastrodonato, M.; Mentino, D.; Lopalco, A.; et al. Spray Dried Chitosan Microparticles for Intravesical Delivery of Celecoxib: Preparation and Characterization. *Pharm. Res.* **2016**, *33*, 2195–2208. [CrossRef]
16. Cutrignelli, A.; Sanarica, F.; Lopalco, A.; Lopedota, A.; Laquintana, V.; Franco, M.; Boccanegra, B.; Mantuano, P.; De Luca, A.; Denora, N. Dasatinib/HP-β-CD inclusion complex based aqueous formulation as a promising tool for the treatment of paediatric neuromuscular disorders. *Int. J. Mol. Sci.* **2019**, *20*, 591. [CrossRef]
17. Devasari, N.; Dora, C.P.; Singh, C.; Paidi, S.R.; Kumar, V.; Sobhia, M.E.; Suresh, S. Inclusion complex of erlotinib with sulfobutyl ether-β-cyclodextrin: Preparation, characterization, in silico, in vitro and in vivo evaluation. *Carbohydr. Polym.* **2015**, *134*, 547–556. [CrossRef]
18. Cutrignelli, A.; Lopedota, A.; Denora, N.; Iacobazzi, R.M.; Fanizza, E.; Laquintana, V.; Perrone, M.; Maggi, V.; Franco, M. A new complex of curcumin with sulfobutylether-β-cyclodextrin: Characterization studies and in vitro evaluation of cytotoxic and antioxidant activity on HepG-2 cells. *J. Pharm. Sci.* **2014**, *103*, 3932–3940. [CrossRef]
19. Challa, R.; Ahuja, A.; Ali, J.; Khar, R.K. Cyclodextrins in drug delivery: An updated review. *AAPS PharmSciTech* **2005**, *6*, 329–357. [CrossRef]
20. Másson, M.; Loftsson, T.; Másson, G.; Stefánsson, E. Cyclodextrins as permeation enhancers: Some theoretical evaluations and in vitro testing. *J. Control. Release* **1999**, *59*, 107–118. [CrossRef]
21. Hammoud, Z.; Khreich, N.; Auezova, L.; Fourmentin, S.; Elaissari, A.; Greige-Gerges, H. Cyclodextrin-membrane interaction in drug delivery and membrane structure maintenance. *Int. J. Pharm.* **2019**, *564*, 59–76. [CrossRef] [PubMed]
22. Rong, W.-T.; Lu, Y.-P.; Tao, Q.; Guo, M.; Lu, Y.; Ren, Y.; Yu, S.-Q. Hydroxypropyl-Sulfobutyl-β-Cyclodextrin Improves the Oral Bioavailability of Edaravone by Modulating Drug Efflux Pump of Enterocytes. *J. Pharm. Sci.* **2014**, *103*, 730–742. [CrossRef] [PubMed]
23. Szente, L.; Szejtli, J. Highly soluble cyclodextrin derivatives: Chemistry, properties, and trends in development. *Adv. Drug Deliv. Rev.* **1999**, *36*, 17–28. [CrossRef]
24. Higuchi, T.; Connors, K.A. Techniques, Phase solubility. *Adv. Anal. Chem. Instrum.* **1965**, *4*, 117–212.
25. Silvente-Poirot, S.; Poirot, M. Cholesterol and Cancer, in the Balance. *Science* **2014**, *343*, 1445–1446. [CrossRef]
26. Brown, A.; Patel, S.; Ward, C.; Lorenz, A.; Ortiz, M.; DuRoss, A.; Wieghardt, F.; Esch, A.; Otten, E.G.; Heiser, L.M.; et al. PEG-lipid micelles enable cholesterol efflux in Niemann-Pick Type C1 disease-based lysosomal storage disorder. *Sci. Rep.* **2016**, *6*, 31750. [CrossRef]
27. Kilsdonk, E.P.C.; Yancey, P.G.; Stoudt, G.W.; Bangerter, F.W.; Johnson, W.J.; Phillips, M.C.; Rothblat, G.H. Cellular cholesterol efflux mediated by cyclodextrins. *J. Biol. Chem.* **1995**, *270*, 17250–17256. [CrossRef]
28. Li, J.; Gu, D.; Lee, S.S.-Y.; Song, B.; Bandyopadhyay, S.; Chen, S.; Konieczny, S.F.; Ratliff, T.L.; Liu, X.; Xie, J.; et al. Abrogating cholesterol esterification suppresses growth and metastasis of pancreatic cancer. *Oncogene* **2016**, *35*, 6378–6388. [CrossRef]
29. Chang, T.-Y.; Chang, C.C.Y.; Ohgami, N.; Yamauchi, Y. Cholesterol Sensing, Trafficking, and Esterification. *Annu. Rev. Cell Dev. Biol.* **2006**, *22*, 129–157. [CrossRef]
30. Cutrignelli, A.; Lopedota, A.; Trapani, A.; Boghetich, G.; Franco, M.; Denora, N.; Laquintana, V.; Trapani, G. Relationship between dissolution efficiency of Oxazepam/carrier blends and drug and carrier molecular descriptors using multivariate regression analysis. *Int. J. Pharm.* **2008**, *358*, 60–68. [CrossRef]
31. Azzariti, A.; Iacobazzi, R.M.; Di Fonte, R.; Porcelli, L.; Gristina, R.; Favia, P.; Fracassi, F.; Trizio, I.; Silvestris, N.; Guida, G.; et al. Plasma-activated medium triggers cell death and the presentation of immune activating danger signals in melanoma and pancreatic cancer cells. *Sci. Rep.* **2019**, *9*, 4099. [CrossRef] [PubMed]

International Journal of
Molecular Sciences

Article

Aminopeptidase N Inhibitors as Pointers for Overcoming Antitumor Treatment Resistance

Oldřich Farsa [1], Veronika Ballayová [1,*], Radka Žáčková [1], Peter Kollar [2], Tereza Kauerová [2] and Peter Zubáč [1]

[1] Department of Chemical Drugs, Faculty of Pharmacy, Masaryk University, Palackého 1946/1, 612 00 Brno, Czech Republic
[2] Department of Pharmacology and Toxicology, Faculty of Pharmacy, Masaryk University, Palackého 1946/1, 612 00 Brno, Czech Republic
* Correspondence: 516259@mail.muni.cz; Tel.: +420-541562936

Abstract: Aminopeptidase N (APN), also known as CD13 antigen or membrane alanyl aminopeptidase, belongs to the M1 family of the MA clan of zinc metallopeptidases. In cancer cells, the inhibition of aminopeptidases including APN causes the phenomenon termed the amino acid deprivation response (AADR), a stress response characterized by the upregulation of amino acid transporters and synthetic enzymes and activation of stress-related pathways such as nuclear factor kB (NFkB) and other pro-apoptotic regulators, which leads to cancer cell death by apoptosis. Recently, APN inhibition has been shown to augment DR4-induced tumor cell death and thus overcome resistance to cancer treatment with DR4-ligand TRAIL, which is available as a recombinant soluble form dulanermin. This implies that APN inhibitors could serve as potential weapons for overcoming cancer treatment resistance. In this study, a series of basically substituted acetamidophenones and the semicarbazones and thiosemicarbazones derived from them were prepared, for which APN inhibitory activity was determined. In addition, a selective anti-proliferative activity against cancer cells expressing APN was demonstrated. Our semicarbazones and thiosemicarbazones are the first compounds of these structural types of Schiff bases that were reported to inhibit not only a zinc-dependent aminopeptidase of the M1 family but also a metalloenzyme.

Keywords: aminopeptidase N; acetamidophenones; Schiff bases; semicarbazones; thiosemicarbazones; inhibition of proliferation

Citation: Farsa, O.; Ballayová, V.; Žáčková, R.; Kollar, P.; Kauerová, T.; Zubáč, P. Aminopeptidase N Inhibitors as Pointers for Overcoming Antitumor Treatment Resistance. *Int. J. Mol. Sci.* **2022**, *23*, 9813. https://doi.org/10.3390/ijms23179813

Academic Editor: Angela Stefanachi

Received: 26 July 2022
Accepted: 27 August 2022
Published: 29 August 2022

Publisher's Note: MDPI stays neutral with regard to jurisdictional claims in published maps and institutional affiliations.

1. Introduction

APN is sometimes called "a moonlighting enzyme". It is a widely expressed ectoenzyme with many functions that do not always depend on its enzymatic activity. The membrane-bound form of APN, which is expressed in the renal and intestinal epithelia, the nervous system, myeloid cells, and fibroblast-like cells, such as synoviocytes, is often referred to as hCD13, whereas the soluble form, which is present in human serum at a concentration of about 4.6 nM, is known as sCD13 [1]. There is a strong correlation between the expression and enzymatic activity of hCD13 and sCD13 and the invasive capacity of numerous tumor cell types. APN also serves as a receptor involved in endocytosis during viral infection such as in the human coronavirus HCoV-229E, among others [2]. As a signaling molecule, it takes part in adhesion, phagocytosis, and angiogenic processes [3]. The plasmatic concentration of sCD13 can be used as a prognostic marker for some cancers such as non-small cell lung cancer (NSCLC) including lung adenocarcinoma [4]. APN is a promising target for anticancer therapy. Newer research suggests that it serves as one of the molecular targets for the anticancer antibiotic actinomycin D and its simplified analogs [5]. Bestatin (3-Amino-2-hydroxy-4-phenylbutanoyl)-L-leucine or ubenimex in INN nomenclature, Figure 1, one of the most known APN inhibitors, was first isolated from the bacteria *Streptomyces olivoreticuli* in 1976. It was used as an anticancer agent for the treatment of lung cancer and acute myeloid leukemia in Japan for several years [6,7]. More recently, it has also been demonstrated to inhibit cell proliferation, migration, and invasion in both renal cell carcinoma and prostate cancer [8,9]. Bestatin has also been shown to be capable of attenuating the acquired resistance of renal cell carcinoma to treatment with sorafenib, which is today a first-line therapy for this cancer [10]. Tosedostat, cyclopentyl (2S)-2-[[(2R)-2-[(1S)-1-hydroxy-2-(hydroxyamino)-2-oxoethyl]-4-methylpentanoyl]amino]-2-phenylacetate, a synthetic dipeptide containing the carbohydroxamic group (Figure 1), is also a known APN inhibitor.

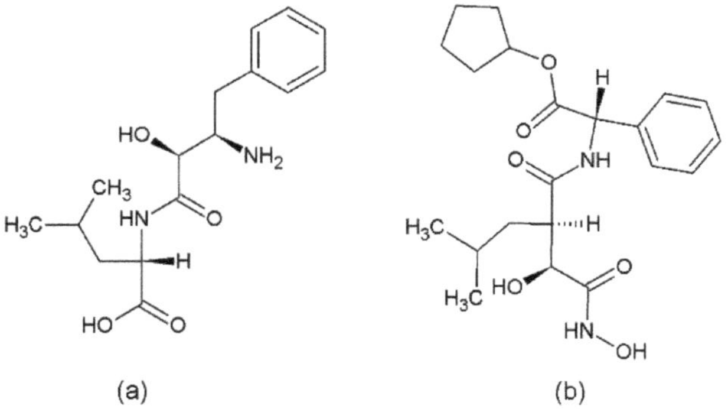

(a)

(b)

Figure 1. The most known dipeptide APN inhibitors: bestatin (**a**) and tosedostat (**b**).

It has undergone more than ten clinical trials of phases 1 or 2 for the treatment of myeloid leukemias and solid tumors. Its anticancer activity is mainly attributed to the inhibition of the cleavage of proteins and peptides by M1 family aminopeptidases including APN. This disrupts normal cellular protein turnover, resulting in both peptide accumulation and a decrease in intracellular free amino acid content, a process that appears to preferentially affect metabolically active cells such as malignant cells. Such an inhibition triggers the phenomenon termed the amino acid deprivation response (AADR), a stress response comprising the upregulation of amino acid transporters and synthetic enzymes and the activation of stress-related pathways such as NFkB and other pro-apoptotic regulators [11]. Death receptor 4 (DR4 or TRAIL-R1), a member of the DR subgroup of the tumor necrosis factor (TNF) receptor superfamily, is overexpressed in various types of tumor cells. DR4 mediates extrinsic apoptotic cascades using binding to TNF-related apoptosis-inducing ligands (TRAIL or Apo2L). Unfortunately, resistance is often observed in the clinical application of TRAIL, which has undergone five clinical trials on its soluble recombinant form dulanermin, and another study for the treatment of peritoneal carcinomatosis continues [12]. In a recent study, bestatin markedly sensitized fibrosarcoma cells previously implanted in athymic nude mice to apoptosis induced by TRAIL [13]. Numerous further APN inhibitors were prepared as potential anti-cancer drugs. Recent progress in this field is summarized in a review article [1]. APN belongs among the zinc metallopeptidases. As far as a mechanism of inhibition is concerned, zinc chelation is frequently mentioned. Many reported inhibitors are attributed to this mechanism of action. Hydroxamic acids with the ureido fragment in their molecules [14–16] or without [17] use their carbohydroxamic group as the coordinating moiety for Zn^{2+}. Vicinal cycloaliphatic amino ketones, specifically 3-amino-2-tetralone derivatives and analogs, use this complexation as their primary amino group together with the oxygen of the adjacent keto group [18]. Semicarbazones and thiosemicarbazones are known zinc chelators [19,20], although this fact has never been used in the design of metalloenzyme inhibitors. In this article, we describe the design, synthesis, and APN inhibition activity of a series of novel, basically substituted acetamidoacetophenone-semicarbazones and -thiosemicarbazones and their starting ketones, with either the dialkylamine group or a saturated nitrogenous heterocycle as a basic substituent in the acetamido part of the molecule.

2. Results

2.1. Synthesis of Target Compounds

Our target compounds were synthesized by a four-stage synthetic sequence that is depicted in Scheme 1.

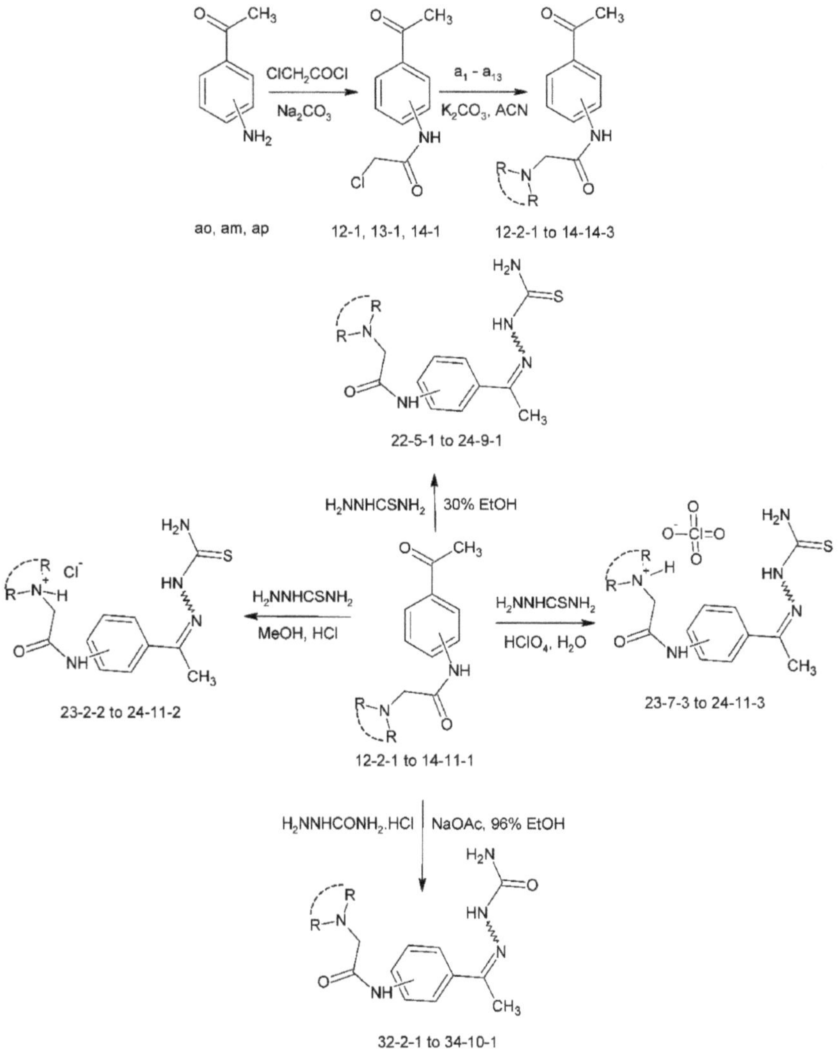

Scheme 1. Synthesis of target compounds. Legend: a_1 to a_{11}: a set of secondary amines; explanation of codes for ketones, thiosemicarbazides, and semicarbazides: XX-X-X: 1st Figure 1 for ketone, 2 for thiosemicarbazone, 3 for semicarbazone; 2nd Figure 2 for ortho substitution, 3 for meta and 4 for para; 3rd Figure 1 for Cl and then in the respective order: 2 diethylamine, 3 dipropyl amine, 4 dibutyl amine, 5 pyrrolidine, 6 piperidine, 7 azepane, 8 morpholine, 9 thiomorpholine, 10 N-benzylpiperazine, 11 N-methyl piperazine.

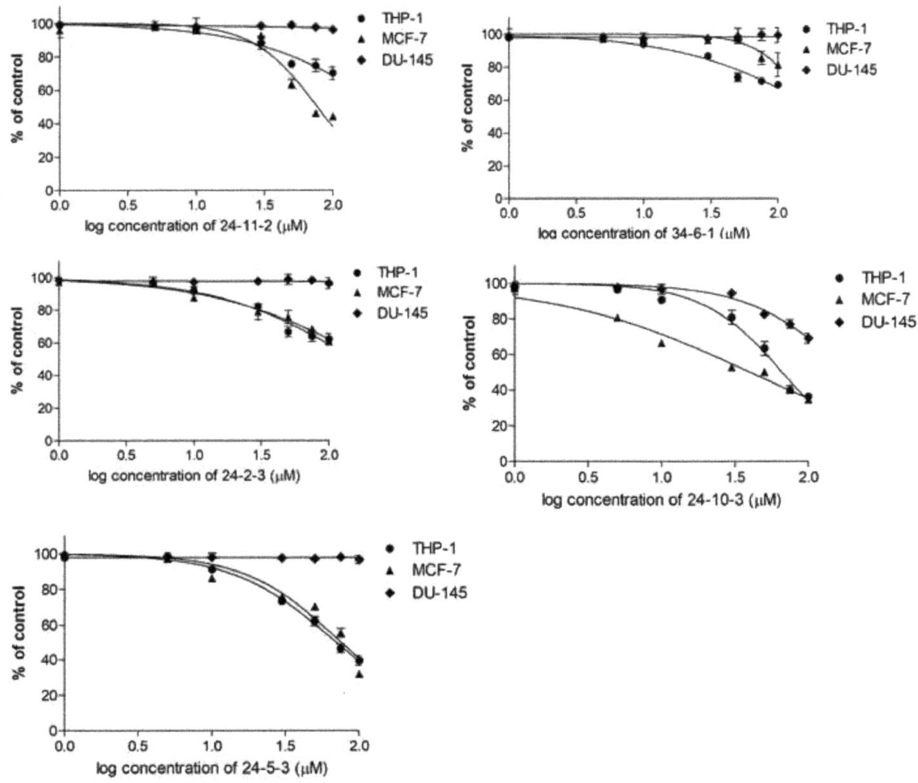

Figure 2. Effect of tested compounds on cell proliferation in THP-1, MCF-7, and DU-145 cell lines after 48 h of incubation. Proliferation was determined using WST-1 assay. The results are shown as the mean ± SD of three independent experiments, each performed in triplicate. Specific values, including statistical analysis, are given in the Supplementary Materials.

The starting 2-,3- or 4-aminoacetophenone **ao, am, ap** reacted with chloroacetyl chloride to give 2-, 3- or 4-(chloroacetamido)acetophenones **12-1**, **13-1**, **14-1**, which were then subjected to the reaction with individual secondary amines to give basically substituted ketones **12-2-1** to **14-11-1** (in case of dialkyl amines as reagents, 2-, 3- or 4-[2-(dialkylamino)acetamido]acetophenones were prepared). The reaction of such ketones with thiosemicarbazide in ethanol without the presence of any strong acid then led to the appropriate thiosemicarbazones as bases (**22-5-1–24-9-1**), whereas the analogous reaction with semicarbazide gave semicarbazones as bases, **32-2-1–34-10-1**. Analogous reactions with thiosemicarbazide in the presence of either perchloric or hydrochloric acid then led to appropriate thiosemicarbazones in the form of perchlorate or hydrochloride, **23-7-3–24-11-3.** The compounds were purified by a simple crystallization from the system ethanol/water with the addition of charcoal in the case of need and characterized with 1H- and 13-C-NMR, IR, and MS spectra. Two-dimensional NMR spectra (H-H-cosy, HMQC, HMBC) were used for the 1D NMR spectra interpretation. NOESY (NMR) spectra were used to determine the geometry of the Schiff bases. They revealed that the prepared products consisted of about equimolar amounts of *E* and *Z* (or *syn*/*anti*) isomers. The structural characteristics of the prepared compounds can be found in the Supplementary Materials as well as the procedure for the determination of their purity, and the yields and values of the purity of the target compounds and key intermediates are summarized in Table 1.

Table 1. Yields and values of purity, determined as chromatography homogeneity by HPLC, of the target compounds and their reaction intermediates.

Substance Code	Yield [%]	Purity [%]
12-1	93	99.7
13-1	95	99.9
14-1	96	99.9
14-2-1	87	95.1
14-6-1	72	91.8
14-8-1	69.4	92.4
14-10-1	70	91.6
22-5-1	49	98.2
22-6-1	54.7	96.4
22-7-1	46	97.7
23-2-1	75	93.1
23-3-1	72	94.2
23-5-1	53	96.6
23-6-1	58.9	97.1
23-8-1	62	95.7
23-11-1	69	98.2
24-3-1	61	95.7
24-4-1	53	95.5
24-5-1	53	98.0
24-6-1	60	94.3
24-7-1	49	97.9
24-8-1	65	99.9
24-9-1	62.6	95.5
24-2-2	59	96.2
24-11-2	71	96.0
23-7-3	59	97.3
24-2-3	68	94.7
24-5-3	71	97.6
24-6-3	74	95.1
24-10-3	58	97.9
24-11-3	63	96.5
34-2-1	45	97.9
34-6-1	36	98.8
34-7-1	21	98.4
34-8-1	19	97.7

2.2. APN Inhibitory Activity and QSAR in It

2.2.1. Determination of APN Inhibition

APN inhibitory activity was determined using a standardized spectrophotometry protocol using L-leucine-p-nitroanilide as a chromogenic substrate for APN. Measurements were performed in triplicate at 405 nm at a Cytation 3 well-plate reader. The results were processed into IC_{50} values using GraFit 5 software and are listed in Table 2 below.

In some compounds, solubility problems occurred, which complicated the inhibition activity determination. We overcame these either by the use of cosolvents or by an alternative HPLC approach [21]. The details are summarized in the Supplementary Materials.

Table 2. Inhibitory activity of basically substituted acetamidoacetophenones and their semicarbazones and thiosemicarbazones expressed as IC_{50} in μmol/L, and their logarithms and values of computed physical parameters used for construction of QSAR Equation (2). Optimized values of log IC_{50} are listed only for compounds that were finally used for the construction of Equation (2) (see the text).

Substance Code	logP	pI	IC_{50} Experimental [μmol/L]	Log IC_{50} Experimental	Log IC_{50} Optimized
14-2-1	1.6	9.67	366 [2]	2.5635	-
14-6-1	2.72	9.46	173	2.6288	-
14-8-1	1.4	9.67	425.4	2.6307	-
14-10-1	2.65	10.34	427.3 [2]	2.8910	-
22-6-1	2.78	9.38	778.0	2.2380	6.9918
22-7-1	3.34	9.68	169.3	2.2287	1.8688
23-7-3	2.57	9.68	35.6	1.5514	1.7158
24-2-2	0.97	9.67	46.4	1.6665	-
24-2-3	0.97	9.67	18.9	1.2765	1.9172
24-3-1	2.03	9.95	1266	3.1024	2.8204
24-4-1	3.1	9.97	1679 [1]	3.2251	-
24-5-1	1.53	9.52	35.6 [1]	1.5514	-
24-5-3	1.53	9.52	23.5 [1]	1.3711	1.1489
24-6-1	2.09	9.46	1549 [1]	3.1901	2.9154
24-6-3	2.09	9.46	4732 [2]	3.6750	-
24-7-1	2.66	9.69	44.1	1.6444	1.6200
24-8-1	0.78	8.67	6.79	0.8319	0.8753
24-10-3	2.03	9.64	22.3	1.3483	1.7886
24-11-2	0.5	9.61	13.3	1.1239	0.8174
24-11-3	0.5	9.61	150.1	2.1764	-
34-6-1	1.49	9.53	18.4	1.2648	1.0857
34-8-1	0.17	8.45	1644	3.2159	-

[1] IC_{50} values determined with the use of NMP as a cosolvent. [2] IC_{50} values determined with the use of DMSO as a cosolvent.

2.2.2. QSAR in APN Inhibitory Activity

The classical Hansch method of regression analysis was used to determine the dependence of the inhibitory expressed as IC_{50} on the important structure descriptors. Typically, the activity of the members of a homologous series of biologically active compounds correlates with their lipophilicity. Furthermore, in our case, a hint of such a correlation was also found. This situation was expressed by Equation (1):

$$\log IC_{50} = 0.6925 \log P + 0.5684 \tag{1}$$

where log P is calculated by an algorithm based on >12,000 experimental logP values using the principle of isolating carbons [22], as a parameter of lipophilicity was used. The low determination coefficient and F-statistic values ($R^2 = 0.4033$, $F = 3.2615$) indicated that the correlation was insufficient and a further structure parameter had to be added. An electronic parameter was then used because it was very probable that dissociation at various sites of the molecule could have an impact on coordination to Zn^{2+} cation as well as on interactions with the acidic or basic parts of the amino acid residues of the enzyme protein. Since our target compounds have two centers of acidity and at least two centers of basicity, it was more advantageous to express the electronic properties of a molecule with one electronic parameter, the isoelectric point pI, than with a set of dissociation constants. The set of pI values computed by the algorithm implemented in the Marvin software [23] was chosen for such a purpose. Ketones **14-2-1, 14-6-1,** and **14-8-1** were preliminarily excluded because there was an assumption of a different Zn^{2+} complexation in them than in the thiosemicarbazones and semicarbazones. Further, **24-4-1** and **34-8-1** were excluded as outliers during the regression analysis. Finally, a regression model with both parameters squared was found. It is expressed by Equation (2):

$$\log IC_{50} = -0.4475(\log P)^2 - 0.1452(pI)^2 + 1.8847(\log P)(pI) - 0.6101(\log P) + 0.6021(pI) + 1.1284 \quad (2)$$

where the multiple correlation coefficient R was 0.9837, the determination coefficient R^2 was 0.9677, and the computed F-statistic was 29.9540. The IC_{50} values for the prepared compounds, for which it was not possible to determine them experimentally due to their poor solubility or a precipitate formation, or those that have not yet been synthesized, were estimated by the calculation from Equation (2). They are listed in Table 3 below together with their calculated values of log P and pI.

Table 3. Calculated inhibitory activity of semicarbazides and thiosemicarbazides of basically substituted acetamidoacetophenones expressed as IC_{50} in µmol/L together with the calculated values of log P and pI used for IC_{50} calculation.

Substance Code	logP	pI	IC_{50} [µmol/L]
22-5-1 [1]	2.21	9.45	149.2
23-6-1 [1]	2.01	9.46	83.1
23-8-1 [1]	0.69	8.37	397.7
23-11-1 [1]	0.42	9.61	7.86
24-9-1 [1]	1.72	8.97	1372
34-7-1 [1]	2.05	9.77	121.9
22-2-1 [2]	1.66	9.62	50.4
22-3-1 [2]	2.72	9.9	70.7
22-4-1 [2]	3.78	9.93	3.4
23-2-1 [1]	0.89	9.67	39.5
23-3-1 [1]	1.95	9.95	686.5
23-4-1 [2]	3.01	9.97	34.6
23-5-1 [1]	1.44	9.51	20.26
23-9-1 [2]	1.64	8.97	714.7
32-2-1 [2]	1.05	9.7	68.3
32-3-1 [2]	2.12	9.97	538.1
32-4-1 [2]	3.18	9.99	19.5
32-5-1 [2]	1.61	9.51	30.0
32-6-1 [2]	2.18	9.46	130.5
32-7-1 [2]	2.74	9.68	92.4
32-8-1 [2]	0.86	8.37	5137
32-11-1 [2]	0.58	9.65	20.6
33-2-1 [2]	0.28	9.75	135.1
33-4-1 [2]	2.41	10.04	413.7
33-5-1 [2]	0.84	9.58	10.4
33-6-1 [2]	1.41	9.52	19.7
33-7-1 [2]	1.97	9.76	121.0
33-10-1 [2]	1.34	9.72	96.8
34-2-1 [2]	0.37	9.75	138.0
34-4-1 [2]	2.49	10.04	305.3
34-5-1 [2]	0.92	9.59	13.6
34-9-1 [2]	1.12	9.04	5.9
34-10-1 [2]	1.43	9.73	109.12

[1] Synthesized compounds in which IC_{50} determination was not possible due to their poor solubility or precipitate formation. [2] Compounds that have not yet been synthesized.

The synthesis and testing of the above-mentioned unsensitized compounds, as well as other structurally related ones, which would serve as a test set for the confirmation of Equation (1), are planned as the continuation of this research.

2.3. Proliferation Inhibitory Effects Induced by Thiosemicarbazides of Basically Substituted Acetamidoacetophenone Compounds in Human Cancer Cell Lines

For the target compounds that effectively inhibited APN activity, the antiproliferative activity in the different human cancer cell lines THP-1, MCF-7, and DU-145 was evaluated as IC_{50} values (50% inhibition concentration). The results are shown in Table 4, Figure 2, and the Supplementary Materials. All five tested compounds significantly decreased the proliferation of THP-1 and MCF-7 cell lines in a concentration-dependent manner. On the contrary, only compound 24-10-3 induced an antiproliferative effect in DU-145 cells as well.

Table 4. Cell proliferation of three human cancer cell lines THP-1, MCF-7, and DU-145 and IC_{50} values were evaluated using WST-1 analysis after 48 h incubation with serial dilutions of tested compounds. The values shown are the mean ± SD from three independent experiments, each performed in triplicate.

Substance Code	$IC_{50} \pm$ SD [μmol/L]		
	THP-1	MCF-7	DU-145
24-11-2	>100	77.11 ± 2.19	>100
34-6-1	>100	>100	>100
24-2-3	>100	>100	>100
24-10-3	66.05 ± 1.71	39.94 ± 2.83	>100
24-5-3	70.18 ±4.51	75.47 ± 3.14	>100

2.4. The Most Active Compounds and SAR

2.4.1. SAR in APN Inhibition

The compound **24-11-2**, which is 4-[2-(4-methyl piperazine-1-yl)acetamido]acetophenone thiosemicarbazone hydrochloride 34 with an IC_{50} of 13.3 μmol/L, was found to be the most active compound, the activity of which was determined, whereas **22-4-1**, 2-[(diethylamino) acetamido]acetophenone thiosemicarbazone with an IC_{50} of 3.4 μmol/L was the most active in the series of compounds with the activities predicted using Equation 2. The overall activity results suggest that there is no significant difference in the activity of semicarbazones and thiosemicarbazones. The terminal basic parts, which seem to improve the activity, are 4-methyl piperazine-1-yl, 4-benzyl piperazine-1-yl, piperidine-1-yl, azepane-1-yl, and pyrrolidine-1-yl in most bulkier substituents with rather greater basicity (except for pyrrolidine-1-yl). As far as the influence of the positional isomerism is concerned, that is, the position of a substituted 2-aminoacetamido substituent regarding a Schiff base-containing group at the benzene ring, the results suggest that *p*-substituted compounds are more active than *m*- and *o*-derivatives.

2.4.2. APN Inhibition vs. Antiproliferative Activity

The five compounds with the best values of IC_{50} for APN (ranging between 13 and 23.5 μmol/L) underwent testing for inhibition of cell proliferation on the three different cell lines, which differ from one another in their levels of APN expression. All five compounds triggered a significant antiproliferative effect in the cell lines expressing APN, THP-1, and MCF-7, whereas in the cell line DU-145 with no APN expression, four out of these five compounds did not affect proliferation at all. The remaining compound also inhibited DU-145 cell proliferation but less than in APN-positive THP-1 or MCF-7 lines (compare Figure 2). These results could suggest that the antiproliferative activity is linked with APN inhibition although other mechanisms that can also participate in it.

3. Discussion

The antiproliferative effect was determined for five selected compounds with the most potent APN inhibitory activity. Proliferation inhibitory effect determination was performed in three human cancer lines that differed from each other in their levels of APN expression. Although APN expression has been proved in the human monocytic leukemia cell line THP-1 [24] and breast carcinoma cell line MCF-7 [25], no APN expression was found in the DU 145 cells [26]. Our tested compounds showed a different effect in the cell lines used for analysis. All tested substances induced a significant antiproliferative effect in the cell lines expressing APN, THP-1, and MCF-7, whereas in the cell line DU-145 with no APN expression, the compounds **24-11-2, 34-6-1, 24-2-3,** and **24-5-3** did not affect proliferation at all. Compound 24-10-3 also decreased DU-145 cells but to a lesser extent than in APN-positive THP-1 or MCF-7 cells. In the case of this substance, we can consider another additional mechanism of antiproliferative action, but based on the data obtained, nothing closer can be stated. Although the obtained IC_{50} values against the APN-positive cell lines are at the level of double-digit micromoles, the substances represent an interesting model structure for the development of potentially therapeutically useful APN inhibitors.

The APN inhibition activity results suggest that the complexation of Zn^{2+} in the catalytic site of APN could be the mechanism underlying the inhibitory activity of basically substituted aminoacetophenones and their semicarbazones and thiosemicarbazones. A significant difference between the median inhibitory activity of the ketones (median IC_{50} = 395.7 μmol/L) and thiosemicarbazones together with the semicarbazones (median IC_{50} = 44.1 μmol/L) in compounds in which the IC_{50} values were experimentally determined, further suggests that the mechanisms of the Zn^{2+}complexation with ketones and the Schiff bases derived from them must be different. This assumption was taken into account in the construction of the QSAR dependence that led to Equation (2). Based on the structure of (T-4)-[2-[1-[5-Acetyl-2-(hydroxy-κO)-4-hydroxyphenyl]ethylidene]hydrazinecarbothioamidato (2-)-κN,κN2]aquazinc [20] (Figure 3a), we proposed a possible mode of the complexation of Zn^{2+} cation with our basic thiosemicarbazones and semicarbazones. A tentative Zn^{2+} complex of **24-8-1** is an example of such a coordination compound, as seen in Figure 3b.

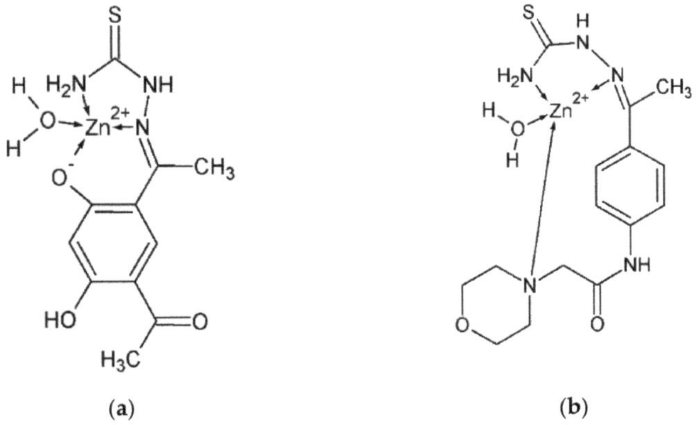

Figure 3. (**a**) Structure of (T-4)-[2-[1-[5-Acetyl-2-(hydroxy-κO)-4-hydroxyphenyl]ethylidene] hydrazinecarbothioamidato(2-)-κN,κN2]aquazinc according to [20], redrawn by ACD/ChemSketch; (**b**) Zn^{2+} complex of **24-8-1**, i.e., 4-[(morpholine-4-yl)acetamido]acetophenone thiosemicarbazone as an example of the possibility of the coordination of the prepared Schiff bases with Zn^{2+}.

The 2^+ charge of such a complex cation must be compensated by the negatively charged carboxylates belonging to the dicarboxylic amino acid residues, i.e., Asp or Glu. Glu320, which together with His297 and His361 takes part in Zn^{2+} complexation in the APN protein, and Glu264 or Glu298, which are nearby in the tertiary structure of the protein [27], can assume this role. Furthermore, when the whole group of Schiff bases with their determined anti-APN activities is separated into two groups—compounds with only one basic nitrogen atom (majority) and those with two basic nitrogens (piperazine derivatives)—the group of piperazines is markedly more active (median IC_{50} = 22.3 μmol/L) than the rest (median IC_{50} = 106.7 μmol/L). This fact could be caused by the possibility of the second nitrogen forming an ionic pair with a free carboxy group of another nearby dicarboxylic amino acid and thus interacting with the protein with a greater affinity. The water molecule, the complexation of which is expected in this model, is also a ligand naturally coordinated to Zn^2 of APN [27] and we suppose that it remains coordinated when the spatial structure of the active site of the enzyme is changed by the binding of our ligand. A comparison of Figure 3a,b also suggests that the activity would benefit from the introduction of a chelating group into the *o*-position and an extension of the linker chain between the carbonyl of the acetamide group and the basic nitrogen to facilitate the formation of a donor–acceptor bond from the nitrogen to the zinc cation. This is the inspiration for our further synthesis of better APN inhibitors.

4. Materials and Methods

4.1. Chemistry

4.1.1. General Information

All chemicals were purchased from commercial suppliers (Sigma-Aldrich, Darmstadt, Germany) and used as supplied without further purification. All reactions were monitored by TLC performed on precoated silica gel 60 F254 plates (Merck, Darmstadt, Germany). For compounds 12-2-1 to 14-14-1, ethyl acetate:hexane:diethylamine = 3:2:1 was used as an eluent, UV light (254 nm). For compounds **22-2-1** to **24-14-1**, petroleum ether: diethylamine = 9:1 was used as eluent, UV light (254 nm), and iodine was used for the detection of spots. NMR spectra were recorded on an FT-NMR ECZR 400 (JEOL, Akishima, Tokyo, Japan) spectrometer using TMS as an internal standard. The FTIR spectra were obtained with a Smart MIRacle™ Nicolet™ Impact 410 FTIR Spectrometer (Thermo Scientific, West Palm. Beach, FL, USA) equipped with the ATR ZnSe module. MS spectra were measured on a Xevo TQ-S triple quadrupole MS spectrometer (Waters, Milford, MA, USA) and analyzed in the positive mode under the formation of $[M-H]^+$ ions. Melting points (uncorrected) were recorded on Kofler's block Büchi Labortechnik AG 535 (BUCHI Labortechnik AG, Flawil, Switzerland). Detailed spectral and other structural data of the prepared compounds can be found in the Supplementary Materials.

4.1.2. General Procedure for the Preparation of **12-1**, **13-1**, and **14-1**

The 2-, 3-, or 4-aminoacetophenone 6.76 g (0.05 mol) was dissolved in 30 mL of acetone. Thereafter, 10 g (0.094 mol) of Na_2CO_3 was added. Then, 7.18 mL (0.09 mol) of 2-chloroacetyl chloride was added to the reaction mixture dropwise. The reaction mixture was stirred for 4 h at room temperature. After the completion of the reaction, 50 mL of 2M, HCl was added to the reaction mixture. The mixture was cooled at 0–5 °C overnight. The precipitate was filtered off, washed with distilled water, and dried to a constant weight.

4.1.3. General Procedure for the Preparation of **12-2-1** to **14-14-1**

The synthetized 2-, 3-, or 4-(chloroacetamido)acetophenone 0.847 g (0.004 mol) was dissolved in 30 mL of acetonitrile. Then, K_2CO_3 1.1 g (0.008 mol) was added to the mixture. Thereafter, 0.0044 mol of appropriate secondary amine was added to the suspension dropwise. The reaction mixture was stirred and refluxed for 4 to 8 h according to the secondary amine used. After the completion of the reaction (monitored by TLC), the reaction mixture was cooled at room temperature, potassium carbonate was filtered off, and the solvent was evaporated under reduced pressure to obtain the crude product. Synthetized 2-, 3-, or 4-[2-(dialkylamino)acetamido]acetophenones were washed with cooled distilled water and a small amount of ethanol and identified.

4.1.4. General Procedure for the Preparation of **22-2-1** to **24-14-1** [28]

2-, 3-, or 4-[2-(dialkylamino)acetamido]acetophenone (0.004 mol) and thiosemicarbazide (0.008 mol) were added to 5 mL of 30% ethanol. The reaction mixture was refluxed for 3 h. After the completion of the reaction (monitored by TLC), the reaction mixture was kept at 0–5 °C overnight. The formed precipitate was then filtered off, washed with cold distilled water, and dried to a constant weight. The final products were recrystallized from ethanol in case of need.

4.1.5. General Procedure for the Preparation of **22-2-2** to **24-14-2** [29]

The thiosemicarbazide (0.01 mol) and the appropriate 2-, 3-, or 4-[2-(dialkylamino) acetamido]acetophenone (0.01 mol) were dissolved in 20 mL of methanol. The reaction mixture was stirred for 10 min at room temperature and then an equivalent amount of 35% hydrochloric acid was added. The reaction mixture was refluxed for 5 h. After the completion of the reaction, the reaction mixture was cooled at 0–5 °C overnight. The precipitate was filtered off and recrystallized from 96% ethanol.

4.1.6. General Procedure for the Preparation of **22-2-3** to **24-14-3** [30]

Thiosemicarbazide (0.01 mol) was added to 15 mL of distilled water and the reaction mixture was stirred and heated at 70 °C for 50 min to dissolve the thiosemicarbazide. Then, 2 mL of 50% perchloric acid was added to the mixture and the mixture was stirred for another 5 min at the same temperature. The solution of 2-, 3-, or 4-[2-(dialkylamino)acetamido]acetophenone (0.011 mol) in 2 mL of ethanol and 13 mL of distilled water was added to the reaction mixture. The mixture was heated and stirred at 85 °C for 3 h. After the completion of the reaction, the reaction mixture was cooled at 0–5 °C overnight. The precipitate was filtered off, washed with a small amount of cooled distilled water, and dried to a constant weight.

4.1.7. General Procedure for the Preparation of **32-2-1** to **34-14-1** [31]

Semicarbazide hydrochloride (0.0025 mol) and sodium acetate (0.0025 mol) were dissolved in 10 mL of 96% ethanol. The mixture was stirred at room temperature for 15 min. Then, 0.0025 mol of synthesized 2-, 3-, or 4-[2-(dialkylamino)acetamido]acetophenone was added to the reaction mixture and the stirring continued at room temperature for the next 12 to 48 h. The reaction progress was monitored by TLC. The final product was filtered off, washed with ethanol, and dried to a constant weight.

4.2. Assessment of APN Inhibitory Activity

IC_{50} values against enzyme APN were determined using L-Leucine-*p*-nitroanilide (Sigma-Aldrich, Darmstadt, Germany) as the substrate and Microsomal Leucine aminopeptidase from porcine kidney EC 3.4.11.2, Type IV-S, ammonium sulfate suspension, 10–40 units/mg protein (Sigma-Aldrich, Darmstadt, Germany). The assay was performed by a Cytation 3 Cell Imaging Multi-Mode Reader (BioTek Instruments, Inc., Winooski, VT, USA) with appropriate Gen-5 software in 96-well plates. A 0.02 mol/L TRIS-HCl buffer solution at pH 7.5 was used as the assay buffer. All tested compounds were dissolved in TRIS-HCl buffer solution and compounds with poor solubility were dissolved in a small amount of DMSO or NMP used as a cosolvent. The assay mixture, which contained a variable amount of inhibitor solution (from 0 up to 80 μL), 10 μL of the enzyme solution, 5 μL of the substrate solution, and the assay buffer adjusted to 200 μL, was incubated at 37 °C for 40 min with short orbital shaking for 10 s. The hydrolysis of the substrate was monitored by the optical photometric method of absorbance in the visible and ultraviolet regions at a wavelength of 405 nm. The enzyme activity inhibitory rate was calculated from the measures of absorbance. The results of the 50% inhibitory activity of the enzyme (IC_{50}) were determined through a regression analysis of the concentration/absorbance data by GraFit 5 software (Erithacus Software Ltd., East Grinstead, UK).

4.3. QSAR Statistic and Parameters Calculations

QCExpert 3.3 (TriloByte Statistical Software, Pardubice, Czech Republic) running onWindows 10 Education was used for the linear and multilinear regression calculations. ACD/ChemSketch was used for the log P values calculation by an algorithm based on >12,000 experimental logP values using the principle of isolating carbons [22]. MarvinSketch 6.2.2 (ChemAxon Ltd., Budapest, Hungary) was the software used for the isoelectric point pI values calculation by the algorithm built into it [23].

4.4. Evaluation of Proliferation Inhibitor Effects

4.4.1. Reagents

All tested compounds were dissolved in dimethyl sulfoxide (DMSO) from Sigma Aldrich. Their fresh solutions were prepared prior to each experiment, whereas the final concentration of DMSO in the assays never exceeded 0.1% (v/v). RPMI 1640 culture medium, phosphate-buffered saline (PBS), fetal bovine serum (FBS), and antibiotics (penicillin and streptomycin) were purchased from HyClone Laboratories, Inc. (GE Healthcare, Logan, UT, USA).

4.4.2. Cell Culture

A human monocytic leukemia cell line (THP-1), human breast adenocarcinoma cells (MCF-7), and human prostate cancer (DU-145) cell line were obtained from ATCC. THP-1 and DU-145 cells were cultivated in an RPMI 1640 culture medium and MCF-7 cells were cultivated in a DMEM medium, both supplemented with the antibiotic solution (100 U/mL of penicillin, 100 μg/mL of streptomycin) and 10% FBS. Cells were maintained in a humidified incubator with 5% CO_2 at 37 °C and were regularly tested for the presence of mycoplasma contamination.

4.4.3. WST-1 Analysis of Cell Proliferation

THP-1, MCF-7, and DU-145 cells were seeded in 96-well plates. Adherent cell lines were allowed to attach to the wells for 24 h. Cells were then treated with various concentrations of tested compounds to reach the final concentrations ranging between 1 μM and 100 μM and were incubated for 48 h. Cell proliferation was determined using Cell Proliferation Reagent WST-1 (2-(4-iodophenyl)-3-(4-nitrophenyl)-5-(2,4-disulfophenyl)-2H-tetrazolium) (Roche Diagnostics, Mannheim, Germany), as previously described [32,33]. WST-1 analysis was performed in three independent experiments, with each condition tested in triplicate. The IC_{50} values were determined using the nonlinear regression four-parameter logistic model using GraphPad Prism 5.00 software (GraphPad Software, San Diego, CA, USA). Statistical significance between the values was assessed by one-way analysis of variance (ANOVA) paired with Dunnett's post hoc test using GraphPad Prism 5.00 software (GraphPad Software, San Diego, CA, USA) at levels of * $p < 0.05$, ** $p < 0.01$, and *** $p < 0.001$.

5. Conclusions

A series of 28 novel compounds based on the structure of 2-, 3-, or 4-[2-amino(acetamido)] acetophenone, where the amino group is a part of either a dialkylamino group or a saturated nitrogenous heterocycle, was synthesized. Twenty-two compounds from this series were tested for APN inhibitory activity. **24-11-2**, 4-[2-(4-methylpiperazine-1-yl)acetamido] acetophenone thiosemicarbazone hydrochloride with an IC_{50} of 13.3 μmol/L was the most active of them. A QSAR study with the semicarbazones and thiosemicarbazones revealed the relationship between the activity on lipophilicity expressed as logP and the acido-basic behavior of the compounds expressed as isoelectric point pI. Equation (1) describing this relationship enabled them to predict the activities of 33 other members of the semicarbazone and thiosemicarbazone series, 6 of which had already been prepared, and the remaining 27 were only proposed for synthesis. The compound with the best-calculated activity was **22-4-1**, 2-[(diethylamino)acetamido]acetophenone thiosemicarbazone with an IC_{50} of 3.4. The complexation of the Zn^{2+} cation of the active site of APN was proposed as the probable mechanism of activity, based on the similarity with other semicarbazones and thiosemicarbazones, which have served as ligands for the synthesis of transition metal complexes including those with Zn^{2+} as the central ion [16,17]. Five compounds with the best values of IC_{50} for APN (ranging between

13 and 23.5 μmol/L) underwent testing for inhibition of cell proliferation on three different cell lines that differ from each other in their levels of APN expression. All five compounds triggered a significant antiproliferative effect in the cell lines expressing APN, THP-1, and MCF-7, whereas in the cell line DU-145 with no APN expression, four of these five compounds, **24-11-2**, 4-[2-(4-methylpiperazine-1-yl)acetamido]acetophenone thiosemicarbazone hydrochloride, **34-6-1**, 4-[2-(piperidine-1-yl)acetamido]acetophenone semicarbazone, **24-2-3**, 4-[2-(diethylamino)acetamido]acetophenone thiosemicarbazone perchlorate, and **24-5-3**, 4-[2-(pyrrolidine-1-yl)acetamido]acetophenone thiosemicarbazone perchlorate, did not affect proliferation at all. The remaining compound **24-10-3**, 4-[2-(4-benzylpiperazine-1-yl)acetamido]acetophenone thiosemicarbazone perchlorate also inhibited DU-145 cell proliferation but less than in the APN-positive THP-1 or MCF-7 lines. These results suggest that the antiproliferative activity is linked with APN inhibition, although other mechanisms can also participate in it. Furthermore, our semicarbazones and thiosemicarbazones are the first compounds of these structural types of Schiff bases that were reported to inhibit not only a zinc-dependent aminopeptidase of the M1 family but also a metalloenzyme. The results, including Equation 2, can enable the proposal and synthesis of highly active APN inhibitors, which could serve as potential anticancer or antiviral drugs, which could contribute to overcoming the resistance of cancers to contemporary treatments.

Supplementary Materials: The following supporting information can be downloaded at: https://www.mdpi.com/article/10.3390/ijms23179813/s1.

Author Contributions: Conceptualization, O.F. and V.B.; methodology, O.F., T.K., P.K. and P.Z.; formal analysis, O.F. and T.K.; investigation, V.B., R.Ž., T.K., O.F. and P.Z.; data curation, T.K.; writing—original draft preparation, O.F., P.K. and V.B.; writing—review and editing, O.F.; visualization, O.F.; supervision, O.F.; project administration, O.F.; funding acquisition, V.B. and O.F. All authors have read and agreed to the published version of the manuscript.

Funding: This research was funded by Masaryk University, grant numbers MUNI/IGA/0932/2021 and MUNI/A/1682/2020.

Institutional Review Board Statement: Not applicable.

Informed Consent Statement: No applicable.

Data Availability Statement: The data presented in this study are only available in this article and its Supplementary Materials.

Acknowledgments: Authors are grateful to Kristián Pršo from the Laboratory of Pharmacokinetics, Jessenius Medical Faculty in Martin, Comenius University in Bratislava, for the measurement and interpretation of MS spectra. The authors would like to thank Karel Souček (Institute of Biophysics, the Czech Academy of Sciences, Brno) for his generous gift of the DU 145 cell lines. The authors are also grateful to Aleš Kroutil, Department of Chemical Drugs, Faculty of Pharmacy, Masaryk University, for consultations concerning metal chelates.

Conflicts of Interest: The authors declare no conflict of interest. The funders had no role in the design of the study; in the collection, analyses, or interpretation of data; in the writing of the manuscript, or in the decision to publish the results.

References

1. Mina-Osorio, P. The moonlighting enzyme CD13: Old and new functions to target. *Trends Mol. Med.* **2008**, *14*, 362–371. [CrossRef]
2. Li, Z.; Tomlinson, A.C.A.; Wong, A.H.M.; Zhou, D.; Desforges, M.; Talbot, P.J.; Benlekbir, S.; Rubinstein, J.L.; Rini, J.M. The human coronavirus HCoV-229E S-protein structure and receptor binding. *eLife* **2019**, *8*, e51230. [CrossRef]
3. Kis, A.; Dénes, N.; Szabó, J.P.; Arató, V.; Beke, L.; Matolay, O.; Enyedi, K.N.; Méhes, G.; Mező, G.; Bai, P.; et al. In Vivo Molecular Imaging of the Efficacy of Aminopeptidase N (APN/CD13) Receptor Inhibitor Treatment on Experimental Tumors Using 68Ga-NODAGA-c(NGR) Peptide. *Bio. Med. Res. Int.* **2021**, *2021*, 6642973. [CrossRef]
4. Zhang, Q.; Wang, J.; Zhang, H.; Zhao, D.; Zhang, Z.; Zhang, S. Expression and clinical significance of aminopeptidase N/CD13 in non-small cell lung cancer. *J. Cancer Res. Ther.* **2015**, *11*, 223–228. [CrossRef]

5. Węglarz-Tomczak, E.; Talma, M.; Giurg, M.; Westerhoff, H.V.; Janowski, R.; Mucha, A. Neutral metalloaminopeptidases APN and MetAP2 as newly discovered anticancer molecular targets of actinomycin D and its simple analogs. *Oncotarget* **2018**, *9*, 29365–29378. [CrossRef]
6. Ichinose, Y.; Genka, K.; Koike, T.; Kato, H.; Watanabe, Y.; Mori, T.; Iioka, S.; Sakuma, A.; Ohta, M. Randomized double-blind placebo-controlled trial of bestatin in patients with resected stage I squamous-cell lung carcinoma. *J. Natl. Cancer Inst.* **2003**, *95*, 605–610. [CrossRef]
7. Scornik, O.A.; Botbol, V. Bestatin as an experimental tool in mammals. *Curr. Drug Metab.* **2001**, *2*, 67–85. [CrossRef] [PubMed]
8. Shuai, L.; Fang, X.; Hafeng, W.; Zheng, L.; Xiaowen, L.; Liang, S.; Zhihong, N. Ubenimex inhibits cell proliferation, migration, and invasion in renal cell carcinoma: The effect is autophagy-associated. *Oncol. Rep.* **2015**, *33*, 1372–1380. [CrossRef]
9. Wang, X.; Niu, Z.; Jia, Y.; Cui, M.; Han, L.; Zhang, Y.; Liu, Z.; Bi, D.; Liu, S. Ubenimex inhibits cell proliferation, migration, and invasion by inhibiting the expression of APN and inducing autophagic cell death in prostate cancer cells. *Oncol. Rep.* **2016**, *35*, 2121–2130. [CrossRef] [PubMed]
10. Liu, S.; Gao, M.; Wang, X.; Ding, S.; Lv, J.; Gao, D.; Wang, Z.; Niu, Z. Ubenimex attenuates acquired sorafenib resistance in renal cell carcinoma by inhibiting Akt signaling in a lipophagy associated mechanism. *Oncotarget* **2016**, *7*, 79141–79153. [CrossRef]
11. DiNardo, C.D.; Cortes, J.E. Tosedostat for the treatment of relapsed and refractory acute myeloid leukemia. *Expert Opin. Investig. Drugs* **2013**, *23*, 265–272. [CrossRef]
12. US National Library of Medicine. ClinicalTrials.gov. Available online: https://clinicaltrials.gov/ct2/results?intr=Apo2L%2FTRAIL+OR+dulanermin&draw=2&rank=5#rowId4 (accessed on 12 January 2022).
13. Ni, J.; Wang, X.; Shang, Y.; Li, Y.; Chen, S. CD13 inhibition augments DR4-induced tumor cell death in a p-ERK1/2-independent manner. *Cancer Biol. Med.* **2021**, *18*, 570–586. [CrossRef]
14. Amin, S.A.; Adhikari, N.; Jha, T. Design of Aminopeptidase N Inhibitors as Anti-cancer Agents. *J. Med. Chem.* **2018**, *61*, 6468–6490. [CrossRef]
15. Ma, C.; Jin, K.; Cao, J.; Zhang, L.; Li, X.; Xu, W. Novel leucine ureido derivatives as inhibitors of aminopeptidase N (APN). *Bioorg. Med. Chem.* **2013**, *21*, 1621–1627. [CrossRef]
16. Ma, C.; Cao, J.; Liang, X.; Huang, Y.; Wu, P.; Li, Y.; Xu, W.; Zhang, Y. Novel leucine ureido derivatives as aminopeptidase N inhibitors. Design, synthesis, and activity evaluation. *Eur. J. Med. Chem.* **2016**, *108*, 21–27. [CrossRef]
17. Zhang, X.; Zhang, L.; Zhang, J.; Feng, J.; Yuan, Y.; Fang, H.; Xu, W. Design, synthesis, and preliminary activity evaluation of novel 3-amino-2-hydroxyl-3-phenylpropanoic acid derivatives as aminopeptidase N/CD13 inhibitors. *J. Enzym. Inhib. Med. Chem.* **2013**, *28*, 545–551. [CrossRef]
18. Schalk, C.; d'Orchymont, H.; Jauch, M.F.; Tarnus, C. 3-Amino-2-tetralone derivatives: Novel potent and selective inhibitors of aminopeptidase-M (EC 3.4.11.2). *Arch. Biochem. Biophys.* **1994**, *311*, 42–46. [CrossRef]
19. Reddy, P.S.; Satyanarayana, B.; Raju, V.J. Synthesis and structural studies on divalent transition metal complexes of 5-acetyl 2,4-dihydroxy acetophenone semicarbazone. *J. Indian Chem. Soc.* **2006**, *83*, 1204–1207.
20. Reddy, P.S.; Satyanarayana, B. Synthesis and structural studies on the transition metal complexes of 5-acetyl 2,4-dihydroxy acetophenone thiosemicarbazone. *Acta Cienc. Indica Chem.* **2006**, *32*, 311–315.
21. Xiong, X.; Barathi, A.; Beuermann, R.W.; Tan, D.T.H. Assay of leucine aminopeptidase activity in vitro using large-pore reversed-phase chromatography with fluorescence detection. *J. Chromatogr. B* **2003**, *796*, 63–70. [CrossRef]
22. Partition Coefficient Calculation with ACD/LogP. Available online: https://www.acdlabs.com/products/percepta/predictors/logP (accessed on 7 December 2021).
23. Isoelectric Point Plugin. Available online: https://docs.chemaxon.com/display/docs/isoelectric-point-plugin.md#src-1806643-isoelectricpointplugin-introduction (accessed on 7 December 2021).
24. Licona-Limón, I.; Garay-Canales, C.A.; Muñoz-Paleta, O.; Ortega, E. CD13 mediates phagocytosis in human monocytic cells. *J. Leukoc. Biol.* **2015**, *98*, 85–98. [CrossRef]
25. Gai, Y.; Jiang, Y.; Long, Y.; Sun, L.; Liu, S.; Qin, C.; Zhang, Y.; Zeng, D.; Lan, X. Evaluation of an Integrin $\alpha v\beta 3$ and Aminopeptidase N Dual-Receptor Targeting Tracer for Breast Cancer Imaging. *Mol. Pharm.* **2020**, *17*, 349–358. [CrossRef]
26. Joshi, S.; Chen, L.; Winter, M.B.; Lin, Y.L.; Yang, Y.; Shapovalova, M.; Smith, P.M.; Liu, C.; Li, F.; LeBeau, A.M. The Rational Design of Therapeutic Peptides for Aminopeptidase N using a Substrate-Based Approach. *Sci. Rep.* **2017**, *7*, 1424. [CrossRef]
27. Ito, K.; Nakajima, Y.; Onohara, Y.; Takeo, M.; Nakashima, K.; Matsubara, F.; Ito, T.; Yoshimoto, T. Crystal structure of aminopeptidase N (proteobacteria alanyl aminopeptidase) from Escherichia coli and conformational change of methionine 260 involved in substrate recognition. *J. Biol. Chem.* **2006**, *281*, 3364–3376. [CrossRef]
28. Zubáč, P.; (Masaryk University, Brno, South Moravia, Czech Republic). Personal communication, 2020.
29. Amishro, C.; Yoji, I.; Junichiro, Y.; Toshiyuki, A.; Ryuchiro, N.; Tomohisa, N. Thiazolidine derivatives. U.S. Patent US2007112044 A1, 17 May 2007.
30. Nakhamovich, A.S.; Elokhina, G.V.; Dolgusin, G.V.; Gushin, A.S.; Poljakov, R.A.; Volkova, K.A.; Punija, V.S. Method for preparing 4-thioureidoiminimethyl-pyridinium perchlorate possessing tuberculostatic activity. RU Patent RU2265014 C1, 27 November 2005.
31. Pizzo, C.; Farral-Tello, P.; Yaluff, G.; Serna, E.; Torres, S.; Vera, N.; Saiz, C.; Robello, C.; Mahler, G. New approach towards the synthesis of selenosemicarbazones, useful compounds for Chagas' disease. *Eur. J. Med. Chem.* **2016**, *109*, 107–113. [CrossRef]

32. Kollar, P.; Barta, T.; Zavalova, V.; Smejkal, K.; Hampl, A. Geranylated flavanone tomentodiplacone B inhibits proliferation of human monocytic leukaemia (THP-1) cells. *Br. J. Pharmacol.* **2011**, *162*, 1534–1541. [CrossRef]
33. Kauerova, T.; Kos, J.; Gonec, T.; Jampilek, J.; Kollar, P. Antiproliferative and Pro-Apoptotic Effect of Novel Nitro-Substituted Hydroxynaphthanilides on Human Cancer Cell Lines. *Int. J. Mol. Sci.* **2016**, *17*, 1219. [CrossRef]

International Journal of *Molecular Sciences*

MDPI

Article

Ketoconazole Reverses Imatinib Resistance in Human Chronic Myelogenous Leukemia K562 Cells

Omar Prado-Carrillo [1,2], Abner Arenas-Ramírez [1], Monserrat Llaguno-Munive [1], Rafael Jurado [1], Jazmin Pérez-Rojas [1], Eduardo Cervera-Ceballos [3] and Patricia Garcia-Lopez [1,*]

1 Laboratorio de Fármaco-Oncología, Subdirección de Investigación Básica, Instituto Nacional de Cancerología, Ciudad de Mexico 14080, Mexico; oprado@live.com.mx (O.P.-C.); alex_abner.ar@ciencias.unam.mx (A.A.-R.); muniv1250@hotmail.com (M.L.-M.); fcojl@yahoo.com (R.J.); jazminmarlen@gmail.com (J.P.-R.)

2 Posgrado en Ciencias Bioquímicas, Universidad Nacional Autónoma de México (UNAM), Ciudad de Mexico 04510, Mexico

3 Dirección de Docencia, Instituto Nacional de Cancerología, Ciudad de Mexico 14080, Mexico; eduardocer@yahoo.com

* Correspondence: pgarcia_lopez@yahoo.com.mx; Tel.: +52-5536935200

Abstract: Chronic myeloid leukemia (CML) is a hematologic disorder characterized by the oncogene BCR-ABL1, which encodes an oncoprotein with tyrosine kinase activity. Imatinib, a BCR-ABL1 tyrosine kinase inhibitor, performs exceptionally well with minimal toxicity in CML chemotherapy. According to clinical trials, however, 20–30% of CML patients develop resistance to imatinib. Although the best studied resistance mechanisms are BCR-ABL1-dependent, P-glycoprotein (P-gp, a drug efflux transporter) may also contribute significantly. This study aimed to establish an imatinib-resistant human CML cell line, evaluate the role of P-gp in drug resistance, and assess the capacity of ketoconazole to reverse resistance by inhibiting P-gp. The following parameters were determined in both cell lines: cell viability (as the IC50) after exposure to imatinib and imatinib + ketoconazole, P-gp expression (by Western blot and immunofluorescence), the intracellular accumulation of a P-gp substrate (doxorubicin) by flow cytometry, and the percentage of apoptosis (by the Annexin method). In the highly resistant CML cell line obtained, P-gp was overexpressed, and the level of intracellular doxorubicin was low, representing high P-gp activity. Imatinib plus a non-toxic concentration of ketoconazole (10 μM) overcame drug resistance, inhibited P-gp overexpression and its efflux function, increased the intracellular accumulation of doxorubicin, and favored greater apoptosis of CML cells. P-gp contributes substantially to imatinib resistance in CML cells. Ketoconazole reversed CML cell resistance to imatinib by targeting P-gp-related pathways. The repurposing of ketoconazole for CML treatment will likely help patients resistant to imatinib.

Keywords: chronic myeloid leukemia; imatinib; tyrosine kinase; ketoconazole; P-glycoprotein; drug efflux transporter

Citation: Prado-Carrillo, O.; Arenas-Ramírez, A.; Llaguno-Munive, M.; Jurado, R.; Pérez-Rojas, J.; Cervera-Ceballos, E.; Garcia-Lopez, P. Ketoconazole Reverses Imatinib Resistance in Human Chronic Myelogenous Leukemia K562 Cells. *Int. J. Mol. Sci.* **2022**, *23*, 7715. https://doi.org/10.3390/ijms23147715

Academic Editor: Angela Stefanachi

Received: 16 May 2022
Accepted: 6 July 2022
Published: 13 July 2022

1. Introduction

Chronic myeloid leukemia (CML), also known as chronic granulocytic leukemia, is a myeloproliferative disorder. It is characterized by neoplastic growth of myeloid cells in the bone marrow, leading to a significant increase of these cells in peripheral blood [1]. CML is traditionally described as a triphasic disease, beginning with the chronic phase and progressing to the accelerated phase, and finally to the blast phase. Chemotherapy given at the early chronic phase usually restores the patient to a normal-like state, which can be sustained for months or years. Nevertheless, without medical treatment or a lack of response to treatment, patients gradually progress to blast crisis [1,2].

The disease has its origins in the formation of the BCR-ABL1 gene, which results from the reciprocal translocation between chromosomes 9 and 22 and the fusion of the ABL1 and BCR genes to create a short chromosome called Philadelphia chromosome. The fused BCR-ABL1 gene codes for an abnormal oncoprotein with tyrosine kinase activity (Bcr-Abl) and with auto-phosphorylation capacity. Bcr-Abl activates multiple signaling pathways that cause the abnormal proliferation of hematopoietic stem cells (HSC) and thus the manifestation of the disease [3,4]. The BCR-ABL gene is present in all cases of CML. It provides a unique biomarker for diagnosis and is targeted during treatment with tyrosine kinase inhibitors (TKIs) to selectively inhibit Bcr-Abl. The first TKI, imatinib mesylate, became the basis of therapy for CML, transforming this disease from a fatal to a chronic one [5].

In 1996, Druker et al. [6] reported the in vitro effects of a specific inhibitor of the BCR-ABL tyrosine kinase on CML cell lines for the first time. This inhibitor was known as signal transduction inhibitor (STI571) but is now called imatinib (Gleevec®). In a phase 1 study of the advanced stage of the disease, STI571 not only controlled blood counts and restored the chronic phase, but 95% of patients achieved a complete hematologic response and a 60% greater cytogenetic response. Despite the short follow-up period existing at that time, imatinib was granted accelerated approval by the FDA in 2001 due to its exceptional efficacy and minimal toxicity [7]. It was established as the first-line treatment for CML [8,9].

The drug occupies the ATP-binding site on the BCR-ABL protein. The resulting conformational change in the tyrosine kinase quaternary structure inhibits autophosphorylation and phosphorylation of tyrosine residues on protein substrates. Thus, imatinib prevents the transduction of signals crucial for the abnormal and uncontrolled cell proliferation caused by the BCR-ABL gene in CML cells [10].

Since the bioavailability of imatinib is around 92% (86–99%) with a half-life of 18 h in healthy volunteers and patients [11], one dose/day seems to be appropriate. The drug is extensively metabolized by cytochrome P450 enzymes (CYP-P450). Additionally, it is a substrate of the drug transporter denominated P-glycoprotein (P-gp or MDR1 [multidrug resistance 1]), an ATP-dependent efflux pump that decreases intracellular drug concentrations [12]. Hence, P-gp influences drug absorption, distribution, metabolism, and excretion (ADME). Even with the high bioavailability of imatinib, its pharmacokinetics show a great variability in the responses of individuals, which are affected by both CYP-P450 and the P-gp efflux transporter [13].

The surprising efficacy of imatinib in CML is attenuated by resistance in a percentage of patients with advanced-stage CML [14–16]. The known mechanisms of resistance to imatinib can be divided into those BCR-ABL-dependent and BCR-ABL-independent. BCR-ABL-dependent mechanisms include mutations in the ABL kinase and/or mutations and amplification of the BCR-ABL gene. Among BCR-ABL-independent mechanisms is drug efflux mediated by ATP-binding cassette (ABC) transporters [17].

Specifically, it has been reported that the P-gp protein (encoded by the ABCB1 gene) may contribute considerably to resistance to imatinib. P-gp is capable of diminishing the intracellular concentration of imatinib by pumping it out of leukemia cells [17,18]. This efflux pump is located in normal human tissue in the liver, kidney, colon, adrenal gland, intestine, placenta, endothelial cells of the blood-brain barrier, and hematopoietic precursor cells. However, its expression is significantly elevated in drug-resistant tumors, making it an obstacle to successful chemotherapy. Therefore, a reduction of the level of P-gp by inhibitors could lead to the sensitization of CML resistant cells to imatinib and therefore the avoidance of resistance.

Numerous studies have identified various competitive substrates and inhibitors of P-gp, allowing for a greater understanding of the regulation of P-gp functions to overcome drug resistance. Several of these compounds are drugs originally approved for clinical indications unrelated to cancer [19], as is the case with ketoconazole. In 2004, Dutreix et al. described an increase in the plasma concentration of imatinib in healthy volunteers who had taken ketoconazole. The mechanisms involved were the blocking of imatinib metabolism by CYP3A4 and the inhibition of its extrusion from gastrointestinal cells by impeding P-gp activity [20]. To date, however, there have been no reports on the capacity of ketoconazole to reverse the resistance of CML cells to imatinib treatment. Thus, the purpose of the current study was first to generate an in vitro model of CML cells resistant to imatinib, and secondly to explore the participation of the drug efflux transporter P-gp in such resistance. Subsequently, an evaluation was made of the capacity of ketoconazole to overcome resistance to imatinib through the inhibition of P-gp. Finally, the possible role of ketoconazole in triggering the apoptosis of imatinib-resistant cells was assessed.

2. Results

2.1. The Development of CML K562 Cell Resistance to Imatinib

By treating the K562 parent cell line with gradually increasing concentrations of imatinib (ranging from 1 to 2500 nM), the resistant phenotype was developed in about 8 months. Drug resistance was confirmed by a cell viability assay based on a 72 h exposure of sensitive (K562) and resistant (K562-RI) cells to imatinib. A clear difference between the two cell viability curves was observed, with the graph displaying a lesser effect of imatinib on the K562-RI cell line.

The IC50 of imatinib was determined for the K562-RI and K562 cell lines, finding the values of 2544 nM and 213 nM, respectively. Hence, there was an approximately 12-fold relative resistance, calculated as the ratio of the IC50 values of resistant and sensitive cells (Figure 1A).

2.2. The Expression of P-Glycoprotein in Resistant and Susceptible CML Cells

After confirming the resistant phenotype, the expression of P-gp was examined by Western blot, finding it to be around 4-fold greater in K562-RI than K562 cells (Figure 1B), correlating with the higher level of P-gp in resistant cells observed by immunofluorescence (Figure 1C). Also indicating a higher level of P-gp in resistant cells was the test with doxorubicin, which showed a lower intracellular accumulation of this compound in resistant than susceptible CML cells, as can be appreciated by the values of relative mean fluorescence intensity (RMFI) (Figure 1D).

2.3. Effect of Ketoconazole on K562 Cells (Sensitive to Imatinib)

The application of imatinib brought about a dose-dependent antiproliferative effect in K562 cells (Figure 2A), which was unmodified by the combination treatment with ketoconazole at 0.1 and 1.0 μM (Figure 2). However, significantly increased antiproliferative activity was detected with imatinib plus ketoconazole at 10 μM (Figure 2A). In this cell line, the IC50 of imatinib monotherapy was 232 nM, while that of imatinib co-incubated with ketoconazole at 10 μM was 150 nM, translating into a 0.65-fold reduction. According to the quantification of intracellular doxorubicin in sensitive cells, no change took place with ketoconazole at 10 or 20 μM (Figure 2B).

2.4. Ketoconazole Induced a Reversal of the Resistance of K562-RI Cells to Imatinib

As with K562 cells, the antiproliferative effect of imatinib on K562-RI cells was not improved by co-treatment with ketoconazole at 0.1 or 1.0 μM. However, a reversal of drug resistance in K562-RI cells was produced by applying imatinib plus ketoconazole at 10 μM (Figure 3A), resulting in an IC50 value of 186 nM. Considering the IC50 value of 1378 nM found after applying imatinib alone to K562-RI cells, the combination treatment afforded an approximately 7.5-fold reversal index.

2.5. Effect of Ketoconazole on P-Glycoprotein Expression in Resistant Cells

The Western blot (Figure 4A) and immunofluorescence assay (Figure 4B) performed with K562-RI cells demonstrated that imatinib alone (at 200 nM) and ketoconazole alone at 10 μM slightly reduced the expression of P-gp, while significantly diminished the level of P-gp, with imatinib plus ketoconazole causing a greater decrease (Figure 4).

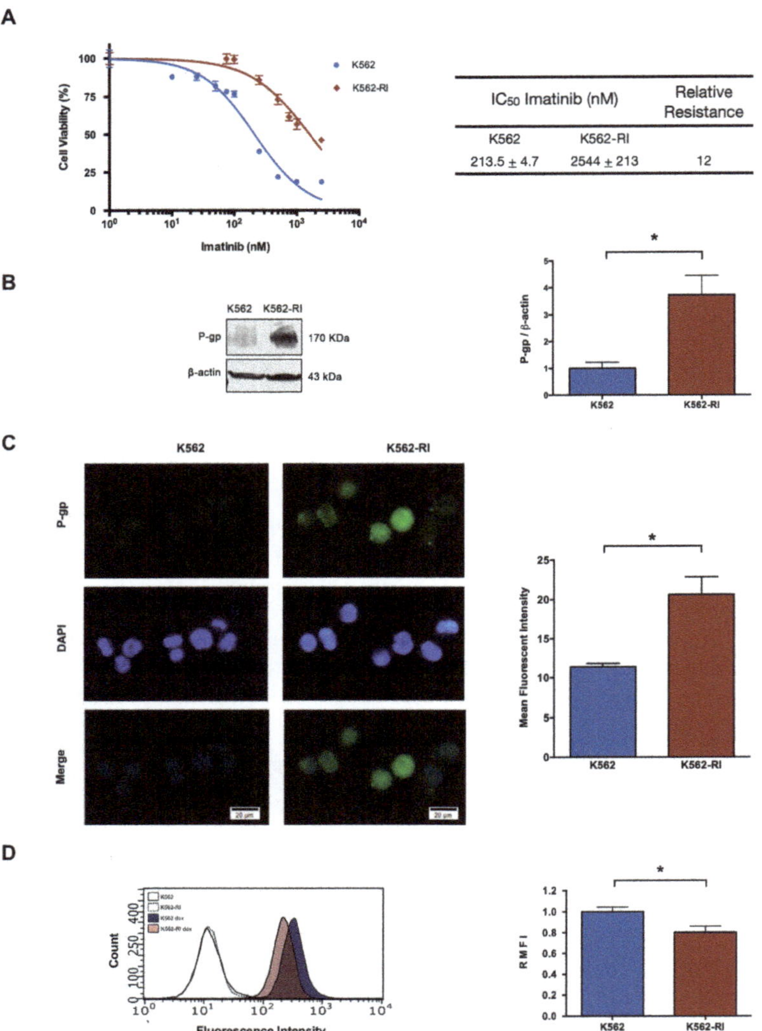

IC50 Imatinib (nM)		Relative Resistance
K562	K562-RI	
213.5 ± 4.7	2544 ± 213	12

Figure 1. The development of a CML cell line with resistance to imatinib treatment. (**A**) Percentage of cell viability after applying imatinib alone (at distinct concentrations), considering CML cells sensitive (K562) and resistant (K562-RI) to this drug. The table denotes the IC50 of imatinib in each cell line and the relative resistance of K562-RI (calculated as the IC50 of K562-RI divided by that of K562). (**B**) P-gp expression levels in the K562 and K562-RI cell lines, based on Western blot and its densitometric analysis (β-actin was the load control); three Western blots from three independent experiments were used for densitometric analysis. (**C**) Analysis of P-gp evaluated by the immunofluorescence assay (the nuclei were visualized with DAPI). The mean fluorescence intensity was quantified by counting P-gp-positive cells from three independent experiment. Scale bars: 20 μm. (**D**) Representative histograms of the fluorescence of uptake of doxorubicin in the K562 and K562-RI cell lines. The empty histograms depict the control cells without doxorubicin. Data are expressed as the relative mean fluorescence intensity (RMFI) ± SEM of three independent experiments. The values of k562-RI were normalized with respect to K562. Statistical analysis was performed by comparing K562-RI to the parental cell line. * $p < 0.05$; Student's *t*-test.

A

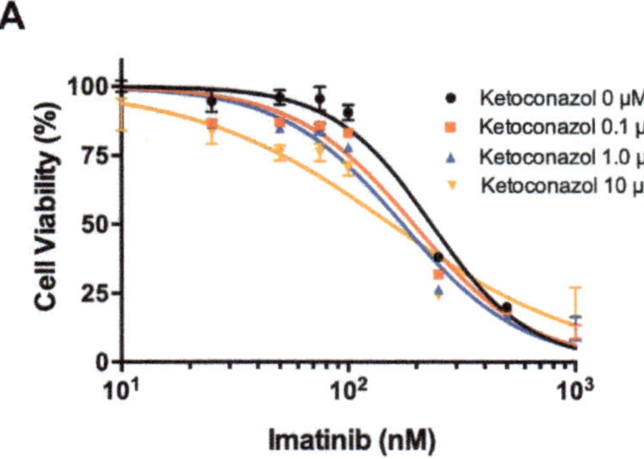

Treatment	IC_{50} Imatinib (nM)
Ktz 0 µM	232.1 ± 9.7
Ktz 0.1 µM	194.5 ± 1.1
Ktz 1.0 µM	173.8 ± 3.6
Ktz 10 µM	150.1 ± 11.6 *

B

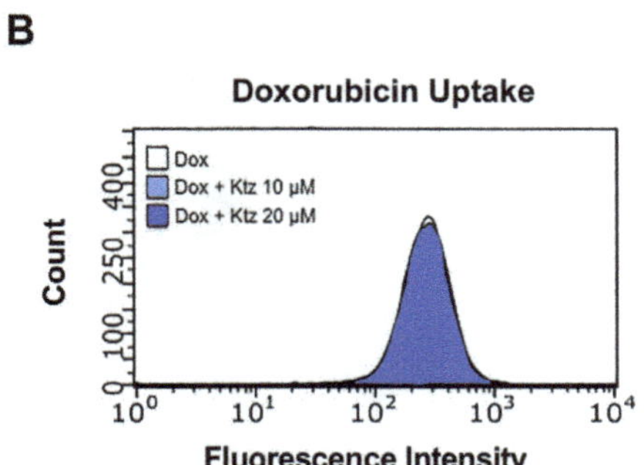

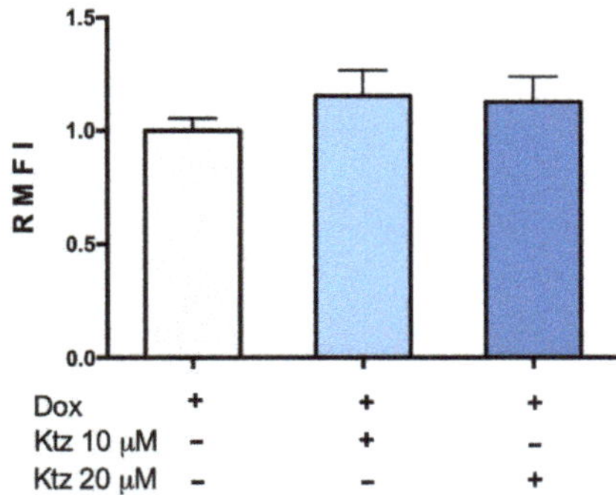

Figure 2. Effect of ketoconazole on K562 cells (sensitive to imatinib). (**A**) Cell viability of the K562 cell line exposed to imatinib in the absence or presence of ketoconazole (at 0.1, 1.0, and 10 µM). The table shows the IC50 of imatinib alone and imatinib plus ketoconazole. (**B**) Effect of ketoconazole (at 10 and 20 µM) on the fluorescence of doxorubicin in K562 cells, representative histograms with raw data and in a bar graph. Data are expressed as the relative mean fluorescence intensity (RMFI) ± SEM of three independent experiments. The cells treated with doxorubicin plus ketoconazole 10 and 20 µM were normalized against doxorubicin alone. Ktz, ketoconazole; Dox, doxorubicin. * Significant difference ($p < 0.05$) by one-way analysis of variance (ANOVA) followed by Tukey's test.

2.6. Effect of Ketoconazole on Apoptosis in Resistant Cells

Since imatinib plus ketoconazole at 10 µM significantly inhibited the viability of K562-RI cells, apoptosis was explored as a possible mechanism of action. Externalization of phosphatidylserine to the outer surface of the plasma membrane is a clear sign of early apoptosis. K562-RI cells were stained with Annexin V-FITC (early apoptosis) and PI (late apoptosis) for flow cytometric analysis, which revealed the lack of any significant change in the percentage of apoptosis produced by imatinib alone or ketoconazole alone (versus the resistant control cells exposed to the vehicle only). However, there was indeed a significantly greater programmed cell death of resistant cells when using imatinib plus ketoconazole (~40%). These data suggest that exposure of cells to ketoconazole triggered apoptosis (Figure 5).

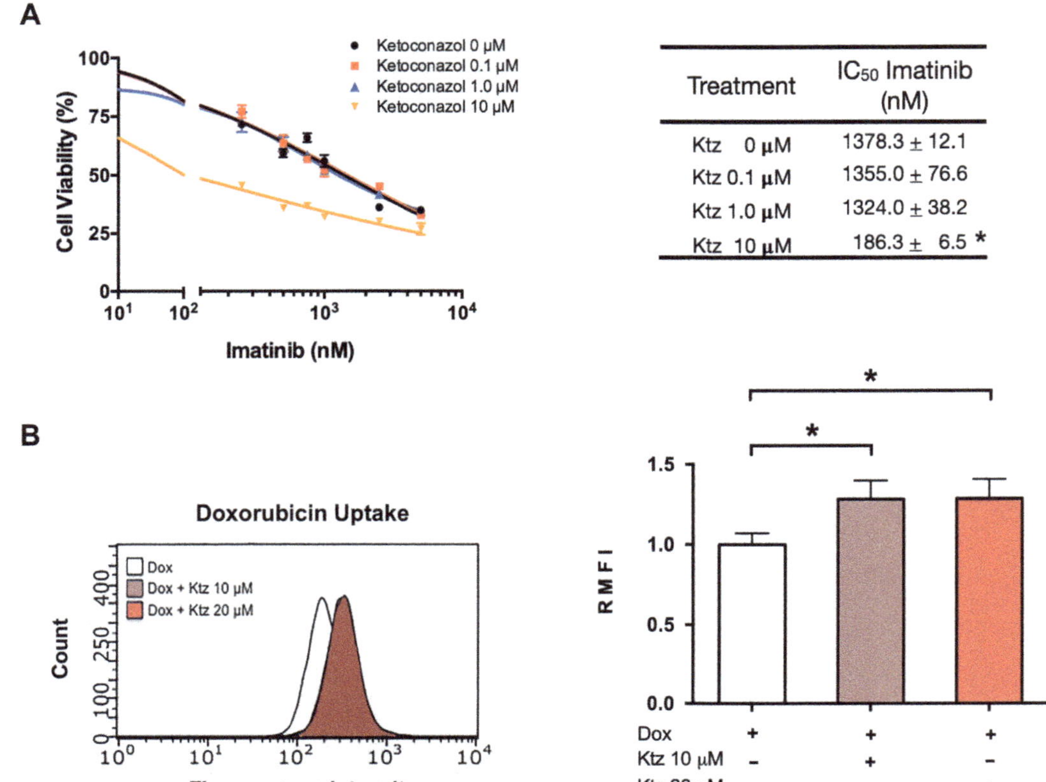

Treatment	IC50 Imatinib (nM)
Ktz 0 μM	1378.3 ± 12.1
Ktz 0.1 μM	1355.0 ± 76.6
Ktz 1.0 μM	1324.0 ± 38.2
Ktz 10 μM	186.3 ± 6.5 *

Figure 3. Ketoconazole-induced reversal of K562-RI cell resistance to imatinib. (**A**) Cell viability of the resistant cell line (K562-RI) exposed to imatinib in the absence and presence of ketoconazole (at 0.1, 1.0 and 10 μM). The table shows the IC50 of imatinib alone and imatinib plus ketoconazole. (**B**) Effect of ketoconazole on the fluorescence of doxorubicin in K562-RI cells. The bar graph portrays the relative fluorescence of doxorubicin in the presence of 10 or 20 μM of ketoconazole. Data are expressed as the relative mean fluorescence intensity (RMFI) ± SEM of three independent experiments. The cells treated with doxorubicin plus ketoconazole 10 and 20 μM were normalized against doxorubicin alone. Statistical analysis was performed by comparing the level of intracellular doxorubicin between the treatment with doxorubicin plus ketoconazole (10 or 20 μM) and doxorubicin alone. * $p < 0.05$; determined with ANOVA followed by Tukey's test. Im, imatinib; Ktz, ketoconazole; Dox, doxorubicin.

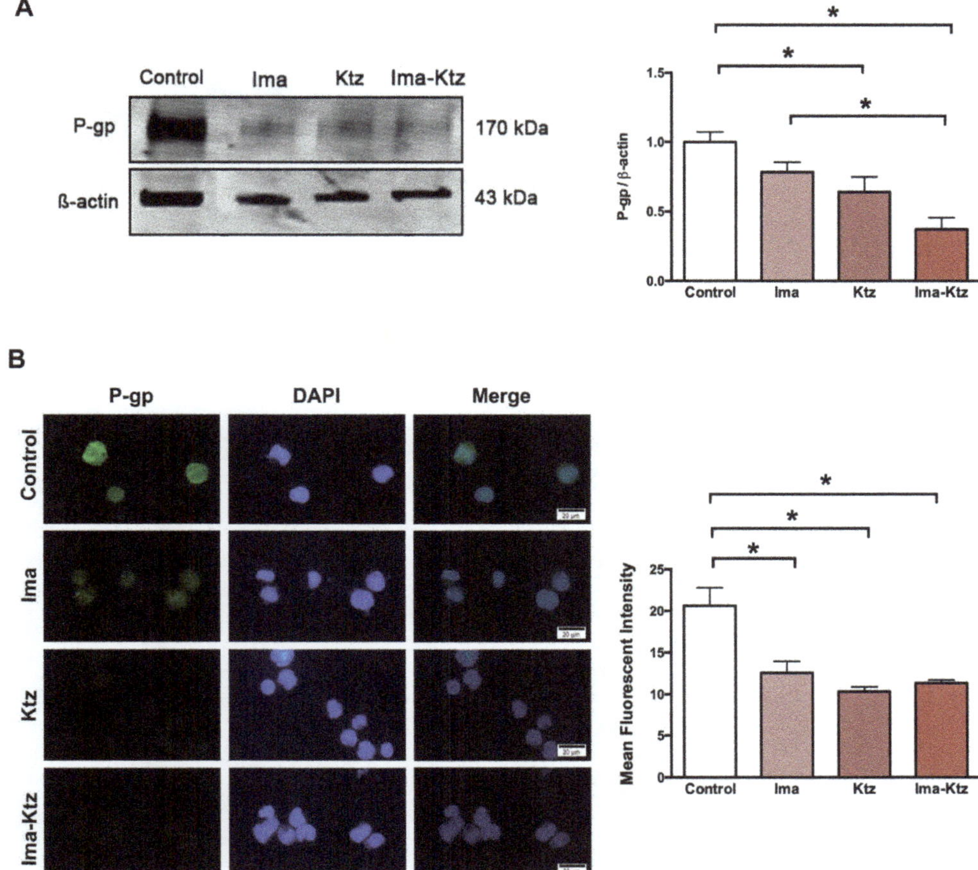

Figure 4. P-gp expression in resistant cells treated with imatinib and ketoconazole. (**A**) P-gp expression levels in K562-RI cells exposed to imatinib in the absence and presence of ketoconazole based on Western blot and its densitometric analysis (β-actin was the load control); three Western blots from three independent experiments were used for densitometric analysis ± SEM. (**B**) analysis of P-gp evaluated by the immunofluorescence assay (DAPI was used to visualize the nuclei). The Mean Fluorescent Intensity was quantified by counting P-gp-positive cells (in 40–45 cells per group counted randomly) from three independent experiments. The cells K562-RI cells were exposed to imatinib alone (200 nM), ketoconazole alone (10 μM), or imatinib plus ketoconazole (200 nM/10 μM). * $p < 0.05$; determined with ANOVA followed by Tukey's test. Scale bars: 20 μm. Ima, imatinib; Ktz, ketoconazole.

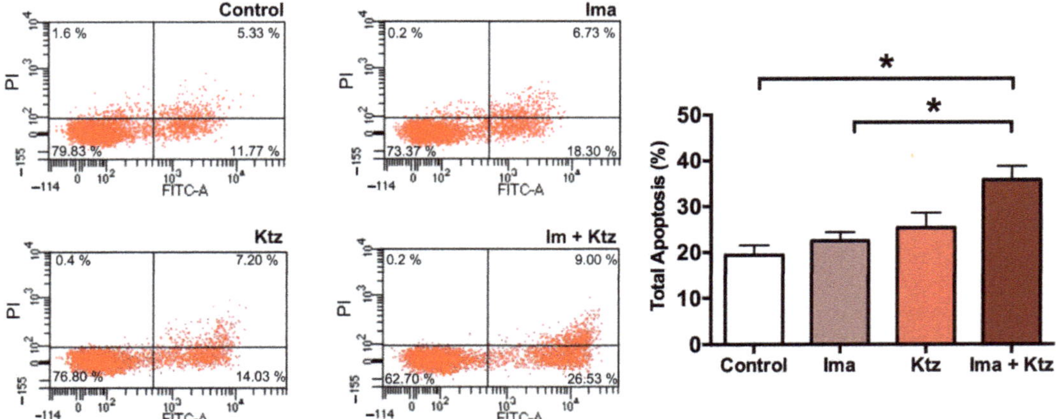

Figure 5. Flow cytometric analysis of apoptosis in K562-RI cells at 96 h post-treatment. Representative histograms depicting cells positive to Annexin V-FITC/IP (early and late apoptosis) by flow cytometric analysis; and percentage of total apoptosis. * $p < 0.05$, examined with one-way analysis of variance (ANOVA) followed by Tukey's test. Data are expressed as the mean ± SEM of three independent experiments. Ima, imatinib (200 nM); Ktz, ketoconazole (10 µM); Ima + Ktz, the combination of imatinib plus ketoconazole (200 nM/10 µM).

3. Discussion

Before 2001, the median survival of patients with CML was 5–7 years. Due to the introduction of TKI therapy, overall survival 5 years after treatment is now 92–95% (1). To date, the Food & Drug Administration (FDA) has approved four TKIs as first-line drugs for CML: imatinib, dasatinib, nilotinib and bosutinib. Imatinib is the first-generation drug, being the first to be approved in 2001 [21]. Whereas annual mortality for CML patients was 10–20% prior to the introduction of imatinib, it is now 1–2% [22].

There are several aspects to be considered in selecting one of the four inhibitors for CML therapy. Imatinib was the first TKI to receive approval by FDA for the treatment of patients with CML in chronic phase, is recommended for older patients with comorbidities, being the safest drug with the fewest side effects [23]. The use of dasatinib or nilotinib, TKI of the second generation, is justified in high-risk patients or young patients in need of a deep response with a first-line treatment. Whereas bosutinib, a third generation TKI, is intended for patients with chronic-, accelerated-, or blast-phase CML who cannot tolerate or are resistant to other therapies [24,25].

Most patients administered imatinib can carry on normal lives as long as they adhere to treatment. Since the dosing is carried out indefinitely, however, adherence is complicated, especially if patients experience side effects. In the latter case, adherence is likely to decline over time, leading a percentage of patients to develop resistance. The ADAGIO study showed that only 14.2% of patients took 100% of the prescribed imatinib doses. Furthermore, non-adherent patients constituted 23.2% of those with a suboptimal response and only 7.3% of those with an optimal response [26]. According to another study conducted on CML patients administered imatinib for a few years, poor adherence may be the predominant reason for a low level of response to imatinib and the development of resistance to the same [27].

Based on clinical trials, around 20–30% of patients eventually develop drug resistance [28], which consists of an absence of the initial desired response to the drug in a certain period of time (primary resistance), or the presence of the desired response followed by its loss (secondary resistance). Among the patients with primary or secondary resistance, many progress to accelerated phase and blast phase [15].

Researchers seeking new strategies to confront the emergence of resistance to imatinib have been obliged to gain new insights into the molecular mechanisms of resistance. Although there are several reports of resistance to imatinib being dependent on BCR-ABL1, the overexpression of P-gp (a drug efflux pump) could also be a crucial resistance mechanism [29–31], because it regulates the intracellular concentration of the drug.

In the current contribution, the first step was to establish drug resistance in the K562 cell line through its constant exposure to imatinib (for about eight months). The resistant cell line exhibited a drastic drop in its response to imatinib, evidenced by a significant increase in the IC_{50} from 213 nM (K562) to 2544 nM. As can be appreciated, an approximately 12-fold greater dose is necessary to attain the same result with the K562-RI cells compared to the K562 cells.

The next step was to examine differences between the two cell lines in relation to certain parameters. In a recent clinical study on CML patients by Ammar et al. (2020), an association was found between elevated levels of P-gp and unresponsiveness to treatment, indicating that the overexpression of P-gp is probably a relevant mechanism in the development of resistance to imatinib [32]. According to the present data from Western blot analysis and immunofluorescence assays, P-gp levels were almost undetectable in the parent cells but were about 4-fold higher in K562-RI cells.

A high level of P-gp implies that its substrates can easily be removed from cells. Like imatinib, doxorubicin is a P-gp substrate and is efficiently pumped out of tumor cells when P-gp is highly expressed, leading to a reduced intracellular accumulation. To evaluate the activity of P-gp, the intracellular accumulation doxorubicin was herein analyzed after incubation of cells with this compound, finding a lower mean fluorescence intensity of intracellular doxorubicin in K562-RI versus K562 cells (Figure 1D). This low uptake of doxorubicin was accompanied by an elevated level of P-gp. Similar results have been reported by other researchers using in vitro models [30,31]. Moreover, K562-RI cells were incubated in doxorubicin only and doxorubicin plus ketoconazole, finding a greater intracellular fluorescence in the latter group with the P-gp inhibitor. Hence, an overexpression of the P-gp drug efflux pump appears to be an important mechanism of resistance of CML cells to imatinib.

The next question to be explored was the effect of applying a P-gp inhibitor to the K562 and K562-RI cell lines. P-gp, the best characterized molecule of the class of efflux pump transporters, is known to produce resistance to treatment by removing drugs from various kinds of cancer cells resistant to drug treatment. Consequently, several studies have evaluated the inhibition of this transporter to try to enhance the anti-proliferative activity of chemotherapy for distinct types of cancer. Some of these studies have tested compounds possibly capable of improving the pharmacokinetics of ITKs. However, there are few reports on the inhibition of P-gp to overcome the resistance of CML cells to imatinib.

Several researchers have demonstrated that ketoconazole is a strong inhibitor not only of CYP-P450 3A4 (CYP3A4) but also of P-gp [33]. On the other hand, imatinib is a substrate of P-gp and is metabolized by CYP-P450, mainly by the CYP3A4 isoenzyme. Thus, imatinib should be susceptible to drug interactions if administered concomitantly with potent inhibitors or inducers of CYP-P450 and/or P-gp. In this sense, the concomitant intake of imatinib and ketoconazole was found to increase the plasma concentration of imatinib in healthy subjects [20].

In the current contribution, imatinib plus 10 μM of ketoconazole was applied to resistant (K562-RI) CML cells with the aim of decreasing the overexpression of P-gp, enhancing the intracellular concentrations of imatinib, and accelerating the rate of programmed cell death. The result was the reversal of resistance to the standard drug. The IC50 of imatinib dropped from 1378 nM (without ketoconazole) to 186 nM (with ketoconazole), reflecting an approximately 7.5-fold reversal index. In K562 cells (with negligible levels of P-gp), this combination regimen only slightly improved the effect of imatinib. Contrarily, K562-RI cells subjected to the combination treatment showed a clear reduction in the expression of P-gp in Western blot and immunofluorescence assays (Figure 4), and a significantly greater relative fluorescence of intracellular doxorubicin after applying ketoconazole (10 and 20 μM) plus doxorubicin (versus doxorubicin alone) (Figure 3). Similarly, Siegsmund et al. found that applying ketoconazole in combination with vinblastine or doxorubicin to treat a highly resistant human cell line (KB-Vl) enhanced the intracellular accumulation of the latter two cytotoxic drugs. Both combinations reversed multidrug resistance with ketoconazole at 1, 3, and 10 μg/mL (~2, 6, and 19 μM, respectively) [34], concentrations comparable to those used in the present work. In another study, it was also shown that ketoconazole was able to inhibit P-gp at a concentration of 6 μM in NIH-3T3-G185 cells that overexpressing human P-gp [35]. It has also been described that ketoconazole at 10 and 20 μM strongly enhanced cell growth inhibition and apoptosis of paclitaxel or cisplatin in ovarian cancer cells through its pregnane X receptor (PXR) antagonism. This nuclear receptor affects drug metabolism/efflux and drug-drug interaction through P-gp expression [36]. The lowest concentration of ketoconazole that presently demonstrated a positive effect was 10 μM, which is within the range of human plasma concentrations observed after a single oral dose of 200–400 mg [37]. Therefore, these non-toxic concentrations of ketoconazole can easily be achieved with clinical pharmacological doses of the drug and may be clinically well-tolerated if administered in combination with imatinib to treat patients with resistance to the standard drug.

Some additional mechanism by which ketoconazole is able to reverse resistance of imatinib could be related to regulation of P-gp/CYP3A4 by PXR, which regulates the expression of metabolic enzymes and transporters [38]. Ketoconazole has been reported as an inhibitor of PXR [39,40]. Therefore, it could regulate the transcription of the P-gp/CYP3A4 gene through disruption of PXR interaction with steroid receptor coactivator (SRC)-1.

A fundamental characteristic of cancer cells is their capacity to avoid apoptotic cell death. The cooperation between P-gp and molecules capable of inhibiting apoptosis-related proteins can generate a more robust drug resistance in cancer cells in general, and CML cells in particular. Among the numerous mechanisms utilized by CML cells to become resistant to imatinib, the avoidance of apoptosis is probably one of the most common.

The present study explored apoptosis as a possible mechanism of growth inhibition by the combination treatment. The Annexin V/PI assay, used for assessing early and late apoptosis, showed that imatinib plus ketoconazole increased apoptosis by 40%, whereas treatment with only imatinib or ketoconazole produced 25% greater apoptosis compared to control cells (Figure 5). Consequently, a plausible mechanism of ketoconazole for improving the efficacy of imatinib is its capacity to trigger apoptosis. Previous studies have reported that ketoconazole produce apoptosis inducing p53 levels and PARP cleavage in breast cancer cells [41], human colorectal and hepatocellular carcinoma cell lines [42]; therefore, a similar mechanism could be involved in the apoptosis of CML

On the other hand, Chen et al. recently reported that ketoconazole accelerates the process of apoptosis in hepatocarcinoma cells by exacerbating mitophagy, activating the PINK1/Parkin (PRKN) signaling pathway and downregulating cyclooxygenase-2 (COX-2), an inducible form of the enzyme that catalyzes the synthesis of prostanoids. The overexpression of COX-2 has been related with resistance to apoptosis [43]. In addition to these results, several reports reveled that the overexpression of COX-2 leads to increased P-gp expression [44]. Likewise, an overexpression of COX-2 and P-gp has been demonstrated in resistant K562 cells, with a decrease in apoptosis involving the Akt/p-Akt signaling pathway, which suggests the participation of COX-2 and P-gp in the development of resistance of CML cells [45]. We propose that ketoconazole stimulates apoptosis through COX-2 inhibition, in addition to P-gp inhibition, in imatinib-resistant cells (K562-RI).

Ketoconazole is an antifungal drug known to impede fungal growth by preventing the synthesis of ergosterol (the fungal equivalent of cholesterol) [46,47]. According to recent reports and the current findings, this drug seems to have great potential for cancer therapy [48–53].

Since its approval by the FDA in 1982, the estimated number of prescriptions of ketoconazole in the United States has been increasing every year. The oral use of ketoconazole is well tolerated in patients with limited toxicity. In the CALGB 9583 trial, 2% of patients who received 400 mg of ketoconazole three times per day, had grade 3 or 4 hepatotoxicity. However, the low ketoconazole dose (200 mg) was less toxic; thus, for patients who cannot tolerate high dose treatment, the low dose would be an option [54]. In another report, Outeiro et al. reported that only 1.7% of patients who received ketoconazole (400 mg/day for 28 days) experienced liver function abnormalities [55]. Several reports in the literature have also described that ketoconazole has mild toxicity and is rarely fatal in comparison with other azoles as voriconazole, fluconazole, itraconazole among others. Ketoconazole toxicity can be reversed upon drug discontinuation [56–58].

Thus, ketoconazole is a safe drug with a relatively low cost compared to the high price of new medications for cancer. The price of new anti-cancer drugs reflects the corresponding research and development costs, generally around a billion dollars or more [59,60]. One strategy now employed more and more frequently is the repositioning of drugs, which involves giving approved drugs new applications. Among the advantages is the known profile of safety and efficiency.

The literature describes a large number of P-gp substrates already approved by the FDA; nevertheless, to date there are few reported studies evaluating P-gp inhibition in CML patients. Cyclosporine was one of the first drugs studied concerning clinical P-gp modulations; despite the results in cells, it also showed high toxicity [61,62]. Recently several in vitro studies, using P-gp inhibitors, supported the role of efflux activity of this protein in CML resistance. Liu et. al, in 2018, demonstrated that non-toxic concentrations of nelfinavir, an anti-HIV drug, reverse the resistance of adriamycin (doxorubicin), colchicine, paclitaxel and imatinib in k562/ADR cells that overexpressed P-gp. Nelfinavir, in addition to being a protease inhibitor drug approved for the treatment of AIDS patients, it has also been proposed as a new antitumor drug for the treatment of CML [63]. Elacridar, a potent P-gp inhibitor, approved recently by FDA, has demonstrated interesting results about to promote a synergic effect with imatinib in resistant cells [64]. Azithromycin, antimicrobial drug, may be another interesting alternative to overcome imatinib resistance according to the results described of its ability to inhibit P-gp function and increase intracellular accumulation of imatinib [65]. Although all these drugs (including ketoconazole), show interesting results as potential drugs for the treatment of resistant CML, clinical trials are needed to demonstrate their effectiveness in reversing resistance by targeting P-gp and prove their low systemic toxicity in CML patients.

4. Materials and Methods

4.1. Cell Lines

Human CML K562 cells were acquired from the American Type Culture Collection (ATTC, Rockville, MD, USA) and they were cultured in Dulbecco's modified Eagle's medium (DMEM) supplemented with 10% fetal bovine serum, at 37 °C in a humidified atmosphere with 5% CO_2.

4.2. Development of the CML Cell Line with Resistance to Imatinib Treatment

The K562 cells were seeded in a culture flask with a surface area of 25 cm^2, they were exposed to gradually increasing concentrations (ranging from 1 nM to 2500 nM) of imatinib over various months, establishing an imatinib-resistant culture. After about eight months, the resistant phenotype, designated as K562-RI, was confirmed with the cell viability assay. To maintain the resistance of the K562-RI cells, 250 nM of imatinib were added to the culture medium.

To evaluate the resistance of K562-RI to imatinib, these cells were seeded into 96-well plates (Costar, Cambridge, MA, USA) at a density of 12×10^3 viable cells per well in 150 µL of culture medium. They were exposed to increasing concentrations of imatinib for 72 h. Cell viability to test cell resistance to imatinib was assessed with the sodium 3′-[1-[(phenylamino)-carbony]-3,4-tetrazolium]-bis(4-methoxy-6-nitro) benzene-sulfonic acid hydrate (XTT) assay (Roche Molecular Biochemicals), which is based on the cleavage of yellow tetrazolium salt XTT to form an orange formazan dye by metabolically active cells.

Briefly, 50 µL of XTT were added to each well with K562-RI cells (for a final concentration 0.3 mg/mL), followed by incubation at 37 °C for 2 h in a humidified atmosphere containing 5% CO_2. The parent K562 cell line was also tested with the XTT assay for comparison. After quantifying the absorbance of the samples from both cell lines spectrophotometrically at 492 nm with an ELISA microplate reader (Thermo Scientific, Waltham, MA, USA), the percentage of viability and relative resistance was calculated. Data are expressed as the mean ± SEM of three independent experiments performed in triplicate. For each experiment, the resistance of the K562-RI cell line was confirmed using the XTT assay to calculate cell viability and corroborate relative resistance.

Additionally, the level of P-gp was determined by Western blot; three independent experiments were performed, and the intracellular accumulation of doxorubicin was evaluated as an indirect measure of P-gp activity (given that doxorubicin is considered a substrate for P-gp transport). Data are expressed as the mean ± SEM of three independent experiments.

4.3. Treatments with Imatinib and Ketoconazole

K562 and K562-RI cells were seeded into 96-well plates (Costar, Cambridge, MA, USA) at a density of 12×10^3 viable cells per well in 150 µL culture medium, and then exposed for 72 h to various amounts of imatinib alone (0–2500 nM) or imatinib plus ketoconazole (0.1, 1.0, and 10 µM). Control cells were only in contact with the vehicle. At the end of the exposure period, cell viability was examined with the XTT assay, as aforementioned. The mean concentration in each set of four wells was determined in triplicate. The dose–response relationship for imatinib applied alone or in combination with ketoconazole was characterized with a sigmoidal function. The percentage of growth inhibition was calculated, and the IC50 values (the concentration of the drug required to afford 50% growth inhibition) were obtained from the survival curve fitted to a non-linear regression using the GraphPad Software, Prism 7.0 (San Diego, CA, USA) The equations used were: "Dose-response curves-Inhibition" and "log(inhibitor) vs. normalized response"; with the following function: Y = 100/(1 + 10^((LogIC50-X) × HillSlope))). Where the HillSlope describes the steepness of curves, and the IC50 is the concentration that provokes a response halfway between the minimum and maximum response [66].

4.4. P-Glycoprotein Expression Analyzed by Western Blotting

To evaluate the P-gp expression levels in resistant cells, K562 and K562-RI cells (1×10^6 cells) were seeded and incubated for 24 h, they were recollected and centrifuged. The resulting pellets were washed three times with PBS, then homogenized with a lysis buffer containing protease inhibitors. To evaluate the P-gp expression levels in K562-RI cells exposed to imatinib in the absence and presence of ketoconazole, K562-RI cells (1×10^6 cells) were cultured overnight, then treated with imatinib (200 nM), ketoconazole (10 µM), or imatinib plus ketoconazole (200 nM/10 µM) during 4 h, then the cells were recollected, washed with PBS and homogenized with a lysis buffer. The proteins were separated by centrifugation at 10,000× g and 4 °C, quantified by the BCA (bicinchoninic acid) assay, and separated electrophoretically on 4–20% gradient gel (Mini-Protean TGX 456-1094, Bio-Rad Laboratories, Inc., Burns, TN, USA). Markers (Bio-Rad, Hercules, CA, USA) were included to establish protein size. Subsequently, the proteins were transferred from the gel onto PVDF membranes (Amersham, UK), which were blocked with 5% non-fat dry milk at room temperature for 2 h. Membranes were incubated overnight at 4 °C with antibodies against P-gp (12683, 1:500, Cell Signaling Technology, Danvers, MA, USA) and β-actin (sc-69879, 1:1000; Santa Cruz Biotechnology, Dallas, TX, USA). The membranes were washed and incubated with IRDye® 800 CW goat anti-mouse or IRDye® 680RD goat anti-rabbit secondary antibodies (1:15,000; LI-COR, Biotechnology, Inc., Lincoln, NE, USA) for 1 h and then scanned on an Odyssey Imaging System. Their intensity of fluorescence was calculated by using Image Studio software. In each figure, representative blot images were selected from the same gel.

4.5. P-Glycoprotein by Immunofluorescence Assay

K562-RI cells (1 × 10^6 cells) were cultured overnight in culture slides (CultureSlides, Falcon Corning, NY, USA), then treated with imatinib (200 nM), ketoconazole (10 μM), or imatinib plus ketoconazole during 4 h. Afterwards, the cells were fixed in 4% paraformaldehyde in PBS (pH 7.4) for 15 min, permeabilized with 0.1% Triton X-100 for another 15 min and blocked with Ultracruz Blocking reagent (sc-516214, Santa Cruz, CA, USA) for 1 h. At the end of this time, they were incubated with monoclonal antibody Mdr-1 conjugated to Alexa Fluor 488 (sc-55510 AF488, 1:200) at 4 °C overnight. Subsequently, the cells were washed three times, and DAPI (ENZ-53003) reagent was used to counterstain the nuclei. Finally, immunofluorescence images were examined through an inverted fluorescence microscope (Olympus XI51).

4.6. Assessment of the Intracellular Accumulation of Doxorubicin

The intracellular accumulation of doxorubicin, a P-gp substrate, was determined in K562 and K562-RI cells to appraise P-gp activity. Briefly, the cells (1 × 10^6 cells) were seeded in triplicate in 25 cm^2 plates and exposed to doxorubicin (10 μg/mL) in the presence or absence of ketoconazole (10 μM) for 1 h. Control cells were only in contact with the vehicle. The cells were centrifuged after incubation, and the pellets were washed three times with ice-cold phosphate buffer solution (PBS). Subsequently, the cells were analyzed by flow cytometry (Guava® easyCyte, Merck Millipore), obtaining data from 10,000 acquired events with InCyte software (Merck Millipore, Darmstadt, DE, USA). The fluorescence of doxorubicin was quantified at 488 nm excitation and 575 nm emission wavelength.

4.7. Determination of Apoptosis by Flow Cytometry

Externalization of phosphatidylserine was evaluated with the Annexin-V-FLUOS Staining Kit. Cells (1 × 10^6) were seeded in Costar® 6-well Clear TC-treated Plates and exposed with imatinib alone (200 nM), ketoconazole alone (10 μM), or imatinib plus ketoconazole for 96 h. At the end of the incubation period, the cells were harvested, centrifuged at 2000 rpm for 5 min, washed once with PBS, and centrifuged again. Then they were resuspended in Annexin-V-FLUOS labeling solution, Annexin V-FITC (FITC), and propidium iodide (PI), to be incubated for 15 min at room temperature in the dark, according to the Annexin-V-FLUOS Staining Kit protocol. Flow cytometry was carried out to obtain 5 × 10^3 cells. The analysis of annexin was conducted with the BD FACS Canto II BD flow cytometry system, and BD FACSDiva software 6.0. The results were expressed as the total percentage of cells undergoing apoptosis. At least three independent experiments were performed for each assay.

5. Conclusions

The current study provides evidence that P-gp, a drug efflux pump, plays an important role in the development of resistance to imatinib in CML cells. In vitro testing presently showed the capacity of an inhibitor of P-gp, ketoconazole, to reverse CML cell resistance to imatinib treatment. In resistant CML cells, the antifungal ketoconazole inhibited the overexpression and efflux function of P-gp. Additionally, it increased the intracellular concentration of doxorubicin (a P-gp substrate) when resistant cells were incubated with doxorubicin plus ketoconazole. Thus, a possible corresponding increment in the intracellular concentration of imatinib may have taken place when resistant cells were incubated with this drug plus ketoconazole. Finally, ketoconazole triggered greater apoptosis of resistant CML cells. According to these findings on the mechanisms of action of ketoconazole in CML cells resistant to imatinib, the administration of imatinib plus ketoconazole is likely to favor reversal of resistance in CML patients treated unsuccessfully with imatinib alone; therefore, we propose the repositioning of ketoconazole for the treatment of CML in patients resistant to imatinib.

Author Contributions: Conceptualization, P.G.-L.; methodology, O.P.-C., A.A.-R., R.J., M.L.-M. and J.P.-R.; software, R.J.; investigation, E.C.-C., O.P.-C. and A.A.-R.; writing, P.G.-L.; supervision, P.G.-L.; funding acquisition, P.G.-L. and E.C.-C. All authors have read and agreed to the published version of the manuscript.

Funding: This research was partially financed by CONACYT (Mexico City, Mexico) through grant number: 142062.

Institutional Review Board Statement: Not applicable.

Informed Consent Statement: Not applicable.

Data Availability Statement: The data that support the findings of this study are available on request to the corresponding author.

Acknowledgments: We thank Bruce Allan Larsen for proofreading the manuscript.

Conflicts of Interest: The authors declare no conflict of interest.

References

1. Cortes, J.; Pavlovsky, C.; Saußele, S. Chronic myeloid leukaemia. *Lancet* **2021**, *398*, 1914–1926. [CrossRef]
2. Apperley, J.F. Chronic myeloid leukaemia. *Lancet* **2015**, *385*, 1447–1459. [CrossRef]
3. Goldman, J.M.; Melo, J.V. BCR-ABL in chronic myelogenous leukemia—how does it work? *Acta Haematol.* **2008**, *119*, 212–217. [CrossRef] [PubMed]
4. Kang, Z.J.; Liu, Y.F.; Xu, L.Z.; Long, Z.J.; Huang, D.; Yang, Y.; Liu, B.; Feng, J.X.; Pan, Y.J.; Yan, J.S.; et al. The Philadelphia chromosome in leukemogenesis. *Chin. J. Cancer* **2016**, *35*, 48. [CrossRef]
5. Di Felice, E.; Roncaglia, F.; Venturelli, F.; Mangone, L.; Luminari, S.; Cirilli, C.; Carrozzi, G.; Giorgi Rossi, P. The impact of introducing tyrosine kinase inhibitors on chronic myeloid leukemia survival: A population-based study. *BMC Cancer* **2018**, *18*, 1069. [CrossRef]
6. Druker, B.J.; Tamura, S.; Buchdunger, E.; Ohno, S.; Segal, G.M.; Fanning, S.; Zimmermann, J.; Lydon, N.B. Effects of a selective inhibitor of the Abl tyrosine kinase on the growth of Bcr-Abl positive cells. *Nat. Med.* **1996**, *2*, 561–566. [CrossRef]
7. Druker, B.J.; Talpaz, M.; Resta, D.J.; Peng, B.; Buchdunger, E.; Ford, J.M.; Lydon, N.B.; Kantarjian, H.; Capdeville, R.; Ohno-Jones, S.; et al. Efficacy and safety of a specific inhibitor of the BCR-ABL tyrosine kinase in chronic myeloid leukemia. *N. Engl. J. Med.* **2001**, *344*, 1031–1037. [CrossRef] [PubMed]
8. Clarkson, B.; Strife, A.; Wisniewski, D.; Lambek, C.L.; Liu, C. Chronic myelogenous leukemia as a paradigm of early cancer and possible curative strategies. *Leukemia* **2003**, *17*, 1211–1262. [CrossRef] [PubMed]
9. Mauro, M.; Deininger, M. Chronic myeloid leukemia in 2006: A perspective. *Chronic. Myeloid. Leuk.* **2006**, *91*, 152. [CrossRef]
10. Maekawa, T.; Ashihara, E.; Kimura, S. The Bcr-Abl tyrosine kinase inhibitor imatinib and promising new agents against Philadelphia chromosome-positive leukemias. *Int. J. Clin. Oncol.* **2007**, *12*, 327–340. [CrossRef] [PubMed]
11. Leveque, D.; Maloisel, F. Clinical pharmacokinetics of imatinib mesylate. *Vivo* **2005**, *19*, 77–84.
12. Deng, J.; Shao, J.; Markowitz, J.S.; An, G. ABC transporters in multi-drug resistance and ADME-Tox of small molecule tyrosine kinase inhibitors. *Pharm. Res.* **2014**, *31*, 2237–2255. [CrossRef] [PubMed]
13. Kivistö, K.T.; Niemi, M.; Fromm, M.F. Functional interaction of intestinal CYP3A4 and P-glycoprotein. *Fundam. Clin. Pharmacol.* **2004**, *18*, 621–626. [CrossRef] [PubMed]
14. Esposito, N.; Colavita, I.; Quintarelli, C.; Sica, A.R.; Peluso, A.L.; Luciano, L.; Picardi, M.; Del Vecchio, L.; Buonomo, T.; Hughes, T.P.; et al. SHP-1 expression accounts for resistance to imatinib treatment in Philadelphia chromosome-positive cells derived from patients with chronic myeloid leukemia. *Blood* **2011**, *118*, 3634–3644. [CrossRef]
15. Quintas-Cardama, A.; Kantarjian, H.M.; Cortes, J.E. Mechanisms of primary and secondary resistance to imatinib in chronic myeloid leukemia. *Cancer Control.* **2009**, *16*, 122–131. [CrossRef] [PubMed]
16. Yang, K.; Fu, L.W. Mechanisms of resistance to BCR-ABL TKIs and the therapeutic strategies: A review. *Crit. Rev. Oncol. Hematol.* **2015**, *93*, 277–292. [CrossRef] [PubMed]
17. Maia, R.C.; Vasconcelos, F.C.; Souza, P.S.; Rumjanek, V.M. Towards Comprehension of the ABCB1/P-Glycoprotein Role in Chronic Myeloid Leukemia. *Molecules* **2018**, *23*, 119. [CrossRef] [PubMed]
18. Thomas, J.; Wang, L.; Clark, R.E.; Pirmohamed, M. Active transport of imatinib into and out of cells: Implications for drug resistance. *Blood* **2004**, *104*, 3739–3745. [CrossRef] [PubMed]
19. Silva, R.; Vilas-Boas, V.; Carmo, H.; Dinis-Oliveira, R.J.; Carvalho, F.; de Lourdes Bastos, M.; Remiao, F. Modulation of P-glycoprotein efflux pump: Induction and activation as a therapeutic strategy. *Pharmacol. Ther.* **2015**, *149*, 1–123. [CrossRef] [PubMed]

20. Dutreix, C.; Peng, B.; Mehring, G.; Hayes, M.; Capdeville, R.; Pokorny, R.; Seiberling, M. Pharmacokinetic interaction between ketoconazole and imatinib mesylate (Glivec) in healthy subjects. *Cancer Chemother. Pharmacol.* **2004**, *54*, 290–294. [CrossRef] [PubMed]
21. Granatowicz, A.; Piatek, C.I.; Moschiano, E.; El-Hemaidi, I.; Armitage, J.D.; Akhtari, M. An Overview and Update of Chronic Myeloid Leukemia for Primary Care Physicians. *Korean J. Fam. Med.* **2015**, *36*, 197–202. [CrossRef]
22. Jabbour, E.; Kantarjian, H. Chronic myeloid leukemia: 2018 update on diagnosis, therapy and monitoring. *Am. J. Hematol.* **2018**, *93*, 442–459. [CrossRef]
23. Flis, S.; Chojnacki, T. Chronic myelogenous leukemia, a still unsolved problem: Pitfalls and new therapeutic possibilities. *Drug Des. Devel. Ther.* **2019**, *13*, 825–843. [CrossRef] [PubMed]
24. Jabbour, E.; Kantarjian, H. Chronic myeloid leukemia: 2020 update on diagnosis, therapy and monitoring. *Am. J. Hematol.* **2020**, *95*, 691–709. [CrossRef] [PubMed]
25. Hill, B.G.; Kota, V.K.; Khoury, H.J. Bosutinib: A third generation tyrosine kinase inhibitor for the treatment of chronic myeloid leukemia. *Expert. Rev. Anticancer. Ther.* **2014**, *14*, 765–770. [CrossRef]
26. Noens, L.; van Lierde, M.A.; De Bock, R.; Verhoef, G.; Zachee, P.; Berneman, Z.; Martiat, P.; Mineur, P.; Van Eygen, K.; MacDonald, K.; et al. Prevalence, determinants, and outcomes of nonadherence to imatinib therapy in patients with chronic myeloid leukemia: The ADAGIO study. *Blood* **2009**, *113*, 5401–5411. [CrossRef]
27. Marin, D.; Bazeos, A.; Mahon, F.X.; Eliasson, L.; Milojkovic, D.; Bua, M.; Apperley, J.F.; Szydlo, R.; Desai, R.; Kozlowski, K.; et al. Adherence is the critical factor for achieving molecular responses in patients with chronic myeloid leukemia who achieve complete cytogenetic responses on imatinib. *J. Clin. Oncol.* **2010**, *28*, 2381–2388. [CrossRef] [PubMed]
28. Druker, B.J.; Sawyers, C.L.; Sawyers, C.L.; Kantarjian, H.; Resta, D.J.; Reese, S.F.; Ford, J.M.; Capdeville, R.; Talpaz, M. Activity of a specific inhibitor of the BCR-ABL tyrosine kinase in the blast crisis of chronic myeloid leukemia and acute lymphoblastic leukemia with the Philadelphia chromosome. *N. Engl. J. Med.* **2001**, *344*, 1038–1042, Erratum in *N. Engl. J. Med.* **2001**, *345*, 232. [CrossRef]
29. Rumjanek, V.M.; Trindade, G.S.; Wagner-Souza, K.; de-Oliveira, M.C.; Marques-Santos, L.F.; Maia, R.C.; Capella, M.A. Multidrug resistance in tumour cells: Characterization of the multidrug resistant cell line K562-Lucena 1. *An. Acad. Bras. Cienc.* **2001**, *73*, 57–69. [CrossRef]
30. Peng, X.X.; Tiwari, A.K.; Wu, H.C.; Chen, Z.S. Overexpression of P-glycoprotein induces acquired resistance to imatinib in chronic myelogenous leukemia cells. *Chin. J. Cancer* **2012**, *31*, 110–118. [CrossRef]
31. Alves, R.; Fonseca, A.R.; Goncalves, A.C.; Ferreira-Teixeira, M.; Lima, J.; Abrantes, A.M.; Alves, V.; Rodrigues-Santos, P.; Jorge, L.; Matoso, E.; et al. Drug transporters play a key role in the complex process of Imatinib resistance in vitro. *Leuk. Res.* **2015**, *39*, 355–360. [CrossRef] [PubMed]
32. Ammar, M.; Louati, N.; Frikha, I.; Medhaffar, M.; Ghozzi, H.; Elloumi, M.; Menif, H.; Zeghal, K.; Ben Mahmoud, L. Overexpression of P-glycoprotein and resistance to Imatinib in chronic myeloid leukemia patients. *J. Clin. Lab. Anal.* **2020**, *34*, e23374. [CrossRef] [PubMed]
33. Kageyama, M.; Namiki, H.; Fukushima, H.; Ito, Y.; Shibata, N.; Takada, K. In vivo effects of cyclosporin A and ketoconazole on the pharmacokinetics of representative substrates for P-glycoprotein and cytochrome P450 (CYP) 3A in rats. *Biol. Pharm. Bull.* **2005**, *28*, 316–322. [CrossRef]
34. Siegsmund, M.J.; Cardarelli, C.; Aksentijevich, I.; Sugimoto, Y.; Pastan, I.; Gottesman, M.M. Ketoconazole effectively reverses multidrug resistance in highly resistant KB cells. *J. Urol.* **1994**, *151*, 485–491. [CrossRef]
35. Wang, E.J.; Lew, K.; Casciano, C.N.; Clement, R.P.; Johnson, W.W. Interaction of common azole antifungals with P glycoprotein. *Antimicrob. Agents Chemother.* **2002**, *46*, 160–165. [CrossRef] [PubMed]
36. Masuyama, H.; Nakamura, K.; Nobumoto, E.; Hiramatsu, Y. Inhibition of pregnane X receptor pathway contributes to the cell growth inhibition and apoptosis of anticancer agents in ovarian cancer cells. *Int. J. Oncol.* **2016**, *49*, 1211–1220. [CrossRef] [PubMed]
37. Huang, Y.C.; Colaizzi, J.L.; Bierman, R.H.; Woestenborghs, R.; Heykants, J. Pharmacokinetics and dose proportionality of ketoconazole in normal volunteers. *Antimicrob. Agents Chemother.* **1986**, *30*, 206–210. [CrossRef] [PubMed]
38. Synold, T.W.; Dussault, I.; Forman, B.M. The orphan nuclear receptor SXR coordinately regulates drug metabolism and efflux. *Nat. Med.* **2001**, *7*, 584–590. [CrossRef]
39. Wang, H.; Huang, H.; Li, H.; Teotico, D.G.; Sinz, M.; Baker, S.D.; Staudinger, J.; Kalpana, G.; Redinbo, M.R.; Mani, S. Activated pregnenolone X-receptor is a target for ketoconazole and its analogs. *Clin. Cancer Res.* **2007**, *13*, 2488–2495. [CrossRef]
40. Svecova, L.; Vrzal, R.; Burysek, L.; Anzenbacherova, E.; Cerveny, L.; Grim, J.; Trejtnar, F.; Kunes, J.; Pour, M.; Staud, F.; et al. Azole antimycotics differentially affect rifampicin-induced pregnane X receptor-mediated CYP3A4 gene expression. *Drug Metab. Dispos.* **2008**, *36*, 339–348. [CrossRef]
41. Bae, S.H.; Park, J.H.; Choi, H.G.; Kim, H.; Kim, S.H. Imidazole Antifungal Drugs Inhibit the Cell Proliferation and Invasion of Human Breast Cancer Cells. *Biomol. Ther.* **2018**, *26*, 494–502. [CrossRef] [PubMed]
42. Ho, Y.S.; Tsai, P.W.; Yu, C.F.; Liu, H.L.; Chen, R.J.; Lin, J.K. Ketoconazole-induced apoptosis through P53-dependent pathway in human colorectal and hepatocellular carcinoma cell lines. *Toxicol. Appl. Pharmacol.* **1998**, *153*, 39–47. [CrossRef] [PubMed]

43. Chen, Y.; Chen, H.N.; Wang, K.; Zhang, L.; Huang, Z.; Liu, J.; Zhang, Z.; Luo, M.; Lei, Y.; Peng, Y.; et al. Ketoconazole exacerbates mitophagy to induce apoptosis by downregulating cyclooxygenase-2 in hepatocellular carcinoma. *J. Hepatol.* **2019**, *70*, 66–77. [CrossRef]
44. Patel, V.A.; Dunn, M.J.; Sorokin, A. Regulation of MDR-1 (P-glycoprotein) by cyclooxygenase-2. *J. Biol. Chem.* **2002**, *277*, 38915–38920. [CrossRef] [PubMed]
45. Arunasree, K.M.; Roy, K.R.; Anilkumar, K.; Aparna, A.; Reddy, G.V.; Reddanna, P. Imatinib-resistant K562 cells are more sensitive to celecoxib, a selective COX-2 inhibitor: Role of COX-2 and MDR-1. *Leuk. Res.* **2008**, *32*, 855–864. [CrossRef] [PubMed]
46. Van Tyle, J.H. Ketoconazole. Mechanism of action, spectrum of activity, pharmacokinetics, drug interactions, adverse reactions and therapeutic use. *Pharmacotherapy* **1984**, *4*, 343–373. [CrossRef] [PubMed]
47. Brunton, L.L.; Hilal-Dandan, R.; Knollmann, B.C. *Goodman & Gilman's: The Pharmacological Basis of Therapeutics*, 13th ed.; McGraw Hill: New York, NY, USA, 2018.
48. Greenberg, J.W.; Kim, H.; Moustafa, A.A.; Datta, A.; Barata, P.C.; Boulares, A.H.; Abdel-Mageed, A.B.; Krane, L.S. Repurposing ketoconazole as an exosome directed adjunct to sunitinib in treating renal cell carcinoma. *Sci. Rep.* **2021**, *11*, 10200. [CrossRef]
49. Tresnanda, R.I.; Pramod, S.V.; Safriadi, F. Ketoconazole for the Treatment of Docetaxel-Naive Metastatic Castration-Resistant Prostate Cancer (mCRPC): A Systematic Review. *Asian Pac. J. Cancer Prev.* **2021**, *22*, 3101–3107. [CrossRef]
50. Lu, C.T.; Leong, P.Y.; Hou, T.Y.; Kang, Y.T.; Chiang, Y.C.; Hsu, C.T.; Lin, Y.D.; Ko, J.L.; Hsiao, Y.P. Inhibition of proliferation and migration of melanoma cells by ketoconazole and Ganoderma immunomodulatory proteins. *Oncol. Lett.* **2019**, *18*, 891–897. [CrossRef]
51. Herrera-Martinez, A.D.; Feelders, R.A.; de Herder, W.W.; Castano, J.P.; Galvez Moreno, M.A.; Dogan, F.; van Dungen, R.; van Koetsveld, P.; Hofland, L.J. Effects of Ketoconazole on ACTH-Producing and Non-ACTH-Producing Neuroendocrine Tumor Cells. *Horm. Cancer* **2019**, *10*, 107–119. [CrossRef]
52. Chen, H.N.; Chen, Y.; Zhou, Z.G.; Wei, Y.; Huang, C. A novel role for ketoconazole in hepatocellular carcinoma treatment: Linking PTGS2 to mitophagy machinery. *Autophagy* **2019**, *15*, 733–734. [CrossRef] [PubMed]
53. Agnihotri, S.; Mansouri, S.; Burrell, K.; Li, M.; Mamatjan, Y.; Liu, J.; Nejad, R.; Kumar, S.; Jalali, S.; Singh, S.K.; et al. Ketoconazole and Posaconazole Selectively Target HK2-expressing Glioblastoma Cells. *Clin. Cancer Res.* **2019**, *25*, 844–855. [CrossRef]
54. Small, E.J.; Halabi, S.; Dawson, N.A.; Stadler, W.M.; Rini, B.I.; Picus, J.; Gable, P.; Torti, F.M.; Kaplan, E.; Vogelzang, N.J. Antiandrogen withdrawal alone or in combination with ketoconazole in androgen-independent prostate cancer patients: A phase III trial (CALGB 9583). *J. Clin. Oncol.* **2004**, *22*, 1025–1033. [CrossRef]
55. Outeiro, N.; Hohmann, N.; Mikus, G. No Increased Risk of Ketoconazole Toxicity in Drug-Drug Interaction Studies. *J. Clin. Pharmacol.* **2016**, *56*, 1203–1211. [CrossRef] [PubMed]
56. Zhou, Z.-X.; Yin, X.-D.; Zhang, Y.; Shao, Q.-H.; Mao, X.-Y.; Hu, W.-J.; Shen, Y.-L.; Zhao, B.; Li, Z.-L. Antifungal Drugs and Drug-Induced Liver Injury: A Real-World Study Leveraging the FDA Adverse Event Reporting System Database. *Front. Pharmacol.* **2022**, *13*, 891336. [CrossRef] [PubMed]
57. Gadour, E.; Kotb, A. Systematic Review of Antifungal-Induced Acute Liver Failure. *Cureus* **2022**, *13*, e18940. [CrossRef] [PubMed]
58. Kyriakidis, I.; Tragiannidis, A.; Munchen, S.; Groll, A.H. Clinical hepatotoxicity associated with antifungal agents. *Expert. Opin. Drug Saf.* **2017**, *16*, 149–165. [CrossRef]
59. Pushpakom, S.; Iorio, F.; Eyers, P.A.; Escott, K.J.; Hopper, S.; Wells, A.; Doig, A.; Guilliams, T.; Latimer, J.; McNamee, C.; et al. Drug repurposing: Progress, challenges and recommendations. *Nat. Rev. Drug Discov.* **2019**, *18*, 41–58. [CrossRef]
60. Hernandez, J.J.; Pryszlak, M.; Smith, L.; Yanchus, C.; Kurji, N.; Shahani, V.M.; Molinski, S.V. Giving Drugs a Second Chance: Overcoming Regulatory and Financial Hurdles in Repurposing Approved Drugs As Cancer Therapeutics. *Front. Oncol.* **2017**, *7*, 273. [CrossRef]
61. Maia, R.C.; Carrico, M.K.; Klumb, C.E.; Noronha, H.; Coelho, A.M.; Vasconcelos, F.C.; Ruimanek, V.M. Clinical approach to circumvention of multidrug resistance in refractory leukemic patients: Association of cyclosporin A with etoposide. *J. Exp. Clin. Cancer Res.* **1997**, *16*, 419–424. [PubMed]
62. List, A.F.; Kopecky, K.J.; Willman, C.L.; Head, D.R.; Slovak, M.L.; Douer, D.; Dakhil, S.R.; Appelbaum, F.R. Cyclosporine inhibition of P-glycoprotein in chronic myeloid leukemia blast phase. *Blood* **2002**, *100*, 1910–1912. [CrossRef] [PubMed]
63. Liu, W.; Meng, Q.; Sun, Y.; Wang, C.; Huo, X.; Liu, Z.; Sun, P.; Sun, H.; Ma, X.; Liu, K. Targeting P-Glycoprotein: Nelfinavir Reverses Adriamycin Resistance in K562/ADR Cells. *Cell Physiol. Biochem.* **2018**, *51*, 1616–1631. [CrossRef] [PubMed]
64. Alves, R.; Gonçalves, A.C.; Jorge, J.; Almeida, A.M.; Sarmento-Ribeiro, A.B. Combination of Elacridar with Imatinib Modulates Resistance Associated with Drug Efflux Transporters in Chronic Myeloid Leukemia. *Biomedicines* **2022**, *10*, 1158. [CrossRef] [PubMed]
65. Ozkan, T.; Hekmatshoar, Y.; Karabay, A.Z.; Koc, A.; Altinok Gunes, B.; Karadag Gurel, A.; Sunguroglu, A. Assessment of azithromycin as an anticancer agent for treatment of imatinib sensitive and resistant CML cells. *Leuk. Res.* **2021**, *102*, 106523. [CrossRef] [PubMed]
66. Motulsky, H.J. GraphPad Curve Fitting Guide. Available online: http://www.graphpad.com/guides/prism/7/curve-fitting/index.htm (accessed on 5 March 2016).

International Journal of *Molecular Sciences*

Article

Isothiocyanates (ITCs) 1-(Isothiocyanatomethyl)-4-phenylbenzene and 1-Isothiocyanato-3,5-bis(trifluoromethyl)benzene—Aldehyde Dehydrogenase (ALDH) Inhibitors, Decreases Cisplatin Tolerance and Migratory Ability of NSCLC

Jolanta Kryczka [1,*], Jakub Kryczka [2], Łukasz Janczewski [3], Anna Gajda [3], Andrzej Frączyk [4], Joanna Boncela [2], Beata Kolesińska [3] and Ewa Brzeziańska-Lasota [1]

1 Department of Biomedicine and Genetics, Medical University of Lodz, 92-213 Lodz, Poland; ewa.brzezianska@umed.lodz.pl
2 Institute of Medical Biology, Polish Academy of Sciences, 93-232 Lodz, Poland; jkryczka@cbm.pan.pl (J.K.); jboncela@cbm.pan.pl (J.B.)
3 Institute of Organic Chemistry, Faculty of Chemistry, Lodz University of Technology, 90-924 Lodz, Poland; lukasz.janczewski@p.lodz.pl (Ł.J.); anna.gajda@p.lodz.pl (A.G.); beata.kolesinska@p.lodz.pl (B.K.)
4 Institute of Applied Computer Science, Lodz University of Technology, 90-537 Lodz, Poland; andrzej.fraczyk@p.lodz.pl
* Correspondence: jolanta.kryczka@umed.lodz.pl

Citation: Kryczka, J.; Kryczka, J.; Janczewski, Ł.; Gajda, A.; Frączyk, A.; Boncela, J.; Kolesińska, B.; Brzeziańska-Lasota, E. Isothiocyanates (ITCs) 1-(Isothiocyanatomethyl)-4-phenylbenzene and 1-Isothiocyanato-3,5-bis(trifluoromethyl)benzene—Aldehyde Dehydrogenase (ALDH) Inhibitors, Decreases Cisplatin Tolerance and Migratory Ability of NSCLC. *Int. J. Mol. Sci.* **2022**, *23*, 8644. https://doi.org/10.3390/ijms23158644

Academic Editor: Angela Stefanachi

Received: 22 July 2022
Accepted: 29 July 2022
Published: 3 August 2022

Publisher's Note: MDPI stays neutral with regard to jurisdictional claims in published maps and institutional affiliations.

Abstract: One of the main treatment modalities for non-small-cell lung cancer (NSCLC) is cisplatin-based chemotherapy. However, the acquisition of cisplatin resistance remains a major problem. Existing chemotherapy regimens are often ineffective against cancer cells expressing aldehyde dehydrogenase (ALDH). As such, there is an urgent need for therapies targeting ALDH-positive cancer cells. The present study compares the anticancer properties of 36 structurally diverse isothiocyanates (ITCs) against NSCLC cells with the ALDH inhibitor disulfiram (DSF). Their potential affinity to ALDH isoforms and ABC proteins was assessed using AutoDockTools, allowing for selection of three compounds presenting the strongest affinity to all tested proteins. The selected ITCs had no impact on NSCLC cell viability (at tested concentrations), but significantly decreased the cisplatin tolerance of cisplatin-resistant variant of A549 (A549CisR) and advanced (stage 4) NSCLC cell line H1581. Furthermore, long-term supplementation with ITC 1-(isothiocyanatomethyl)-4-phenylbenzene reverses the EMT phenotype and migratory potential of A549CisR to the level presented by parental A549 cells, increasing E-Cadherin expression, followed by decreased expression of ABCC1 and ALDH3A1. Our data indicates that the ALDH inhibitors DSF and ITCs are potential adjuvants of cisplatin chemotherapy.

Keywords: non-small-cell lung cancer; cisplatin resistance; aldehyde dehydrogenase; isothiocyanates; disulfiram; epithelial to mesenchymal transition

1. Introduction

Since its introduction into clinical trials in 1971 and subsequent Food and Drug Administration approval in 1978, cisplatin represents a major landmark in the history of successful anti-cancer therapeutics. It has changed the management of several solid malignancies, including lung cancer, which remains the second-most-common cancer globally and the leading cause of cancer death [1]. Approximately 2.2 million of new cases of lung cancer are estimated to occur each year worldwide, with a mean 5-year survival rate of 22% [2,3].

One of the factors that contributes to such a low survival rate is the development of cisplatin resistance [4,5]. Cisplatin resistance is very complicated because of the multifactorial composition of several mechanisms including DNA repair, induction of anti-apoptotic signals, and the active efflux of drugs from the cell cytoplasm. Additionally, cisplatin-resistant cells were proven to undergo Epithelial to Mesenchymal Transition (EMT), which plays a substantial role in cancer progression and metastasis, increasing cancer cell motility and invasiveness [5]. EMT renders cancer cells virtually impervious to the majority of anticancer drugs, decreasing cell proliferation and directly influencing the expression of ABC-family transporters, which are involved in multidrug resistance [6,7].

Recently, aldehyde dehydrogenase (ALDH) was confirmed to play a role in drug resistance in lung cancer [8]. A bioinformatic analysis of metabolic enzymes found cisplatin-resistant NSCLC to express ALDH [9]. ALDH family members are cytosolic or mitochondrial isoenzymes that are responsible for oxidizing intracellular aldehydes. They play a role in the oxidation of retinol to retinoic acid in early stem cell differentiation [10]. Several of the 19 genes known to encode the ALDH family, such as ALDH1A1, ALDH1A2, ALDH1A3, ALDH1A7, ALDH3A1, ALDH4A1, ALDH5A1, ALDH6A1, and ALDH9A1, are considered the cancer stem cells (CSC) markers involved in drug resistance [11–13]. Notably, currently used chemotherapy regimens were shown to be ineffective against ALDH-positive, cisplatin-resistant cancer cells [8]. Importantly, while ALDH1 activity has been reported in a number of NSCLC cell lines and tumor samples, its role in chemotherapy resistance remains unclear [4]. Therefore, it may be beneficial to investigate therapies targeting drug-resistant cancer cells expressing ALDH [8]. One group of small-molecule ALDH inhibitors that may be promising candidates for anti-cancer therapy is the isothiocyanates (ITCs).

Isothiocyanates are commonly known for their anti-cancer properties. They are low-molecular-weight, natural, organic composites characterized by a pungent odor [14], with the general formula R–NCS. They are found in cruciferous vegetables, such as radish, horseradish, wasabi, broccoli, or Brussels sprouts [15,16] and are produced by the reaction of glucosinolates with myrosinases [17–20]. In addition to natural ITCs such as benzyl isothiocyanate [21,22], phenethyl isothiocyanate [23,24] or best-tested sulforaphane (SFN) [25–30], synthetic counterparts of ITCs have been produced by modification with a fluorine atom [31,32], phosphorous group [33,34], or other functional groups [35]. ITCs are used in organic chemistry as substrates to synthesize inter alia heterocyclic compounds or thioamides [36]. However, these compounds are best-known for their anticancer activity [27,37–41] and antibacterial activity [42–44].

In prostate cancer, ITCs decreased the concentration of the anti-apoptotic proteins Bcl-2 (B-cell lymphoma 2) and Bcl-xl (B-cell lymphoma-extra large) and increased the expression of pro-apoptotic proteins Bax (BCL2 Associated X, Apoptosis Regulator) and activate caspases [45]. Additionally, they decreased the activity of apoptosis inhibitors such as cIAP1 (cellular inhibitor of apoptosis protein-1), cIAP2 (cellular inhibitor of apoptosis 2), and XIAP (X-linked inhibitor of apoptosis protein) and induced Apaf1 (Apoptotic Peptidase Activating Factor 1) protein activity [46]. In addition, in human embryonic kidney cell line HEK293, ITCs were found to suppress transcription of histone deacetylases (HDACs), thus deregulating apoptosis- and differentiation-controlling mechanisms [47]. Furthermore, several studies have confirmed that ITCs suppress both the angiogenesis of human umbilical vein endothelial cells (HUVECs) [48] and metastasis of B16F-10 melanoma [33,49]. Additionally, benzyl isothiocyanates BITCs were proven to inhibit the phosphorylation activity of three major mitogen-activated protein kinases (MAPKs), ERK1/2, p38, and p-JNK1/2, thus presenting direct anti-metastatic activities in SK-Hep1 cells [50].

To improve the current therapeutic efficacy of this cisplatin, alternative strategies are needed to overcome resistance. In the present study, 36 structurally different ITCs (**1–36**) were synthesized and tested in vitro using lines of lung cancer cells and their cisplatin-resistant variants A549, A549CisR, and NCI-H1581. The protein level and function of two isoforms of ALDH were noted, these being potential markers of cisplatin resistance. Furthermore, the effects of targeting ALDH3A1 and ALDH7A1 by chemical inhibition were assessed in terms of their ability to re-sensitize resistant lung tumor cells to the cytotoxic effects of cisplatin. Moreover, as multiple ITCs were proven to present many, ostensibly not related, anticancer properties, due to different ITCs cellular targets, ITCs 1-(isothiocyanatomethyl)-4-phenylbenzene (named **19**) was screened in regards to possible antimetastatic abilities. ITCs 19 supplementation significantly decreases proteolytic abilities and in the aftermath the invasiveness of A549CisR, thus highlighting multifactorial anticancer properties.

2. Results

2.1. Characteristics of the Cisplatin-Resistant A549 Cell Line

A cisplatin-resistant variant of A549 (named A549CisR) was created by constant culturing in increasing cisplatin concentrations (1–10 μM), resulting in an IC_{50} value of 150 μM, compared to 75 μM for A549 (data not shown). A549CisR acquired a mesenchymal-like phenotype manifested by upregulation of mesenchymal marker N-cadherin, with simultaneous repression of epithelial marker E-cadherin (Figure 1A), as noted previously [51,52]. The degree of mesenchymal properties acquired via EMT varied between cells from an epithelial-like status, through a mixed epithelial/mesenchymal (E/M hybrid) form to a strongly mesenchymal phenotype. The hybrid and mesenchymal cells exhibited increased invasive features and circulating tumor cell (CTC) characteristics, suggesting that EMT plays an important role during metastatic dissemination [53]. The A549CisR EMT phenotype was followed by changes in ALDH3A1, ALDH7A1, and ABC protein expression (Figure 1A). A549CisR presented significantly higher ALDH3A1 (stem cell marker [10]), ALDH7A1, ABCC1, and ABCC4 expression and lower expression of ABCG2 compared to A549. Interestingly, ALDH1A1 expression did not differ (data not shown). Furthermore, we noticed changes in cell morphology, A549CisR cells became larger, spindle-shaped, and less densely packed than the parental A549 (Figure 1B). The cisplatin-induced EMT model mimics a natural shift toward higher aggression, and increased the migratory ability and metastasis obtained by chemo-resistant cancer cells during cancer progression [5,54].

Regarding the effect of cisplatin resistance related changes on NSCLC migration, A549CisR presented a higher 2D migration rate (observed in wound healing assay) than the parental sensitive variant (A549) (Figure 1C), even though this variant is considered to be a fairly aggressive NSCLC model [55]. This further demonstrates the increased metastatic potential associated with cisplatin resistance.

2.2. ALDH Inhibitor—Disulfiram Impact on Suppression of Cisplatin Resistance

The study also analyzed the impact of disulfiram (DSF), a well-known ALDH inhibitor, on A549 and A549CisR. DSF has been used to control alcohol abuse for many decades; however, it has recently been found to have strong anticancer activity both in vitro and in cancer xenografts [56]. DSF treatment (3 μM, 48 h) slightly reduces the mesenchymal phenotype of the cisplatin-resistant variant of A549, restoring the expression of E-cadherin, an epithelial marker (Figure 1D), to the level observed in parental A549. However, the phenotype was not fully restored, as the expression of N-cadherin, a mesenchymal marker, remained unchanged. Importantly, DSF partially re-sensitized A549CisR cells to cisplatin treatment (Figure 1E). Supplementation with 3 μM DSF followed by 75 μM cisplatin (24 h) significantly reduced the tolerance of A549CisR to cisplatin treatment. DSF impact on A549CisR viability is presented in Supplementary Materials (Figure S1).

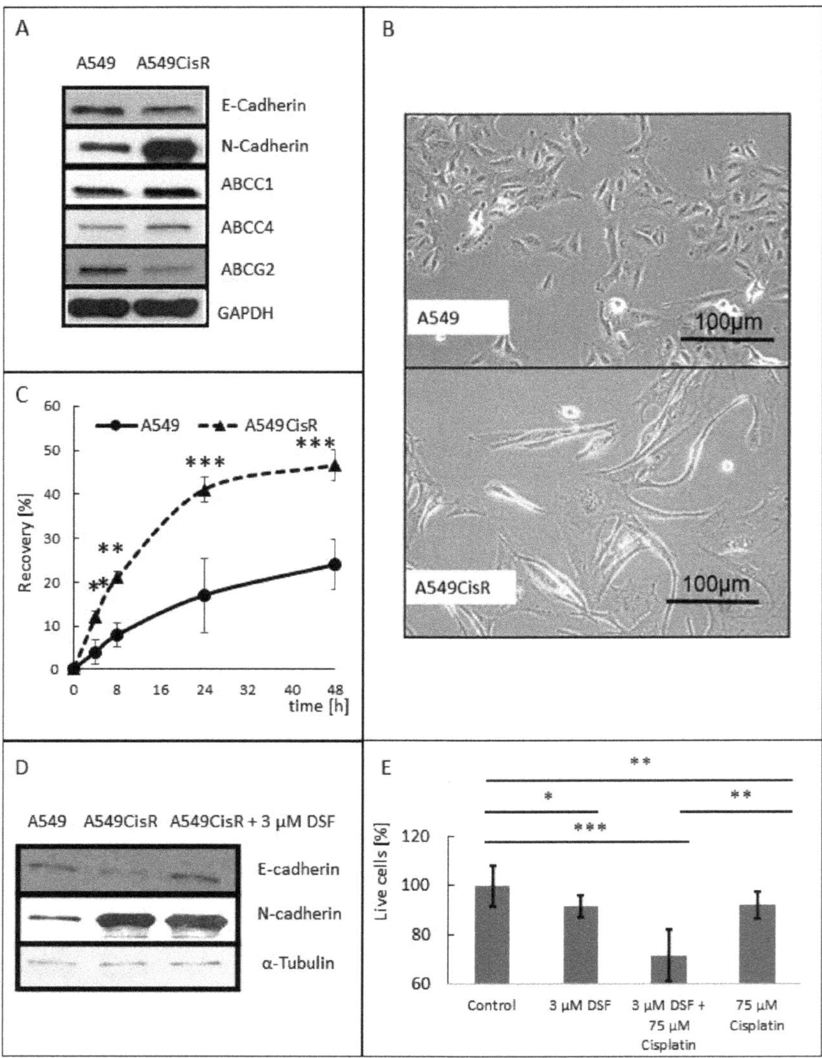

Figure 1. Characteristics of the cisplatin-resistant A549 cell line and DSF reversion of cisplatin resistance. (**A**) Western blot analysis of EMT markers and ABC proteins, performed in standard reducing SDS PAGE conditions. (**B**) Cell morphology (phase contrast microscopy). (**C**) Wound healing such as analysis of resistant-cell migration. A549 and A549CisR cells were seeded on a six-well plate and grown to confluence. Next, a wound was created and rinsed twice with PBS. New full medium was added. Wounded area was visualized after 0, 4, 8, 24, and 48 h using an OLYMPUS IX53 microscope and calculated by ImageJ software, $n = 4$, * $p < 0.05$; ** $p < 0.005$; *** $p < 0.001$, NS—not statistically significant. (**D**) DSF treatment reverses cisplatin resistance. Standard SDS-PAGE Western blot analysis of EMT markers. The A549CisR cells were treated with 3 µM DSF (48 h), and cell lysates were obtained using M-PER Mammalian Protein Extraction Reagent #78501 as described in Materials and Methods. (**E**) To determine the impact of DSF on cell viability, A549CisR cells were seeded on 96-well plates and treated with 75 µM cisplatin, 3 µM DSF, and 3 µM DSF 75 µM cisplatin. After 48 h, cell viability was tested using WST-1 assay (ScienCell, Research Lab., Carlsbad, CA, USA); $n = 3$, * $p < 0.05$; ** $p < 0.005$; *** $p < 0.001$, NS—not statistically significant.

2.3. Analysis of Cisplatin Resistance in NSCLC Cells Based on Changes in mRNA Expression of ALDH Family Proteins, EMT Marker, and ABC Proteins According to GEO Data

Delivery of cisplatin-resistant NSCLC cells is considered the standard approach (Materials and Methods section). However, to confirm whether the phenotypical changes occurring in the A549CisR variant are universal, the differences in mRNA expression between A549 and A549CisR were compared using the Gene Expression Omnibus GSE108214 database (Figure 2). A549CisR demonstrated significantly higher ALDH3A1 (stem cell marker) [10] and ALDH7A1 mRNA expression compared to parental A549, similarly to those obtained by our Western blot results. Interestingly, no statistically significant changes were observed in the mRNA expression of the most well-known stem cell marker, ALDH1A1 (Figure 2A) [10].

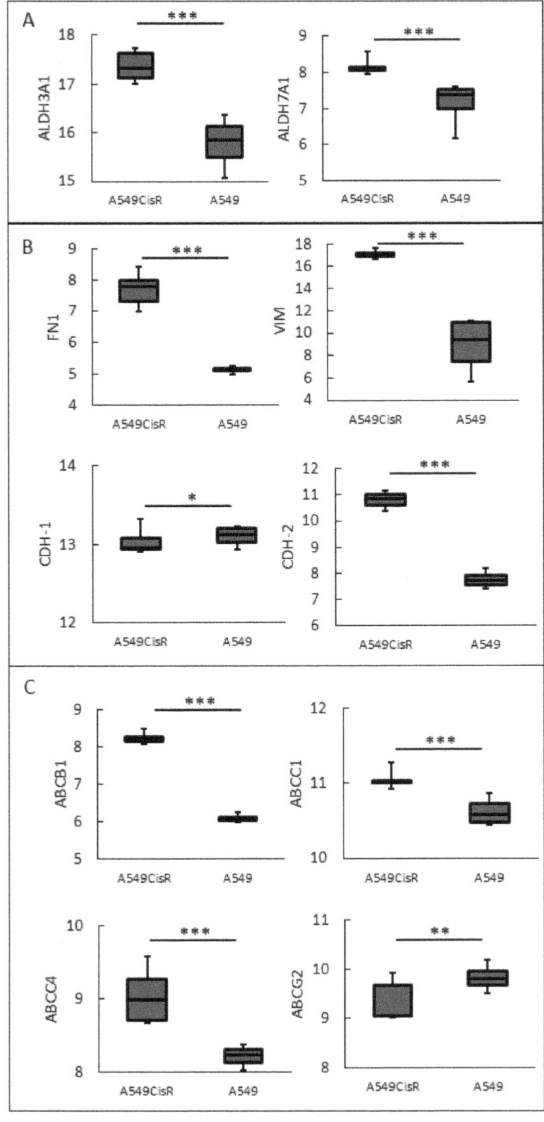

Figure 2. Cisplatin-resistance-related changes in mRNA expression of ALDH family proteins (**A**),

EMT markers (**B**), and ABC proteins (**C**). NSCLC cell line A549 and its cisplatin-resistant variant A549CisR mRNA levels were analyzed using microarray dataset GSE108214, acquired from the public Gene Expression Omnibus (GEO) databases [59]. Normality test (Shapiro–Wilk) was performed, followed by the Mann–Whitney U test; * $p < 0.05$; ** $p < 0.005$; *** $p < 0.001$, not statistically significant—no indicator.

In addition, A549 demonstrated significant upregulation of all major EMT markers, including vimentin (VIM), fibronectin (FN1), and N-Cadherin (CDH-2) (Figure 2B), with no repression of epithelial E-Cadherin (CDH-1). This suggests that acquisition of cisplatin resistance is accompanied by the development of an advanced (not yet fully completed) hybrid EMT phenotype, with strong migratory potential. Recently, EMT was confirmed to be an important regulator of several ABC proteins, as the promoters of ABC transporters contain several binding sites for EMT-inducing transcription factors such as: Twist, Snail, and ZEB. Thus, this leads to increased ABC protein mRNA expression and increased broad, multidrug resistance [7,57,58]. Analysis of the GSE108214 data set confirmed that the mRNA expression of ABCB1, ABCC1, and ABCC4 by cisplatin-resistant variants of A549 (A549CisR) was significantly upregulated, whereas ABCG2 was downregulated (Figure 2C).

2.4. Assessment of Affinity of Isothiocyanates to Chosen ADHD and ABC Family Protein

In vivo, DSF is rapidly metabolized to diethyldithiocarbamate (DDTC), which is further converted to *S*-methyl-*N*,*N*-diethyldithiocarbamate (Me-DDTC) and *S*-methyl-*N*,*N*-diethylthiocarbamate (DETC). Subsequently, P450 catalyzes the oxidation of DETC, and Me-DDTC produces DETC-sulfoxide (DETC-SO) and *S*-methyl-*N*,*N*-diethyldithiocarbamate-sulfoxide (Me-DDTC-SO) and -sulfone (Me-DTC-SO_2), metabolites that are most likely directly involved in ALDH inhibition. Importantly, when downstream steps of DSF metabolism are blocked by a chemical P450 inhibitor, liver ALDH remains uninhibited, confirming that it is the metabolites of DSF that are the true inhibitors of ALDH in vivo [60].

Therefore, the study examined the effects of isothiocyanate DSF analogues (ITCs) synthesized for the purpose of the study. In silico molecular modeling allowed the selection of the three most promising compounds: (2-isothiocyanatoethane-1,1-diyl)dibenzene named **18**, 1-(isothiocyanatomethyl)-4-phenylbenzene named **19**, and 1-isothiocyanato-3,5-bis(trifluoromethyl)benzene named **36** (Figure 3A); these presented the highest affinity to the several available models (RCSB Protein Data Bank) of ALDH isoforms, including ALDH3A1 and ALDH7A1, which were upregulated in the cisplatin-resistant variant of A549 (Figure 3B), and the ABC proteins involved in drug resistance (ABCB1, ABCC1, and ABCG2) (Figure 3C). Among the tested ITCs, the highest average affinity to all selected proteins was demonstrated by **18**, **19**, and **36** (Figure 3D).

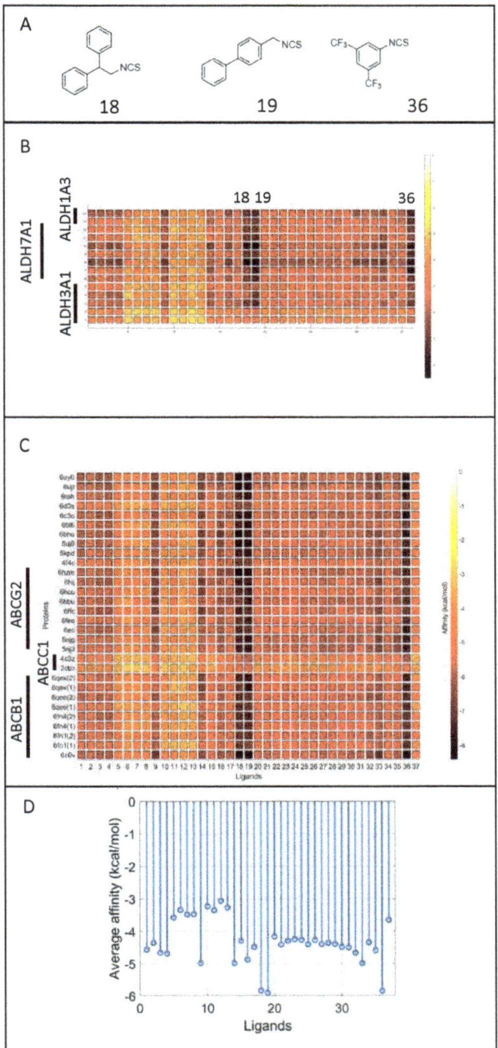

Figure 3. Molecular modeling of isothiocyanate (ICT) affinity to the tested ALDH and ABC proteins. (**A**) Chemical structures of three chosen ICTs: **18**, **19**, and **36**. (**B**) ICT affinity to the models of ALDH isoforms and (**C**) ABC proteins obtained from RCSB Protein Data Bank. (**D**) Average affinity of ICTs to all selected proteins, calculated as average of all affinity value to all structural models of each protein.

2.5. Impact of ITCs on the Reversion of Cisplatin Resistance, EMT Phenotype and Cell Migration

The chosen isothiocyanates are well-tolerated by the A549 cell line across a broad concentration range for 24 h: the IC_{50} values were approximately 80 μM for ITC **18** but are not present for ITC **19** and are 360 μM for ITC **36**. The ITCs demonstrated no cytotoxic effect against A549 or A549CisR at concentrations of 0.1–5 μM (Figure 4A). However, **19** and **36** significantly repressed the cisplatin resistance of A549CisR cells at 10-fold lower concentrations (0.3 μM), similar to DSF itself (3 μM) (Figure 4B). No statistically significant changes in cisplatin resistance were observed for combined 0.3 μM ITC **18** and 75 μM cisplatin (Figure 4B). Additionally, ITCs 19 impact on cisplatin resistance repression was tested in 3 μM concentration, presenting nearly 50% higher sensitivity compared to the untreated, resistant variant A549 (Figure 4B).

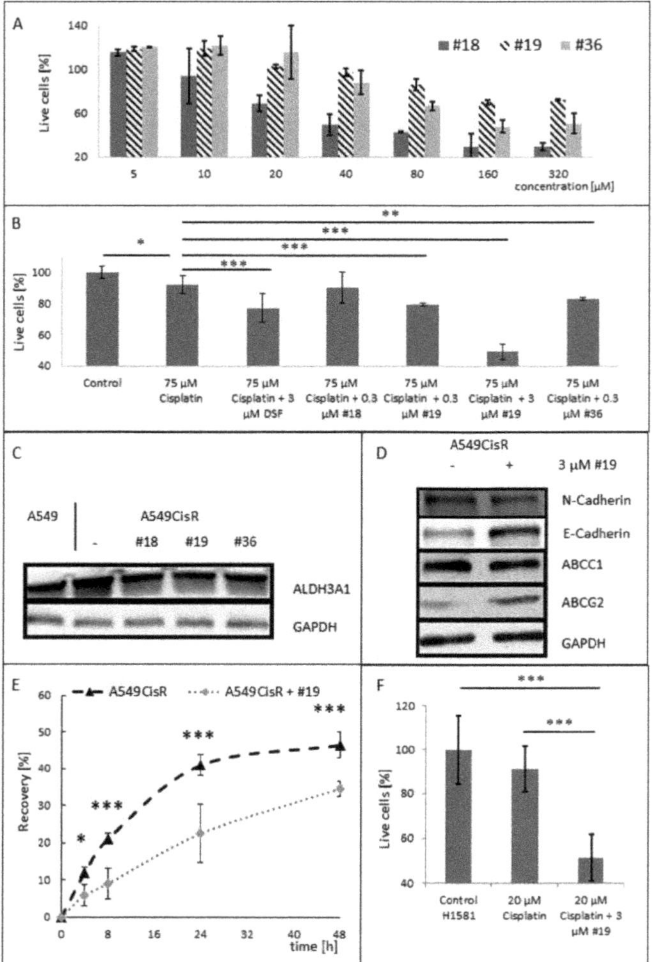

Figure 4. Induction of cisplatin-resistance reversion by isothiocyanates. (**A**) The impact of isothiocyanate concentration on A549 cell viability—WST-1 assay after 24 h supplementation with compounds **18**, **19**, or **36** (ScienCell, Research Lab., Carlsbad, CA, USA). (**B**) Reversion of cisplatin resistance (WST-1 assay). (**C**) Downregulation of ALDH3A1 by ITCs. Long-term (3 μM, 15 days) impact of compound **19** on EMT marker and ABC protein levels. (**D**) Standard SDS-PAGE Western blot. (**E**) Wound-healing-like analysis of the impact of compound **19** on cisplatin-resistant NSCLC cell migration. A549CisR and A549CisR supplemented with compound **19** (3 μM, 15 days) cells were seeded on a six-well plate, grown to confluence. A wound was made and rinsed twice with PBS. New full medium was added. The wounded area was visualized after 0, 4, 8, 24, and 48 h using an OLYMPUS IX53 microscope and calculated by ImageJ software, $n = 3$. (**F**) ITC **19** increases cisplatin sensitization of advanced NSCLC cells NCI-H1581. * $p < 0.05$; ** $p < 0.005$; *** $p < 0.001$, not statistically significant—no marker.

As compound **19** presented the strongest cisplatin-resistance reversion abilities and the highest tolerability in both tested cell lines, it was chosen for further study of its long-term impact on cisplatin-resistant NSCLC cells when applied at the same concentration as DSF itself (3 μM). Fifteen-day supplementation with 3 μM ITC **19** resulted in significant repression of ALDH3A1 expression (similarly to **18** and **36**) (Figure 4C) suggesting the reversion of a stem-cell-like phenotype. Furthermore, compound **19** was found to partially reverse EMT phenotype, presenting increased expression of the epithelial marker E-cadherin, with no significant changes in N-cadherin expression (Figure 4D) (similar to DSF). Phenotype reversion was also accompanied by the acquisition of an ABC protein-expression pattern, characteristic of the non-resistant A549 parental variant (Figure 4D), i.e., a decrease in ABCC1 and an increase in ABCG2 expression.

Furthermore, acquisition of cisplatin resistance increased the migratory abilities of A549CisR cells, rendering them highly metastatic. However, treatment with ITC **19** (15 days, 3 μM) significantly lowered 2D migration rate, as indicated by wound healing assay (Figure 4E).

2.6. Isothiocyanate #19 Increases Cisplatin Sensitization of Advanced NSCLC Cells

As ITC **19** significantly reduces cisplatin resistance in A549CisR, it was also tested against NCI-H1581, a stage 4 NSCLC cell line (CRL-5878) (Figure 4F). NSCLC predominantly consists of adenocarcinoma (AC) and squamous cell carcinoma (SCC). H1581 represents the smallest subfraction (10%) of NSCLC: a large cell carcinoma (LCC) that tends to grow rapidly and spread more aggressively than some other forms of lung cancer. H1581 is characterized by high focal amplification of FGFR1 [61] overexpression, which is related to increased aggressiveness, metastasis, and poor prognosis in various cancer types (especially in NSCLC) [62]. Activation of FGFR1 was reported to initiate EMT in several cancer types, including primary or secondary drug resistant lung cancer and lung cancer cell lines such as H1581 [63]. Importantly, H1581 cell line presents an EMT-derived, highly drug-resistant, cancer stem-cell-like phenotype, with increased ALDH activity. Inhibition of FGFR1 and ALDH activity suppress the growth, viability, and stem-cell-like phenotype of H1581 [64]. Thus, H1581 may be considered a well indicator, partially proving ALDH importance in drug resistance. Supplementation of H1581 with 20 μM cisplatin (24 h) had no significant impact on cell viability; however, the combination of 3 μM **19** and 20 μM cisplatin significantly decreased cell viability, as observed using a standard WST-1 assay (Figure 4F). ITCs 19′s impact on H1581 viability is presented in Supplementary Materials (Figure S2).

2.7. Isothiocyanate #19 Decreases Proteolytical Abilities and Invasive Properties of Cisplatin-Resistant NSCLC Cells

The obtained cisplatin-resistant variant of A549 presents a significantly higher migration rate as observed in the wound healing assay. However, migration has two main types: "path finding" (amoeboid migration type) and "path generating" (mesenchymal type of migration). Amoeboid migration is characterized by rounded cell morphology, low adhesion, high migration velocity, extensive cell body deformations caused by actin protrusions or hydrostatic membrane blebs, and its independence of extracellular matrix (ECM) degradation. Thus, amoeboid migration is based on cells' abilities to find and fit into existing "paths". On the other hand mesenchymal type of migration is acquired in non-mesenchymal cells via EMT and strongly depends on proteolytic degradation of ECM components (mainly via matrix metalloproteinases—MMPs), which enables crossing of the anatomical boundaries and in-aftermath metastasis [65,66]. Cisplatin resistance is often accompanied by increased metastatic potential. A549CisR presents higher proteolytic abilities than parental A549, as visualized by confocal microscopy imaging (zymography in situ assay—white arrows point increased gelatinolytic effect) (Figure 5A) and calculated using fluorescent dequenching (DQ) gelatin assay (Figure 5B). A549CisR supplemented with 3 μM **19** decreases gelatin degradation to the level observed for parental A549 and A549CisR treated with MMP2 inhibitor ARP101 (24 h, 10 μM). Interestingly, gelatin degradation presented by A549CisR supplemented with both 10 μM ARP101 and 3 μM **19** is slightly, yet statistically significant, lower than the one observed for either of the compounds alone, whereas the MMP2 protein level remains unchanged by **19** (Figure 5C). An increased ability to cleave ECM components such as collagen (or its degraded form—gelatin) is required by cancer cells during invasion and metastasis, as it provides physical disintegration of anatomical boundaries allowing for invasion of surrounding as well as distant metastasis [66]. Thus, invasive properties of A549 and A549CisR were tested (Figure 5D,E). Cells were treated with or without 3 μM **19** for 24 h, and next were transferred to gelatin coated 8 μm pore size upper chamber of Nunc Cell Culture Inserts in starving medium (with or without 3 μM **19**). A full medium was used in lower chamber as chemoattractant. Cells were allowed to enter the membrane pores thru gelatin layer for 3 h. Next, the medium and the gelatin from the top surface of the membrane were removed, the invaded cells on the bottom surface of the membrane were washed 2× with PBS, and then fixed, stained with Hoechst 33342, and counted in five random spots. The cisplatin-resistant variant presents a significantly higher invasion rate than the parental A549 cells. Furthermore, supplementation with **19** significantly decreases invasion of A549CisR, presenting no statistically significant effect on A549.

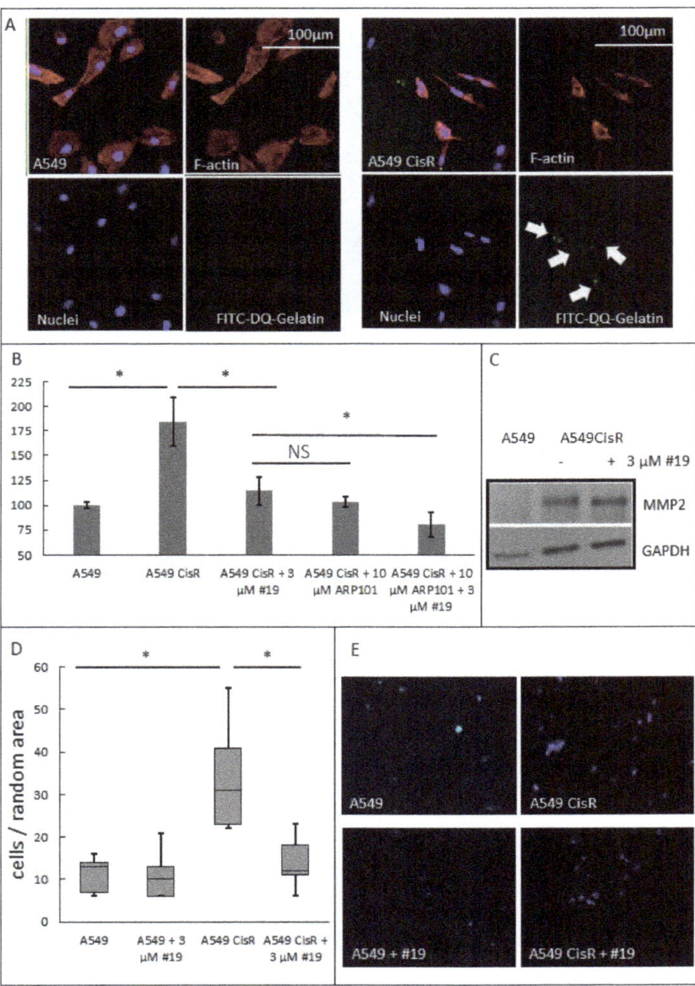

Figure 5. Repression of proteolytical and invasive abilities of A549CisR by isothiocyanates. (**A**) Confocal

microscopy image of A549 and A549CisR proteolytic degradation of FITC-conjugated DQ gelatin (Life Technologies, Waltham, MA, USA) (zymography in situ). Cleaved DQ gelatin becomes fluorescently active as highlighted by white arrows. F-actine stained with Texas Red-X Phalloidin, nuclei stained with Hoechst 33342. (**B**) Gelatinolytic activity of A549 and A549CisR calculated by measuring fluorescence intensity of cleaved DQ gelatin. MMP2 inhibitor ARP101 used as an additional control to verify ITCs impact on DG gelatin degradation. Calculated fluorescence intensity in arbitrary units of fluorescence were et as control—100% for A549. * $p < 0.05$, not statistically significant—NS. (**C**) Standard SDS-PAGE Western blot analysis of MMP2 expression in A549, A549CisR, and A549CisR supplementation with compound **19** (24 h 3 μM). (**D**) Gelatin transwell invasion assay based on [57,67]. A549 and A49CisR were treated with or without compound **19** (24 h 3 μM) and transferred to upper chamber in starving medium supplemented with or without **19** 3 μM. Full medium was used as chemoattractant in lower chamber. Invasion was visualized and calculated after 3 h in 5 random spots. * $p < 0.05$. (**E**) Gelatin transwell invasion assay visualization by fluorescence microscopy—random representative image.

3. Discussion

Currently, the best treatment for most types of carcinomas is surgical excision of the early primary tumor with proper histopathologic margins [68]. However, often due to late diagnosis (advanced stage of tumor), cancer cells are able to increase their mass and invade surrounding/distant tissue, thus preventing surgical removal. In such cases, chemotherapy remains one of the basic treatment modality [5,54]. This fact is extremely important in case of lung cancer: due to the lack of effective early-detection methods, it has one of the highest mortality rates among cancers [1].

Cisplatin is commonly used in many lung cancer types, including squamous cell carcinoma (SCC), large cell carcinoma (LCC), and adenocarcinoma (AC) [69]. Although cisplatin shows remarkable effects during initial treatment, a large majority of patients develop resistance as treatment proceeds, presenting a higher number of secondary tumors after period of remission [5,54]; for example, 30–55% of non-small-cell lung cancer (NSCLC) patients (both adenocarcinoma and squamous cell carcinoma) [70] suffer from cisplatin-resistant cancer reoccurrence within one year of surgery and associated chemotherapy. There are many mechanisms responsible for cisplatin therapy failure such as DNA-damage repair, cell-death inhibition, drugs efflux and inactivation, drug-target alteration, and metabolic shift. However, one of the most prominent mechanisms is EMT induction, which allows an epithelial cell to obtain a mesenchymal phenotype, resulting in an increased ability to migrate [5]. Importantly, cisplatin-related EMT leads to the acquisition of a migratory phenotype, which allows passage across anatomical boundaries and the invasion of both local and distant tissue [71]. Interestingly, inhibition of Ataxia Telangiectasia Mutated (ATM) results in reversion of the EMT phenotype in cisplatin-resistant NSCLC cells, inhibiting cell invasion and tumor metastasis [72]. Although cisplatin resistance is driven by multiple, often ostensibly unrelated mechanisms, the Western blot analysis of our present cisplatin-resistant A549CisR found it to correspond on the protein level with the mRNA profile of a previously described A549CisR variant [73], which is publicly available in the GEO database as the GSE108214 dataset.

Furthermore, in our A549CisR cell line, cisplatin resistance manifests as ongoing EMT, indicated by upregulation of N-cadherin and repression of E-cadherin, which substantially increased migratory potential. Previous studies have also noted a similar co-existence between cisplatin resistance and EMT induction [72,74]. Importantly, our derived A549CisR cell line, representing an EMT phenotype, demonstrated ABCC1 and ABCC4 upregulation and ABCG2 downregulation; this relationship was observed for GSE108214, while ABCG2 repression has also been noted in advanced colorectal cancer cells undergoing EMT and patient samples [57]. Several EMT-inducing transcription factors, such as Twist, Snail, and ZEB, were recently confirmed to be important regulators of certain ABC proteins, directly interacting with E-Box sites in their corresponding promoter regions [7,57]. This mechanism may be utilized by cisplatin-resistant cells to increase overall multidrug resistance, as ABC proteins have broad spectrum of transported chemotherapeutic agents such as 5-fluorouracil, irinotecan, doxorubicin, mitoxantrone, or vinblastine, to name a few [75,76].

Currently, potential supplements to cisplatin-based chemotherapy are being sought [77–79]. One such candidate is Disulfiram (DSF), which is an aldehyde dehydrogenase (ADH) inhibitor that has been used as a first-line anti-alcoholism Drug The drug has been reported to cause cell-cycle arrest in the G2/M phase and enhance cisplatin sensitivity in NSCLC lines [80]. Recently, disulfiram has been called "a novel repurposed drug for cancer therapy", with an anti-cancer effect noted in several cancer types, including liver, breast, prostate, pancreatic, and NSCLC [81]. Therefore, the present study examines the potential of ADH inhibitors for improving the treatment of cisplatin-resistant NSCLC. Moreover, interestingly, aldehyde dehydrogenase (ALDH) is considered to be marker of NSCLC circulating tumor cells (CTC), which indicate an advanced EMT phenotype [82].

Our data confirm that DSF triggers re-sensitization of a cisplatin-resistant NSCLC cell line (A549CisR). Although DSF is known to inhibit NF-kB signaling, proteasome activity, and aldehyde dehydrogenase (ALDH) activity and to induce endoplasmic reticulum (ER) stress and autophagy, the exact mechanisms of its anti-cancer properties remain unclear [81].

Our findings indicate that DSF reverses an acquired mesenchymal phenotype to a certain extent, repressing the expression of mesenchymal marker N-cadherin. Interestingly, DSF is rapidly metabolized to diethyldithiocarbamate (DDTC), which is further converted to *S*-methyl-*N*,*N*-diethyldithiocarbamate (DETC) and *S*-methyl-*N*,*N*-diethyldithiocarbamate (Me-DDTC), and it is these metabolites of DSF that are the true inhibitors of ALDH in vivo [60].

The present study evaluates the potential of de novo synthesized isothiocyanates (ICTs) that resemble DSF. Out of 36 tested ITCs, the three most promising compounds were chosen, viz. **18**, **19**, and **36**: these were found to manifest strong affinity to ALDH isoforms and ABC proteins based on in silico analysis. Interestingly, **18** and **19** have also presented strong anti-tumoral properties in a xenograft zebrafish model [83]. The chosen ITCs appear to be strong anti-cisplatin-resistance agents: they were found to significantly repress cisplatin tolerance in both A549CisR and in the stage 4 NSCLC cell line NCI-H1581 at concentrations 10-fold lower than DSF. ALDH family and/or ABC proteins most probably are not the only important target for ITCs, that are beneficial during anticancer therapy. Benzyl isothiocyanates BITCs are one of the most extensively studied ITCs with regard to cancer chemoprevention, which was proven to inhibit the phosphorylation activities of three major mitogen-activated protein kinases (MAPKs): ERK1/2, p38, and p-JNK1/2 [50].

In a previous study, ITCs were found to demonstrate anti-proliferative activity in vitro in the human colon, uterus, mammary gland, and lung carcinoma cell lines. ITC treatment led to cell-cycle arrest and cell death. ITCs have also been found to be effective against a murine mammary gland carcinoma 4T1 model in vivo, with administration resulting in reduced tumor mass [33]. Furthermore, 3 μM ITC analogs inhibited the motility of three highly malignant cell lines derived from cervical (HeLa), glioblastoma (U87), and breast (MDA-MB-231) carcinomas [83]. This finding is consistent with our present observations.

DSF has been used to treat alcohol abuse for about 70 years. Typically, patients receiving DSF-based therapy are exposed to high doses for a long period of time, i.e., six or more months, during which time the drug appears to possess non-lethal properties, if not mixed with alcohol [84].

In the present study, 15-day supplementation with **19** was found to partially reverse the EMT phenotype (restoration of E-cadherin expression) stem-cell-like phenotype (down-regulation of ALDH3A1) of the tested on A549CisR cells, and repress their migratory potential to the level observed in parental A549 cells.

Cell migration is a complicated process consisting of cell-body polarization, followed by the formation and extension of cell protrusions; these protrusions adhere to the substratum, and cell contraction moves the cell body forward toward the leading edge. The migration cycle is completed by deadhesion of the attachments at the rear of the cell [85]. This cascade of events requires substantial energy expense in the form of ATP [86]. Thus, since ALDH mediates the production of NADH, which is used as an energy source for ATP production during oxidative phosphorylation (OXPHOS) [87], long-term ALDH inhibition may slow the migration rate. Importantly, while OXPHOS uses NADH supplied from ALDH in cancer cells, OXPHOS obtains NADH via the tricarboxylic acid cycle (TCA cycle) in normal cells. Thus, targeting cancer cell OXPHOS by inhibiting ALDH could selectively reduce their ATP levels, significantly repressing many of the mechanisms of cancer cells [88]. High ALDH3A1 expression and activity correlates with cell proliferation and resistance against drug toxicity, such as cyclophosphamide, ifosfamide and trofosfamide. Repression of cells proliferation and drug resistance can be observed upon ALDH3A1 directly inhibition by the administration of specific synthetic inhibitors, antisense oligonucleotides, or siRNA [89,90]. Thus, even though ALDH3A1 expression was downregulated in a discrete manner by the tested ITCs in A549CisR cells, inhibition of its activity is the most important factor influencing cell response. Furthermore, mitochondria are able to interact with the nucleus through the retrograde signaling mechanism; this results in the activation of diverse nuclear responses that regulate survival rate, metastasis, and drug resistance [91]. Additionally, mesenchymal type of migration that strongly relies on proteolytical degradation of ECM components is acquired in non-mesenchymal cells via EMT [65,66]. Thus, often cisplatin-resistant cells, presenting EMT phenotype, are characterized by increased proteolytical abilities and a path-generating type of migration [71]. In this study, tested A549CisR presented enhanced gelatinolytic and invasive properties in comparison to the parental A549. Both invasion and gelationlysis were significantly suppressed by 24 h supplementation with 3 μM ITCs **19**. Gelatin is mainly degraded by matrix metalloproteinases 2 and 9 (MMP2 and MMP9), thus, inhibition of their activity results in altered invasion and metastasis. Importantly, **19** in the tested concentration decreases gelatin proteolysis to the level presented by the well-known MMP2 inhibitor ARP101 (10 μM, 24 h), with no changes observed in the MMP2 protein level. Importantly, MMP2 is produced as an inactive proenzyme that requires activation, canonically performed by MMP14 (non-canonically performed by other enzymes, such as MMP2 itself or Cathepsins); thus, the unchanged protein level upon **19** supplementation is less relevant in terms of MMP2 activity and involvement in cell migration [67]. This effect was presented also by benzyl isothiocyanates (BITCs), which was proven to downregulate both MMP2 and MMP9 but, more importantly, to increase the mRNA level of tissue inhibitor of matrix metalloproteinases-2 (TIMP-2) [50]. Thus, we can strongly assume that compound **19** acts as indirect suppressor of MMP2 activity, however, the exact mechanism is yet to be discovered.

Furthermore, in the A549CisR cell line, the ABC protein levels were restored to the parental non-resistant variant: ABCC1 was downregulated whereas ABCG2 was upregulated. Interestingly, neither ABCC1 nor ABCG2 are cisplatin exporters [75,76]; hence, it was unclear why cisplatin resistance regulates ABC protein levels, while not being a substrate for particular transporters itself, and why compound **19** reverses their level to the one observed in parental A549 cells [92]. However, cisplatin-resistant cells exhibit an EMT phenotype, which has been proposed to be the main cause of the primary and acquired drug resistance in several cancer types [7,93]. The observed downregulation of ABCG2 in the more mesenchymal A549CisR cell line may be difficult to explain, but our data correspond well with previous findings, indicating that the downregulation of ABCG2, at both the mRNA and protein level, reverses the correlation with mesenchymal markers and acquisition of advance EMT in CRC cells (in vivo and in vitro) [57]. Furthermore, long-term supplementation with compound **19** partially represses and reverses EMT, i.e., restores E-Cadherin expression, and EMT triggers the metastatic and drug-resistance properties of cisplatin-resistant NSCLC cells.

Importantly, compound **19** presents high affinity to ABC proteins, including ABCB1, ABCC1, and ABCG2, which are present in various pharmaceuticals, such as Doxorubicin, Paclitaxel, Vinblastine, Methotreaxate (MTX), Irinotecan, and Topotecan [75,76]. This can potentially increase their cytosolic accumulation, leading to the repression of cisplatin-induced multidrug resistance and the resensitization of cells; however, this needs further investigation. Compound **19** appears to possess strong anti-cisplatin/anti-multidrug resistance and anti-metastatic properties, and its supplementation during chemotherapy may be highly beneficial for NSCLC patients.

4. Materials and Methods

4.1. Cell Culturing and Induction of Drug Resistance

Lung cancer cell lines A549 and NCI-H1581 were obtained from the ATCC (Manassas, VA, USA). The A549 cisplatin-resistant sub-line A549CisR was established by growing A549 cells in the presence of increasing concentrations of cisplatin to a final concentration of 10μM over approximately six months. Both cell lines (A549/A549CisR) were cultured in Ham's F12-K medium (Corning, Manassas, VA, USA), H1581 were cultured in DNEM/HAM F-12 (Corning, Manassas, VA, USA) in a 90–95% humidified atmosphere of 5% CO_2; the media were supplemented with 10% heat-inactivated fetal bovine serum (FBS) (Biowest, Nuaillé, France) and the antibiotics streptomycin, penicillin (Biowest), and primocin (Invivogen, San Diego, CA, USA). The cells were plated in 25 cm^2 cell culture flasks and sub-cultured before reaching confluency using Accutase (Biowest). The culture medium was changed every two days. The cells were split 1:10 during each passage.

4.2. Reagents

Cisplatin [cis-diammineplatinum(II) dichloride], was obtained from Sigma-Aldrich (St. Louis, MO, USA) and dissolved in DMSO. Disulfiram was obtained from Cayman Chemical and dissolved in DMSO. Aliquots were stored at −20 °C for up to a maximum of three months and thawed immediately before use. WST-1 [2-(4-Iodophenyl)-3-(4-nitrophenyl)-5-(2,4-disulfophenyl)-2H-tetrazolium] was purchased from ScienCell (Research Lab., Carlsbad, CA, USA) freshly made, #8038; RIPA lysis buffer was purchased from VWR Chemicals, #N653-100 mL; protease inhibitor was purchased from Thermo Fisher Scientific (Waltham, MA, USA), #PIER87785; BCA purchased from Thermo Fisher Scientific, #PIER23225. Primary antibodies: E cadherin #3195P cell signaling, N cadherin #13116P cell signaling, ABCC1 #VMA00330 BioRad (Hercules, CA, USA), ABCC4 #PA5-18315 Thermo Fisher Scientific, ABCG2 #BRB155559 Biorbyt (Cambridge, UK), GAPDH #sc-32233 Santa Cruz Biotechnology (Dallas, TX, USA), α-Tubulin #NB100-690H Novus Biologicals (Littleton, CO, USA), ALDH3A1 #PA5-80332 Thermo Fisher Scientific, ALDH7A1 #MA5-29028 Thermo Fisher Scientific. Secondary HRP-conjugated antibodies were purchased from Santa Cruz Biotechnology.

4.3. Western Immunoblotting

Total protein was extracted from cells using ice-cold M-PER Mammalian Protein Extraction Reagent #78501 supplemented with the Halt protease inhibitor cocktail (Thermo Scientific, Waltham, MA, USA), and the soluble protein fraction was collected through centrifugation. The protein concentrations in the cell lysates were measured with the BCA method (Pierce/Thermo Scientific, Waltham, MA, USA) and equalized between samples. Protein (40 μg) from whole cell lysates was fractionated on SDS-PAGE gels and transferred to a PVDF or nitrocellulose membrane (BioRad, Hercules, CA, USA). Transfer efficiency and loading were confirmed by reversible staining of the membrane with Ponseau S solution (Sigma-Aldrich, UK) following protein transfer. Membranes were blocked at room temperature with blocking buffer (BioRad, Hercules, CA, USA). Primary antibodies were added in 1:1000–1:5000 dilution and incubated for one hour at RT (Materials and Methods Section 4.2). Membranes were washed 3 × 15′ with TBST and incubated with a secondary horseradish peroxidase (HRP)—labelled antibody for 1 h RT (1:2000). Membranes were washed in 3 × 15′ with TBST following incubation with secondary antibodies. Bound antibody complexes were detected and visualized using Clarity Western ECL Substrate (BioRad, Hercules, CA, USA). Densitometric analysis was carried out using ImageJ software, and percentage expression was represented relative to controls (100%) [94].

4.4. Cell Viability Assay

The in vitro cell viability effects of DSF, analogs of disulfiram, cisplatin, DSF/cisplatin, and analogs of disulfiram/cisplatin were determined by WST-1 assay. In brief, the cells (2×10^5 cells/mL) were seeded on 96-well culture plates and left for 24 h. Next, parental and resistant tumor cells were incubated in 100 μL of fresh medium containing different drug concentrations. Next, after 48 h of incubation, 10 μL of WST-1 reagent (ScienCell, Research Lab., Carlsbad, CA, USA), freshly made, #8038, was added for 2 h. Calculation of cell viability was done by $OD_{450\ nm}$–$OD_{630\ nm}$ using the BioTek ELx800 multimode microplate reader.

4.5. Wound Healing (Scratch) Assay

The 2D migration was tested using wound healing assay. Treated or untreated with ITC **19** (15 days, 3 μM), A549 and A549CisR cells were seeded on a six-well plate and grown to confluence. Wounds were created by scraping monolayer cells using a 20 μL pipet tip, and non-adherent cells were rinsed off twice with PBS. Fresh medium was added with or without 3 μM ITC **19**. The scratch and surrounding cells were imaged at 0 h (immediately after scratching). The area of the wound was visualized (OLYMPUS IX53 microscope, magnification, 100×) and measured 4, 8, 10, and 24 h after scratching. The wound area was calculated by ImageJ software. Cell motility was estimated by quantification of percentage recovery using the equation: R (%) = [1 − (wound area at Tt/wound area at T0)] × 100, where T0 is the wounded area at 0 h, and Tt is the wounded area after th. The assays were replicated three times.

4.6. GEO Database Analysis

Microarray profiles and the GSE108214 dataset of the NSCLC cell line A549 and its cisplatin-resistant variant (A549CisR) were acquired from the public Gene Expression Omnibus (GEO) databases—National Center for Biotechnology Information (NCBI), USA National Library of Medicine 8600 Rockville Pike, Bethesda, MD 20894, USA (https://www.ncbi.nlm.nih.gov/geo/, accessed on 14 April 2021) [59]; these were described previously [5]. The data were analyzed and presented as box charts, with the median depicted, as described previously [57]. Statistical analysis was performed using Jasp software (https://jasp-stats.org/, accessed on 6 May 2022) [95].

4.7. Synthesis of ITCs—Similar Compounds to DSF

The tested isothiocyanates **1–36** (Table 1) had previously been synthesized using four known methods [96–98] (Schemes 1–4, Methods A–D).

Table 1. Structure of tested isothiocyanates **1–36** [a].

Substrate		Isothiocyanate (Product)		Yield %	Method
Symbol	Structure	Symbol	Structure		
37	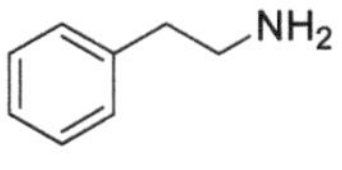 NH_2	1	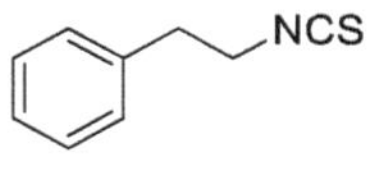 NCS	87	A
38	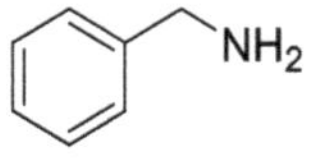 NH_2	2	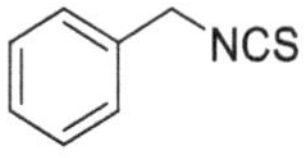 NCS	83	A
39	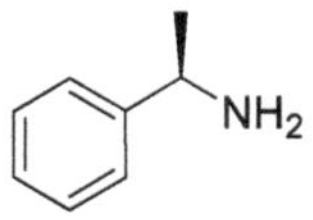 NH_2	3	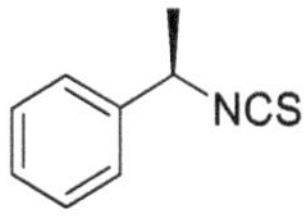 NCS	83	A

Table 1. *Cont.*

Substrate		Isothiocyanate (Product)		Yield %	Method
Symbol	Structure	Symbol	Structure		
40	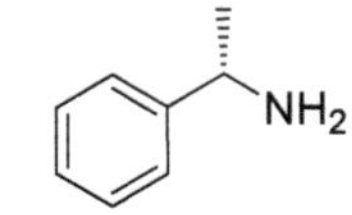	4	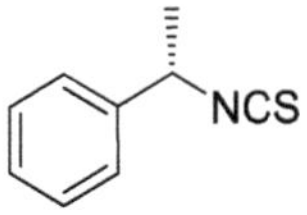	85	A
41	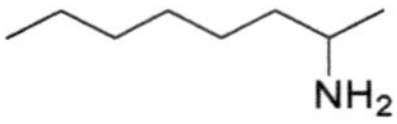	5	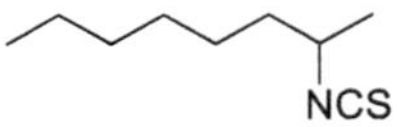	83	A
42	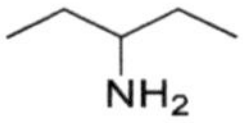	6	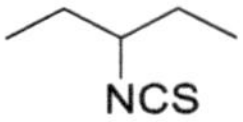	76	A
43	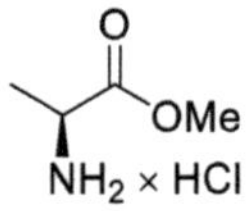	7	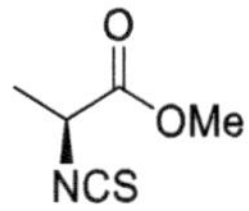	50	C
44	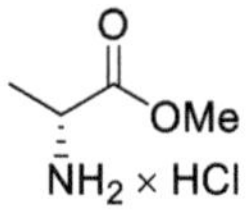	8	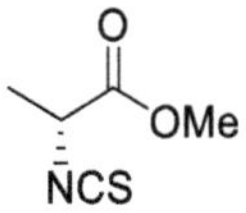	51	C
45	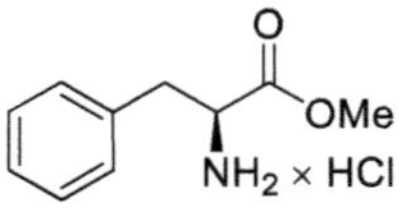	9	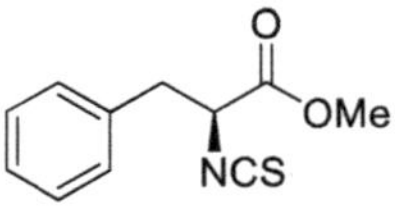	30	C
46	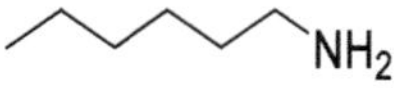	10		82	A
47	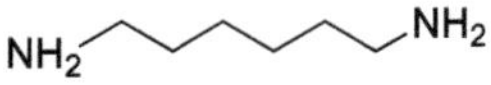	11	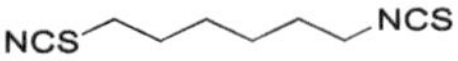	60	D
48		12		55	D
49	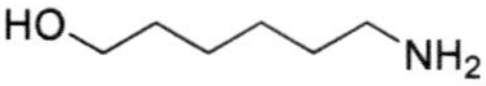	13	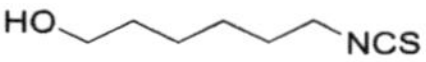	75	A
50	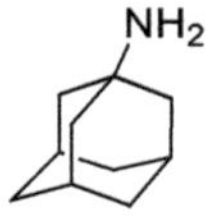	14	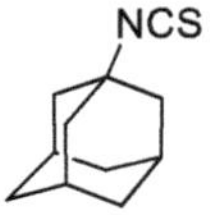	82	A
51	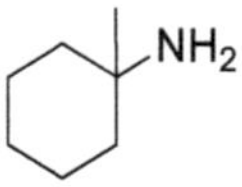	15	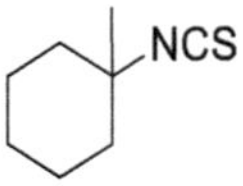	54	A
52	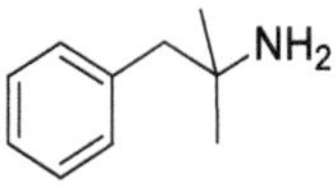	16	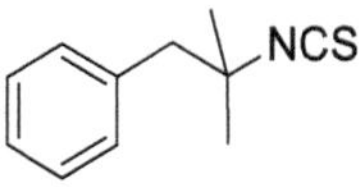	82	A
53	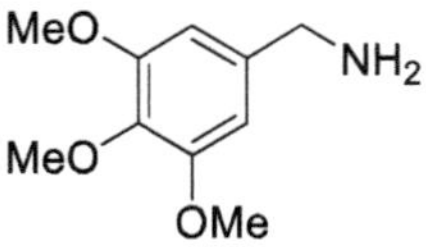	17	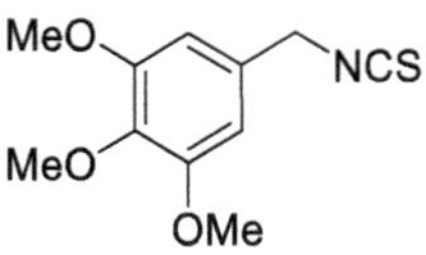	53	A

Table 1. *Cont.*

Substrate		Isothiocyanate (Product)		Yield %	Method
Symbol	Structure	Symbol	Structure		
54	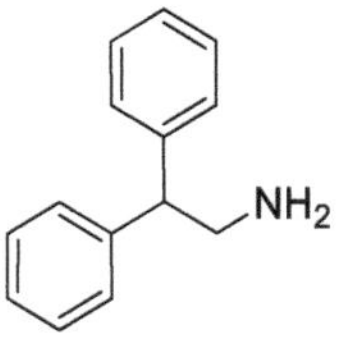 NH_2	**18**	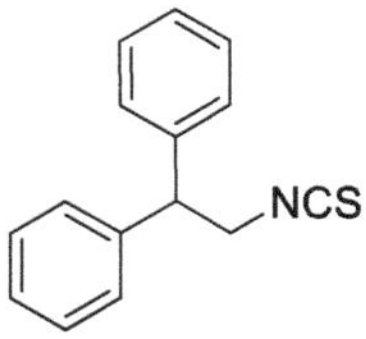 NCS	79	A
55	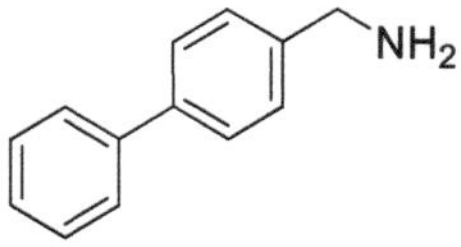 NH_2	**19**	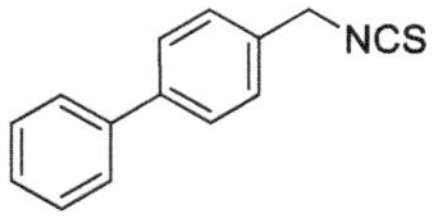 NCS	61	A
56	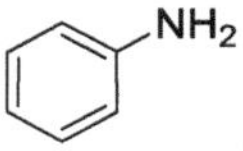 NH_2	**20**	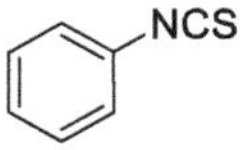 NCS	77	A
57	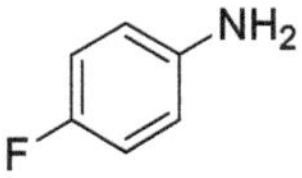 NH_2, F	**21**	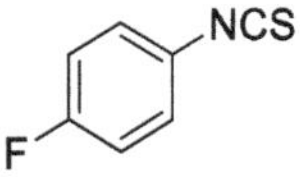 NCS, F	61	A
58	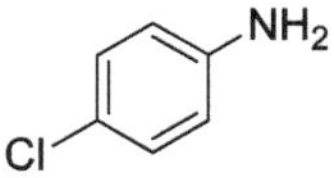 NH_2, Cl	**22**	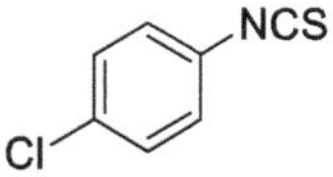 NCS, Cl	75	A
59	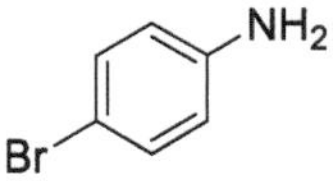 NH_2, Br	**23**	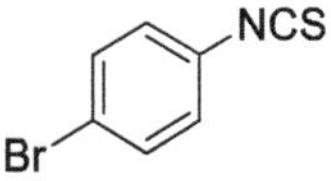 NCS, Br	95	A
60	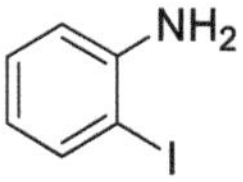 NH_2, I	**24**	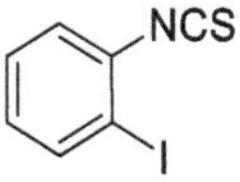 NCS, I	54	A
61	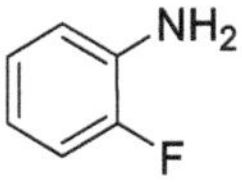 NH_2, F	**25**	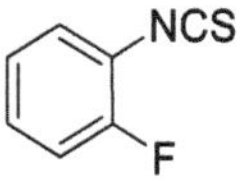 NCS, F	62	A
62	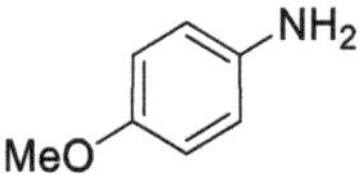 NH_2, MeO	**26**	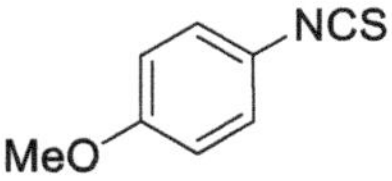 NCS, MeO	75	A
63	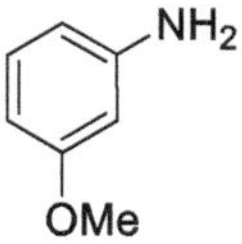 NH_2, OMe	**27**	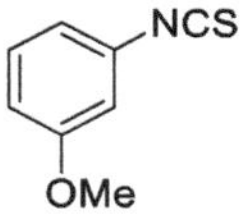 NCS, OMe	80	A
64	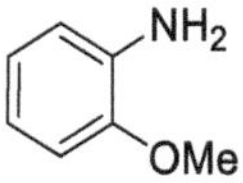 NH_2, OMe	**28**	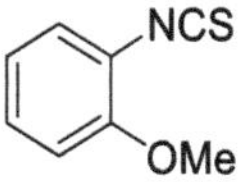 NCS, OMe	87	A
65	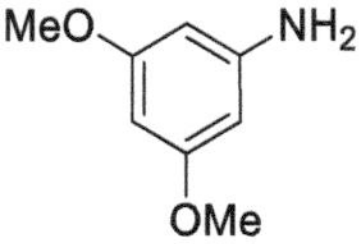 MeO, NH_2, OMe	**29**	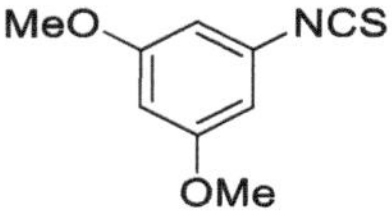 MeO, NCS, OMe	67	B

Table 1. *Cont.*

Substrate		Isothiocyanate (Product)		Yield %	Method
Symbol	Structure	Symbol	Structure		
66	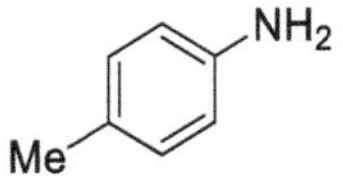	30	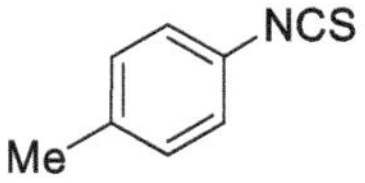	87	B
67	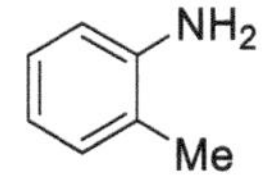	31	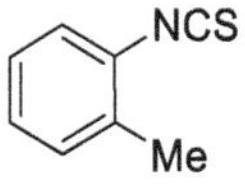	90	A
68	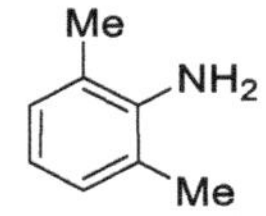	32	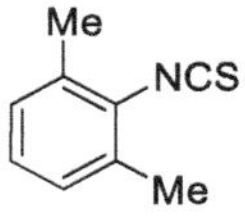	98	A
69	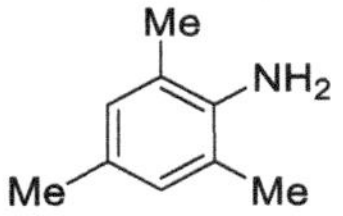	33	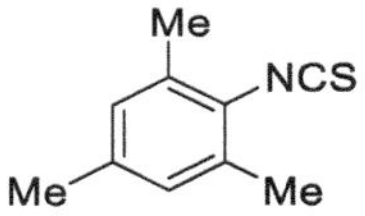	94	A
70	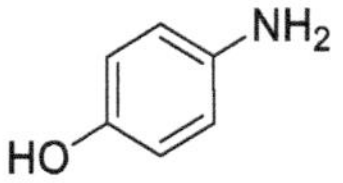	34	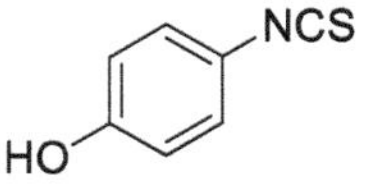	72	A
71	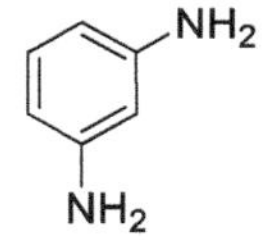	35	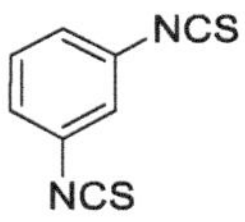	57	A
72	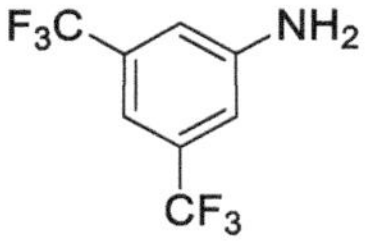	36	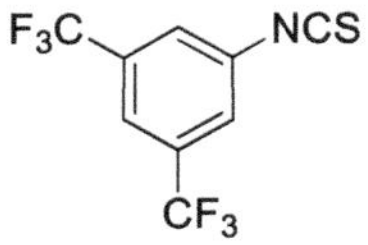	25	A

[a] Method A [96]; method B and method C [97]; method D [98].

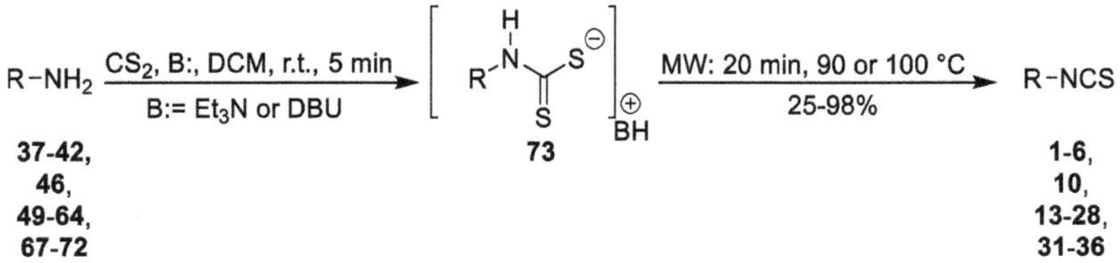

$R{-}NH_2$ → (CS_2, B:, DCM, r.t., 5 min; B:= Et_3N or DBU) → [73] → (MW: 20 min, 90 or 100 °C; 25-98%) → R-NCS

37-42, 46, 49-64, 67-72 → 1-6, 10, 13-28, 31-36

Scheme 1. Method A: microwave-assisted synthesis of isothiocyanates **1–6**, **10**, **13–28**, and **31–36**.

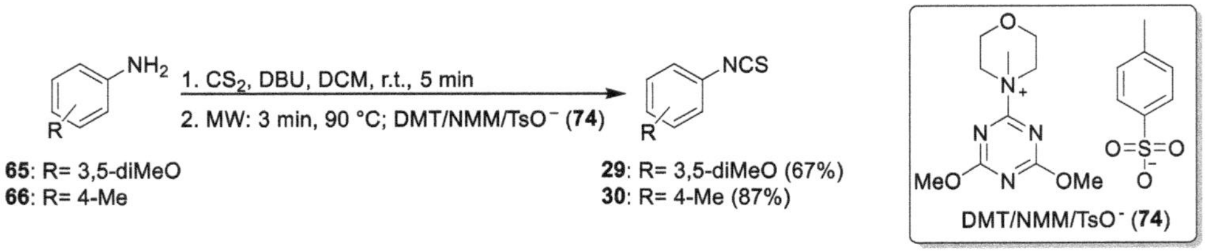

1. CS_2, DBU, DCM, r.t., 5 min
2. MW: 3 min, 90 °C; DMT/NMM/TsO⁻ (**74**)

65: R= 3,5-diMeO
66: R= 4-Me

29: R= 3,5-diMeO (67%)
30: R= 4-Me (87%)

DMT/NMM/TsO⁻ (**74**)

Scheme 2. Method B: microwave-assisted synthesis of isothiocyanates **29** and **30** using DMT/NMM/TsO$^-$ (**74**) as desulfurating reagent.

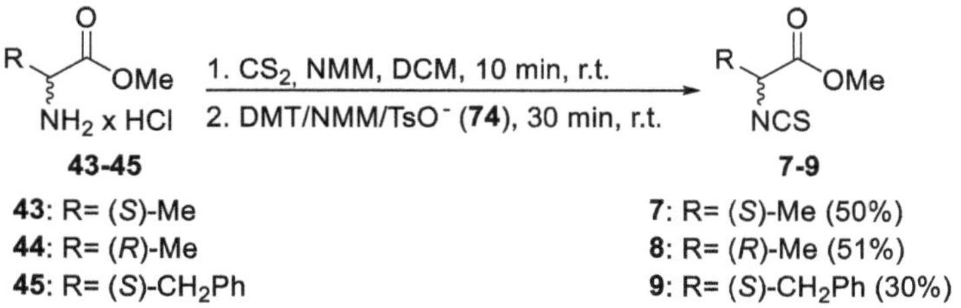

Scheme 3. Method C: synthesis of isothiocyanate derivatives of methyl ester amino acids **7–9** using DMT/NMM/TsO$^-$ (**74**) as desulfurating reagent.

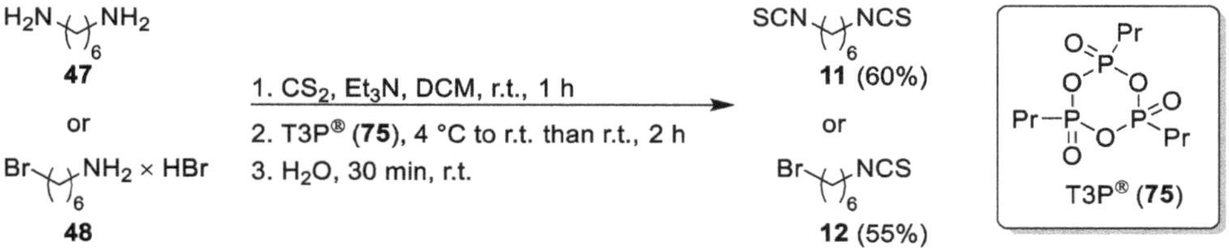

Scheme 4. Method D: synthesis of isothiocyanates **11** and **12** using propane phosphonic acid anhydride (T3P) (**75**).

The isothiocyanates **1–6**, **10**, **13–28**, and **31–36** were obtained by one-pot, two-step, microwave-assisted (MW) synthesis (Method A). Briefly, a mixture of aliphatic primary amines **37–42**, **46**, and **49–55** or aromatic amines **56–64** and **67–72**, carbon disulfide (CS_2) and triethylamine (Et_3N) (for amines **37–42**, **46**, and **49–55**), or DBU (for amines **56–64** and **67–72**) were transformed under normal conditions at room temperature into the intermediate dithiocarbamates **73**. Next, dithiocarbamates **73** were converted without any desulfurating reagent under microwave-assisted reaction (MW: 20 min, 90 °C for **37–42**, **46**, and **49–55** or 100 °C for **56–64** and **67–72**) to ITCs **1–6**, **10**, **13–28**, and **31–36** in good yields (25–98%) (Scheme 1, Table 1) [96].

Isothiocyanates **29–30** were also synthesized by one-pot, two-step microwave-assisted synthesis using CS_2, DBU, and aromatic amines **65** and **66** as substrates; however, a different MW method was used (Method B). The intermediate dithiocarbamates were synthesized, as shown in Method A. The second step however, performed under microwave-assisted conditions, was accomplished in a shorter time (MW: 3 min and 90 °C) and in the presence of 4-(4,6-dimethoxy-1,3,5-triazin-2-yl)-4-methylmorpholinium toluene-4-sulfonate (DMT/NMM/TsO$^-$, **74**) as a desulfurating reagent. Isothiocyanates **29** and **30** were isolated at high yields (67–87%) (Scheme 2) (Table 1) [97].

Isothiocyanate derivatives of methyl ester amino acids **7–9** were obtained under normal conditions in a one-pot, two-step procedure in the presence of DMT/NMM/TsO$^-$ (**74**) as desulfurating agent; however, the reaction was performed in normal conditions (Method C). The first step was performed in the presence of CS_2, NMM at room temperature in 10 min, using hydrochloride **43–45** as substrates; the second step was performed with desulfurating reagent **74** at room temperature and in 30 min. ITCs **7–9** were isolated with satisfactory yields (30–51%) after flash chromatography (Scheme 3) (Table 1) [97].

The last two isothiocyanates **11** and **12** were synthesized in a two-step process with propane phosphonic acid anhydride (T3P) used as a desulfurating reagent (**75**) (Method D). Diamine **47** or hydrobromide **48** was reacted with CS_2 in the presence of Et_3N at room temperature for one hour. Next, the reactions were cooled to 4 °C, T3P (**75**) was added in two portions, and the mixture was stirred for another two hours at room temperature. Products **11** and **12** were isolated in good yields (55–60%) using flash chromatography (Scheme 4) (Table 1) [98].

4.8. Molecular Modeling

Affinity calculations were performed for 26 ABC proteins and 27 ligands using software dedicated to molecular docking: AutoDockTools v.1.5.6 (La Jolla, California, USA), included in the MGLTools 1.5.6 (La Jolla, California, USA) package, and AutoDock Vina 1.1.2 (La Jolla, California, USA). Files describing proteins were downloaded in pdb format from the RCSB Protein Data Bank [99] and then converted using AutoDockTools to the pdbqt format required by AutoDock Vina.

The structures of ITCs **1–37** were drawn in ACD/ChemSketch (Toronto, Ontario, Canada) (Freeware) and then 3D Structure Optimization was performed. The structures were saved in .mol (MDL MOL) format. Next, the structures in .mol format were converted to .pdb (Protein Data Bank) format using OpenBabelGUI. The .pdb format files were used to model docking in the Vina package.

AutoDockTools was also used to determine the size of the search area and its center. For protein molecules with size that exceeded the maximum size of the search area, search sub-areas totaling the entire protein molecule were defined. For example, two search sub-areas covering the entire molecule were created for the 6qex protein, and these were treated as separate cases in the calculations: 6qex(1) and 6qex(2).

As a non-deterministic search algorithm was implemented in AutoDock Vina, each docking variant was calculated 10-fold. As a result, it was necessary to perform 9620 calculations (37 ligands × 26 proteins × 10 searches). As it would be difficult to run AutoDock Vina manually so many times, the calculation process was automated using a script written in Matlab 2021a. This script, in addition to running individual calculation cases, aggregated the results generated by AutoDock Vina. The calculation process is given in the flow diagram presented in Figure 6.

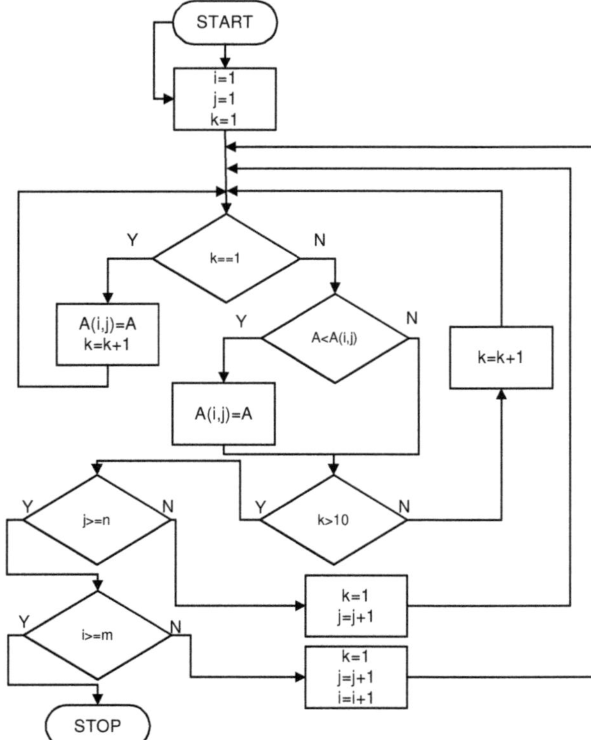

Figure 6. Flow diagram of the algorithm controlling the calculations, where i—protein index; j—ligand

index; k—search variant index; m—number of proteins; n—number of ligands; A—affinity for the current protein and ligand; A_{set}—set of affinity values.

The calculations were carried out on a PC equipped with a 12-core/24-thread AMD Ryzen 9 3900X processor. In order to fully utilize all processor cores, three instances of AutoDock Vina were run simultaneously. In order to make a quantitative assessment of the affinity of the studied ligands to selected proteins, the average affinity values of the analyzed ligands for selected proteins, expressed by the relationship (1), were calculated:

$$AA(j) = \frac{\sum_{i=1}^{i=m} A(i,j)}{m} \quad (1)$$

where *i*—protein index; *j*—ligand index; k—search variant index; *m*—number of proteins; n—number of ligands; *A*—affinity; *AA*—average affinity.

4.9. Fluorescent Dequenching (DQ) Gelatin Assay

The surface of 96-well plates was coated with 75 µL 0.1 mg/mL DQ gelatin (Life Technologies, Waltham, MA, USA) overnight at 4 °C and then washed 3× with PBS. Then, 25 × 10^5 cells/mL were added for 24 h to earlier prepared DQ gelatin-coated dishes in full medium. Additionally, as a control medium supplemented with 10 µM MMP-2 inhibitor ARP101 was used. FITC fluorescence generated by the cleavage of DQ gelatin was measured using a Thermo Labsystem Fluoroscan Ascent reader (ThermoFisher Scientific, Waltham, MA, USA), fit with FITC excitation and emission filters. Data are presented as the percent of increase above background fluorescence (100%) observed in the control A549 cell line [67].

Visualization of DQ gelatin assay (zymography in situ assay) was performed as described by us in [67]. Briefly, cells were grown on FITC gelatine-coated chamber slides (Life Technologies, Waltham, MA, USA) until 60–70% confluency and were subsequently incubated with Hoechst 33342 (Molecular Probes/Life Technologies, Waltham, MA, USA) for 15 min in the incubator. The cells were washed with PBS (3×), fixed for 10 min in CellFIX™ (1% formaldehyde, 0.35% methanol, 0.09% sodium azide) from BD Biosciences (cat no. 340181) for 10 min, washed with PBS (3 × 5 min), and blocked with 3% BSA/PBS at RT for 1 h. After washing with PBS, the slides were incubated with F-actin probe—Texas Red-X Phalloidin (ThermoFisher Scientific # T7471) at RT for 20 min in the dark. The slides were washed with PBS and mounted with Mowiol (Sigma-Aldrich, St. Louis, MO, USA), and the cells were visualized under a confocal microscope (Nikon D-Eclipse C1; Nikon, Tokyo, Japan) with a 40× objective and were analyzed with EZ-C1 version 3.6 software (Tokyo, Japan).

4.10. Transwell Invasion Assay

Transwell invasion assay was performed using our modified protocol [57,67]. Briefly: Nunc Cell Culture Inserts (transwell) with 8.0µm pore diameter (#141006) were covered with 50 µL 0.2% gelatin 1 h 37 °C. Next, gelatin was carefully removed. A549 or A549CisR cells were treated with or without 3 µM #19 for 24 h. Then, they were trypsinized, washed 2× with medium, and transferred (2 × 10^5 cells/chamber) to upper chamber of Nunc Cell Culture Inserts in 0.1% BSA medium—supplemented with or without 3 µM #19. Full medium in lower chamber was used as chemoattractant. Next, medium and the gelatin from the top surface of the membrane were removed, invaded cells on the bottom surface of the membrane were washed 2× with PBS and then fixed for 5 min with CellFIX™. Cells were dyed at RT 15 min with Hoechst 33,342 (Molecular Probes/Life Technologies, Waltham, MA, USA). Finally, membranes were cut out from chambers and placed on microscope glass, and number of cells that migrate into the membrane was counted in 5 random spots.

5. Conclusions

The presence of significant drug resistance, as well as the combination of cisplatin resistance and increased metastatic potential in the case of NSCLC, remains an obstacle during anticancer therapy. Chemotherapy can result in the selection of highly resistant and metastatic cancer cells, and this may be one of the reasons why patients present a high number of secondary tumors after apparent post-chemotherapeutic reemission [54]. Thus, long-term therapeutic strategies should focus on adjuvant "adaptive therapies" before resistance emerges [100]. Our research indicates that DSF and the tested ITCs have no impact on NSCLC cell viability at the tested concentrations, but significantly repress the cisplatin resistance of both cisplatin-resistant and metastatic NSCLC cells. Furthermore, supplementation with the tested ITCs significantly decreased the metastatic potential of all tested NSCLC models, reversing EMT toward an epithelial phenotype and decreasing the migratory potential as well as the proteolytic and invasive abilities. Therefore, the tested ITCs, especially ITC **19**, possess anti-drug-resistant and anti-metastatic potential, which may be of value during cisplatin-based chemotherapy.

Supplementary Materials: The supporting information can be downloaded at: https://www.mdpi.com/article/10.3390/ijms23158644/s1.

Author Contributions: J.K. (Jolanta Kryczka) generated the A549CisR cell line, designed the research, carried out all experiments, prepared the first draft of the manuscript; J.K. (Jakub Kryczka) participated in A549CisR cell-line generation, performed GEO database analysis, participated in data evaluation and interpretation of migration assay; J.B. critically revived the manuscript; Ł.J. synthesized ITCs; A.G. synthesized ITCs; B.K. participated in the substantive evaluation and discussion of the results; A.F. performed molecular modeling; E.B.-L. designed the research, mentored and developed the protocols, provided data evaluation and interpretation. All authors have read and agreed to the published version of the manuscript.

Funding: This research was funded by the National Center for Research and Development (Warsaw, Poland) within an InterChemMed grant (POWR.03.02.00–00-I029/16). This research was funded by the Medical University of Lodz (Statute No. 503/1-013-02/503-11-001-19-00).

Institutional Review Board Statement: Not applicable.

Informed Consent Statement: Not applicable.

Data Availability Statement: Gene Expression Omnibus dataset #GSE108214 is available at https://www.ncbi.nlm.nih.gov/geo/query/acc.cgi?acc=GSE108214, accessed on 21 July 2022.

Acknowledgments: The authors Ł.J. and A.G. express their gratitude to Tadeusz Gajda for their helpful discussions.

Conflicts of Interest: The authors declare no conflict of interest.

References

1. Kryczka, J.; Migdalska-Sęk, M.; Kordiak, J.; Kiszałkiewicz, J.M.; Pastuszak-Lewandoska, D.; Antczak, A.; Brzeziańska-Lasota, E. Serum Extracellular Vesicle-Derived MiRNAs in Patients with Non-Small Cell Lung Cancer-Search for Non-Invasive Diagnostic Biomarkers. *Diagnostics* **2021**, *11*, 425. [CrossRef]
2. Cancer Today. Available online: http://gco.iarc.fr/today/home (accessed on 15 January 2020).
3. Lung Cancer—Non-Small Cell—Statistics. Available online: https://www.cancer.net/cancer-types/lung-cancer-non-small-cell/statistics (accessed on 9 May 2022).
4. MacDonagh, L.; Gallagher, M.F.; Ffrench, B.; Gasch, C.; Breen, E.; Gray, S.G.; Nicholson, S.; Leonard, N.; Ryan, R.; Young, V.; et al. Targeting the Cancer Stem Cell Marker, Aldehyde Dehydrogenase 1, to Circumvent Cisplatin Resistance in NSCLC. *Oncotarget* **2017**, *8*, 72544–72563. [CrossRef] [PubMed]
5. Kryczka, J.; Kryczka, J.; Czarnecka-Chrebelska, K.H.; Brzeziańska-Lasota, E. Molecular Mechanisms of Chemoresistance Induced by Cisplatin in NSCLC Cancer Therapy. *Int. J. Mol. Sci.* **2021**, *22*, 8885. [CrossRef]
6. Sobczak, M.; Strachowska, M.; Gronkowska, K.; Robaszkiewicz, A. Activation of ABCC Genes by Cisplatin Depends on the CoREST Occurrence at Their Promoters in A549 and MDA-MB-231 Cell Lines. *Cancers* **2022**, *14*, 894. [CrossRef] [PubMed]
7. Jiang, Z.-S.; Sun, Y.-Z.; Wang, S.-M.; Ruan, J.-S. Epithelial-Mesenchymal Transition: Potential Regulator of ABC Transporters in Tumor Progression. *J. Cancer* **2017**, *8*, 2319–2327. [CrossRef]

8. Januchowski, R.; Wojtowicz, K.; Zabel, M. The Role of Aldehyde Dehydrogenase (ALDH) in Cancer Drug Resistance. *Biomed. Pharmacother.* **2013**, *67*, 669–680. [CrossRef]
9. Kang, J.H.; Lee, S.-H.; Lee, J.-S.; Nam, B.; Seong, T.W.; Son, J.; Jang, H.; Hong, K.M.; Lee, C.; Kim, S.-Y. Aldehyde Dehydrogenase Inhibition Combined with Phenformin Treatment Reversed NSCLC through ATP Depletion. *Oncotarget* **2016**, *7*, 49397–49410. [CrossRef]
10. Tomita, H.; Tanaka, K.; Tanaka, T.; Hara, A. Aldehyde Dehydrogenase 1A1 in Stem Cells and Cancer. *Oncotarget* **2016**, *7*, 11018–11032. [CrossRef] [PubMed]
11. Wang, N.-N.; Wang, L.-H.; Li, Y.; Fu, S.-Y.; Xue, X.; Jia, L.-N.; Yuan, X.-Z.; Wang, Y.-T.; Tang, X.; Yang, J.-Y.; et al. Targeting ALDH2 with Disulfiram/Copper Reverses the Resistance of Cancer Cells to Microtubule Inhibitors. *Exp. Cell Res.* **2018**, *362*, 72–82. [CrossRef]
12. Yokoyama, A.; Muramatsu, T.; Omori, T.; Yokoyama, T.; Matsushita, S.; Higuchi, S.; Maruyama, K.; Ishii, H. Alcohol and Aldehyde Dehydrogenase Gene Polymorphisms and Oropharyngolaryngeal, Esophageal and Stomach Cancers in Japanese Alcoholics. *Carcinogenesis* **2001**, *22*, 433–439. [CrossRef]
13. Marchitti, S.A.; Brocker, C.; Stagos, D.; Vasiliou, V. Non-P450 Aldehyde Oxidizing Enzymes: The Aldehyde Dehydrogenase Superfamily. *Expert Opin. Drug Metab. Toxicol.* **2008**, *4*, 697–720. [CrossRef]
14. Bell, L.; Oloyede, O.O.; Lignou, S.; Wagstaff, C.; Methven, L. Taste and Flavor Perceptions of Glucosinolates, Isothiocyanates, and Related Compounds. *Mol. Nutr. Food Res.* **2018**, *62*, e1700990. [CrossRef] [PubMed]
15. Avato, P.; Argentieri, M.P. Brassicaceae: A Rich Source of Health Improving Phytochemicals. *Phytochem. Rev.* **2015**, *14*, 1019–1033. [CrossRef]
16. Verhoeven, D.T.; Verhagen, H.; Goldbohm, R.A.; van den Brandt, P.A.; van Poppel, G. A Review of Mechanisms Underlying Anticarcinogenicity by Brassica Vegetables. *Chem. Biol. Interact.* **1997**, *103*, 79–129. [CrossRef]
17. Rollin, P.; Tatibouët, A. Glucosinolates: The Synthetic Approach. *C. R. Chim.* **2011**, *14*, 194–210. [CrossRef]
18. Bell, L.; Wagstaff, C. Glucosinolates, Myrosinase Hydrolysis Products, and Flavonols Found in Rocket (Eruca Sativa and Diplotaxis Tenuifolia). *J. Agric. Food Chem.* **2014**, *62*, 4481–4492. [CrossRef]
19. Soundararajan, P.; Kim, J.S. Anti-Carcinogenic Glucosinolates in Cruciferous Vegetables and Their Antagonistic Effects on Prevention of Cancers. *Molecules* **2018**, *23*, 2983. [CrossRef]
20. Maina, S.; Misinzo, G.; Bakari, G.; Kim, H.-Y. Human, Animal and Plant Health Benefits of Glucosinolates and Strategies for Enhanced Bioactivity: A Systematic Review. *Molecules* **2020**, *25*, 3682. [CrossRef]
21. Rao, C.V. Benzyl Isothiocyanate: Double Trouble for Breast Cancer Cells. *Cancer Prev. Res.* **2013**, *6*, 760–763. [CrossRef]
22. Dinh, T.N.; Parat, M.-O.; Ong, Y.S.; Khaw, K.Y. Anticancer Activities of Dietary Benzyl Isothiocyanate: A Comprehensive Review. *Pharmacol. Res.* **2021**, *169*, 105666. [CrossRef]
23. Coscueta, E.R.; Sousa, A.S.; Reis, C.A.; Pintado, M.M. Phenylethyl Isothiocyanate: A Bioactive Agent for Gastrointestinal Health. *Molecules* **2022**, *27*, 794. [CrossRef] [PubMed]
24. Qin, C.-Z.; Zhang, X.; Wu, L.-X.; Wen, C.-J.; Hu, L.; Lv, Q.-L.; Shen, D.-Y.; Zhou, H.-H. Advances in Molecular Signaling Mechanisms of β-Phenethyl Isothiocyanate Antitumor Effects. *J. Agric. Food Chem.* **2015**, *63*, 3311–3322. [CrossRef] [PubMed]
25. Briones-Herrera, A.; Eugenio-Pérez, D.; Reyes-Ocampo, J.G.; Rivera-Mancía, S.; Pedraza-Chaverri, J. New Highlights on the Health-Improving Effects of Sulforaphane. *Food Funct.* **2018**, *9*, 2589–2606. [CrossRef] [PubMed]
26. Jiang, X.; Liu, Y.; Ma, L.; Ji, R.; Qu, Y.; Xin, Y.; Lv, G. Chemopreventive Activity of Sulforaphane. *Drug Des. Devel. Ther.* **2018**, *12*, 2905–2913. [CrossRef] [PubMed]
27. Leone, A.; Diorio, G.; Sexton, W.; Schell, M.; Alexandrow, M.; Fahey, J.W.; Kumar, N.B. Sulforaphane for the Chemoprevention of Bladder Cancer: Molecular Mechanism Targeted Approach. *Oncotarget* **2017**, *8*, 35412–35424. [CrossRef] [PubMed]
28. Khan, S.; Awan, K.A.; Iqbal, M.J. Sulforaphane as a Potential Remedy against Cancer: Comprehensive Mechanistic Review. *J. Food Biochem.* **2022**, *46*, e13886. [CrossRef]
29. Kaiser, A.E.; Baniasadi, M.; Giansiracusa, D.; Giansiracusa, M.; Garcia, M.; Fryda, Z.; Wong, T.L.; Bishayee, A. Sulforaphane: A Broccoli Bioactive Phytocompound with Cancer Preventive Potential. *Cancers* **2021**, *13*, 4796. [CrossRef]
30. Mahn, A.; Castillo, A. Potential of Sulforaphane as a Natural Immune System Enhancer: A Review. *Molecules* **2021**, *26*, 752. [CrossRef]
31. Kiełbasiński, P.; Łuczak, J.; Cierpiał, T.; Błaszczyk, J.; Sieroń, L.; Wiktorska, K.; Lubelska, K.; Milczarek, M.; Chilmończyk, Z. New Enantiomeric Fluorine-Containing Derivatives of Sulforaphane: Synthesis, Absolute Configurations and Biological Activity. *Eur. J. Med. Chem.* **2014**, *76*, 332–342. [CrossRef]
32. Cierpiał, T.; Kiełbasiński, P.; Kwiatkowska, M.; Łyżwa, P.; Lubelska, K.; Kuran, D.; Dąbrowska, A.; Kruszewska, H.; Mielczarek, L.; Chilmonczyk, Z.; et al. Fluoroaryl Analogs of Sulforaphane—A Group of Compounds of Anticancer and Antimicrobial Activity. *Bioorg. Chem.* **2020**, *94*, 103454. [CrossRef] [PubMed]
33. Psurski, M.; Janczewski, Ł.; Świtalska, M.; Gajda, A.; Goszczyński, T.M.; Oleksyszyn, J.; Wietrzyk, J.; Gajda, T. Novel Phosphonate Analogs of Sulforaphane: Synthesis, in Vitro and in Vivo Anticancer Activity. *Eur. J. Med. Chem.* **2017**, *132*, 63–80. [CrossRef] [PubMed]
34. Gajda, T.; Janczewski, Ł.; Psurski, M.; Świtalska, M.; Gajda, A.; Goszczyński, T.; Oleksyszyn, J.; Wietrzyk, J. Design, Synthesis, and Evaluation of ω-(Isothiocyanato)Alkylphosphinates and Phosphine Oxides as Antiproliferative Agents. *ChemMedChem* **2017**, *13*, 105–115. [CrossRef]

35. Janczewski, Ł. Sulforaphane and Its Bifunctional Analogs: Synthesis and Biological Activity. *Molecules* **2022**, *27*, 1750. [CrossRef]
36. Pace, V.; Monticelli, S.; de la Vega-Hernández, K.; Castoldi, L. Isocyanates and Isothiocyanates as Versatile Platforms for Accessing (Thio)Amide-Type Compounds. *Org. Biomol. Chem.* **2016**, *14*, 7848–7854. [CrossRef] [PubMed]
37. Milelli, A.; Fimognari, C.; Ticchi, N.; Neviani, P.; Minarini, A.; Tumiatti, V. Isothiocyanate Synthetic Analogs: Biological Activities, Structure-Activity Relationships and Synthetic Strategies. *Mini. Rev. Med. Chem.* **2014**, *14*, 963–977. [CrossRef] [PubMed]
38. Gründemann, C.; Huber, R. Chemoprevention with Isothiocyanates—From Bench to Bedside. *Cancer Lett.* **2018**, *414*, 26–33. [CrossRef] [PubMed]
39. Mastuo, T.; Miyata, Y.; Yuno, T.; Mukae, Y.; Otsubo, A.; Mitsunari, K.; Ohba, K.; Sakai, H. Molecular Mechanisms of the Anti-Cancer Effects of Isothiocyanates from Cruciferous Vegetables in Bladder Cancer. *Molecules* **2020**, *25*, 575. [CrossRef]
40. Mitsiogianni, M.; Koutsidis, G.; Mavroudis, N.; Trafalis, D.T.; Botaitis, S.; Franco, R.; Zoumpourlis, V.; Amery, T.; Galanis, A.; Pappa, A.; et al. The Role of Isothiocyanates as Cancer Chemo-Preventive, Chemo-Therapeutic and Anti-Melanoma Agents. *Antioxidants* **2019**, *8*, 106. [CrossRef] [PubMed]
41. Palliyaguru, D.L.; Yuan, J.-M.; Kensler, T.W.; Fahey, J.W. Isothiocyanates: Translating the Power of Plants to People. *Mol. Nutr. Food Res.* **2018**, *62*, e1700965. [CrossRef]
42. Dufour, V.; Stahl, M.; Baysse, C. The Antibacterial Properties of Isothiocyanates. *Microbiology* **2015**, *161*, 229–243. [CrossRef]
43. Romeo, L.; Iori, R.; Rollin, P.; Bramanti, P.; Mazzon, E. Isothiocyanates: An Overview of Their Antimicrobial Activity against Human Infections. *Molecules* **2018**, *23*, 624. [CrossRef] [PubMed]
44. Janczewski, Ł.; Burchacka, E.; Psurski, M.; Ciekot, J.; Gajda, A.; Gajda, T. New Diaryl ω-(Isothiocyanato)Alkylphosphonates and Their Mercapturic Acids as Potential Antibacterial Agents. *Life Sci.* **2019**, *219*, 264–271. [CrossRef] [PubMed]
45. Myzak, M.C.; Hardin, K.; Wang, R.; Dashwood, R.H.; Ho, E. Sulforaphane Inhibits Histone Deacetylase Activity in BPH-1, LnCaP and PC-3 Prostate Epithelial Cells. *Carcinogenesis* **2006**, *27*, 811–819. [CrossRef]
46. Choi, S.; Lew, K.L.; Xiao, H.; Herman-Antosiewicz, A.; Xiao, D.; Brown, C.K.; Singh, S.V. D,L-Sulforaphane-Induced Cell Death in Human Prostate Cancer Cells Is Regulated by Inhibitor of Apoptosis Family Proteins and Apaf-1. *Carcinogenesis* **2007**, *28*, 151–162. [CrossRef] [PubMed]
47. Myzak, M.C.; Karplus, P.A.; Chung, F.-L.; Dashwood, R.H. A Novel Mechanism of Chemoprotection by Sulforaphane: Inhibition of Histone Deacetylase. *Cancer Res.* **2004**, *64*, 5767–5774. [CrossRef]
48. Asakage, M.; Tsuno, N.H.; Kitayama, J.; Tsuchiya, T.; Yoneyama, S.; Yamada, J.; Okaji, Y.; Kaisaki, S.; Osada, T.; Takahashi, K.; et al. Sulforaphane Induces Inhibition of Human Umbilical Vein Endothelial Cells Proliferation by Apoptosis. *Angiogenesis* **2006**, *9*, 83–91. [CrossRef]
49. Thejass, P.; Kuttan, G. Antimetastatic Activity of Sulforaphane. *Life Sci.* **2006**, *78*, 3043–3050. [CrossRef]
50. Hwang, E.-S.; Lee, H.J. Benzyl Isothiocyanate Inhibits Metalloproteinase-2/-9 Expression by Suppressing the Mitogen-Activated Protein Kinase in SK-Hep1 Human Hepatoma Cells. *Food Chem. Toxicol.* **2008**, *46*, 2358–2364. [CrossRef] [PubMed]
51. Yu, M.; Zhang, C.; Li, L.; Dong, S.; Zhang, N.; Tong, X. Cx43 Reverses the Resistance of A549 Lung Adenocarcinoma Cells to Cisplatin by Inhibiting EMT. *Oncol. Rep.* **2014**, *31*, 2751–2758. [CrossRef]
52. Zhao, Z.; Zhang, L.; Yao, Q.; Tao, Z. MiR-15b Regulates Cisplatin Resistance and Metastasis by Targeting PEBP4 in Human Lung Adenocarcinoma Cells. *Cancer Gene. Ther.* **2015**, *22*, 108–114. [CrossRef]
53. Sinha, D.; Saha, P.; Samanta, A.; Bishayee, A. Emerging Concepts of Hybrid Epithelial-to-Mesenchymal Transition in Cancer Progression. *Biomolecules* **2020**, *10*, 1561. [CrossRef] [PubMed]
54. Kryczka, J.; Boncela, J. Cell Migration Related to MDR—Another Impediment to Effective Chemotherapy? *Molecules* **2018**, *23*, 331. [CrossRef] [PubMed]
55. Lundholm, L.; Hååg, P.; Zong, D.; Juntti, T.; Mörk, B.; Lewensohn, R.; Viktorsson, K. Resistance to DNA-Damaging Treatment in Non-Small Cell Lung Cancer Tumor-Initiating Cells Involves Reduced DNA-PK/ATM Activation and Diminished Cell Cycle Arrest. *Cell. Death Dis.* **2013**, *4*, e478. [CrossRef] [PubMed]
56. Yang, Z.; Guo, F.; Albers, A.E.; Sehouli, J.; Kaufmann, A.M. Disulfiram Modulates ROS Accumulation and Overcomes Synergistically Cisplatin Resistance in Breast Cancer Cell Lines. *Biomed. Pharmacother.* **2019**, *113*, 108727. [CrossRef]
57. Kryczka, J.; Sochacka, E.; Papiewska-Pająk, I.; Boncela, J. Implications of ABCC4–Mediated CAMP Efflux for CRC Migration. *Cancers* **2020**, *12*, 3547. [CrossRef]
58. Saxena, M.; Stephens, M.A.; Pathak, H.; Rangarajan, A. Transcription Factors That Mediate Epithelial–Mesenchymal Transition Lead to Multidrug Resistance by Upregulating ABC Transporters. *Cell. Death Dis.* **2011**, *2*, e179. [CrossRef]
59. Home—GEO—NCBI. Available online: https://www.ncbi.nlm.nih.gov/geo/ (accessed on 14 April 2021).
60. Skrott, Z.; Majera, D.; Gursky, J.; Buchtova, T.; Hajduch, M.; Mistrik, M.; Bartek, J. Disulfiram's Anti-Cancer Activity Reflects Targeting NPL4, Not Inhibition of Aldehyde Dehydrogenase. *Oncogene* **2019**, *38*, 6711–6722. [CrossRef]
61. Fumarola, C.; Bozza, N.; Castelli, R.; Ferlenghi, F.; Marseglia, G.; Lodola, A.; Bonelli, M.; La Monica, S.; Cretella, D.; Alfieri, R.; et al. Expanding the Arsenal of FGFR Inhibitors: A Novel Chloroacetamide Derivative as a New Irreversible Agent with Anti-Proliferative Activity Against FGFR1-Amplified Lung Cancer Cell Lines. *Front. Oncol.* **2019**, *9*, 179. [CrossRef]
62. Deng, C.; Xiong, J.; Gu, X.; Chen, X.; Wu, S.; Wang, Z.; Wang, D.; Tu, J.; Xie, J. Novel Recombinant Immunotoxin of EGFR Specific Nanobody Fused with Cucurmosin, Construction and Antitumor Efficiency in Vitro. *Oncotarget* **2017**, *8*, 38568–38580. [CrossRef]
63. Wang, K.; Ji, W.; Yu, Y.; Li, Z.; Niu, X.; Xia, W.; Lu, S. FGFR1-ERK1/2-SOX2 Axis Promotes Cell Proliferation, Epithelial-Mesenchymal Transition, and Metastasis in FGFR1-Amplified Lung Cancer. *Oncogene* **2018**, *37*, 5340–5354. [CrossRef]

64. Ji, W.; Yu, Y.; Li, Z.; Wang, G.; Li, F.; Xia, W.; Lu, S. FGFR1 Promotes the Stem Cell-like Phenotype of FGFR1-Amplified Non-Small Cell Lung Cancer Cells through the Hedgehog Pathway. *Oncotarget* **2016**, *7*, 15118–15134. [CrossRef] [PubMed]
65. Kryczka, J.; Przygodzka, P.; Bogusz, H.; Boncela, J. HMEC-1 Adopt the Mixed Amoeboid-Mesenchymal Migration Type during EndMT. *Eur. J. Cell Biol.* **2017**, *96*, 289–300. [CrossRef] [PubMed]
66. Yamada, K.M.; Sixt, M. Mechanisms of 3D Cell Migration. *Nat. Rev. Mol. Cell Biol.* **2019**, *20*, 738–752. [CrossRef] [PubMed]
67. Kryczka, J.; Papiewska-Pajak, I.; Kowalska, M.A.; Boncela, J. Cathepsin B Is Upregulated and Mediates ECM Degradation in Colon Adenocarcinoma HT29 Cells Overexpressing Snail. *Cells* **2019**, *8*, 203. [CrossRef] [PubMed]
68. Noordzij, I.C.; Curvers, W.L.; Schoon, E.J. Endoscopic Resection for Early Esophageal Carcinoma. *J. Thorac. Dis.* **2019**, *11*, S713–S722. [CrossRef]
69. Siddik, Z.H. Cisplatin: Mode of Cytotoxic Action and Molecular Basis of Resistance. *Oncogene* **2003**, *22*, 7265–7279. [CrossRef] [PubMed]
70. Uramoto, H.; Tanaka, F. Recurrence after Surgery in Patients with NSCLC. *Transl. Lung. Cancer Res.* **2014**, *3*, 242–249. [CrossRef] [PubMed]
71. Han, M.-L.; Zhao, Y.-F.; Tan, C.-H.; Xiong, Y.-J.; Wang, W.-J.; Wu, F.; Fei, Y.; Wang, L.; Liang, Z.-Q. Cathepsin L Upregulation-Induced EMT Phenotype Is Associated with the Acquisition of Cisplatin or Paclitaxel Resistance in A549 Cells. *Acta Pharmacol. Sin.* **2016**, *37*, 1606–1622. [CrossRef]
72. Shen, M.; Xu, Z.; Xu, W.; Jiang, K.; Zhang, F.; Ding, Q.; Xu, Z.; Chen, Y. Inhibition of ATM Reverses EMT and Decreases Metastatic Potential of Cisplatin-Resistant Lung Cancer Cells through JAK/STAT3/PD-L1 Pathway. *J. Exp. Clin. Cancer Res.* **2019**, *38*, 149. [CrossRef]
73. Sarin, N.; Engel, F.; Rothweiler, F.; Cinatl, J.; Michaelis, M.; Frötschl, R.; Fröhlich, H.; Kalayda, G.V. Key Players of Cisplatin Resistance: Towards a Systems Pharmacology Approach. *Int. J. Mol. Sci.* **2018**, *19*, 767. [CrossRef]
74. Lima de Oliveira, J.; Moré Milan, T.; Longo Bighetti-Trevisan, R.; Fernandes, R.R.; Machado Leopoldino, A.; Oliveira de Almeida, L. Epithelial–Mesenchymal Transition and Cancer Stem Cells: A Route to Acquired Cisplatin Resistance through Epigenetics in HNSCC. *Oral Dis.* **2022**. [CrossRef]
75. Yin, J.; Zhang, J. Multidrug Resistance-Associated Protein 1 (MRP1/ABCC1) Polymorphism: From Discovery to Clinical Application. *Zhong Nan Da Xue Xue Bao Yi Xue Ban* **2011**, *36*, 927–938. [CrossRef] [PubMed]
76. Ween, M.P.; Armstrong, M.A.; Oehler, M.K.; Ricciardelli, C. The Role of ABC Transporters in Ovarian Cancer Progression and Chemoresistance. *Crit. Rev. Oncol. Hematol.* **2015**, *96*, 220–256. [CrossRef] [PubMed]
77. Khan, P.; Bhattacharya, A.; Sengupta, D.; Banerjee, S.; Adhikary, A.; Das, T. Aspirin Enhances Cisplatin Sensitivity of Resistant Non-Small Cell Lung Carcinoma Stem-like Cells by Targeting MTOR-Akt Axis to Repress Migration. *Sci. Rep.* **2019**, *9*, 16913. [CrossRef] [PubMed]
78. Peng, S.; Wang, J.; Lu, C.; Xu, Z.; Chai, J.-J.; Ke, Q.; Deng, X.-Z. Emodin Enhances Cisplatin Sensitivity in Non-Small Cell Lung Cancer through Pgp Downregulation. *Oncol. Lett.* **2021**, *21*, 1. [CrossRef] [PubMed]
79. Morelli, A.P.; Tortelli Jr, T.C.; Pavan, I.C.B.; Silva, F.R.; Granato, D.C.; Peruca, G.F.; Pauletti, B.A.; Domingues, R.R.; Bezerra, R.M.N.; De Moura, L.P.; et al. Metformin Impairs Cisplatin Resistance Effects in A549 Lung Cancer Cells through MTOR Signaling and Other Metabolic Pathways. *Int. J. Oncol.* **2021**, *58*, 1–15. [CrossRef]
80. Duan, L.; Shen, H.; Zhao, G.; Yang, R.; Cai, X.; Zhang, L.; Jin, C.; Huang, Y. Inhibitory Effect of Disulfiram/Copper Complex on Non-Small Cell Lung Cancer Cells. *Biochem. Biophys. Res. Commun.* **2014**, *446*, 1010–1016. [CrossRef] [PubMed]
81. Lu, C.; Li, X.; Ren, Y.; Zhang, X. Disulfiram: A Novel Repurposed Drug for Cancer Therapy. *Cancer Chemother. Pharmacol.* **2021**, *87*, 159–172. [CrossRef] [PubMed]
82. Ntzifa, A.; Strati, A.; Kallergi, G.; Kotsakis, A.; Georgoulias, V.; Lianidou, E. Gene Expression in Circulating Tumor Cells Reveals a Dynamic Role of EMT and PD-L1 during Osimertinib Treatment in NSCLC Patients. *Sci. Rep.* **2021**, *11*, 2313. [CrossRef]
83. Rudzinska-Radecka, M.; Janczewski, Ł.; Gajda, A.; Godlewska, M.; Chmielewska-Krzesinska, M.; Wasowicz, K.; Podlasz, P. The Anti-Tumoral Potential of Phosphonate Analog of Sulforaphane in Zebrafish Xenograft Model. *Cells* **2021**, *10*, 3219. [CrossRef]
84. Wright, C.; Moore, R.D. Disulfiram Treatment of Alcoholism. *Am. J. Med.* **1990**, *88*, 647–655. [CrossRef]
85. Horwitz, R.; Webb, D. Cell Migration. *Curr. Biol.* **2003**, *13*, R756–R759. [CrossRef] [PubMed]
86. Zanotelli, M.R.; Goldblatt, Z.E.; Miller, J.P.; Bordeleau, F.; Li, J.; VanderBurgh, J.A.; Lampi, M.C.; King, M.R.; Reinhart-King, C.A. Regulation of ATP Utilization during Metastatic Cell Migration by Collagen Architecture. *MBoC* **2018**, *29*, 1–9. [CrossRef] [PubMed]
87. Kang, J.H.; Lee, S.-H.; Hong, D.; Lee, J.-S.; Ahn, H.-S.; Ahn, J.-H.; Seong, T.W.; Lee, C.-H.; Jang, H.; Hong, K.M.; et al. Aldehyde Dehydrogenase Is Used by Cancer Cells for Energy Metabolism. *Exp. Mol. Med.* **2016**, *48*, e272. [CrossRef]
88. Lee, J.-S.; Lee, H.; Jang, H.; Woo, S.M.; Park, J.B.; Lee, S.-H.; Kang, J.H.; Kim, H.Y.; Song, J.; Kim, S.-Y. Targeting Oxidative Phosphorylation Reverses Drug Resistance in Cancer Cells by Blocking Autophagy Recycling. *Cells* **2020**, *9*, 2013. [CrossRef] [PubMed]
89. Muzio, G.; Maggiora, M.; Paiuzzi, E.; Oraldi, M.; Canuto, R.A. Aldehyde Dehydrogenases and Cell Proliferation. *Free Radic. Biol. Med.* **2012**, *52*, 735–746. [CrossRef] [PubMed]
90. Parajuli, B.; Fishel, M.L.; Hurley, T.D. Selective ALDH3A1 Inhibition by Benzimidazole Analogues Increase Mafosfamide Sensitivity in Cancer Cells. *J. Med. Chem.* **2014**, *57*, 449–461. [CrossRef]

91. Cocetta, V.; Ragazzi, E.; Montopoli, M. Mitochondrial Involvement in Cisplatin Resistance. *Int. J. Mol. Sci.* **2019**, *20*, 3384. [CrossRef] [PubMed]
92. Vesel, M.; Rapp, J.; Feller, D.; Kiss, E.; Jaromi, L.; Meggyes, M.; Miskei, G.; Duga, B.; Smuk, G.; Laszlo, T.; et al. ABCB1 and ABCG2 Drug Transporters Are Differentially Expressed in Non-Small Cell Lung Cancers (NSCLC) and Expression Is Modified by Cisplatin Treatment via Altered Wnt Signaling. *Respir. Res.* **2017**, *18*, 52. [CrossRef]
93. Jolly, M.K.; Somarelli, J.A.; Sheth, M.; Biddle, A.; Tripathi, S.C.; Armstrong, A.J.; Hanash, S.M.; Bapat, S.A.; Rangarajan, A.; Levine, H. Hybrid Epithelial/Mesenchymal Phenotypes Promote Metastasis and Therapy Resistance across Carcinomas. *Pharmacol. Ther.* **2019**, *194*, 161–184. [CrossRef]
94. Barr, M.P.; Gray, S.G.; Hoffmann, A.C.; Hilger, R.A.; Thomale, J.; O'Flaherty, J.D.; Fennell, D.A.; Richard, D.; O'Leary, J.J.; O'Byrne, K.J. Generation and Characterisation of Cisplatin-Resistant Non-Small Cell Lung Cancer Cell Lines Displaying a Stem-like Signature. *PLoS ONE* **2013**, *8*, e54193. [CrossRef] [PubMed]
95. JASP—A Fresh Way to Do Statistics. Available online: https://jasp-stats.org/ (accessed on 6 May 2022).
96. Janczewski, Ł.; Gajda, A.; Gajda, T. Direct, Microwave-Assisted Synthesis of Isothiocyanates. *Eur. J. Org. Chem.* **2019**, *2019*, 2528–2532. [CrossRef]
97. Janczewski, Ł.; Kręgiel, D.; Kolesińska, B. Synthesis of Isothiocyanates Using DMT/NMM/TsO− as a New Desulfurization Reagent. *Molecules* **2021**, *26*, 2740. [CrossRef] [PubMed]
98. Janczewski, Ł.; Gajda, A.; Frankowski, S.; Goszczyński, T.M.; Gajda, T. T3P®—A Benign Desulfurating Reagent in the Synthesis of Isothiocyanates. *Synthesis* **2018**, *50*, 1141–1151. [CrossRef]
99. Bank, R.P.D. RCSB PDB: Homepage. Available online: https://www.rcsb.org/ (accessed on 6 May 2022).
100. Fillon, M. Cancer and Natural Selection. *J. Natl. Cancer Inst.* **2012**, *104*, 1773–1774. [CrossRef]

International Journal of
Molecular Sciences

Article

The Multidirectional Effect of Azelastine Hydrochloride on Cervical Cancer Cells

Ewa Trybus *, Teodora Król * and Wojciech Trybus

Department of Medical Biology, The Jan Kochanowski University, Uniwersytecka 7, 25-406 Kielce, Poland; wojciech.trybus@ujk.edu.pl
* Correspondence: ewa.trybus@ujk.edu.pl (E.T.); teodora.krol@ujk.edu.pl (T.K.)

Abstract: A major cause of cancer cell resistance to chemotherapeutics is the blocking of apoptosis and induction of autophagy in the context of cell adaptation and survival. Therefore, new compounds are being sought, also among drugs that are commonly used in other therapies. Due to the involvement of histamine in the regulation of processes occurring during the development of many types of cancer, antihistamines are now receiving special attention. Our study concerned the identification of new mechanisms of action of azelastine hydrochloride, used in antiallergic treatment. The study was performed on HeLa cells treated with different concentrations of azelastine (15–90 μM). Cell cycle, level of autophagy (LC3 protein activity) and apoptosis (annexin V assay), activity of caspase 3/7, anti-apoptotic protein of Bcl-2 family, ROS concentration, measurement of mitochondrial membrane potential (Δψm), and level of phosphorylated H2A.X in response to DSB were evaluated by cytometric method. Cellular changes were also demonstrated at the level of transmission electron microscopy and optical and fluorescence microscopy. Lysosomal enzyme activities-cathepsin D and L and cell viability (MTT assay) were assessed spectrophotometrically. Results: Azelastine in concentrations of 15–25 μM induced degradation processes, vacuolization, increase in cathepsin D and L activity, and LC3 protein activation. By increasing ROS, it also caused DNA damage and blocked cells in the S phase of the cell cycle. At the concentrations of 45–90 μM, azelastine clearly promoted apoptosis by activation of caspase 3/7 and inactivation of Bcl-2 protein. Fragmentation of cell nucleus was confirmed by DAPI staining. Changes were also found in the endoplasmic reticulum and mitochondria, whose damage was confirmed by staining with rhodamine 123 and in the MTT test. Azelastine decreased the mitotic index and induced mitotic catastrophe. Studies demonstrated the multidirectional effects of azelastine on HeLa cells, including anti-proliferative, cytotoxic, autophagic, and apoptotic properties, which were the predominant mechanism of death. The revealed novel properties of azelastine may be practically used in anti-cancer therapy in the future.

Keywords: azelastine; oxidative stress; autophagy; apoptosis; mitotic catastrophe

Citation: Trybus, E.; Król, T.; Trybus, W. The Multidirectional Effect of Azelastine Hydrochloride on Cervical Cancer Cells. *Int. J. Mol. Sci.* **2022**, *23*, 5890. https://doi.org/10.3390/ijms23115890

Academic Editors: Angela Stefanachi and Marcin Majka

Received: 9 April 2022
Accepted: 23 May 2022
Published: 24 May 2022

Publisher's Note: MDPI stays neutral with regard to jurisdictional claims in published maps and institutional affiliations.

1. Introduction

Drug resistance is a very big problem in most advanced cancers [1,2]. The biggest obstacle in cancer chemotherapy, including the treatment of cervical cancer, is resistance to cisplatin, among others, resulting from the induction of autophagy and inhibition of tumor cell apoptosis [3,4]. The process of programmed cell death can also be inhibited during oncogenesis. Cancer cells with multiple genetic and epigenetic alterations avoid apoptosis, which is initially triggered by the transformation process itself and then by the unfavorable tumor environment and the implemented therapy [5,6]. The resulting limitations in cancer therapy contribute to increased mortality. Therefore, recently, a new trend in worldwide research has become the search for alternative treatments, also inducing other types of cell death, especially among compounds already used in other therapies [5,7–9], an example of which may be antihistamines (AHs).

Antihistamines (AHs), due to their proven strong anti-inflammatory and anti-allergenic properties, are widely used worldwide as first-line drugs in the treatment of numerous allergic diseases [10]. Their mechanism of action involves stabilization of the inactive form of histamine H1 receptor, thereby blocking the action of histamine [11–13], which, as a major mediator of inflammatory response, not only underlies many allergic diseases [14], but is also directly involved in the regulation of biological processes during the development of various types of cancer, including cervical cancer [1,15,16]. Hence, in recent years, attention has been focused on the potential antitumor properties of antihistamines, both among the long-used and new second-generation representatives. Compounds have been identified, that alone or in combination with other drugs show significant activity against various types of cancer cells, confirmed both in vitro and in clinical trials. An example is astemizole, a second-generation drug, that has been described as an inhibitor of hepatocellular carcinoma proliferation [17] as well as an inducer of apoptotic death in various human melanoma cell lines [18,19]. In the case of terfenadine, the ability to induce apoptosis in prostate cancer cells has been demonstrated [20]. In turn, new representatives of AHs, more often used in practice, i.e., loratadine and its active metabolite desloratadine, improve survival in breast cancer [21–23] and skin melanoma [24]. Additionally, desloratadine has properties to induce apoptosis of T-cell lymphoma cells [25], and loratadine interferes with cell cycle progression of human colon cancer cells by increasing their sensitivity to radiation [26], and improves survival for ovarian cancer [7].

Azelastine hydrochloride (a phthalazinone derivative) is commonly used especially in the topical treatment of respiratory diseases, i.e., in allergic rhinitis (also in the course of asthma and COPD), vasomotor rhinitis, and as part of the prophylaxis and therapy of allergic conjunctivitis [27–30]. Furthermore, recent in vitro studies have demonstrated the ability of this compound to prevent and inhibit SARS-CoV-2 infection in nasal tissue [31]. Azelastine was also included in the list of compounds that exhibit lysosomotropic properties and have the ability to accumulate in the lungs when administered systemically, which creates the potential to achieve an effective drug concentration and was therefore recommended for use in patients with SARS-CoV-2 [32]. It should be emphasized that azelastine is a new representative of the second generation of H1 receptor antagonists, characterized by a different chemical structure than other preparations from this group [33] and high selectivity to the receptor, and thus low risk of side effects and very good tolerance both in adults and children [34–38]. It was also found that this group has an equivalent or faster onset of action compared to the first generation AHs [39]. Numerous scientific studies confirm that the biological properties of H1 receptor antagonists, including azelastine, also result from the possibility of non-receptor activity [13,40–42], which offers a broad perspective for the discovery of new therapeutic properties of these compounds.

In recent years, azelastine has also been tested for anti-inflammatory [43], antibacterial [44], and antiparasitic properties [45]. In turn, little attention has been paid to research into the potential anticancer mechanisms of this compound. So far, the property of azelastine to induce apoptosis in human colorectal adenocarcinoma cells (HT-29 line) has been described, where the tested compound at concentrations of 10 μM-20 μM, independently of the receptor, decreased the expression of Bcl-2 protein and caused significant changes in mitochondria [46]. In another study [47], azelastine at a concentration of 5 μM sensitized KBV20C cells to the effects of vincristine (in combination administration), leading to decreased cell viability, arrest in G2 phase, and increased apoptosis. The results of the cited studies inspired the present study.

Therefore, due to the well-known resistance of HeLa cells to chemotherapy, which manifests itself by induction of autophagy and blockade of apoptosis, we decided to study the changes occurring in these cells under the influence of azelastine hydrochloride in the context of induction of apoptosis as well as other types of cell death as potential anticancer mechanisms of action of this compound.

2. Results

2.1. *Azelastine Induces Apoptosis in HeLa Cells*

Exposure of cells to azelastine resulted in an increase in the frequency of both cells in early (Annexin V-PE+/7-AAD−) and late apoptosis (Annexin V-PE+/7-AAD+).

At a concentration of 15 μM, apoptotic cells were over 26% ($p \leq 0.0001$) and at 25 μM over 34% ($p \leq 0.0001$) (Figure 1A,C). Azelastine at a concentration of 45 μM further increased the number of apoptotic cells to 60.13% ($p \leq 0.0001$). Subsequent concentrations (60 μM and 90 μM) significantly increased the number of apoptotic cells, which were more than 93% and 98%, respectively ($p \leq 0.0001$), with a clear predominance of cells with a late apoptotic phenotype.

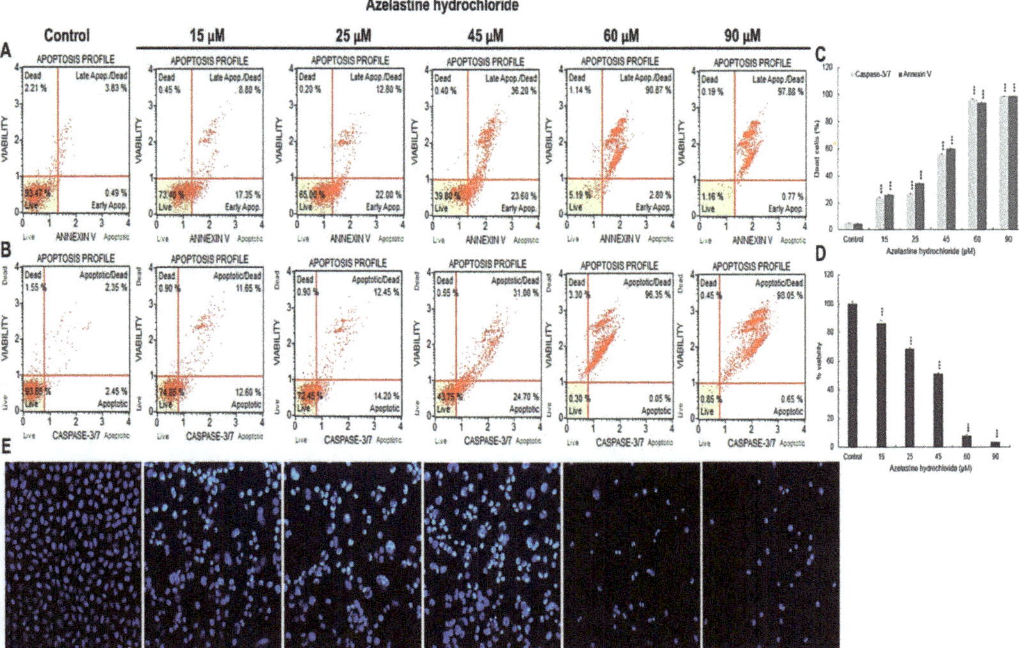

Figure 1. Proapoptotic effects of azelastine hydrochloride. Control cells and treated for 48 h with azelastine at concentrations of 15 μM, 25 μM, 45 μM, 60 μM, and 90 μM. (**A**) Level of apoptosis determined by annexin V-PE/ 7-AAD staining. Live cells (annexin V-PE−/7-AAD−), cells in early (annexin V-PE+/7-AAD−) and late-stage apoptosis (annexin V-PE+/7-AAD+), and necrotic cells (V-PE−/7-AAD+). (**B**) Changes in 3/7 caspase activity. Live cells (caspase 3/7-/7-AAD−), cells in early (caspase 3/7+/7-AAD−) and late apoptosis (caspase 3/7+/7-AAD+), dead cells (caspase 3/7-/7-AAD+). (**C**) Percentage of apoptotic cells dependent on azelastine concentration. (**D**) Cell viability as determined by the MTT assay. (**E**) Changes in nuclei of cells labeled with 4′,6-diamidino-2-phenylindole (DAPI). Control cells showing normal cell nuclei morphology. Cells treated with azelastine showing changes typical of apoptosis, i.e., marked condensation of chromatin and fragmentation of cell nucleus. Images were taken at 4000× magnification. Data representative of three parallel experiments correspond to mean values ± standard error (SE). Differences were statistically confirmed at: *** $p < 0.001$.

Moreover, microscopic analysis (DAPI staining) showed that azelastine induces typical nuclear changes for apoptosis i.e., chromatin condensation and nuclear fragmentation, especially at concentrations of 60 μM and 90 μM (Figure 1E). The obtained results were dependent on the concentration of the test compound.

2.2. Azelastine Induces Caspase 3/7-Dependent Apoptosis

As shown in Figure 1B,C, caspase 3/7 activity increased significantly ($p \leq 0.0001$) under the azelastine concentrations used. At 15 μM and 25 μM, cells with active caspase were more than 24% and 26%, respectively, and more than 55% at 45 μM concentration. The highest caspase 3/7 activity was shown for concentrations of 60 μM (96.4%) and 90 μM (98.7%). These results indicate a caspase 3/7 activation-dependent proapoptotic effect of azelastine.

2.3. Azelastine Inhibits the Viability of HeLa Cells

The MTT assay showed a highly statistically significant ($p \leq 0.0001$) reduction in the ability of the cells to reduce the dye compared to the control, which was taken as 100% (Figure 1D). Already at the lowest concentration of 15 μM, the cell viability was 86% and at subsequent concentrations of 25 μM and 45 μM, it decreased significantly to 68.33% and 51.33%. However, the lowest percentage of living cells was obtained as a consequence of the highest concentrations of the test compound, i.e., 60 μM (8%) and 90 μM (about 4%). Azelastine inhibited mitochondrial metabolic activity to a concentration-dependent degree, which was also indicative of mitochondrial membrane damage.

2.4. Azelastine Generates ROS Inducing Changes in Mitochondrial Structure and Induces Endoplasmic Reticulum Stress

Compared to the control image (Figure 2(A1)), mitochondria with clear matrix and an irregular arrangement of mitochondrial cristae were observed already at the lowest azelastine concentration (15 μM) (Figure 2(A2)). As a result of 25 μM concentration, mitochondria showed a significant enlargement, a highly clear matrix, and a reduction in the mitochondrial cristae. Swollen channels of the rough endoplasmic reticulum were also visible in their close proximity (Figure 2A(3,3a)). In turn, the cytoplasm of cells exposed to 45 μM azelastine (Figure 2A(4,4a)) was dominated by swollen mitochondria with a strongly clear matrix, with disorganization of the inner mitochondrial membrane and pronounced damage to the cristae. Mitochondria were also characterized by disruption of the mitochondrial membrane, resulting in leakage of matrix contents into the cytoplasm (Figure 2(A4)). In contrast, visible in the microphotographs the rough endoplasmic reticulum appeared as dilated channels (Figure 2A(4,4a)). At subsequent azelastine concentrations of 60 μM and 90 μM (Figure 2A(5–6a)), the mitochondria were characterized by increased structure disorganization indicating significant damage, and in the vicinity of these structures remained the rough endoplasmic reticulum in the form of strongly widened and swollen cisterns. Compared with the control group, in which ROS (+) cells constituted 3.29%, treatment of HeLa cell with azelastine resulted in concentration-dependent intracellular ROS production (Figure 2C,D). The concentrations of 15 μM and 25 μM showed a small but statistically significant increase in ROS (+) cells to 12.4% and 24.8%, respectively ($p \leq 0.0001$). Increasing the azelastine concentration to 45 μM resulted in increased generation of reactive oxygen species, ROS (+) cells accounted for more than 45% ($p \leq 0.0001$). Significant levels of cellular ROS (+) were observed following azelastine treatment at concentrations of 60 μM (48.93%) and 90 μM (49.99%) ($p \leq 0.0001$). The induction of reactive oxygen species generation correlated with a progressive decrease in mitochondrial membrane potential (Figure 2E,F).The lowest percentage of cells with mitochondrial membrane depolarization was shown at a concentration of 15 μM (9.76%) ($p \leq 0.0001$). At 25 μM and 45 μM, cells with reduced mitochondrial membrane potential were 14.05% and 15.89%, respectively ($p \leq 0.0001$). The highest percentage of cells with mitochondrial membrane depolarization (being more than 50%; $p \leq 0.0001$) was found for concentrations of 60 μM and 90 μM. These results were confirmed in the imaging of rhodamine 123-labeled mitochondria, as it was shown that with increasing azelastine concentration, there is a gradual extinction of green fluorescence emission, significant in the range 45–90 μM (Figure 2B). The azelastine-induced increase in the level of reactive oxygen species, contributed to increased oxidative stress and stress of the rough endoplasmic reticulum and to the induction of apoptotic changes.

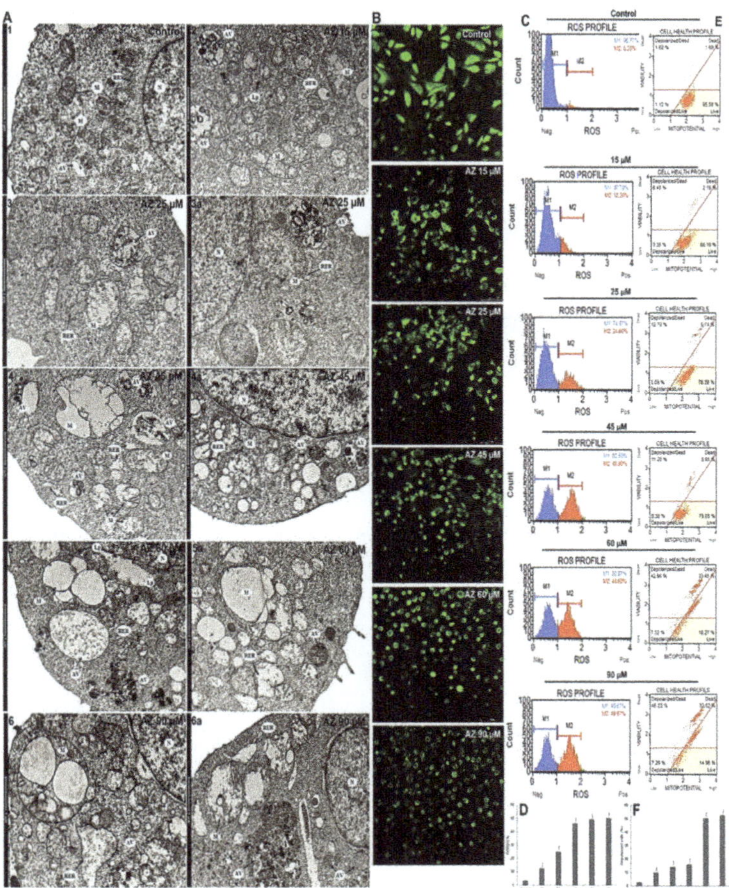

Figure 2. Changes in mitochondria, induction of oxidative stress, and endoplasmic reticulum stress in HeLa cells caused by the action of azelastine hydrochloride. Control cells (**A1**). Azelastine concentration-dependent ultrastructural changes indicative of apoptosis (**A2–6a**); brightening of the matrix and irregular arrangement of cristae in the mitochondria of cells subjected to the 15 µM concentration (**2**); at 25 µM, visible enlarged mitochondria with reduction of mitochondrial cristae remaining in close proximity to the dilated channels of the rough endoplasmic reticulum (**3,3a**); at a concentration of 45 µM, mitochondria with enhanced damage characteristics are present, i.e., strongly enlarged with disruption of the mitochondrial membrane (**4**) and damaged mitochondrial cristae (**4a**) and altered rough endoplasmic reticulum in the form of dilated channels (**4,4a**); at concentrations of 60 µM (**5,5a**) and 90 µM (**6,6a**), visible mitochondria with severe disorganization of the structure indicating damage, and rough endoplasmic reticulum located in their vicinity with strongly enlarged and swollen cisterns. Explanation of abbreviations: N—nucleus, M—mitochondria, AG—Golgi apparatus, RER—rough endoplasmic reticulum, AV—autophagic vacuoles, Lp—primary lysosomes, Ls—secondary lysosomes. Images were taken at 11,500× magnification. (**B**) Gradual and azelastine concentration-dependent loss of green fluorescence derived from rhodamine 123-labeled mitochondria. (**C**) Generation of reactive oxygen species and (**D**) percentage of ROS (+) cells as a result of azelastine. (**E**) Changes of mitochondrial membrane potential ($\Delta\psi m$) and the percentage of cells with mitochondrial membrane depolarization (**F**) at different azelastine concentrations. Each sample was analyzed in triplicate. The differences were statistically confirmed at: *** $p < 0.001$.

2.5. Azelastine Induces DNA Damage

The 48 h effect of azelastine induced a concentration-dependent increase in phosphorylated H2A.X in response to DNA double strand breaks (DSBs) (Figure 3B,C). The increase in DNA damage was as follows for the subsequent concentrations: 15 µM (12.9%, $p \leq 0.0001$), 25 µM (18.38%, $p \leq 0.0001$), 45 µM (24.44%, $p \leq 0.0001$), and 60 µM (28.92%, $p \leq 0.0001$). At the highest concentration of 90 µM, cells with DBS accounted for more than 30% ($p \leq 0.0001$) of all analyzed cells compared to the control (2.79%). DNA damage in azelastine-treated cells could have led to apoptosis.

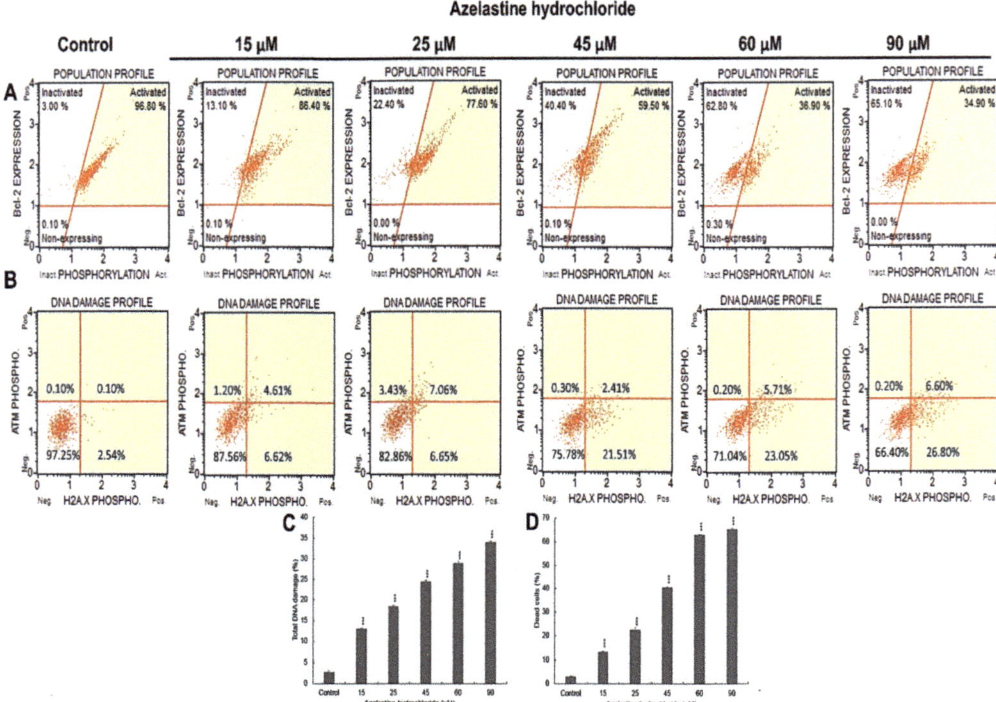

Figure 3. Bcl-2 protein inactivation and DNA damage demonstrated in cells exposed to 48 h action of azelastine hydrochloride. (**A**) Cells expressing Bcl-2 are clustered in the top two quadrants of the dot plot (inactivated and activated). Over 60% are dephosphorylated after treatment with azelastine at 60 µM and 90 µM, confirming inactivation of the Bcl-2 signaling pathway. (**B**) Azelastine at concentrations of 15–90 µM generates DNA damage that induces DNA repair mechanisms such as γH2AX. (**C**) Azelastine concentration-dependent percentage of cells with double-stranded DNA damage and (**D**) percentage of cells with Bcl-2 protein inactivation. Data representative of three parallel experiments correspond to mean values ± standard error (SE). Differences were statistically confirmed at: *** $p < 0.001$.

2.6. Azelastine Inactivates the Bcl-2 Protein

Azelastine induced inactivation of the anti-apoptotic protein Bcl-2 in HeLa cells (Figure 3A,D). At concentrations of 15 µM and 25 µM, cells with inactivated Bcl-2 protein represented 13.17% and 22.47% ($p \leq 0.0001$), respectively, relative to the control (3.1%). Progressive changes in the expression of the test protein were shown at higher concentrations, i.e., 45 µM (40.56%), 60 µM (62.82%), and 90 µM (65%) ($p \leq 0.0001$), which indicates a mechanism of proapoptotic action of azelastine involving the mitochondrial pathway.

2.7. Azelastine Enhances Vacuolization and Apoptotic Changes in HeLa Cells—Morphological Evaluation

In cells exposed to 48 h of azelastine, a significant concentration-dependent increase in the number of cells with vacuolization changes in the cytoplasm was observed (Figure 4D). Compared to control values (16 cells), the highest number of cells with enhanced vacuolization was observed at 15 μM (2119 cells) and 25 μM (2010 cells) ($p \leq 0.0001$). Within the vacuole, a strong eosinophilic material destined for degradation was visible (Figure 4A(2,2a,3)). In contrast, at higher concentrations of the test compound, vacuolization changes showed a decreasing trend (Figure 4D). A lower but equally highly statistically significant result (1282 cells) was shown at a concentration of 45 μM (Figure 4A(4,4a)). On the other hand, at concentrations of 60 μM and 90 μM (Figure 4A(5–6a)), the presence of the lowest number of vacuolized cells was confirmed.

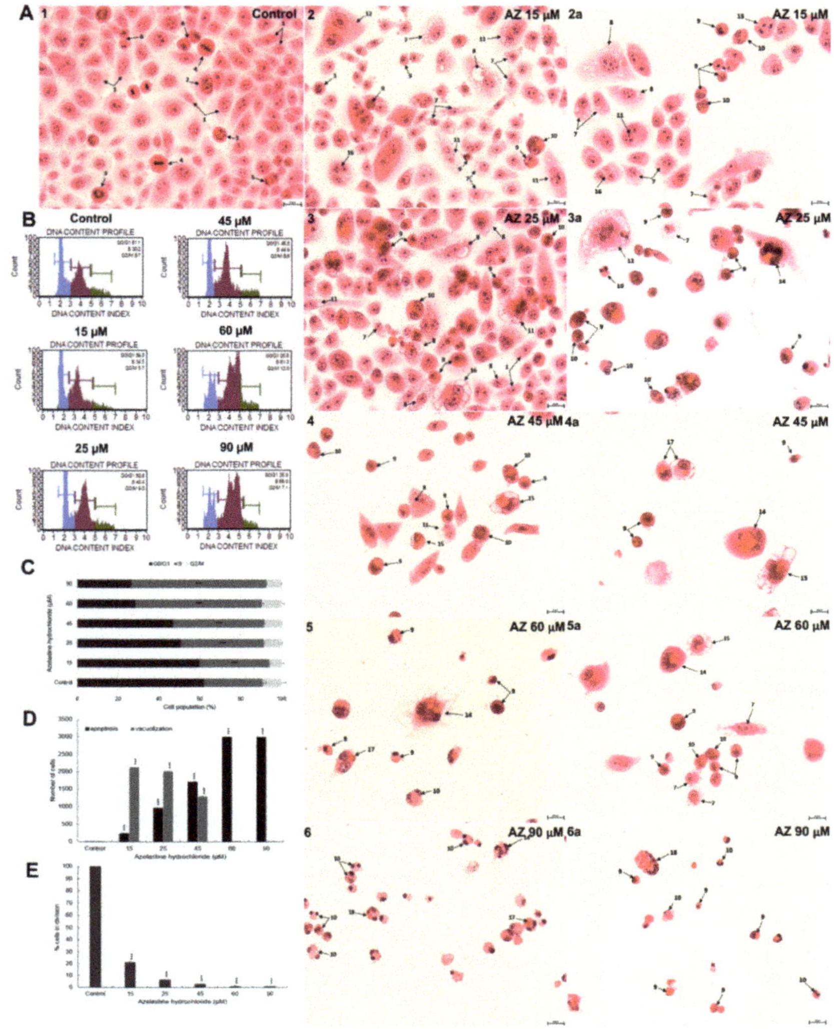

Figure 4. Morphological changes indicating the induction of vacuolating and apoptotic changes and a decrease in the dividing capacity of Hela cells as a consequence of 48 h treatment with azelastine hydrochloride. (**A1**) Control cells with normal morphology, including interphase cells and cells with

multiple mitotic figures. (**A2–6a**) Cells treated with azelastine at concentrations of 15–90 μM; (**A2–3a**) cells with numerous vacuolization changes in the cytoplasm, strongly eosinophilic material is visible within the vacuoles, which is destined for degradation, indicating the process of autophagy; (**A4,4a**) cells with intensive cytoplasmic vacuolization and a pyknotic cell nucleus with visible partial chromatin condensation; (**5–6a**) strong pro-apoptotic effect of azelastine at concentrations of 60 μM and 90 μM, expressed by the presence of numerous apoptotic cells and cells with efferocytosis. Explanation of markings: 1—interphase, 2—prophase, 3—prometaphase, 4—metaphase, 5—telophase, 6—cytokinesis, 7—vacuolization of cytoplasm, 8—vacuoles with visible digestion material, 9—apoptotic cells, 10—binucleated cells in apoptosis, 1—-binucleated cells with vacuolization, 12—giant cells, 13—abnormal segregation of chromosomes, 14—multinucleated cells in apoptosis, 15—cells with features of vacuolization and apoptosis, 16—multinucleated cells with vacuolization, 17—efferocytosis, 18—cells with phagocytosed material (by efferocytosis), which were directed toward the apoptotic pathway. Hematoxylin and eosin staining. Images were taken at 4000× magnification. (**B**) Cell cycle changes of HeLa line cells treated for 48 h with azelastine at concentrations of 15–90 μM analyzed by flow cytometry. (**C**) Percentage of cells in different phases of the cell cycle indicating blocking of cells in S phase. (**D**) Azelastine concentration-dependent number of vacuolated and apoptotic cells; at concentrations of 15–25 μM, azelastine induced vacuolization changes; at a concentration of 45 μM, there was a reduction in vacuolization changes in favor of apoptotic cells, while at concentrations of 60–90 μM, azelastine promoted apoptosis. (**E**) Changes in the mitotic index indicating an inhibition of the dividing capacity of azelastine. Average values from three independent experiments. The differences were statistically confirmed at: *** $p < 0.001$.

The action of azelastine on tested cells resulted in the simultaneous appearance of cells with apoptotic changes such as reduced size, increased cytoplasm staining, pyknotic nucleus with chromatin condensation, and the presence of apoptotic bodies. At concentrations of 15 μM and 25 μM, apoptotic cells accounted for 236 and 963 ($p \leq 0.0001$), respectively (Figure 4A(2–3a),D), compared to the control (11 cells). However, at 45 μM, the number increased significantly to 1708 ($p \leq 0.0001$) (Figure 4A(4,4a),D). The highest value was observed at 60 μM (2988 cells) and 90 μM (2992 cells) ($p \leq 0.0001$) (Figure 4A(5–6a),D). It should be noted that at a concentration of 45 μM (Figure 4A(4,4a)), there were cells with simultaneously observed features such as enhanced vacuolization of the cytoplasm and a pyknotic cell nucleus with partial chromatin condensation, indicating a gradual switch from vacuolization to apoptotic changes. The presence of phagocytosed apoptotic cells was observed within the cytoplasm of living cells (Figure 4A(4a,5a)) and those that were directed into the apoptosis pathway (Figure 4A(6,6a)), indicating induction of the efferocytosis process.

2.8. Azelastine Blocks Cells in S Phase and Reduces Mitotic Index

Cytometric analysis (Figure 4B,C) showed a statistically significant ($p \leq 0.0001$) increase in the number of cells arrested in the S phase of the cell cycle, progressing with azelastine concentration. At the concentration of 15 μM, these cells accounted for 34.23%. Slightly higher results were obtained at a concentration of 25 μM (40.64%) and at 45 μM (44.47%). However, at 60 μM and 90 μM, there was a 2-fold increase in the number of cells in the above-mentioned phase as compared to the control (28.77%). At the same time, in the concentration range of 25–90 μM, there was a significant reduction in the number of cells in the G0/G1 phase ($p \leq 0.0001$).

Comparison of cells incubated with azelastine at all concentrations used (15–90 μM) with cells from the control group (considered as 100%) showed statistically significant ($p \leq 0.0001$) decrease in mitotic index (Figure 4E). Already at a concentration of 15 μM azelastine, the dividing capacity of the cells decreased significantly to 22% and this was also the highest recorded result, because at the other concentrations of the test compound, i.e., 25 μM, 45 μM, 60 μM and 90 μM, the mitotic index decreased to, respectively, 7%, 3%, 2%, and 1%. These changes demonstrate the antiproliferative properties of azelastine.

2.9. Azelastine Induces Mitotic Catastrophe

Morphological analysis showed that azelastine at 15 μM resulted in changes considered morphological markers of mitotic catastrophe (Figure 5). These included multiple abnormalities occurring during mitotic division, such as the presence of anaphase bridges (Figure 5(A2)), tripolar metaphase (Figure 5(A1)), and pentapolar anaphase (Figure 5(A3)). Azelastine also induced the formation of micronuclei (micronucleation) (Figure 5B), which were present in the highest and also statistically significant numbers at a concentration of 15 μM (Figure 5A(3–6)). Furthermore, the data obtained indicated clear multinucleation due to the action of the test compound (Figure 5B). The highest results were recorded at a concentration of 15 μM with 372 binucleated cells, 132 multinucleated cells, and 23 giant cells (at $p \leq 0.0001$). At the next concentration of 25 μM, the results remained high in the range of statistically significant values, there were 267 binucleated cells, 90 multinucleated cells, and 12 giant cells found. However, at 45 μM azelastine, the number of binucleated, multinucleated, and giant cells significantly decreased to 47, 20, and 9, respectively, while at high concentrations of 60 μM and 90 μM, it was further reduced to levels below control values (Figure 5B).

Of note are the vacuolization (Figure 5C) and apoptotic (Figure 5D) changes observed simultaneously in cells with multinucleation. At low concentrations of azelastine (15 μM and 25 μM), vacuolization changes predominated over apoptotic ones, whereas at 45–90 μM, bi- and multinucleated cells were directed towards the apoptotic pathway. The results indicate that azelastine induces mitotic catastrophe, which precedes the onset of apoptosis.

2.10. Azelastine Enhances Degradation Processes

Analysis of changes at the ultrastructural level revealed numerous autophagic vacuoles in the cytoplasm of cells with azelastine at a concentration of 15 μM (Figure 6A(1–1b)); vacuoles were differentiated in size and content indicating different stages of degradation. In the studied cells, expanded Golgi apparatuses and dilated channels of the rough endoplasmic reticulum were present; these changes indicated the intensification of the process of synthesis of proteins crucial for subsequent stages of intracellular digestion. The presence of numerous mitochondria (Figure 6(A1a)) in the tested cells may result from the increased demand for ATP necessary for the macroautophagy process. Also at 25 μM concentration (Figure 6A(2–2b)), numerous and highly enlarged autophagic vacuoles containing material at different stages of degradation were shown, and vacuoles at the formation stage were also present (Figure 6(A2b)). In the lumen of these structures, large fragments of the cytosol with organelles were visible (Figure 6A(2,2b)), and some vacuoles took the form of emptiness and clearly demarcated from the cytoplasm spaces (Figure 6(A2a)). Swollen mitochondria (Figure 6(A2a)), dilated channels of rough endoplasmic reticulum, and reduced Golgi apparatus (Figure 6(A2)), whose membranes could be used for vacuole formation, were also observed in the cells. In contrast, in cells exposed to azelastine at 45 μM concentration (Figure 6A(3–3b)), the number of autophagic vacuoles was reduced; however, they had different shapes and covered a large area of the cytosol. In addition, altered mitochondria and single, slightly dilated channels of rough endoplasmic reticulum were present within the cytoplasm of these cells. When cells were treated with high concentrations of azelastine 60 μM and 90 μM (Figure 6A(4–5b)), the presence of secondary lysosomes was clearly marked alongside altered cell nuclei with local chromatin condensation (Figure 6(A4a), an expanded nuclear envelope (Figure 6A(4a,5,5a)), often with features of fragmentation (Figure 6(A5b)). There were also single, damaged mitochondria (Figure 6A(5a,5b)) and dilated channels of rough endoplasmic reticulum (Figure 6A(4–5a)). The demonstrated changes were dependent on the concentration of azelastine and indicated the intensification of the degradation processes. The progressive degradation observed at high concentrations may indicate a switch of cellular metabolism with the possibility of triggering programmed cell death.

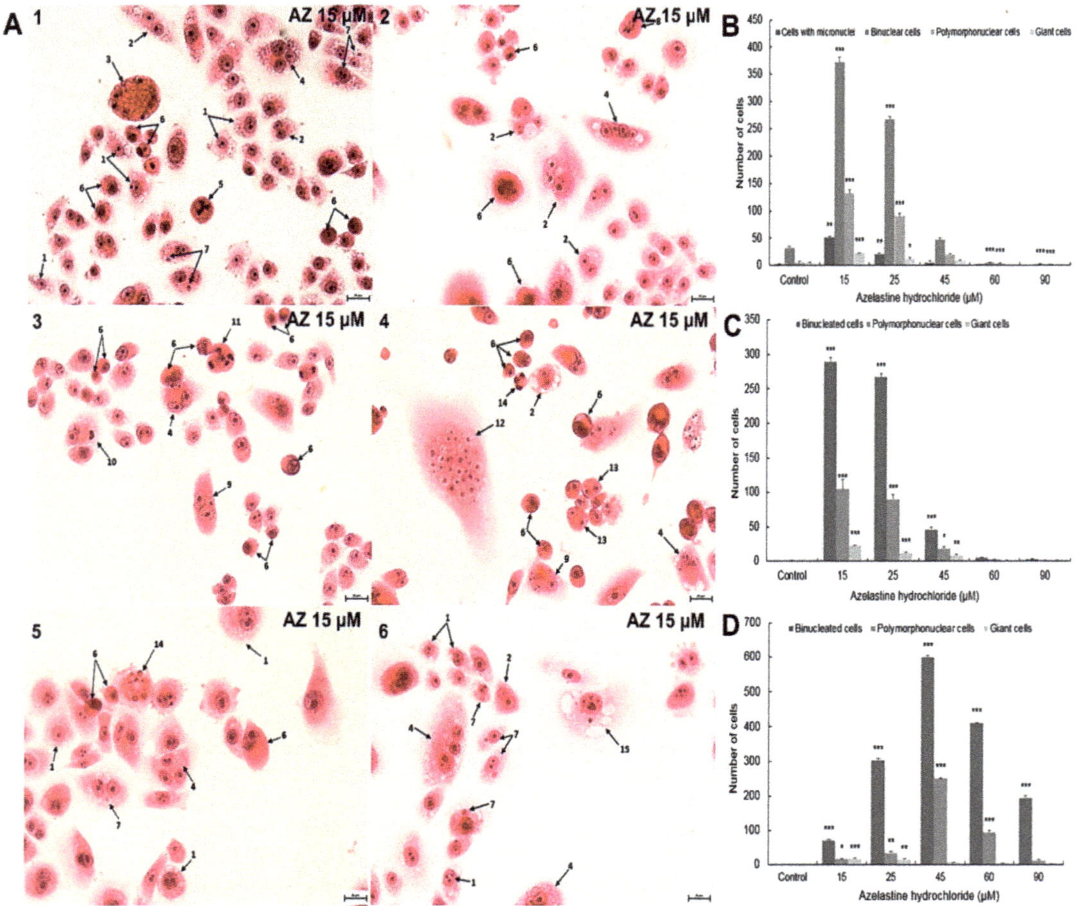

Figure 5. Morphological markers of mitotic catastrophe induced in HeLa cells exposed to 48 h treatment of azelastine hydrochloride. (**A**) Intensified changes (multipolar mitosis, micronucleation, and multinucleation) in cells treated with 15 μM azelastine. Explanation of markings: 1—cells with vacuolization, 2—binucleated cells with vacuolization, 3—apoptotic giant cell, 4—multinucleated cells with vacuolization, 5—tripolar metaphase, 6—apoptotic cells, 7—cells with micronuclei, 8—anaphase bridges, 9—multinucleated cells with micronuclei, 10—efferocytosis, 11—pentapolar anaphase, 12—giant cells, 13—binucleated cells in apoptosis, 14—multinucleated cells in apoptosis, 15—giant cells with vacuolization and micronuclei. Hematoxylin and eosin staining. Images were taken at a magnification of 4000×. (**B**) Distribution of cells with micronuclei, bi-, multinucleated cells, and giant cells at different concentrations of azelastine 15–90 μM. Azelastine concentration-dependent induction of vacuolization (**C**) and apoptotic (**D**) changes in bi-, multinucleated, and giant cells. Average values from three independent experiments. The differences were statistically confirmed at: * $p < 0.05$, ** $p < 0.01$, *** $p < 0.001$.

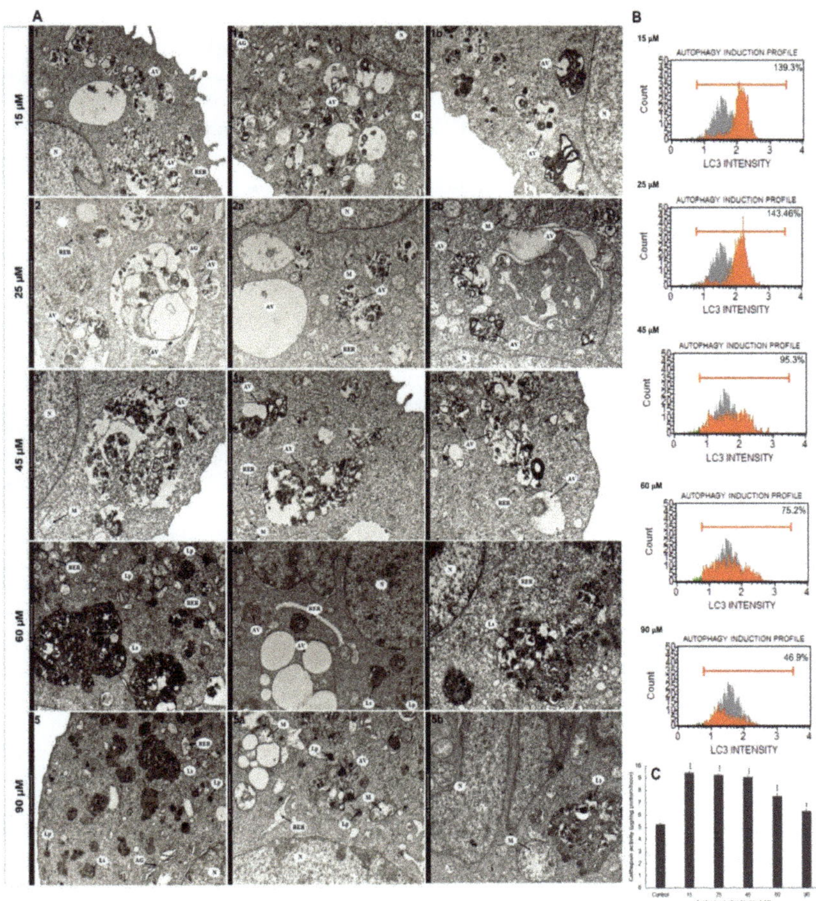

Figure 6. The intensification of degradation processes in HeLa cells with azelastine hydrochloride. (**A**) Ultrastructural changes; (**1–1b**) in the cytoplasm of cells very numerous autophagic vacuoles filled with material at various stages of degradation can be seen, and an extensive Golgi apparatus and dilated channels of the rough endoplasmic reticulum; (**2–2b**) clear changes in cells indicating a progressive degradation process, i.e., multiple vacuoles at different stages of digestion; (**3–3b**), the image shows vacuoles differentiated in terms of shape, covering large areas of the cytosol, indicating intensive degradation; (**4–5b**) dominant changes in the form of numerous primary and secondary lysosomes and changes at the level of the cell nucleus (local chromatin condensation, enlarged nuclear envelope and fragmentation). The changes obtained in the concentrations of 60 μM and 90 μM of azelastine indicate a progressive degradation as well as the initiation of cell death by apoptosis. (**B**) Histograms of cells indicating azelastine-induced autophagy as manifested by an increase in fluorochrome fluorescence intensity (red area) indicating LC3 protein activation. In the concentration range of 45–90 μM, the fluorescence intensity decreased, which argued for a switch of autophagy to apoptosis. Cells were stained with anti-LC3/Alexa Fluor® 555 conjugated antibody and the fluorescence intensity was measured cytometrically. (**C**) Changes in cathepsin D and L activities (mean ± SE) in the lysosomal fraction of HeLa cells after 48 h of exposure to different concentrations of azelastine. Results are the average of three independent experiments. Differences were statistically confirmed at: * $p < 0.05$, ** $p < 0.01$, *** $p < 0.001$.

2.11. Azelastine Activates Cathepsin D and L

As shown in the study, the effect of azelastine compared to control (which was taken as 100%) resulted in concentration-dependent changes in cathepsin D and L activity ($p \leq 0.0001$) (Figure 6C). The highest increase in the activity of enzymes to 179.96% and 177.54% occurred at the concentrations of 15 μM and 25 μM, respectively. At 45 μM, the enzymes activity was found to be 173.89%. Further increase in the concentration of the test compound to 60 μM and 90 μM resulted in reduction of cathepsin D and L activities to 144.33% and 120.53%, respectively. The behavior of lysosomal enzymes is a reflection of the degradation processes activated by azelastine.

2.12. Azelastine Induces Autophagy by Increasing LC3 Protein Levels

According to the principle of the assay used, LC3 is a cytoplasmic protein involved in autophagosome formation during autophagy, which is translocated from the cytoplasm to the interior of autophagosomes, and its fluorescence is monitored cytometrically. According to the studies performed, azelastine induced autophagy depending on the concentration (Figure 6B). The highest intensity of fluorescence in cells was observed at concentrations of 15 μM and 25 μM, it was 139.3% and 143.36%, respectively, compared to the cells of the control group (gray area) (48.9%). With increasing azelastine concentrations, a gradual reduction in dye emission in labeled cells was observed to 95.3% at 45 μM and to 75.2% at 60 μM. At the highest concentration used (90 μM), a further reduction of the fluorescence intensity to 46.9% was demonstrated.

3. Discussion

Despite continuous advances in anticancer therapy, low treatment efficacy with concomitant high side effects is still a major problem [16]. Therefore, in the search for potential chemotherapeutic agents, particular attention is paid to the safety of the drug and its good tolerability [18]. Such features may be possessed by the well-studied new-generation H1 antihistamines, which have almost completely displaced the old-generation drugs used in anti-allergic treatment [48]. Another important aspect in the search for new oncological treatment options is the complexity of the oncological disease. The success of cancer therapy is also influenced by the possibility of modulating molecular and cellular factors found in the tumor and its microenvironment [16]. Thus, the identification of compounds with multidirectional mechanisms of action is crucial for the further development of anticancer therapies [6], and azelastine, used in anti-allergy treatment, may be such a drug.

The results obtained from our study allow us to conclude that the studied compound induced in HeLa cells two important processes for anticancer therapy, namely autophagy and apoptosis (Figure 4D).

At low concentrations (15 μM and 25 μM), azelastine clearly promoted autophagy while apoptosis remained low. The induction of autophagy is indicated by an increased number of cells with intensified vacuolization of the cytoplasm (Figure 4A(2–3a),D). An important role in this process is played by the maintenance of an acidic pH inside the vacuole, which was documented by the presence of a strongly eosinophilic content within the large vacuoles of the studied cells (Figure 4A(2,3)). Adequate pH is necessary for the activity of lysosomal enzymes required to digest cellular material [49,50]. In our study, we showed that azelastine treatment caused a marked increase in the activity of lysosomal enzymes, i.e., cathepsin D and L (Figure 6C). The revealed concentration-dependent increase in lysosomal hydrolases activity was correlated with ultrastructural changes of studied cells, indicating an increase in degradative processes. The numerous autophagic vacuoles seen in the microphotographs (Figure 6A(1–2b)), which are very large and contain fragments of cytosol with organelles, indicate the presence of macroautophagy. This was also confirmed by examining the autophagy-specific marker, LC3 protein, where the highest fluorescence intensity (i.e., 139.3% and 143.36%) was found at the lowest concentrations of azelastine (15 μM and 25 μM, respectively) (Figure 6B).

As shown in Figure 4D, cells with morphological features typical of apoptosis clearly gained advantage at 45 μM azelastine, and with increasing concentrations of 60 μM and 90 μM, they constituted more than 90% of all analyzed cells. The switch of autophagy to apoptosis is documented in Figure 4A(4,4a), where cells with characteristics of both types of cell death are seen. Such a condition can be associated with progressive degradation of organelles, confirmed by the presence of giant autophagic vacuoles in cells loaded with 45 μM of azelastine (Figure 6A(3,3a)), as well as the presence of increased numbers of primary and secondary lysosomes at high concentrations (60 μM and 90 μM) of the test compound (Figure 6A(4–5b)). In the studied cells, enlarged mitochondria were visible next to the vacuoles (Figure 6A(5a,5b)), which according to the literature could be related to the increasing demand for ATP, necessary for enhanced autophagy as well as for triggering programmed cell death [51]. The nuclei of the cells also showed altered morphology, including chromatin condensation and fragmentation (Figure 6A(4a,5b)), which was confirmed by DAPI staining (Figure 1E). The pro-apoptotic effect was additionally confirmed by the cytometric method; it was shown that azelastine, depending on the concentration, significantly induced the number of apoptotic cells with the dominance of the late-apoptotic phenotype. These values increased as follows: up to 60% at a concentration of 45 μM, 93% at 60 μM, and 98% at 90 μM (Figure 1A,C).

Autophagy and apoptosis are interconnected and can occur in the same cell in response to a given stimulus simultaneously or separately [52,53]. According to Fimia and Piacentini [54], induction of apoptosis is often associated with increased autophagy. In the presence of apoptotic stimuli, autophagy may be an adaptive response or a distinct type of cell death [55]. The tendency to change the regulation of both processes demonstrated in our studies was dependent on the concentration of azelastine. The targeting of cells to the apoptotic pathway was likely the result of a failed attempt to restore cellular homeostasis as a consequence of increased cellular stress [56]. During excessive autophagy, mitochondria tend to show accelerated production of reactive oxygen species due to increased oxidative phosphorylation. A slight, but statistically significant increase in the ROS level (Figure 2C,D), was noted already at the lowest concentrations of the tested compound (15 μM and 25 μM). The ROS values increased significantly at the concentration of 45 μM, which could have triggered the apoptosis process in the tested cells. Similar results were obtained by the team of Nicolau-Galmés [55] in a study on terfenadine, an old generation antihistamine, which enhanced autophagy and consequently led to the induction of apoptosis.

The oxidative stress activated by azelastine in HeLa cells was correlated with the simultaneous increase in the level of phosphorylated H2A.X (Figure 3B,C). The results obtained in this study indicate the participation of ROS in inducing DNA damage, which could have been a signal to trigger apoptosis. The significantly reduced division capacity of HeLa cells (Figure 4E) and their arrest in the S phase of the cell cycle (Figure 4B,C) may be associated with the DNA damage response. As shown in the literature, cell proliferation may be crucial for tumor development and progression, and histamine may be the main mediator of this process in various types of cancer [16]. On the other hand, DNA damage and inhibition of cell proliferation are among the important mechanisms of action of anticancer drugs [57]. The various properties of azelastine demonstrated in this study can also be used in anticancer therapy.

The antiproliferative properties of azelastine are also confirmed by the mitotic catastrophe induction capacity demonstrated in studies, documented in Figure 5A showing the abnormal course of mitosis. This process was most likely induced by DNA damage, and resulted in demonstrated changes such as multinucleation and micronucleation (Figure 5A(1–6),B). Of particular note is the presence of giant, multinucleated cells (Figure 5(A4)). At high concentrations of the tested compound, cells with significant nuclear changes eventually underwent apoptosis (Figure 5D), which is considered one of the necessary final steps in the course of mitotic catastrophe. The mitotic catastrophe shows a strong mechanistic relationship with the cellular and molecular changes accompanying carcinogenesis and therefore seems to be a preferentially stimulated process in cancer cells [58–60]. Compounds promoting mitotic death, such as azelastine, may be a promising therapeutic alternative for apoptosis-resistant cancer cells.

In cells, the functions of "stress sensors" are performed by mitochondria and they are the central executors of apoptosis [61] as well as the course of mitotic catastrophe [59]. As our results showed, the induction of apoptosis by azelastine was also associated with the mitochondrial pathway. At the level of submicroscopic studies, already at low concentrations of azelastine, mitochondria were enlarged with a clear matrix (Figure 2A(2,3,3a)). However, under the influence of high concentrations, enhanced changes were demonstrated, with significant mitochondrial damage (Figure 2A(4–6a)) and disorganization of the inner membrane. At the same time, the cytometric analysis determined the highest percentage of cells with depolarization of the mitochondrial membrane (over 50%) for the concentrations of 60 μM and 90 μM (Figure 2E,F). The violation of the mitochondrial membrane integrity was confirmed by the concentration-dependent gradual quenching of green fluorescence emissions from labeled mitochondria (Figure 2B), which was also associated with the demonstrated inactivation of the Bcl-2 protein (Figure 3A,D) and activation of the executive caspases (Figure 1B,C). We also demonstrated the cytotoxic effect of azelastine related to the reduction of metabolic activity of mitochondria using the MTT test. Depending on the concentration, this compound reduced the viability of HeLa cells (Figure 1D), which at the highest concentration (90 μM) was only 4%. According to the studies by Cornet-Masana [62] conducted on leukemia lines, mitochondria in cancer cells are characterized by numerous changes, which, according to Pathania's team [63], makes them more susceptible to therapies aimed at the metabolism of cancer cells.

The analysis performed at the level of submicroscopic changes revealed that mitochondrial disorganization is also accompanied by significant changes in the profile of the rough endoplasmic reticulum. It has been reported that even at low concentrations of azelastine, there is significant dilatation of reticulum channels (Figure 2A(3,3a)). These changes intensified as a consequence of the action of increasing concentrations of the tested compound, and at its high concentrations, they became significantly swollen (Figure 2A(5–6a)), which can be explained by the stress of the reticulum. The revealed changes in the endoplasmic reticulum homeostasis may be induced by an increased level of ROS [64,65], which is also confirmed by the results of our research.

Cao and Kaufman [66] emphasize in their works the importance of spatial and functional distribution in cells of organelles such as mitochondria and endoplasmic reticulum. Also in the analyzed electronograms, close proximity of altered mitochondria and expanded channels of rough endoplasmic reticulum was demonstrated (Figure 2A(5a–6a)), indicating a functional relationship between these organelles and which may be relevant to the processes regulating apoptotic cell death. Pro-apoptotic factors derived from mitochondria induce signals from the rough endoplasmic reticulum, which in turn cause changes in the mitochondria. On the other hand, reticulum stress can lead to mitochondrial dysfunction and consequent oxidative stress, followed by impaired homeostasis and apoptosis [67–69]. Apoptosis involving endoplasmic reticulum stress has attracted a lot of attention in recent years [64]. Mild stress of cancer cells can lead to the activation of adaptive mechanisms, however, therapeutic benefits of compounds that induce endoplasmic reticulum stress and put cells on the apoptosis pathway have been confirmed for certain types of cancer cells [70,71]. In our studies, azelastine induced in HeLa cells oxidative stress, stress of rough endoplasmic reticulum, and mitochondrial dysfunction, which by reinforcing each other, disrupted cellular functions and activated proapoptotic signals [65,66,68]. A similar mechanism of action has been reported for terfenadine, an old generation antihistamine in relation to A375, HT144, and Hs294T cell lines [55]. However, in studies on the action of rupatadine, ebastine, and loratadine in relation to acute myeloblastic leukemia cells, the cytotoxicity of these compounds consisted of bidirectional, mitochondrial-lysosomal action, ROS generation, and reduction of mitochondrial metabolic activity, which led to the activation of caspase 3 and 7 and induction of the apoptosis pathway [62].

Efferocytosis-phagocytosis of dead cells was also observed under the action of azelastine (Figure 4A(4a,5,6,6a)). According to the literature, this process under certain conditions can be performed by "non-professional phagocytes" [72–74]. In the context of carcinogenesis, efferocytosis suppresses the body's natural immune response, then facilitates the immune escape of tumor cells while promoting the tumor microenvironment [50]. This process not only affects the proliferation, invasion, metastasis, and angiogenesis of cancer cells, but also regulates adaptive responses and decreases the positive response to radiotherapy and to many commonly used anticancer antibodies and chemotherapeutic agents [75]. The data obtained in our study indicate, that initially azelastine at low concentrations (Figure 4(A4a)) induced efferocytosis in the context of an adaptive response of HeLa cells, and then cells with phagocytosed apoptotic cells were directed to the apoptotic pathway (Figure 4A(6,6a)). Such action is in contrast to that of traditional therapies, which induce apoptosis of tumor cells and increase subsequent efferocytosis, which suppresses the inflammatory response [76]. Thus, the demonstrated property of azelastine indicates an additional possibility of the interference of the compound in the tumorigenesis, and at the same time, fits in the current view of combining traditional therapies with therapies targeting the efferocytosis process in order to improve their effectiveness [50].

4. Material and Methods

4.1. In Vitro Culture Conditions

Human cervical adenocarcinoma cells (HeLa cells) were cultured in a Direct Heat incubator (Thermo Scientific, Waltham, MA, USA), under standard culture conditions, i.e., 37 °C and 5% CO_2, on a modified DMEM medium (GIBCO, New York, NY, USA), containing 10% fetal calf serum (Biowest, Nuaillé, France) and a mixture of antibiotics (penicillin G, streptomycin, amphotericin B) (Corning, Manassas, VA, USA). The HeLa cells were purchased from the American Type Tissue Culture Collection (Rockville, MD, USA). Cells were treated for 48 h with azelastine hydrochloride ($\geq$98% HPLC), (4-[(4-chlorophenyl)methyl]-2-(1-methylazepan-4-yl) phthalazin-1-one hydrochloride), which was purchased from Sigma Aldrich (St. Louis, MO, USA). The following concentrations of the test compound were used in the experiment: 15 μM, 25 μM, 45 μM, 60 μM, and 90 μM.

According to the literature data, the tested concentrations are used in research on antihistamine drugs conducted on cancer cell lines. Control cells were cultured in complete maintenance medium without the addition of the test compound.

4.2. Assessment of Cell Viability—MTT Test

The level of cytotoxicity of azelastine against HeLa cells was determined using MTT (3-(4,5-dimethyl-2-yl)-2,5-diphenyltetrazolium bromide) reduction assay. Cells seeded in Falcon 96-well plates (Fisher Scientific, Waltham, MA, USA) after azelastine treatment were stained with MTT solution (1 mg/mL) (Sigma Aldrich, St. Louis, MO, USA). After 2 h of incubation of the cells with the dye, dimethylsulfoxide (DMSO) was applied to solubilize the formed formazan crystals. Optical density was measured at 570 nm using a Synergy 2 multi-detector microplate reader (BioTek, Winooski, VT, USA). Cell viability was calculated in comparison with the control group using Gen5 software.

4.3. Visualization of Apoptotic Cells under A Fluorescence Microscope

Morphological evaluation of nuclei of control and tested cells was performed using 4′,6-diamidino-2-phenylindole (DAPI) staining. Cells cultured in dishes (Falcon, Fisher Scientific, Waltham, MA, USA) were stained with 2.5 μg/mL DAPI solution (Sigma Aldrich, St. Louis, MO, USA) for 15 min and then washed with PBS. The preparations were analyzed using a Nikon Eclipse Ti epi-fluorescence inverted microscope (Nikon Instruments Inc., Melville, NY, USA).

4.4. Detection of Apoptosis

Phosphatidylserine externalization in azelastine-exposed cells was assessed using Annexin V and Dead Cell test kit (Merck Millipore, Burlington, MA, USA). Control and azelastine treated cells were detached using 0.25% trypsin-EDTA (Corning, New York, NY, USA), centrifuged and washed with PBS. Then, cells were stained with annexin V (100 μL) for 20 min at room temperature in the dark. The fluorescence intensity was analyzed using a Muse analyzer (Merc Millipore, Burlington, MA, USA).

4.5. Activity of Caspase-3/7

The level of caspase-3/7 activation was measured using a caspase-3/7 assay kit (Merck-Millipore, Burlington, MA, USA). After 48 h of incubation with azelastine, cells were harvested by trypsinization and incubated at 37 °C with 5 μL of Caspase-3/7 working solution (as per protocol). Then, 150 μL of Caspase 7-AAD working solution was added to the cells. Detection of caspase-positive cells was performed using a Muse analyzer (Merck-Millipore, Burlington, MA, USA).

4.6. Analysis of Ultrastructural Changes

Cells for electron microscopy were fixed in 3% glutaraldehyde (Serva Electrophoresis GmbH, Heidelberg, Germany) followed by 2% OsO4 (Spi, West Chester, PA, USA) in cacodyl buffer. The material was then dehydrated in an ascending series of ethanol solutions (10–99.8%) and embedded in Epon 812 epoxy resin (Serva Electrophoresis GmbH, Heidelberg, Germany), followed by polymerization at 40 °C and 60 °C. The epoxy blocks were cut into ultra-thin sections on a Leica EM UC7 ultramicrotome (Leica Biosystems, Wetzlar, Germany), and the obtained preparations were further contrasted with uranyl acetate and lead citrate. Analysis was performed using a Tecnai G2 Spirit transmission electron microscope (FEI Company, Hillsboro, OR, USA) equipped with a Morada camera (Olympus, Soft Imagine Solutions, Münster, Germany). The interpretation of the changes in azelastine-exposed cells was based on the image of control cells.

4.7. Measurement of the Mitochondrial Membrane Potential (Δψm)

The decrease in Δψm was analyzed using the Muse MitoPotential Assay kit (Merck Millipore). Cells after incubation with azelastine were resuspended in 95 μL of Muse MitoPotential working solution and incubated at 37 °C for 20 min. The cells were then stained with 7-AAD dead cell marker (5 μL) at room temperature for 5 min, and the cell suspension was analyzed by flow cytometry.

4.8. Microscopic Evaluation of Changes in the Potential of Mitochondrial Membrane

After 48 h incubation with azelastine, cells were fixed in 4% paraformaldehyde and then incubated for 30 min with rhodamine 123 (Sigma Aldrich, St. Louis, MO, USA) at a concentration of 5 μg/mL ethanol. The used fluorochrome binds to metabolically active mitochondria, so the fading of fluorescence is proportional to the decrease in mitochondrial membrane potential. The cells were then washed with PBS and analyzed under a Nikon A1 confocal microscope based on a Nikon Eclipse Ti inverted microscope (Nikon Instruments Inc., Melville, NY, USA) and equipped with Nikon Nis Elements AR software (Nikon Instruments Inc., Melville, NY, USA).

4.9. Oxidative Stress Analysis

The Muse Oxidative Stress Assay kit (Merck Millipore, Burlington, MA, USA) based on intracellular detection of superoxide radicals was used to investigate the level of reactive oxygen species. As according to the manufacturer's instructions, cells were treated with Muse Oxidative Stress Reagent working solution (190 μL) after 48 h incubation with azelastine. Samples were then incubated at 37 °C for 30 min and the percentage of gated ROS (−) and ROS (+) cells with ROS activity were analyzed.

4.10. Assessment of Bcl-2 Protein Phosphorylation

Changes in Bcl-2 phosphorylation in HeLa cells were assessed using the Muse™ Bcl-2 Activation Dual Detection Assay kit (Merck-Millipore, Guyancourt, France) according to the manufacturer's instructions. Two direct conjugated antibodies were used in the kit, i.e., phospho-specific anti-phospho-Bcl-2 (Ser70)-Alexa Fluor® 555 and a conjugated anti-Bcl-2-PECy5 antibody to measure total Bcl-2 expression levels. The degree of activation of the Bcl-2 pathway was assessed by measuring Bcl-2 phosphorylation relative to total Bcl-2 expression in the tested cells.

4.11. DNA Damage Assessment

To determine whether azelastine causes DNA damage, cells were fixed and permeabilized with Muse Fixation Buffer and Permeabilization Buffer reagents, followed by staining with anti-phospho-Histone H2A.X (Ser139) and anti-phospho-ATM (Ser1981) antibodies according to the instructions for the Muse H2A.X Activation Dual Detection kit (Millipore, Darmstadt, Germany).

4.12. Cell Cycle Analysis

Cells were analyzed using the Muse Cell Cycle Assay Kit (Merck Millipore, Burlington, MA, USA). Cells were trypsinized and centrifuged, and the obtained cell pellet was fixed in 70% ice-cold ethanol. Cells were then treated with Muse Cell Cycle Reagent (Merck Millipore, Burlington, MA, USA) for 30 min and then analyzed with a Muse analyzer (Merck Millipore, Burlington, MA, USA).

4.13. Visualization of Morphological Changes and Assessment of the Dividing Capacity of HeLa Cells

Cells were cultured on sterile coverslips in Falcon dishes (Fisher Scientific, Waltham, MA, USA) in DMEM medium supplemented with azelastine (test cells) or without test compound (control cells). Methanol-fixed cells were stained with Harris hematoxylin and eosin, then dehydrated in an ascending series of ethanol solutions and immersed in xylene. Each preparation was analyzed based on a control image, taking into account changes mainly concerning cell nucleus (presence of bi- and multinucleated cells, giant cells, cells with micronuclei, with chromatin condensation, with pyknotic nucleus), cytoplasm (increased or decreased pigmentation, vacuolization changes, presence of apoptotic bodies), and mitotic division (presence of cells in particular phases of division and abnormal mitotic figures). Quantitative and qualitative analysis of morphological changes in the studied cells and photographic documentation were performed using a Nikon Eclipse 80i microscope with Nikon NIS Elements D 3.10 software (Nikon Instruments Inc., Melville, NY, USA). The mitotic index was evaluated by determining the number of cells in each phase of mitotic division, and the result was expressed as a percentage. In preparations, 3000 cells each were analyzed in three independent experiments (9000 cells/concentration), and the final score for a given trait was the mean value.

4.14. Evaluation of Cathepsin D and L Activity Levels

After 48 h of incubation with azelastine, cells were trypsinized, resuspended in 0.25 M sucrose solution and homogenized using a Potter S homogenizer (Sartorius, Gottingen, Germany). The homogenate was initially centrifuged at an overload of 700× g, for 10 min. The extranuclear supernatant was then centrifuged at 20,000× g overload for 20 min, and the obtained lysosomal pellet was resuspended in Triton X-100 (Sigma-Aldrich, St. Louis, MO, USA). The activities of degradative enzymes, cathepsin D and L, were determined in the lysosomal fraction according to the modified Langner's method. According to the procedure, 2% azocasein (Sigma-Aldrich, St. Louis, MO, USA), 0.2 M acetate buffer pH = 5.0, and 10% TCA (+4 °C) were used. After incubation at 37 °C, samples were centrifuged, and enzyme activity was measured by colorimetric method at 366 nm using a Spekol 1500 spectrophotometer (Analytik Jena GmbH, Jena, Germany). Simultaneously, the total protein content (at 680 nm) was determined using the Lowry's method modified by Kirschke and Wiederanders. Enzyme activity was expressed as μmol/mg protein/hour.

4.15. LC3-Antibody Detection

The level of azelastine-induced autophagy was assessed by cytometric assay using Autophagy LC3 antibody (Merck Millipore, Burlington, MA, USA). The kit includes reagent to selective membrane permeabilization (Autophagy Reagent A) that allows to distinguish between cytosolic and autophagic LC3. This is accomplished by extracting the cytosolic protein while protecting the LC3 that is translocated to and remains intact in autophagosomes. Addition of Anti-LC3 Alexa Fluor® 555 and Autophagy Reagent B to the cells allows quantification of LC3 by measuring fluorescence using flow cytometry. According to the protocol, cells were seeded in Falcon 96-well plates. Autophagy A reagent in EBSS medium (Corning, Corning, NY, USA) was then added to the cells and incubated for 4 h under CO_2 atmosphere, followed by washing with HBSS (Corning, Corning, NY, USA), trypsinization, and centrifugation. The supernatant was removed and anti-LC3 Alexa Fluor® 555 and Autophagy Reagent B were added to the cells and incubated on ice for 30 min in the dark. The samples were then analyzed using flow cytometry technique. Cells that were treated with serum-free medium for 4 h were used as a positive control.

4.16. Statistical Analysis

Statistical analysis of the study results was performed using one-way analysis of variance (ANOVA) with multiple post-hoc comparisons using Tukey's test. Differences were considered statistically significant at $p < 0.05$. Statistica 10.0 software (StatSoft, Krakow, Poland) was used for data analysis.

5. Conclusions

In our study we demonstrated potential anticancer properties of azelastine based on autophagic, proapoptotic, cytotoxic, or antiproliferative activity, which, taking into account safety of its application and potent anti-inflammatory properties, can be regarded as features of a compound that is part of the current canon of fight against cancer. Azelastine may be therefore an alternative method of oncological treatment, which requires further research.

Author Contributions: Conceptualization, E.T., W.T. and T.K.; methodology, E.T. and W.T.; software, E.T. and W.T.; validation, E.T.; formal analysis, E.T., W.T. and T.K.; investigation, E.T., W.T. and T.K.; resources, E.T., W.T. and T.K.; data curation, E.T., W.T. and T.K.; writing—original draft preparation, E.T.; writing—review and editing, E.T., W.T. and T.K.; visualization, E.T. and W.T.; supervision, E.T.; project administration, E.T., W.T. and T.K.; funding acquisition, E.T., W.T. and T.K. All authors have read and agreed to the published version of the manuscript.

Funding: This work was supported by the statutory activity of the Jan Kochanowski University in Kielce, grant No. SUPB.RN.21.248.

Institutional Review Board Statement: Not applicable.

Informed Consent Statement: Informed consent was obtained from all subjects involved in the study.

Data Availability Statement: The data that support the findings of this study are available from the corresponding author upon reasonable request.

Acknowledgments: We would like to thank Anna Lankoff from the Center of Radiobiology and Biological Dosimetry of the Institute of Nuclear Chemistry and Technology, Warsaw, Poland for making the cell line available for research.

Conflicts of Interest: The authors declare no conflict of interest.

References

1. Matsumoto, N.; Ebihara, M.; Oishi, S.; Fujimoto, Y.; Okada, T.; Imamura, T. Histamine H1 receptor antagonists selectively kill cisplatin-resistant human cancer cells. *Sci. Rep.* **2021**, *11*, 1492. [CrossRef] [PubMed]
2. Chang, A. Chemotherapy, chemoresistance and the changing treatment landscape for NSCLC. *Lung Cancer* **2011**, *71*, 3–10. [CrossRef] [PubMed]
3. Xia, J.; Yu, X.; Song, X.; Li, G.; Mao, X.; Zhang, Y. Inhibiting the cytoplasmic location of HMGB1 reverses cisplatin resistance in human cervical cancer cells. *Mol. Med. Rep.* **2017**, *15*, 488–494. [CrossRef] [PubMed]
4. Chang, H.; Zou, Z. Targeting autophagy to overcome drug resistance: Further developments. *J. Hematol. Oncol.* **2020**, *13*, 159. [CrossRef] [PubMed]
5. Ellegaard, A.M.; Dehlendorff, C.; Vind, A.C.; Anand, A.; Cederkvist, L.; Petersen, N.H.T.; Nylandsted, J.; Stenvang, J.; Mellemgaard, A.; Osterlind, K.; et al. Repurposing Cationic Amphiphilic Antihistamines for Cancer Treatment. *EBioMedicine* **2016**, *9*, 130–139. [CrossRef]
6. Greten, F.R.; Grivennikov, S.I. Inflammation and Cancer: Triggers, Mechanisms, and Consequences. *Immunity* **2019**, *51*, 27–41. [CrossRef]
7. Verdoodt, F.; Dehlendorff, C.; Jaattela, M.; Strauss, R.; Pottegard, A.; Hallas, J.; Friis, S.; Kjaer, S.K. Antihistamines and Ovarian Cancer Survival: Nationwide Cohort Study and in Vitro Cell Viability Assay. *J. Natl. Cancer Inst.* **2020**, *112*, 964–967. [CrossRef]
8. Kuzu, O.F.; Toprak, M.; Noory, M.A.; Robertson, G.P. Effect of lysosomotropic molecules on cellular homeostasis. *Pharmacol. Res.* **2017**, *117*, 177–184. [CrossRef]
9. Serrano-Puebla, A.; Boya, P. Lysosomal membrane permeabilization as a cell death mechanism in cancer cells. *Biochem. Soc. Trans.* **2018**, *46*, 207–215. [CrossRef]
10. Tatarkiewicz, J.; Rzodkiewicz, P.; Zochowska, M.; Staniszewska, A.; Bujalska-Zadrozny, M. New antihistamines—Perspectives in the treatment of some allergic and inflammatory disorders. *Arch. Med. Sci. AMS* **2019**, *15*, 537–553. [CrossRef]
11. Bens, A.; Dehlendorff, C.; Friis, S.; Cronin-Fenton, D.; Jensen, M.B.; Ejlertsen, B.; Lash, T.L.; Kroman, N.; Mellemkjaer, L. The role of H1 antihistamines in contralateral breast cancer: A Danish nationwide cohort study. *Br. J. Cancer* **2020**, *122*, 1102–1108. [CrossRef] [PubMed]
12. Mandola, A.; Nozawa, A.; Eiwegger, T. Histamine, histamine receptors, and anti-histamines in the context of allergic responses. *LymphoSign J.* **2019**, *6*, 35–51. [CrossRef]
13. Criado, P.R.; Criado, R.F.J.; Maruta, C.W.; Fiho, C. Histamine, histamine receptors and antihistamines: New concepts. *An. Bras. De Dermatol.* **2010**, *85*, 195–210. [CrossRef] [PubMed]
14. Thangam, E.B.; Jemima, E.A.; Singh, H.; Baig, M.S.; Khan, M.; Mathias, C.B.; Church, M.K.; Saluja, R. The Role of Histamine and Histamine Receptors in Mast Cell-Mediated Allergy and Inflammation: The Hunt for New Therapeutic Targets. *Front. Immunol.* **2018**, *9*, 1873. [CrossRef] [PubMed]
15. Zhang, E.; Zhang, Y.; Fan, Z.; Cheng, L.; Han, S.; Che, H. Apigenin Inhibits Histamine-Induced Cervical Cancer Tumor Growth by Regulating Estrogen Receptor Expression. *Molecules* **2020**, *25*, 1960. [CrossRef] [PubMed]
16. Massari, N.A.; Nicoud, M.B.; Medina, V.A. Histamine receptors and cancer pharmacology: An update. *Br. J. Pharmacol.* **2020**, *177*, 516–538. [CrossRef] [PubMed]
17. de Guadalupe Chavez-Lopez, M.; Perez-Carreon, J.I.; Zuniga-Garcia, V.; Diaz-Chavez, J.; Herrera, L.A.; Caro-Sanchez, C.H.; Acuna-Macias, I.; Gariglio, P.; Hernandez-Gallegos, E.; Chiliquinga, A.J.; et al. Astemizole-based anticancer therapy for hepatocellular carcinoma (HCC), and Eag1 channels as potential early-stage markers of HCC. *Tumour Biol. J. Int. Soc. Oncodev. Biol. Med.* **2015**, *36*, 6149–6158. [CrossRef]
18. Garcia-Quiroz, J.; Camacho, J. Astemizole: An old anti-histamine as a new promising anti-cancer drug. *Anti-Cancer Agents Med. Chem.* **2011**, *11*, 307–314. [CrossRef]
19. Jangi, S.M.; Diaz-Perez, J.L.; Ochoa-Lizarralde, B.; Martin-Ruiz, I.; Asumendi, A.; Perez-Yarza, G.; Gardeazabal, J.; Diaz-Ramon, J.L.; Boyano, M.D. H1 histamine receptor antagonists induce genotoxic and caspase-2-dependent apoptosis in human melanoma cells. *Carcinogenesis* **2006**, *27*, 1787–1796. [CrossRef]
20. Wang, W.T.; Chen, Y.H.; Hsu, J.L.; Leu, W.J.; Yu, C.C.; Chan, S.H.; Ho, Y.F.; Hsu, L.C.; Guh, J.H. Terfenadine induces anti-proliferative and apoptotic activities in human hormone-refractory prostate cancer through histamine receptor-independent Mcl-1 cleavage and Bak up-regulation. *Naunyn-Schmiedeberg's Arch. Pharmacol.* **2014**, *387*, 33–45. [CrossRef]
21. Fritz, I.; Wagner, P.; Olsson, H. Improved survival in several cancers with use of H1-antihistamines desloratadine and loratadine. *Transl. Oncol.* **2021**, *14*, 101029. [CrossRef] [PubMed]
22. Fritz, I.; Wagner, P.; Broberg, P.; Einefors, R.; Olsson, H. Desloratadine and loratadine stand out among common H1-antihistamines for association with improved breast cancer survival. *Acta Oncol.* **2020**, *59*, 1103–1109. [CrossRef] [PubMed]
23. Olsson, H.; Einefors, R.; Broberg, P. Second generation antihistmines after breast cancer diagnosis to improve prognosis both in patients with ER+ and ER- breast cancer. *J. Clin. Oncol.* **2015**, *33*, 3062. [CrossRef]
24. Fritz, I.; Wagner, P.; Bottai, M.; Eriksson, H.; Ingvar, C.; Krakowski, I.; Nielsen, K.; Olsson, H. Desloratadine and loratadine use associated with improved melanoma survival. *Allergy* **2020**, *75*, 2096–2099. [CrossRef]
25. Dobbeling, U.; Waeckerle-Men, Y.; Zabel, F.; Graf, N.; Kundig, T.M.; Johansen, P. The antihistamines clemastine and desloratadine inhibit STAT3 and c-Myc activities and induce apoptosis in cutaneous T-cell lymphoma cell lines. *Exp. Dermatol.* **2013**, *22*, 119–124. [CrossRef]

26. Soule, B.P.; Simone, N.L.; DeGraff, W.G.; Choudhuri, R.; Cook, J.A.; Mitchell, J.B. Loratadine dysregulates cell cycle progression and enhances the effect of radiation in human tumor cell lines. *Radiat. Oncol.* **2010**, *5*, 8. [CrossRef]
27. Procopiou, P.A.; Ford, A.J.; Gore, P.M.; Looker, B.E.; Hodgson, S.T.; Holmes, D.S.; Vile, S.; Clark, K.L.; Saunders, K.A.; Slack, R.J.; et al. Design of Phthalazinone Amide Histamine H1 Receptor Antagonists for Use in Rhinitis. *ACS Med. Chem. Lett.* **2017**, *8*, 577–581. [CrossRef]
28. Lee, C.; Corren, J. Review of azelastine nasal spray in the treatment of allergic and non-allergic rhinitis. *Expert Opin. Pharmacother.* **2007**, *8*, 701–709. [CrossRef]
29. Ellis, A.K.; Zhu, Y.; Steacy, L.M.; Walker, T.; Day, J.H. A four-way, double-blind, randomized, placebo controlled study to determine the efficacy and speed of azelastine nasal spray, versus loratadine, and cetirizine in adult subjects with allergen-induced seasonal allergic rhinitis. *Allergy Asthma Clin. Immunol. Off. J. Can. Soc. Allergy Clin. Immunol.* **2013**, *9*, 16. [CrossRef]
30. Jonathan, M.; Horbal, J.M.; Bernstein, J.A. Azelastine HCl: A Review of the Old and New Formulations. *Clin. Med. Insights Ther.* **2010**, *2*, 427–437.
31. Konrat, R.; Papp, H.; Szijártó, V.; Gesell, T.; Gábor, N.; Madai, M.; Zeghbibc, S.; Kuczmog, A.; Lanszki, Z.; Helyes, Z.; et al. The Anti-histamine Azelastine, Identified by Computational Drug Repurposing, Inhibits SARS-CoV-2 Infection in Reconstituted Human Nasal Tissue In Vitro. *Biorxiv* **2020**. [CrossRef]
32. Blaess, M.; Kaiser, L.; Sommerfeld, O.; Csuk, R.; Deigner, H.P. Drugs, Metabolites, and Lung Accumulating Small Lysosomotropic Molecules: Multiple Targeting Impedes SARS-CoV-2 Infection and Progress to COVID-19. *Int. J. Mol. Sci.* **2021**, *22*, 1797. [CrossRef]
33. Mourad, A.K.; Makhlouf, A.A.; Soliman, A.Y.; Mohamed, S.A. Phthalazines and phthalazine hybrids as antimicrobial agents: Synthesis and biological evaluation. *J. Chem. Res.* **2019**, *44*, 31–41. [CrossRef]
34. Papadopoulos, N.G.; Aggelides, X.; Stamataki, S.; Prokopakis, E.; Katotomichelakis, M.; Xepapadaki, P. *New Concepts in Pediatric Rhinitis*; European Society of Pediatric Allergy and Immunology: Zurich, Switzerland, 2021; Volume 32, pp. 635–646. [CrossRef]
35. Berger, W.; Sher, E.; Gawchik, S.; Fineman, S. Safety of a novel intranasal formulation of azelastine hydrochloride and fluticasone propionate in children: A randomized clinical trial. *Allergy Asthma Proc.* **2018**, *39*, 110–116. [CrossRef]
36. Church, M.K. Allergy, Histamine and Antihistamines. In *Handbook of Experimental Pharmacology*; Springer: Berlin/Heidelberg, Germany, 2017; Volume 241, pp. 321–331. [CrossRef]
37. Horak, F. Effectiveness of twice daily azelastine nasal spray in patients with seasonal allergic rhinitis. *Ther. Clin. Risk Manag.* **2008**, *4*, 1009–1022. [CrossRef]
38. Saito, K.; Abe, N.; Toyama, H.; Ejima, Y.; Yamauchi, M.; Mushiake, H.; Kazama, I. Second-Generation Histamine H1 Receptor Antagonists Suppress Delayed Rectifier K(+)-Channel Currents in Murine Thymocytes. *BioMed Res. Int.* **2019**, *2019*, 6261951. [CrossRef]
39. Fein, M.N.; Fischer, D.A.; O'Keefe, A.W.; Sussman, G.L. CSACI position statement: Newer generation H1-antihistamines are safer than first-generation H1-antihistamines and should be the first-line antihistamines for the treatment of allergic rhinitis and urticaria. *Allergy Asthma Clin. Immunol. Off. J. Can. Soc. Allergy Clin. Immunol.* **2019**, *15*, 61. [CrossRef]
40. Marshal, G.D. Therapeutic options in allergic disease: Antihistamines as systemic antiallergic agents. *J. Allergy Clin. Immunol.* **2000**, *106*, 303–309. [CrossRef]
41. Watts, A.M.; Cripps, A.W.; West, N.P.; Cox, A.J. Modulation of Allergic Inflammation in the Nasal Mucosa of Allergic Rhinitis Sufferers With Topical Pharmaceutical Agents. *Front. Pharmacol.* **2019**, *10*, 294. [CrossRef]
42. Hill, S.J.; Ganellin, C.R.; Timmerman, H.; Schwartz, J.C.; Shankley, N.P.; Young, J.M.; Schunack, W.; Levi, R.; Haas, H.L. International Union of Pharmacology. XIII. Classification of histamine receptors. *Phamacol. Rev.* **1997**, *49*, 253–278.
43. McNeely, W.; Wiseman, L.R. Intranasal azelastine. A review of its efficacy in the management of allergic rhinitis. *Drugs* **1998**, *56*, 91–114. [CrossRef]
44. El-Nakeeb, M.A.; Abou-Shleib, H.M.; Khalil, A.M.; Omar, H.G.; El-Halfawy, O.M. In vitro antibacterial activity of some antihistaminics belonging to different groups against multi-drug resistant clinical isolates. *Braz. J. Microbiol.* **2011**, *42*, 980–991. [CrossRef]
45. Peniche, A.G.; Osorio, E.Y.; Melby, P.C.; Travi, B.L. Efficacy of histamine H1 receptor antagonists azelastine and fexofenadine against cutaneous Leishmania major infection. *PLoS Negl. Trop. Dis.* **2020**, *14*, e0008482. [CrossRef]
46. Hu, H.F.; Xu, W.W.; Li, Y.J.; He, Y.; Zhang, W.X.; Liao, L.; Zhang, Q.H.; Han, L.; Yin, X.F.; Zhao, X.X.; et al. Anti-allergic drug azelastine suppresses colon tumorigenesis by directly targeting ARF1 to inhibit IQGAP1-ERK-Drp1-mediated mitochondrial fission. *Theranostics* **2021**, *11*, 1828–1844. [CrossRef]
47. Kim, J.Y.; Kim, K.S.; Kim, I.S.; Yoon, S. Histamine Receptor Antagonists, Loratadine and Azelastine, Sensitize P-gp-overexpressing Antimitotic Drug-resistant KBV20C Cells Through Different Molecular Mechanisms. *Anticancer. Res.* **2019**, *39*, 3767–3775. [CrossRef]
48. Church, D.S.; Church, M.K. Pharmacology of antihistamines. *World Allergy Organ. J.* **2011**, *4*, 22–27. [CrossRef]
49. Johnson, D.E.; Ostrowski, P.; Jaumouille, V.; Grinstein, S. The position of lysosomes within the cell determines their luminal pH. *J. Cell Biol.* **2016**, *212*, 677–692. [CrossRef]
50. Zhou, J.; Tan, S.H.; Nicolas, V.; Bauvy, C.; Yang, N.D.; Zhang, J.; Xue, Y.; Codogno, P.; Shen, H.M. Activation of lysosomal function in the course of autophagy via mTORC1 suppression and autophagosome-lysosome fusion. *Cell Res.* **2013**, *23*, 508–523. [CrossRef]

51. Bursch, W.; Hochegger, K.; Torok, L.; Marian, B.; Ellinger, A.; Hermann, R.S. Autophagic and apoptotic types of programmed cell death exhibit different fates of cytoskeletal filaments. *J. Cell Sci.* **2000**, *113*, 1189–1198. [CrossRef]
52. Chen, Q.; Kang, J.; Fu, C. The independence of and associations among apoptosis, autophagy, and necrosis. *Signal Transduct. Target. Ther.* **2018**, *3*, 18. [CrossRef]
53. Condello, M.; Pellegrini, E.; Caraglia, M.; Meschini, S. Targeting Autophagy to Overcome Human Diseases. *Int. J. Mol. Sci.* **2019**, *20*, 725. [CrossRef]
54. Fimia, G.M.; Piacentini, M. Regulation of autophagy in mammals and its interplay with apoptosis. *Cell. Mol. Life Sci. CMLS* **2010**, *67*, 1581–1588. [CrossRef]
55. Nicolau-Galmes, F.; Asumendi, A.; Alonso-Tejerina, E.; Perez-Yarza, G.; Jangi, S.M.; Gardeazabal, J.; Arroyo-Berdugo, Y.; Careaga, J.M.; Diaz-Ramon, J.L.; Apraiz, A.; et al. Terfenadine induces apoptosis and autophagy in melanoma cells through ROS-dependent and -independent mechanisms. *Apoptosis Int. J. Program. Cell Death* **2011**, *16*, 1253–1267. [CrossRef]
56. Shubin, A.V.; Demidyuk, I.V.; Komissarov, A.A.; Rafieva, L.M.; Kostrov, S.V. Cytoplasmic vacuolization in cell death and survival. *Oncotarget* **2016**, *7*, 55863–55889. [CrossRef]
57. Espinosa, E.; Zamora, P.; Feliu, J.; Gonzalez Baron, M. Classification of anticancer drugs—A new system based on therapeutic targets. *Cancer Treat. Rev.* **2003**, *29*, 515–523. [CrossRef]
58. Mc Gee, M.M. Targeting the Mitotic Catastrophe Signaling Pathway in Cancer. *Mediat. Inflamm.* **2015**, *2015*, 146282. [CrossRef]
59. Vakifahmetoglu, H.; Olsson, M.; Zhivotovsky, B. Death through a tragedy: Mitotic catastrophe. *Cell Death Differ.* **2008**, *15*, 1153–1162. [CrossRef]
60. Gu, J.J.; Kaufman, G.P.; Mavis, C.; Czuczman, M.S.; Hernandez-Ilizaliturri, F.J. Mitotic catastrophe and cell cycle arrest are alternative cell death pathways executed by bortezomib in rituximab resistant B-cell lymphoma cells. *Oncotarget* **2017**, *8*, 12741–12753. [CrossRef]
61. Chan, D.C. Mitochondria: Dynamic organelles in disease, aging, and development. *Cell* **2006**, *125*, 1241–1252. [CrossRef]
62. Cornet-Masana, J.M.; Banus-Mulet, A.; Carbo, J.M.; Torrente, M.A.; Guijarro, F.; Cuesta-Casanovas, L.; Esteve, J.; Risueno, R.M. Dual lysosomal-mitochondrial targeting by antihistamines to eradicate leukaemic cells. *EBioMedicine* **2019**, *47*, 221–234. [CrossRef]
63. Pathania, D.; Millard, M.; Neamati, N. Opportunities in discovery and delivery of anticancer drugs targeting mitochondria and cancer cell metabolism. *Adv. Drug Deliv. Rev.* **2009**, *61*, 1250–1275. [CrossRef] [PubMed]
64. Hasnain, S.Z.; Prins, J.B.; McGuckin, M.A. Oxidative and endoplasmic reticulum stress in beta-cell dysfunction in diabetes. *J. Mol. Endocrinol.* **2016**, *56*, 33–54. [CrossRef]
65. Malhotra, J.D.; Kaufman, R.J. Endoplasmic reticulum stress and oxidative stress: A vicious cycle or a double-edged sword? *Antioxid. Redox Signal.* **2007**, *9*, 2277–2293. [CrossRef] [PubMed]
66. Cao, S.S.; Kaufman, R.J. Endoplasmic reticulum stress and oxidative stress in cell fate decision and human disease. *Antioxid. Redox Signal.* **2014**, *21*, 396–413. [CrossRef]
67. Bolisetty, S.; Jaimes, E.A. Mitochondria and reactive oxygen species: Physiology and pathophysiology. *Int. J. Mol. Sci.* **2013**, *14*, 6306–6344. [CrossRef]
68. Xiong, Y.; Yin, Q.; Li, J.; He, S. Oxidative Stress and Endoplasmic Reticulum Stress Are Involved in the Protective Effect of Alpha Lipoic Acid Against Heat Damage in Chicken Testes. *Anim. Open Access J. MDPI* **2020**, *10*, 384. [CrossRef]
69. Sena, L.A.; Chandel, N.S. Physiological roles of mitochondrial reactive oxygen species. *Mol. Cell* **2012**, *48*, 158–167. [CrossRef]
70. Rahmani, M.; Davis, E.M.; Crabtree, T.R.; Habibi, J.R.; Nguyen, T.K.; Dent, P.; Grant, S. The kinase inhibitor sorafenib induces cell death through a process involving induction of endoplasmic reticulum stress. *Mol. Cell. Biol.* **2007**, *27*, 5499–5513. [CrossRef]
71. Schardt, J.A.; Weber, D.; Eyholzer, M.; Mueller, B.U.; Pabst, T. Activation of the unfolded protein response is associated with favorable prognosis in acute myeloid leukemia. *Clin. Cancer Res. Off. J. Am. Assoc. Cancer Res.* **2009**, *15*, 3834–3841. [CrossRef]
72. Doran, A.C.; Yurdagul, A., Jr.; Tabas, I. Efferocytosis in health and disease. *Nat. Rev. Immunol.* **2020**, *20*, 254–267. [CrossRef]
73. Abdolmaleki, F.; Farahani, N.; Gheibi Hayat, S.M.; Pirro, M.; Bianconi, V.; Barreto, G.E.; Sahebkar, A. The Role of Efferocytosis in Autoimmune Diseases. *Front. Immunol.* **2018**, *9*, 1645. [CrossRef] [PubMed]
74. Boada-Romero, E.; Martinez, J.; Heckmann, B.L.; Green, D.R. The clearance of dead cells by efferocytosis. *Nat. Rev. Mol. Cell Biol.* **2020**, *21*, 398–414. [CrossRef]
75. Rajput, S.; Wilber, A. Roles of inflammation in cancer initiation, progression, and metastasis. *Front. Biosci.* **2010**, *2*, 176–183. [CrossRef]
76. Gheibi Hayat, S.M.; Bianconi, V.; Pirro, M.; Sahebkar, A. Efferocytosis: Molecular mechanisms and pathophysiological perspectives. *Immunol. Cell Biol.* **2019**, *97*, 124–133. [CrossRef]

International Journal of
Molecular Sciences

Article

Arylquin 1 (Potent Par-4 Secretagogue) Inhibits Tumor Progression and Induces Apoptosis in Colon Cancer Cells

Yi-Ting Chen [1,2], Tzu-Ting Tseng [1], Hung-Pei Tsai [3] and Ming-Yii Huang [4,5,6,*]

1 Department of Pathology, Kaohsiung Medical University Hospital, Kaohsiung 80708, Taiwan; yt0728@gmail.com (Y.-T.C.); cawaii7992@gmail.com (T.-T.T.)
2 Department of Pathology, Faculty of Medicine, College of Medicine, Kaohsiung Medical University, Kaohsiung 80708, Taiwan
3 Department of Surgery, Division of Neurosurgery, Kaohsiung Medical University Hospital, Kaohsiung 80708, Taiwan; carbugino@gmail.com
4 Department of Radiation Oncology, Kaohsiung Medical University Hospital, Kaohsiung 80708, Taiwan
5 Cancer Center, Kaohsiung Medical University Hospital, Kaohsiung 80708, Taiwan
6 Department of Radiation Oncology, Faculty of Medicine, College of Medicine, Kaohsiung Medical University, Kaohsiung 80708, Taiwan
* Correspondence: miyihu@gmail.com

Abstract: Colorectal cancer (CRC) is one of the most common gastrointestinal cancers worldwide. Current therapeutic strategies mainly involve surgery and chemoradiotherapy; however, novel antitumor compounds are needed to avoid drug resistance in CRC, as well as the severe side effects of current treatments. In this study, we investigated the anticancer effects and underlying mechanisms of Arylquin 1 in CRC. The MTT assay was used to detect the viability of SW620 and HCT116 cancer cells treated with Arylquin 1 in a dose-dependent manner in vitro. Further, wound-healing and transwell migration assays were used to evaluate the migration and invasion abilities of cultured cells, and Annexin V was used to detect apoptotic cells. Additionally, Western blot was used to identify the expression levels of N-cadherin, caspase-3, cyclin D1, p-extracellular signal-regulated kinase (ERK), p-c-JUN N-terminal kinase (JNK), and phospho-p38, related to key signaling proteins, after administration of Arylquin 1. Xenograft experiments further confirmed the effects of Arylquin 1 on CRC cells in vivo. Arylquin 1 exhibited a dose-dependent reduction in cell viability in cultured CRC cells. It also inhibited cell proliferation, migration, and invasion, and induced apoptosis. Mechanistic analysis demonstrated that Arylquin 1 increased phosphorylation levels of ERK, JNK, and p38. In a mouse xenograft model, Arylquin 1 treatment diminished the growth of colon tumors after injection of cultured cancer cells. Arylquin 1 may have potential anticancer effects and translational significance in the treatment of CRC.

Keywords: Arylquin 1; colon cancer; tumor progression; apoptosis

Citation: Chen, Y.-T.; Tseng, T.-T.; Tsai, H.-P.; Huang, M.-Y. Arylquin 1 (Potent Par-4 Secretagogue) Inhibits Tumor Progression and Induces Apoptosis in Colon Cancer Cells. *Int. J. Mol. Sci.* **2022**, 23, 5645. https://doi.org/10.3390/ijms23105645

Academic Editor: Angela Stefanachi

Received: 26 April 2022
Accepted: 17 May 2022
Published: 18 May 2022

Publisher's Note: MDPI stays neutral with regard to jurisdictional claims in published maps and institutional affiliations.

1. Introduction

Colorectal cancer (CRC) is the third most common malignancy in men and the second in women globally. In 2020, there were almost two million newly diagnosed cases of CRC, and it caused nearly one million deaths [1]. Wide excision for resectable cancer combined with postoperative chemotherapy remains the first choice of therapy in cases of CRC. Combined radiotherapy is now regarded as the standard treatment for advanced rectal cancer. However, not only chemotherapy but also radiation may cause unbearable side effects and toxicities to lead treatment failure due to early drug withdrawal. The development of innovative drugs that induce tumor apoptosis is the main focus of research into antineoplastic therapeutic agents. To find selective cancer-targeted therapeutics remains one of the greatest challenges.

Arylquin 1, a new molecule first identified by scientists at the University of Kentucky, is regarded as a secretagogue of prostate apoptosis response-4 (Par-4) in healthy cells. Par-4, regarded a tumor specific suppressor protein, is identified in normal cells, but it is often decreased or absent in neoplastic cells to evade the destiny of apoptosis [2]. Extracellular Par-4 can induce apoptosis through the binding of the 78 kDa glucose-regulated protein (GRP78) on tumor cell membranes [3]. Intracellularly, as seen in prostate cancers, Par-4 mainly activates the signaling pathway of the Fas death receptor and inhibits cellular pro-survival mechanisms [4]. Downregulation of Par-4 protein during tumorigenesis has been proposed associated with poor prognosis and treatment resistance in many solid tumors to highlight the importance as a critical event of therapeutic target agent. For example, Par-4 level was decreased in human renal cell carcinoma compared with normal renal tubular cells [5]. Low level of Par-4 was found correlated with low survival period in glioblastoma patients, and Tamoxifen-induced cell death was alleviated by Par-4 specific siRNA in vitro [6]. In breast cancer, knockdown of Par-4 increased the proliferation and reduced the chemosensitization [7]. However, gene therapy such as transfection with plasmids or adenoviral vectors is very complex and hundreds of challenges remain to be overcome. Therefore, replenishment of Par-4 by Arylquin 1 via sensitizing tumor cells to apoptosis is promising. Previously, Arylquin 1 has been reported to promote morphology changes and decrease viability in various cancer cells [8]; however, any anticancer effects of Arylquin 1 in CRC have yet to be determined.

The present study aimed to investigate the cytotoxic effects and mechanisms of Arylquin-1-induced cell death in CRC. To our knowledge, this is the first study to explore the therapeutic value of Arylquin 1 specifically in CRC.

2. Results

2.1. *Arylquin-1-Attenuated Cell Viability in CRC Cell Lines*

To explore the effects of Arylquin 1 on the proliferation of cultured CRC cells, cells were incubated with Arylquin 1 at different concentrations for 72 h, after which cell viability was assessed using the MTT assay. SW620 and HCT116 cells demonstrated reduced viability at doses of 0.25, 0.5, 1, 1.5, 2, 2.5, and 3 μM Arylquin 1 relative to the 0 μM control (Figure 1). These data showed a dose-dependent reduction in proliferation after Arylquin 1 treatment. The IC_{50} concentrations of Arylquin 1 were 1.8 and 2.3 μM for SW620 and HCT116 cells, respectively.

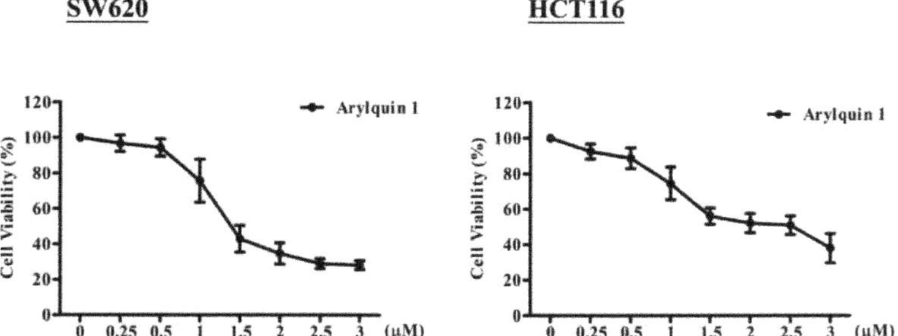

Figure 1. Viability of SW620 and HCT116 cells treated with different doses of Arylquin 1.

2.2. *Arylquin-1-Inhibited Cell Migration, Invasion, and Epithelial–Mesenchymal Transition (EMT) in CRC Cells*

To investigate the effects of cell migration after treatment with Arylquin 1, we used a wound-healing assay and compared the results between a control group and the Arylquin-1-treated groups. In both SW620 and HCT116 cells, 1, 1.5, and 2 μM Arylquin 1 markedly inhibited cell migration at 16, 24, and 48 h after administration (Figure 2). These data also indicated the dose-dependent effects of Arylquin 1 on cell migration in CRC cells.

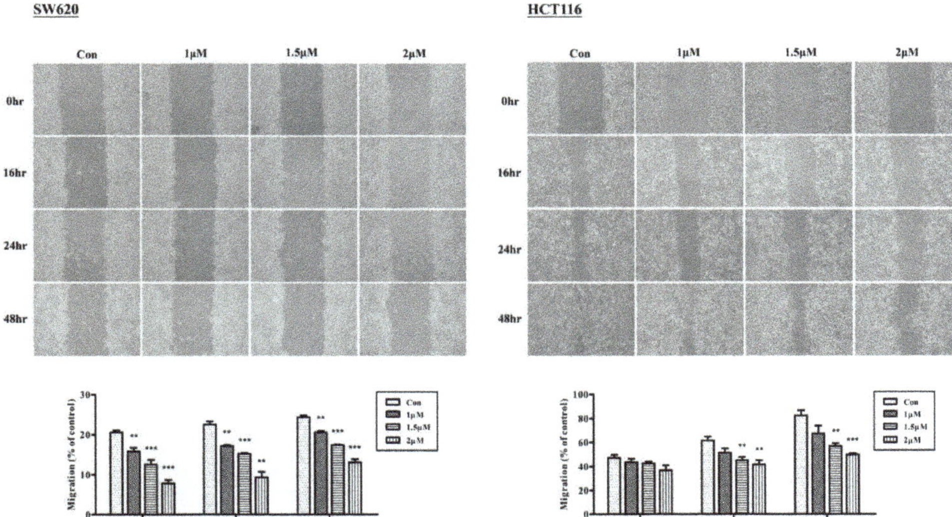

Figure 2. Wound-healing assay and percentage of cell migration in cultured SW620 and HCT116 cells 0, 16, 24, and 48 h after treatment with Arylquin 1; quantitative data are expressed as mean ± SEM. ** $p < 0.01$ and *** $p < 0.001$ compared to the control group.

To evaluate cell invasion, we used a Matrigel invasion assay. In both SW620 and HCT116 cells, 1, 1.5, and 2 μM Arylquin 1 significantly inhibited cell invasion 24 h after administration (Figure 3). As in the wound-healing and proliferation assays, Arylquin 1 inhibited the cell migration in CRC cells in a dose-dependent manner.

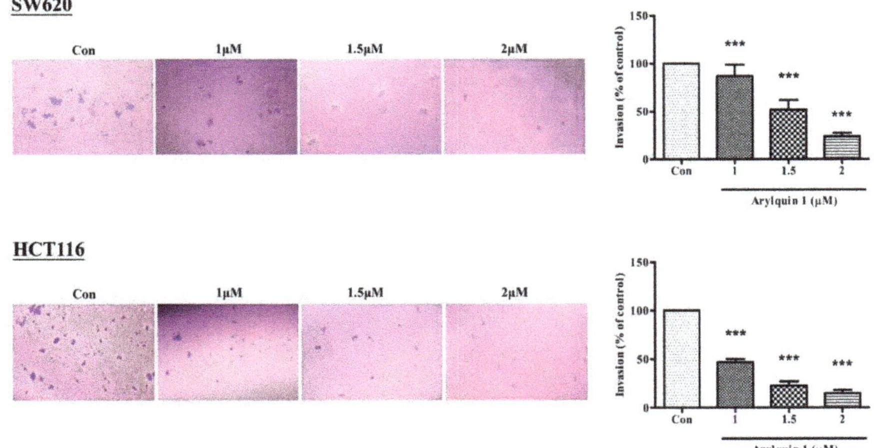

Figure 3. Wound-healing assay and percentage of cell migration in cultured SW620 and HCT116 cells 24 h after treatment with Arylquin 1; quantitative data are expressed as mean ± SEM. *** $p < 0.001$ compared to the control group.

Cadherins are markers of metastasis used to evaluate the ability of cells to undergo an epithelial–mesenchymal transition (EMT). SW620 and HCT116 cells treated with Arylquin 1 were subjected to Western blot (Figure 4), showing that N-cadherin levels were significantly lower in Arylquin-1-treated CRC cells in a dose-dependent manner, indicating that Arylquin 1 attenuates EMT in CRC.

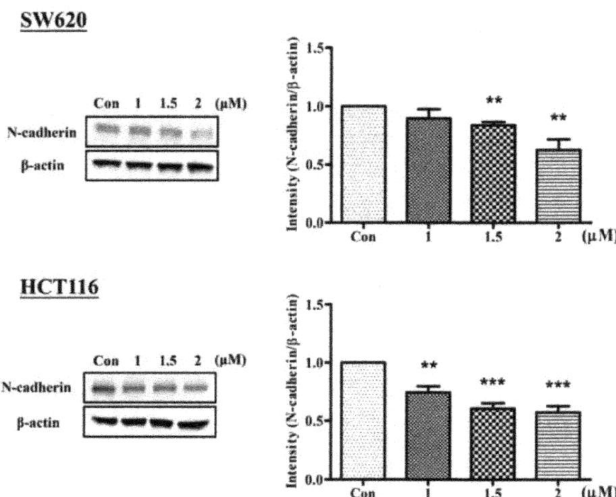

Figure 4. Western blot for N-cadherin, with β-actin-loading control, in SW620 and HCT116 cells treated with Arylquin 1; quantitative data are expressed as mean ± SEM; ** $p < 0.01$ and *** $p < 0.001$ compared to the control group.

2.3. Arylquin 1 Promotes Apoptosis in Cultured CRC Cells

As shown in Figure 5, the highest level of cleaved caspase-3 and the lowest level of Cyclin-D1 were detected 72 h after treatment with a dose of 2 μM Arylquin 1, indicating that apoptotic activity is positively associated with Arylquin 1 dose strength in both SW620 and HCT116 cells. Moreover, BCl2 levels showed a tendency to significantly decrease after treatment with Arylquin 1 in HCT116 cells.

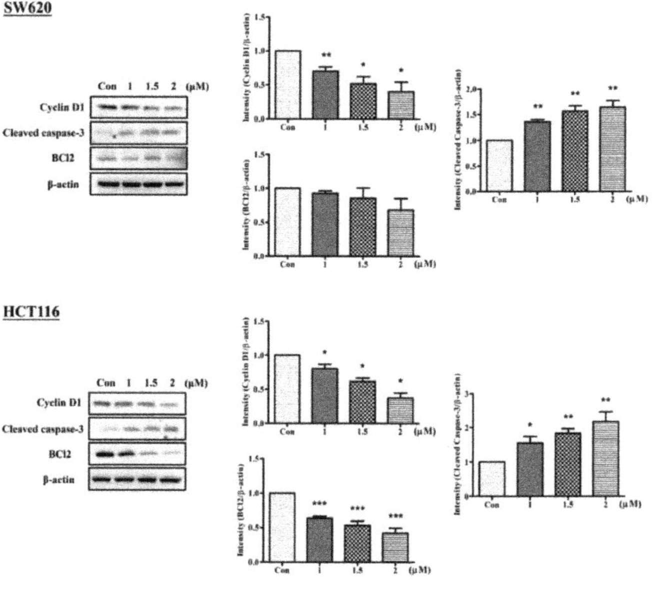

Figure 5. Western blot for Cyclin D1, cleaved caspase-3, and BCl2, with a β-actin-loading control, in SW620 and HCT116 cells 72 h after treated with Arylquin 1; quantitative data are expressed as mean ± SEM. * $p < 0.05$, ** $p < 0.01$, and *** $p < 0.001$ compared to the control group.

The Annexin V/FITC assay was performed to identify apoptotic cells. A high proportion of viable cells was observed in the untreated control group in both SW620 (93.52 ± 3.1%) and HCT116 (83.65 ± 5.37%) cells (Figure 6). In both cell lines, a pattern of the cell population shifting from viable cells to early apoptotic stages to late apoptotic stages was observed, and this was associated with the Arylquin 1 dosage. However, a significant increase in early and late apoptotic/necrotic cells was observed after treatment with 2 μM Arylquin 1 in SW620 cells (19.68 ± 5%). These results revealed that Arylquin 1 can induce apoptosis and necrosis in a dose-dependent manner.

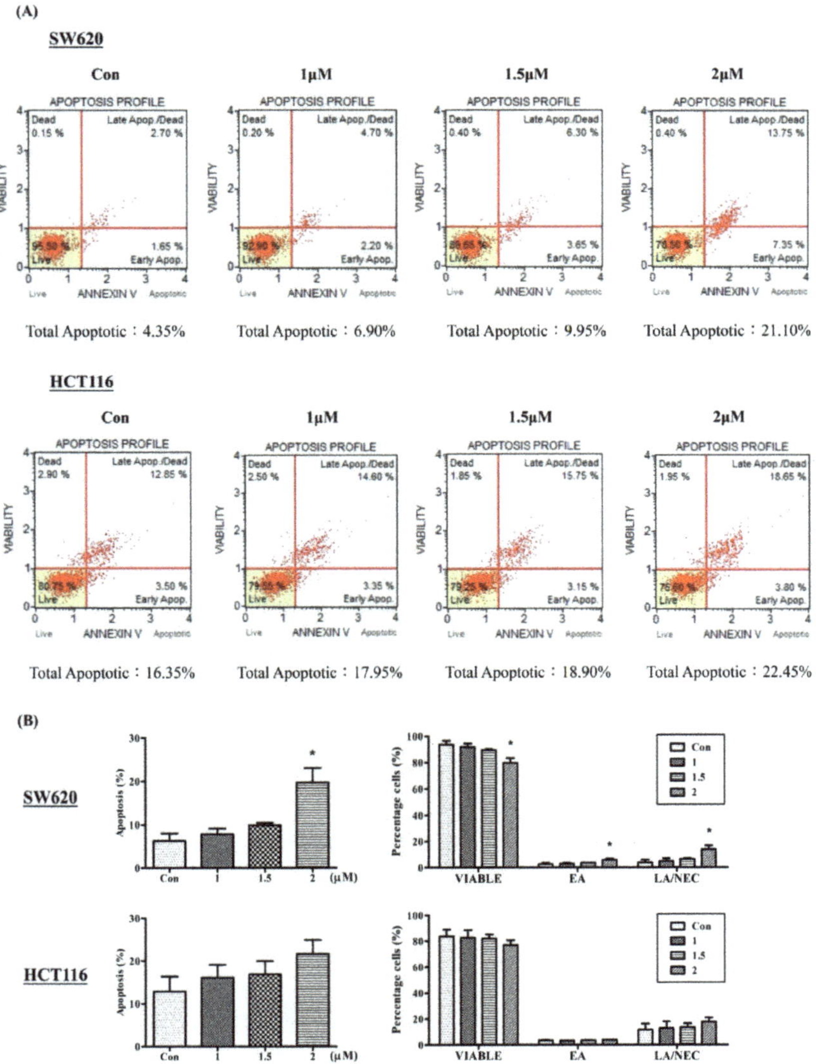

Figure 6. Annexin V/FITC Assay: (**A**) dot plot of Annexin V/FITC in SW620 and HCT116 cells after treatment with Arylquin 1; (**B**) quantification of Annexin V/FITC in SW620 and HCT116 cells. EA—early apoptosis; LA/NEC—late apoptosis/necrosis. All data are expressed as mean ± SEM. * $p < 0.05$ compared with corresponding controls.

2.4. Arylquin 1 Regulated Apoptosis via the MAPK Pathway in CRC Cells

To further clarify the signaling pathways involved and whether CRC cell viability decreased as a consequence of Arylquin-1-induced apoptosis, SW620 and HCT116 cells treated with Arylquin 1 were subjected to Western blot for proteins downstream of the mitogen-activated protein kinase (MAPK), Akt, and signal transducer and activator of transcription 3 (STAT3) pathways. The MAPK family consists of three major subfamilies of related proteins (extracellular-signal-regulated kinases [ERKs], c-Jun N-terminal kinase [JNK], and p38). In both SW620 and HCT116 cells, Arylquin 1 administration led to the upregulation of ERK, p38, and JNK expression (Figure 7).

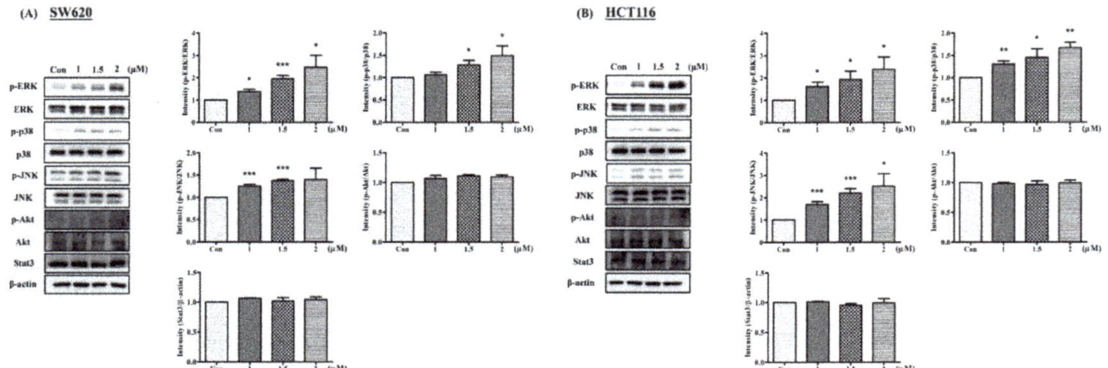

Figure 7. Western blot for p-ERK, ERK, p-p38, p38, p-JNK, JNK, p-Akt, Akt, and Stat3, with a β-actin-loading control, in (**A**) SW620 and (**B**) HCT116 cells treated with Arylquin 1; quantitative data are expressed as mean ± SEM; * $p < 0.05$, ** $p < 0.01$, and *** $p < 0.001$ compared to the control group.

2.5. Arylquin 1 Suppressed the Growth of CRC Cells in Mice

Having validated our cell model in vitro, we next evaluated the in vivo effects of Arylquin 1 on HCT116 tumor xenograft growth. Tumor or control (empty) cell suspensions were injected subcutaneously into the flanks of 6-week-old mice, and tumor growth was evaluated and registered periodically (Figure 8). Arylquin 1 was injected intraperitoneally on day 7 after cell implantation. Our results demonstrate a significant and marked reduction in tumor volume in the Arylquin 1 treatment group, from day 14 after cell injection until the end of the experiment at day 42 (all $p < 0.01$). The Arylquin 1 treatment group showed significantly slower tumor growth compared with the untreated control group, displaying final tumor volumes of 127 ± 69 and 1042 ± 157 mm^3, respectively.

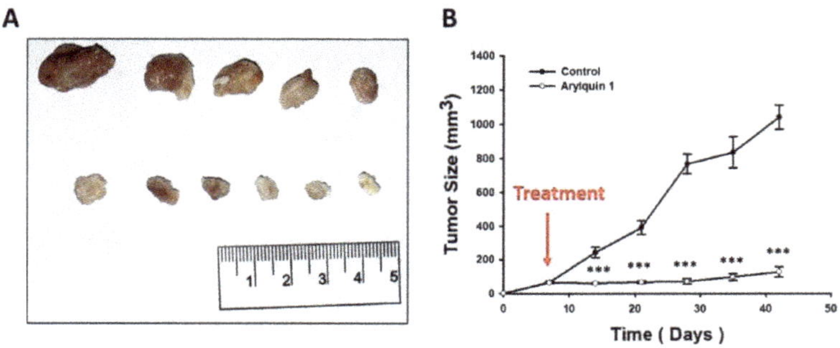

Figure 8. (**A**) Gross examination and (**B**) tumor growth curves of mouse xenografts of human colorectal cancer with and without Arylquin 1 treatment; quantitative data are expressed as mean ± SEM; *** $p < 0.001$ compared to the control group.

3. Discussion

CRC is one of the most commonly diagnosed cancers and is a leading cause of cancer-related death. Currently, the chemotherapeutic strategy for the treatment of advanced CRC is based on a variety of reagents, sometimes accompanied by lethal and dose-dependent adverse side effects. Thus, it is worth exploring potential drug substitutions to increase treatment effectiveness and cause fewer severe side effects in patients with CRC.

In the present study, various assays were used to investigate the antitumor effects of Arylquin 1 in human CRC. We first identified that treatment with Arylquin 1 inhibited growth in cultured SW620 and HCT116 cells with an IC_{50} of 1.8 and 2.3 μM, respectively. In mice injected with cultured human CRC cells, treatment with Arylquin 1 decreased tumor size. Arylquin 1 was also found to inhibit cell migration and invasion. Western blot showed that Arylquin 1 treatment led to downregulated N-cadherin expression.

Arylquin 1, a molecular compound regarded as a Par-4 secretagogue in healthy cells, was identified in 2014 by Rangnekar et al. [9]. This structure–activity study has been confirmed, with results showing that Arylquin 1 targets vimentin and causes the release of Par-4 [10]. Par-4, a 38 kDa tumor suppressor protein, is expressed in healthy cells, but it is usually downregulated or silenced in cancer cells to elude the induction of apoptosis [2]. Recently, the intracellular apoptotic effects of Par-4 and the ability of cancer cells to inhibit Par-4 release were carefully investigated [4]. Apoptosis induced by interactions between Par-4 and the tumor cell surface receptor GRP78 leads to programmed cell death [3,10]. Moreover, Min et al. reported that Arylquin 1 can induce lysosomal membrane permeabilization, with useful antitumor effects [8]. Chloroquine, another secretagogue of Par-4, has also been identified to induce apoptosis and inhibit tumor metastasis. Arylquin 1 and Chloroquine share a common pharmacophore [11]. Therefore, the potential cytotoxic activity of Arylquin 1 to bolster the release of Par-4 is a promising therapeutic advance.

In our study, the ability of inhibiting cell migration and invasion by Arylquin 1 treatment was noted. We found that Arylquin 1 attenuated N-cadherin expression in CRC. EMT has been identified to play an important role in tumor progression and metastasis. Upregulation of vimentin and N-cadherin regulated by a complex signaling pathway is the hallmark of EMT [12]. In previous studies, vimentin, which is highly involved in EMT and metastasis, was reported as the primary target of Arylquin 1 to inhibit the spread of lung cancer [9,13]. Silencing endogenous Par-4 using siRNA has also been shown to promote tumor growth and metastasis resulted in increased N-cadherin expression in pancreatic cancers [14]. Upregulation of Vimentin and N-cadherin were also detected associated with the interaction of Par-4 in cervical cancer during transforming growth factor (TGF)-β-induced EMT [15]. Furthermore, overexpression of Par-4 was proved to reduce EMT which was later promote apoptosis in pancreatic cancer [16]. It seems that our results may provide evidence that Aryqulin-1 may enhance the anti-EMT effects via Par-4 in tumor cells.

From the Annexin V/FITC assay in the present study, it was observed that both SW620 and HCT116 cells underwent apoptosis after Arylquin 1 treatment. The proportion of cells observed undergoing apoptosis increased as the dose increased. This shows that Arylquin 1 can induce apoptosis in a dose-dependent manner in both SW620 and HCT116 cells. It seems that CRC cell viability decreased is a consequence of Arylquin-1-induced apoptosis. There are two main pathways involved in apoptosis, the mitochondria-related intrinsic pathway and the death-receptor-related extrinsic pathway [17,18]. In the present study, downregulation of BCl2 and upregulation of caspase-3, which is involved in both intrinsic and extrinsic apoptosis pathways, were identified in HCT116 cells treated by Arylquin 1. Rangnekar et al. have previously shown that Par-4 upregulation induces apoptotic death in prostate cancer [19]. Wang et al. further reported a direct correlation between Par-4 levels and apoptotic activity induced by 5-fluorouracil in CRC [20]. Moreover, CRC tumor growth—namely the tumor growth rate as reflected by size after 42 days—was also suppressed by Arylquin 1 in mice in the present study, confirming the apoptosis-inducing effects of Arylquin 1.

Higher apoptosis percentage with a significant decrease in BCl2 expression was identified in HCT116 cells than SW620 cells in our data. As we know, the programmed cell death leaded by the interactions between Par-4 and the tumor cell surface receptor GRP78 has been investigated [2]. GRP78 is regarded as the central regulator of endoplasmic reticulum stress in apoptosis. Upregulation of GRP78 in tumor cells has been identified associated with reduce tumorigenicity and increase sensitivity to DNA crosslinking agents due to promote its localization in cell surface [21,22]. The oncogenic role of GRP78 draws attention to its value as a prognostic and drug response marker to mediate the therapeutic efficiency [23]. Therefore, the variable surface and cellular location of GRP78 expressions may alter and result in the different behaviors of SW620 and HCT116 cells.

In the present study, we found that both promoted apoptosis and increased expression of phosphorylated ERK in CRC treated with Arylquin 1 were observed, in addition to JNK and p38 upregulation. The MAPK signaling pathway, including ERK1/2, JNK/SAPK, and p38, is involved in cell proliferation, apoptosis, and metastasis of cancer cells, depending on the cell type and stimulus [24]. Therefore, modulating this pathway can provide a potential approach to treating CRC. ERK has been reported to promote cell growth and control migration [25,26]. On the other hand, tumor-necrosis-factor-associated apoptosis-inducing ligand and TGF-β were identified to induce apoptosis via ERK-mediated upregulation of death receptors in colon cancer cells [27,28]. Activation of JNK has also been shown to be involved in cell survival, proliferation, migration, invasion, and cell death [29,30]. Previously, ERK and JNK were reported to participate in the regulation of apoptosis (intrinsic and extrinsic; [30,31]). JNK can activate apoptotic pathway by regulating the activities of mitochondria directly or upregulating proapoptotic genes through transcriptional factors [32]. Activation of JNK pathway of apoptosis induced by Vernodalin or TRAIL was found in colon cancer cells [33,34]. p38 is also involved in the signal integration of migration, metastasis, and apoptosis in cancer [35]. In HeLa cells, Lee et al. reported that TRAIL induced apoptosis via p38 activation stimulated by reactive oxygen species [36]. Our results support the crucial role of MAPK in the regulation of proliferation, migration, metastasis, and apoptotic processes in CRC cells. Since these proteins act on tumor behavior, their dysregulation by Arylquin 1 highlights the novel mechanisms that may produce its antiproliferative, anti-invasive, and apoptotic functions.

4. Materials and Methods

4.1. Cell Culture

SW620 and HCT116 cells were purchased from the Bioresource Collection and Research Center (Taiwan) and cultured in Dulbecco's Modified Eagle Medium (DMEM) (Gibco; 12800-017, Waltham, MA; USA) with 10% fetal bovine serum (FBS) at 37 °C in an atmosphere of 5% CO_2.

4.2. Cell Viability

Cell lines were seeded on a 24-well plate at a density of about 3×10^4 cells in 500 μL of DMEM with 10% FBS in each well. The viable cell counts were performed following the MTT assay after culturing with different doses of Arylquin 1 (0, 0.25, 0.5, 1, 1.5, 2, 2.5, and 3 μM) for 72 h to identify the IC_{50} dose.

4.3. Migration and Invasion Assays In Vitro

Cell migration was investigated using the wound-healing assay (Cat. Nr. 80209; Ibidi GmbH, Graefelfing, Germany) using two-sided wound gaps. The wound-healing assay was performed in 24-well plates seeded with 3×10^5 cells per insert and cultured at 37 °C for 24 h, after which 0, 1, 1.5, or 2 μM Arylquin 1 was added to the wells. After 24 h, these cells were washed twice with phosphate-buffered saline (PBS) and photographed at 16, 24, and 48 h after washing. The transwell migration assay (COR3452; Corning Inc., Corning, NY, USA) was used to measure cell invasion in vitro. Incubated cells treated with 0, 1, 1.5, or 2 μM Arylquin 1 were seeded at 5×10^3 cells per insert, and the lower chamber of the transwell assay was filled with 0.5 mL DMEM with 10% FBS. After 24 h, cells that remained on the upper surface of the transwell membrane were removed by a cotton swab. Cells that had passed through the transwell membrane to the bottom of the insert were fixed using formalin, stained using 0.5% crystal violet, and quantified via manual counts from photographs of six randomly selected fields.

4.4. Protein Extraction and Western Blot

All samples were prepared in 100 μL of RIPA lysis buffer, and 30 μg protein from each sample was loaded in the wells of a sodium dodecyl sulfate–polyacrylamide gel electrophoresis gel; gels were run at 80 V for 2 h. The separated proteins were then transferred from the gel to a polyvinyl difluoride membrane with a 400 mA current for 2 h. After 1 h in blocking buffer, the membranes were incubated with primary antibodies overnight, followed by incubation with secondary antibodies for 1 h. The primary antibodies used were anti-β-actin (MAB1501R; Sigma-Aldrich; Burlington, MA, USA), anti-p-p38 (9211S; Cell Signaling; Danvers, MA, USA), anti-p38 (9212S; Cell Signaling; Danvers, MA, USA), anti-p-JNK (4668S; Cell Signaling; Danvers, MA, USA), anti-JNK (9252S; Cell Signaling; Danvers, MA, USA), anti-p-ERK (9101S; Cell Signaling; Danvers, MA, USA), anti-ERK (9102S; Cell Signaling; Danvers, MA, USA), anti-p-Akt (9271S; Cell Signaling; Danvers, MA, USA), anti-Akt (9272S; Cell Signaling; Danvers, MA, USA), anti-Stat3 (sc-8019; Santa Cruz, CA, USA), anti-N-cadherin (22018-1-AP; Proteintech; Rosemont, IL, USA), anti-Cyclin D1 (60186-1-Ig; Proteintech; Rosemont, IL, USA), anti-cleaved caspase-3 (9661S; Cell Signaling; Danvers, MA, USA), and anti-BCl2 (4223S; Cell Signaling; Danvers, MA, USA). Horseradish peroxidase (HRP)-conjugated secondary antibodies were goat anti-rabbit IgG (H + L)-HRP (C04003; Croyez, Taipei, Taiwan) and goat anti-mouse IgG (H + L)-HRP (C04001; Croyez, Taipei, Taiwan). Signals were detected using an enhanced chemiluminescent solution (Western Lightning Plus; PerkinElmer, Waltham, MA, USA) with a chemiluminescence imager (Minichemi; Thermo Fisher Scientific, Waltham, MA, USA).

4.5. Flow Cytometry

Cells were transfected to 6-well plates and treated with 0, 1, 1.5, or 2 μM Arylquin 1 for 72 h. Both detached and attached cells were centrifuged at 1000 rpm for 5 min, washed once with 1× PBS, then treated according to the manufacturer's instructions using the Muse® Annexin V and Dead Cell Kit (Cat. No. MCH100105; MilliporeSigma, Burlington, MA, USA).

4.6. Animal Model

Six-week-old male NU/NU nude mice were obtained from BioLasco Taiwan (Taipei, Taiwan). All animal experiments followed the protocols of the Institutional Animal Care and Use Committee of Kaohsiung Medical University (IACUC Approval No: 110266) and were performed according to the Guiding Principles for the Care and Use of Laboratory Animals. Mice were acclimatized to a 12:12 h light/dark cycle at 24 ± 1 °C with ad libitum access to food and water. Human HCT116 cells were chosen for tumor xenograft to evaluate the effects of Arylquin 1 on tumor growth, invasion, and metastasis of colon cancer according to previous studies [37–39]. HCT116 cells were subcutaneously injected at a density of 1×10^7 in NU/NU mice. One week after injection, 150 uM/kg Arylquin 1 was injected intraperitoneally. Tumor volume (mm^3) was measured three times a week, calculated as (length × width^2)/2 on days 7, 10, 14, 21, 28, 35, and 42. Mice were killed 42 days after the injection of tumor cells.

4.7. Statistical Analysis

All statistical analyses were performed with SPSS 19.0 (IBM Corp., Armonk, NY, USA). Intensities of Western blot bands were digitally analyzed using ImageJ software. All tests were two-sided, and a p-value less than 0.05 was considered statistically significant.

5. Conclusions

In summary, the present study demonstrates for the first time that Arylquin 1, as a secretagogue of Par-4, can inhibit the proliferation of CRC cells in vitro and can decrease migration, invasion, and metastasis. Moreover, promoting apoptosis in vitro and suppressing tumor size in vivo were also observed. In conclusion, Arylquin 1 may be used as a potential anticancer drug with strong translational significance in the treatment of CRC.

Author Contributions: Y.-T.C. conducted data analysis and manuscript drafting; the cell and animal studies were finished with the assistance of T.-T.T. and H.-P.T.; M.-Y.H. was involved in conceptualization and manuscript review editing and led the project. All authors have read and agreed to the published version of the manuscript.

Funding: This work was supported by grants from the Ministry of Science and Technology (MOST 107-2320-B-037-018, MOST 110-2314-B-037-075-MY2), Kaohsiung Medical University Chung-Ho Memorial Hospital (KMUH107-7R87, KMUH108-8M66, KMUH109-9M78, KMUH110-0R72, KMUH-DK(B)110005-2), and Kaohsiung Medical University (KMU-Q108003).

Institutional Review Board Statement: Not applicable.

Informed Consent Statement: Not applicable.

Data Availability Statement: The datasets used and/or analyzed during the current study are available from the corresponding author on reasonable request.

Conflicts of Interest: All authors declare no conflict of interest.

References

1. FIAfRoC, L. Global Cancer Observatory: Cancer Today. Available online: https://gcoiarcfr/today/home (accessed on 17 August 2020).
2. Hebbar, N.; Wang, C.; Rangnekar, V.M. Mechanisms of apoptosis by the tumor suppressor Par-4. *J. Cell Physiol.* **2012**, *227*, 3715–3721. [CrossRef] [PubMed]
3. Burikhanov, R.; Zhao, Y.; Goswami, A.; Qiu, S.; Schwarze, S.R.; Rangnekar, V.M. The tumor suppressor Par-4 activates an extrinsic pathway for apoptosis. *Cell* **2009**, *138*, 377–388. [CrossRef] [PubMed]
4. Chakraborty, M.; Qiu, S.G.; Vasudevan, K.M.; Rangnekar, V.M. Par-4 drives trafficking and activation of Fas and Fasl to induce prostate cancer cell apoptosis and tumor regression. *Cancer Res.* **2001**, *61*, 7255–7263.
5. Cook, J.; Krishnan, S.; Ananth, S.; Sells, S.F.; Shi, Y.; Walther, M.M.; Linehan, W.M.; Sukhatme, V.P.; Weinstein, M.H.; Rangnekar, V.M. Decreased expression of the pro-apoptotic protein Par-4 in renal cell carcinoma. *Oncogene* **1999**, *18*, 1205–1208. [CrossRef] [PubMed]
6. Jagtap, J.C.; Parveen, D.; Shah, R.D.; Desai, A.; Bhosale, D.; Chugh, A.; Ranade, D.; Karnik, S.; Khedkar, B.; Mathur, A.; et al. Secretory prostate apoptosis response (Par)-4 sensitizes multicellular spheroids (MCS) of glioblastoma multiforme cells to tamoxifen-induced cell death. *FEBS Open Bio* **2015**, *5*, 8–19. [CrossRef] [PubMed]
7. Pereira, M.C.; de Bessa-Garcia, S.A.; Burikhanov, R.; Pavanelli, A.C.; Antunes, L.; Rangnekar, V.M.; Nagai, M.A. Prostate apoptosis response-4 is involved in the apoptosis response to docetaxel in MCF-7 breast cancer cells. *Int. J. Oncol.* **2013**, *43*, 531–538. [CrossRef] [PubMed]
8. Min, K.J.; Shahriyar, S.A.; Kwon, T.K. Arylquin 1, a potent Par-4 secretagogue, induces lysosomal membrane permeabilization-mediated non-apoptotic cell death in cancer cells. *Toxicol. Res.* **2020**, *36*, 167–173. [CrossRef]
9. Burikhanov, R.; Sviripa, V.M.; Hebbar, N.; Zhang, W.; Layton, W.J.; Hamza, A.; Zhan, C.G.; Watt, D.S.; Liu, C.; Rangnekar, V.M. Arylquins target vimentin to trigger Par-4 secretion for tumor cell apoptosis. *Nat. Chem. Biol.* **2014**, *10*, 924–926. [CrossRef]
10. Sviripa, V.M.; Burikhanov, R.; Obiero, J.M.; Yuan, Y.; Nickell, J.R.; Dwoskin, L.P.; Zhan, C.G.; Liu, C.; Tsodikov, O.V.; Rangnekar, V.M.; et al. Par-4 secretion: Stoichiometry of 3-arylquinoline binding to vimentin. *Org. Biomol. Chem.* **2016**, *14*, 74–84. [CrossRef]
11. Burikhanov, R.; Hebbar, N.; Noothi, S.K.; Shukla, N.; Sledziona, J.; Araujo, N.; Kudrimoti, M.; Wang, Q.J.; Watt, D.S.; Welch, D.R.; et al. Chloroquine-Inducible Par-4 Secretion Is Essential for Tumor Cell Apoptosis and Inhibition of Metastasis. *Cell Rep.* **2017**, *18*, 508–519. [CrossRef]
12. Kang, Y.; Massagué, J. Epithelial-mesenchymal transitions: Twist in development and metastasis. *Cell* **2004**, *118*, 277–279. [CrossRef] [PubMed]
13. Usman, S.; Waseem, N.H.; Nguyen, T.K.N.; Mohsin, S.; Jamal, A.; Teh, M.T.; Waseem, A. Vimentin Is at the Heart of Epithelial Mesenchymal Transition (EMT) Mediated Metastasis. *Cancers* **2021**, *13*, 4985. [CrossRef] [PubMed]
14. Katoch, A.; Jamwal, V.L.; Faheem, M.M.; Kumar, S.; Senapati, S.; Yadav, G.; Gandhi, S.G.; Goswami, A. Overlapping targets exist between the Par-4 and miR-200c axis which regulate EMT and proliferation of pancreatic cancer cells. *Transl. Oncol.* **2021**, *14*, 100879. [CrossRef]

15. Chaudhry, P.; Fabi, F.; Singh, M.; Parent, S.; Leblanc, V.; Asselin, E. Prostate apoptosis response-4 mediates TGF-beta-induced epithelial-to-mesenchymal transition. *Cell Death Dis.* **2014**, *5*, e1044. [CrossRef] [PubMed]
16. Mohd Faheem, M.; Rasool, R.U.; Ahmad, S.M.; Jamwal, V.L.; Chakraborty, S.; Katoch, A.; Gandhi, S.G.; Bhagat, M.; Goswami, A. Par-4 mediated Smad4 induction in PDAC cells restores canonical TGF-β/Smad4 axis driving the cells towards lethal EMT. *Eur. J. Cell Biol.* **2020**, *99*, 151076. [CrossRef] [PubMed]
17. Obeng, E. Apoptosis (programmed cell death) and its signals—A review. *Braz. J. Biol.* **2021**, *81*, 1133–1143. [CrossRef]
18. Elmore, S. Apoptosis: A review of programmed cell death. *Toxicol. Pathol.* **2007**, *35*, 495–516. [CrossRef] [PubMed]
19. Sells, S.F.; Wood, D.P., Jr.; Joshi-Barve, S.S.; Muthukumar, S.; Jacob, R.J.; Crist, S.A.; Humphreys, S.; Rangnekar, V.M. Commonality of the gene programs induced by effectors of apoptosis in androgen-dependent and -independent prostate cells. *Cell Growth Differ.* **1994**, *5*, 457–466.
20. Wang, B.D.; Kline, C.L.; Pastor, D.M.; Olson, T.L.; Frank, B.; Luu, T.; Sharma, A.K.; Robertson, G.; Weirauch, M.T.; Patierno, S.R.; et al. Prostate apoptosis response protein 4 sensitizes human colon cancer cells to chemotherapeutic 5-FU through mediation of an NF kappaB and microRNA network. *Mol. Cancer* **2010**, *9*, 98. [CrossRef]
21. Hardy, B.; Raiter, A.; Yakimov, M.; Vilkin, A.; Niv, Y. Colon cancer cells expressing cell surface GRP78 as a marker for reduced tumorigenicity. *Cell Oncol.* **2012**, *35*, 345–354. [CrossRef]
22. Belfi, C.A.; Chatterjee, S.; Gosky, D.M.; Berger, S.J.; Berger, N.A. Increased sensitivity of human colon cancer cells to DNA cross-linking agents after GRP78 up-regulation. *Biochem. Biophys. Res. Commun.* **1999**, *257*, 361–368. [CrossRef] [PubMed]
23. Gifford, J.B.; Hill, R. GRP78 Influences Chemoresistance and Prognosis in Cancer. *Curr. Drug Targets* **2018**, *19*, 701–708. [CrossRef] [PubMed]
24. Burotto, M.; Chiou, V.L.; Lee, J.M.; Kohn, E.C. The MAPK pathway across different malignancies: A new perspective. *Cancer* **2014**, *120*, 3446–3456. [CrossRef] [PubMed]
25. Park, J.I. Growth arrest signaling of the Raf/MEK/ERK pathway in cancer. *Front. Biol.* **2014**, *9*, 95–103. [CrossRef] [PubMed]
26. Wu, W.S.; Wu, J.R.; Hu, C.T. Signal cross talks for sustained MAPK activation and cell migration: The potential role of reactive oxygen species. *Cancer Metastasis Rev.* **2008**, *27*, 303–314. [CrossRef]
27. Kim, B.; Seo, J.H.; Lee, K.Y.; Park, B. Icariin sensitizes human colon cancer cells to TRAIL-induced apoptosis via ERK-mediated upregulation of death receptors. *Int. J. Oncol.* **2020**, *56*, 821–834. [CrossRef]
28. Zhao, Y.; Xia, S.; Cao, C.; Du, X. Effect of TGF-β1 on Apoptosis of Colon Cancer Cells Via the ERK Signaling Pathway. *J. BUON* **2019**, *24*, 449–455.
29. Tournier, C. The 2 Faces of JNK Signaling in Cancer. *Genes Cancer* **2013**, *4*, 397–400. [CrossRef]
30. Dhanasekaran, D.N.; Reddy, E.P. JNK-signaling: A multiplexing hub in programmed cell death. *Genes Cancer* **2017**, *8*, 682–694. [CrossRef]
31. Cagnol, S.; Chambard, J.C. ERK and cell death: Mechanisms of ERK-induced cell death—Apoptosis, autophagy and senescence. *FEBS J.* **2010**, *277*, 2–21. [CrossRef]
32. Dhanasekaran, D.N.; Reddy, E.P. JNK signaling in apoptosis. *Oncogene* **2008**, *27*, 6245–6251. [CrossRef] [PubMed]
33. Mohebali, N.; Pandurangan, A.K.; Mustafa, M.R.; Anandasadagopan, S.K.; Alagumuthu, T. Vernodalin induces apoptosis through the activation of ROS/JNK pathway in human colon cancer cells. *J. Biochem. Mol. Toxicol.* **2020**, *34*, e22587. [CrossRef] [PubMed]
34. Guo, X.; Meng, Y.; Sheng, X.; Guan, Y.; Zhang, F.; Han, Z.; Kang, Y.; Tai, G.; Zhou, Y.; Cheng, H. Tunicamycin enhances human colon cancer cells to TRAIL-induced apoptosis by JNK-CHOP-mediated DR5 upregulation and the inhibition of the EGFR pathway. *Anticancer Drugs* **2017**, *28*, 66–74. [CrossRef] [PubMed]
35. Wagner, E.F.; Nebreda, A.R. Signal integration by JNK and p38 MAPK pathways in cancer development. *Nat. Rev. Cancer* **2009**, *9*, 537–549. [CrossRef]
36. Lee, M.W.; Park, S.C.; Yang, Y.G.; Yim, S.O.; Chae, H.S.; Bach, J.H.; Lee, H.J.; Kim, K.Y.; Lee, W.B.; Kim, S.S. The involvement of reactive oxygen species (ROS) and p38 mitogen-activated protein (MAP) kinase in TRAIL/Apo2L-induced apoptosis. *FEBS Lett.* **2002**, *512*, 313–318. [CrossRef]
37. Yang, M.; Li, W.Y.; Xie, J.; Wang, Z.L.; Wen, Y.L.; Zhao, C.C.; Tao, L.; Li, L.F.; Tian, Y.; Sheng, J. Astragalin Inhibits the Proliferation and Migration of Human Colon Cancer HCT116 Cells by Regulating the NF-κB Signaling Pathway. *Front. Pharmacol.* **2021**, *12*, 639256. [CrossRef]
38. Rajput, A.; San Martin, I.D.; Rose, R.; Beko, A.; Levea, C.; Sharratt, E.; Mazurchuk, R.; Hoffman, R.M.; Brattain, M.G.; Wang, J. Characterization of HCT116 human colon cancer cells in an orthotopic model. *J. Surg. Res.* **2008**, *147*, 276–281. [CrossRef]
39. Phang, C.W.; Abd Malek, S.N.; Karsani, S.A. Flavokawain C exhibits antitumor effects on in vivo HCT 116 xenograft and identification of its apoptosis-linked serum biomarkers via proteomic analysis. *Biomed. Pharmacother.* **2021**, *137*, 110846. [CrossRef]

International Journal of
Molecular Sciences

Article

Sulforaphane Suppresses the Nicotine-Induced Expression of the Matrix Metalloproteinase-9 via Inhibiting ROS-Mediated AP-1 and NF-κB Signaling in Human Gastric Cancer Cells

Shinan Li [1,†], Pham Ngoc Khoi [2,†], Hong Yin [3], Dhiraj Kumar Sah [1], Nam-Ho Kim [1], Sen Lian [3,*] and Young-Do Jung [1,4,*]

1 Research Institute of Medical Sciences, Chonnam National University Medical School, Gwangju 61469, Korea; 156103@chonnam.edu (S.L.); 197784@chonnam.edu (D.K.S.); nhk111@jnu.ac.kr (N.-H.K.)
2 Faculty of Basic Medical Sciences, Pham Ngoc Thach University of Medicine, Ho Chi Minh City 740500, Vietnam; khoicnsh@gmail.com
3 Department of Biochemistry and Molecular Biology, School of Basic Medical Sciences, Southern Medical University, Guangzhou 510515, China; graceyinh@126.com
4 Department of Biochemistry, Chonnam National University Medical School, Hwasun 58128, Korea
* Correspondence: senlian@i.smu.edu.cn (S.L.); ydjung@chonnam.ac.kr (Y.-D.J.); Tel.: +82-61-379-2772 (S.L.); +86-20-6278-9385 (Y.-D.J.); Fax: +82-81-379-2781 (S.L.); +86-20-62-789-385 (Y.-D.J.)
† These authors contributed equally to this work.

Abstract: Sulforaphane, a natural phytochemical compound found in various cruciferous vegetables, has been discovered to present anti-cancer properties. Matrix metalloproteinase-9 (MMP-9) plays a crucial role in gastric cancer metastasis. However, the role of sulforaphane in MMP-9 expression in gastric cancer is not yet defined. Nicotine, a psychoactive alkaloid found in tobacco, is associated with the development of gastric cancer. Here, we found that sulforaphane suppresses the nicotine-mediated induction of MMP-9 in human gastric cancer cells. We discovered that reactive oxygen species (ROS) and MAPKs (p38 MAPK, Erk1/2) are involved in nicotine-induced MMP-9 expression. AP-1 and NF-κB are the critical transcription factors in MMP-9 expression. ROS/MAPK (p38 MAPK, Erk1/2) and ROS functioned as upstream signaling of AP-1 and NF-κB, respectively. Sulforaphane suppresses the nicotine-induced MMP-9 by inhibiting ROS-mediated MAPK (p38 MAPK, Erk1/2)/AP-1 and ROS-mediated NF-κB signaling axes, which in turn inhibit cell invasion in human gastric cancer AGS cells. Therefore, the current study provides valuable evidence for developing sulforaphane as a new anti-invasion strategy for human gastric cancer therapy.

Keywords: sulforaphane; nicotine; metalloproteinase-9; gastric cancer; cell invasion

Citation: Li, S.; Khoi, P.N.; Yin, H.; Sah, D.K.; Kim, N.-H.; Lian, S.; Jung, Y.-D. Sulforaphane Suppresses the Nicotine-Induced Expression of the Matrix Metalloproteinase-9 via Inhibiting ROS-Mediated AP-1 and NF-κB Signaling in Human Gastric Cancer Cells. *Int. J. Mol. Sci.* **2022**, *23*, 5172. https://doi.org/10.3390/ijms23095172

Academic Editor: Angela Stefanachi

Received: 11 April 2022
Accepted: 2 May 2022
Published: 5 May 2022

Publisher's Note: MDPI stays neutral with regard to jurisdictional claims in published maps and institutional affiliations.

1. Introduction

Gastric cancer is the most common gastrointestinal cancer with high mortality worldwide. The vast majority of gastric cancer cases detected are in the advanced stage [1]. The poor prognosis and treatment of gastric cancer are highly relative to metastasis [2]. Therefore, exact studies on the mechanisms underlying gastric cancer may contribute to the development of improved treatments. Tobacco abuse is strongly associated with gastric cancer progression [3]. Nicotine, a psychoactive alkaloid contained in tobacco, is a known carcinogen, related to cancer metastasis, including gastric cancer [4]. It has been reported that the pathophysiological roles of nicotine are mediated by nicotinic acetylcholine receptors [5]. In addition, nicotine promotes invasion and angiogenesis by the induction of cyclooxygenase-2 and vascular endothelial growth factor receptor-2 in gastric cancer [6].

Tumor metastasis includes the cancer progression of proliferation, migration, invasion, and the following progression of adhesion and angiogenesis in a distant tissue [2]. Most cancer-related mortalities are due to tumor metastasis. Studies indicate that cell invasion is one of the fundamental properties of malignant cancer cells [7]. The breakdown of the extracellular matrix by proteinases is an essential step in cancer cell invasion [8]. Matrix metalloproteinases (MMPs), a family of extracellular-matrix-degrading proteinases, induce cancer cell invasion through the degradation of the extracellular matrix and the basal membrane [9]. Among the MMPs, MMP-2, and MMP-9 play crucial roles in cancer metastasis [10]. Normally, MMP-2 is constitutively expressed in highly malignant tumors, whereas MMP-9 is induced by growth factors [11]. High MMP-9 expression is observed in tumor extracts in gastric cancer; furthermore, aberrant expression of MMP-9 can increase tumor cell detachment and metastasis, which are related to malignancy and poor prognosis in gastric cancer [12]. Therefore, agents with the ability to suppress MMP-9 expression may be useful for the development of treatment strategies for gastric cancer.

Sulforaphane, a natural compound that is abundant in cruciferous vegetables, exhibits several beneficial properties, including anti-inflammatory, antioxidant, and anticancer activities. Recently, the anticancer effect of sulforaphane has attracted much attention [13]. Intake of cruciferous vegetables has been proven to prevent gastric cancer. It has been reported that sulforaphane induces cell cycle arrest and apoptosis in human colorectal cancer cells [14]. Sulforaphane was also suggested to sensitize hepatoma cancer cells to TRAIL-mediated apoptosis by reactive oxygen species (ROS)-mediatedDR5 expression [15]. Our recent studies show that sulforaphane inhibits bladder cancer cell proliferation via suppression of HIF-1α-mediated glycolysis in hypoxia [16]. Moreover, sulforaphane shows a protective effect on gastric mucosa via the Nrf2 mechanism [17]. The effects of MMP-9 inhibitors on the treatment of gastric cancer have been widely reported. However, the potential mechanisms by which sulforaphane inhibits MMP-9 expression are not fully understood in gastric cancer.

In this study, we examined the effect of sulforaphane on nicotine-mediated induction of MMP-9 and explored the underlying mechanisms. Based on our results, we reported that sulforaphane suppresses the nicotine-induced MMP-9 by inhibiting ROS-mediated MAPK (p38 MAPK, Erk1/2)/AP-1 and ROS-mediated NF-κB signaling axes, which in turn inhibit cell invasion in human gastric cancer AGS cells.

2. Results

2.1. Sulforaphane Suppresses Nicotine-Induced MMP-9 Expression in AGS Cells

To determine the effect of sulforaphane on nicotine-induced MMP-9 expression, AGS cells were pretreated with sulforaphane and treated with nicotine. The expression levels of MMP-9 mRNA and protein were measured, respectively. As shown in Figure 1A,B, treatment with nicotine significantly induced MMP-9 expression at both the mRNA and protein levels, which can be abrogated by sulforaphane. In addition, suppression of nicotine-induced MMP-9 promoter activity was also observed in sulforaphane-pretreated AGS cells (Figure 1C). These results showed that sulforaphane inhibits nicotine-induced MMP-9 expression in human gastric cancer AGS cells.

A

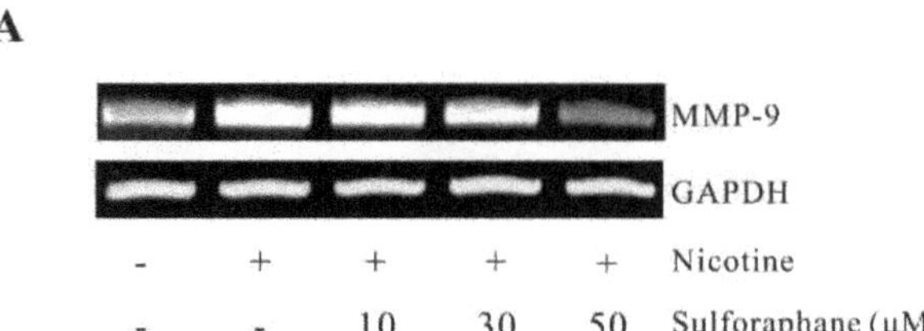

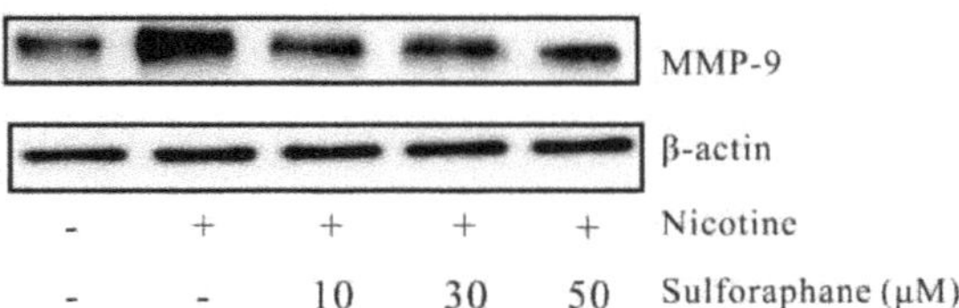

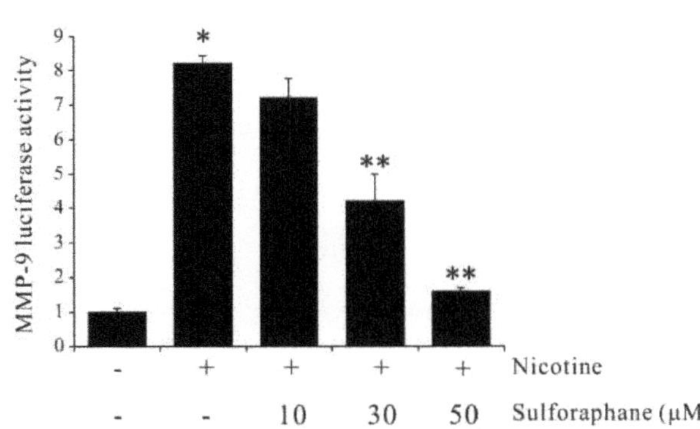

Figure 1. Sulforaphane inhibits nicotine-induced MMP-9 expression in human gastric AGS cells. AGS cells were pretreated with sulforaphane (10, 30, 50 μM) for 1 h, followed by treatment with 100 μg/mL nicotine for 4 h or 12 h, and MMP-9 expression was analyzed by performing RT-PCR (**A**), Western blot (**B**), and luciferase activity assay (**C**), respectively. The data represent the mean ± SEM from three experimental trials. * $p < 0.05$ in comparison with the control; ** $p < 0.05$ in comparison with nicotine alone.

2.2. Sulforaphane Suppresses Nicotine-Induced MMP-9 Expression by Inhibiting ROS Generation

Due to the important role of oxidative stress in the pathogenesis of cancer [18], the ROS production levels were determined by H_2DCFDA treated with nicotine in the presence or absence of sulforaphane. As shown in Figure 2A,B, sulforaphane suppressed the nicotine-induced ROS production levels. N-Acetylcysteine (NAC) was used as a positive control. NAC treatment abrogated nicotine-induced MMP-9 expression (Figure 2C). These results indicate that sulforaphane suppressed the nicotine-induced MMP-9 via regulating ROS generation in human gastric cancer AGS cells.

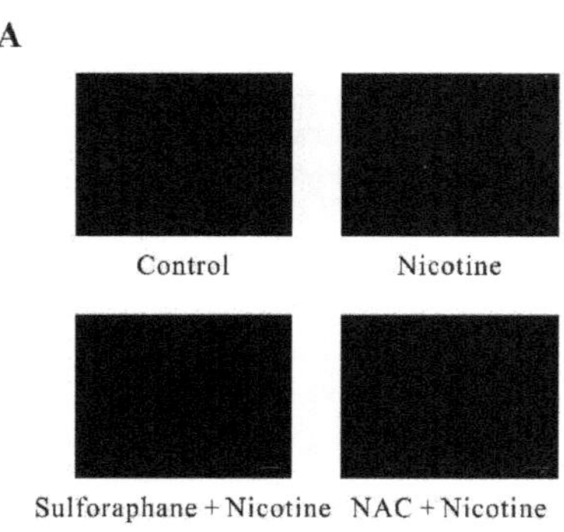

B

Relative fluorescence intensity (fold)

0 1 2 3 4 5

*

** **

-	-	30	-	Sulforaphane (μM)
-	-	-	5	NAC (mM)
-	+	+	+	Nicotine

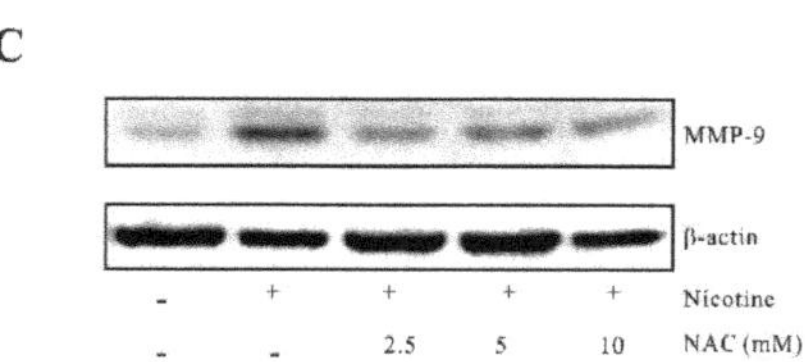

Figure 2. Sulforaphane inhibits nicotine-induced ROS in human gastric cancer AGS cells. (**A**) AGS cells were pretreated with 30 μM sulforaphane and 5 mM NAC for 1 h prior to nicotine treatment for 30 min. The cells were then incubated in the dark for 10 min with 10 μM H_2DCFDA. The H_2DCFDA fluorescence was imaged using a confocal laser scanning fluorescence microscope. (**B**) Relative fluorescence intensities of the ROS production level. (**C**) Cells pretreated with NAC (2.5, 5, 10 mM) for 1 h were incubated with nicotine (100 μg/mL) for 12 h. After incubation, extracted proteins were analyzed for the induction of MMP-9 expression by Western blot. The data represent the mean ± SEM from three experimental trials. * $p < 0.05$ in comparison with the control; ** $p < 0.05$ in comparison with nicotine alone.

2.3. Sulforaphane Suppresses Nicotine-Induced MMP-9 Expression by Inhibiting p38 MAPK and Erk1/2 Activation

MAPKs have well-established roles in the progression of human cancers [19,20]. To determine the role of MAPKs on nicotine-induced MMP-9 expression, pharmacological inhibitors of MAPKs, SB203580 (a p38 MAPK inhibitor), and PD98059 (a MEK inhibitor) were used along with nicotine treatment in AGS cells. As shown in Figure 3A, both SB203580 and PD98059 inhibited the nicotine-induced MMP-9 expression at the transcriptional level. Transfection of dominant-negative mutant constructs mP38 (p38 MAPK) or K97M (MEK-1) attenuated nicotine-induced MMP-9 promoter activity (Figure 3B). Additionally, we found that sulforaphane suppressed nicotine-induced p38 MAPK and Erk1/2 (Figure 3C,D).

These results suggest that sulforaphane suppressed nicotine-induced MMP-9 expression by inhibiting p38 MAPK and Erk1/2 activation in AGS cells.

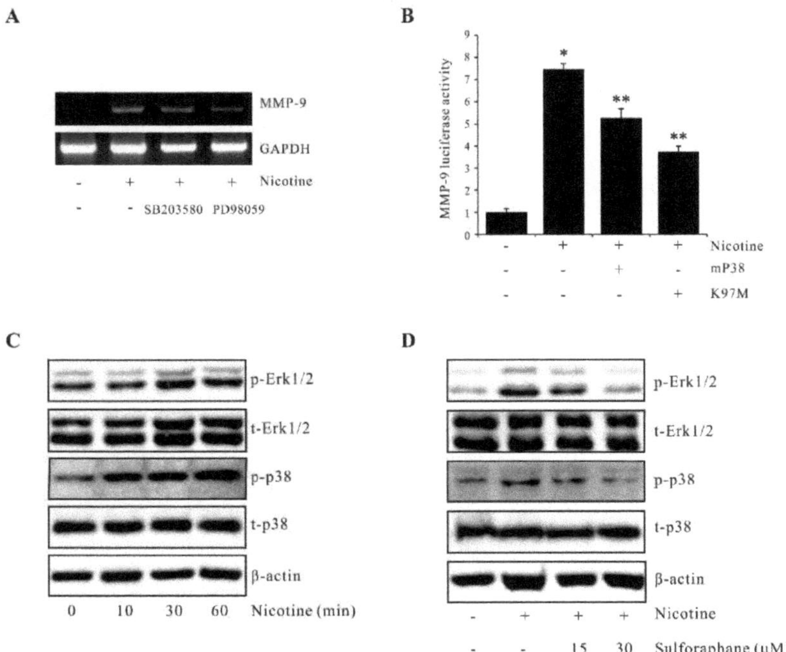

Figure 3. Sulforaphane inhibits nicotine-induced MMP-9 expression by suppressing p38 MAPK and Erk1/2 signaling pathways. (**A**) AGS cells were pretreated with 20 μM SB203580 and 20 μM PD 98059 for 1 h and incubated with 100 μg/mL nicotine for 4 h. After incubation, extracted mRNA was analyzed for the induction of MMP-9 expression by RT-PCR. (**B**) AGS cells were cotransfected with dominant-negative mutants of p38 MAPK (mP38) or MEK-1 (K97M) and the pGL4-MMP-9 promoter-reporter construct. The luciferase activity was determined using a luminometer after incubating the cells with 100 μg/mL nicotine for 4 h. (**C**) Cells were treated with 100 μg/mL nicotine for 0–60 min, and extracted proteins were analyzed by Western blot. (**D**) Cells were pretreated with sulforaphane (15, 30 μM) followed by 100 μg/mL nicotine treatment for 30 min, and extracted proteins were analyzed by Western blot. The data represent the mean ± SEM from three experimental trials. * $p < 0.05$ in comparison with the control; ** $p < 0.05$ in comparison with nicotine alone.

2.4. *Sulforaphane Suppresses Nicotine-Induced MMP-9 Expression by Inhibiting Reporter Activities of AP-1 and NF-κB*

Studies showed that AP-1 plays a pivotal role in tumor carcinogenesis [21]. Curcumin (an AP-1 inhibitor) pretreatment significantly suppressed the nicotine-induction MMP-9 protein expression and promoter activity (Figure 4A,B). Furthermore, sulforaphane treatment resulted in significant inhibition of nicotine-induced c-fos and c-jun phosphorylation. (Figure 4C). Moreover, NF-κB is also a key transcription factor in tumor carcinogenesis [22]. BAY 11-7082 (an NF-κB inhibitor) pretreatment decreased the nicotine-induced MMP-9 protein expression and promoter activity (Figure 4D,E). It is observed that sulforaphane suppressed the nicotine-enhanced phosphorylation of NF-κB and IκBα (Figure 4F). These results demonstrated that sulforaphane inhibited nicotine-induced MMP-9 expression via suppressing AP-1 and NF-κB activation.

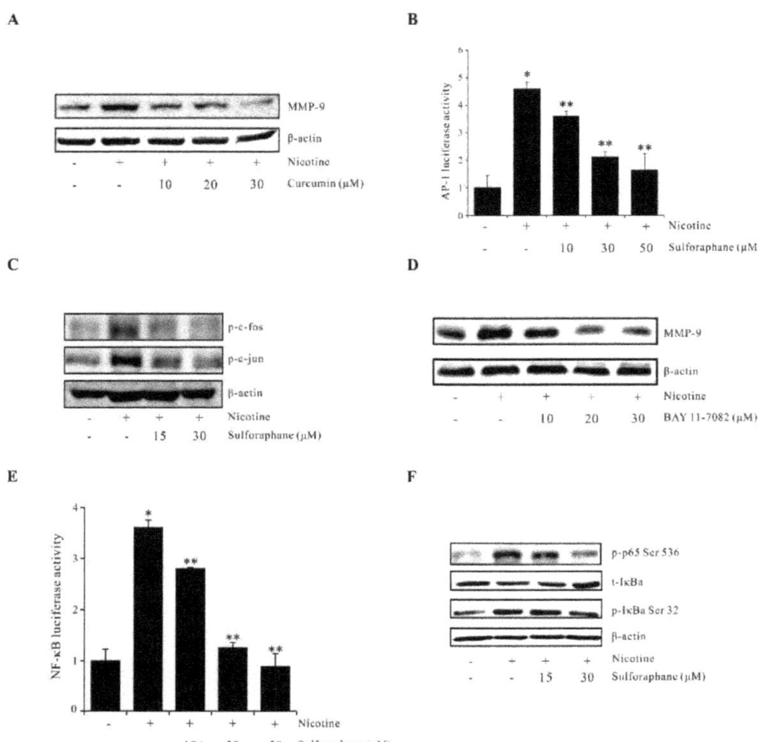

Figure 4. Sulforaphane inhibits nicotine-induced MMP-9 expression by suppressing the transcription factors of AP-1 and NF-κB. (**A**) AGS cells were pretreated with the indicated concentration of curcumin and treated with 100 μg/mL nicotine for 12 h, and extracted proteins were analyzed for the induction of MMP-9 expression by Western blot. (**B**) Cells were transfected with the AP-1 luciferase reporter. The luciferase activity was determined using a luminometer after incubating the cells with sulforaphane for 1 h prior to nicotine treatment for 4 h. (**C**) Cells were pretreated with sulforaphane and treated with 100 μg/mL nicotine for 1 h; the expression of phos-c-fos and phos-c-jun were analyzed by Western blot. (**D**) Cells were pretreated with the indicated concentration of BAY11-7082 and treated with 100 μg/mL nicotine for 12 h, and extracted proteins were analyzed for the induction of MMP-9 expression by Western blot. (**E**) Cells were transfected with the NF-κB luciferase reporter. The luciferase activity was determined using a luminometer after incubating the cells with sulforaphane for 1 h prior to nicotine treatment for 4 h. (**F**) Cells were pretreated with sulforaphane and treated with 100 μg/mL nicotine for 1 h, the expression of phos-p65 (Ser536 and phos-IκBα (Ser32) were analyzed by Western blot. The data represent the mean ± SEM from three experimental trials. * $p < 0.05$ in comparison with the control; ** $p < 0.05$ in comparison with nicotine alone.

2.5. ROS/(p38 MAPK, Erk1/2) and ROS Functioned as Upstream Regulators of AP-1 and NF-κB Respectively

To dissect the relevant signaling pathways contributing to AP-1 activation induced by nicotine, we performed inhibitor studies with luciferase activity assay and Western blot. As shown in Figure 5A, SB203580 partially suppressed the AP-1 transcription, while both PD98059 and NAC significantly blocked the AP-1 transcription. Similar results are shown at protein levels (Figure 5B). In addition, the data presented in Figure 5C indicated that the ROS inhibitor, NAC, decreases p38 MAPK and Erk1/2 activation. These results indicate that ROS/(p38 MAPK, Erk1/2) is the upstream regulator of AP-1 in nicotine-induced MMP-9 expression in AGS cells. Next, we examined which relevant regulator contributed to

AP-1 activation induced by nicotine. To determine whether the ROS contributed to NF-κB activation induced by nicotine, the effects of an inhibitor of ROS on nicotine-induced NF-κB activation were examined. The inhibitor of ROS, NAC, inhibited nicotine-mediated NF-κB reporter activity (Figure 5D). Pretreatment of NAC attenuated nicotine-media activation of p65 (Figure 5E). These findings supported that ROS functioned as upstream signaling of NF-κB in nicotine-induced MMP-9 expression in AGS cells.

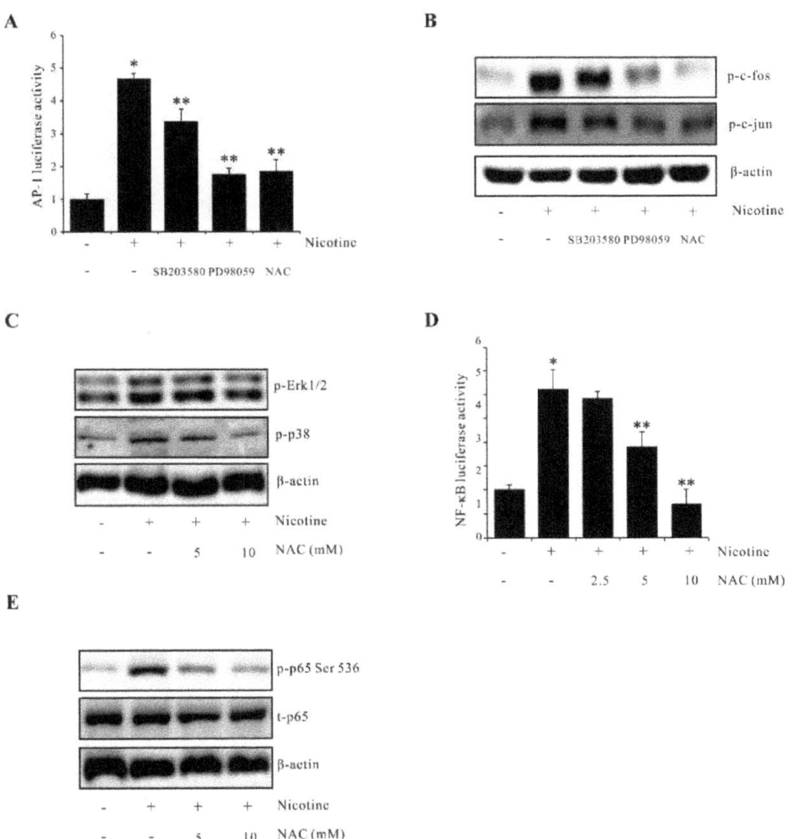

Figure 5. ROS/(p38 MAPK, Erk1/2) and ROS functioned as upstream regulators of AP-1 and NF-κB, respectively. (**A**) AGS cells were transfected with the AP-1 luciferase reporter. The luciferase activity was determined using a luminometer after incubating the cells with 20 μM SB203580, 20 μM PD 98059, or 5 mM NAC for 1 h prior to nicotine treatment for 4 h. (**B**) AGS cells were pretreated with 20 μM SB203580, 20 μM PD 98059, or 5 mM NAC and treated with 100 μg/mL nicotine for 30 min; the expression of phos-c-fos and phos-c-jun were analyzed by Western blot. (**C**) Cells were pretreated with the indicated concentration of NAC and treated with 100 μg/mL nicotine for 30 min; the expression of phos-Erk1/2 and phos-p38 were analyzed by Western blot. (**D**) Cells were transfected with the NF-κB luciferase reporter. The luciferase activity was determined using a luminometer after incubating the cells with NAC for 1 h prior to nicotine treatment for 4 h. (**E**) Cells were pretreated with the indicated concentration of NAC and treated with 100 μg/mL nicotine for 30 min; the expression of phos-p65 (Ser5360 and phos-IκBα (Ser32) were analyzed by Western blot. The data represent the mean ± SEM from three experimental trials. * $p < 0.05$ in comparison with the control; ** $p < 0.05$ in comparison with nicotine alone.

2.6. Sulforaphane Attenuates the Invasiveness of AGS Cells by Suppressing MMP-9 Expression

It is well known that high expression of MMP-9 is important for the invasive phenotype of cancer cells. The effect of sulforaphane on nicotine-induced cell invasion was examined by performing a matrigel invasion assay. AGS cells incubated in nicotine resulted in increased activity of the cell invasive phenotype. However, in the presence of sulforaphane or an MMP-9 antibody, the number of invaded cells decreased, suggesting that sulforaphane suppressed the nicotine-induced invasive phenotype by inhibiting MMP-9 expression (Figure 6A). We further counted the invading cells and the cell invasion results showed with statistically significant values that the sulforaphane pretreatment significantly reduced the nicotine-induced cell invasive activity as well as the neutralizer, MMP-9 antibody (Figure 6B). These results further indicated that sulforaphane inhibited the AGS cell invasive activity by downregulating the MMP-9 expression.

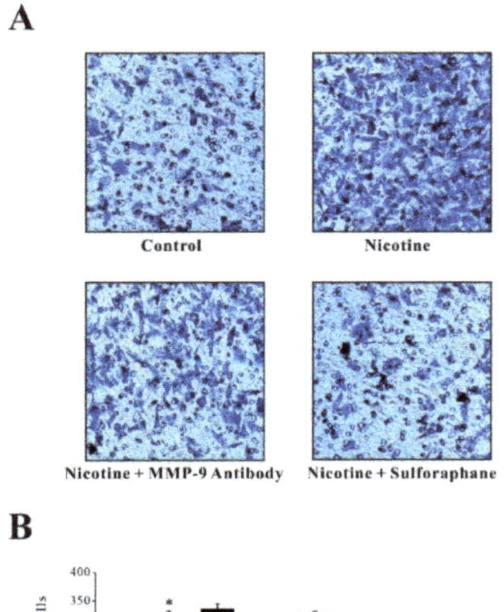

Figure 6. Sulforaphane inhibits nicotine-induced cell invasion in AGS cells. (**A**) AGS cells (3×10^5) were incubated with nicotine (100 μg/mL) in the presence or absence of sulforaphane (30 μM) or anti-MMP-9 antibody (200 ng/mL) in a Corning Matrigel matrix for 24 h. (**B**) AGS cells (3×10^5) were incubated with nicotine (100 μg/mL) in the presence or absence of nonspecific IgG, anti-MMP-9 antibody (200 ng/mL), or sulforaphane (10, 30, 50 μM) in a Corning Matrigel matrix for 24 h. Cells invading the undersurface of the membrane were stained using a Diff-Quick stain kit and counted under a phase-contrast light microscope. The data represent the mean ± SEM from three experimental trials. * $p < 0.05$ in comparison with the control; ** $p < 0.05$ in comparison with nicotine alone.

3. Discussion

Gastric cancer ranks as the fourth most common cancer and is one of the leading causes of cancer-related death worldwide [23]. Phytochemicals, derived from plants, have become an important source of anticancer medicines, with antioxidant activities [24]. Sulforaphane, 1-isothiocyanato-4-(methylsulfinyl)butane, a natural compound that includes the isothiocyanate group of organosulfur compounds, is one of the major phytochemicals found in cruciferous vegetables [25]. Many studies have been directed at defining the role of sulforaphane as an anticancer medicine in humans, due to various reasons. Firstly, cruciferous vegetables, particularly broccoli, are rich in sulforaphane, which can prevent cancer risk [26]. Sulforaphane may protect against various types of cancer. In breast cancer, combination therapy with sulforaphane has been shown to improve the outcome [27]. Sulforaphane can inhibit breast cancer stem cells via downregulation of the Wnt/β-catenin self-renewal pathway in the xenograft mice model [28]. In colorectal cancer, sulforaphane inhibits the stemness of cancer stem cells both in vitro and in vivo by targeting TAp63α [29]. Rutz et al. reported that sulforaphane acts as a histone deacetylase (HDAC) inhibitor to prostate cancer cell progression [30]. In addition, sulforaphane has a potential therapeutic application in the treatment and prevention of gastric cancer by induction of apoptosis of gastric cancer cells [31]. Our earlier studies indicated that sulforaphane decreased glycolytic metabolism in a hypoxia microenvironment by inhibiting hypoxia-induced HIF-1α and HIF-1α trans-localization in non-muscle-invasive bladder cancer cell lines [16]. Moreover, it has been known that sulforaphane has many health benefits. Sulforaphane could prevent memory dysfunction and improve cognitive function [32]. Sulforaphane prevents type 2 diabetes-induced cardiomyopathy by activating the lipid metabolic pathway and enhancing NRF2 activation [33]. Sulforaphane presents anti-inflammation properties by suppression of cyclooxygenase-2 expression [34].

Clinical and epidemiological research has revealed that smokers are more likely to develop cancer progression as compared to non-smokers [35]. Cigarette smoke caused many diseases and cancers, and nicotine is a major poison in cigarette smoke [36]. Nicotine caused more harm to human organs and tissues than other compounds of cigarette smoke [37]. Recently, we demonstrated that nicotine promotes gastrointestinal cancer progression through IL-8 upregulation [38]. Aberrant processes of wound healing contribute to cancer progression [39]. Matrix metalloproteinases (MMPs) have been identified as the main factors in both acute and chronic wounds and the excess protease activity can lead to a chronic nonhealing wound [40]. Reiss et al. reported that when MMP-9 is expressed at excessive levels, it prevents the reestablishment of the dermal/epidermal junction and, thereby, limits epithelial migration and wound closure in a murine wound model [41].

In the present study, we attempted to explore the role and potential mechanisms of sulforaphane in nicotine-challenged gastric cancer cells. We revealed that nicotine induces MMP-9 expression and cell invasiveness in gastric cancer AGS cells. Sulforaphane effectively suppressed ROS, p38 MAPK, Erk1/2, AP-1, and NF-κB activation by inhibiting MMP-9 expression in gastric cancer AGS cells.

Healthy bodies and tissues are often subjected to sublethal doses of various oxidants [42]. There is considerable evidence suggesting oxidative stress has been associated with the development of cancer [18]. Increased ROS generation was observed in cancer cells compared with normal cells [43]. ROS function as secondary messengers and control various signaling cascades in cells [44]. Nicotine promotes atherosclerosis by the induction of ROS in endothelial cells [45]. The present study suggested that nicotine induces ROS generation in gastric cancer AGS cells, and NAC abrogated nicotine-induced MMP-9 expression. Sulforaphane suppresses ROS production to inhibit nicotine-induced MMP-9 expression. NADPH oxidases were identified as upstream signal molecules of ROS in AGS cells. NADPH oxidase activation is regulated by several processes such as phosphorylation of its components, exchange of GDP/GTP on Rac2, and binding of p47^{phox} and p40^{phox} to phospholipids [46]. Membrane translocation of p47^{phox} plays a critical role in the activation of NADPH oxidase [47]. Nicotine can trigger the generation of ROS through NADPH oxidase [48]. Sulforaphane decreases ROS and inhibits carcinogenesis by the activation of NRF2 [49]. Sulforaphane was also reported to induce HO-1 in microglia [50]. In T24 bladder cancer cells, sulforaphane upregulates ROS to induce cell apoptosis [51]. Sulforaphane induces ROS generation to promote tumor necrosis factor-related apoptosis-inducing ligand sensitivity [52]. In this respect, the mechanisms involved in sulforaphane inhibited nicotine-activated ROS are revealed in this study.

MAPK cascade plays a vital role in various cancer progression [20]. MAPK-regulated MMP-9 in cancer cells has been reported in many studies [7]. Here, nicotine stimulated the phosphorylation of p38 MAPK and Erk1/2 to induce MMP-9 expression in AGS cells. The aberrant activation of EGFR has been implicated in tumor growth [53]. Previously, we observed that EGFR is involved in MMP-9 expression in human endothelial cells [54]. One study reported that Akt and PKCδ are associated with TPA-induced MMP-9 expression [55]. MAPKs are studied as the downstream of PKCα/β [56,57]. Experiments on colorectal tumor cells documented that MAPK signaling may directly depend on ROS [58]. Treatment with sulforaphane significantly reduced the amount of phosphorylated Akt and phosphorylation of the mTOR subunit [59]. In this study, sulforaphane inhibited p38 MAPK and Erk1/2 activation to suppress nicotine-induced MMP-9 expression. Thus, many additional signaling modulators should be explored to define sulforaphane suppression of nicotine-induced MMP-9 expression in AGS gastric cancer cells.

Our previous study revealed the important role of AP-1 and NF-κB in regulating MMP-9 by cadmium in endothelial cells [54]. AP-1 is composed of members of the c-fos and c-jun families, which have been shown to regulate the expression of several genes involved in tumor development. Here, enhanced phosphorylation of c-fos and c-jun was observed in nicotine-treated cells. Our results showing that AP-1 inhibitor ameliorated MMP-9 expression indicated that AP-1 contributed to nicotine-mediated induction of MMP-9. Sulforaphane's inhibition of c-fos and c-jun phosphorylation accompanied by a reduction in AP-1 transcription factor activity, therefore, suppressed MMP-9 expression. To further determine the underlying mechanisms, we treated AGS cells with the inhibitors of ROS, p38 MAPK, and Erk1/2. We found that inhibitors of ROS, p38 MAPK, and Erk1/2 suppressed the nicotine-mediated c-fos and c-jun activation and the AP-1 reporter activity. ROS inhibitor reduced the nicotine-induced activation of p38 MAPK and Erk1/2. Our results indicated that ROS/MAPK (p38 MAPK, Erk1/2) functioned as the upstream signaling molecules in the nicotine-activated AP-1 pathway. Src tyrosine was reported to be upstream of AP-1 [60]. We observed that JNK1/2 mediates AP-1 activation in AGS cells [61]. Increased NF-κB translocation is usually associated with its phosphorylation and IκB proteasomal degradation in many types of cancer progression. In our study, we found that IκBα and p65, the two subunit elements of NF-κB, play vital roles in the nicotine-mediated induction of MMP-9 in AGS cells. It was observed that sulforaphane inhibited the nicotine-induced NF-κB p65 and IκBα in a dose-dependent manner. ROS contribute to the upstream signaling of NF-κB [62]. Our results showed that ROS are the upstream molecules of NF-κB in nicotine-induced MMP-9 in AGS cells. A prior study suggested that the EGFR signaling activates NF-κB via mTORC2 [63]. We found that Erk1/2 and JNK were critical for NF-κB in bladder cancer cells [64].

4. Conclusion

Figure 7 illustrates that sulforaphane suppresses the nicotine-induced MMP-9 by inhibiting ROS-mediated MAPK (Erk1/2, p38 MAPK)/AP-1 and ROS-mediated NF-κB signaling axes, which in turn inhibit cell invasion in human gastric cancer AGS cells. These findings demonstrate that sulforaphane might be a potential functional food ingredient with the feature of gastric cancer therapy.

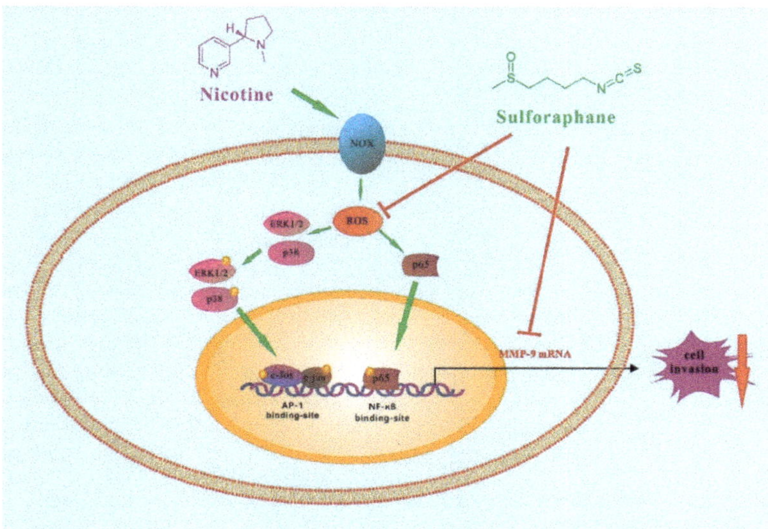

Figure 7. Schematic representation of the mechanism underlying the inhibition of nicotine-induced MMP-9 expression by sulforaphane in AGS cells. Sulforaphane inhibits nicotine-induced MMP-9 expression via suppression of the ROS/MAPKs(p38 MAPK, Erk1/2)/AP-1 and ROS/NF-κB signaling pathways, which in turn attenuate AGS cell invasiveness.

5. Materials and Methods

5.1. Reagents

RPMI-1640, OPTI-modified Eagle's medium, fetal bovine serum (FBS), phosphate-buffered saline, and penicillin–streptomycin solution were obtained from HyClone (Logan, UT, USA). TrypLE™ Express was obtained from Gibco (Grand Island, NY, USA). Sulforaphane, nicotine, DMSO, curcumin, and all other chemicals were purchased from Sigma-Aldrich (St. Louis, MO, USA). BAY11-7082, PD98059, and SB203580 were purchased from Calbiochem (San Diego, CA, USA). Antibodies against MMP-9, phos-Erk1/2, Erk1/2, phos-p38, p38, phos-c-jun, phos-c-fos, phos-p65 (Ser536), phos-IκBα (Ser32), and IκBα were purchased from Cell Signaling Technology (Danvers, MA, USA).

5.2. Cell Culture

The AGS human gastric cancer cell line was obtained from American Type Culture Collection (Manassas, VA, USA) and cultured in RPMI-1640 medium supplemented with 10% fetal bovine serum (FBS) and 0.6% penicillin–streptomycin at 37 °C in a 5% CO_2 humidified incubator. In these experiments, stimulants such as nicotine were added to serum-free media for the indicated time intervals. When the inhibitors were used, they were added 1 h before the nicotine treatment.

5.3. Reverse Transcription PCR

AGS cells were treated with 100 μg/mL nicotine for 4 h. Then, total RNA was extracted from the AGS cells using TRIzol reagent (Invitrogen). One μg of total RNA was used for first-strand complementary DNA synthesis using random primers and M-MLV transcriptase (Promega). The complementary DNA was subjected to PCR amplification with primer sets for GAPDH and MMP-9 using a PCR master mix solution (iNtRON, Korea). The specific primer sequences were as follows: GAPDH sense, 5′-TTG TTG CCA TCA ATG ACC CC-3′, and GAPDH antisense, 5′-TGA CAA AGT GGT CGT TGA GG-3′ (836 bp); and MMP-9 sense, 5′- AAG TGG CAC CAC CAC AAC AT -3′ and MMP-9 anti-sense, 5′-TTT CCC ATC AGC ATT GCC GT-3′ (497 bp). The PCR conditions were as follows:

denaturation at 94 °C for 30 s, annealing at 52 °C for 20 s, and extension at 72 °C for 30 s, 28 cycles.

5.4. Western Blot Analysis

AGS cells were treated with 100 μg/mL nicotine for 12 h to detect the MMP-9 changes and were treated with 100 μg/mL nicotine for 30–60 min to detect the signal molecule changes. After each experiment, cells were washed twice with cold PBS and were harvested in 100 μL of protein extraction solution (iNtRON, Seongnam, Korea). Cell homogenates were centrifuged at 10,000× *g* for 20 min at 4 °C. Equal amounts of total cellular protein (50 μg) were electrophoresed in sodium dodecyl sulfate (SDS)-polyacrylamide gels, and the protein was then transferred to polyvinylidene difluoride membranes (Millipore, Billerica, MA, USA). Nonspecific binding sites on the membranes were blocked with 5% nonfat dry milk in 15 mM Tris/150 mM NaCl buffer (pH 7.4) at room temperature for 2 h. Membranes were incubated with the target antibody. The membranes were then probed with a secondary antibody labeled with horseradish peroxidase. The bands were visualized using an enhanced chemiluminescence kit (Millipore, Billerica, MA, USA) and were scanned by a luminescence image analyzer (Vilber Lourmat, Collégien, France).

5.5. Transient Transfection with Dominant Negative Mutants

The plasmids encoding dominant-negative mutants of MEK-1 (pMCL-K97M) and p38 MAPK (pMCL-mP38) were kindly provided by Dr. N. G. Ahn (University of Colorado, Boulder, CO, USA) and Dr. J. Han (Scripps Research Institute, CA, USA), respectively. All mutants were prepared by using TIANGEN (Beijing, China) plasmid DNA preparation kits. Dominant-negative mutants (1 μg) were carried out using Lipofectamine 3000 from Invitrogen (Carlsbad, CA, USA).

5.6. Measurement of MMP-9, AP-1 and NF-κB Luciferase Activity

The transcriptional regulation of MMP-9 was examined by the transient transfection of an MMP-9 promoter–luciferase reporter construct (pGL4-MMP-9). The plasmid pGL4-MMP-9 promoter (spanning nucleotides from −925 to +13) was kindly provided by Dr. Young-Han Lee (Konkuk University, Korea). The NF-κB and AP-1 luciferase reporter plasmid were purchased from Clontech (Palo Alto, CA, USA). The effects of sulforaphane on MMP-9 promoter activity were determined by pretreating cells with sulforaphane for 1 h prior to the nicotine treatment. Cells were collected with cell culture lysis reagent (Promega, Madison, WI, USA) and the luciferase activity was determined using a luminometer (Centro XS lb960 microplate luminometer, Berthold Technologies, Oak Ridge, TN, USA) according to the manufacturer's protocol.

5.7. Detection of ROS by H_2DCFDA

ROS production levels were performed by modifying the method described by our previous study [65]. Briefly, H_2DCFDA (MCE, Romulu, MI, USA), a cell-permeable probe, was used to detect changes in intracellular ROS produced by AGC cells in the sulforaphane and nicotine treatment group (30 min) and control, which were incubated with H_2DCFDA at 37 °C with 5% CO_2 for 30 min, digested with trypsin, and suspended in PBS. Images were acquired using the Laser Scanning Microscope 5 PASCAL program (Carl Zeiss) and a confocal microscope. DCF fluorescence was excited at 488 nm with an argon laser, and the evoked emission was filtered with a 515 nm long-pass filter.

5.8. Matrigel Invasion Assay

The cell invasion assay was carried out according to our previous study [66] using 10-well chemotaxis chambers (Neuro Probe, Gaithersburg, Maryland, USA) with an 8-μM pore membrane (Neuro Probe) in RPMI-1640 with 10% FBS as the chemoattractant in the lower chamber. Briefly, AGS cells were added to the upper chamber with nicotine for 24 h, the non-invading cells on the upper surface of each membrane were removed from the chamber by using cotton swabs, and the invading cells on the lower surface of each membrane were stained using the Quick-Diff stain kit (Becton-Dickinson, Franklin Lakes, NJ, USA). After two washes with water, the chambers were allowed to air dry. The number of invading cells was counted using a phase-contrast microscope.

5.9. Statistics Analysis

Quantitative data were analyzed using a one-way analysis of variance followed by Tukey's honestly significant difference tests between individual groups. Data are expressed as the mean ± SEM. A value of $p < 0.05$ was considered to be significant. The statistical software package Prism 5.0 (GraphPad Software, La Jolla, CA, USA) was used for analysis.

Author Contributions: Conceptualization: S.L. (Shinan Li), S.L. (Sen Lian) and Y.-D.J.; investigation and methodology: S.L. (Shinan Li) and P.N.K.; software and formal analysis: P.N.K., H.Y. and D.K.S.; resources: H.Y. and N.-H.K.; validation: P.N.K. and Y.-D.J.; data curation: S.L. (Shinan Li) and Y.-D.J.; writing—original draft: S.L. (Sen Lian); writing—review and editing: S.L. (Shinan Li) and Y.-D.J.; project administration: Y.-D.J.; funding acquisition: S.L. (Sen Lian) and Y.-D.J. All authors have read and agreed to the published version of the manuscript.

Funding: This study was supported by the Basic Science Research Program grant through the National Research Foundation of Korea funded by the Ministry of Education, Science, and Technology (2018R1D1A1B07049918); the Guangdong Basic and Applied Basic Research Foundation (2020A1515011433); the Guangzhou Basic and Applied Basic Research Foundation (202102021052).

Institutional Review Board Statement: Not applicable.

Informed Consent Statement: Not applicable.

Data Availability Statement: The data presented in this study are available in the article.

Conflicts of Interest: The authors declare that they have no conflict of interest.

References

1. Ganesh, S.; Sier, C.; Heerding, M.M.; Griffioen, G.; Lamers, C.; Verspaget, H.W. Prognostic relevonce of the plasminogen activation system in tissue of patients with gastric cancer. *Neth. J. Med.* **1995**, *47*, A38. [CrossRef]
2. Jin, X.; Zhu, Z.; Yi, S. Metastasis mechanism and gene/protein expression in gastric cancer with distant organs metastasis. *Bull Cancer* **2014**, *101*, 1–12. [CrossRef] [PubMed]
3. Sasazuki, S.; Sasaki, S.; Tsugane, S. Cigarette smoking, alcohol consumption and subsequent gastric cancer risk by subsite and histologic type. *Int. J. Cancer* **2002**, *101*, 560–566. [CrossRef] [PubMed]
4. Lien, Y.C.; Wang, W.; Kuo, L.J.; Liu, J.J.; Wei, P.L.; Ho, Y.S.; Ting, W.C.; Wu, C.H.; Chang, Y.J. Nicotine Promotes Cell Migration Through Alpha7 Nicotinic Acetylcholine Receptor in Gastric Cancer Cells. *Ann. Surg. Oncol.* **2011**, *18*, 2671–2679. [CrossRef]
5. Lindstrom, J. Nicotinic acetylcholine receptors in health and disease. *Mol. Neurobiol.* **1997**, *15*, 193–222. [CrossRef]
6. Shin, V.Y. Nicotine induces cyclooxygenase-2 and vascular endothelial growth factor receptor-2 in association with tumor-associated invasion and angiogenesis in gastric cancer. *Mol. Cancer Res.* **2005**, *3*, 607. [CrossRef]
7. Xia, Y.; Lian, S.; Ngoc, K.P.; Joong, Y.H.; Eun, J.Y.; Oh, C.K.; Keun, K.K.; Young, D.J.; Singh, P.K. Chrysin Inhibits Tumor Promoter-Induced MMP-9 Expression by Blocking AP-1 via Suppression of ERK and JNK Pathways in Gastric Cancer Cells. *PLoS ONE* **2015**, *10*, e0124007. [CrossRef]
8. Curran, S.; Murray, G.I. Matrix metalloproteinases in tumour invasion and metastasis. *J. Pathol.* **1999**, *189*, 300–308. [CrossRef]
9. Velinov, N.; Poptodorov, G.; Gabrovski, N.; Gabrovski, S. The role of matrixmetalloproteinases in the tumor growth and metastasis. *Khirurgiia* **2010**, *1*, 44–49.
10. Zhao, X.L.; Tao, S.; Na, C.; Sun, D.; Nan, Z.; Dong, X.Y.; Gu, Q.; Yao, Z.; Sun, B.C. Promotion of hepatocellular carcinoma metastasis through matrix metalloproteinase activation by epithelial-mesenchymal transition regulator Twist1. *J. Cell. Mol. Med.* **2011**, *15*, 691–700. [CrossRef]
11. Ramírez, G.; Hagood, J.S.; Sanders, Y.; Ramírez, R.; Becerril, C.; Segura, L.; Barrera, L.; Selman, M.; Pardo, A. Absence of Thy-1 results in TGF- | [beta] | induced MMP-9 expression and confers a profibrotic phenotype to human lung fibroblasts. *Lab. Investig.* **2011**, *91*, 1206–1218. [CrossRef] [PubMed]
12. Shan, Y.Q.; Ying, R.C.; Zhou, C.H.; Zhu, A.K.; Ye, J.; Zhu, W.; Ju, T.F.; Jin, H.C. MMP-9 is increased in the pathogenesis of gastric cancer by the mediation of HER2. *Cancer Gene Ther.* **2015**, *22*, 101–107. [CrossRef] [PubMed]
13. Fimognari, C.; Hrelia, P. Sulforaphane as a promising molecule for fighting cancer. *Mutat. Res.* **2007**, *635*, 90–104. [CrossRef] [PubMed]

14. Gamet-Payrastre, L.; Li, P.; Lumeau, S.; Cassar, G.; Tercé, F. Sulforaphane, a Naturally Occurring Isothiocyanate, Induces Cell Cycle Arrest and Apoptosis in HT29 Human Colon Cancer Cells. *Cancer Res.* **2000**, *60*, 1426–1433. [PubMed]
15. Kim, H. Sulforaphane Sensitizes Tumor Necrosis Factor–Related Apoptosis-Inducing Ligand (TRAIL)–Resistant Hepatoma Cells to TRAIL-Induced Apoptosis through Reactive Oxygen Species–Mediated Up-regulation of DR5. *Cancer Res.* **2016**, *66*, 1740–1750. [CrossRef] [PubMed]
16. Xia, Y.; Kang, T.W.; Jung, D.Y.; Zhang, C.; Lian, S. Sulforaphane Inhibits Nonmuscle Invasive Bladder Cancer Cells Proliferation through Suppression of HIF-1α-Mediated Glycolysis in Hypoxia. *J. Agric. Food Chem.* **2019**, *67*, 7844–7854. [CrossRef]
17. Yanaka, A. Sulforaphane Enhances Protection and Repair of Gastric Mucosa against Oxidative Stress via Nrf2-Dependent Mechanisms. *Front. Gastrointest. Res.* **2012**, *30*, 170–180.
18. Kumar, B.; Koul, S.; Khandrika, L.; Meacham, R.B.; Koul, H.K. Oxidative stress is inherent in prostate cancer cells and is required for aggressive phenotype. *Cancer Res.* **2008**, *68*, 1777–1785. [CrossRef]
19. Schmitt, D.C.; Silva, L.M.D.; Zhang, W.; Liu, Z.; Tan, M. ErbB2-intronic MicroRNA-4728: A novel tumor suppressor and antagonist of oncogenic MAPK signaling. *Cell Death Dis.* **2015**, *6*, e1742. [CrossRef]
20. Dhillon, A.S.; Hagan, S.; Rath, O.; Kolch, W. MAP kinase signalling pathways in cancer. *Oncogene* **2007**, *26*, 3279–3290. [CrossRef]
21. Matthews, C.P.; Colburn, N.H.; Young, M.R. AP-1 a target for cancer prevention. *Curr. Cancer Drug Targets* **2007**, *7*, 317–324. [CrossRef] [PubMed]
22. Baeuerle, P.A.; Baltimore, D. NF-κB: Ten Years After. *Cell* **1996**, *87*, 13–20. [CrossRef]
23. Hartgrink, H.H.; Jansen, E.P.M.; Grieken, N.C.T.V.; Velde, C.J.H.V.D. Gastric cancer. *Lancet* **2009**, *374*, 477–490. [CrossRef]
24. Ahmad, R.; Khan, M.A.; Srivastava, A.N.; Gupta, A.; Srivastava, A.K. Anticancer Potential of Dietary Natural Products: A Comprehensive Review. *Anti-Cancer Agents Med. Chem. (Former. Curr. Med. Chem. Anti-Cancer Agents)* **2019**, *19*, 122–236. [CrossRef] [PubMed]
25. Vanduchova, A.; Anzenbacher, P.; Anzenbacherova, E. Isothiocyanate from Broccoli, Sulforaphane, and Its Properties. *J. Med. Food* **2018**, *22*, 121–126. [CrossRef] [PubMed]
26. Marion, N. Broccoli Sprouts in Cancer Prevention. *Nutr. Rev.* **2010**, *4*, 127–130.
27. Aumeeruddy, M.Z.; Mahomoodally, M.F. Combating Breast Cancer using Combination Therapy of three phytochemicals: Piperine, Sulforaphane, and Thymoquinone. *Cancer* **2019**, *125*, 1600–1611. [CrossRef]
28. Li, Y.; Zhang, T.; Korkaya, H.; Liu, S.; Lee, H.F.; Newman, B.; Yu, Y.; Clouthier, S.G.; Schwartz, S.J.; Wicha, M.S. Sulforaphane, a Dietary Component of Broccoli/Broccoli Sprouts, Inhibits Breast Cancer Stem Cells. *Clin. Cancer Res.* **2010**, *16*, 2580. [CrossRef]
29. Chen, Y.; Wang, M.H.; Zhu, J.Y.; Xie, C.F.; Zhong, C.Y. TAp63α targeting of Lgr5 mediates colorectal cancer stem cell properties and sulforaphane inhibition. *Oncogenesis* **2020**, *9*, 89. [CrossRef]
30. Rutz, J.; Thaler, S.; Maxeiner, S.; Chun, F.K.; Blaheta, R.A. Sulforaphane Reduces Prostate Cancer Cell Growth and Proliferation In Vitro by Modulating the Cdk-Cyclin Axis and Expression of the CD44 Variants 4, 5, and 7. *Int. J. Mol. Sci.* **2020**, *21*, 8724. [CrossRef]
31. Wang, Y.; Wu, H.; Dong, N.; Su, X.; Duan, M.; Wei, Y.; Wei, J.; Liu, G.; Peng, Q.; Zhao, Y. Sulforaphane induces S-phase arrest and apoptosis via p53-dependent manner in gastric cancer cells. *Sci. Rep.* **2021**, *11*, 2504. [CrossRef] [PubMed]
32. Dash, P.K.; Zhao, J.; Orsi, S.A.; Min, Z.; Moore, A.N. Sulforaphane improves cognitive function administered following traumatic brain injury. *Neurosci. Lett.* **2009**, *460*, 103–107. [CrossRef] [PubMed]
33. Li, Z.; Guo, H.; Li, J.; Ma, T.; Cai, L. Sulforaphane prevents type 2 diabetes-induced nephropathy via AMPK-mediated activation of lipid metabolic pathways and Nrf2 anti-oxidative function. *Clin. Sci.* **2020**, *134*, 2469–2487. [CrossRef] [PubMed]
34. Woo, K.J.; Kwon, T.K. Sulforaphane suppresses lipopolysaccharide-induced cyclooxygenase-2 (COX-2) expression through the modulation of multiple targets in COX-2 gene promoter. *Int. Immunopharmacol.* **2007**, *7*, 1776–1783. [CrossRef]
35. Vineis, P.; Alavanja, M.; Buffler, P.; Fontham, E.; Franceschi, S.; Gao, Y.T.; Gupta, P.C.; Hackshaw, A.; Matos, E.; Samet, J. Tobacco and Cancer: Recent Epidemiological Evidence. *J. Natl. Cancer Inst.* **2004**, *96*, 99–106. [CrossRef]
36. Prochaska, J.J.; Benowitz, N.L. Current advances in research in treatment and recovery: Nicotine addiction. *Sci. Adv.* **2019**, *5*, eaay9763. [CrossRef]
37. Benowitz, N.L. Pharmacology of nicotine: Addiction, smoking-induced disease, and therapeutics. *Annu. Rev. Pharmacol.* **2009**, *49*, 57–71. [CrossRef]
38. Lian, S.; Li, S.; Zhu, J.; Xia, Y.; Do Jung, Y. Nicotine stimulates IL-8 expression via ROS/NF-κB and ROS/MAPK/AP-1 axis in human gastric cancer cells. *Toxicology* **2022**, *466*, 153062. [CrossRef]
39. Sundaram, G.M.; Quah, S.; Sampath, P. Cancer: The dark side of wound healing. *FEBS J.* **2018**, *285*, 4516–4534. [CrossRef]
40. Caley, M.P.; Martins, V.L.; O'Toole, E.A. Metalloproteinases and Wound Healing. *Adv. Wound Care* **2015**, *4*, 225–234. [CrossRef]
41. Reiss, M.J.; Han, Y.P.; Garcia, E.; Goldberg, M.; Yu, H.; Garner, W.L. Matrix metalloproteinase-9 delays wound healing in a murine wound model. *Surgery* **2010**, *147*, 295–302. [CrossRef] [PubMed]
42. Gopalakrishna, R.; Jaken, S. Protein kinase C signaling and oxidative stress. *Free Radic. Biol. Med.* **2000**, *28*, 1349–1361. [CrossRef]
43. Joo, J.H.; Oh, H.; Kim, M.; An, E.J.; Kim, R.K.; Lee, S.Y.; Kang, D.H.; Kang, S.W.; Park, C.K.; Kim, H. NADPH Oxidase 1 Activity and ROS Generation Are Regulated by Grb2/Cbl-Mediated Proteasomal Degradation of NoxO1 in Colon Cancer Cells. *Cancer Res.* **2016**, *76*, 855–865. [CrossRef] [PubMed]
44. Ewald, C. Redox Signaling of NADPH Oxidases Regulates Oxidative Stress Responses, Immunity and Aging. *Antioxidants* **2018**, *7*, 130. [CrossRef] [PubMed]

45. Wu, X.; Zhang, H.; Qi, W.; Zhang, Y.; Li, J.; Li, Z.; Lin, Y.; Bai, X.; Liu, X.; Chen, X. Nicotine promotes atherosclerosis via ROS-NLRP3-mediated endothelial cell pyroptosis. *Cell Death Dis.* **2018**, *9*, 171. [CrossRef] [PubMed]
46. Belambri, S.A.; Rolas, L.; Raad, H.; Hurtado-Nedelec, M.; El-Benna, J. NADPH oxidase activation in neutrophils: Role of the Phosphorylation of its subunits. *Eur. J. Clin. Investig.* **2018**, *48* (Suppl. 2), e12951. [CrossRef]
47. Chakraborti, S.; Sarkar, J.; Chakraborti, T. Role of PLD-PKCζ signaling axis in p47phox phosphorylation for activation of NADPH oxidase by angiotensin II in pulmonary artery smooth muscle cells. *Cell Biol. Int.* **2019**, *43*, 678–694. [CrossRef]
48. Asano, H.; Horinouchi, T.; Mai, Y.; Sawada, O.; Fujii, S.; Nishiya, T.; Minami, M.; Katayama, T.; Iwanaga, T.; Terada, K. Nicotine- and tar-free cigarette smoke induces cell damage through reactive oxygen species newly generated by PKC-dependent activation of NADPH oxidase. *J. Pharmacol. Sci.* **2012**, *118*, 275. [CrossRef]
49. Yuting, W.; Kumar, M.A.; Young-Ok, S.; Poyil, P.; Wise, J.T.F.; Lei, W.; Zhuo, Z.; Xianglin, S.; Zhimin, C. Roles of ROS, Nrf2, and autophagy in cadmium-carcinogenesis and its prevention by sulforaphane. *Toxicol. Appl. Pharmacol.* **2018**, *353*, 23–30.
50. Subedi, L.; Lee, J.H.; Yumnam, S.; Ji, E.; Kim, S.Y. Anti-Inflammatory Effect of Sulforaphane on LPS-Activated Microglia Potentially through JNK/AP-1/NF-κB Inhibition and Nrf2/HO-1 Activation. *Cells* **2019**, *8*, 194. [CrossRef]
51. Shan, Y.; Zhang, L.; Bao, Y.; Li, B.; He, C.; Gao, M.; Feng, X.; Xu, W.; Zhang, X.; Wang, S. Epithelial-mesenchymal transition, a novel target of sulforaphane via COX-2/MMP2, 9/Snail, ZEB1 and miR-200c/ZEB1 pathways in human bladder cancer cells. *J. Nutr. Biochem.* **2013**, *24*, 1062–1069. [CrossRef] [PubMed]
52. Jin, C.Y.; Molagoda, I.M.N.; Karunarathne, W.A.H.M.; Kang, S.H.; Park, C.; Kim, G.Y.; Choi, Y.H. TRAIL attenuates sulforaphane-mediated Nrf2 and sustains ROS generation, leading to apoptosis of TRAIL-resistant human bladder cancer cells. *Toxicol. Appl. Pharmacol.* **2018**, *132*, 132–141. [CrossRef]
53. Sweeney, K. EGF receptor activation by heterologous mechanisms. *Cancer Cell* **2002**, *1*, 405–406.
54. Lian, S.; Xia, Y.; Khoi, P.N.; Ung, T.T.; Yoon, H.J.; Kim, N.H.; Kim, K.K.; Jung, Y.D. Cadmium induces matrix metalloproteinase-9 expression via ROS-dependent EGFR, NF-small ka, CyrillicB, and AP-1 pathways in human endothelial cells. *Toxicology* **2015**, *338*, 104–116. [CrossRef] [PubMed]
55. Chen, H.W.; Chao, C.Y.; Lin, L.L.; Lu, C.Y.; Liu, K.L.; Lii, C.K.; Li, C.C. Inhibition of matrix metalloproteinase-9 expression by docosahexaenoic acid mediated by heme oxygenase 1 in 12-O-tetradecanoylphorbol-13-acetate-induced MCF-7 human breast cancer cells. *Arch. Toxicol.* **2013**, *87*, 857–869. [CrossRef] [PubMed]
56. Hwang, Y.P.; Yun, H.J.; Kim, H.G.; Han, E.H.; Lee, G.W.; Jeong, H.G. Suppression of PMA-induced tumor cell invasion by dihydroartemisinin via inhibition of PKCα/Raf/MAPKs and NF-κB/AP-1-dependent mechanisms. *Biochem. Pharmacol.* **2010**, *79*, 1714–1726. [CrossRef] [PubMed]
57. Lin, C.-W.; Hou, W.-C.; Shen, S.-C.; Juan, S.-H.; Ko, C.-H.; Wang, L.-M.; Chen, Y.-C. Quercetin inhibition of tumor invasion via suppressing PKCδ/ERK/AP-1-dependent matrix metalloproteinase-9 activation in breast carcinoma cells. *Carcinogenesis* **2008**, *29*, 1807–1815. [CrossRef]
58. Lan, H.; Yuan, H.; Lin, C. Sulforaphane induces p53deficient SW480 cell apoptosis via the ROSMAPK signaling pathway. *Mol. Med. Rep.* **2017**, *16*, 7796–7804. [CrossRef]
59. Rai, R.; Essel, K.; Benbrook, D.M.; Garland, J.; Chandra, V. Preclinical Efficacy and Involvement of AKT, mTOR, and ERK Kinases in the Mechanism of Sulforaphane against Endometrial Cancer. *Cancers* **2020**, *12*, 1273. [CrossRef] [PubMed]
60. Yuan, M.; Meng, W.; Liao, W.; Lian, S. Andrographolide Antagonizes TNF-α-induced IL-8 via Inhibition of NADPH Oxidase/ROS/NF-κB and Src/MAPKs/AP-1 Axis in Human Colorectal Cancer HCT116 Cells. *J. Agric. Food Chem.* **2018**, *66*, 5139–5148. [CrossRef]
61. Lian, S.; Xia, Y.; Ung, T.T.; Khoi, P.N.; Yoon, H.J.; Lee, S.G.; Kim, K.K.; Jung, Y.D. Prostaglandin E2 stimulates urokinase-type plasminogen activator receptor via EP2 receptor-dependent signaling pathways in human AGS gastric cancer cells. *Mol. Carcinog.* **2017**, *56*, 664–680. [CrossRef] [PubMed]
62. Khoi, P.N.; Park, J.S.; Kim, N.H.; Jung, Y.D. Nicotine stimulates urokinase-type plasminogen activator receptor expression and cell invasiveness through mitogen-activated protein kinase and reactive oxygen species signaling in ECV304 endothelial cells. *Toxicol. Appl. Pharmacol.* **2012**, *259*, 248–256. [CrossRef] [PubMed]
63. Tanaka, K.; Babic, I.; Nathanson, D.; Akhavan, D.; Guo, D. Oncogenic EGFR signaling activates an mTORC2-NF-κB pathway that promotes chemotherapy resistance. *Cancer Discov.* **2011**, *1*, 524–538. [CrossRef] [PubMed]
64. Xia, Y.; Yuan, M.; Li, S.; Trong, T.U.; Thinh, N.T.; Kang, T.W.; Liao, W.; Lian, S.; Young, J.D. Apigenin Suppresses the IL-1β-Induced Expression of the Urokinase-Type Plasminogen Activator Receptor by Inhibiting MAPK-Mediated AP-1 and NF-κB Signaling in Human Bladder Cancer T24 Cells. *J. Agric. Food Chem.* **2018**, *66*, 7663–7673. [CrossRef]
65. Li, S.; Nguyen, T.T.; Ung, T.T.; Sah, D.K.; Park, S.Y.; Lakshmanan, V.K.; Jung, Y.D. Piperine Attenuates Lithocholic Acid-Stimulated Interleukin-8 by Suppressing Src/EGFR and Reactive Oxygen Species in Human Colorectal Cancer Cells. *Antioxidants* **2022**, *11*, 530. [CrossRef]
66. Li, S.; Ung, T.T.; Nguyen, T.T.; Sah, D.K.; Park, S.Y.; Jung, Y.D. Cholic Acid Stimulates MMP-9 in Human Colon Cancer Cells via Activation of MAPK, AP-1, and NF-kappaB Activity. *Int. J. Mol. Sci.* **2020**, *21*, 3420. [CrossRef]

International Journal of
Molecular Sciences

Article

Pancreatic Ductal Cell-Derived Extracellular Vesicles Are Effective Drug Carriers to Enhance Paclitaxel's Efficacy in Pancreatic Cancer Cells through Clathrin-Mediated Endocytosis

Haoyao Sun [1,2], Kritisha Bhandari [2], Stephanie Burrola [2], Jinchang Wu [1,*] and Wei-Qun Ding [2,*]

1 Department of Radiation Oncology, The Affiliated Suzhou Hospital of Nanjing Medical University, Suzhou 215001, China; sun464209459@gmail.com
2 Department of Pathology, University of Oklahoma Health Sciences Center, Oklahoma City, OK 73104, USA; kritisha-bhandari@ouhsc.edu (K.B.); stephaine-burrola@ouhsc.edu (S.B.)
* Correspondence: wjinchang@sina.com (J.W.); weiqun-ding@ouhsc.edu (W.-Q.D.); Tel.: +86-1377-604-8328 (J.W.); +1-405-271-1605 (W.-Q.D.)

Abstract: Chemo-resistance challenges the clinical management of pancreatic ductal adenocarcinoma (PDAC). A limited admittance of chemotherapeutics to PDAC tissues is a key obstacle in chemotherapy of the malignancy. An enhanced uptake of drugs into PDAC cells is required for a more effective treatment. Extracellular vesicles (EVs), especially small EVs (sEVs), have emerged as drug carriers for delivering chemotherapeutics due to their low immunogenicity and propensity for homing toward tumor cells. The present study evaluated sEVs derived from six different human cell lines as carriers for paclitaxel (PTX). The encapsulation of the chemotherapeutics was achieved using incubation, sonication and electroporation. The cytotoxicity of the EV drugs was evaluated by MTS assay. While sonication led to a higher efficiency of drug loading than incubation and electroporation, PTX loaded through incubation with HPNE-derived sEVs (HI-PTX) was the most efficacious in killing PDAC cells. Furthermore, HI-PTX was taken up by PDAC cells more efficiently than other EV drugs, implying that the efficacy of HI-PTX is associated with its efficient uptake. This was supported by the observation that the cytotoxicity and uptake of HI-PTX is mediated via the clathrin-dependent endocytosis. Our results indicate that the hTERT-HPNE cell-derived EVs are effective drug carriers to enhance paclitaxel's efficacy in PDAC cells.

Keywords: extracellular vesicles; pancreatic cancer; paclitaxel; clathrin; endocytosis

Citation: Sun, H.; Bhandari, K.; Burrola, S.; Wu, J.; Ding, W.-Q. Pancreatic Ductal Cell-Derived Extracellular Vesicles Are Effective Drug Carriers to Enhance Paclitaxel's Efficacy in Pancreatic Cancer Cells through Clathrin-Mediated Endocytosis. *Int. J. Mol. Sci.* **2022**, *23*, 4773. https://doi.org/10.3390/ijms23094773

Academic Editor: Angela Stefanachi

Received: 6 April 2022
Accepted: 22 April 2022
Published: 26 April 2022

1. Introduction

Pancreatic ductal adenocarcinoma (PDAC) is a devastating disease, with its 5-year overall survival rate being less than 10% [1]. The high mortality rate of PDAC is related to the fact that the majority of pancreatic cancer patients have their tumor already metastasized at the time of diagnosis [2], making systemic therapy the mainstay of treatment. For those with advanced or metastasized tumors, chemotherapy is one of the most effective regimens recommended by the National Comprehensive Cancer Network guidelines [3]. Traditional chemotherapeutics are mainly small-molecule cytotoxic drugs that have distinct pharmacological profiles [4]. Upon intravenous administration, these drugs are passively distributed in the body via the bloodstream. Most of the drugs will gather in the liver and few can reach the tumor sites. Furthermore, drug distribution in the body mainly depends on the passive diffusion of the concentration gradient in the tumor microenvironment. It has been reported in the three-dimensional cell model that the penetration rate of small-molecule anti-tumor drugs in tumor tissue is only about 5% [5]. The limited admittance of chemotherapeutics to PDAC tissues in vivo is even more evident due to the unique stroma composition of PDAC that is histologically manifested as desmoplasia [6]. This pharmacological kinetic feature of chemotherapeutics causes traditional chemotherapy drugs to have serious side effects with limited potential of dose elevation in PDAC patients.

To overcome this challenge, new chemo drug carriers have been developed to improve the pharmacokinetics, increase tumor-targeting efficacy and reduce the side effects of chemotherapeutics [7–11]. One of the recently studied natural Nano drug-carriers are extracellular vesicles (EVs), which are small lipid bilayer-delimited particles generated through various cellular processes and released from all types of cells investigated. They are able to transfer genetic and cellular materials between different cell types to mediate intercellular communication [12,13]. The most attractive properties of human cell-derived EVs as drug carriers include their lower immunogenicity and higher capacity of homing toward tumor cells [14–17]. Results from 12 recent clinical trials testing small EVs (sEVs) as therapeutic carriers or potential cancer therapeutics demonstrated safe profiles of sEVs delivered in humans, supporting the development of human cell-line-derived sEVs as chemotherapeutic carriers [18]. However, questions remain to be answered as to the choice of EV sources, the strategies of drug encapsulation, and the mechanisms of the cellular uptake of the delivered EV drugs.

The present study was driven by the above-mentioned questions. We compared the drug loading efficiency of sEVs derived from six different human cell lines, including the human normal pancreatic duct cell line hTERT-HPNE, the human embryonic kidney cell line HEK-293T, the cancer-associated fibroblast cell line CAF19 and three human PDAC cell lines, PANC-1, MIA PaCa-2, and BxPC-3. Direct incubation, sonication and electroporation of sEVs were applied to incorporate paclitaxel (PTX) or gemcitabine (GEM), two commonly used chemotherapeutics. The sEV-drug efficacy was tested in the PDAC cell lines PANC-1, MIA PaCa-2, and BxPC-3. Our results show that while sonication leads to a higher efficiency of drug loading than incubation and electroporation, sEV derived from hTERT-HPNE cells and incubated with PTX (HI-PTX) was the most efficacious in killing PDAC cells. By using specific endocytosis pathway inhibitors and gene manipulation techniques, we demonstrated that the increased cellular cytotoxicity of HI-PTX is associated with enhanced cellular uptake of the sEV-drug complex through clathrin-mediated endocytosis.

2. Results

2.1. Characterization of sEVs and Drug Encapsulation

Small EVs were isolated from culture medium of the six human cell lines as we described [19]. Protein analysis of isolated sEVs showed that hTERT-HPNE sEV has the lowest protein concentration, while MIA PaCa-2 sEV the highest. Cancer-cell-derived sEVs seemed to have a higher EV concentration than normal-cell-derived sEVs (Figure 1a). EV particle quantity derived from the six cell lines also differed, but not as much as the protein concentrations (Figure 1b). Two sEV-positive markers, flotillin-1 and CD63, and one negative marker, calnexin, were detected by Western blot to verify the isolated sEVs (Figure 1c). All sEVs had iconic peaks around 100–200 nm, based on nanoparticle tracking analysis (Figure 1d). To quantify EV-drug concentration, both EV drugs and sEVs were analyzed by Nanodrop in the UV–Vis spectrum. Standard curves of free PTX, GEM and sEVs derived from the six cell lines were generated separately at 230 or 275 nm (Figure S1). The quantity of loaded drugs was expressed as ng of drug/μg of sEV. Our data showed that sonication leads to a higher concentration of drug loading than incubation and electroporation (Figure 1e).

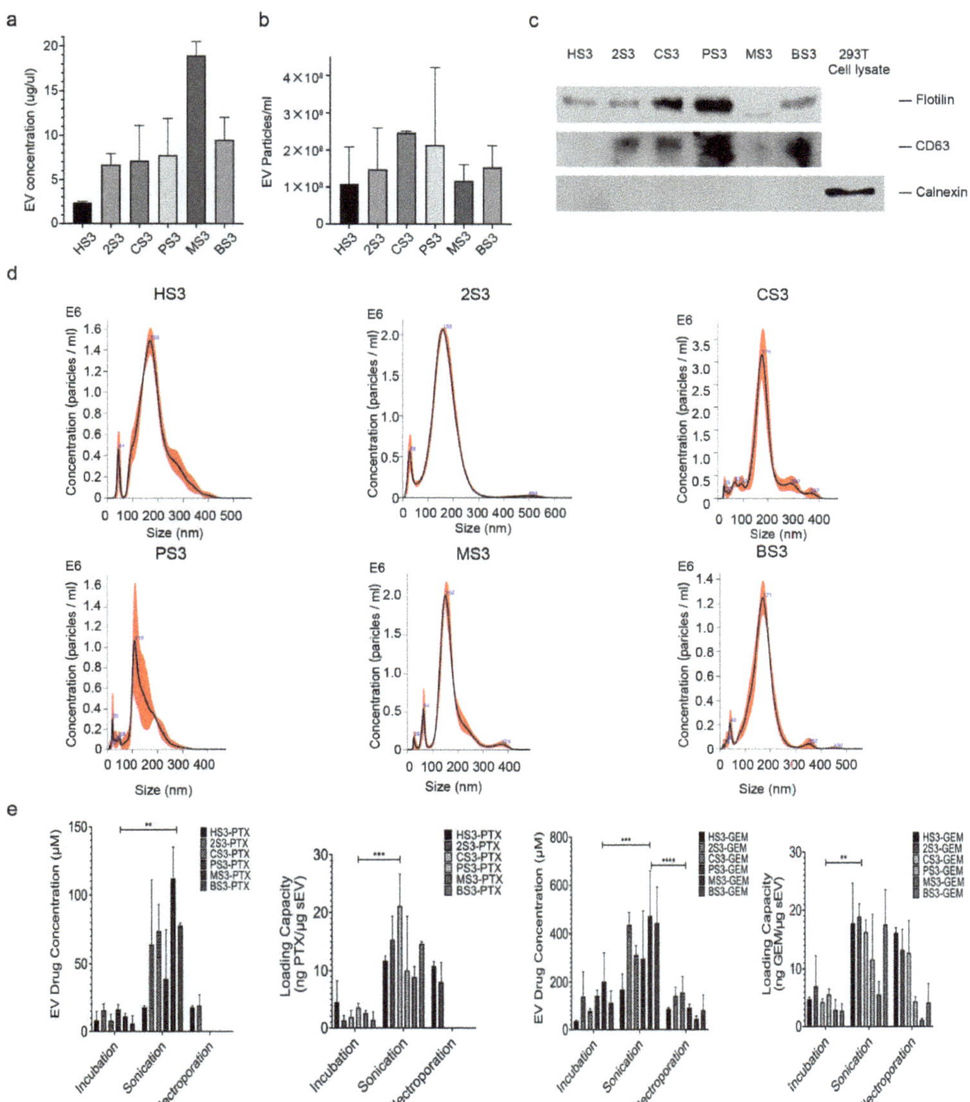

Figure 1. Validation of extracellular vesicles and quantification of EV drugs. (**a**) EVs were derived from 20 mL medium of 6 different human cell lines, and EV concentrations were analyzed by BCA assay; (**b**) EV particle analysis by NanoSight (1:1000 dilution); (**c**) Small EV-positive markers, CD63 and Flotillin-1, and a negative marker, Calnexin, detected by Western blot; (**d**) Representative EV size distribution and particle numbers analyzed by Nanosight; (**e**) Sonication leads to higher efficiency of EV drug loading than incubation and electroporation. **** $p < 0.0001$, *** $p < 0.001$, ** $p < 0.01$, one-way ANOVA (n = 3–5, comparison of Incubation, Sonication, and Electroporation). HS3: HPNE sEV; 2S3: HEK-293T sEV; CS3: CAF19 sEV; PS3: PANC-1 sEV; MS3: MIA PaCa-2 sEV; BS3: BxPC-3 sEV; PTX: Paclitaxel; GEM: Gemcitabine.

2.2. HPNE sEV-PTX Derived from Incubation (HI-PTX) Is most Efficacious in Killing Pancreatic Cancer Cells

To test whether the EV-PTX are cytotoxic toward cancer cells, PANC-1, MIA PaCa-2 and BxPC-3 cells were seeded in 96-well plates and treated with free or EV drugs for 24–72 h. MTS assay was applied to assess cell cytotoxicity as we previously described [20,21]. As expected, PTX and GEM suppressed cell viability in a time- and concentration-dependent manner, with IC_{50} values ranging from 10 to 100 nM in the three cancer cell lines (data not shown). We then treated the cells using EV drugs or free drugs with the same drug concentrations ranging from 1 to 1000 nM for 72 h. While five EV-PTX drugs showed lower IC_{50} values than that of free PTX, only HI-PTX showed consistent cytotoxicity in all three cancer cell lines (Figure 2a–f). In contrast to EV-PTX, by following the same protocol and procedures, no EV-GEM drugs were more cytotoxic than free GEM in PANC-1 cells (Figure 2g,h). These results indicate that the efficacy of EV-encapsulated chemotherapeutics is associated with the loading methods, EV sources, and the drug of interest.

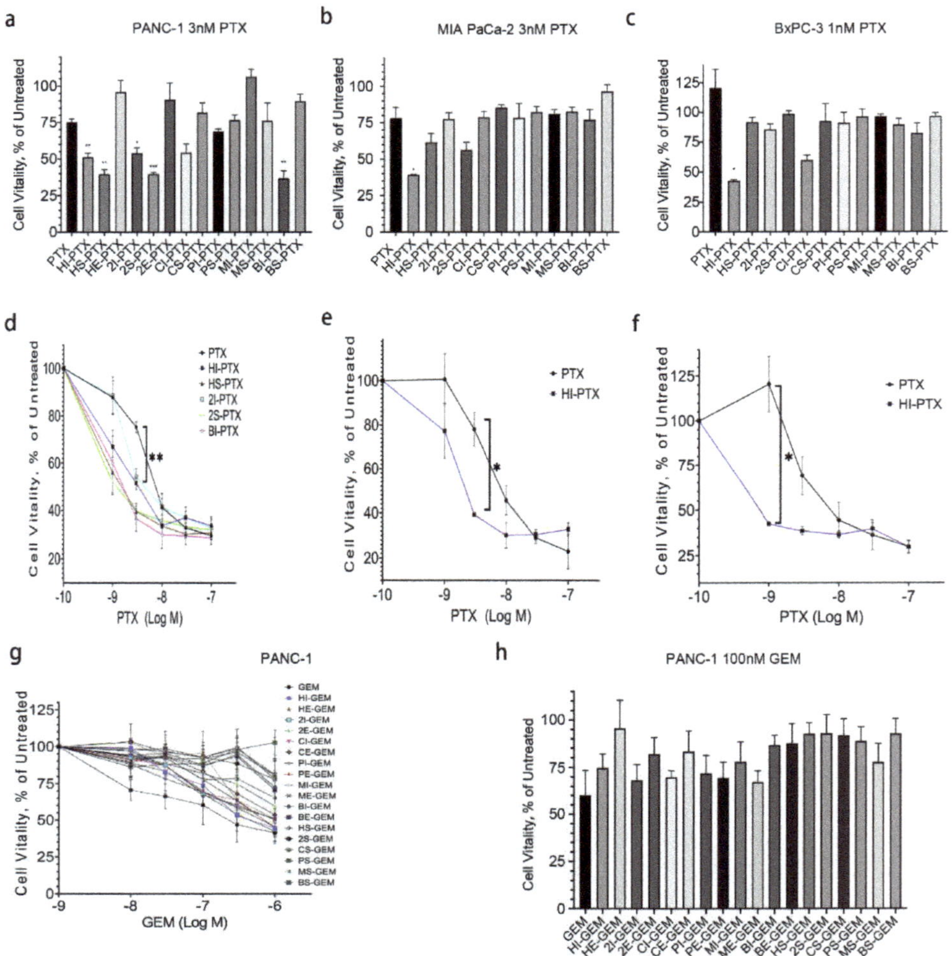

Figure 2. HI-PTX is more cytotoxic than other EV drugs. (**a**,**d**) Cell vitality of PANC−1 cells treated with equivalent of 3 nM PTX and 1 to 100 nM EV−PTX for 72 h; (**b**,**e**) Cell vitality of MIA PaCa−2 cells

treated with equivalent of 3 nM PTX and 1 to 100 nM EV−PTX for 72 h; (**c**,**f**) Cell vitality in BxPC−3 cells treated with equivalent of 1 nM PTX or 1 to 100 nM EV−PTX for 72 h; (**g**,**h**) Cell vitality of PANC−1 cells treated with equivalent of 100 nM GEM or 10 nM to 1 μM EV−GEM for 72 h. The cytotoxicity of all EV−GEM was lower than that of free GEM. Statistical analysis was performed using one-way ANOVA followed by Dunnett's post-test for (**a**) and two-way ANOVA for (**d**–**f**). *** $p < 0.001$, ** $p < 0.01$, * $p < 0.05$ (data from three individual determinations). HI-PTX: Incubation of HPNE sEV with paclitaxel; HS−PTX: Sonication of HPNE sEV with paclitaxel; HE−PTX: Electroporation of HPNE sEV with paclitaxel; 2I-PTX: Incubation of HEK−293T sEV with paclitaxel; 2S−PTX: Sonication of HEK−293T sEV with paclitaxel; 2E−PTX: Electroporation of HEK−293T sEV with paclitaxel; CI-PTX: Incubation of CAF19 sEV with paclitaxel; CS−PTX: Sonication of CAF19 sEV with paclitaxel; PI−PTX: Incubation of PANC−1 sEV with paclitaxel; PS−PTX: Sonication of PANC−1 sEV with paclitaxel; MI−PTX: Incubation of MIA PaCa−2 sEV with paclitaxel; MS−PTX: Sonication of MIA PaCa−2 sEV with paclitaxel; BI−PTX: Incubation of BxPC−3 sEV with paclitaxel; BS−PTX: Sonication of BxPC−3 sEV with paclitaxel.

2.3. Uptake of HI-PTX in PANC-1 Cells

To understand why cytotoxicity differs among the EV drugs, we examined EV-drug uptake using the PKH67 dye under a fluorescent microscope. We found that both HPNE sEVs or HI-PTX uptake by PANC-1 cells were time-dependent, with the highest uptake at 10 h post sEV addition. Interestingly, HI-PTX uptake was more pronounced than HPNE sEV uptake at each time point (Figure 3a). To exclude the potential effects of PTX on fluorescent imaging, we compared the uptake of HPNE sEV versus HPNE sEV plus free PTX and found no differences in any of the three cancer cell lines between the two groups of sEVs (Figure 3b). To determine whether the method of drug encapsulation affects the uptake, we compared the uptake of sEVs derived from HPNE and HEK-293T cells and prepared with PTX via sonication, electroporation, and incubation. It turned out that HI-PTX had the highest uptake of the EV drugs in PANC-1 and BxPC-3 cells (Figure 3c), consistent with its more pronounced cytotoxicity. While there was also a tendency for increased uptake of HI-PTX in MIA PaCa-2 cells, statistical significance could not be reached. These results suggest a connection between cell cytotoxicity and uptake of the EV drugs.

2.4. HI-PTX'S Uptake and Cytotoxicity Is Associated with Clathrin-Mediated Endocytosis

To explore the mechanism of HI-PTX uptake and cytotoxicity in pancreatic cancer cells, several endocytic pathway inhibitors were applied. The higher uptake of HI-PTX was diminished by the inhibitors Monesin, Bafilomycin A1 (BFA), and clathrin-mediated endocytosis inhibitor Pitstop2. To the contrary, the caveolin-mediated endocytosis inhibitor Genistein had no effect on HI-PTX uptake (Figure 4a). However, among the inhibitors tested, only Pitstop2 could reverse the cytotoxicity of HI-PTX in PANC-1 cells when 3 nM PTX equivalent concentration of HI-PTX was applied (Figure 4b), indicating that the clathrin-mediated endocytosis is primarily involved in HI-PTX's uptake and cytotoxicity. To further confirm the contribution of clathrin-mediated endocytosis to HI-PTX uptake, overexpression and siRNA knockdown of clathrin light chain and caveolin was achieved in PANC-1 cells (Figure 4c,d). The most successful knockdown of clathrin light chain was obtained using Si-CLTB-93, which was used for subsequent experiments (Figure 4e,f). As shown in Figure 4e, EV uptake was associated with the expression levels of clathrin, not that of caveolin, for both HPNE sEVs and the HI-PTX. These observations support the conclusion that HI-PTX uptake is facilitated, at least in part, by clathrin-mediated endocytosis in pancreatic cancer cells. To make sure the knockdown of clathrin light chain impairs clathrin-mediated endocytosis, an RFP-tagged transferrin receptor construct (TfR-pHuji plasmid [22] (Addgene Plasmid #61505) was used to monitor cellular localization of the transferrin receptor during siRNA knockdown (Supplemental Figure S2). It confirmed that the knockdown leads to more transferrin receptor on the cell surface. This is consistent with previous reports showing that the depletion of clathrin light chain effectively inhibits the clathrin-mediated internalization of cargos such as bacteria and virus particles that are too large for conventional endocytosis [23,24] and that exosomes or small EVs share physical properties and size ranges with viral particles [25].

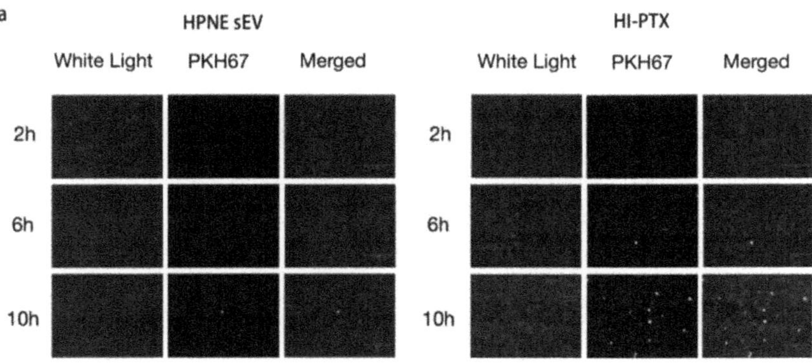

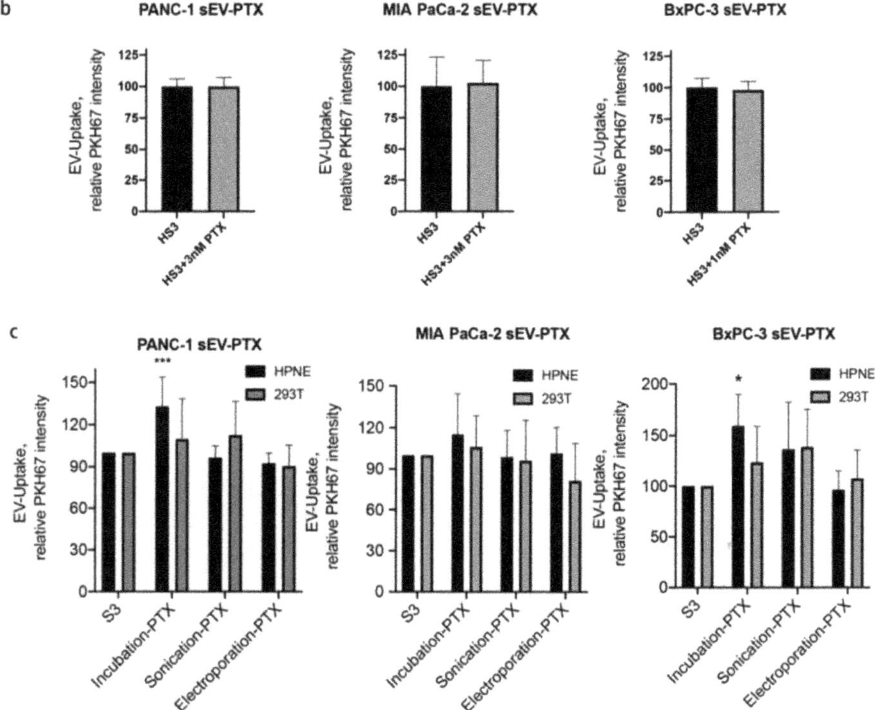

Figure 3. HI-PTX is taken up more effectively by pancreatic cancer cells. (**a**) Equal amounts of HPNE-sEV and HI-PTX were labeled with PKH67 and loaded onto PANC-1 cells for 2 to 10 h. Cellular uptake of HPNE-sEV and HI-PTX was monitored by fluorescent microscopy (excitation 488 nm, detection, 510 nm; Nikon TE2000-E, 10× magnification); (**b**) Adding HPNE-sEV and PTX simultaneously to PANC-1, MIA PaCa-2, and BxPC-3 cells did not enhance EV uptake. The fluorescence was detected by fluorescent microscopy and the fluorescence intensity was semi quantified, and presented as relative levels; (**c**) The uptake of HPNE- and 293T-derived EV, when loaded with PTX by incubation, sonication and electroporation, in the three pancreatic cancer cell lines. *** $p < 0.001$, * $p < 0.05$, one-way ANOVA followed by Dunnett's post-test (data from three individual determinations).

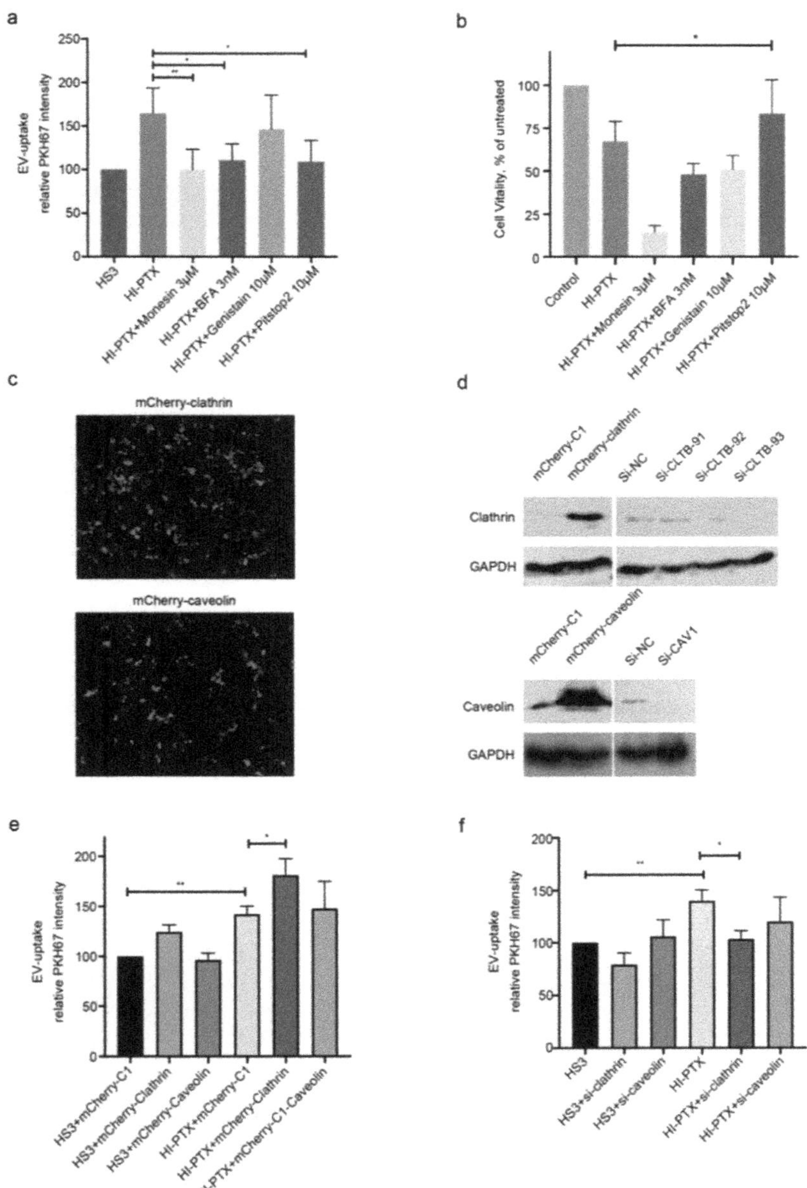

Figure 4. HI-PTX's uptake and cytotoxicity is associated with clathrin-mediated endocytosis. (**a**) The clathrin-mediated endocytosis inhibitor Pitstop2 inhibited the uptake of HI-PTX by PANC-1 cells; (**b**) Treatment with Pitstop2 attenuated HI-PTX's cytotoxicity in PANC-1 cells; (**c**) Overexpression of clathrin and caveolin (Nikon TE2000-E, 10× magnification); (**d**) Knockdown of clathrin and caveolin expression in PANC-1 cells, evidenced by fluorescent images and Western blot analysis, mCherry-clathrin, 63 kDa; clathrin, 31 kDa; mCherry-caveolin, 53 kDa; caveolin, 22 kDa; (**e**,**f**) The uptake efficiency of HI-PTX was altered by overexpression or knockdown of clathrin in pancreatic cancer cells. ** $p < 0.01$, * $p < 0.05$, one-way ANOVA followed by Dunnett's post-test (data from three individual determinations).

3. Discussion

Although the experimental evidence showing the effectiveness of EV drugs against cancer has been abundant, there has been no clear consensus regarding the choice of methods for EV encapsulation of drugs, the source of EVs as carriers, and the mechanisms of EV-drug internalization to achieve the best therapeutic effects. The results from the present study demonstrate that, while sonication leads to higher efficiency of PTX loading than incubation and electroporation, HI-PTX prepared by incubation is most efficacious in killing pancreatic cancer cells, observations in line with a recent report using EV-encapsulated doxorubicin (DOX) [26]. Furthermore, we demonstrated that the uptake and cytotoxicity of HI-PTX are associated with clathrin-mediated endocytosis in pancreatic cancer cells, implicating endocytosis pathways in EV-drug efficacy.

Cellular uptake of molecules larger than one kilo Dalton, such as proteins or nanoparticles, is usually facilitated by endocytic pathways [4]. It has been reported that EV uptake is mediated through various endocytic pathways, including clathrin-dependent endocytosis and clathrin-independent endocytosis, such as caveolin-mediated uptake and lipid raft-mediated internalization. Because EV populations are often heterogeneous, more than one route of uptake is generally involved during EVs internalization into cells [27]. For example, clathrin- and caveolin-dependent endocytosis and macropinocytosis are the predominant routes of sEV-mediated communication between bone marrow stromal cells and multiple myeloma cells, and the knocking down of calveolin-1 and clathrin heavy chain in multiple myeloma cells significantly suppressed sEV uptake and chemo sensitivity to bortezomib [28]. However, the endocytosis pathways involved in the uptake of drug-loaded EVs has not been previously established. Our experiment results showed that the cellular uptake of HI-PTX (PTX encapsulated by HPNE sEVs via incubation) is enhanced when compared to the uptake of HPNE sEVs in all three pancreatic cancer cell lines. This enhanced uptake of HI-PTX is attributed to clathrin-mediated endocytosis, since modulation of this process using the inhibitor Pitstop2 or by expression manipulation of clathrin altered the uptake of HI-PTX and its cytotoxicity, whereas the caveolin-dependent endocytosis seemed to be irrelevant in this process. Our observations thus provide novel information in the understanding of EV drug uptake and efficacy in cancer cells. The higher uptake of HI-PTX, when compared to EV-PTX prepared via sonication and electroporation, may be explained by the possibility that, compared to incubation, both sonication and electroporation are likely to trigger a harsh process to sEVs that cause damage of sEV membranes, thereby compromising their ability for cellular internalization [29]. Nonetheless, the mechanisms responsible for a higher uptake of HI-PTX, when compared to the uptake of HPNE sEVs, remain to be explored in the future.

Various sources of sEVs have been tested for their potential as therapeutic carriers, and each type of sEV may have pros and cons when used for drug delivery [30]. This is most likely due to their differences in the EV cargos in which each may have unique lipid, protein and RNA profiles that may directly influence sEVs' ability to interact with receiving cells [31]. In particular, tumor-cell-derived EVs were considered drug carriers for selective targeting and enhanced immune response, yet may ironically promote tumor growth and invasion, due to their cargo compositions [30]. In this context, a strategy using exosomes coated with magnetic nanoparticles to deliver chemotherapeutics specifically to tumor cells has been described [32–34]. However, the best EV sources have yet to be identified in the development of EVs as therapeutic carriers. In this study, we tested sEVs derived from six human cell lines, including cancer lines and non-cancer lines. We found that the sEVs derived from the human pancreatic ductal cell line HPNE (a non-cancer line) are the most efficacious when used to encapsulate PTX via incubation, suggesting that this group of sEVs is a promising candidate for further development as cancer therapeutic carriers. In vivo testing is warranted for the efficacy and safety of HI-PTX.

An interesting finding from this study was that in contrast to HI-PTX, no EV-GEM drugs showed superior cytotoxicity compared with free GEM in pancreatic cancer cells, suggesting that individual chemotherapeutics entails a different fate when encapsulated by sEVs. Specific to GEM, two research groups have reported that GEM is successfully loaded into autologous exosomes that suppress tumor growth [35,36]; however, others have also reported that GEM is inefficiently entrapped by nanoparticles due to leakage or hydrophilic property [37,38]. These inconsistent observations, along with ours, are most likely due to the nature of the sEVs used, and the procedures applied for sEV isolation and drug encapsulation.

In summary, we have demonstrated that HI-PTX is the most efficacious EV-PTX in suppressing pancreatic cancer cell viability, which is primarily mediated via the clathrin-dependent endocytosis pathway. Our data indicate that the efficacy of EV-encapsulated chemotherapeutics is associated with the loading methods, EV sources, and EV uptake efficiency.

4. Materials and Methods

Cell culture. The human pancreatic cancer cell lines PANC-1, MIA PaCa-2 and BxPC-3, the immortalized human pancreatic duct cell line HPNE, and the human embryonic kidney cell line HEK-293T were obtained from the American Type Culture Collection (ATCC, Manassas, VA, USA). The cancer-associated fibroblast cell line CAF19 was kindly provided by Dr. Priyabrata Mukherjee, University of Oklahoma Health Sciences Center. Cells were cultured following ATCC's instructions, and CAF19 was cultured in DMEM supplemented with 10% FBS. Exosome-depleted FBS and horse serum were prepared by pelleting the serum EVs at 100,000× g for 1.5 h at 4 °C and then filtered through a 0.22 µm PVDF centrifuge filter. Cells were routinely incubated in a humidified environment at 37 °C and 5% CO_2.

EV isolation and validation. EVs from 20 mL cell culture medium were isolated following the protocol we previously described, with minor modifications [19,39]. Briefly, after being pre-cleared by 10,000× g centrifugation for 30 min at 4 °C, the resulting supernatant was transferred to a 100 kDa cut-off centrifugal column (Merckmillipore, Burlington, MA, USA) and centrifuged for 15 min at 4000× g. The concentrated supernatant was then filtered through PVDF centrifuge filters (Merckmillipore, Burlington, MA, USA) as we described [39]. Small EVs (exosomes) were recovered using the Total Exosome Isolation Reagent (Thermofisher, Waltham, MA, USA) following the manufacturer's instructions. The isolated sEVs dissolved in PBS were verified by Western blot detecting positive and negative exosome proteins and nanoparticle tracking analysis (Nanosight NS300 System, Malvern Instruments, Malvern, UK) measuring sEV sizes and concentrations.

Drug encapsulation and quantification. PTX (Sigma-Aldrich, St. Louis, MO, USA, 0.01 µmol, CAS 33069-62-4) or GEM (Sigma-Aldrich, St. Louis, MO, USA, 0.1 µmol, CAS 122111-03-9) was mixed with the purified sEVs (around 50 µg) in 1 mL PBS. Three loading methods, including incubation, sonication and electroporation, were applied. For the incubation method, the sEV-drug mixture was incubated at 37 °C for 1 h. For the sonication method, the mixture was sonicated using a FB505 sonicator (Fisher Scientific, Pittsburgh, PA, USA) with the following settings: 20% amplitude, 6 cycles of 30 s on and off, followed by 2 min cooling period. After sonication, the EV-drug mixture was incubated at 37 °C for 1 h. For the electroporation method, the sEV-drug mixture was electroporated using the P3 Primary Cell 4D-NucleofectorTM X Kit L (Lonza, Basel, Switzerland) with DN-100 program of the 4D-NucleofectorTM Core Unit. After electroporation, the EV-drug mixture was also incubated at 37 °C for 1 h. The EV-drug mixture was then pelleted by the total exosome isolation reagent and dissolved in 200 µL PBS to remove unbounded drugs. UV absorbance of the EV-drug mixtures at 230 nm (PTX) or 275 nm (GEM) was recorded by Nanodrop (Denovix, Wilmington, DE, USA) to determine the drug concentrations as previously described [40,41]. A standard curve of free PTX or GEM was established, and drug concentration in the EV-drug mixture was calculated as the following: EV-drug UV absorbance minus sEV UV absorbance, and the resulting absorbance was fitting to the free drug standard curve. The drug loading efficiency was presented as drugs (ng)/sEVs (µg).

Western blot analysis. Western blot was performed as we recently described [39]. Primary antibody raised against Clathrin-LC was obtained from Santa Cruz Biotechnology

(Dallas, TX, USA), and those against Caveolin 1 and GAPDH were purchased from Cell Signaling Technology (Danvers, MA, USA). Antibodies used for sEV marker detection include: CD63 (Santa Cruz, Dallas, TX, USA), Flotillin-1 and Calnexin (Cell Signaling Technology, Danvers, MA, USA). The Li-Cor Odyssey® Fc Imaging System was used to visualize and image the blots.

Cell proliferation (MTS) assay. Cells were seeded onto 96-well plates at a density of 6000–8000 cells/well in triplicate and drugs were added the next morning. After 72 h incubation at 37 °C with 5% CO_2, the medium was replaced with 100 μL fresh medium supplemented with 20 μL CellTiter 96® AQueous One Solution (Promega, Madison, WI, USA). After 1 h of incubation, the absorbance value at 490 nm was recorded using a spectrometer (VWR SpectraMax® 190, Radnor, PA, USA). The data were expressed as percentage of the absorbance value detected in untreated control cells.

EV uptake analysis. Small EVs were labelled using the PKH67 Green Fluorescent Cell Linker kit (Sigma-Aldrich, St. Louis, MO, USA) following the manufacture's protocol. Same amount of sEV or EV-drug (around 20 μg) was added to 500 μL diluent before adding 1 μL PKH67. The mixture was incubated at room temperature for 20 min and 500 μL EV-depleted FBS was added to stop the labeling. The labeled sEVs were pelleted by ultracentrifuge at 100,000× g, 4 °C for 1.5 h. The pellet was resolved in EV-depleted medium. Cancer cells (5×10^4 cells/well) were plated on a 24-well plate. The labeled sEVs or sEV-drug mixtures (around 20 μg) were added to the cell culture 24 h post seeding and incubated for 2 to 10 h. In some experiments, endocytosis inhibitors were added 6 h prior to addition of the labeled sEVs or EV drugs. Cells were washed by PBS, fixed by paraformaldehyde and mounted using Prolong Antifade Reagents (Thermofisher, Waltham, MA, USA). Fluorescent signal in cells was detected by a fluorescent microscope (Thermofisher, Waltham, MA, USA) and analyzed using the ImageJ software [42].

Manipulation of clathrin and caveolin expression. The mCherry-clathrin-LC and mCherry-caveolin1 plasmids were purchased from Addgene (Watertown, MA, USA). The mCherry-C1 was purchased from Takara Bio USA (Mountain View, CA, USA) as control. DNA transfection to PANC-1 cells was performed using Lipofectamine 3000 (Fisher Scientific, Pittsburgh, PA, USA) and clathrin, and caveolin overexpression was verified by fluorescent microscopy and Western blot. For clathrin light chain and caveolin knockdown, three clathrin siRNA—s3191 (siCLTB91), s3192 (siCLTB92), and s3193 (siCLTB93)—one caveolin siRNA—s2446 (siCAV1)—and one negative control siRNA were synthesized (Table 1, Thermofisher, Waltham, MA, USA). Transfection of the siRNAs to PANC-1 cells was performed using Lipofectamine 3000 (Fisher Scientific, Pittsburgh, PA, USA) and the knockdown of clathrin and caveolin was verified by Western blot.

Table 1. siRNA sequences for knockdown of clathrin light chain and caveolin.

Name	Sequence	Target
si_CLTB91	sense: GCCCAGCUAUGUGACUUCATT antisense: UGAAGUCACAUAGCUGGGCCA	Clathrin light chain
si_CLTB92	sense: CCUCCUCUCAGUCUACUCATT antisense: UGAGUAGACUGAGAGGAGGCG	Clathrin light chain
si_CLTB93	sense: GAACAAGUAGAGAAGAACATT antisense: UGUUCUUCUCUACUUGUUCAC	Clathrin light chain
si_CAV1	sense: GCUUCCUGAUUGAGAUUCATT antisense: UGAAUCUCAAUCAGGAAGCTC	Caveolin

Statistics. Statistical analyses were performed using GraphPad Prism software (La Jolla, CA, USA). One-way ANOVA was used to determine p-values among experimental groups and a p-value of ≤ 0.05 was considered statistically significant.

Supplementary Materials: The following are available online at https://www.mdpi.com/article/10.3390/ijms23094773/s1.

Author Contributions: Conceptualization, W.-Q.D. and J.W.; Investigation, H.S., K.B., W.-Q.D. and K.B.; Methodology, H.S. and S.B.; Resources, W.-Q.D. and J.W. All authors have read and agreed to the published version of the manuscript.

Funding: This research was funded by the National Cancer Institute (CA235208-01), Suzhou Science and Technology Development: applied basic research (SKJY2021118), the Presbyterian Health Foundation, the Peggy and Charles Stephenson Cancer Center, and the Affiliated Suzhou Hospital of Nanjing Medical University.

Institutional Review Board Statement: Not applicable.

Informed Consent Statement: Not applicable.

Data Availability Statement: The data presented in this study are contained within the article and Supplementary Materials.

Acknowledgments: We thank the Department of Pathology at the University of Oklahoma Health Sciences Center for administrative support, and the core facility at the Peggy and Charles Stephenson Cancer Center for support in nanoparticle analysis. We thank Jundong Zhou at the Department of Radiation Oncology, the Affiliated Suzhou Hospital of Nanjing Medical University, for his participation in initiating this study.

Conflicts of Interest: The authors declare no conflict of interest. The funders had no role in the design of the study; in the collection, analyses, or interpretation of data; in the writing of the manuscript, or in the decision to publish the results.

References

1. Siegel, R.L.; Miller, K.D.; Jemal, A. Cancer statistics, 2020. *CA Cancer J. Clin.* **2020**, *70*, 7–30. [CrossRef] [PubMed]
2. Tas, F.; Sen, F.; Keskin, S.; Kilic, L.; Yildiz, I. Prognostic factors in metastatic pancreatic cancer: Older patients are associated with reduced overall survival. *Mol. Clin. Oncol.* **2013**, *1*, 788–792. [CrossRef]
3. National Comprehensive Cancer Network Pancreatic Cancer (Version 1, 2021). Available online: https://www.nccn.org/professionals/physician_gls/pdf/pancreatic.pdf (accessed on 6 August 2021).
4. Mosquera, J.; Garcia, I.; Liz-Marzan, L.M. Cellular Uptake of Nanoparticles versus Small Molecules: A Matter of Size. *ACC Chem. Res.* **2018**, *51*, 2305–2313. [CrossRef]
5. Grantab, R.; Sivananthan, S.; Tannock, I.F. The penetration of anticancer drugs through tumor tissue as a function of cellular adhesion and packing density of tumor cells. *Cancer Res.* **2006**, *66*, 1033–1039. [CrossRef] [PubMed]
6. Hosein, A.N.; Brekken, R.A.; Maitra, A. Pancreatic cancer stroma: An update on therapeutic targeting strategies. *Nat. Rev. Gastroenterol. Hepatol.* **2020**, *17*, 487–505. [CrossRef] [PubMed]
7. Wicki, A.; Witzigmann, D.; Balasubramanian, V.; Huwyler, J. Nanomedicine in cancer therapy: Challenges, opportunities, and clinical applications. *J. Control Release* **2015**, *200*, 138–157. [CrossRef]
8. Vishnu, P.; Roy, V. Safety and Efficacy of nab-Paclitaxel in the Treatment of Patients with Breast Cancer. *Breast Cancer* **2011**, *5*, 53–65. [CrossRef]
9. Dobrovolskaia, M.A. Pre-clinical immunotoxicity studies of nanotechnology-formulated drugs: Challenges, considerations and strategy. *J. Control Release* **2015**, *220*, 571–583. [CrossRef]
10. Giannakou, C.; Park, M.V.; de Jong, W.H.; van Loveren, H.; Vandebriel, R.J.; Geertsma, R.E. A comparison of immunotoxic effects of nanomedicinal products with regulatory immunotoxicity testing requirements. *Int. J. Nanomed.* **2016**, *11*, 2935–2952. [CrossRef]
11. Hannon, G.; Lysaght, J.; Liptrott, N.J.; Prina-Mello, A. Immunotoxicity Considerations for Next Generation Cancer Nanomedicines. *Adv. Sci.* **2019**, *6*, 1900133. [CrossRef]
12. Baj-Krzyworzeka, M.; Szatanek, R.; Weglarczyk, K.; Baran, J.; Urbanowicz, B.; Branski, P.; Ratajczak, M.Z.; Zembala, M. Tumour-derived microvesicles carry several surface determinants and mRNA of tumour cells and transfer some of these determinants to monocytes. *Cancer Immunol. Immunother.* **2006**, *55*, 808–818. [CrossRef] [PubMed]
13. Ratajczak, J.; Miekus, K.; Kucia, M.; Zhang, J.; Reca, R.; Dvorak, P.; Ratajczak, M.Z. Embryonic stem cell-derived microvesicles reprogram hematopoietic progenitors: Evidence for horizontal transfer of mRNA and protein delivery. *Leukemia* **2006**, *20*, 847–856. [CrossRef]
14. Bastos, N.; Ruivo, C.F.; da Silva, S.; Melo, S.A. Exosomes in cancer: Use them or target them? *Semin. Cell Dev. Biol.* **2018**, *78*, 13–21. [CrossRef] [PubMed]
15. Ohno, S.; Drummen, G.P.; Kuroda, M. Focus on Extracellular Vesicles: Development of Extracellular Vesicle-Based Therapeutic Systems. *Int. J. Mol. Sci.* **2016**, *17*, 172. [CrossRef] [PubMed]

16. Hoshino, A.; Costa-Silva, B.; Shen, T.L.; Rodrigues, G.; Hashimoto, A.; Tesic Mark, M.; Molina, H.; Kohsaka, S.; Di Giannatale, A.; Ceder, S.; et al. Tumour exosome integrins determine organotropic metastasis. *Nature* **2015**, *527*, 329–335. [CrossRef] [PubMed]
17. Tian, Y.; Li, S.; Song, J.; Ji, T.; Zhu, M.; Anderson, G.J.; Wei, J.; Nie, G. A doxorubicin delivery platform using engineered natural membrane vesicle exosomes for targeted tumor therapy. *Biomaterials* **2014**, *35*, 2383–2390. [CrossRef]
18. Sun, H.; Burrola, S.; Wu, J.; Ding, W.Q. Extracellular Vesicles in the Development of Cancer Therapeutics. *Int. J. Mol. Sci.* **2020**, *21*, 6097. [CrossRef]
19. Xu, Y.F.; Xu, X.; Gin, A.; Nshimiyimana, J.D.; Mooers, B.H.M.; Caputi, M.; Hannafon, B.N.; Ding, W.Q. SRSF1 regulates exosome microRNA enrichment in human cancer cells. *Cell Commun. Signal* **2020**, *18*, 130. [CrossRef]
20. Ding, W.Q.; Liu, B.; Vaught, J.L.; Palmiter, R.D.; Lind, S.E. Clioquinol and docosahexaenoic acid act synergistically to kill tumor cells. *Mol. Cancer* **2006**, *5*, 1864–1872. [CrossRef]
21. Hannafon, B.N.; Carpenter, K.J.; Berry, W.L.; Janknecht, R.; Dooley, W.C.; Ding, W.Q. Exosome-mediated microRNA signaling from breast cancer cells is altered by the anti-angiogenesis agent docosahexaenoic acid (DHA). *Mol. Cancer* **2015**, *14*, 133. [CrossRef]
22. Shen, Y.; Rosendale, M.; Campbell, R.E.; Perrais, D. pHuji, a pH-sensitive red fluorescent protein for imaging of exo- and endocytosis. *J. Cell Biol.* **2014**, *207*, 419–432. [CrossRef] [PubMed]
23. Cureton, D.K.; Massol, R.H.; Whelan, S.P.; Kirchhausen, T. The length of vesicular stomatitis virus particles dictates a need for actin assembly during clathrin-dependent endocytosis. *PLoS Pathog.* **2010**, *6*, e1001127. [CrossRef] [PubMed]
24. Majeed, S.R.; Vasudevan, L.; Chen, C.Y.; Luo, Y.; Torres, J.A.; Evans, T.M.; Sharkey, A.; Foraker, A.B.; Wong, N.M.; Esk, C.; et al. Clathrin light chains are required for the gyrating-clathrin recycling pathway and thereby promote cell migration. *Nat. Commun.* **2014**, *5*, 3891. [CrossRef] [PubMed]
25. Nolte-'t Hoen, E.; Cremer, T.; Gallo, R.C.; Margolis, L.B. Extracellular vesicles and viruses: Are they close relatives? *Proc. Natl. Acad. Sci. USA* **2016**, *113*, 9155–9161. [CrossRef] [PubMed]
26. Kanchanapally, R.; Deshmukh, S.K.; Chavva, S.R.; Tyagi, N.; Srivastava, S.K.; Patel, G.K.; Singh, A.P.; Singh, S. Drug-loaded exosomal preparations from different cell types exhibit distinctive loading capability, yield, and antitumor efficacies: A comparative analysis. *Int. J. Nanomed.* **2019**, *14*, 531–541. [CrossRef] [PubMed]
27. Mulcahy, L.A.; Pink, R.C.; Carter, D.R. Routes and mechanisms of extracellular vesicle uptake. *J. Extracell. Vesicles* **2014**, *3*, 24641. [CrossRef] [PubMed]
28. Tu, C.; Du, Z.; Zhang, H.; Feng, Y.; Qi, Y.; Zheng, Y.; Liu, J.; Wang, J. Endocytic pathway inhibition attenuates extracellular vesicle-induced reduction of chemosensitivity to bortezomib in multiple myeloma cells. *Theranostics* **2021**, *11*, 2364–2380. [CrossRef]
29. Schindler, C.; Collinson, A.; Matthews, C.; Pointon, A.; Jenkinson, L.; Minter, R.R.; Vaughan, T.J.; Tigue, N.J. Exosomal delivery of doxorubicin enables rapid cell entry and enhanced in vitro potency. *PLoS ONE* **2019**, *14*, e0214545. [CrossRef]
30. Nedeva, C.; Mathivanan, S. Engineering Extracellular Vesicles for Cancer Therapy. *Subcell. Biochem.* **2021**, *97*, 375–392.
31. Thery, C.; Witwer, K.W.; Aikawa, E.; Alcaraz, M.J.; Anderson, J.D.; Andriantsitohaina, R.; Antoniou, A.; Arab, T.; Archer, F.; Atkin-Smith, G.K.; et al. Minimal information for studies of extracellular vesicles 2018 (MISEV2018): A position statement of the International Society for Extracellular Vesicles and update of the MISEV2014 guidelines. *J. Extracell. Vesicles* **2018**, *7*, 1535750. [CrossRef] [PubMed]
32. Wang, J.; Chen, P.; Dong, Y.; Xie, H.; Wang, Y.; Soto, F.; Ma, P.; Feng, X.; Du, W.; Liu, B.F. Designer exosomes enabling tumor targeted efficient chemo/gene/photothermal therapy. *Biomaterials* **2021**, *276*, 121056. [CrossRef] [PubMed]
33. Wang, J.; Li, W.; Zhang, L.; Ban, L.; Chen, P.; Du, W.; Feng, X.; Liu, B.F. Chemically Edited Exosomes with Dual Ligand Purified by Microfluidic Device for Active Targeted Drug Delivery to Tumor Cells. *ACS Appl. Mater. Interfaces* **2017**, *9*, 27441–27452. [CrossRef] [PubMed]
34. Wang, J.; Dong, Y.; Li, Y.W.; Li, W.; Cheng, K.; Qian, Y.; Xu, G.Q.; Zhang, X.S.; Hu, L.; Chen, P.; et al. Designer Exosomes for Active Targeted Chemo-Photothermal Synergistic Tumor Therapy. *Adv. Funct. Mater.* **2018**, *28*, 1707360. [CrossRef]
35. Zhou, Y.; Zhou, W.; Chen, X.; Wang, Q.; Li, C.; Chen, Q.; Zhang, Y.; Lu, Y.; Ding, X.; Jiang, C. Bone marrow mesenchymal stem cells-derived exosomes for penetrating and targeted chemotherapy of pancreatic cancer. *Acta Pharm. Sin. B* **2020**, *10*, 1563–1575. [CrossRef] [PubMed]
36. Li, Y.J.; Wu, J.Y.; Wang, J.M.; Hu, X.B.; Cai, J.X.; Xiang, D.X. Gemcitabine loaded autologous exosomes for effective and safe chemotherapy of pancreatic cancer. *Acta Biomater.* **2020**, *101*, 519–530. [CrossRef] [PubMed]
37. Meng, H.; Wang, M.; Liu, H.; Liu, X.; Situ, A.; Wu, B.; Ji, Z.; Chang, C.H.; Nel, A.E. Use of a lipid-coated mesoporous silica nanoparticle platform for synergistic gemcitabine and paclitaxel delivery to human pancreatic cancer in mice. *ACS Nano* **2015**, *9*, 3540–3557. [CrossRef]
38. Miao, L.; Guo, S.; Zhang, J.; Kim, W.Y.; Huang, L. Nanoparticles with Precise Ratiometric Co-Loading and Co-Delivery of Gemcitabine Monophosphate and Cisplatin for Treatment of Bladder Cancer. *Adv. Funct. Mater.* **2014**, *24*, 6601–6611. [CrossRef]
39. Xu, Y.F.; Xu, X.; Bhandari, K.; Gin, A.; Rao, C.V.; Morris, K.T.; Hannafon, B.N.; Ding, W.Q. Isolation of extra-cellular vesicles in the context of pancreatic adenocarcinomas: Addition of one stringent filtration step improves recovery of specific microRNAs. *PLoS ONE* **2021**, *16*, e0259563. [CrossRef]
40. Singh, R.; Shakya, A.K.; Naik, R.; Shalan, N. Stability-indicating HPLC determination of gemcitabine in pharmaceutical formulations. *Int. J. Anal. Chem.* **2015**, *2015*, 862592. [CrossRef]

41. Pascucci, L.; Cocce, V.; Bonomi, A.; Ami, D.; Ceccarelli, P.; Ciusani, E.; Vigano, L.; Locatelli, A.; Sisto, F.; Doglia, S.M.; et al. Paclitaxel is incorporated by mesenchymal stromal cells and released in exosomes that inhibit in vitro tumor growth: A new approach for drug delivery. *J. Control Release* **2014**, *192*, 262–270. [CrossRef]
42. Schneider, C.A.; Rasband, W.S.; Eliceiri, K.W. NIH Image to ImageJ: 25 years of image analysis. *Nat. Methods* **2012**, *9*, 671–675. [CrossRef] [PubMed]

International Journal of
Molecular Sciences

Review

Delivering Therapeutics to Glioblastoma: Overcoming Biological Constraints

Elza N. Mathew, Bethany C. Berry, Hong Wei Yang, Rona S. Carroll and Mark D. Johnson *

Department of Neurological Surgery, University of Massachusetts Medical School, 55Lake Avenue North, Worcester, MA 01655, USA; elzaneelima.mathew@gmail.com (E.N.M.); bethany.berry@umassmed.edu (B.C.B.); hongwei.yang@umassmed.edu (H.W.Y.); rona.carroll@umassmed.edu (R.S.C.)
* Correspondence: mark.johnson3@umassmemorial.org; Tel.: +1-508-334-0605

Abstract: Glioblastoma multiforme is the most lethal intrinsic brain tumor. Even with the existing treatment regimen of surgery, radiation, and chemotherapy, the median survival time is only 15–23 months. The invasive nature of this tumor makes its complete removal very difficult, leading to a high recurrence rate of over 90%. Drug delivery to glioblastoma is challenging because of the molecular and cellular heterogeneity of the tumor, its infiltrative nature, and the blood–brain barrier. Understanding the critical characteristics that restrict drug delivery to the tumor is necessary to develop platforms for the enhanced delivery of effective treatments. In this review, we address the impact of tumor invasion, the molecular and cellular heterogeneity of the tumor, and the blood–brain barrier on the delivery and distribution of drugs using potential therapeutic delivery options such as convection-enhanced delivery, controlled release systems, nanomaterial systems, peptide-based systems, and focused ultrasound.

Keywords: glioblastoma; brain tumor; drug delivery

Citation: Mathew, E.N.; Berry, B.C.; Yang, H.W.; Carroll, R.S.; Johnson, M.D. Delivering Therapeutics to Glioblastoma: Overcoming Biological Constraints. *Int. J. Mol. Sci.* **2022**, *23*, 1711. https://doi.org/10.3390/ijms23031711

Academic Editor: Angela Stefanachi

Received: 30 December 2021
Accepted: 31 January 2022
Published: 2 February 2022

Publisher's Note: MDPI stays neutral with regard to jurisdictional claims in published maps and institutional affiliations.

1. Introduction

Glioblastoma multiforme (GBM) is the most common primary brain malignancy in adults, accounting for more than 50% of intrinsic brain tumors [1]. According to the World Health Organization (WHO) classification, GBM is a grade IV glioma, highly invasive with a five-year survival rate of less than 5% [2,3]. This highly aggressive disease presents a very poor prognosis. The median survival time from diagnosis of the tumor is approximately 15–23 months. The incidence of GBM is 3.19 per 100,000 persons in the United States with a median age of 64 years [4–8].

Widely recognized risk factors associated with GBM occurrence are exposure to high-dose ionizing radiation and certain genetic syndromes, including neurofibromatosis type 1 and Li-Fraumeni syndrome [2,9,10]. The occurrence of GBM is higher in males and among individuals 50 years of age and older [5].

GBM is characterized by rapid cell proliferation and extensive invasion of tumor cells into the surrounding brain, making complete removal of the tumor impossible [11]. These features lead to a high recurrence rate, even with the current treatment regimen of maximum safe surgical removal, radiation therapy and temozolomide chemotherapy. Intratumoral molecular heterogeneity, presence of the blood–brain barrier, and tumor immune evasion via local immunosuppression limits the success of existing therapies [12]. Numerous drug therapies directed against GBM have shown promising results in in vitro assays, but all have had limited success in vivo. This is due in part to the diffuse, heterogeneous nature of the tumor, and the blood–brain barrier that limits the ability of many drugs to enter the brain parenchyma. Understanding the critical factors that restrict drug delivery to the brain is necessary to develop platforms for the enhanced delivery of effective drugs.

In this review, we discuss the impact that molecular and cellular tumor heterogeneity, tumor dispersion and the blood–brain barrier have on the delivery and distribution of pharmacological agents to GBM. The currently available chemotherapeutic delivery systems designed to overcome these constraints are also discussed (Figure 1). We will focus on the use of convection-enhanced delivery, controlled release systems, nanomaterial systems, peptide-based therapeutics and focused ultrasound (FUS) for the treatment of this aggressive tumor.

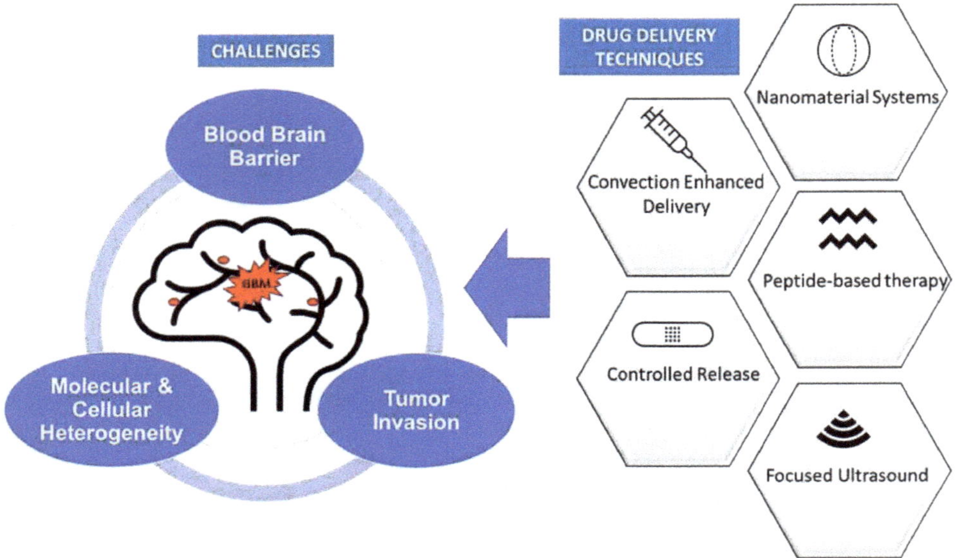

Figure 1. Challenges to delivery of therapeutics to glioblastoma and the currently available drug delivery techniques.

1.1. Molecular Heterogeneity

GBM is characterized by molecular heterogeneity within a single tumor in addition to inter-patient heterogeneity. Molecular heterogeneity underlies the cellular heterogeneity in GBM, i.e., the differences between cell types within a tumor [13]. This molecular heterogeneity may be responsible for differences in individual patient responses to therapy and prognosis as well as the failure of targeted therapies [14,15]. Common gene mutations include IDH1/2 mutations, O6-methylguanine DNA methyltransferase (MGMT) promoter methylation, co-deletion of 1p and 19q, and EGFR amplification/truncation. Primary GBM frequently exhibits epidermal growth factor receptor (EGFR) over expression, PTEN (MMC-I) mutation or deletion, CDKN2A (p16) deletion, and MDM2 amplification. Secondary GBM predominantly has IDH1, ATRX and TP53 mutations [16]. Based on differences in genetic alterations and the expression of EGFR, NF1, PDGFRA/IDH1, PI3K and other key genes, primary GBM can be classified into four subtypes: proneural, neural, classical, and mesenchymal, with each subtype varying in its gene expression signature [17]. Secondary GBMs predominantly have IDH1 (or less commonly IDH2) mutations that have a proneural gene expression signature and a better prognosis than GBMs with wild-type IDH1/IDH2.

In addition to the molecular heterogeneity described above, GBM cells are morphologically and functionally heterogeneous. Single-cell RNA-Seq studies of individual tumor cells have revealed that GBM tumor cells exist in multiple stages of differentiation [18]. Other studies have shown that astrocyte-like GBM cells can transdifferentiate to become endothelial-like cells [19,20]. As a result of this extraordinary cellular heterogeneity, individual GBM cells may be more or less replicative, invasive, or sensitive to radiation or chemotherapy [21].

1.2. Blood–Brain Barrier

The blood–brain barrier is composed of a highly specialized circuit of blood vessels that are lined by brain microvascular endothelial cells (BMEC), the cell–cell junctions between which restrict the entry of potentially harmful substances (including systemically administered therapeutics) into the brain. The BMECs are surrounded by pericytes, astrocytes and the basal membrane [22,23]. This is indeed a protective layer safeguarding the brain from damaging agents in the systemic circulation and keeping CNS homeostasis to allow proper neuronal function [24]. The two major features of the blood–brain barrier are: (1) the presence of tight junctions limiting paracellular transport, and (2) reduced fenestrations and transport vesicles limiting transcellular transport. Small molecules need to be less than 400 Da in size and lipid soluble in order to cross the barrier [25].

Disruption of the blood–brain barrier (BBB) is observed in glioma tumor regions. Unfortunately, this degree of BBB disruption is not sufficient to allow for the ability of therapeutics to reach this diffusely infiltrating tumor [26,27]. Biochemical and physical methods can be used to further increase the permeability of this biological barrier. Ion channel activators such as ATP-sensitive and calcium-activated potassium channel activators, phosphodiesterase 5 (PDE5) inhibitors, bradykinin type 2 receptor activators, adenosine 2A receptor (A2AR) agonists, papaverine, and certain microRNAs represent biochemical approaches to biochemical modulation of the BBB. PDE5 inhibitors decrease cGMP degradation and increase vesicular transport, thereby increasing BBB permeability within the tumor region. Potassium channel activators downregulate tight junction protein expression and increase the formation of pinocytotic vesicles. Bradykinin activators increase transcytosis and modulate tight junction protein expression and cGMP synthesis. A2AR agonists and papaverine downregulate tight junction protein expression. MicroRNAs such as miR-132-3p increase blood–brain barrier permeability by increasing transcytosis. Mannitol, an osmotic agent, is widely used to disrupt the blood–brain barrier by vasodilatation and shrinkage of endothelial cells [28,29]. Physical strategies for modulating BBB permeability include the application of electromagnetic pulses, laser-induced thermal therapy, radiotherapy, and focused ultrasound (FUS). All these methods downregulate the expression of tight junction proteins. FUS is also involved in increasing transcytosis [22,30].

2. Methods of Drug Delivery to Glioblastoma

In this review, we will discuss the application of controlled release systems, convection-enhanced delivery, nanomaterial systems, peptide-based therapeutics, and focused ultrasound for drug delivery to glioblastoma, overcoming the obstacles posed by GBM molecular and cellular heterogeneity, GBM cell invasion/dispersal and the blood–brain barrier.

2.1. Controlled-Release Systems

Implantable drug release systems enable the direct delivery of therapeutic agents to the tumor site, circumventing the need to cross the blood–brain barrier. Drug-loaded biocompatible materials such as drug-impregnated gels can be designed so that they release low doses of drugs at the tumor site over a prolonged period of time.

Implantable controlled-release delivery systems can be constructed with degradable or non-degradable polymers. Between these two options, biodegradable polymers are more commonly used [31]. One such biodegradable polymer delivery system that has been used clinically for GBM is the Gliadel® wafer. This is a biodegradable polymer wafer loaded with the chemotherapeutic drug BCNU. It is the only FDA-approved drug delivery implant for treating GBM [11]. Gliadel was approved by the FDA in 1996 for recurrent GBM and later in 2003 for upfront treatment of malignant glioma [31–33]. In another study, biodegradable wafers were created for the combined delivery of temozolomide and carmustine in a rat glioma model. This approach increased the median survival of the animals significantly, with 25% of the animals living long term >120 days [34]. Biodegradable polymer implants releasing rapamycin were found to increase survival both in the presence and absence of radiation therapy in a rat malignant glioma model [35]. The rigid structure of these systems has limited drug loading capacity and they can be dislodged from the original site of implantation. In addition, the wafer systems also led to the occurrence of seizures, intracranial hypertension, meningitis, cerebral edema, and impaired wound healing in neurosurgical patients [11].

A pH-responsive carboxymethylcellulose biopolymer system was used to deliver the chemotherapeutic drug, rhodamine B encapsulated in multiple emulsions to glioblastoma cells in vitro [36]. In a mouse GBM tumor resection model, hydrogel-based co-delivery of the chemotherapeutic agents paclitaxel and temozolomide enhanced survival [37]. Hydrogels can adapt their shape while nevertheless retaining sufficient drug capacity [11]. The use of multiple emulsions and hydrogels avoided the limitations seen with the rigid structure of wafer implants. However, the intrinsic hydrophilic nature of hydrogels does not allow for the effective delivery of hydrophobic therapeutic agents [38].

With sustainable release of therapeutic agents over time, implantable controlled-release systems limit the local GBM recurrence after resection by interacting with the tumor cells at the resection site. However, these systems are not only limited by a variety of side effects as listed above, but also by poor drug distribution to distant tumor cells that have migrated in the normal brain parenchyma. Consequently, these implantable local delivery systems have limited ability to affect infiltrating GBM cells distant to the site of implantation.

2.2. Convection-Enhanced Delivery

Convection-enhanced delivery (CED) is a catheter-based drug delivery method that depends on pressure gradients rather than diffusion to deliver therapeutic agents into brain tumors. This technique involves the stereotactic insertion of one or more catheters into the tumor. The catheters are connected to a syringe/drug delivery pump that maintains a positive pressure gradient that promotes the distribution of higher volumes of drug over a larger area. Ensuring direct intratumoral drug delivery independent of drug concentration or diffusivity of the infusate, CED allows the use of large volumes of the drug at lower and less-toxic doses [39]. The ratio between the volume of distribution and the volume of infusion is a key factor determining the optimum flow rate and success of this technique.

CED of temozolomide, the standard cytostatic drug for the treatment of GBM, combined with subcutaneous immunizations with irradiated cancer cells had a synergistic effect on tumor growth and overall survival in a mouse-GL261 glioma model. However, the same protocol did not display synergy in the KR158 mouse glioma model, which is resistant to radiation and chemotherapy. In the latter model, CED of temozolomide increased survival [40]. By initiating antitumor immune response, irradiated cells may help to overcome intratumoral immunosuppression and enhance the antitumor activity of temozolomide.

In a phase I clinical trial, CED was used to infuse carboplatin to patients with WHO grade III astrocytomas or grade IV gliomas (GBM). This approach increased median overall survival without any systemic toxicity [41]. This was the first clinical trial to demonstrate that CED of carboplatin into the brain is safe. Oral or intravenous administration of platinated drugs does not produce effective concentrations in the brain and has been associated with systemic toxicity [42].

MRI-guided CED of iron oxide nanoparticles (IONP) conjugated to epidermal growth factor receptor deletion mutant III antibody (EGFRVIIIAb) showed significant increase in survival in a mouse GBM model [43]. Binding of EGFRvIIIAb-IONP conjugate to target cells was evaluated by changes in the MRI signal. IONP is a theranostic nanoparticle which has imaging properties as well as anticancer activity. EGFRvIIIAb is a tumor-specific cell surface protein. The experimental animal groups treated with EGFRvIIIAb-IONP conjugate as well as EGFRvIIIAb alone had a single survivor after 120 days. This study demonstrated the feasibility of conjugating biological agents or ligands that bind to tumor-specific proteins along with concomitant imaging to ensure targeted distribution of the therapeutic agent.

Direct delivery of anticancer agents in large volumes and at less toxic doses makes CED an attractive drug delivery method for GBM. Implantable catheters can be used to repeat CED at required intervals, but the efficacy of this approach may be limited by the highly invasive nature of the tumor with GBM cells dispersed to distant sites [44]. Challenges in maintaining the ratio between the volumes of distribution and infusion, flow rate, formation of air bubbles, infusate reflux along the catheter, and the variability in the tumor tissue composition can limit the establishment of the required pressure gradient, thereby limiting the efficacy of CED [45].

2.3. Nanomaterial Systems

Nanomedicine is an emerging candidate for cancer therapy that uses a variety of different types of nanocarriers such as lipid-based, polymer-based, inorganic, viral, and drug-conjugated nanoparticles [46]. Small size, high surface-to-volume ratio and other physico-chemical parameters make nanoparticles unique [47]. Accentuated drug delivery along with potential applications in diagnosis and imaging makes nanomedicine an appealing approach to cancer therapeutics [48].

Nanomaterial systems for drug delivery improve GBM tumor targeting mainly by promoting drug diffusion through the blood–brain barrier and by the enhanced permeability and retention (EPR) effect. The EPR effect is dependent on the highly angiogenic nature of GBM where leaky vasculature is commonly present [49]. Because of this phenomenon, nanoparticles can passively modulate the biodistribution of loaded molecules and increase their accumulation in cancer tissues with leaky vasculature [50–52]. The efficacy of antitumor drug delivery via nanoparticles is also enhanced by several additional mechanisms, including increased cellular internalization, activation of immune cells, and reactive oxygen species (ROS) production [53]. Surface functionalization of a nanoparticle with an active principle, a targeting agent, and a compound for detection will increase the functional range of the nanomaterial drug delivery targeting biomolecules, thereby enabling both drug delivery and disease diagnosis [54]. The long half-life and improved cellular uptake and biodistribution increase the appeal of nanomaterial systems for tumor therapy [49].

The delivery of tumor antigens and adjuvants using nanoparticles as a vehicle has been shown to increase the efficacy of immune therapy against GBM [55–57]. Different types of nanomaterials, including polymeric nanoparticles or lipid-based nanoparticles, may also be used to deliver nucleic acids intended for tumor therapy [49]. A poly(ε-caprolactone) (PCL)-based nanoparticle system was used to deliver the natural growth modulating tripeptide GHK (glycyl-L-histidyl-L-lysine) to human GBM cells in vitro, thereby reducing GBM cell viability to nearly 65%. However, this approach did not show anticancer activity at concentrations less than 20 mg/mL [58]. GHK is a natural tripeptide with a varied array of biological activities including anticancer, antioxidant, anti-inflammatory, skin remodeling, wound healing, etc. [59]. The anti-GBM activity of GHK demonstrated in this study in vitro requires further validation in vivo.

A nanobubble-based theranostic system consisting of intravenously administered iron-platinum nanoparticles loaded with doxorubicin and surface-functionalized with transferrin (to allow for tumor targeting) reduced glioma growth in a mouse model by almost 70%. The nanobubbles were burst by exposing to high intensity focused ultrasound (HIFU) to bypass the blood–brain barrier by generating a cavitation effect. This single nanocomposite, combining nanomaterials, chemotherapeutic agents and MRI contrast agents all together for the first time, was able to cross the blood–brain barrier, target the tumor cells and allow imaging of the brain tumor. This multimodal system is an example of combining several strategies to enhance the success rate of therapy [60].

The physicochemical properties of nanomaterials enable them to deliver therapeutic agents to the brain via drug encapsulation or by surface modification. The clinical use of the nanomaterial drug delivery systems is limited by the poorly controlled accumulation and distribution of particles in and around the specific target site [61,62]. The use of multiple therapeutic strategies in combination with nanocarriers may improve the success rate of targeted tumor therapies. However, the potential toxic effects of nanoparticle systems due to aggregation upon introduction into biological systems remains poorly understood. The effect of protein corona formation around nanoparticles and its relevance regarding their biological activity is also being extensively investigated [63].

2.4. Peptide Based Therapeutics

There are three main types of peptide-based therapeutics—tumor homing peptides, peptides targeting aberrant cellular signaling pathways and cell-penetrating peptides [64]. GBM cells have increased expression of membrane proteins that are responsible for cellular function and maintenance, protein synthesis, intercellular signaling, cell movement, and antigen presentation [8,65]. Tumor homing peptides can bind to specific molecular targets on the surface of GBM cells and will be taken up by the cells by endocytosis [66]. The binding process of these peptides is faster than antibodies. These can also be used for in vivo tumor imaging. Binding of some of these peptides can also enhance or antagonize signal transduction pathways in cancer cells or tumor tissues. Peptides and their derivatives targeting aberrant cellular signaling pathways can improve the efficacy of tumor therapy by increased selectivity in their interaction with the oncogenic pathways [64]. Cell-penetrating peptides (CPPs) are small, basic, positively charged peptide derivatives that can pass through the cell membrane [64,67].

Highly selective tumor-targeting peptides obtained using a biopanning phage display library directed against GBM cells were able to cross the BBB and deliver the oncolytic virus VSVΔM51 to GBM in a mouse model in vivo [68]. These peptides, when delivered in combination with gadolinium, also enabled the visualization of the tumors via MRI. The use of peptides directed against multiple targets provides a mechanism for addressing the heterogeneity of the tumor while nevertheless allowing for tumor specific oncolytic virus delivery.

Self-assembled spherical nanoparticles containing a peptide probe (Cy5.5-SAPD-99mTc) that combines tumor homing ability with mitochondria targeting was found to have promising theranostic possibilities owing to the enhanced apoptosis in GBM cells coupled with imaging functionality [69]. Incorporating both tumor-homing and mitochondria-targeting components helps to increase the specificity of drug delivery.

Peptide derivatives of rabies virus glycoproteins, RVG29 and RVG15-liposome, were used to deliver anticancer chemotherapeutic docetaxel nanoparticles and paclitaxel-cholesterol to glioma-bearing ICR mice with a positive effect on animal survival [70,71]. The RVG peptides target the nicotinic acetylcholine receptor, the increased expression of which is noted in the hypoxic and ischemic conditions within the tumor microenvironment. Administration of RVG peptides thus aids in tumor-specific chemotherapeutic delivery.

WSW (also called PhrCACET1) is a tumor-targeting peptide (derived from Clostridium acetobutylicum) that was fused to paclitaxel nanosuspensions and used to target GBM cell membranes in a glioma mouse model. The use of WSW induced apoptosis and prolonged survival of the animals [72]. By combining BBB penetration with tumor targeting, this biomimetic drug delivery system has enhanced tumor-targeting specificity.

The use of polydopamine (PDA)-coated zein-curcumin nanoparticles functionalized with the peptide G23 inhibited cell proliferation and migration in glioma cells in vitro [73]. Here, the peptide G23 binds to ganglioside GM-1 and enables crossing of the BBB. The anti-inflammatory, antimicrobial, and anticancer activities of curcumin have been widely demonstrated [74–76].

A dual peptide nanocomplex created by combining SynB3 (a cell penetration peptide) with PVGLIG (an MMP-2 sensitive peptide) and paclitaxel inhibited cell migration and invasion in multiple GBM cell lines, suppressed GBM tumor growth in vivo, and increased overall survival in a mouse model of GBM [77]. The aberrant expression of matrix metalloproteinases (MMPs) has been widely reported in tumors, and the addition of an MMP-sensitive peptide increased the tumor specificity of the drug cargo in this system.

To enhance the membrane permeability of peptides, membrane receptors such as low-density lipoprotein receptor, IL-4 receptor, and transferrin receptor which are abundantly expressed on GBM cells have been used to direct the delivery of tumor-homing peptides to brain malignancies, utilizing receptor-mediated transcytosis [78–80]. Peptide-mediated drug delivery is limited by the poor in vivo stability due to the proteolytic degradation of peptides in the circulation when administered systemically. In addition, the short half-life of peptides results in limited bioavailability. This can be overcome by chemical modification or conjugation with macromolecules or nanocarriers with greater target specificity [64]. Identifying additional GBM-targeting peptides is needed to further exploit the benefits of this mode of drug delivery.

2.5. Focused Ultrasound

Focused ultrasound (FUS) is an image-guided, noninvasive method to transiently open the blood barrier, thereby enhancing the efficacy of therapeutic delivery to GBM. FUS can be used to reversibly disrupt the BBB without irreversible tissue damage [81]. The use of FUS in combination with circulating microbubbles works by creating mechanical cavitation effects that transiently disrupt the BBB [82]. In addition to the cavitation effects and thermal ablation, FUS may also work via immunomodulation [83]. Future concurrent application of noninvasive FUS along with other modes of therapeutic drug delivery to GBM is promising.

By increasing BBB permeability, FUS can enhance the delivery of varied therapeutic agents to the tumor. For example, FUS-induced disruption of the BBB enhanced the local delivery of temozolomide to tumors and increased the overall survival of rats harboring experimentally induced gliomas [84]. FUS application followed by BCNU administration resulted in a reduced rate of tumor growth and improved survival in rats [85]. In a mouse model bearing temozolomide-resistant glioma, low-intensity fluorescent ultrasound (LIFU) was used to deliver a liposomal O6-(4-bromothenyl)guanine (O6BTG) derivative that inactivates MGMT [86]. Because MGMT enhances DNA repair in tumor cells, MGMT silencing has been associated with more favorable outcomes after temozolomide treatment [87].

The use of imaging techniques along with FUS increases the rate of drug delivery to specifically identified tumor tissues. MRI guided FUS (MRgFUS) was used to achieve a higher tissue delivery of liposome-encapsulated doxorubicin in rats, temozolomide in mice, and cisplatin-conjugated gold nanoparticles in mice [84,88,89]. In an additional study, MRgFUS was used to deliver the intravenously administered monoclonal antibody, trastzumab, to Her2-positive intracranial metastases in breast cancer patients without any adverse events [90].

Using this noninvasive technique, relatively low systemic doses of therapeutic agents can be used, thereby reducing systemic toxic effects [91]. Transient application of FUS only results in a transient opening of the BBB and does not lead to long-term BBB defects [25]. FUS has the potential to enhance therapeutic drug efficacy against GBM because it is noninvasive and provides reproducible enhancements in drug delivery in early investigational studies. Clinical trials using FUS in GBM are currently underway [81,92,93]. Nevertheless, FUS is not immune to side effects which include edema, intracerebral hemorrhage, and uncontrolled thermal injury in brain [94].

3. Conclusions

Molecular and cellular heterogeneity, GBM cell dispersal and the BBB are critical constraints limiting the efficacy of anti-GBM drug therapy. Applications of CED, controlled-release systems, nanomaterial systems, peptide-based therapeutics and focused ultrasound for drug delivery to tumor enhance survival with reduced toxicity in animal studies (Table 1). Despite currently available treatments, the highly invasive GBM continues to be a deadly disease without cure in patients. Therefore, clinical trials that combine currently available therapies with the novel drug delivery approaches discussed here may enhance the effectiveness of molecular therapeutics in GBM.

Table 1. Summary of the methods of drug delivery to GBM with the examples discussed in the text.

Method of Drug Delivery	Specific Examples
Controlled release systems	Gliadel [11,31–33]
	Biodegradable wafers for the combined delivery of temozolomide and carmustine [34]
	Biodegradable polymer implants releasing rapamycin [35]
	Carboxymethylcellulose biopolymer system delivering rhodamine B [36]
	Hydrogel based co-delivery of paclitaxel and temozolomide [37]
Convection enhanced delivery	Temozolomide [40]
	Carboplatin [41]
	Iron oxide nanoparticles conjugated to epidermal growth factor receptor deletion mutant III antibody (EG-FRVIIIAb)/MRI-guided [43]
Nanomaterial Systems	Poly(ε-caprolactone) (PCL) based nanoparticle system to deliver the natural growth modulating tripeptide GHK (glycyl-L-histidyl-L-lysine) [58]
	Nanobubble-based theranostic system consisting of intravenously administered iron-platinum nanoparticles loaded with doxorubicin and surface-functionalized with transferrin [60]
Peptide based therapeutics	Tumor targeting peptides delivering deliver the oncolytic virus VSVΔM51, in combination with gadolinium [68]
	Self-assembled spherical nanoparticles containing a peptide probe (Cy5.5-SAPD-99mTc) with mitochondria targeting [69]
	Peptide derivatives of rabies virus glycoproteins, RVG29 and RVG15-liposome, delivering anticancer chemotherapeutic docetaxel nanoparticles and paclitaxel-cholesterol [70,71]
	WSW (also called PhrCACET1) peptide fused to paclitaxel nanosuspensions [72]
	Use of polydopamine (PDA)-coated zein-curcumin nanoparticles functionalized with the peptide G23 [73]
	Dual peptide nanocomplex created by combining SynB3 (a cell penetration peptide) with PVGLIG (an MMP-2 sensitive peptide) and paclitaxel [77]
Focused ultrasound	Temozolomide [84]
	BCNU [85]
	Liposomal O6-(4-bromothenyl)guanine (O6BTG) [86]
	Liposome-encapsulated doxorubicin [88]
	Cisplatin conjugated gold nanoparticles [89]
	Trastzumab [90]

Author Contributions: E.N.M. prepared the manuscript. B.C.B., H.W.Y., R.S.C. and M.D.J. critically reviewed the manuscript. All authors have read and agreed to the published version of the manuscript.

Funding: The Maroun Semaan Professorship and the University of Massachusetts Chan Medical School.

Institutional Review Board Statement: Not applicable.

Informed Consent Statement: Not applicable.

Data Availability Statement: Not applicable.

Conflicts of Interest: The authors declare no conflict of interest.

Abbreviation

GBM	Glioblastoma multiforme
WHO	World Health Organization
IDH	Isocitrate dehydrogenase
EGFR	Epidermal growth factor receptor
MGMT	O6-methylguanine DNA methyltransferase
PTN	Pleiotrophin
CDKN2A	Cyclin Dependent Kinase Inhibitor 2A
MDM2	Mouse double minute 2 homolog
NF1	Neurofibromatosis type 1
PDGFRA	Platelet Derived Growth Factor Receptor Alpha
GSC	Glioblastoma stem cells
NSC	Neural stem cell
BEMC	Brain microvascular endothelial cells
ATP	Adenosine triphosphate
PDE5	Phosphodiesterase 5
FUS	Focused ultrasound
CED	Convection enhanced delivery
EPR	Enhanced permeability and retention
ROS	Reactive oxygen species
PCL	Poly(ε-caprolactone)
GHK	glycyl-L-histidyl-L-lysine
CPP	Cell penetrating peptide
PDA	Polydopamine
MMP	Matrix Metalloproteinase
MRI	Magnetic resonance imaging
MRgFUS	Magnetic resonance guided focused ultrasound
LIFU	Low intensity fluorescent ultrasound
HIFU	High intensity fluorescent ultrasound

References

1. Schiffer, D.; Annovazzi, L.; Casalone, C.; Corona, C.; Mellai, M. Glioblastoma: Microenvironment and Niche Concept. *Cancers* **2018**, *11*, 5. [CrossRef] [PubMed]
2. Hanif, F.; Muzaffar, K.; Perveen, K.; Malhi, S.M.; Simjee, S.U. Glioblastoma Multiforme: A Review of its Epidemiology and Pathogenesis through Clinical Presentation and Treatment. *Asian Pac. J. Cancer Prev.* **2017**, *18*, 3–9. [CrossRef] [PubMed]
3. Louis, D.N.; Perry, A.; Reifenberger, G.; von Deimling, A.; Cavenee, W.K.; Ohgaki, H.; Wiestler O., D.; Kleihues, P.; Ellison, D.W. The 2016 World Health Organization Classification of Tumors of the Central Nervous System: A summary. *Acta Neuropathol.* **2016**, *131*, 803–820. [CrossRef] [PubMed]
4. SEER*Explorer Application. Available online: https://seer.cancer.gov/explorer/application.html?site=661&data_type=4&graph_type=5&compareBy=sex&chk_sex_3=3&chk_sex_2=2&series=9&race=1&age_range=1&stage=101&advopt_precision=1&advopt_show_ci=on&advopt_display=2 (accessed on 13 October 2021).
5. Tamimi, A.F.; Juweid, M. Epidemiology and Outcome of Glioblastoma. In *Glioblastoma*; De Vleeschouwer, S., Ed.; Codon Publications: Brisbane, Australia, 2017. Available online: http://www.ncbi.nlm.nih.gov/books/NBK470003/ (accessed on 29 April 2021).
6. Thakkar, J.P.; Dolecek, T.A.; Horbinski, C.; Ostrom, Q.T.; Lightner, D.D.; Barnholtz-Sloan, J.S.; Villano, J.L. Epidemiologic and Molecular Prognostic Review of Glioblastoma. *Cancer Epidemiol. Prev Biomark.* **2014**, *23*, 1985–1996. [CrossRef] [PubMed]
7. Koshy, M.; Villano, J.L.; Dolecek, T.A.; Howard, A.; Mahmood, U.; Chmura, S.J.; Weichselbaum, R.R.; McCarthy, B.J. Improved survival time trends for glioblastoma using the SEER 17 population-based registries. *J. Neurooncol.* **2012**, *107*, 207–212. [CrossRef] [PubMed]
8. Shergalis, A.; Bankhead, A.; Luesakul, U.; Muangsin, N.; Neamati, N. Current Challenges and Opportunities in Treating Glioblastoma. *Pharm. Rev.* **2018**, *70*, 412–445. [CrossRef]
9. Orr, B.A.; Clay, M.R.; Pinto, E.M.; Kesserwan, C. An update on the central nervous system manifestations of Li–Fraumeni syndrome. *Acta Neuropathol* **2020**, *139*, 669–687. [CrossRef]
10. Han, S.; Liu, Y.; Cai, S.J.; Qian, M.; Ding, J.; Larion, M.; Gilbert, M.R.; Yang, C. IDH mutation in glioma: Molecular mechanisms and potential therapeutic targets. *Br. J. Cancer* **2020**, *122*, 1580–1589. [CrossRef]
11. Bastiancich, C.; Danhier, P.; Préat, V.; Danhier, F. Anticancer drug-loaded hydrogels as drug delivery systems for the local treatment of glioblastoma. *J. Control. Release* **2016**, *243*, 29–42. [CrossRef]

12. Janjua, T.I.; Rewatkar, P.; Ahmed-Cox, A.; Saeed, I.; Mansfeld, F.M.; Kulshreshtha, R.; Kumeria, T.; Ziegler, D.S.; Kavallaris, M.; Mazzieri, R.; et al. Frontiers in the treatment of glioblastoma: Past, present and emerging. *Adv. Drug Deliv Rev.* **2021**, *171*, 108–138. [CrossRef]
13. Piraino, S.W.; Thomas, V.; O'Donovan, P.; Furney, S.J. Mutations: Driver Versus Passenger. In *Encyclopedia of Cancer*, 3rd ed.; Boffetta, P., Hainaut, P., Eds.; Academic Press: Salt Lake City, UT, USA, 2019; pp. 551–562. [CrossRef]
14. Friedmann-Morvinski, D. Glioblastoma Heterogeneity and Cancer Cell Plasticity. *Crit Rev. Oncog.* **2014**, *19*, 327–336. [CrossRef] [PubMed]
15. Aum, D.J.; Kim, D.H.; Beaumont, T.L.; Leuthardt, E.C.; Dunn, G.P.; Kim, A.H. Molecular and cellular heterogeneity: The hallmark of glioblastoma. *Neurosurg. Focus* **2014**, *37*, E11. [CrossRef] [PubMed]
16. Kleihues, P.; Ohgaki, H. Primary and secondary glioblastomas: From concept to clinical diagnosis. *Neuro-Oncology* **1999**, *1*, 44–51. [CrossRef] [PubMed]
17. Verhaak, R.G.W.; Hoadley, K.A.; Purdom, E.; Wang, V.; Qi, Y.; Wilkerson, M.D.; Miller, C.R.; Ding, L.; Golub, T.; Mesirov, J.P.; et al. An integrated genomic analysis identifies clinically relevant subtypes of glioblastoma characterized by abnormalities in PDGFRA, IDH1, EGFR and NF1. *Cancer Cell.* **2010**, *17*, 98. [CrossRef] [PubMed]
18. Couturier, C.P.; Ayyadhury, S.; Le, P.U.; Nadaf, J.; Monlong, J.; Riva, G.; Allache, R.; Baig, S.; Yan, X.; Bourgey, M.; et al. Single-cell RNA-seq reveals that glioblastoma recapitulates a normal neurodevelopmental hierarchy. *Nat. Commun.* **2020**, *11*, 3406. [CrossRef]
19. Guelfi, S.; Duffau, H.; Bauchet, L.; Rothhut, B.; Hugnot, J.P. Vascular Transdifferentiation in the CNS: A Focus on Neural and Glioblastoma Stem-Like Cells. *Stem Cells Int.* **2016**, *2016*, e2759403. [CrossRef]
20. Matarredona, E.R.; Pastor, A.M. Neural Stem Cells of the Subventricular Zone as the Origin of Human Glioblastoma Stem Cells. Therapeutic Implications. *Front. Oncol.* **2019**, *9*, 779. [CrossRef]
21. Auffinger, B.; Spencer, D.; Pytel, P.; Ahmed, A.U.; Lesniak, M.S. The role of glioma stem cells in chemotherapy resistance and glioblastoma multiforme recurrence. *Expert Rev. Neurother.* **2015**, *15*, 741–752. [CrossRef]
22. Luo, H.; Shusta, E.V. Blood-Brain Barrier Modulation to Improve Glioma Drug Delivery. *Pharmaceutics* **2020**, *12*, 1085. [CrossRef]
23. Dong, X. Current Strategies for Brain Drug Delivery. *Theranostics* **2018**, *8*, 1481–1493. [CrossRef]
24. Daneman, R.; Prat, A. The Blood–Brain Barrier. *Cold Spring Harb. Perspect Biol.* **2015**, *7*, a020412. [CrossRef] [PubMed]
25. Burgess, A.; Shah, K.; Hough, O.; Hynynen, K. Focused ultrasound-mediated drug delivery through the blood-brain barrier. *Expert Rev. Neurother.* **2015**, *15*, 477–491. [CrossRef] [PubMed]
26. Cao, Y.; Sundgren, P.C.; Tsien, C.I.; Chenevert, T.T.; Junck, L. Physiologic and metabolic magnetic resonance imaging in gliomas. *J. Clin. Oncol.* **2006**, *24*, 1228–1235. [CrossRef]
27. Abbott, N.J.; Patabendige, A.A.K.; Dolman, D.E.M.; Yusof, S.R.; Begley, D.J. Structure and function of the blood-brain barrier. *Neurobiol. Dis.* **2010**, *37*, 13–25. [CrossRef] [PubMed]
28. Li, J.; Zheng, M.; Shimoni, O.; Banks, W.A.; Bush, A.I.; Gamble, J.R.; Shi, B. Development of Novel Therapeutics Targeting the Blood–Brain Barrier: From Barrier to Carrier. *Adv. Sci.* **2021**, *8*, 2101090. [CrossRef]
29. Rapoport, S.I. Osmotic opening of the blood-brain barrier: Principles, mechanism, and therapeutic applications. *Cell. Mol. Neurobiol.* **2000**, *20*, 217–230. [CrossRef] [PubMed]
30. Wala, K.; Szlasa, W.; Saczko, J.; Rudno-Rudzińska, J.; Kulbacka, J. Modulation of Blood–Brain Barrier Permeability by Activating Adenosine A2 Receptors in Oncological Treatment. *Biomolecules* **2021**, *11*, 633. [CrossRef]
31. Sawyer, A.J.; Piepmeier, J.M.; Saltzman, W.M. New Methods for Direct Delivery of Chemotherapy for Treating Brain Tumors. *Yale J. Biol. Med.* **2006**, *79*, 141–152.
32. Zhou, J.; Atsina, K.B.; Himes, B.T.; Strohbehn, G.W.; Saltzman, W.M. Novel Delivery Strategies for Glioblastoma. *Cancer J. Sudbury Mass.* **2012**, *18*, 89–99. [CrossRef]
33. Yeini, E.; Ofek, P.; Albeck, N.; Ajamil, D.R.; Neufeld, L.; Eldar-Boock, A.; Kleiner, R.; Vaskovich, D.; Koshrovski-Michael, S.; Dangoor, S.I.; et al. Targeting Glioblastoma: Advances in Drug Delivery and Novel Therapeutic Approaches. *Adv. Ther.* **2021**, *4*, 2000124. [CrossRef]
34. Shapira-Furman, T.; Serra, R.; Gorelick, N.; Doglioli, M.; Tagliaferri, V.; Cecia, A.; Peters, M.; Kumar, A.; Rottenberg, Y.; Langer, R.; et al. Biodegradable wafers releasing Temozolomide and Carmustine for the treatment of brain cancer. *J. Control. Release* **2019**, *295*, 93–101. [CrossRef] [PubMed]
35. Tyler, B.; Wadsworth, S.; Recinos, V.; Mehta, V.; Vellimana, A.; Li, K.; Rosenblatt, J.; Do, H.; Gallia, G.L.; Siu, I.M.; et al. Local delivery of rapamycin: A toxicity and efficacy study in an experimental malignant glioma model in rats. *Neuro-Oncology* **2011**, *13*, 700–709. [CrossRef] [PubMed]
36. Dluska, E.; Markowska-Radomska, A.; Metera, A.; Ordak, M. Multiple Emulsions as a Biomaterial-Based Delivery System for the Controlled Release of an Anti-cancer Drug. *J. Phys. Conf. Ser.* **2020**, *1681*, 012021. [CrossRef]
37. Zhao, M.; Bozzato, E.; Joudiou, N.; Ghiassinejad, S.; Danhier, F.; Gallez, B.; Préat, V. Codelivery of paclitaxel and temozolomide through a photopolymerizable hydrogel prevents glioblastoma recurrence after surgical resection. *J. Control. Release* **2019**, *309*, 72–81. [CrossRef]
38. Basso, J.; Miranda, A.; Nunes, S.; Cova, T.; Sousa, J.; Vitorino, C.; Pais, A. Hydrogel-Based Drug Delivery Nanosystems for the Treatment of Brain Tumors. *Gels* **2018**, *4*, 62. [CrossRef]
39. Jahangiri, A.; Chin, A.T.; Flanigan, P.M.; Chen, R.; Bankiewicz, K.; Aghi, M.K. Convection-enhanced delivery in glioblastoma: A review of preclinical and clinical studies. *J. Neurosurg.* **2017**, *126*, 191–200. [CrossRef]

40. Enríquez Pérez, J.; Kopecky, J.; Visse, E.; Darabi, A.; Siesjö, P. Convection-enhanced delivery of temozolomide and whole cell tumor immunizations in GL261 and KR158 experimental mouse gliomas. *BMC Cancer* **2020**, *20*, 7. [CrossRef]
41. Wang, J.L.; Barth, R.F.; Cavaliere, R.; Puduvalli, V.K.; Giglio, P.; Lonser, R.R.; Elder, J.B. Phase I trial of intracerebral convection-enhanced delivery of carboplatin for treatment of recurrent high-grade gliomas. *PLoS ONE* **2020**, *15*, e0244383. [CrossRef]
42. Muldoon, L.L.; Soussain, C.; Jahnke, K.; Johanson, C.; Siegal, T.; Smith, Q.R.; Hall, W.A.; Hynynen, K.; Senter, P.D.; Peereboom, D.M.; et al. Chemotherapy Delivery Issues in Central Nervous System Malignancy: A Reality Check. *J. Clin. Oncol.* **2007**, *25*, 2295–2305. [CrossRef]
43. Hadjipanayis, C.G.; Machaidze, R.; Kaluzova, M.; Wang, L.; Schuette, A.J.; Chen, H.; Wu, X.; Mao, H. EGFRvIII Antibody–Conjugated Iron Oxide Nanoparticles for Magnetic Resonance Imaging–Guided Convection-Enhanced Delivery and Targeted Therapy of Glioblastoma. *Cancer Res.* **2010**, *70*, 6303–6312. [CrossRef]
44. Nwagwu, C.D.; Immidisetti, A.V.; Jiang, M.Y.; Adeagbo, O.; Adamson, D.C.; Carbonell, A.M. Convection Enhanced Delivery in the Setting of High-Grade Gliomas. *Pharmaceutics* **2021**, *13*, 561. [CrossRef] [PubMed]
45. Mehta, A.M.; Sonabend, A.M.; Bruce, J.N. Convection-Enhanced Delivery. *Neurotherapeutics* **2017**, *14*, 358–371. [CrossRef] [PubMed]
46. Tran, S.; DeGiovanni, P.J.; Piel, B.; Rai, P. Cancer nanomedicine: A review of recent success in drug delivery. *Clin. Transl. Med.* **2017**, *6*, e44. [CrossRef] [PubMed]
47. Wicki, A.; Witzigmann, D.; Balasubramanian, V.; Huwyler, J. Nanomedicine in cancer therapy: Challenges, opportunities, and clinical applications. *J. Control. Release* **2015**, *200*, 138–157. [CrossRef]
48. Shi, J.; Kantoff, P.W.; Wooster, R.; Farokhzad, O.C. Cancer nanomedicine: Progress, challenges and opportunities. *Nat. Rev. Cancer* **2017**, *17*, 20–37. [CrossRef]
49. Michael, J.S.; Lee, B.S.; Zhang, M.; Yu, J.S. Nanotechnology for Treatment of Glioblastoma Multiforme. *J. Transl. Intern. Med.* **2018**, *6*, 128–133. [CrossRef]
50. Kalyane, D.; Raval, N.; Maheshwari, R.; Tambe, V.; Kalia, K.; Tekade, R.K. Employment of enhanced permeability and retention effect (EPR): Nanoparticle-based precision tools for targeting of therapeutic and diagnostic agent in cancer. *Mater. Sci. Eng. C.* **2019**, *98*, 1252–1276. [CrossRef]
51. Greish, K. Enhanced Permeability and Retention (EPR) Effect for Anticancer Nanomedicine Drug Targeting. In *Cancer Nanotechnology: Methods and Protocols. Methods in Molecular Biology*; Grobmyer, S.R., Moudgil, B.M., Eds.; Humana Press: New York, NY, USA, 2010; pp. 25–37. [CrossRef]
52. Brannon-Peppas, L.; Blanchette, J.O. Nanoparticle and targeted systems for cancer therapy. *Adv. Drug Deliv. Rev.* **2004**, *56*, 1649–1659. [CrossRef]
53. Alphandéry, E. Nano-Therapies for Glioblastoma Treatment. *Cancers* **2020**, *12*, 242. [CrossRef]
54. Mout, R.; Moyano, D.F.; Rana, S.; Rotello, V.M. Surface functionalization of nanoparticles for nanomedicine. *Chem. Soc. Rev.* **2012**, *41*, 2539–2544. [CrossRef]
55. Taiarol, L.; Formicola, B.; Magro, R.D.; Sesana, S.; Re, F. An update of nanoparticle-based approaches for glioblastoma multiforme immunotherapy. *Nanomedicine* **2020**, *15*, 1861–1871. [CrossRef] [PubMed]
56. Ngobili, T.A.; Daniele, M.A. Nanoparticles and direct immunosuppression. *Exp. Biol. Med.* **2016**, *241*, 1064–1073. [CrossRef] [PubMed]
57. Blank, F.; Gerber, P.; Rothen-Rutishauser, B.; Sakulkhu, U.; Salaklang, J.; De Peyer, K.; Gehr, P.; Nicod, L.P.; Hofmann, H.; Geiser, T.; et al. Biomedical nanoparticles modulate specific CD4+ T cell stimulation by inhibition of antigen processing in dendritic cells. *Nanotoxicology* **2011**, *5*, 606–621. [CrossRef] [PubMed]
58. Budama-Kilinc, Y.; Kecel-Gunduz, S.; Cakir-Koc, R.; Aslan, B.; Bicak, B.; Kokcu, Y.; Ozel, A.E.; Akyuz, S. Structural Characterization and Drug Delivery System of Natural Growth-Modulating Peptide Against Glioblastoma Cancer. *Int. J. Pept. Res. Ther.* **2021**, *27*, 2015–2028. [CrossRef]
59. Pickart, L.; Vasquez-Soltero, J.M.; Margolina, A. The Effect of the Human Peptide GHK on Gene Expression Relevant to Nervous System Function and Cognitive Decline. *Brain Sci.* **2017**, *7*, 20. [CrossRef]
60. Chan, M.H.; Chen, W.; Li, C.H.; Fang, C.Y.; Chang, Y.C.; Wei, D.H.; Liu, R.S.; Hsiao, M. An Advanced In Situ Magnetic Resonance Imaging and Ultrasonic Theranostics Nanocomposite Platform: Crossing the Blood–Brain Barrier and Improving the Suppression of Glioblastoma Using Iron-Platinum Nanoparticles in Nanobubbles. *ACS Appl. Mater. Interfaces* **2021**, *13*, 26759–26769. [CrossRef]
61. Pasut, G. Grand Challenges in Nano-Based Drug Delivery. *Front. Med. Technol.* **2019**, *1*, 1. Available online: https://www.frontiersin.org/article/10.3389/fmedt.2019.00001 (accessed on 28 January 2022). [CrossRef]
62. Sharma, D.; Sharma, N.; Pathak, M.; Agrawala, P.K.; Basu, M.; Ojha, H. Chapter 2—Nanotechnology-based drug delivery systems: Challenges and opportunities. In *Drug Targeting and Stimuli Sensitive Drug Delivery Systems*; Grumezescu, A.M., Ed.; William Andrew Publishing: Norwich, NY, USA, 2018; pp. 39–79. [CrossRef]
63. Rampado, R.; Crotti, S.; Caliceti, P.; Pucciarelli, S.; Agostini, M. Recent Advances in Understanding the Protein Corona of Nanoparticles and in the Formulation of "Stealthy" Nanomaterials. *Front. Bioeng. Biotechnol.* **2020**, *8*, 166. Available online: https://www.frontiersin.org/article/10.3389/fbioe.2020.00166 (accessed on 26 January 2022). [CrossRef]
64. Raucher, D. Tumor targeting peptides: Novel therapeutic strategies in glioblastoma. *Curr. Opin. Pharmacol.* **2019**, *47*, 14–19. [CrossRef]

65. Polisetty, R.V.; Gautam, P.; Sharma, R.; Harsha, H.C.; Nair, S.C.; Gupta, M.K.; Uppin, M.S.; Challa, S.; Puligopu, A.K.; Ankathi, P.; et al. LC-MS/MS analysis of differentially expressed glioblastoma membrane proteome reveals altered calcium signaling and other protein groups of regulatory functions. *Mol. Cell. Proteom.* **2012**, *11*, M111.013565. [CrossRef]
66. Kondo, E.; Iioka, H.; Saito, K. Tumor-homing peptide and its utility for advanced cancer medicine. *Cancer Sci.* **2021**, *112*, 2118–2125. [CrossRef] [PubMed]
67. Audrey, G.; Claire, L.C.; Joel, E. Effect of the NFL-TBS.40-63 peptide on canine glioblastoma cells. *Int. J. Pharm.* **2021**, *605*, 120811. [CrossRef] [PubMed]
68. Rahn, J.J.; Lun, X.; Jorch, S.K.; Hao, X.; Venugopal, C.; Vora, P.; Ahn, B.Y.; Babes, L.; Alshehri, M.M.; Cairncross, J.; et al. Development of a peptide-based delivery platform for targeting malignant brain tumors. *Biomaterials* **2020**, *252*, 120105. [CrossRef] [PubMed]
69. Rizvi, S.F.A.; Shahid, S.; Mu, S.; Zhang, H. Hybridization of tumor homing and mitochondria-targeting peptide domains to design novel dual-imaging self-assembled peptide nanoparticles for theranostic applications. *Drug Deliv. Transl. Res.* **2021**, 1–12. [CrossRef] [PubMed]
70. Hua, H.; Zhang, X.; Mu, H.; Meng, Q.; Jiang, Y.; Wang, Y.; Lu, X.; Wang, A.; Liu, S.; Zhang, Y.; et al. RVG29-modified docetaxel-loaded nanoparticles for brain-targeted glioma therapy. *Int. J. Pharm.* **2018**, *543*, 179–189. [CrossRef] [PubMed]
71. Xin, X.; Liu, W.; Zhang, Z.A.; Han, Y.; Qi, L.-L.; Zhang, Y.-Y.; Zhang, X.-T.; Duan, H.-X.; Chen, L.-Q.; Jin, M.-J.; et al. Efficient Anti-Glioma Therapy Through the Brain-Targeted RVG15-Modified Liposomes Loading Paclitaxel-Cholesterol Complex. *Int. J. Nanomed.* **2021**, *16*, 5755–5776. [CrossRef]
72. Fan, Y.; Cui, Y.; Hao, W.; Chen, M.; Liu, Q.; Wang, Y.; Yang, M.; Li, Z.; Gong, W.; Song, S.; et al. Carrier-free highly drug-loaded biomimetic nanosuspensions encapsulated by cancer cell membrane based on homology and active targeting for the treatment of glioma. *Bioact. Mater.* **2021**, *6*, 4402–4414. [CrossRef] [PubMed]
73. Zhang, H.; van Os, W.L.; Tian, X.; Zu, G.; Ribovski, L.; Bron, R.; Bussmann, J.; Kros, A.; Liu, Y.S.; Zuhorn, I. Development of curcumin-loaded zein nanoparticles for transport across the blood–brain barrier and inhibition of glioblastoma cell growth. *Biomater. Sci.* **2021**, *9*, 7092–7103. [CrossRef]
74. Kunnumakkara, A.B.; Bordoloi, D.; Padmavathi, G.; Monisha, J.; Roy, N.K.; Prasad, S.; Aggarwal, B.B. Curcumin, the golden nutraceutical: Multitargeting for multiple chronic diseases. *Br. J. Pharmacol.* **2017**, *174*, 1325–1348. [CrossRef]
75. Shabaninejad, Z.; Pourhanifeh, M.H.; Movahedpour, A.; Mottaghi, R.; Nickdasti, A.; Mortezapour, E.; Shafiee, A.; Hajighadimi, S.; Moradizarmehri, S.; Sadeghian, M.; et al. Therapeutic potentials of curcumin in the treatment of glioblstoma. *Eur. J. Med. Chem.* **2020**, *188*, 112040. [CrossRef]
76. Zorofchian Moghadamtousi, S.; Abdul Kadir, H.; Hassandarvish, P.; Tajik, H.; Abubakar, S.; Zandi, K. A Review on Antibacterial, Antiviral, and Antifungal Activity of Curcumin. *BioMed Res. Int.* **2014**, *2014*, e186864. [CrossRef] [PubMed]
77. Hua, D.; Tang, L.; Wang, W.; Tang, S.; Yu, L.; Zhou, X.; Wang, Q.; Sun, C.; Shi, C.; Luo, W.; et al. Improved Antiglioblastoma Activity and BBB Permeability by Conjugation of Paclitaxel to a Cell-Penetrative MMP-2-Cleavable Peptide. *Adv. Sci.* **2021**, *8*, 2001960. [CrossRef] [PubMed]
78. Ayo, A.; Laakkonen, P. Peptide-Based Strategies for Targeted Tumor Treatment and Imaging. *Pharmaceutics* **2021**, *13*, 481. [CrossRef] [PubMed]
79. Demeule, M.; Currie, J.C.; Bertrand, Y.; Ché, C.; Nguyen, T.; Régina, A.; Gabathuler, R.; Castaigne, J.P.; Béliveau, R. Involvement of the low-density lipoprotein receptor-related protein in the transcytosis of the brain delivery vector angiopep-2. *J. Neurochem.* **2008**, *106*, 1534–1544. [CrossRef]
80. Hong, H.-Y.; Lee, H.Y.; Kwak, W.; Yoo, J.; Na, M.-H.; So, I.S.; Kwon, T.-H.; Park, H.-S.; Huh, S.; Oh, G.T.; et al. Phage display selection of peptides that home to atherosclerotic plaques: IL-4 receptor as a candidate target in atherosclerosis. *J. Cell. Mol. Med.* **2008**, *12*, 2003–2014. [CrossRef] [PubMed]
81. Bunevicius, A.; McDannold, N.J.; Golby, A.J. Focused Ultrasound Strategies for Brain Tumor Therapy. *Oper. Neurosurg.* **2020**, *19*, 9–18. [CrossRef]
82. Wu, S.K.; Tsai, C.L.; Huang, Y.; Hynynen, K. Focused Ultrasound and Microbubbles-Mediated Drug Delivery to Brain Tumor. *Pharmaceutics* **2020**, *13*, 15. [CrossRef]
83. Cohen-Inbar, O.; Xu, Z.; Sheehan, J.P. Focused ultrasound-aided immunomodulation in glioblastoma multiforme: A therapeutic concept. *J. Ther. Ultrasound.* **2016**, *4*, 2. [CrossRef]
84. Wei, K.C.; Chu, P.C.; Wang, H.Y.J.; Huang, C.Y.; Chen, P.Y.; Tsai, H.C.; Lu, Y.J.; Lee, P.I.; Tseng, I.C.; Feng, L.Y.; et al. Focused ultrasound-induced blood-brain barrier opening to enhance temozolomide delivery for glioblastoma treatment: A preclinical study. *PLoS ONE* **2013**, *8*, e58995. [CrossRef]
85. Liu, H.L.; Hua, M.Y.; Chen, P.Y.; Chu, P.C.; Pan, C.H.; Yang, H.W.; Huang, C.Y.; Wang, J.J.; Yen, T.C.; Wei, K.C. Blood-Brain Barrier Disruption with Focused Ultrasound Enhances Delivery of Chemotherapeutic Drugs for Glioblastoma Treatment. *Radiology* **2010**, *255*, 415–425. [CrossRef]
86. Papachristodoulou, A.; Signorell, R.D.; Werner, B.; Brambilla, D.; Luciani, P.; Cavusoglu, M.; Grandjean, J.; Silginer, M.; Rudin, M.; Martin, E.; et al. Chemotherapy sensitization of glioblastoma by focused ultrasound-mediated delivery of therapeutic liposomes. *J. Control. Release* **2019**, *295*, 130–139. [CrossRef] [PubMed]

87. Hegi, M.E.; Diserens, A.C.; Gorlia, T.; Hamou, M.F.; de Tribolet, N.; Weller, M.; Kros, J.M.; Hainfellner, J.A.; Mason, W.; Mariani, L.; et al. MGMT Gene Silencing and Benefit from Temozolomide in Glioblastoma. *N. Engl. J. Med.* **2005**, *352*, 997–1003. [CrossRef] [PubMed]
88. Treat, L.H.; McDannold, N.; Vykhodtseva, N.; Zhang, Y.; Tam, K.; Hynynen, K. Targeted delivery of doxorubicin to the rat brain at therapeutic levels using MRI-guided focused ultrasound. *Int. J. Cancer* **2007**, *121*, 901–907. [CrossRef] [PubMed]
89. Coluccia, D.; Figueiredo, C.A.; Wu, M.Y.; Riemenschneider, A.N.; Diaz, R.; Luck, A.; Smith, C.; Das, S.; Ackerley, C.; O'Reilly, M.; et al. Enhancing glioblastoma treatment using cisplatin-gold-nanoparticle conjugates and targeted delivery with magnetic resonance-guided focused ultrasound. *Nanomed. Nanotechnol. Biol. Med.* **2018**, *14*, 1137–1148. [CrossRef]
90. Meng, Y.; Reilly, R.M.; Pezo, R.C.; Trudeau, M.; Sahgal, A.; Singnurkar, A.; Perry, J.; Myrehaug, S.; People, C.B.; Davidson, B.; et al. MR-guided focused ultrasound enhances delivery of trastuzumab to Her2-positive brain metastases. *Sci. Transl. Med.* **2021**, *13*, eabj4011. [CrossRef]
91. Timbie, K.F.; Mead, B.P.; Price, R.J. Drug and gene delivery across the blood–brain barrier with focused ultrasound. *J. Control. Release* **2015**, *219*, 61–75. [CrossRef]
92. ClinicalTrials.gov [Internet]. Bethesda (MD): National Library of Medicine (US).. Assessment of Safety and Feasibility of ExAblate Blood-Brain Barrier Disruption for the Treatment of High Grade Glioma in Patients Undergoing Standard Chemotherapy. 2021. Available online: https://clinicaltrials.gov/ct2/show/NCT03551249 (accessed on 6 December 2021).
93. ClinicalTrials.gov [Internet]. Bethesda (MD): National Library of Medicine (US).. A Study to Evaluate the Safety and Feasibility of Exablate Model 4000 Type-2 to Temporarily Mediate Blood-Brain Barrier Disruption (BBBD) in Patients with Suspected Infiltrating Glioma in the Setting of Planned Surgical Interventions. 2021. Available online: https://clinicaltrials.gov/ct2/show/NCT03322813 (accessed on 6 December 2021).
94. Wang, D.; Wang, C.; Wang, L.; Chen, Y. A comprehensive review in improving delivery of small-molecule chemotherapeutic agents overcoming the blood-brain/brain tumor barriers for glioblastoma treatment. *Drug Deliv.* **2019**, *26*, 551–565. [CrossRef]

International Journal of
Molecular Sciences

Article

Indomethacin and Diclofenac Hybrids with Oleanolic Acid Oximes Modulate Key Signaling Pathways in Pancreatic Cancer Cells

Maria Narożna [1,2], Violetta Krajka-Kuźniak [1], Robert Kleszcz [1] and Wanda Baer-Dubowska [1,*]

1 Department of Pharmaceutical Biochemistry, Poznan University of Medical Sciences, 4, Święcicki Street, 60-781 Poznań, Poland; maria.narozna@ump.edu.pl (M.N.); vkrajka@ump.edu.pl (V.K.-K.); kleszcz@ump.edu.pl (R.K.)
2 Program in Cell Cycle and Cancer Biology, Oklahoma Medical Research Foundation, 825, NE 13th Street, Oklahoma City, OK 73104, USA
* Correspondence: baerw@ump.edu.pl

Abstract: Our earlier studies showed that coupling nonsteroidal anti-inflammatory drugs (NSAIDs) with oleanolic acid derivatives increased their anti-inflammatory activity in human hepatoma cells. The aim of this study was to evaluate their effect on the signaling pathways involved in inflammation processes in human pancreatic cancer (PC) cells. Cultured PSN-1 cells were exposed for 24 h (30 μM) to OA oxime (OAO) derivatives substituted with benzyl or morpholide groups and their conjugates with indomethacin (IND) or diclofenac (DCL). The activation of NF-κB and Nrf2 was assessed by the evaluation of the translocation of their active forms into the nucleus and their binding to specific DNA sequences via the ELISA assay. The expression of NF-κB and Nrf2 target genes was evaluated by R-T PCR and Western blot analysis. The conjugation of IND or DCL with OAO derivatives increased cytotoxicity and their effect on the tested signaling pathways. The most effective compound was the DCL hybrid with OAO morpholide (4d). This compound significantly reduced the activation and expression of NF-κB and enhanced the activation and expression of Nrf2. Increased expression of Nrf2 target genes led to reduced ROS production. Moreover, MAPKs and the related pathways were also affected. Therefore, conjugate 4d deserves more comprehensive studies as a potential PC therapeutic agent.

Keywords: diclofenac; indomethacin; oleanolic acid derivative conjugates; NF-κB; Nrf2; MAPKs; PSN-1 cells; inflammation; reactive oxygen species

Citation: Narożna, M.; Krajka-Kuźniak, V.; Kleszcz, R.; Baer-Dubowska, W. Indomethacin and Diclofenac Hybrids with Oleanolic Acid Oximes Modulate Key Signaling Pathways in Pancreatic Cancer Cells. *Int. J. Mol. Sci.* **2022**, 23, 1230. https://doi.org/10.3390/ijms23031230

Academic Editor: Angela Stefanachi

Received: 21 December 2021
Accepted: 19 January 2022
Published: 22 January 2022

1. Introduction

Pancreatic cancer (PC) is a major cause of cancer-associated mortality worldwide, with a dismal overall prognosis that has remained practically unchanged for many decades. Prevention or early diagnosis at a curable stage is still exceedingly difficult; patients rarely exhibit symptoms, and tumors do not display sensitive and specific markers to aid detection [1].

Chronic inflammation is considered a significant risk factor of many cancers, and pancreatic cancer is one of them. During this process, upregulation of pro-inflammatory molecules and pathways provides a favorable microenvironment for the exponential growth of malignant cells. Therefore, inflammation may provide both the critical mutations resulting from the DNA damage, e.g., by reactive oxygen species (ROS), and create the proper environment to foster tumor growth [2].

The NF-κB and Nrf2 signaling pathways play key roles in regulating the body's responses to stress and inflammation. Moreover, several lines of evidence suggest that these pathways play a crucial role in PC development, progression and resistance, and both transcription factors are often overexpressed in pancreatic cancer cells [3,4].

NF-κB comprises a family of conserved and structurally related proteins, including RelA/p65, Rel/cRel, RelB, NF-κB1/p50, and NF-B2/p52. When inactivated, NF-κB is found in the cytosol bound to an IκB inhibitor protein (IκBα). Through the involvement of membrane receptors, a variety of extracellular signals can activate the IκB kinase (IKK) enzyme. IKK, in turn, phosphorylates the IκBα protein, leading to its ubiquitination and degradation by the proteasome [5].

Nrf2 is associated with the Keap1 protein and resides in cytosol. In normal conditions, Keap1 targets Nrf2 for ubiquitylation, leading to its proteasomal degradation [6]. In response to chemical or oxidative stress, the interaction between Nrf2 and Keap1 is perturbed, resulting in the stabilization and nuclear accumulation of Nrf2 [7–9]. The Nrf2 in the nucleus interacts with antioxidant response elements (ARE) in the promoter regions of a plethora of genes coding for phase 2 detoxifying enzymes (e.g., glutathione-S-transferases and NAD(P)H quinone oxidoreductase), antioxidant enzymes (e.g., SOD and GPx) and several other cytoprotective proteins [10–13].

For a long time, Nrf2 was considered an attractive and effective target for chemopreventive strategies [14,15] to maintain electrophilic and cellular redox balance. Through the induction of antioxidative target genes, Nrf2 could protect non-transformed cells against DNA damage and, thereby, may prevent mutagenesis. However, there is growing evidence that Nrf2 activation can lead to tumor development and enhanced chemoresistance once a malignant transformation occurs. Stable overexpression of Nrf2 results in the enhanced resistance of cancer cells to chemotherapeutics [16].

The interplay between these pathways occurs through a range of complex molecular interactions and can often depend on the cell type and tissue context. These interactions operate via both transcriptional and posttranscriptional mechanisms, allowing dynamic responses to ever-changing environmental cues. Despite convincing evidence for functional interactions between the Nrf2 and NF-κB pathways, many aspects of the conditional and dynamic nature of crosstalk remain unknown [16]. However, often, an inverse relationship between these pathways is observed. In this regard, anti-inflammatory agents that suppress NF-κB signaling may activate Nrf2. In addition, Nrf2 may reverse the cellular damage of inflammation-exposed cells and indirectly prevent the sustained activation of NF-κB [16].

Apart from NF-κB and Nrf2, several other signaling pathways might be involved in inflammation and, thus, pancreatic cancer induction. These include MAPKs and the Jak/STAT, CREB, and PI3K/Akt pathways [17]. In this regard, up to 60% of pancreatic cancer tissues and most pancreatic cancer cell lines exhibit increased Akt activity [18]. Moreover, crosstalk between these pathways is possible [17]. Therefore, suppressing multiple signaling pathways involved in inflammation responses may be a promising approach for the treatment of pancreatic cancer.

Several studies have shown that the use of nonsteroidal anti-inflammatory drugs (NSAIDs) is associated with decreased cancer incidence and recurrence. The specific COX-2 inhibitor, celecoxib, was approved by the FDA for use in the adjuvant therapy of familial adenomatous polyposis [19]. Beneficial effects after treatment with several COXIBs, e.g., celecoxib and 2,5-dimethylcelecoxib, were also observed in glioblastoma cells [20]. Moreover, early studies showed that long-term use (>2 years) of NSAIDs may protect against pancreatic cancer, but replication of these findings was recommended [21]. However, in general, anti-inflammatory agents, even those that obtained FDA approval, have limited use in cancer prophylaxis or therapy because of the presence of off-target effects and toxicities [19]. Therefore, it is important to develop safe and effective agents that reduce cancer risk. Coupling NSAIDs with natural compounds exhibiting anti-inflammatory potential might enhance their anti-inflammatory activity and prevent the unfavorable side effects of the long-term use of NSAIDs. One such compound is the naturally occurring triterpenoid oleanolic acid (OA) and its synthetic derivatives. Our earlier studies showed that conjugates of NSAIDs, namely, aspirin, indomethacin (IND), or diclofenac (DCL) with novel synthetic oxime derivatives of OA (OAO) enhanced their modulating effect on Nrf2-ARE and NF-κB signaling pathways in hepatoma cancer cells (HCC), resulting in increased apoptosis rate and reduced cell proliferation. Moreover, the DCL hybrid with OAO morpholide derivative reduced the tumor volume in mice bearing xenografts [22].

The aim of our present study was to evaluate the effect of the IND and DCL conjugates (3c, 3d, 4c, 4d) with two OAO derivatives, namely, benzyl ester and morpholide (2c, 2d) (Figure 1), on the signaling pathways involved in the inflammation process in pancreatic cancer cells, and their possible effects on cell cycle distribution and proliferation.

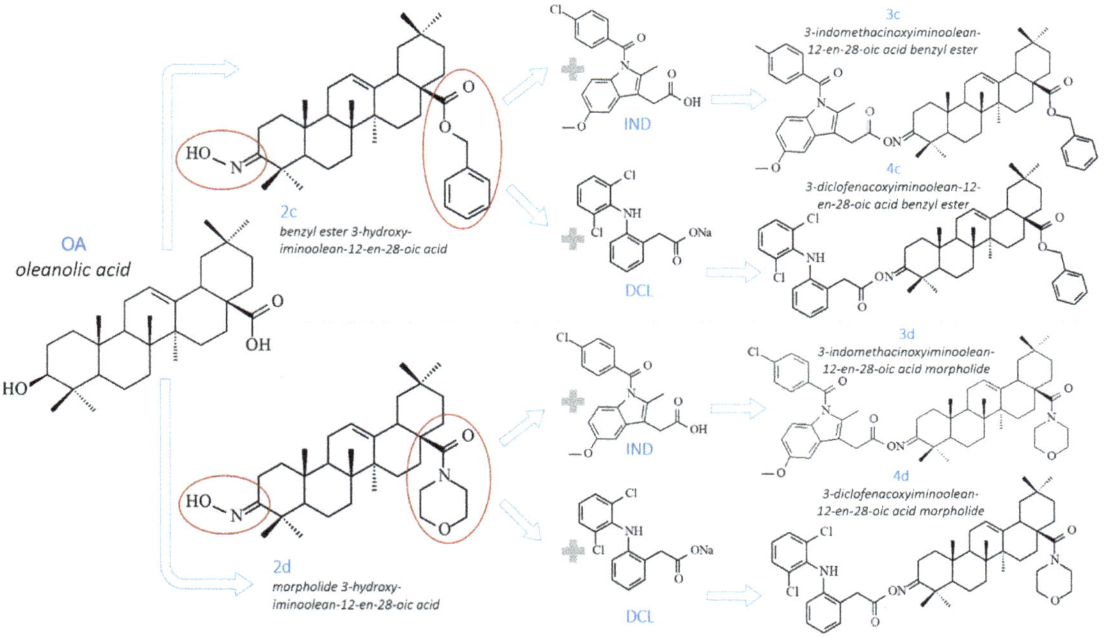

Figure 1. The chemical structures of oleanolic acid (OA), oleanolic acid oximes derivatives: benzyl ester, and morpholide (2c, 2d), and their conjugates with IND (3c, 3d) and DCL (4c, 4d).

2. Results

2.1. Conjugation with OAO Derivatives Increases the Cytotoxicity of Indomethacin and Diclofenac

The viability of pancreatic adenocarcinoma-derived PSN-1 cells and immortalized pancreatic endothelial cells MS-1 was assessed after treatment with OA, OAO, DCL, IND, and their conjugates at the concentration range of 1–150 μM. As shown in Figure 2A,B, and Table 1, the conjugation of IND and DCL with OAO significantly increased their cytotoxicity, reaching even 100% in PSN-1 cells at the highest tested concentration. Comparing the IC_{50} values obtained on both cell lines, slightly higher cytotoxicity toward the cancer PSN-1 cells was noticed.

Based on the MTT assay results, the concentrations of 30 μM assuring 70% viability were applied in the further assays.

Table 1. Comparison of the cytotoxicity of the tested compounds in cancer PSN-1 cells and immortalized MS-1 cells *.

Compound	PSN-1 *	MS-1 *
OA	>150	>150
2c	>150	>150
2d	126.56 ± 1.22	>150
IND	>150	>150
3c	87.55 ± 4.87	97.50 ± 1.49
3d	71.89 ± 2.98	93.75 ± 0.85
DCL	>150	>150
4c	103.13 ± 0.28	125.00 ± 2.44
4d	55.21 ± 1.97	69.50 ± 1.83

* IC_{50} values ± SEM (μM).

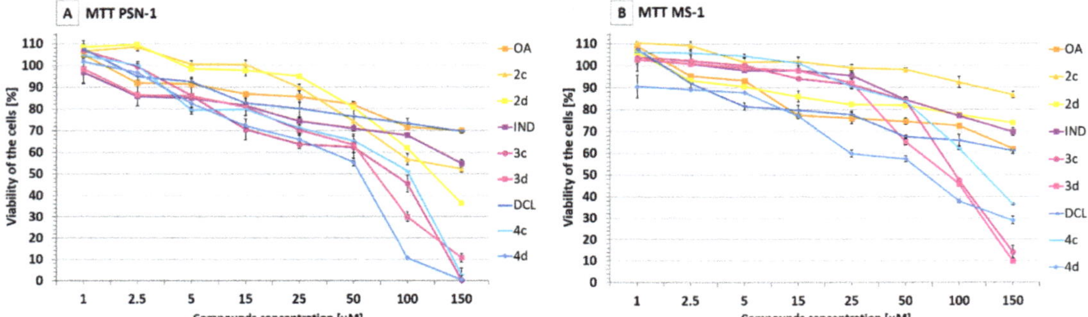

Figure 2. The effect of the OA, OAO derivatives (2c, 2d), and OAO conjugates with IND (3c, 3d) and DCL (4c, 4d) on the viability of PSN-1 (**A**) and MS-1 cells (**B**). Data are expressed as means ± SEM from three separate measurements.

2.2. IND and DCL-OAO Conjugates Diminish the Activation of NF-κB and COX-2 Expression in Pancreatic Cancer PSN-1 Cells

The active NF-κB generally refers to a p50-p65 heterodimer, depicting the major Rel/NF-κB complex in cells [23]. Translocation of the active heterodimer into the nucleus, where it binds with specific sequences of the target genes, results from phosphorylation of the IκB by the IκB kinase beta (IKK), leading to a degradation of the inhibitory subunit [24]. Therefore, the effect of OAO and their conjugates with IND and DCL was assessed, based on the binding of their p50 and p65 subunits to their immobilized consensus site and their nuclear level.

As shown in Figure 3A,B, hybrids 3d and 4d, i.e., conjugates of IND and DCL with OAO morpholide, were the most efficient at reducing the binding level of p50 and p65. However, the effect was more pronounced for the p65 subunit. The latter was also reduced by the conjugate of DCL with OAO benzyl derivative (4c) and OAO morpholide itself (2d). The reduced binding of the p65 subunit to its consensus site after treatment with compounds 4d and 2d correlated with the level of p65 nuclear protein (Figure 3F) and total cytosolic protein (Figure 3E).

The reduced activation of NF-κB in the case of the p65 subunit might also be related to a diminished expression of its gene, as indicated by the reduced level of p65 mRNA (Figure 3D) after treatment with compounds 4c and 4d, 3c and 3d and OAO morpholide (2d). The 4d derivative caused ~60% diminished expression of p65 mRNA, while the expression of p50 was affected only by this derivative, although it merely caused a ~30% decrease (Figure 3C). One of the NF-κB target genes is COX-2, which similarly to NF-κB, is often overexpressed in pancreatic cancer cells [25]. Thus, as a result of the diminished activation and expression of NF-κB, reduced COX-2 transcript and protein levels occurred as a result of treatment with compounds 2c, 2d, 3c, 3d, and 4c (Figure 3H,I). Hybrid 4d significantly reduced the COX-2 transcript level but did not affect its protein level.

These data indicate that IND and DCL conjugates with OAO morpholide are the most effective inhibitors of NF-kB in pancreatic cancer cells, unlike IND and DCL and the OAO derivatives, which on their own showed a negligible or no effect.

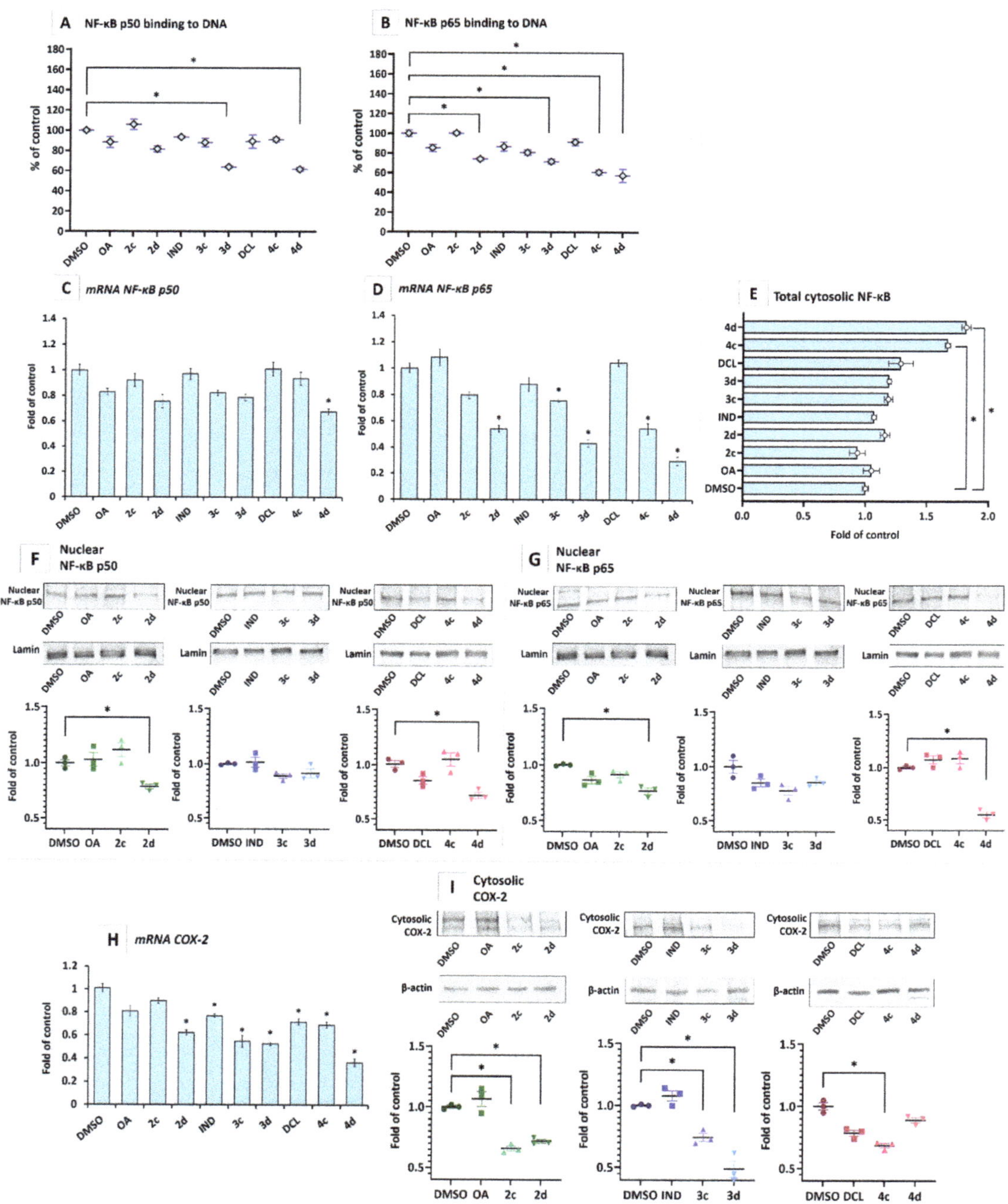

Figure 3. The effect of the OA, OAO derivatives (2c, 2d), and OAO conjugates with IND (3c, 3d) and DCL (4c, 4d) in the PSN-1 cells treated with these compounds at concentrations of 30 μM for 24h on NF-κB activation and expression. Panels (**A**,**B**) show NF-κB p50 and p65 subunits binding to DNA, its cytosolic and nuclear protein levels panels (**E**–**G**) and mRNA transcripts panels (**C**,**D**). Panels (**H**,**I**)

show the COX-2 gene mRNA transcript and protein. Activated p50 and p65 subunits were assessed regarding the amount of NF-κB p50 (**A**) and NF-κB p65 (**B**) contained in the DNA-binding complexes extracted from the nuclear fraction. The values (mean ± SEM) from three separate experiments are presented in comparison to control cells, set to 100%. Representative Western immunoblots of the nuclear NF-κB p50 (**F**), NF-κB p65 (**G**), and cytosolic COX-2 (**I**) are shown above the graphs. The sequence of the bands corresponds to the sequence of the bars in the graph. Lamin and β-actin were used as a loading control. The values were calculated as protein levels compared to control cells (expression equal to 1). The total cytosolic NF-κB protein level (**E**) was assessed using the MAGPIX® system, and the results from three independent measurements are presented as a fold of fluorescence intensity obtained in control cells. The values (mean ± SEM) of mRNA levels (**C**,**D**,**H**) were calculated as the relative change in comparison to control cells (expression equal to 1). * Significantly different from control: DMSO-treated cells, $p < 0.05$. Student's t-test was used to assess the statistical significance.

2.3. *Conjugates with OAO Increased Activation and Expression of Nrf2 in PSN-1 Cells*

The activation of Nrf2 requires its translocation from the cytosol into the nucleus. As shown in Figure 4A,B, treatment with the DCL conjugate with OAO benzyl ester (4c) and morpholide (4d) significantly increased the Nrf2 translocation and its level in the nucleus. The effect of the same conjugates with IND (3c, 3d) was slightly less pronounced. An increased level of Nrf2 in the nucleus resulted in its enhanced binding to the ARE sequence, measured based on the amount of Nrf2 contained in the DNA binding complex to the ARE sequence. The oligonucleotides containing ARE consensus-binding site (5′-GTCACAGTGACTCAGCAGAATCTG-3′) for Nrf2 were immobilized on microplates as bait (Figure 4C). An increased level of Nrf2 transcript was observed after treatment with OAO morpholide and its conjugates with IND and DCL. Increased levels of Nrf2 transcript were also observed after treatment with OA and OAO morpholide (Figure 4D).

The activation of Nrf2 leads to the elevated mRNA levels of its target genes *SOD-1*, *NQO1*, and *GSTA* encoding superoxide dismutase, NAD(P)H: quinone oxidoreductase, and glutathione-S-transferase A, respectively.

While this increased transcript level after treatment with 4c and 4d was confirmed for SOD-1 protein level (Figure 5A,B) and, in the case of 4d, on NQO1 protein (Figure 5C,D), those conjugates did not affect the GSTA protein (Figure 5F). In contrast, conjugates 3c and 3d even showed a tendency to reduce the expression of this gene (Figure 5E). Since increased expression of GSTA is often associated with an increased risk of cancer, this effect might be considered to be desirable [26]. The transcript of *HO-1* and *GPx* genes encoding heme oxygenase-1 and glutathione peroxidase was elevated by about 30–50% after treatment with 2d, 3c, and 3d derivatives (Figure 5G,H). DCL conjugate with OAO morpholide derivative was the most potent inducer of the GPx gene, leading to a more than a two-fold increase in its transcript level (Figure 5H).

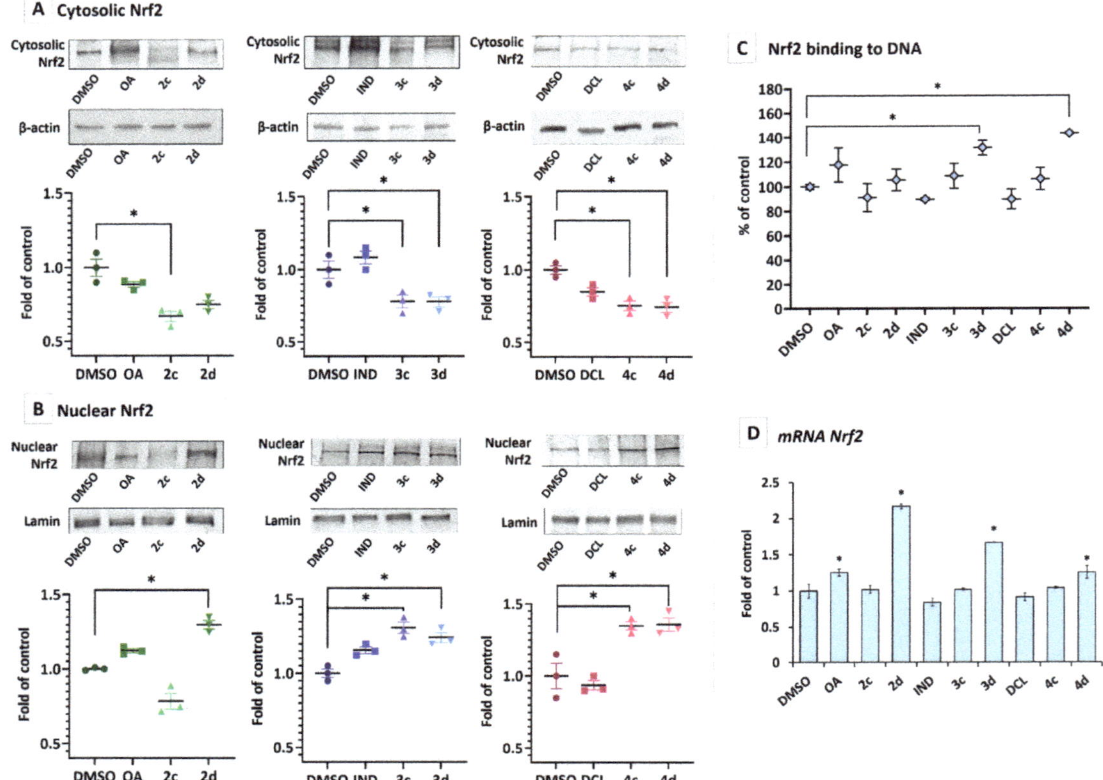

Figure 4. The effect of the OA, OAO derivatives (2c, 2d), and OAO conjugates with IND (3c, 3d) and DCL (4c, 4d) in the PSN-1 cells treated with these compounds at concentrations of 30 μM for 24 h on Nrf2 activation and expression. Panels (**A**,**B**) present cytosolic and nuclear protein levels of Nrf2, panel (**C**) Nrf2 binding to DNA, and mRNA transcript panels (**D**). Representative Western immunoblots are presented above the graphs. The sequence of the bands corresponds to the sequence of the bars in the graph. Lamin and β-actin were used as a loading control. The values were calculated as protein levels compared to control cells (expression equal to 1). Activated Nrf2 (**C**) was assessed in terms of the amount of Nrf2 contained in the DNA-binding complexes extracted from the nuclei isolated from the cells and calculated compared to control cells set to 100%. The values (means ± SEM) of mRNA level (**D**) were calculated as the relative change in comparison to control cells (expression equal to 1). * Significantly different from control: DMSO-treated cells, $p < 0.05$. Student's t-test was used to assess the statistical significance.

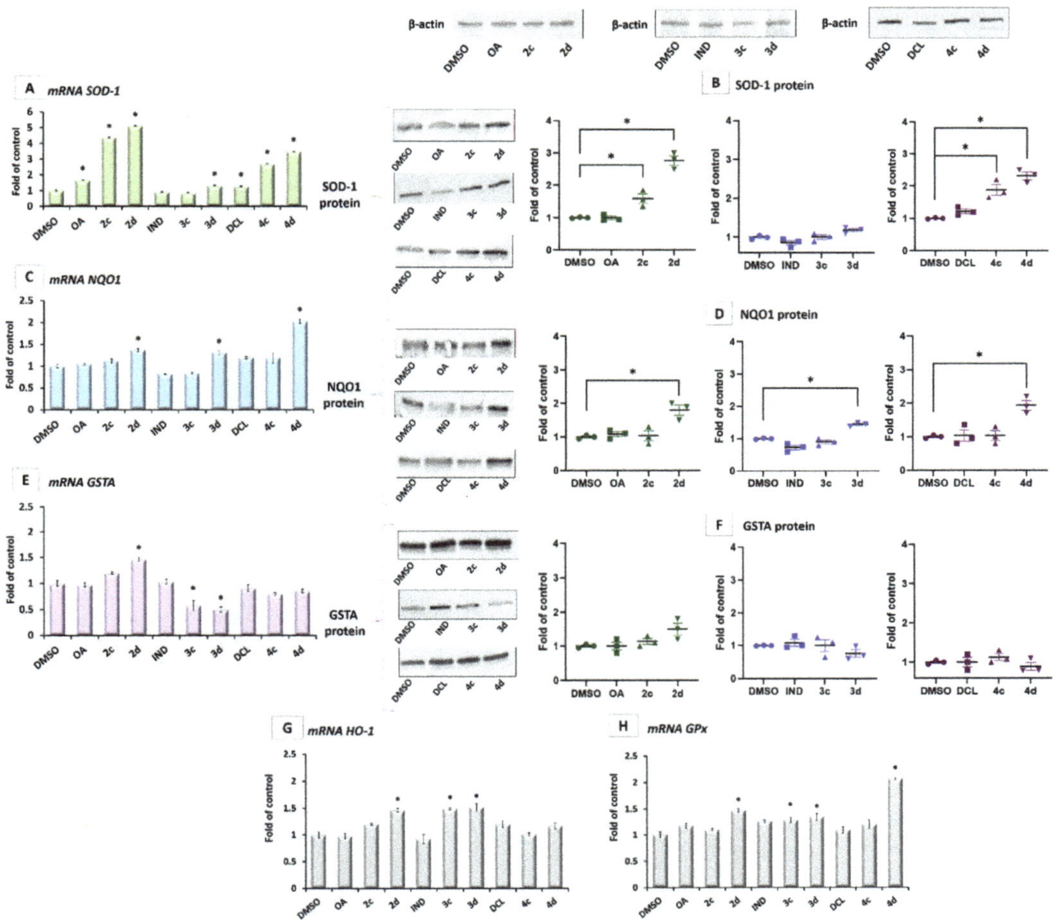

Figure 5. The effect of the OA, OAO derivatives (2c, 2d), and OAO conjugates with IND (3c, 3d) and DCL (4c, 4d) on the SOD-1 (**A**,**B**), NQO1 (**C**,**D**), and GSTA (**E**,**F**) mRNA expression and protein levels, as well as HO-1 (**G**) and GPx (**H**) mRNA expression levels, measured in PSN-1 cells after 24 h incubation with tested compounds at the concentrations of 30 μM. The values (mean ± SEM) for the mRNA expression levels (**A**,**C**,**E**,**G**,**H**) were calculated from three separate experiments in comparison with control cells (expression of control was set as 1). Representative immunoblots for the analysis of the cytosolic levels of SOD-1 (**B**), NQO1 (**D**), and GSTA (**F**) are shown. The sequence of the bands corresponds to the sequence of the bars in the graph. β-actin was used as a loading control. The values were calculated as protein levels compared to control cells (expression equal to 1). * Significantly different from control: DMSO-treated cells, $p < 0.05$. Student's *t*-test was used to assess the statistical significance.

These results show that conjugates of OAO derivatives, particularly OAO morpholide with DCL, in concert with inhibition of the NF-κB pathway, activate Nrf2 signaling, leading to the increased expression of antioxidant enzymes.

The increased expression of SOD-1 may be the reason for the observed reduced production of ROS (namely, superoxide radicals) as a result of treatment with 2d, 3d, 4c, and 4d (Figure 6).

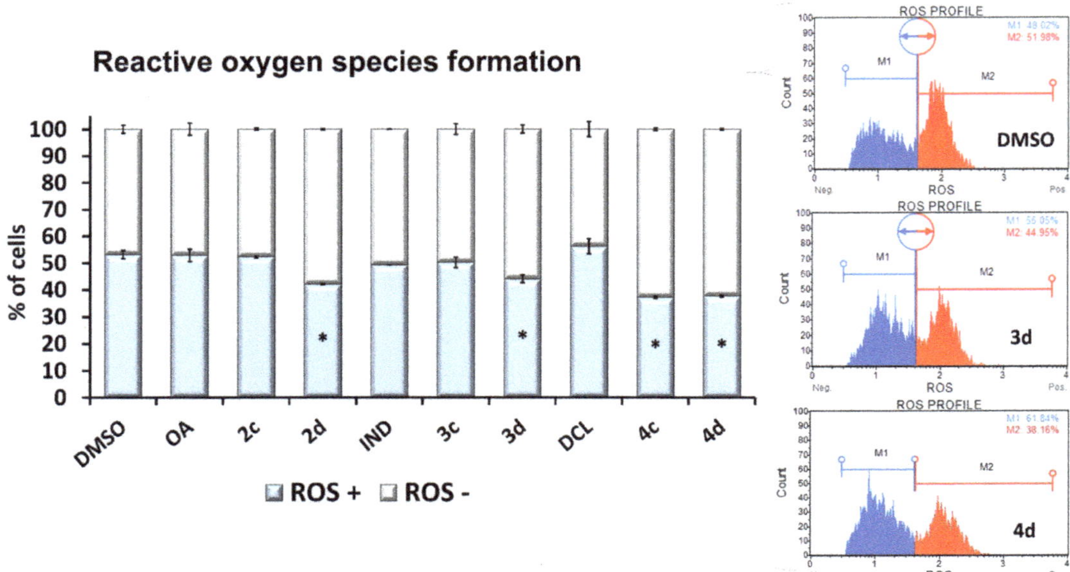

Figure 6. The effect of the OA, OAO derivatives (2c, 2d), and OAO conjugates with IND (3c, 3d) and DCL (4c, 4d) on the generation of reactive oxygen species in PSN-1 cells after 24 h incubation with tested compounds at concentrations of 30 μM. Graphs represent the relative percentage of cells that are ROS-negative and -positive (percentage of cells undergoing oxidative stress, based on the intracellular detection of superoxide radicals). The mean values (mean ± SEM) from three independent experiments are shown. * Significantly different from control: DMSO-treated cells, $p < 0.05$. Student's t-test was used to assess the statistical significance.

2.4. Bead-Based Multiplex Immunoassay Revealed Possible Modulation of Protein Regulating Several Signaling Pathways by OAO-IND and OAO-DCL Conjugates

A bead-based multiplex immunoassay was applied to evaluate the effect of tested compounds on cytosolic levels of kinases from the MAPK family, p38, JNK, AKT, ERK 1/2, and p70S6K. Additionally, CREB, STAT3, and STAT5 protein levels were assessed. As shown in Figure 7, significant decreases of cytosolic AKT and p70S6K protein levels, as a result of treatment with 4d, were observed, reduced by ~50% and ~90%, respectively. In contrast, the same compound most efficiently increased the levels of certain proteins: CREB (by ~80%), JNK (by ~70%), and STAT3 (three-fold increase). ERK1/2 and STAT5 were relatively unaffected. A decreased level of p70S6K was observed as a result of treatment with compounds 4c, 3d, and 2d, while an increase in the p38 protein level was detected after treatment with 4c, DCL, 3d, and 2d. STAT3 protein level was affected by all tested compounds except OA and IND.

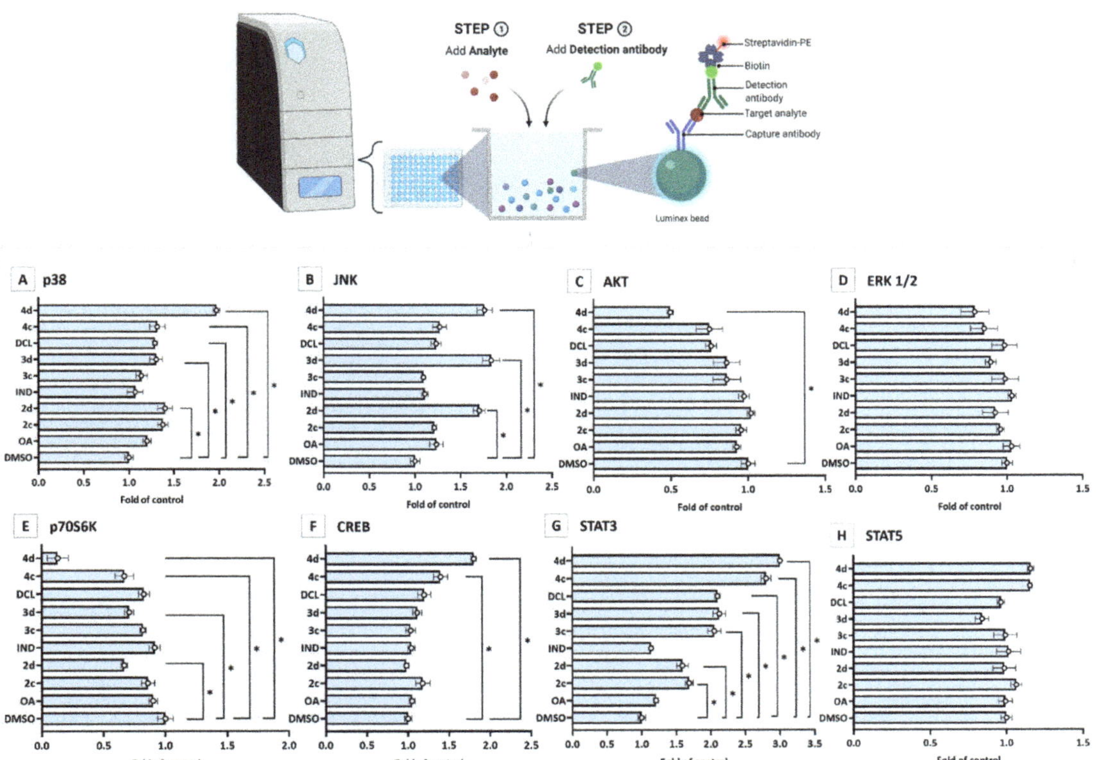

Figure 7. The effect of the OA, OAO derivatives (2c, 2d), and OAO conjugates with IND (3c, 3d) and DCL (4c, 4d) on the cytosolic level of p38 (**A**), JNK (**B**), AKT (**C**), ERK1/2 (**D**), p70S6K (**E**), CREB (**F**), STAT3 (**G**) and STAT5 (**H**), measured and calculated on the MAGPIX® system in PSN-1 cells, after 24 h incubation with tested compounds at concentrations of 30 μM. The mean values (mean± SEM) of fluorescence intensity are presented as the fold of control from three independent measurements. * Significantly different from control: DMSO-treated cells, $p < 0.05$. Student's t-test was used to assess the statistical significance.

2.5. Conjugates of IND, DCL, and OAO Derivatives Do Not Change the Cell Cycle Distribution and Have No Effect on Cell Proliferation in Pancreatic Cancer Cells

Figure 8A shows the effect of OA, its oximes derivatives, and their conjugates with IND or DCL on cell cycle distribution in the PSN-1 cells. While topotecan, used as a reference compound, significantly increased the percentage of the cells in the S phase and G2/M, none of the tested compounds changed the cell-cycle distribution. The effect of the tested OA derivatives and conjugates on cell proliferation was evaluated, based on Ki67 protein expression. Similar to effects on cell cycle distribution, the compounds did not affect PSN-1 cell proliferation (Figure 8B). Although the reduced activation of NF-κB may enhance proapoptotic gene expression, the enhanced activation of Nrf2 and reduced ROS production may be responsible for the lack of induction of cell cycle arrest or inhibition of cell proliferation by the tested compounds. This concept supports the fact that the proliferation and differentiation of diverse cell types are often influenced and modulated by the cellular redox balance. Therefore, Nrf2 regulating the cellular levels of ROS may control these cellular processes.

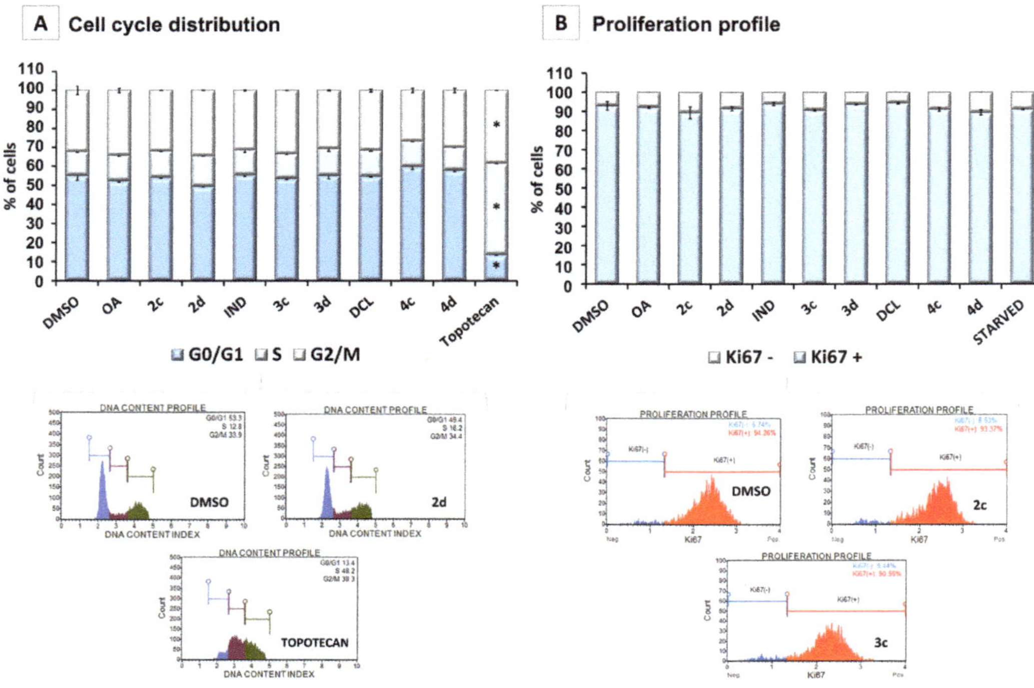

Figure 8. The effect of the OA, OAO derivatives (2c, 2d), and OAO conjugates with IND (3c, 3d) and DCL (4c, 4d) on the cell cycle distribution (**A**) and the proliferation level (**B**) measured in PSN-1 cells after 24 h incubation with tested compounds at concentrations of 30 µM. (**A**) Graphs represent the percentage of cells in the G1/G0, S, and G2/M phases measured by flow cytometry after propidium iodide staining. Topotecan (100 nM)-treated cells were used as a positive control for the cell cycle arrest. (**B**) Graphs representing the percentage of proliferating [Ki67(+)] and nonproliferating [Ki67(−)] cells measured by flow cytometry. Starved cells (cultured in an FBS-free medium) were used as a reference for the antiproliferative conditions. Representative plots are presented. The mean values ± SEM from three independent experiments are shown for both panels. * Significantly different from control: DMSO-treated cells, $p < 0.05$. Student's *t*-test was used to assess the statistical significance.

3. Discussion

Chronic inflammation is considered one of the major risk factors of PC. During chronic pancreatitis, inflammatory cells and macrophages enhance stroma formation, which increases the risk of PC development. Moreover, enhanced activation of the NF-κB signaling pathway, in concert with an increased level of cytokines and chemokines, was observed in this type of cancer [27]. Thus, the NF-κB pathway is emerging as an important element in PC cancer development [28]. Consequently, NSAIDs, along with natural compounds with anti-inflammatory potential, might be considered as therapeutic agents targeting this signaling pathway.

Our earlier studies showed that conjugates of OAO derivatives, particularly morpholides and benzyl esters with ASP [29,30], IND [31], or DCL [22], effectively inhibited NF-κB activation and expression in HCC cells, resulting in a reduced rate of proliferation. The focus of this study was to evaluate whether the same effect might be observed in pancreatic cancer cells.

While the highest activity of IND and DCL conjugates with OAO morpholide was confirmed, this effect was generally less pronounced in PSN-1 cells than in HCC cells, requiring higher concentrations of the compound. Initial screening of the effects of these compounds, i.e., their impact on the PSN-1 cells at the concentrations applied in HCC cells

(10 and 20 μM), in most cases, did not show significant effects. Therefore, a concentration of 30 μM was used in further experiments. At this dose, significant changes in the activation and expression of NF-κB and related signaling pathways were observed. The activation of NF-κB was measured in terms of binding its active subunits, p65 and p50, to DNA, and their nuclear levels were the most reduced as a result of treating PSN-1 cells with OAO-DCL morpholide hybrid (4d), although the conjugate of DCL with OAO-benzyl ester (4c) as well as OAO-morpholide itself (2d) were also effective inhibitors of this pathway. These derivatives also affected NF-κB expression, as indicated by the reduced level of p65 transcript, while p50 mRNA was diminished only by OAO benzyl and morpholide derivatives.

The effect of tested compounds, similar to our previous study in HCC cells, was more explicit for the p65 NF-κB subunit than for p50. The p65 subunit is basically responsible for the transcription initiation, while the p50 subunit serves only as a helper in NF-κB DNA binding [32,33]. Excessively activated p65 subunit and, ultimately, effector molecules might be related to PC pathogenesis [5]. One such effector molecule, often overexpressed in PC, is COX-2 [25]. All compounds tested in this study, except OA and 2c, reduced the expression of *COX-2* at the mRNA level, but this effect was not confirmed for the protein level in the case of conjugate 4d, which reduced COX-2 mRNA to the greatest extent. There are many processes between transcription and translation that affect the protein level and contribute to this discrepancy. The half-life of different proteins can vary from minutes to days, whereas the degradation rate of mRNA would fall within a much tighter range of 2–7 h for mammalian mRNAs, vs. 48 h for protein [34].

NF-κB negatively interferes with the Nrf2 signaling pathway, and, often, agents that suppress NF-κB signaling will activate Nrf2 [35]. The activity of Nrf2 is regulated by the Keap1 protein, which promotes the ubiquitination and subsequent degradation of Nrf2 under normal conditions while releasing the active Nrf2 upon exposure to stress [36]. The activation of protein kinases, such as extracellular signal-regulated kinase (ERK), induces Nrf2 phosphorylation, which may also stimulate the dissociation of Nrf2 from its repressor Keap1 and its subsequent translocation into the nucleus and binding to the ARE sequence [37]. This study showed enhanced activation of the Nrf2 pathway, as indicated by the translocation of its active form into the nucleus and binding to the ARE sequence, particularly by treatment with IND and DCL hybrids with OAO morpholide (3d, 4d) in PSN-1 cells. These compounds, along with OAO morpholide (2d) itself, also increased Nrf2 expression at the mRNA level. The same effect in the case of OAO morpholide and the other OAO derivatives was observed in human hepatoma HepG2 cells. However, in HCC cells, it was less pronounced as a result of treatment with their conjugates with ASP [29]. Moreover, the association between cytotoxicity and Nrf2 activation was observed in these cells, i.e., the most cytotoxic derivatives activated Nrf2 to the greatest extent [26,27]. Basically, the same effect was observed in this study in pancreatic cancer cells.

The contribution of the Nrf2-ARE signaling pathway to protection against reactive oxygen species (ROS) and electrophilic insults has been well established and occurs through the induction of genes encoding antioxidant enzymes. Activation of Nrf2 stimulated by OAO derivatives 2c and 2d in PSN-1 cells increased the expression of SOD-1, GPx, HO-1, and NQO1 genes, similar to the effects seen in HepG2 cells.

However, in contrast to a conjugate with ASP, a hybrid of DCL and, to a lesser extent, the IND hybrid were the most potent inducers of these genes. The proteins encoded by these genes play an important role in protecting against reactive electrophiles produced as a result of xenobiotic exposure and ROS, which are involved in PC pathogenesis [38]. Consequently, reduced production of ROS under the influence of these compounds was found. In this way, enhanced activation of Nrf2 may play protective roles in pancreatic inflammation. Thus, the tested compounds might be considered chemopreventive agents. On the other hand, accumulating evidence suggests that an amplified activity of Nrf2 may promote cancer growth and increase chemoresistance [39]. In this regard, constitutive overexpression of Nrf2 in PC models such as the K-Ras (G12D), B-Raf (V619E), Myc (ERT2), and human samples was described [16].

However, it was demonstrated using synthetic triterpenoid RTA 405 as an example that the pharmacological activation of Nrf2 is distinct from genetic activation and does not provide a growth or survival advantage to several cancer cells, including PC cells. Moreover, it was shown that pretreatment with RTA 405 did not protect cancer cells from doxorubicin- or cisplatin-mediated growth inhibition [40].

MAPKs such as ERK, JNK, and p38 are key molecules at the convergence of several signal transduction pathways. They transduce the converged signals into the nucleus, resulting in various cellular responses, including proliferation, differentiation, and apoptosis [41]. Several studies have correlated the phosphorylation of ERK and Nrf2 activation [42]. Interestingly, it was shown that the activation of MAPK–Akt and ERK is required for OA-induced activation of Nrf2, followed by HO-1 expression, in primary rat vascular smooth muscle cells [43]. Using a bead-based multiplex immunoassay in our study, we observed decreased levels of cytosolic AKT and increased levels of unphosphorylated JNK and p38 [44,45]. The accumulation of p38 and JNK in the cytosol could result from decreased ROS production and may be related to an increase in cytosolic CREB protein level [46]. The latter is often overexpressed and constitutively phosphorylated in many types of cancer, including pancreatic cancer [47]. Therefore, its reduced activation is desirable. Additionally, increased STAT3 cytosolic level may suggest the inhibition of mTOR signaling as the crosstalk between these pathways has been implicated in several malignant diseases, e.g., osteosarcoma [48]. Moreover, a reduced level of cytosolic p70S6 kinase, which is activated by mTOR, further supports this concept [49].

Overall, our current study indicates that the conjugation of NSAIDs with OAO derivatives may enhance their modifying effect on key signaling pathways, not only in HCC but also in PC cells. Moreover, in the PC cells, the conjugation of DCL with OAO morpholide (4d) shows greater efficiency than OAO morpholide (2d) alone. However, in contrast to HCC cells, reduced activation of NF-κB led to enhanced activation of Nrf2, which may be responsible for the lack of effect on cell-cycle distribution and proliferation. Such an interpretation of the data presented in this study is further supported by the fact that the analyses of the genome-wide distribution of Nrf2 have identified new sets of Nrf2 target genes whose products are indeed involved in cell proliferation and differentiation. Moreover, it was shown that Nrf2 is required for the transition from the G2 phase to the M phase in the cell cycle [50].

Importantly, IND or DCL alone did not affect the parameters evaluated in PC cells. Since inflammation can predispose to PC, NSAIDs were proposed for cancer prevention and therapy [19,21]. However, the output of these attempts is not consistent. Therefore, the results of this study, indicating that DCL conjugate with OAO morpholide is a more potent modulator of the key signaling pathways in PC cells than parent compounds, may pave the way to designing more intensive in vitro and subsequently in vivo studies on these types of hybrids for prophylaxis and/or PC treatment.

4. Materials and Methods

4.1. Chemistry

OAO derivatives (2c, 2d) and OAO conjugates with IND (3c, 3d) and DCL (4c, 4d) were synthesized as described previously [22,30,31]. Their chemical structures were elucidated based on IR, ^{1}H NMR, ^{13}C NMR, MS, and HRMS elemental analysis. The NMR spectra and the detailed description of synthesized compounds' identity were presented in our previous papers, along with the synthesis overview and detailed purity data.

4.2. Biological Assays

4.2.1. Cell Culture and Viability Assay

Pancreatic carcinoma cell lines, PSN-1, and immortalized pancreatic MS-1 cell lines were purchased from American Type Culture Collection (Rockville, MD, USA). The PSN-1 cells were maintained in RPMI-1640 medium (Sigma-Aldrich, St. Louis, MO, USA) and MS-1 cells in Dulbecco's Modified Eagle's Medium (Sigma-Aldrich, St. Louis, MO, USA).

Both cell lines were maintained with 10% FBS (EURx, Gdańsk, Poland), 1% antibiotic solution (Sigma-Aldrich, St. Louis, MO, USA), and were grown under standard conditions (37 °C and 5% CO_2). After 24 h of initial incubation, the cells were treated with 0.1% of vehicle control and tested compounds at a concentration of 30 μM for a further 24 h and then harvested. The concentration of OAO derivatives and their conjugates was selected based on the MTT viability assay.

The MTT assay was performed following the standard protocol. Briefly, PSN-1 and MS-1 cells were seeded at 10^4 cells per well in 96-well plates. After 24 h of preincubation in a complete medium, compounds were added, and cells were incubated for 24 h. Later, the cells were washed twice with phosphate-buffered saline (PBS) and incubated for 4 h with a medium containing 0.5 mg/mL 3-(4,5-dimethylthiazol-2-yl)-2,5-diphenyl-2H-tetrazolium bromide (MTT). Then, the formazan crystals were dissolved in acidic isopropanol, and the absorbance was measured at 570 nm and 690 nm. All experiments were repeated three times.

4.2.2. Total RNA Isolation, cDNA Synthesis, and Quantitative Real-Time PCR (R-T PCR)

The total RNA extraction was performed using the GeneMatrix Universal DNA/RNA/Protein Purification Kit (EURx, Gdańsk, Poland). According to the manufacturer's instructions, samples were subjected to reverse transcription via the RevertAid First Strand cDNA Synthesis Kit (Thermo Fisher Scientific, Waltham, MA, USA).

The Maxima SYBR Green Kit (Thermo Fisher Scientific, Waltham, MA, USA) and the BioRad Chromo4 thermal cycler (BioRad Laboratories, Hercules, CA, USA) were used for the quantitative R-T PCR analyses. Protocol started with 5 min enzyme activation at 95 °C, followed by 40 cycles of 95 °C for 15 s, 56 °C for 20 s and 72 °C for 40 s, and the final elongation at 72 °C for 5 min. Melting curve analysis was used for amplicon verification. The expression of *TBP* (*TATA box binding protein*) and *PBGD* (*porphobilinogen deaminase*) was used to normalize data. The Pfaffl comparative method was used for fold-change quantification. Primers were designed using the Beacon Designer software and subjected to a BLAST search to minimize unspecific binding. Only the primer pairs that generated intron-spanning amplicons were selected. The primers' sequences used to analyze *NF- B p50, NF-B p65, COX-2, Nrf2, SOD-1, NQO1, GSTA, HO-1, GPx, PBGD,* and *TBP* genes are listed in Table 2.

Table 2. Primers used in the R-T PCR.

Gene	Forward Primer	Reverse Primer
NF-kB p50	5′ATCATCCACCTTCATTCTCAA	5′AATCCTCCACCACATCTTCC
NF-kB p65	5′CGCCTGTCCTTTCTCATC	5′ACCTCAATGTCCTCTTTCTG
COX-2	5′CCTGTGCCTGATGATTGC	5′CAGCCCGTTGGTGAAAGC
Nrf2	5′ATTGCTACTAATCAGGCTCAG	5′GTTTGGCTTCTGGACTTGG
SOD-1	5′CGACAGAAGGAAAGTAATG	5′TGGATAGAGGATTAAAGTGAGG
NQO1	5′CAATTCAGAGTGGCATTC	5′GAAGTTTAGGTCAAAGAGG
GSTA	5′GGGAAAGACATAAAGGAGAGAG	5′TCAAAGGCAGGGAAGTAGC
HO-1	5′CAGGCAGAGGGTGATAGAAGAG	5′GGAGCGGGTGTTGAGTGG
GPx	5′CAACCAGTTTGGGCATCAG	5′TTCACCTCGCACTTCTCG
PBGD	5′TCAGATAGCATACAAGAGACC	5′TGGAATGTTACGAGCAGTG
TBP	5′GGCACCACTCCACTGTATC	5′GGGATTATATTCGGCGTTTCG

4.2.3. Nuclear and Cytosolic Fractions Preparation

According to the manufacturer's protocol, the subcellular extracts from PSN-1 cells were prepared using the Nuclear/Cytosol Fractionation Kit (BioVision Research, Milpitas, CA, USA). Protein concentration was assessed, and the samples were stored at −80 °C for further analysis.

4.2.4. Western Blot Analysis

Cytosolic extracts for Nrf2, SOD-1, NQO1, GSTA, COX-2, and β-actin, or nuclear extracts for Nrf2, NF-B p65, NF-B p50, and lamin protein detection, were separated on 12%

or 10% SDS-PAGE slab gels. The β-actin and lamin were used as a loading control. Then, 100 μg of cytosolic and nuclear fractions were added per well for the SDS-PAGE. Proteins were transferred to the nitrocellulose Immobilon P membrane. After blocking for 2 h with 10% skimmed milk, proteins were probed with primary antibodies against Nrf2, NF-B p65, NF-B p50, SOD-1, NQO1, GSTA, COX-2, β-actin, and lamin. Alkaline phosphatase AP-labeled anti-rabbit IgG, anti-goat IgG, and anti-mouse IgG secondary antibodies (BioRad Laboratories, Hercules, CA, USA) were used in the staining reaction. Bands were visualized with the AP Conjugate Substrate Kit NBT/BCIP (BioRad Laboratories, USA). The number of immunoreactive products in each lane was determined using the ChemiDoc Imaging System (BioRad Laboratories, Hercules, CA, USA). Values were calculated as relative absorbance units (RQ) per mg of protein and expressed as a percentage of control.

4.2.5. Nrf2 and NF-kB Binding Assay

Nrf2, NF-κB p50, and NF-κB p65 activation were assessed by the enzymatic immunoassay (Transcription Factor ELISA Assay Kit; Active Motif, La Hulpe, Belgium) according to the manufacturer's instructions. The activated Nrf2 was evaluated based on the amount of Nrf2 contained in the DNA-binding complex to the ARE sequence. The oligonucleotides containing the ARE consensus-binding site (5′-GTCACAGTGACTCAGCAGAATCTG-3′) for Nrf2 were immobilized on microplates as bait. However, the activated NF-kB was measured using the number of p65 and p50 subunits held in the DNA-binding complex. The oligonucleotides containing (5′-GGGACTTTCC-3′), a consensus site for NF-κB, were immobilized on microplates as bait. Nuclear fractions were incubated with oligonucleotides for 1 h, then the wells were washed, and DNA-bound subunits were detected by the specific primary antibody and a secondary antibody conjugated with the HRP. The results were expressed as the normalized level of absorbance (OD450 nm per mg of protein).

4.2.6. Cell Cycle Distribution

The analysis of the cell cycle distribution was performed using the Muse Cell Cycle Kit (Merck, Darmstadt, Germany) according to the manufacturer's protocol. Briefly, cells (3×10^5 per well) were seeded and grown on 6-well plates for 24 h before stimulation. Then, the tested compounds were added and cells were incubated for 24 h. Topotecan (100 nM)-treated cells were used as a positive control for the cell cycle arrest. Subsequently, cells were harvested by trypsinization, washed with PBS buffer, fixed in ice-cold 70% ethanol, and frozen until staining at −20 °C. After overnight storage, cells were washed with cold PBS buffer, stained with propidium iodide in the presence of RNAase A, and, after 30 min incubation at room temperature and in the dark, the fluorescence signal was analyzed by flow cytometry on the Muse Cell Analyzer (Merck, Darmstadt, Germany). Data were evaluated using Muse 1.5 analysis software (Merck, Darmstadt, Germany).

4.2.7. Proliferation Profile

According to the manufacturer's protocol, the Ki67 protein was used as a marker of proliferating cells, and its expression was detected by fluorochrome-labeled antibody using the Muse Ki67 Proliferation Kit (Merck, Darmstadt, Germany). Briefly, cells (3×10^5 per well) were seeded on the 6-well plates. After 24 h, the tested compounds were added, and cells were then incubated for 24 h. Cells cultured in a medium without FBS (starved cells) were used as a positive control of antiproliferative conditions. After the incubation, cells were collected by trypsinization, washed with PBS buffer, instantly fixed, and exposed to a permeabilization buffer. After 30 min incubation at room temperature and dark with the Muse Hu Ki67-PE antibody, cells were analyzed on Muse Cell Analyzer (Merck, Darmstadt, Germany) by flow cytometry. Data were evaluated using Muse 1.5 Analysis Software (Merck, Darmstadt, Germany).

4.2.8. ROS Generation

Dihydroethidium (DHE) reacts with superoxide radicals, generating fluorophores with a high affinity to DNA. The Muse Oxidative Stress Kit (Merck, Darmstadt, Germany) containing DHE was used according to the manufacturer's protocol for the quantitative measurement of superoxide radical positive cells. Briefly, cells (3×10^5 per well) were seeded in the 6-well plates. After 24 h, the tested compounds were added, and cells were incubated for 24 h. Subsequently, cells were collected by trypsinization, washed with PBS buffer, resuspended in 1X Assay Buffer, mixed with Muse Oxidative Stress Reagent working solution, and subjected to a 30 min incubation (37 °C, 95% humidity, 5% CO_2). Cells were analyzed by flow cytometry on the Muse Cell Analyzer (Merck, Darmstadt, Germany), and data were evaluated using the Muse 1.5 Analysis Software (Merck, Darmstadt, Germany).

4.2.9. Bead-Based Immunoassay on the Luminex MAGPIX Instrument

A magnetic bead-based immunoassay was performed on the Luminex-MAGPIX multiplex immunoassay system. Data were analyzed with MILLIPLEX® Analyst 5.1 software (EMD Millipore, Burlington, MA, USA). The panel that we performed quantitated the following proteins in the cytosolic fractions of PSN-1 cells, NF-κB, AKT, p70S6, p38, JNK, ERK 1/2, CREB, STAT3, and STAT5, according to the manufacturer's instructions. The magnetic bead panel, with high-sensitivity antibodies, was obtained from Merck (Darmstadt, Germany). A multiplex test based on microspheres, using Luminex xMAP technology with different fluorescent colors, was detected on a compatible MAGPIX camera. Cytosolic fractions were suspended in MILLIPLEX® MAP buffer. The bead suspension was added to each well of a 96-well plate, then samples were added into the wells and incubated overnight at 2–8 °C on a shaker protected from light. The plate was washed with 2× buffer, and then 1X MILLIPLEX® MAP detection antibodies were added. After shaking for 1 h at room temperature, the antibodies were removed, and 1X MILLIPLEX® MAP Streptavidin/Phycoerythrin (SAPE) was added. Then the MILLIPLEX® MAP Amplification Buffer was added to each well and shaken for 15 min. The beads were suspended in MILLIPLEX® MAP buffer, and each microsphere was identified with a MAGPIX Luminex Analyzer (Merck, Darmstadt, Germany); the results were calculated from the reporters' fluorescent signals. Mean fluorescence intensities were quantified with the xPonent 4.2 software (Luminex Corporation, Austin, TX, USA). The raw MFI results for the tested protein levels were converted, relative to the control of DMSO-treated cells.

4.3. Statistical Analysis

Statistical analysis and graphs were calculated and prepared using GraphPad Prism 9.2.0 (GraphPad Software, San Diego, CA, USA), assuming the significance level of changes as $p < 0.05$. Student's *t*-test was used to assess the statistical significance between the experimental and control groups.

5. Conclusions

We show here that the conjugation of NSAIDs with OAO derivatives enhances their effect on key signaling pathways that are involved in the inflammation process. Our studies indicating that DCL conjugate with OAO morpholide is a more potent modulator of the key signaling pathways in PC cells than parent compounds should pave the way to the design of more intensive studies on these types of hybrids for PC treatment and/or adjuvant therapy.

Author Contributions: M.N., investigation, biological data collection and interpretation, formal analysis, and preparation of a draft of the manuscript; V.K.-K., experiment planning and designing, methodology, investigation, biological data collection, formal analysis, and data interpretation; R.K., flow cytometry methodology and data interpretation; W.B.-D., supervision, funding acquisition, project administration, data interpretation, and final writing. All authors have read and agreed to the published version of the manuscript.

Funding: This study was supported by the Polish National Science Centre, grant no. 2016/21/B/NZ7/01758.

Data Availability Statement: Data will be available if requested.

Acknowledgments: The authors express their gratitude to Barbara Bednarczyk-Cwynar for providing compounds for this study. Graphical abstract and Figure 7 cartoon were created with BioRender.com.

Conflicts of Interest: The authors declare no conflict of interest, financial or otherwise.

References

1. Nihr, J.K.; Biomedical, P.; Kleeff, J.; Korc, M.; Apte, M.; La Vecchia, C.; Johnson, C.D.; Biankin, A.V.; Neale, R.E.; Tempero, M.; et al. Pancreatic cancer. *Nat. Rev. Dis. Primers* **2016**, *1*, 16022. [CrossRef]
2. Rayburn, E.R.; Ezell, S.J.; Zhang, R. Anti-Inflammatory Agents for Cancer Therapy. *Mol. Cell. Pharmacol.* **2009**, *1*, 29. [CrossRef] [PubMed]
3. Taniguchi, K.; Karin, M. NF-κB, inflammation, immunity and cancer: Coming of age. *Nat. Rev. Immunol.* **2018**, *18*, 309–324. [CrossRef] [PubMed]
4. Qin, J.J.; Cheng, X.D.; Zhang, J.; Zhang, W.D. Dual roles and therapeutic potential of Keap1-Nrf2 pathway in pancreatic cancer: A systematic review. *Cell. Commun. Signal.* **2019**, *17*, 1–15. [CrossRef] [PubMed]
5. Pramanik, K.C.; Makena, M.R.; Bhowmick, K.; Pandey, M.K. Advancement of NF-κB Signaling Pathway: A Novel Target in Pancreatic Cancer. *Int. J. Mol. Sci.* **2018**, *19*, 3890. [CrossRef]
6. McMahon, M.; Thomas, N.; Itoh, K.; Yamamoto, M.; Hayes, J.D. Dimerization of substrate adaptors can facilitate cullin-mediated ubiquitylation of proteins by a "tethering" mechanism: A two-site interaction model for the Nrf2-Keap1 complex. *J. Biol. Chem.* **2006**, *281*, 24756–24768. [CrossRef]
7. Tong, K.I.; Kobayashi, A.; Katsuoka, F.; Yamamoto, M. Two-site substrate recognition model for the Keap1-Nrf2 system: A hinge and latch mechanism. *Biol. Chem.* **2006**, *387*, 1311–1320. [CrossRef] [PubMed]
8. Copple, I.M.; Goldring, C.E.; Kitteringham, N.R.; Park, B.K. The Keap1-Nrf2 Cellular Defense Pathway: Mechanisms of Regulation and Role in Protection Against Drug-Induced Toxicity. *Handb. Exp. Pharmacol.* **2010**, *196*, 233–266. [CrossRef]
9. Lister, A.; Nedjadi, T.; Kitteringham, N.R.; Campbell, F.; Costello, E.; Lloyd, B.; Copple, I.M.; Williams, S.; Owen, A.; Neoptolemos, J.P.; et al. Nrf2 is overexpressed in pancreatic cancer: Implications for cell proliferation and therapy. *Mol. Cancer* **2011**, *10*, 37. [CrossRef] [PubMed]
10. Sasaki, H.; Sato, H.; Kuriyama-Matsumura, K.; Sato, K.; Maebara, K.; Wang, H.; Tamba, M.; Itoh, K.; Yamamoto, M.; Bannai, S. Electrophile Response Element-mediated Induction of the Cystine/Glutamate Exchange Transporter Gene Expression. *J. Biol. Chem.* **2002**, *277*, 44765–44771. [CrossRef] [PubMed]
11. Maher, J.M.; Dieter, M.Z.; Aleksunes, L.M.; Slitt, A.L.; Guo, G.; Tanaka, Y.; Scheffer, G.L.; Chan, J.Y.; Manautou, J.E.; Chen, Y.; et al. Oxidative and electrophilic stress induces multidrug resistance–associated protein transporters via the nuclear factor-E2–related factor-2 transcriptional pathway. *Hepatology* **2007**, *46*, 1597–1610. [CrossRef] [PubMed]
12. Chanas, S.A.; Jiang, Q.; McMahon, M.; McWalter, G.K.; McLellan, L.I.; Elcombe, C.R.; Henderson, C.J.; Wolf, C.R.; Moffat, G.J.; Itoh, K.; et al. Loss of the Nrf2 transcription factor causes a marked reduction in constitutive and inducible expression of the glutathione S-transferase *GSTA1, GSTA2, GSTM1, GSTM2, GSTM3* and *GSTM4* genes in the livers of male and female mice. *Biochem. J.* **2002**, *365*, 405. [CrossRef] [PubMed]
13. Itoh, K.; Chiba, T.; Takahashi, S.; Ishii, T.; Igarashi, K.; Katoh, Y.; Oyake, T.; Hayashi, N.; Satoh, K.; Hatayama, I.; et al. An Nrf2/Small Maf Heterodimer Mediates the Induction of Phase II Detoxifying Enzyme Genes through Antioxidant Response Elements. *Biochem. Biophys. Res. Commun.* **1997**, *236*, 313–322. [CrossRef] [PubMed]
14. Hayes, J.D.; McMahon, M. NRF2 and KEAP1 mutations: Permanent activation of an adaptive response in cancer. *Trends Biochem. Sci.* **2009**, *34*, 176–188. [CrossRef] [PubMed]
15. Osburn, W.O.; Kensler, T.W. Nrf2 signaling: An adaptive response pathway for protection against environmental toxic insults. *Mutat. Res.* **2008**, *659*, 31. [CrossRef] [PubMed]
16. Arlt, A.; Schäfer, H.; Kalthoff, H. The 'N-factors' in pancreatic cancer: Functional relevance of NF-κB, NFAT and Nrf2 in pancreatic cancer. *Oncogenesis* **2012**, *1*, e35. [CrossRef]
17. Lee, S.; Rauch, J.; Kolch, W. Targeting MAPK Signaling in Cancer: Mechanisms of Drug Resistance and Sensitivity. *Int. J. Mol. Sci.* **2020**, *21*, 1102. [CrossRef]
18. Arlt, A.; Müerköster, S.S.; Schäfer, H. Targeting apoptosis pathways in pancreatic cancer. *Cancer Lett.* **2013**, *332*, 346–358. [CrossRef]
19. Zappavigna, S.; Cossu, A.M.; Grimaldi, A.; Bocchetti, M.; Ferraro, G.A.; Nicoletti, G.F.; Filosa, R.; Caraglia, M. Anti-Inflammatory Drugs as Anticancer Agents. *Int. J. Mol. Sci.* **2020**, *21*, 2605. [CrossRef] [PubMed]
20. Majchrzak-Celińska, A.; Misiorek, J.O.; Kruhlenia, N.; Przybyl, L.; Kleszcz, R.; Rolle, K.; Krajka-Kuźniak, V. COXIBs and 2,5-dimethylcelecoxib counteract the hyperactivated Wnt/β-catenin pathway and COX-2/PGE2/EP4 signaling in glioblastoma cells. *BMC Cancer* **2021**, *21*, 1–18. [CrossRef] [PubMed]

21. Bradley, M.C.; Hughes, C.M.; Cantwell, M.M.; Napolitano, G.; Murray, L.J. Non-steroidal anti-inflammatory drugs and pancreatic cancer risk: A nested case–control study. *Br. J. Cancer* **2010**, *102*, 1415. [CrossRef]
22. Narożna, M.; Krajka-Kuźniak, V.; Bednarczyk-Cwynar, B.; Kucińska, M.; Kleszcz, R.; Kujawski, J.; Piotrowska-Kempisty, H.; Plewiński, A.; Murias, M.; Baer-Dubowska, W. Conjugation of diclofenac with novel oleanolic acid derivatives modulate Nrf2 and NF-κB activity in hepatic cancer cells and normal hepatocytes leading to enhancement of its therapeutic and chemopreventive potential. *Pharmaceuticals* **2021**, *14*, 688. [CrossRef] [PubMed]
23. Pires, B.R.B.; Silva, R.C.M.C.; Ferreira, G.M.; Abdelhay, E. NF-kappaB: Two Sides of the Same Coin. *Genes* **2018**, *9*, 24. [CrossRef] [PubMed]
24. Christian, F.; Smith, E.L.; Carmody, R.J. The Regulation of NF-κB Subunits by Phosphorylation. *Cells* **2016**, *5*, 12. [CrossRef] [PubMed]
25. Okami, J.; Yamamoto, H.; Fujiwara, Y.; Tsujie, M.; Kondo, M.; Noura, S.; Oshima, S.; Nagano, H.; Dono, K.; Umeshita, K.; et al. Overexpression of Cyclooxygenase-2 in Carcinoma of the Pancreas. *Clin. Cancer Res.* **1999**, *5*, 2018–2024. [PubMed]
26. Liu, H.; Yang, Z.; Zang, L.; Wang, G.; Zhou, S.; Jin, G.; Yang, Z.; Pan, X. Downregulation of glutathione S-transferase A1 suppressed tumor growth and induced cell apoptosis in A549 cell line. *Oncol. Lett.* **2018**, *16*, 467–474. [CrossRef] [PubMed]
27. Prabhu, L.; Mundade, R.; Korc, M.; Loehrer, P.J.; Lu, T. Critical role of NF-κB in pancreatic cancer. *Oncotarget* **2014**, *5*, 10969. [CrossRef] [PubMed]
28. Nomura, A.; Majumder, K.; Giri, B.; Dauer, P.; Dudeja, V.; Roy, S.; Banerjee, S.; Saluja, A.K. Inhibition of NF-kappa B pathway leads to deregulation of epithelial–mesenchymal transition and neural invasion in pancreatic cancer. *Lab. Investig.* **2016**, *96*, 1268–1278. [CrossRef] [PubMed]
29. Narożna, M.; Krajka-Kuźniak, V.; Kleszcz, R.; Bednarczyk-Cwynar, B.; Szaefer, H.; Baer-Dubowska, W. Activation of the Nrf2 response by oleanolic acid oxime morpholide (3-hydroxyiminoolean-12-en-28-oic acid morpholide) is associated with its ability to induce apoptosis and inhibit proliferation in HepG2 hepatoma cells. *Eur. J. Pharmacol.* **2020**, *883*, 173307. [CrossRef]
30. Krajka-Kuźniak, V.; Bednarczyk-Cwynar, B.; Paluszczak, J.; Szaefer, H.; Narożna, M.; Zaprutko, L.; Baer-Dubowska, W. Oleanolic acid oxime derivatives and their conjugates with aspirin modulate the NF-κB-mediated transcription in HepG2 hepatoma cells. *Bioorg. Chem.* **2019**, *93*, 103326. [CrossRef]
31. Narożna, M.; Krajka-Kuźniak, V.; Bednarczyk-Cwynar, B.; Kleszcz, R.; Baer-Dubowska, W. The Effect of Novel Oleanolic Acid Oximes Conjugated with Indomethacin on the Nrf2-ARE And NF-κB Signaling Pathways in Normal Hepatocytes and Human Hepatocellular Cancer Cells. *Pharmaceuticals* **2021**, *14*, 32. [CrossRef] [PubMed]
32. Yu, Y.; Wan, Y.; Huang, C. The Biological Functions of NF-κB1 (p50) and its Potential as an Anti-Cancer Target. *Curr. Cancer Drug Targets* **2009**, *9*, 566. [CrossRef] [PubMed]
33. Giridharan, S.; Srinivasan, M. Mechanisms of NF-κB p65 and strategies for therapeutic manipulation. *J. Inflamm. Res.* **2018**, *11*, 407. [CrossRef] [PubMed]
34. Vogel, C.; Marcotte, E.M. Insights into the regulation of protein abundance from proteomic and transcriptomic analyses. *Nat. Rev. Genet.* **2012**, *13*, 227–232. [CrossRef]
35. Saha, S.; Buttari, B.; Panieri, E.; Profumo, E.; Saso, L. An Overview of Nrf2 Signaling Pathway and Its Role in Inflammation. *Molecules* **2020**, *25*, 5474. [CrossRef]
36. Baird, L.; Yamamoto, M. The Molecular Mechanisms Regulating the KEAP1-NRF2 Pathway. *Mol. Cell. Biol.* **2020**, *40*, e00099-20. [CrossRef]
37. Surh, Y.J.; Na, H.K. NF-κB and Nrf2 as prime molecular targets for chemoprevention and cytoprotection with anti-inflammatory and antioxidant phytochemicals. *Genes Nutr.* **2008**, *2*, 313–317. [CrossRef] [PubMed]
38. Kurutas, E.B. The importance of antioxidants which play the role in cellular response against oxidative/nitrosative stress: Current state. *Nutr. J.* **2016**, *15*, 71. [CrossRef]
39. Geismann, C.; Grohmann, F.; Sebens, S.; Wirths, G.; Dreher, A.; Häsler, R.; Rosenstiel, P.; Hauser, C.; Egberts, J.H.; Trauzold, A.; et al. c-Rel is a critical mediator of NF-κB-dependent TRAIL resistance of pancreatic cancer cells. *Cell Death Dis.* **2014**, *5*, e1455. [CrossRef]
40. Probst, B.L.; McCauley, L.; Trevino, I.; Wigley, W.C.; Ferguson, D.A. Cancer Cell Growth Is Differentially Affected by Constitutive Activation of NRF2 by KEAP1 Deletion and Pharmacological Activation of NRF2 by the Synthetic Triterpenoid, RTA 405. *PLoS ONE* **2015**, *10*, e0135257. [CrossRef]
41. Rodríguez-Carballo, E.; Gámez, B.; Ventura, F. p38 MAPK signaling in osteoblast differentiation. *Front. Cell Dev. Biol.* **2016**, *4*, 40. [CrossRef]
42. Yang, S.Y.; Pyo, M.C.; Nam, M.H.; Lee, K.W. ERK/Nrf2 pathway activation by caffeic acid in HepG2 cells alleviates its hepatocellular damage caused by t-butylhydroperoxide-induced oxidative stress. *BMC Complement. Altern. Med.* **2019**, *19*, 139. [CrossRef]
43. Feng, J.; Zhang, P.; Chen, X.; He, G. PI3K and ERK/Nrf2 pathways are involved in oleanolic acid-induced heme oxygenase-1 expression in rat vascular smooth muscle cells. *J. Cell. Biochem.* **2011**, *112*, 1524–1531. [CrossRef] [PubMed]
44. Fernandes, J.V.; Cobucci, R.N.O.; Jatobá, C.A.N.; de Medeiros Fernandes, T.A.A.; de Azevedo, J.W.V.; de Araújo, J.M.G. The role of the mediators of inflammation in cancer development. *Pathol. Oncol. Res.* **2015**, *21*, 527–534. [CrossRef] [PubMed]
45. Huang, G.; Shi, L.Z.; Chi, H. Regulation of JNK and p38 MAPK in the immune system: Signal integration, propagation and termination. *Cytokine* **2009**, *48*, 161. [CrossRef] [PubMed]

46. Liou, G.Y.; Storz, P. Reactive oxygen species in cancer. *Free Radic. Res.* **2010**, *44*, 479–496. [CrossRef]
47. Arensman, M.D.; Telesca, D.; Lay, A.R.; Kershaw, K.M.; Wu, N.; Donahue, T.R.; Dawson, D.W. The CREB-Binding Protein Inhibitor ICG-001 Suppresses Pancreatic Cancer Growth. *Mol. Cancer Ther.* **2014**, *13*, 2303–2314. [CrossRef] [PubMed]
48. Wang, Y.-T.; Tang, F.; Hu, X.; Zheng, C.-X.; Gong, T.; Zhou, Y.; Luo, Y.; Min, L. Role of crosstalk between STAT3 and mTOR signaling in driving sensitivity to chemotherapy in osteosarcoma cell lines. *IUBMB Life* **2020**, *72*, 2146–2153. [CrossRef]
49. Xiao, L.; Wang, Y.C.; Li, W.S.; Du, Y. The role of mTOR and phospho-p70S6K in pathogenesis and progression of gastric carcinomas: An immunohistochemical study on tissue microarray. *J. Exp. Clin. Cancer Res.* **2009**, *28*, 152. [CrossRef]
50. Murakami, S.; Motohashi, H. Roles of Nrf2 in cell proliferation and differentiation. *Free Radic. Biol. Med.* **2015**, *88*, 168–178. [CrossRef]

International Journal of
Molecular Sciences

Article

Biotin Transport-Targeting Polysaccharide-Modified PAMAM G3 Dendrimer as System Delivering α-Mangostin into Cancer Cells and *C. elegans* Worms

Joanna Markowicz [1,*], Łukasz Uram [1], Stanisław Wołowiec [2] and Wojciech Rode [3,*]

1. Faculty of Chemistry, Rzeszow University of Technology, 6 Powstancow Warszawy Ave., 35-959 Rzeszow, Poland; luram@prz.edu.pl
2. Medical College, Rzeszow University, 1a Warzywna Str., 35-310 Rzeszow, Poland; swolowiec@ur.edu.pl
3. Nencki Institute of Experimental Biology, 3 Pasteur Street, 02-093 Warsaw, Poland

* Correspondence: jmarkowicz@stud.prz.edu.pl (J.M.); w.rode@nencki.edu.pl (W.R.)

Abstract: The natural xanthone α-mangostin (αM) exhibits a wide range of pharmacological activities, including antineoplastic and anti-nematode properties, but low water solubility and poor selectivity of the drug prevent its potential clinical use. Therefore, the targeted third-generation poly(amidoamine) dendrimer (PAMAM G3) delivery system was proposed, based on hyperbranched polymer showing good solubility, high biocompatibility and low immunogenicity. A multifunctional nanocarrier was prepared by attaching αM to the surface amine groups of dendrimer via amide bond in the ratio 5 ($G3^{2B12gh5M}$) or 17 ($G3^{2B10gh17M}$) residues per one dendrimer molecule. Twelve or ten remaining amine groups were modified by conjugation with D-glucoheptono-1,4-lactone (gh) to block the amine groups, and two biotin (B) residues as targeting moieties. The biological activity of the obtained conjugates was studied in vitro on glioma U-118 MG and squamous cell carcinoma SCC-15 cancer cells compared to normal fibroblasts (BJ), and in vivo on a model organism *Caenorhabditis elegans*. Dendrimer vehicle $G3^{2B12gh}$ at concentrations up to 20 μM showed no anti-proliferative effect against tested cell lines, with a feeble cytotoxicity of the highest concentration seen only with SCC-15 cells. The attachment of αM to the vehicle significantly increased cytotoxic effect of the drug, even by 4- and 25-fold for $G3^{2B12gh5M}$ and $G3^{2B10gh17M}$, respectively. A stronger inhibition of cells viability and influence on other metabolic parameters (proliferation, adhesion, ATP level and Caspase-3/7 activity) was observed for $G3^{2B10gh17M}$ than for $G3^{2B12gh5M}$. Both bioconjugates were internalized efficiently into the cells. Similarly, the attachment of αM to the dendrimer vehicle increased its toxicity for *C. elegans*. Thus, the proposed α-mangostin delivery system allowed the drug to be more effective in the dendrimer-bound as compared to free state against both cultured the cancer cells and model organism, suggesting that this treatment is promising for anticancer as well as anti-nematode chemotherapy.

Keywords: α-mangostin; poly(amidoamine) dendrimer; targeted drug delivery; biotin targeting; glioblastoma multiforme; squamous cell carcinoma; antiparasitic therapy

Citation: Markowicz, J.; Uram, Ł.; Wołowiec, S.; Rode, W. Biotin Transport-Targeting Polysaccharide-Modified PAMAM G3 Dendrimer as System Delivering α-Mangostin into Cancer Cells and *C. elegans* Worms. *Int. J. Mol. Sci.* **2021**, *22*, 12925. https://doi.org/10.3390/ijms222312925

Academic Editor: Angela Stefanachi

Received: 25 October 2021
Accepted: 26 November 2021
Published: 29 November 2021

Publisher's Note: MDPI stays neutral with regard to jurisdictional claims in published maps and institutional affiliations.

1. Introduction

The isoprenylated xanthone α-mangostin (αM) was firstly isolated from the pericarp of mangosteen tree (*Garcinia mangostana* L., *Clusiaceae*) that has long been used in traditional medicine in Southeast Asia to treat inflammation, ulcer, skin infection, and wounds [1]. Nowadays, numerous in vitro and in vivo studies have proved that mangostins possess diverse pharmacological activities, such as antioxidant, anti-inflammatory, antibacterial, antifungal, antimalarial, anticancer and anthelmintic [2–5].

Considering the potential pharmacological application of αM, it is significantly limited by its low water solubility (2.03×10^{-4} mg/L at 25 °C) [6], a major obstacle in achieving maximum drug bioavailability and accumulation in the target organs. The latter, together with insufficient target selectivity in the human body, resulted in the fact that αM has not been approved for clinical use.

One strategy increasing the concentration of biologically active substances in target locations and cells is to take advantage of a drug delivery agent [7], e.g., to bind a drug with nanoparticles, in order to improve its physicochemical properties and transport. Among the αM delivery systems tested so far, nanolipids, nanopolymers, nanomicelles, nanoliposomes, nanoemulsion, nanofibers and metal nanoparticles have been studied [8], e.g., αM-loaded fibroin nanoparticles (FNPs) crosslinked with EDC or PEI, compared to the free αM, confirmed the crosslinked FNPs to increase the drug's solubility by about 3-fold [9], as well as to offer sustained drug release and reduced hematotoxicity. Anticancer studies of these nanoparticles with Caco-2 colorectal and MCF-7 breast adenocarcinoma cell lines showed αM to cause apoptosis with cytotoxicity exceeding that of free drug. Additionally, cyclodextrin-based hyperbranched polymer nanoparticles (CDNPs) were used [10] to solubilize αM encapsulated in the CD cavity. They assumed that release of αM in the slow mode is essential for its retention until the cancerous region is reached. Samprasit et al. [11] used αM-loaded mucoadhesive nanoparticles as colon-targeted drug delivery. They proved that chitosan and thiolated chitosan nanoparticles crosslinked by genipin (GP) and modified by Eudragit® L100 increased αM loading limited the release of the drug in the upper gastrointestinal tract, and enhanced its delivery to the colon.

The vast majority of the proposed carriers encapsulate αM; moreover, there are many potential carriers that allow binding it via covalent bonds and significantly increase the stability of such a system. An interesting example presents poly(amidoamine) dendrimers (PAMAM) that are spheroidal or ellipsoidal three-dimensional polymers with the peripheral functional groups and internal cavities. PAMAM dendrimers may be favorably used in drug delivery due to their hydrophilic, biocompatible, and non-immunogenic nature [12], having been applied as targeting drug or gene delivery systems and as diagnostic agents. Drugs or biologically active molecules can be encapsulated in the interior space of dendrimers, attracted by electrostatic interactions or linked via surface groups [13], but covalently attached drugs, e.g., methotrexate, are more stable compared to non-covalent drugs [14]. Additionally, polyvalency of PAMAMs has a potential role in intracellular targeted drug delivery by surface modification [13]. Especially in anticancer therapy, drug targeting is important in order to maximize the therapeutic potential in tumor area with minimizing side effects in normal tissues [15]. The PAMAM dendrimers present the platform for surface modification with numerous ligand moieties that can improve active targeting to the cancer cells and increase tumor specificity with minimum systemic toxicity [12]. Examples of such ligands include monoclonal antibodies, polypeptides, antibody–drug conjugates, nucleic acids and signal transduction inhibitors [16]. Additionally, vitamins, such as folic acid and biotin, are promising molecules targeting nanoparticles and potential drugs conjugates into cancer cells overexpressing folate and biotin receptors, respectively, e.g., breast, ovarian, lung, and colon cancer cells [17–19]. Notably, biotinylated dendrimers have been demonstrated to be absorbed into cells stronger than non-biotinylated [20,21].

The aim of this study was to design, synthesize, characterize, and investigate the biological, anti-cancer activity of biotin-targeted, polysaccharide modified PAMAM G3 dendrimers substituted covalently by αM ($G3^{2B12gh5M}$ and $G3^{2B10gh17M}$, differing by the substitution level) against human grade IV glioma U-118 MG cells, human squamous cell carcinoma (SCC-15), compared to normal human fibroblasts (BJ), with the three cell lines found previously to accumulate biotin [22]. The effects of the above-mentioned conjugates or their vehicle ($G3^{2B12gh}$) on viability, proliferation, adhesion, apoptosis, as well as the intracellular ATP level were determined. Cellular accumulation and localization of fluorescently labeled analogues were also evaluated. Additionally, the in vivo toxicity of dendrimer conjugates and αM was tested with model nematode *Caenorhabditis elegans*, which has been used extensively in toxicological studies of many nanoparticles and as a model organism providing insights into cancer cells metabolism, also allowing to assess the anti-parasitic activity.

2. Materials and Methods

2.1. Reagents

α-Mangostin (αM, purity ≥ 98% (HPLC)) was purchased from Aktin Chemicals, Inc. (Chengdu, China). Ethylenediamine, methyl acrylate, D-glucoheptono-1,4-lactone (GHL), biotin *N*-hydroxysuccinimide ester (NHS-Biotin), 4-nitrophenyl chloroformate (NPCF), 2-Chloro-1-methylpyridinium iodide (Mukaiyama reagent), 4-dimethylaminopyridine (DMAP), 6-[fluorescein-5(6)-carboxamido]hexanoic acid (FCH), dimethyl sulfoxide (dmso) and other reagents used in syntheses were obtained from Merck KGaA (Darmstadt, Germany). Spectra/Por® 3 RC dialysis membrane (cellulose, MW_{cutoff}–3.5 kD) was provided by Carl Roth GmbH & Co. KG (Karlsruhe, Germany).

Human glioblastoma (U-118 MG, ATCC® HTB-15), human squamous cell carcinoma (SCC-15, ATCC® CRL-1623), and human normal fibroblast (BJ, ATCC® CRL-2522) cell lines were purchased from the American Type Culture Collection (ATCC, Manassas, VA, USA). Dulbecco's Modified Eagle's Media (DMEM and DMEM/ F-12), Eagle's Minimum Essential Medium (EMEM), and fetal bovine serum (FBS) were obtained from Corning Inc. (New York, NY, USA). Penicillin and streptomycin solution, phosphate-buffered saline (PBS) with and without magnesium and calcium ions, and DAPI (4′,6-diamidino-2-phenylindole, dihydrochloride), Hoechst 33342, and MitoTracker™ Deep Red FM were provided by Thermo Fisher Scientific Inc. (Waltham, MA, USA). Trypsin-EDTA solution, hydrocortisone, 0.33% neutral red solution (3-amino-m-dimethylamino-2-methyl-phenazine hydrochloride), XTT sodium salt (2,3-bis[2-methoxy-4-nitro-5-sulfophenyl]-2H-tetrazolium-5-carboxanilide inner salt), phenazinemethosulfate (PMS), crystal violet, 0.4% trypan blue solution, dimethylsulfoxide (dmso) for molecular biology, 5-Fluoro-2′-deoxy-uridine (FUdR), and other chemicals and buffers were purchased from Merck KGaA (Darmstadt, Germany). CellTiter-Glo® Luminescent Cell Viability Assay and Apo-ONE® Homogenous Caspase-3/7 Assay were obtained from Promega Corporation (Madison, WI, USA). Cell cultures dishes and materials were from Corning Incorporated (Corning, NY, USA), Greiner (Kremsmünster, Austria), or Nunc (Roskilde, Denmark). All reagents used for *C. elegans* culture and synchronization were supplied by Sigma-Aldrich (Saint Louis, MO, USA) or Carl Roth GmbH & Co., KG (Karlsruhe, Germany).

2.2. Chemical Syntheses and Purification Methods

2.2.1. PAMAM G3 Conjugation with Biotin

PAMAM G3 dendrimer was synthesized starting from an ethylenediamine core by repeatable two-step procedure according to Tomalia's protocol [23] and stored as 20.1 mM solution in methanol for further use. Then PAMAM G3, after prior methanol evaporation, was substituted with two equivalents of biotin by stepwise addition of 99 mg (290 μmoles) of solid NHS-Biotin to 695 mg of G3 (145 μmoles) dissolved in 4 mL of dmso with vigorous stirring. The solution was left at 50 °C for 4 h, then transferred into dialytic tube (cellulose, MW_{cutoff} = 3.5 kD) and dialyzed for 3 days against water (7 × 3 L). Afterwards, the solvent was removed by vacuum rotary evaporation and dried under high vacuum overnight. The product was identified by ^{1}H NMR spectroscopy as G3 substituted with average two residues of amide-bonded biotin $G3^{2B}$, as described before [24,25]. The isolated yield of the product was 47% (501 mg; 68 μmoles; MW_{calc} = 7362 g mol^{-1}).

2.2.2. Activation of α-Mangostin with NPCF and Attachment to $G3^{2B}$

α-Mangostin (αM) was activated with 4-nitrophenyl chloroformate (NPCF; 10% molar excess), which further provided one carbon linker between hydroxyl group of αM and amino group of PAMAM G3. Thus, 223.3 mg (544 μmoles) of αM was dissolved in 25 mL of chloroform, then 120.6 mg (598 μmoles) of NPCF was added in portions as solid with vigorous stirring. To this solution, 146.6 mg (1200 μmoles) of 4-dimethylaminopyridine (DMAP) was added and the mixture was heated under reflux for 4 h. Then, chloroform was removed by vacuum rotary evaporation and remained solid dried under high vacuum. The solid product, α-mangostin 4-nitrophenylcarbonate (αM-NPC) was dissolved in 10 mL dmso to obtain 54.4 mM stock solution. This solution (6 mL, 326.4 μmoles of αM-NPC, 143 mg of αM) was added dropwise into 162 mg of $G3^{2B}$ (22 μmoles) in 1.6 mL dmso to obtain highly substituted conjugate [A]. In another synthesis process, stock solution of αM-NPC (2 mL, 109 μmoles) was added to 22 μmoles of $G3^{2B}$ in 1.6 mL dmso [B]. In both cases, the mixtures were heated at 50 °C for 3 h, then dialyzed and isolated solids were dried in vacuo. Yields: 273.2 mg, 91.4% for [A], identified by ^{1}H NMR spectroscopy as $G3^{2B17M}$ and 188.7 mg, 90% for [B], identified by ^{1}H NMR spectroscopy as $G3^{2B5M}$. Long-lasting dialysis of synthesized conjugates, especially $G3^{2B17M}$ (4 days, 12 times 5 L of receiving water), resulted in loss of low-substituted fraction of $G3^{2BxM}$ (x < 5). During the dialytic process, immediate precipitation of highly substituted derivatives was observed. The solids inside dialytic tubes were irritated throughout dialysis in order to remove solid-entrapped low molecular weight DMAP, *p*-nitrophenol, dmso, and other reagents. The products [A] and [B] were further converted by reaction with GHL (vide infra).

2.2.3. $G3^{2B}$ and $G3^{2BM}$ Supplementation with D-Glucoheptono-1,4-lactone

To the solutions of $G3^{2B17M}$ and $G3^{2B5M}$ obtained in the previous step according to protocol [A] and [B] (ca 20 μmoles each), in 5 mL dmso, 264 μmoles of solid GHL (54.9 mg) were added in portions with vigorous stirring. The mixtures were left for 16 h at 50 °C. Thereafter, purification was performed according to the procedure described earlier (3 days of dialysis followed by 24 h drying under high vacuum).

The steps of synthesis of dendrimer PAMAM G3 conjugates with α-mangostin, biotin and D-glucoheptono-1,4-lactone is presented in the Scheme 1.

The obtained conjugates were characterized by the ^{1}H NMR (Figures 1 and 2), and 2-D ^{1}H-^{1}H COSY spectroscopy in dmso-d_6 (Figure S1), which revealed the level of PAMAM $G3^{2B}$ substitution by αM and gh residues: $G3^{2B12gh}$, $G3^{2B12gh5M}$, and $G3^{2B10gh17M}$, with the isolated yield 94.5% (20.8 μmoles, MW_{calc} = 9859 g mol^{-1}), 86% (19 μmoles, MW_{calc} = 11,660 g mol^{-1}), and 91% (20 μmoles, MW_{calc} = 16,896 g mol^{-1}), respectively.

Analytical Data

^{1}H NMR (dmso-d_6; for atom numbering see Scheme 1): chemical shift [ppm] (intensity, multiplicity, assignment):

α-Mangostin (Figure 1A): 13.72 ([1H], s, $1'^{M}$); 11.01 ([1H], bs, $3'^{M}$); 10.82 ([1H], bs, $6'^{M}$); 6.80 ([1H], s, 5^{M}); 6.34 ([1H], s, 4^{M}); 5.16 ([2H], t overlapped, $12,17^{M}$); 4.01 ([2H], d, 11^{M}); 3.70 ([3H], s, $7'^{M}$); 3.34 (s, HDO); 3.21 ([2H], d, 16^{M}); 1.70 ([12H], m, $14,15,19,20^{M}$).

$G3^{2B12gh5M}$ (Figure 1B): 7.10 ([5H], s, NH(carbamate)); 6.51 ([5H], s, 5^{M}); 6.26 ([5H], s, 4^{M}); 5.17 ([10H], t overlapped, $12,17^{M}$); 3.95 ([10H], m, 11^{M}); 3.68 ([15H], s, $7'^{M}$); 1.69 ([60H], overlapped s, $14,15,19,20^{M}$); PAMAM G3 CH_2 broad resonances: 3.12, 2.64, 2.42, 2.19 [484H]; PAMAM G3 NH: 8.0 ppm, [73H]; Biotin resonances: 6.43 ([4H], bs, 10^{B} and 11^{B}); 4.31 and 4.14 ([2H] and [2H], 8^{B} and 9^{B}); 2.04 ([4H], 2^{B}); 1.49–1.22 ([12H], 3^{B},4^{B},5^{B}); gh resonances: 3.87 and 3.58–3.36 ([84H], s and m, CH^{gh}).

$G3^{2B10gh17M}$ (Fig 1C): 6.65 ([17H], s, 5^{M}); 6.29 ([17H], s, 4^{M}); 5.16 ([34H], t overlapped, $12,17^{M}$); 3.99 ([34H]; d, 11^{M}); 3.70 ([51H], s, $7'^{M}$); 3.19 ([34H], d, 16^{M}); 1.70 ([204H], four overlapped singlets, $14,15,19,20^{M}$); PAMAM G3 CH_2 broad resonances: 3.14–2.20 [484H]; PAMAM G3 NH: 8.02 ppm, [60H]; biotin resonances: 6.51 and 6.44 ([4H], s and s, 10^{B} and 11^{B}); 4.31 and 4.13 ([4H], s and s, 8^{B} and 9^{B}); 2.06 ([4H], 2^{B}); 1.51–1.28 ([12H], 3^{B},4^{B},5^{B}); gh resonances: 3.87 and 3.59–3.37 ([70H], s and m, CH^{gh}).

$G3^{2B12gh}$ (Figure 2A): 6.47 ([4H], d, $10,11^{B}$); 4.36 (overlapped singlets, OH^{gh}); 3.66 ([84H], m, CH^{gh}); 1.39 ([12H], m, $3,4,5^{B}$); PAMAM G3 CH_2 broad resonances: 3.10, 2.63, 2.41, 2.19 [480H]; PAMAM G3 NH broad resonance is centered at 8.00 ppm, [78H].

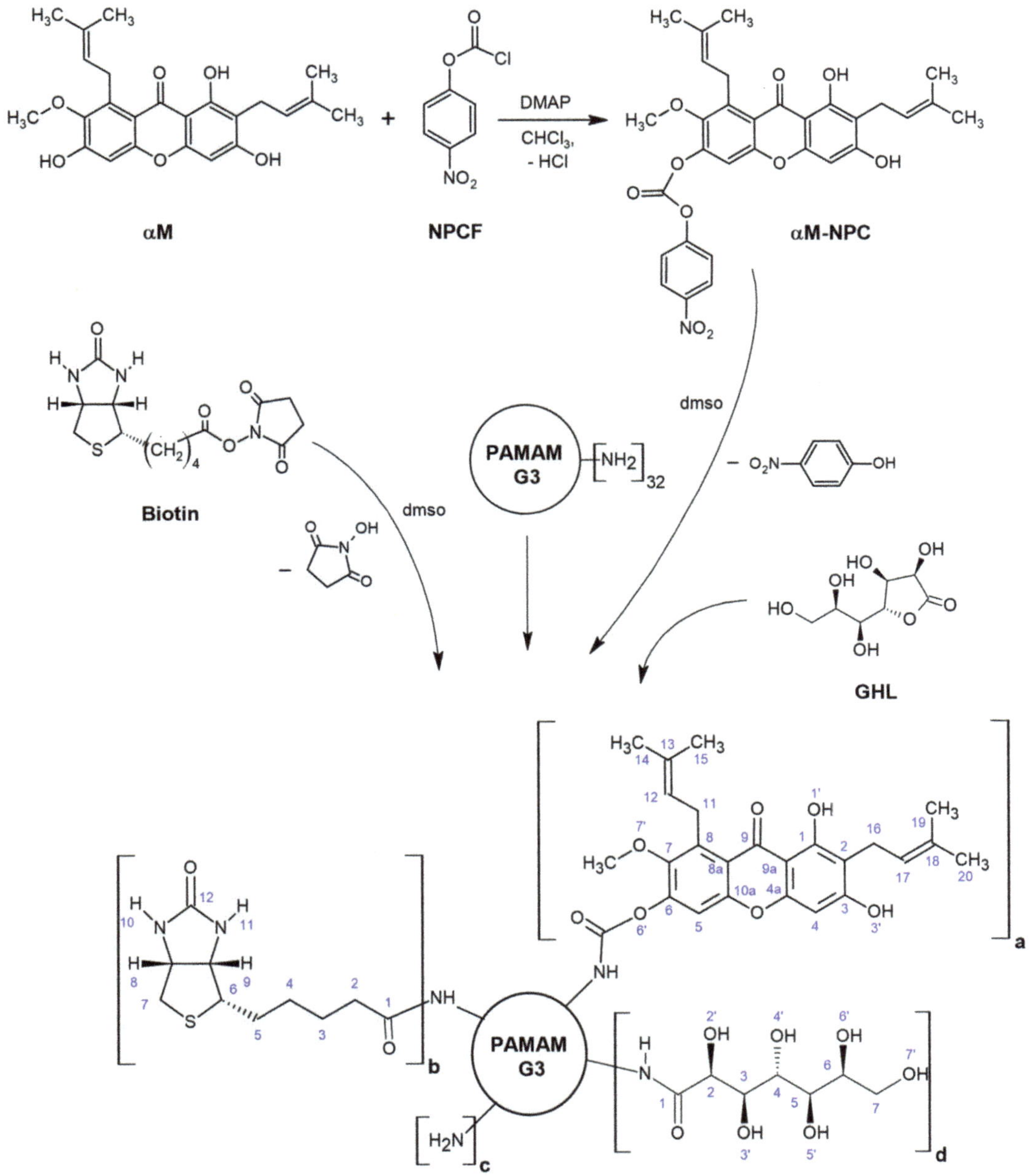

Scheme 1. The scheme of synthesis of dendrimer PAMAM G3 conjugates with αmangostin, biotin and D-glucoheptono-1,4-lactone attached via amide bond. The following conjugates were obtained: G3[2B12gh5M] (a = 5, b = 2, c = 13, d = 12), G3[2B10gh17M] (a = 17, b = 2, c = 3, d = 10), and G3[2B12gh] (a = 0, b = 2, c = 18, d = 12). Abbreviations: NPCF—4-nitrophenyl chloroformate, αM-NPC—p-nitrophenyl) carbonate derivative of αM, DMAP—4-dimethylaminopyridine.

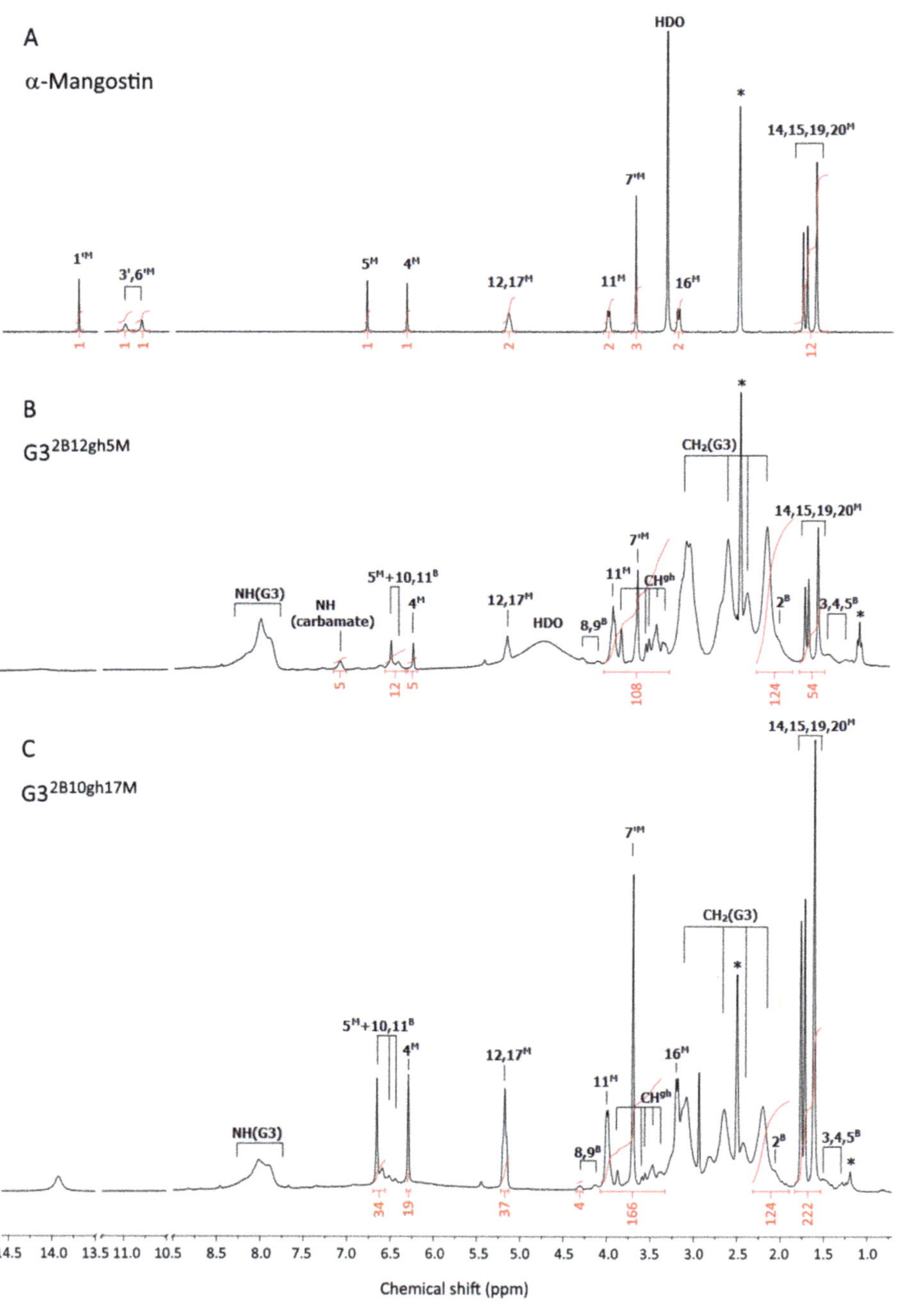

Figure 1. The ^{1}H-NMR spectra of α-mangostin (**A**), G3^{2B12gh5M} (**B**), and G3^{2B10gh17M} (**C**) conjugates in dmso-d$_6$. The residual solvent peak at 2.5 and impurity resonances are marked with asterisks *. The PAMAM G3 core dendrimer resonances are labeled as CH$_2$(G3) and NH(G3) in spectrum B and C. The resonances of α-mangostin, biotin, and glucoheptoamide are labeled by locants with ^{M}, ^{B}, and ^{gh} upper indexes, respectively, according to the atom numbering in Scheme 1.

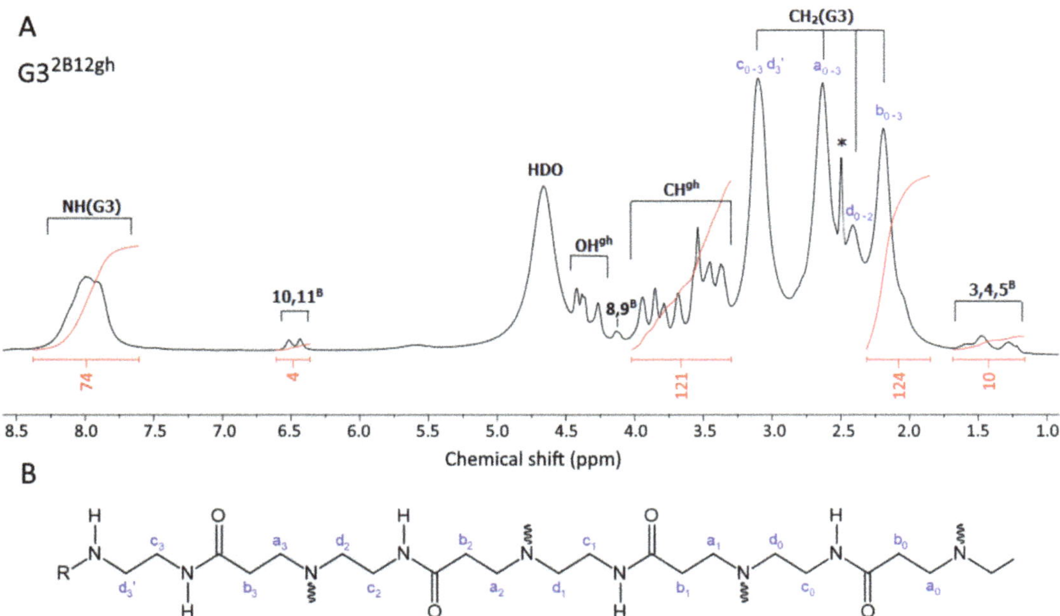

Figure 2. (**A**) The ^{1}H-NMR spectrum of G3^{2B12gh} vehicle in dmso-d$_6$. The PAMAM G3 core dendrimer resonances are labeled as CH$_2$(G3) and NH(G3). The resonances of biotin and D-glucoheptoamide residues are labeled with locants with ^{B} and ^{gh}. The residual solvent peak is labeled with an asterisk *. For atom numbering, see Scheme 1. (**B**) The selected chain of the PAMAM G3 dendrimer core with carbon atom numbering. Branch points at ternary nitrogen atoms are marked with wavy lines. R = H, biotin or gh residues.

2.2.4. Fluorescent Labeling of the Conjugates

Fluorescently labeled analogues of G3^{2B12gh5M} and G3^{2B10gh17M} were synthesized by attachment of one equivalent of 6-[fluorescein-5(6)-carboxamido]hexanoic acid (FCH) via an ester bond. The carboxyl groups of FCH were activated by Mukaiyama reagent (2-Chloro-1-methylpyridinium iodide). Specifically, 9.0 mg (18 µmoles) of solid FCH was dissolved in 1 mL of dmso with vigorous stirring and protection from light. Then, 6.2 mg (27 µmoles) of Mukaiyama reagent and 6.6 mg (54 µmoles) of DMAP were added with stirring and the mixture was heated at 45 °C for 0.5 h. The activated FCH was divided into two parts and added dropwise to 0.8 mL (7.1 µmoles) of G3^{2B12gh5M} and 2 mL (7.8 µmoles) of G3^{2B10gh17M} in dmso in equimolar amount. Both reaction mixtures were kept on the heating block at 45 °C overnight followed by purification as before. The ^{1}H NMR spectra in dmso-d$_6$ were taken and confirmed the attachment of one equivalent of FCH to the dendrimer conjugates. The isolated yield for G3^{2B12gh5MF} was 59% (4.19 µmoles, MW$_{calc}$ = 12,150 g mol^{-1}) and for G3^{2B10gh17MF} was 53% (4.1 µmoles, MW$_{calc}$ = 17,386 g mol^{-1}). These fluorescent-labeled compounds were used for cellular internalization studies by CLSM.

2.3. NMR Spectroscopy

The 1D ^{1}H and ^{13}C NMR spectra and 2D ^{1}H-^{1}H correlations spectroscopy (COSY), ^{1}H-^{13}C heteronuclear single quantum correlation (HSQC), and heteronuclear multiple bond correlation (HMBC) spectra were recorded in dmso-d$_6$ using Bruker 300 MHz instrument (Rheinstetten, Germany) at College of Natural Sciences, University of Rzeszów.

2.4. Conjugate Size and ζ Potential Measurements

Size and ζ potential of G3^{2B12gh}, G3^{2B12gh5M}, and G3^{2B10gh17M} conjugates were measured using the dynamic light scattering technique at pH 5 (0.05 M acetate buffer) and in water using the Zetasizer Nano instrument (Malvern, UK) for 1 mg/mL samples (0.7–1.0 mM solutions).

2.5. Biological Studies

2.5.1. Cell Cultures

Human glioblastoma cells (U-118 MG, doubling time—37 h) were grown in DMEM, human squamous carcinoma cells (SCC-15, doubling time—48 h) were cultured in DMEM F-12 supplemented with hydrocortisone (400 ng/mL), and normal human skin fibroblasts (BJ, doubling time—1.9 day) were cultured in EMEM. Culture media were supplemented with heat-inactivated 10% FBS and 100 U/mL penicillin and 1% streptomycin solution. All cell lines were cultured at 37 °C in a humidified atmosphere of 95% air with 5% CO_2. Growth media were changed every 2–3 days and cells were passaged at 80–85% confluence with 0.25% trypsin-0.03% EDTA in PBS (calcium and magnesium ions free). Cell morphology was observed with Nikon TE2000S Inverted Microscope (Tokyo, Japan) with phase contrast. Viability and cell density were estimated by trypan blue exclusion test using Automatic Cell Counter TC20TM (BioRad Laboratories, Hercules, CA, USA). All assays were performed in triplicates in three independent experiments. The working solutions of synthesized dendrimer conjugates were prepared from stock solutions in a corresponding cell culture media by adjusting the dmso concentration to 0.05–0.2% (depending on the type of conjugate), which had no effect on treated cells. Control samples with non-treated cells in complete culture medium with adjusted dmso concentration were included in all assays.

2.5.2. Cytotoxicity (NR and XTT Assays)

The cytotoxicity of PAMAM G3 conjugates with αM ($G3^{2B12gh5M}$ or $G3^{2B10gh17M}$) and the $G3^{2B12gh}$ vehicle was assessed with neutral red uptake (NR) assay and XTT reduction assay. Cells were seeded in flat, clear bottom 96-well culture plates in triplicate (100 μL cell suspension/well) at a density of 1×10^4 cells/well and allowed to attach for 24 h. Then, cells were incubated with working solutions of dendrimer conjugates and vehicle in complete culture medium for 48 h. After that, the NR assay and XTT reduction assay were performed as described in Uram et al. [20]. Additionally, to present cells morphology and the level of accumulation of neutral red dye in lysosomes, microscopic images were collected with a Delta Optical IB-100 microscope with contrast phase under 200× magnification.

2.5.3. Fluorescently Labeled $G3^{2B12gh5M}$ or $G3^{2B10gh17M}$ Cellular Accumulation and Distribution

Cellular accumulation as well as colocalization with nuclei and mitochondria of fluorescently labeled dendrimer conjugates were analyzed using confocal microscopy. Cells were seeded into an 8-well Lab-tek™ Chambered Coverglass (Nunc, Denmark) with a borosilicate glass bottom at a density of 7×10^4 cells/well in 400 μL and placed in an incubator for 24 h. Then, BJ, U118 MG or SCC-15 cells were incubated with non-toxic concentrations of FCH-labeled $G3^{2B12gh5M}$ (1 μM concentration) and $G3^{2B10gh17M}$ (0.1 μM concentration) conjugates, respectively, for 4 or 48 h. After incubation, the dendrimer solution was replaced with an 8 μM Hoechst 33,342 and 50 nM MitoTracker Deep Red FM mixture in growth medium without FBS and incubated at 37 °C for 15 min. Following nuclei and mitochondria staining, cells were washed with PBS (three times) and fixed with 3.7% formaldehyde in PBS. The images were collected in three channels with a confocal microscope (Olympus Fluoview FV10i, Tokyo, Japan) at 361/497 nm for Hoechst, 491/515 nm for FCH, and 644/665 nm for MitoTracker. Images were collected using an objective with water immersion, under a total magnification of 240×. The obtained images had an optical section thickness of app. 1.2 μm. Image analysis was performed with the ImageJ software.

2.5.4. Apoptosis and Intracellular ATP Level

The activity of caspase-3 and -7, the apoptosis marker, was measured to check the ability of the synthesized compounds to induce apoptosis in studied cells. In addition, the intracellular ATP level was assessed. For both parameters, commercially available kits Apo-ONE® Homogenous Caspase-3/7 Assay (Promega) and CellTiter-Glo® Luminescent Cell Viability Assay (Promega) were used. Cells were plated at a density of 1×10^4 per well into a flat, black bottom 96-microplates for apoptosis assay and into a clear 96-microplate for ATP level assay. After 24 h of incubation, cells were treated with the $G3^{2B12gh5M}$ or $G3^{2B10gh17M}$ solutions (100 μL/well) for a further 48 h. Both assays were carried out as described by Uram et al. [26], who used Hoechst 33342 staining to determine the number of cells.

2.5.5. Proliferation

Cell proliferation was estimated using DAPI staining. The 5×10^3 cells/well were seeded into a flat, clear bottom 96-well plates and incubated for 24 h at 37 °C to attach. After removing the growth media, cells were treated with working solutions of $G3^{2B12gh5M}$ or $G3^{2B10gh17M}$ conjugates for 72 h in increasing concentrations. In the next step, plates were centrifuged (5 min, 700 g) and the medium was gently removed. The assay was performed according to Uram et al. [26].

2.5.6. Adhesion

Evaluation of cell adhesion was performed with crystal violet (CV) assay. Cells were seeded into a 96-well clear bottom plates at a density of 1×10^4 cells/well and left 24 h at 37 °C to attach. Afterwards, cells were incubated with working solutions of studied compounds for 48 h and the subsequent operations were carried out as described in Markowicz et al. [5]. Images of the studied cells stained with crystal violet were collected with a Delta Optical IB-100 microscope with contrast phase, with a 10× objective magnification.

2.5.7. Toxicity to Caenorhabditis Elegans and the Worm Survival Analysis

The model organism *Caenorhabditis elegans* was used to estimate in vivo activity of the synthesized dendrimer conjugates and α-mangostin alone in the multicellular system. The wild type *C. elegans* strain N2, variety Bristol, was cultured at 20 °C on NGM agar plates with *E. coli* OP50 lawn as a food source [27]. Survival assay was based on the protocol by Bischof et al. [28] and was adapted to a 96-well plate. Worms at L4 stage were used in the assay and obtained from eggs by prior synchronization of all-stages-nematode population by treatment with hypochlorite. The obtained eggs were left in M9 buffer overnight at room temperature for hatching. The next day, nematodes at L1 stage were transferred to NGM plates with bacterial lawn and grown at 20 °C until the L4 stage (approx. 44 h). After that, the L4 worms were washed twice with distilled water and centrifuged at 250 g followed by resuspension in complete S medium ([29] and centrifugation. Then, the density of nematode suspension was estimated according to Scanlan et al. [30]. Nematodes were suspended in complete S medium supplemented with E. coli OP50 (1:1000), 0.08% cholesterol (5 mg/mL in Et-OH), 1% penicillin-streptomycin, 1% nystatin, and 100 mM FUdR (in final concentration 200 μM) to obtain 20 nematodes in 50 μL transferred to each well. FUdR was added to sterilize nematodes. The working solutions of tested compounds were prepared in the complete S medium with dmso (adjusted to 0.1%, 0.36%, and 0.41% for αM, $G3^{2B12gh5M}$, and $G3^{2B10gh17M}$, respectively) and pipetted (50 μL/well) to a 96-well plate with previously seeded worms. Then, worms were incubated at 20 °C for 7 days. Every day live (moving and curling) and dead nematodes were counted under an inverted microscope. Additionally, images of some morphological changes in nematodes after incubation with $G3^{2B12gh5M}$ were collected using Delta Optical IB-100 microscope under 10× and 20× magnification. The assay was performed in triplicates in three independent experiments.

2.5.8. Statistical Analysis

Due to the lack of a normal distribution of the data in the experimental groups, the non-parametric Kruskal–Wallis test was used to evaluate the differences between dendrimer-treated cells and non-treated control in each cell line. To determine the statistically significant differences between G3^{2B12gh5M} and G3^{2B10gh17M} treated groups, a Mann–Whitney *U* test was performed. $p \leq 0.05$ was considered statistically significant. The survival curves of *C. elegans* were presented in a plot of the Kaplan–Meier estimator. Statistically significant differences (with $p \leq 0.05$) between treated and non-treated control groups were indicated using Gehan's Wilcoxon test. All analyses and calculations were performed using Statistica 13.3 software (StatSoft Poland, Cracow).

3. Results and Discussion

3.1. Dendrimer Conjugates Synthesis and Characterization

The bioavailability of α-mangostin (αM) is low due to its poor solubility in aqueous solutions, below 0.5 μM at ambient temperature. On the other hand, it has been evidenced that αM shows a wide range of pharmacological activities, including anticancer, antioxidant, anti-inflammatory [31], and antibacterial [32]. In vitro and in vivo studies on various cancer models have proven that the anticancer activity of αM is manifested by, i.a., inhibition of cell proliferation, induction of apoptosis [33,34], and inhibition of metastasis [35]. In order to overcome limitations of αM, we performed several attempts to functionalize it by attachment it to PAMAM dendrimers, which are known to act both as solubilizers for hydrophobic drug molecules and as macromolecular carriers for covalently attached (pro)drugs [36]. Thus, we have used poly(amidoamine) dendrimer of the third generation, G3, as macromolecular carrier, and attempted to bind αM via short link provided by *p*-nitrophenyl chloroformate (NPCF). αM was functionalized with NPCF to obtain *p*-nitrophenylcarbonate derivative, αM-NPC which was further used to bind it to terminal primary amine groups of G3. Although we were able to isolate G3-αM with NPCF-derived carbonyl linker, the water solubility of obtained derivatives was too low to progress with biological tests. Therefore, we converted G3-αM conjugates into more soluble compounds by reaction of remained primary amine groups with D-glucoheptono-1,4-lactone (GHL) to obtain amide-attached polyhydroxyalkyl chains. Two G3-αM derivatives were prepared (Scheme 1) differing by αM substitution levels amounting on average to 17 and 5 residues per dendrimer molecule. These were additionally furnished with two equivalents of amide-attached biotin per dendrimer molecule and exhaustively or semi-exhaustively glucoheptoamidated with GHL. The macromolecular G3 substrate was equipped with 2 biotin to obtain G3^{2B} as before, reacted with αM-NPC, and further converted with excess of GHL, as described before [24]. Additionally, in order to obtain the control carrier, void of αM, G3 substrate was substituted with two biotin residues per dendrimer molecule and further glucoheptoamidated with excess of GHL to obtain G3^{2B12gh}. The conjugates were well soluble in dmso and characterized by NMR spectroscopy in this solvent, while molecular size was determined by DLS measurements in diluted aqueous solutions.

The ^{1}H NMR spectra of the final products, G3^{2B12gh5M} and G3^{2B10gh17M}, are presented in Figure 1B,C. For comparison, the ^{1}H-NMR spectrum of αM in dmso-d$_6$ was recorded (Figure 1A) and ^{1}H resonances were assigned in accordance with literature data [37,38]. Common features of the ^{1}H NMR spectra were the PAMAM G3 core methylene proton (a,b,c,d) resonances within 2.19–3.10 ppm region, while protons of amide groups gave broad signal at 8.0 ppm (NH(G3)). In the case of biotin, resonances of aliphatic chain protons (2^{B} and 3,4,5^{B}), ureido ring protons (10,11^{B}), and thiophane ring protons (8,9^{B}) were separated from gh and G3 core resonances (Figure 1B,C). The resonances of gh C-H protons were observed in the 3.3–3.9 ppm region.

For both dendrimer conjugates G3^{2B12gh5M} and G3^{2B10gh17M}, the characteristic resonances of αM in the ^{1}H NMR spectra are observed (Figure 1): partially overlapped singlets within 1.61–1.77 ppm from protons of four methyl groups, and methoxy group proton singlet resonance at 3.70 ppm. Although prenyl methylene proton (11 and 16) doublets remain non-equivalent in both G3^{2B12gh5M} and G3^{2B10gh17M}, the 11-H doublet was found at 3.99 ppm (for $G3^{2B10gh17M}$) and at 3.95 (for $G3^{2B12gh5M}$). The 12 and 17-H triplets are overlapped at 5.16 ppm for αM, $G3^{2B12gh5M}$, and $G3^{2B10gh17M}$.

The resonances of two αM aromatic protons in both conjugated spectra are considerably shifted upfield, due to the shielding effect of the hyperbranched dendrimer with additional polyhydroxyalkylamide chains. The larger shift of 5^{M} resonance in both conjugates related to αM (0.30 ppm upfield) in comparison with the 4^{M} signal (0.08 ppm upfiled) were observed. The ^{1}H and ^{13}C resonances of αM in $G3^{2B12gh5M}$ and $G3^{2B10gh17M}$ were assigned based on 2-D ^{1}H-^{1}H COSY (Figure S1) and heteronuclear ^{1}H-^{13}C HSQC and HMBC experiments (for combined HSQC/HMBC map see Figure S2) and are listed in Table S1.

The considerable shift of H5 proton resonance (0.30 ppm upfield) together with 4.4 ppm downfield shift of C6 and 1.5 ppm upfield shift of C7 resonance in relation to αM alone spectra in dmso-d_6 strongly suggest that 6′ oxygen is involved in carbamide linker. In such a bonding mode, both prenyl proton multiplets remain unshifted in the conjugates.

We were not able to isolate an αM-NPC intermediate single crystal to verify unambiguously which hydroxyl group was substituted with *para*-nitrophenylcarbonate and further involved in a bonding with dendrimer.

Additionally, $G3^{2B12gh}$ conjugate was synthesized to examine the vehicle cytotoxicity. All of these conjugates were extensively purified by dialysis against water, in order to remove low molecular reagents or side products, such as *p*-nitrophenol and DMAP, and dried under high vacuum. The ^{1}H NMR spectrum of $G3^{2B12gh}$ conjugate is shown in Figure 2A. The ^{1}H resonances were assigned based on COSY spectrum as described before [24]. The number of attached residues of αM, biotin and gh was estimated based upon the integral intensity of ^{1}H resonances in relation to the reference intensity of $b_{0\text{-}3}$ protons resonance from the dendrimer core, namely [120H].

Because no free amine groups of G3 were readily available in $G3^{2B10gh17M}$, all conjugates were single-labeled with fluorescein using FCH, which was activated with Mukaiyama reagent and attached to gh hydroxyl groups via ester bond. The average 1:1 stoichiometry was confirmed by ^{1}H NMR spectra (not shown) in $G3^{2B12gh5MF}$ and $G3^{2B10gh17MF}$.

3.2. Size and Zeta Potential of Conjugates

Unlike PAMAM G3, the $G3^{2B12gh}$ derivative has smaller dynamic diameter in water, which increases upon acidification of the solution to pH 5 about four times (Figure 3; for numbers, see Table 1) due to protonation of internal tertiary amine groups, as observed before within a series of $G3^{gh}$ derivatives [39]. The highest value of diameter was observed already for half-substituted $G3^{16gh}$, namely 4.60 nm (averaged by volume) or 3.80 nm (averaged by number); the biotin-containing conjugate $G3^{2B12gh}$ has a comparable size.

Although we have prepared two relatively well water-soluble αM conjugates, namely $G3^{2B12gh5M}$ and $G3^{2B10gh17M}$, in contrary to non-glucoheptoamidated semiproducts, the size measurements of aqueous solutions evidenced high association of each $G3^{2B12gh5M}$ and $G3^{2B10gh17M}$. DLS measurements indicated the number-averaged size for $G3^{2B12gh5M}$, amounting to 1262 nm with major contribution from large size particles (d(V) = 1367 > d(N)). These associates were destabilized at pH 5, resulting in a decrease of particle size within the <200,100> nm region. The $G3^{2B10gh17M}$ conjugate did not show an additional association in neutral aqueous solution, and finally, the size of particles was within <230,100> nm regardless of the pH (7 or 5). Nonetheless, both conjugates showed major contribution of larger associates in acidic solution, which is illustrated by d(V) > d(N) in Figure 3B.

The values of the zeta potential are positive for all conjugates studied (Figure 4). The vehicle $G3^{2B12gh}$ has a ζ of circa 11 mV, the same as the $G3^{16gh}$ studied in a regular series of derivatives [39]. αM-containing conjugates showed the same ζ, circa 22 mV, which is twice that for the vehicle itself, despite the 10-fold size difference between $G3^{2B12gh5M}$ and $G3^{2B10gh17M}$ in water. On the other hand, the potential increased considerably upon a two log increase of hydrogen cation concentration, presumably due to protonation of core PAMAM and considerable increase of cationic charge of nanoparticles of the conjugates. For the size-dependent biological behavior of conjugates vide infra.

Table 1. Size and zeta potential values ± standard deviation determined by DLS analysis.

Compound	Size [nm]				Zeta Potential [mV]	
	pH 7		pH 5		pH 7	pH 5
	d(V)	d(N)	d(V)	d(N)		
$G3^{2B12gh}$	1.0 ± 0.24	0.9 ± 0.22	4.5 ± 0.14	3.8 ± 0.18	11.1 ± 2.89	17.4 ± 1.87
$G3^{2B12gh5M}$	1367 ± 245.7	1262 ± 196.4	178.3 ± 5.73	113.8 ± 7.34	22.7 ± 1.01	37.5 ± 2.32
$G3^{2B10gh17M}$	149 ± 33.9	111 ± 14.1	230.6 ± 12.4	130.8 ± 12.25	22.1 ± 0.61	39.5 ± 3.67

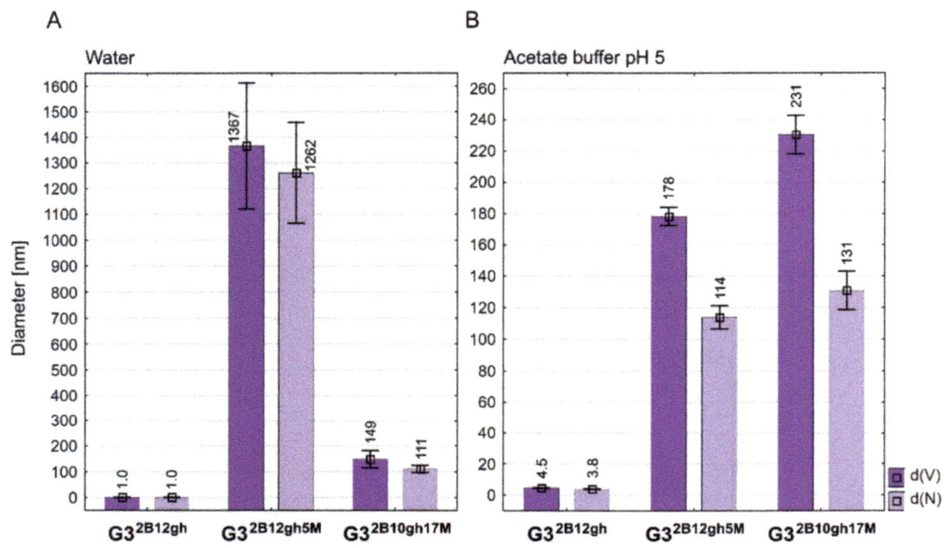

Figure 3. Diameter of conjugates averaged by volume (d(V)) and by number of molecules (d(N)) measured in water (**A**) and in acetate buffer pH 5 (**B**). Data are presented as mean ± standard deviation with mean values marked above the columns.

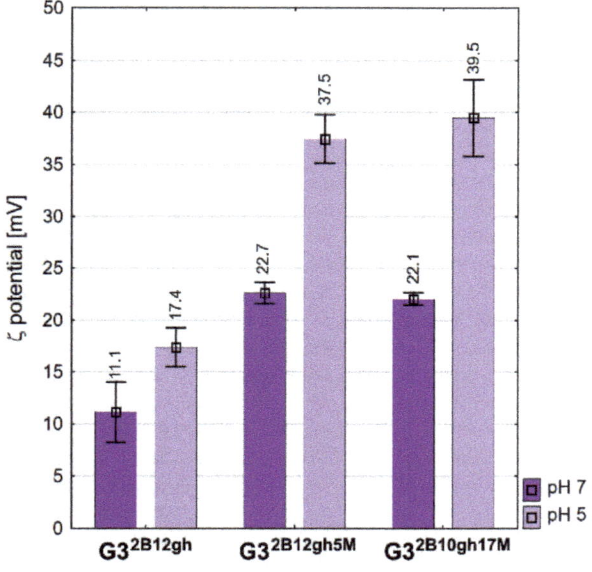

Figure 4. Zeta potential values of conjugates measured in water and acetate buffer pH 5. Data are presented as mean ± standard deviation with mean values marked above the columns.

3.3. Cytotoxicity

The main goal of this study was to design an efficient αM delivery vehicle, based on the biotinylated and glucoheptoamidated (gh) PAMAM G3 dendrimer, increasing the drug's toxicity against cancer cells and *C. elegans*. While substitution by biotin was supposed to increase the intracellular accumulation of the conjugates, modification by gh residues was expected to reduce the toxicity of the surface groups of the αM carrier (PAMAM G3 dendrimer).

The results of the NR assay (Figure 5) revealed that both studied αM conjugates were active at low, micromolar concentrations, with the effect depending on the level of dendrimer substitution by the drug. Thus, $G3^{2B10gh17M}$ killed all cells more efficiently than $G3^{2B12gh5M}$. Surprisingly, normal cells were as sensitive as SCC-15 cells, with glioma cells being less sensitive. The results were confirmed by microscopic analysis. At toxic or higher concentrations, the studied compounds cells formed aggregates, shrank, lost their adhesion, and absorbed a lower amount of neutral red dye (Figure 5B).

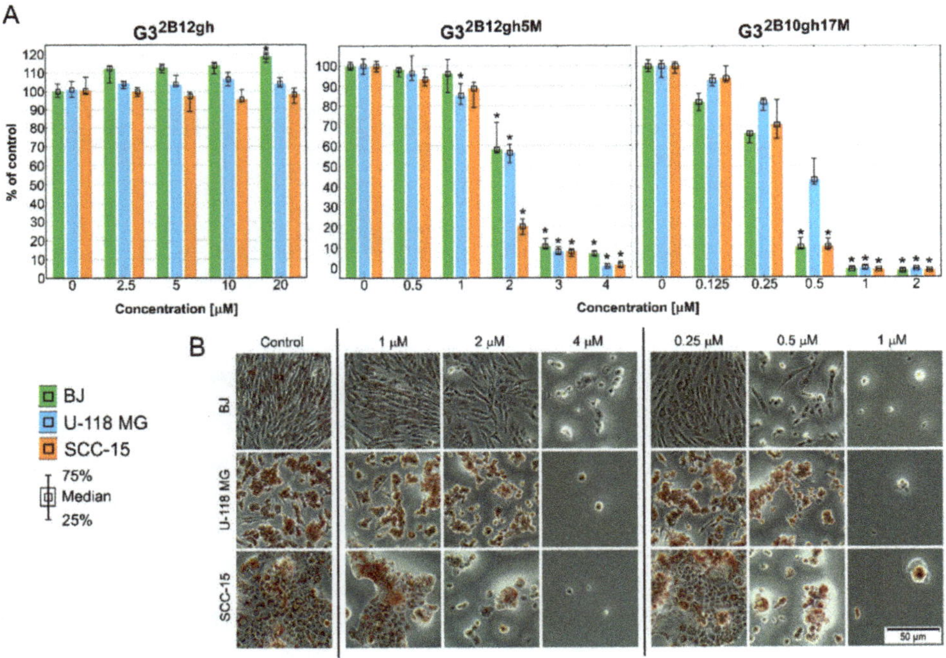

Figure 5. Cytotoxicity of $G3^{2B12gh5M}$, $G3^{2B10gh17M}$, and the control vehicle $G3^{2B12gh}$ against human cells: normal fibroblasts (BJ), glioma cells (U-118 MG), and squamous carcinoma cells (SCC-15) after 48 h of incubation estimated with an NR assay. (**A**) Cell viability expressed as medians of a percent against non-treated control (control expressed as 100%). The whiskers are the lower (25%) and upper (75%) quartile ranges. * $p \leq 0.05$; Kruskal–Wallis test (against non-treated control). (**B**) Morphology of the cells incubated with chosen concentrations of $G3^{2B12gh5M}$ and $G3^{2B10gh17M}$. Red signal on microscopic images represents a neutral red dye.

The results of XTT assay, detecting mainly mitochondria damages, showed a weaker response than those of NR test (Figure 6). Of note is that despite the effect of the $G3^{2B12gh5M}$ compound, compared to that of $G3^{2B10gh17M}$ compound, the former seems to be a better candidate for the destruction of neoplastic cells, in view of a strong toxicity of the latter towards normal human fibroblasts (Figure 6, Table 2).

The results of the NR and XTT assays allowed to determine the half maximal inhibitory concentration IC_{50} values, demonstrating both drug conjugates to lack cytotoxic selectivity against BJ vs. squamous carcinoma SCC-15/glioma U-118 MG cells (Table 2, Figures 5 and 6).

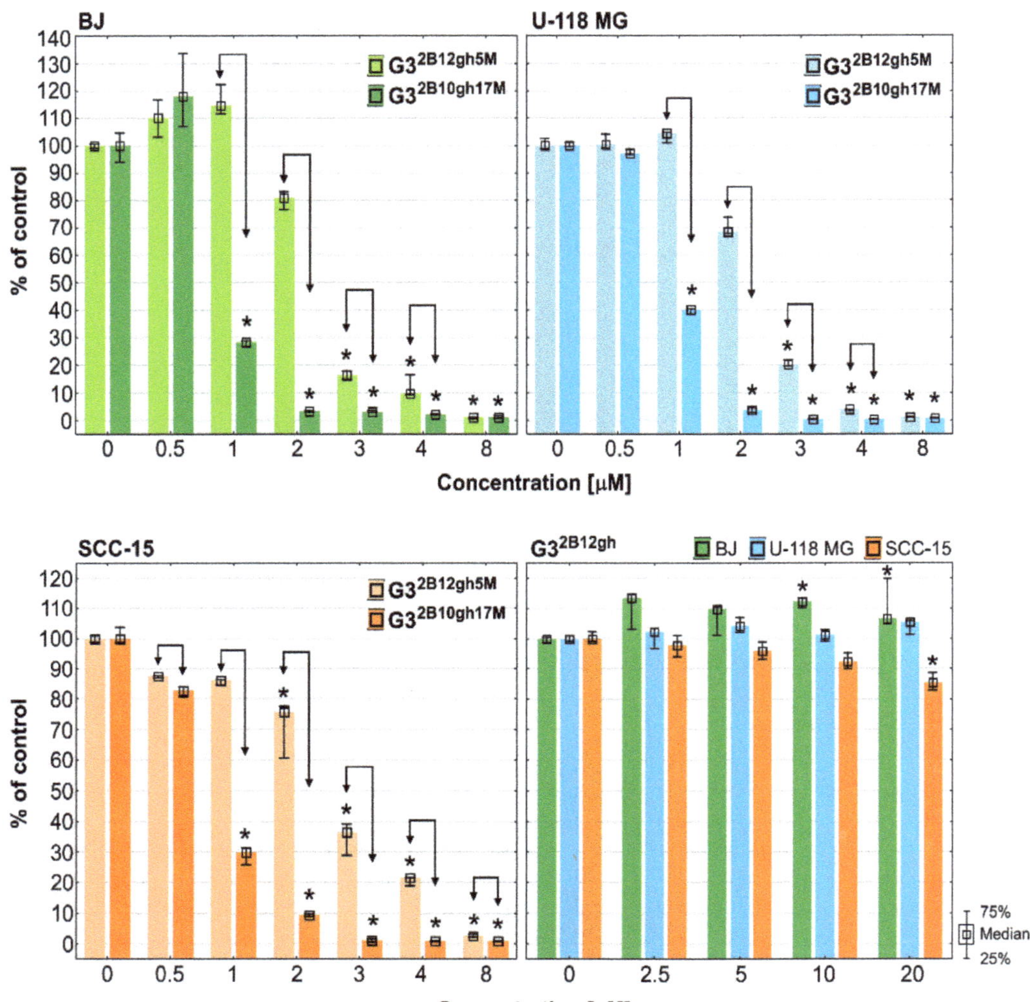

Figure 6. The results of XTT assay performed on BJ, U-118 MG and SCC-15 cells after 48 h of incubation with G3^{2B12gh5M} and G3^{2B10gh17M} conjugates and the control vehicle G3^{2B12gh}. Cells viability expressed as medians of a percent against non-treated control (control expressed as 100%). The whiskers are lower (25%) and upper (75%) quartile ranges. * $p \leq 0.05$; Kruskal–Wallis test (against non-treated control), ↕ $p \leq 0.05$; Mann–Whitney *U* test (the G3^{2B12gh5M}-treated group against the G3^{2B10gh17M}-treated group).

Of note is that with the control dendrimer vehicle G3^{2B12gh}, void of αM, at concentrations up to 20 μM (the highest allowed in view of dmso in the dendrimer preparation) following 48 h of treatment, a feeble cytotoxicity of the highest concentration was seen only for SCC-15 cells with the XTT assay (Figures 5 and 6). These results are in agreement with our earlier results, indicating that a similar G3 PAMAM dendrimer substituted with 16 gh residues was non-toxic up to 100 μM [39]. It confirms that biotinylated and partially glucoheptoamidated PAMAM G3 dendrimer is useful as a drug nanocarrier. The conjugation of αM with non-toxic dendrimeric carrier (G3^{2B12gh}) resulted in the drug's toxicity increase, as compared to that of the free αM [5]. The increase of toxicity was dependent on the level of substitution of the carrier by the drug, observed not only against SCC-15 cells (about 4.5- or 21-fold), but also for U-118 MG (about 5.2- or 25-fold) and BJ cells (4.5- or 11.5-fold for G3^{2B12gh5M} and G3^{2B10gh17M}, respectively). It should be noted that the conjugate with five residues of αM enhanced its toxicity on average by ca. 5-fold, which may be related to the polyvalency or multivalency phenomenon [40]. Of interest is that the conjugate with 17 residues enhanced αM activity by about 21–25-fold in cancer cells but by only 11.5-fold in normal cells. This suggests that addition of more xanthone residues increase its activity in cancer cells with simultaneous moderation of this effect in normal cells, resulting in an apparent enhancing of antineoplastic efficacy [41].

Table 2. The half maximal inhibitory concentration (IC_{50}) values determined following 48 h of treatment of BJ, U-118 MG, or SCC-15 cells with $G3^{2B12gh5M}$ or $G3^{2B10gh17M}$.

	IC_{50} [μM] NR Assay		
	BJ	U-118 MG	SCC-15
$G3^{2B12gh5M}$	2	1.83	1.41
$G3^{2B10gh17M}$	0.28	0.39	0.31
α-Mangostin [1]	8.97	9.59	6.43
	IC_{50} [μM] XTT Assay		
	BJ	U-118 MG	SCC-15
$G3^{2B12gh5M}$	2.37	2.05	2.52
$G3^{2B10gh17M}$	0.78	1.01	0.85
α-Mangostin [1]	18.58	18.15	7.72

[1] IC_{50} values for α-mangostin quoted from our earlier work [5].

It should also be mentioned that U-118 MG cells, compared to the remaining cell lines, did not present higher sensitivity to the drug bound to both biotinylated vehicles, which was unexpected considering that glioma cells tend to overexpress biotin receptors and in consequence demonstrate increased biotin uptake [42,43].

3.4. Cellular Accumulation and Distribution of Fluorescently Labeled $G3^{2B12gh5M}$ or $G3^{2B10gh17M}$

Drug vehicles, including dendrimers, may change cellular uptake and accumulation of delivered substances. Based on our earlier experience [22] and considering higher cellular uptake of biotinylated vs. non-biotinylated dendrimers [20,21] the glucoheptoamidated dendrimers used in this study were substituted with two biotin residues per dendrimer molecule, in order to enhance their accumulation and, in consequence, also αM accumulation in the studied cancer cells. It should be added that an increased uptake of biotinylated dendrimer-drug conjugates by cancer cells could be expected, in view of biotin receptors and transporters being frequently overexpressed by such cells [19].

Cellular uptake, accumulation and toxicity are dependent on molecule size, charge and charge-related zeta potential of nanoparticles. Increased cytotoxicity of nanoparticles in non-phagocytic cells correlates with their small size [44]. At physiological pH, $G3^{2B12gh5M}$ and $G3^{2B10gh17M}$ conjugates indicated high association with diameter determined by DLS, equal to about 1300 nm for $G3^{2B12gh5M}$ and 130 nm for $G3^{2B10gh17M}$ compared to an αM-free carrier (about 1 nm diameter). Probably for this reason, microscopic observations of cells after 4 h of incubation with both FCH-labeled $G3^{2B12gh5M}$ and $G3^{2B10gh17M}$ revealed no visible fluorescence signal inside them. Only 48 h of incubation of fluorescently labeled $G3^{2B12gh5M}$ or $G3^{2B10gh17M}$ indicated that both compounds were taken up by all cell lines with the highest accumulation in normal human fibroblasts, while in cancer cells, their amount was significantly lower (Figure 7). This phenomenon may be due to the ability of fibroblasts to more efficiently uptake biotin compared to other cells as we described [22]. It is also possible that the stronger accumulation of studied compound in these cells may be a result of their ability to non-professional phagocytosis or phagocytosis [45,46]. The formation of associates by αM conjugates resulted not only in slower uptake of these nanoparticles—the toxicity was also caused by the presence of αM itself rather than by the interaction of positively charged dendrimer vehicles and negatively charged cell membranes. Microscopic observations revealed that absorbed conjugates remained in endocytic vesicles, with only a small part of the dendrimer present in the cytoplasm of the cells. It is known that αM induces caspase-3 and -9 activation causes swelling, loss of membrane potential (delta psi), decrease in intracellular ATP, ROS accumulation, and cytochrome c/AIF release in human leukemia HL60 cells [47] and also provokes the release of cytochrome C, increase of Bax, decrease of Bcl-2, and activation of caspase-9/caspase-3 cascade in cervical cancer cells [33]. Therefore, mitochondria seem to be a natural goal for the anticancer activity of αM. In this study, both conjugates penetrated into mitochondria of SCC-15 cells and normal fibroblasts, but very poorly into mitochondria of glioma U-118 MG cells (Figure 7), suggesting a reason of glioblastoma cells being resistant to αM conjugates (see Figures 5 and 6).

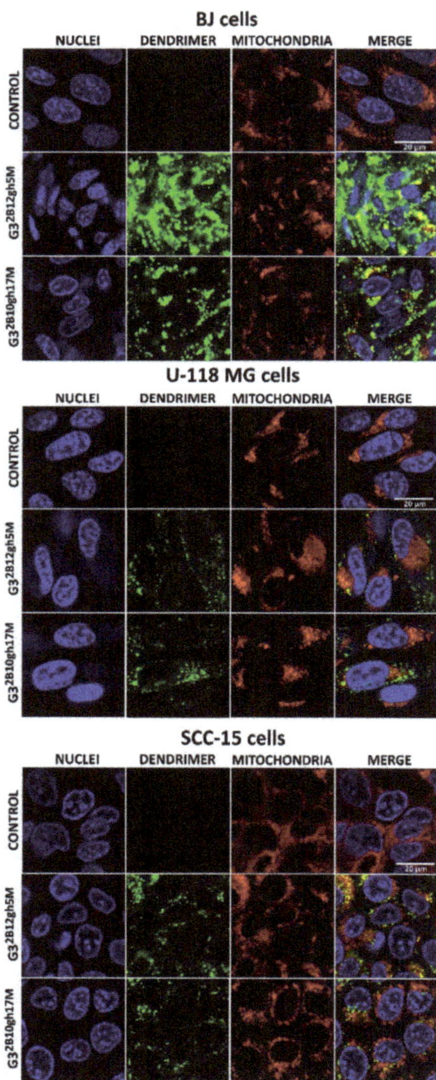

Figure 7. Images obtained with confocal microscopy presenting cellular accumulation and localization of fluorescently labeled $G3^{2B12gh5M}$ or $G3^{2B10gh17M}$ conjugates after 48 h of incubation with 1 µM or 0.1 µM non-toxic concentrations. Green signal: FCH-labeled dendrimer, blue signal: cell nuclei stained with Hoechst 33342, red signal: MitoTracker Deep Red FM labeled mitochondria, yellow signal: colocalization of FCH labeled dendrimer and MitoTracker labeled mitochondria. Scale bar is equal 20 µm.

3.5. Caspase-3/7 and Intracellular ATP Level

Another sign of advanced cytotoxicity is the entry of cells into the programmed pathway of death, apoptosis. Regardless of whether this is an extrinsic or intrinsic mechanism, executive caspases (caspase 3,6,7) are produced to proteolytically degrade components of damaged or malfunctioning cells [48]. In this experiment, the potential of dendrimer conjugates with 5 or 17 residues of αM to induce apoptosis in cancer cells compared to normal cells was studied. In addition, the intracellular ATP level was also estimated. The results are shown in Figure 8 as a relationship between the activity of executioner caspases and intracellular ATP level in studied cell lines. Both conjugates $G3^{2B12gh5M}$ and $G3^{2B10gh17M}$ significantly increased activity of caspase-3/7 in all tested cell lines at concentrations higher than IC_{50}. The most active was the conjugate with 17 residues of αM, inducing Cas-3/7 activity growth with a concentration of 0.5 μM, while the conjugate with 5 αM residues induced this effect with a concentration of 3 μM.

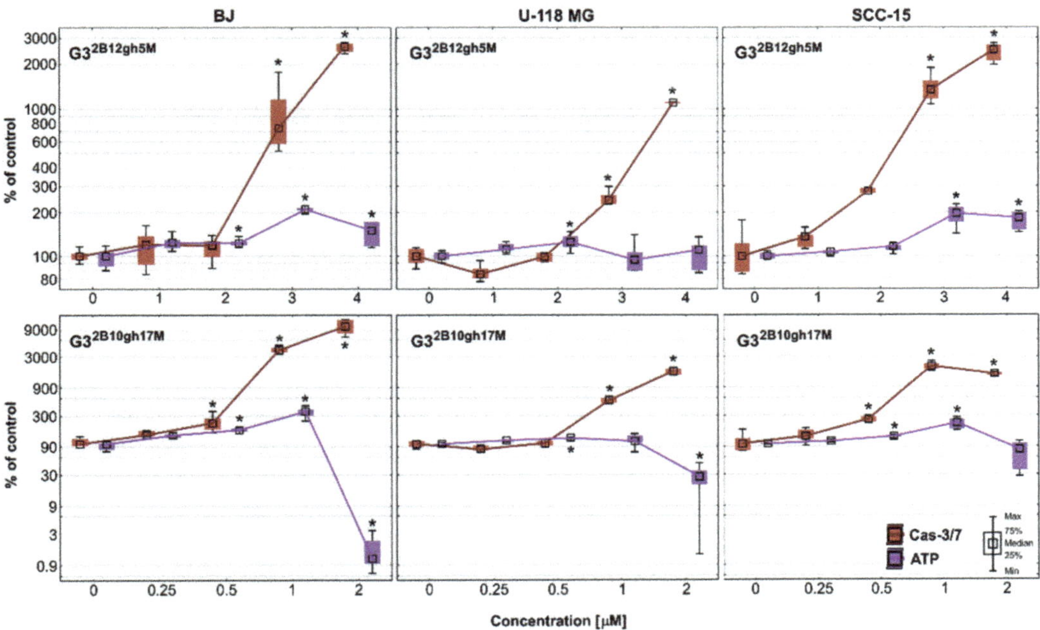

Figure 8. Activity of caspase-3/7 and intracellular ATP level in BJ, U-118 MG, and SCC-15 cells after 48 h of incubation with $G3^{2B12gh5M}$ (1–4 μM) and $G3^{2B10gh17M}$ (0.25–2 μM) conjugates. The results from three independent experiments are medians expressed as a percentage of non-treated control, where data for caspase-3/7 level are marked in red, while data for ATP level are marked in violet. Boxes represent the first (25%) and third (75%) quartile; whiskers are the minimum and maximum range. * $p \leq 0.05$, Kruskal–Wallis test (against non-treated control).

As the cytotoxicity of αM significantly increases upon the conjugation with the dendrimer, the same effect is observed for apoptosis. In our earlier studies [5], αM alone induced apoptosis from 10 μM in cancer and from 20 μM concentration in normal cells. Thus, lower concentrations of αM affected cancer than normal cells. Two other similar xanthones, namely alvaxanthone and rheediaxanthone B, induced higher activity of caspases-3/7 in cancer cell lines comparing to normal cells [49]. For αM conjugated with PAMAM G3 dendrimer, this effect was not seen—the most sensitive were fibroblasts. Referring to non-treated control, at the highest concentration of $G3^{2B12gh5M}$, there was a 11-, 25- and 26-fold increase in caspases activity in U-118 MG, SCC-15, and BJ cells, respectively. In case of $G3^{2B10gh17M}$, there was 17-fold growth in glioblastoma cells and 21-fold growth in the SCC-15 cell line, while fibroblasts obtained a 104-fold increase of caspases activity.

This result may be due to the presence of biotin in conjugates and its higher uptake by fibroblasts compared to remaining cancer cells, as previously observed by Uram et al. [22]. The results of the accumulation studies of fluorescently labeled $G3^{2B12gh5M}$ and $G3^{2B10gh17M}$ conjugates are consistent with this, demonstrating higher uptake of biotinylated conjugates by fibroblasts (Figure 7). On the other hand, the lowest level of caspases activity in glioblastoma cells may be caused by their intrinsic mechanisms of decreased apoptosis, which makes GBM resistant to apoptosis inhibitors [50] and causes the propensity of astrocytic glioma to necrosis [51]. Considering this fact, the obtained statistically significant induction of apoptosis by αM dendrimer conjugates at low concentrations makes them interesting in the context of anticancer properties.

Since ATP plays a key role in the decision of cell death fate and the ATP-dependent nature of apoptosis [48,52], intracellular ATP level after 48 h of incubation with $G3^{2B12gh5M}$ and $G3^{2B10gh17M}$ was assessed. After treatment with 1–3 μM $G3^{2B12gh5M}$, the ATP amount has grown to 195% in SCC-15 cells and to 210% in BJ cells, and then, at the highest concentration, this was 183% and 150%, respectively. In the case of U-118 MG cells, the level of ATP remained mostly unchanged. In the same range of concentrations, activity of caspases has grown significantly in all three cell lines. Meanwhile, incubation of both cancer and normal cells with $G3^{2B10gh17M}$ led to an increase in ATP level at 1 μM, followed by a reduction at 2 μM concentration. The biggest rise and fall of ATP level was observed in fibroblasts (363% and 1.2% of control), while in SCC-15 cells it was 225% and 80%. In glioblastoma cells, the ATP level increased only to 112%, but then decreased to 27%. These ATP changes were accompanied by growth in caspases-3/7 activity in all tested cells, alike for $G3^{2B12gh5M}$. As a high level of intracellular ATP often favors apoptosis [52], and caspases 3 and 7 are activated in the advanced stage of this process, it can be concluded that the studied conjugates induced ATP-dependent apoptosis in a concentration-dependent manner. However, considering the fact that various forms of cell death (autophagy, apoptosis, and necrosis) can be induced simultaneously or successively and that low level of ATP often promotes necrosis [52], the tested conjugates at highly toxic concentrations (4 μM of $G3^{2B12gh5M}$ and 2 μM of $G3^{2B10gh17M}$) could also induce necrosis in both cancer and normal cells. A similar effect was observed by Uram et al. [53] in studies regarding biotinylated dendrimer PAMAM G3 conjugated with celecoxib and Fmoc-L-Leucine (G3-BCL). In fibroblasts, the apoptotic pathway was dominant up to 2 μM of G3-BCL conjugate with a constant ATP supply, while at 4 μM concentration, a significant increase of late apoptosis/necrosis was observed with simultaneous extensive depletion of ATP level. The same pattern of cell death fate was revealed in glioblastoma cells but still with a high level of ATP observed at 4 μM of G3-BCL.

3.6. Proliferation

It is known that αM exhibits a strong anti-proliferative effect in human colon HCT116 carcinoma cells with inhibition of the activity of DNA topoisomerases I and II, and blockade of the cell cycle in the G2/M phase [54], and also in human breast cancer T47D cells [55] and several others [56]. $G3^{2B12gh5M}$ inhibited BJ and SCC-15 cell proliferation from 2 μM concentration and U-118 MG glioma cells from 3 μM, while at higher concentrations cell proliferation was lower than 20% and close to zero. A stronger effect was observed in case of $G3^{2B10gh17M}$, which significantly suppressed proliferation from 1 μM concentration for BJ and U-118 MG or even 0.5 μM for SCC-15 cells. Moreover, activity of conjugate with 17 αM residues was always stronger than $G3^{2B12gh5M}$. It is worth noting that the control dendrimer vehicle $G3^{2B12gh}$ at concentrations up to 20 μM showed no anti-proliferative effect against tested cell lines and even increased the proliferation of fibroblasts by 40% (Figure 9).

Comparing this effect with the toxicity results, it can be concluded that the anti-proliferative effect coincides with the toxicity pattern (Figures 5 and 6). It is also visible that conjugation of αM with PAMAM G3 dendrimers enhanced its antiproliferative effect since our previous studies showed activity of αM alone against BJ, U-118 MG and SCC-15 cells in range of 7.5–20 μM concentrations [5]. Finally, the use of the biotinylated carrier for 5 αM residues led to a reduction of cell proliferation about three-fold, while $G3^{2B10gh17M}$ inhibited cell divisions by 7-, 15-, and 10-fold in BJ, U-118 MG, and SCC-15 cells, respectively, compared to αM alone. This phenomenon may be the result of multi/polyvalency effect or higher amount of αM accumulation inside the studied cells and showed that the studied compounds may play an important role as new anticancer agents.

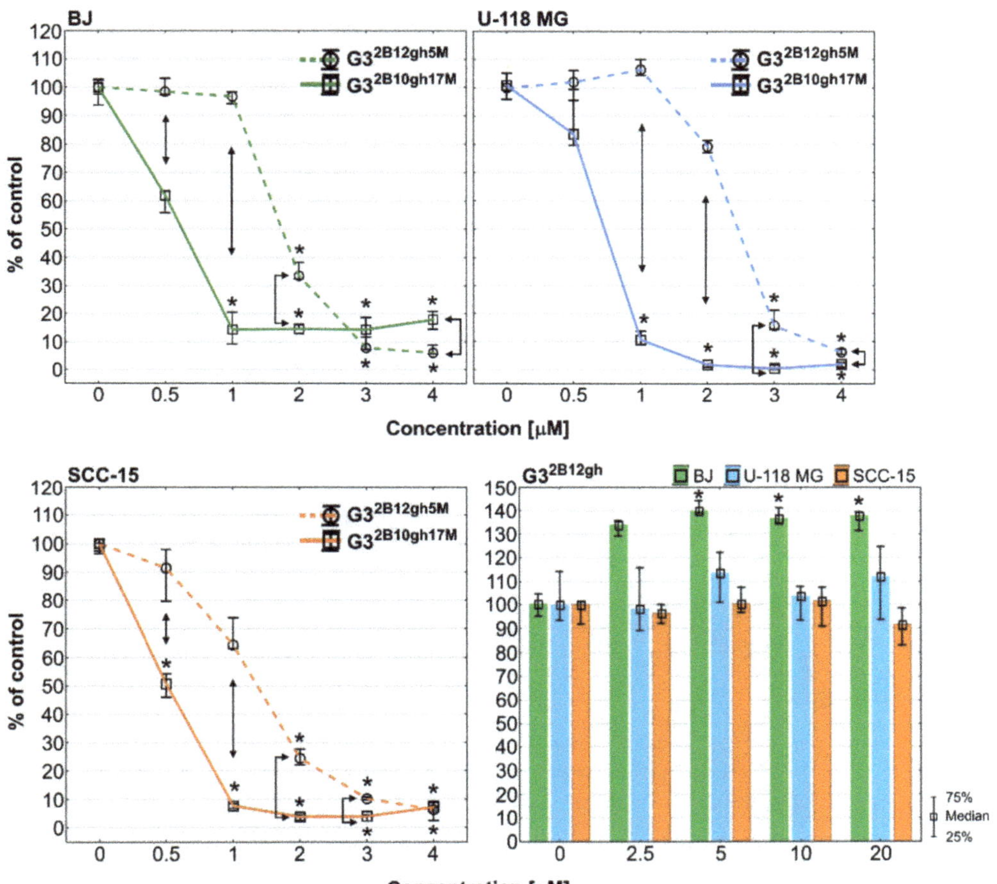

Figure 9. Anti-proliferative action of $G3^{2B12gh5M}$ and $G3^{2B10gh17M}$ on BJ, U-118 MG, and SCC-15 cells after 72 h of incubation, determined with Hoechst staining. Results are from three independent experiments, performed in triplicates and presented as medians (percentage of non-treated control). Whiskers indicate the lower (25%) and upper (75%) quartile ranges. * $p \leq 0.05$, Kruskal–Wallis test (against non-treated control). ↕ $p \leq 0.05$; Mann–Whitney *U* test (the $G3^{2B12gh5M}$-treated group against the $G3^{2B10gh17M}$-treated group).

3.7. Adhesion

Among the ten hallmarks of cancer development and progression, there is the activation of invasion and metastatic processes [57]. Malignant cells during migration adhere to extracellular matrix (ECM) proteins, providing a path to the new metastasis site [58,59]. As cell-to-cell and cell-to-ECM adhesion plays a very important role in tumor metastasis, the molecules with anti-adhesion properties may contribute to suppress the cancer expansion to other tissues. The ability of αM dendrimer conjugates to affect cells adhesion was evaluated by the assay with crystal violet dye, which stains DNA of intact adherent cells. The obtained results are presented in Figure 10.

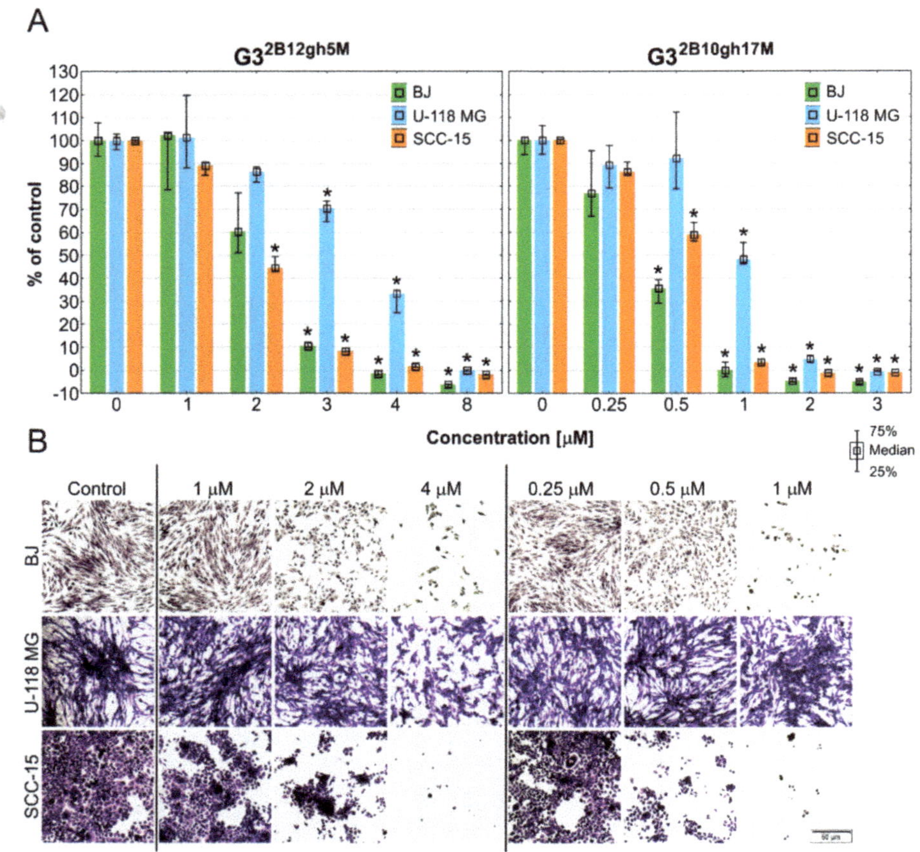

Figure 10. The ability of BJ, U-118 MG, and SCC-15 cells to adhere determined by crystal violet assay after 48 h of treatment with $G3^{2B12gh5M}$ or $G3^{2B10gh17M}$ at different concentrations. (**A**) Results are medians with first (25%) and third (75%) quartile obtained in three independent experiments and expressed as a percentage of non-treated control. * $p \leq 0.05$, Kruskal-Wallis test (against control group). (**B**) Cells attached to the bottom of 96-well plate and stained with crystal violet after 48 h of incubation with studied dendrimer conjugates. Violet signal represents a crystal violet dye inside the cells. Scale bar = 50 μm.

The SCC-15 cells were the most susceptible to loss of adhesion after incubation with $G3^{2B12gh5M}$, having 45% of attached cells at 2 μM and 1.7% at 4 μM concentration. Fibroblasts and glioblastoma cells achieved a statistically significant decrease of adhesion at 3 μM of $G3^{2B12gh5M}$, but the difference between these two cell lines was 60% in favor of U-118 MG cells. PAMAM G3 dendrimer with 17 residues of αM influenced cells attachment stronger, significantly decreasing adhesion of BJ and SCC-15 cells from 0.5 μM concentration. For this conjugate, U-118 MG cells were also the least affected, having 48% of adherent cells at 1 μM, which is 12 times more than for BJ and SCC-15 cells. These results are in line with the microscopic observations shown in the Figure 10B, which present the number of adherent cells stained with crystal violet dye after incubation with tested conjugates in selected concentrations. U-118 MG cells retained relatively high adhesiveness even after treatment with conjugates at concentrations higher than the IC_{50} values. Similar results were obtained for αM and other xanthones [5,49]. In these studies, the adhesion of squamous carcinoma cells was slightly more disturbed compared to fibroblasts, while the weakest effect was observed in glioblastoma cells. Considering the cytotoxicity of the tested conjugates, the $G3^{2B12gh5M}$ conjugate seems to be more potent in anti-adhesion properties in squamous cell carcinoma treatment. In case of GBM, extracranial metastases are extremely rare, affecting 0.4–0.5% of all patients with GBM [60], but in the last decade, clinical reported cases of extraneural dissemination have become more frequent [61]. In addition, GBM is suggested to disseminate not only via cerebrospinal fluid, but also through bloodstream and lymphatic vessels. Therefore, the ability of $G3^{2B10gh17M}$ to reduce adhesion of U-118 MG cells by 52% at 1 µM concentration is noteworthy.

3.8. Toxicity to C. elegans and the Effect on the Worm Survival

Caenorhabditis elegans nematode was used as a model organism to examine in vivo effect of synthesized αM dendrimer conjugates on multicellular system. It has been shown that *C. elegans* is valuable animal model for studies on toxicity and biocompatibility of various nanoparticles [62,63]. Synchronized population of L4 stage-worms was incubated for 7 days in the medium (see Section 2.5.7) containing different concentrations of αM, $G3^{2B12gh5M}$ and $G3^{2B10gh17M}$, as well as $G3^{2B12gh}$ (control vehicle). The results are presented in Figure 11 as survival curves determined by the Kaplan–Meier method.

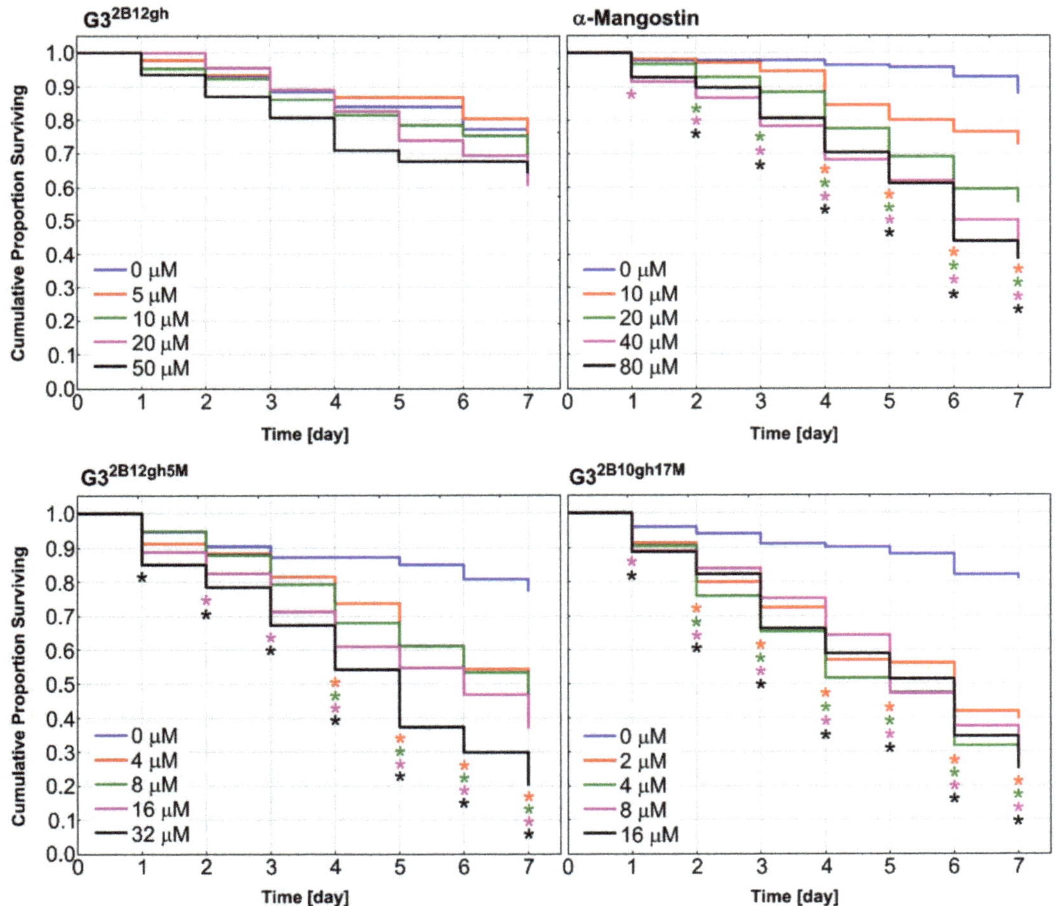

Figure 11. Kaplan–Meier survival curves of *C. elegans* after 7 days of incubation with α-mangostin, $G3^{2B12gh5M}$, $G3^{2B10gh17M}$, and $G3^{2B12gh}$ vehicle at different concentrations. Results are presented as cumulative proportion surviving. Statistically significant differences against non-treated control obtained in Gehan's Wilcoxon test are marked with asterisks ($p \leq 0.05$) in the colors corresponding to the tested concentrations.

Similar to the results of the in vitro studies, attachment of αM to dendrimer vehicle increased its toxicity against *C. elegans* organism, but to a slightly lesser extent. The concentrations in which the proportion of live nematodes was 0.4 after 7 days of incubation were 40 μM of αM, 16 μM of $G3^{2B12gh5M}$, and 2 μM of $G3^{2B10gh17M}$ (Figure 11), pointing to the activity of $G3^{2B12gh5M}$ and $G3^{2B10gh17M}$, compared to free αM, being 2.5- and 20-fold stronger, respectively.

Based on the survival curves after 7 days of incubation, the LC_{50} values were also calculated for αM, $G3^{2B12gh5M}$, $G3^{2B10gh17M}$, and control vehicle $G3^{2B12gh}$ (Table 3, Figure 12), the latter showing a moderate toxicity level [64].

Table 3. The LC_{50} values against *C. elegans* for free and dendrimer-bound αM, and a control dendrimer vehicle, determined following 7 days of incubation with αM, $G3^{2B12gh5M}$, $G3^{2B10gh17M}$, or $G3^{2B12gh}$ presented as medians and the first and third quartile. The LC_{50} values were determined using an online calculator (AAT Bioquest, Inc. Quest Graph™ IC_{50} Calculator. Retrieved from https://www.aatbio.com/tools/ic50-calculator (accessed on 14 October 2021).

	LC50 [μM]	1st Quartile	3rd Quartile
$G3^{2B12gh}$	49.88	40.58	57.98
α-Mangostin	18.74	18.64	20.23
$G3^{2B12gh5M}$	7.87	4.95	8.09
$G3^{2B10gh17M}$	1.38	1.03	1.42

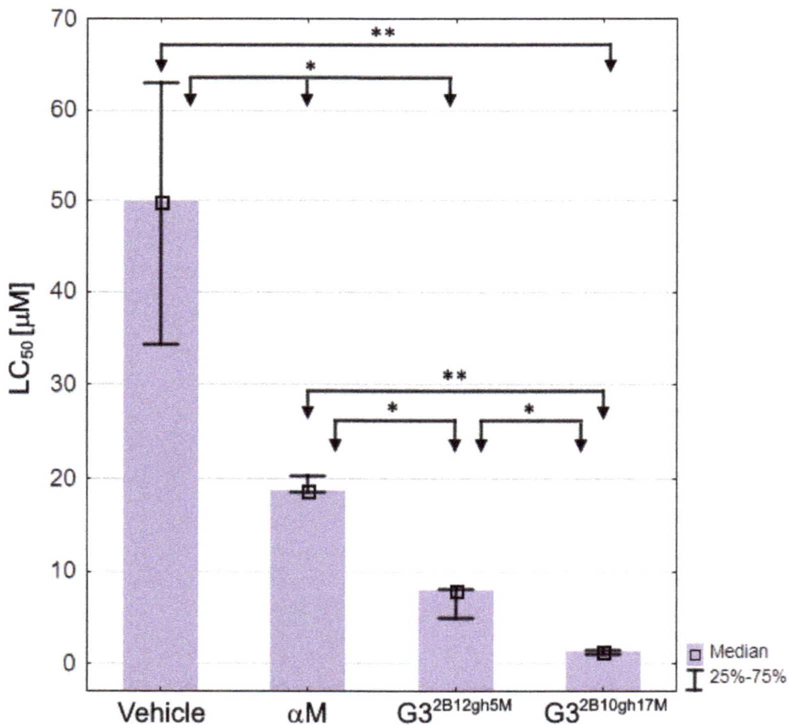

Figure 12. The LC_{50} values against *C. elegans* for αM, dendrimer vehicle $G3^{2B12gh}$ and $G3^{2B12gh5M}$ or $G3^{2B10gh17M}$ conjugates presented as medians with the first and third quartile. Statistically significant differences estimated by the Kruskal–Wallis test are marked with ** and by U Mann–Whitney test with * $p \leq 0.05$.

As expected, each of the $G3^{2B12gh5M}$ and $G3^{2B10gh17M}$ conjugates was more toxic than the $G3^{2B12gh}$ control vehicle (by 6.3- and 36-fold, respectively) or free αM, with the activities of $G3^{2B12gh5M}$ and $G3^{2B10gh17M}$, compared to free αM, being stronger by 2.4- and 13.6-fold, respectively (Table 3).

The present results obtained for free αM, tested against the L4 stage nematodes (Table 3), differ from those obtained earlier in toxicity assay performed by a different method and against a mixed *C. elegans* population [5]. The LC_{50} value for αM was found 3.8 ± 0.5 μM, thus the nematode at L4 stage (LC_{50} = 18.74) appears distinctly less sensitive to the drug than the other developmental forms.

The toxic effects of αM and its conjugates with biotinylated PAMAM G3 dendrimer was accompanied by morphological and locomotion changes of the treated nematodes. Already after 24 h of incubation with the highest concentrations of $G3^{2B12gh5M}$ and $G3^{2B10gh17M}$, slower and vibrating movements were observed. As the toxic effect progressed, the wavy motion declined, leading to straightening and stiffening of the *C. elegans* worm, until it became motionless. Moreover, the creases of the cuticula and degradation of this tissue were seen (Figure 13).

It should be noted that the increase of αM toxicity, resulting from the drug binding with the dendrimer molecules in $G3^{2B12gh5M}$ and further enhancement in consequence of the higher level of dendrimer substitution by αM in $G3^{2B10gh17M}$, was similar when tested in vivo with *C. elegans* (by 14-fold in terms of the LC_{50} values; Table 3) and in vitro with cells (by 9–32-fold considering the IC_{50} values; Table 2). Thus, the free drug and its dendrimer-bound forms appear to interact with cells and the model organism following a common mechanism. Further studies are needed to follow the dendrimer-bound αM uptake by *C.elegans* and compare it with that documented for certain other nanomolecules [65–67], such a study being impeded by a certain level of the worm autofluorescence.

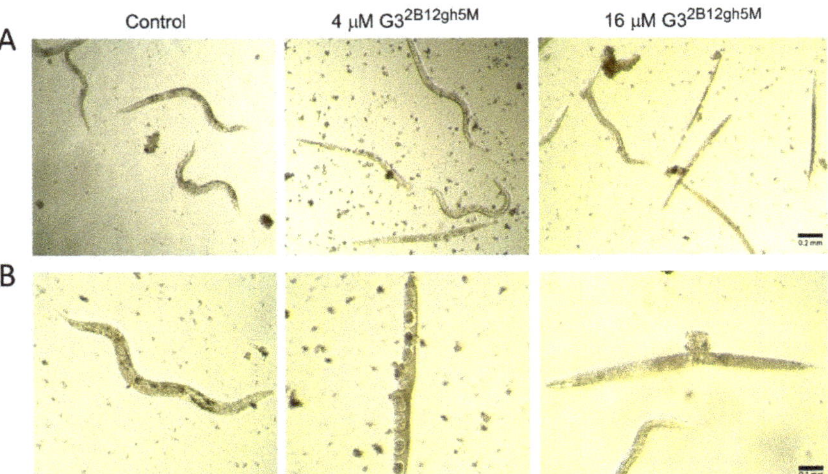

Figure 13. Morphology of *C. elegans* nematodes after 7 days of incubation with $G3^{2B12gh5M}$ dendrimer at a concentration of 4 and 16 μM in comparison to the control group representative, presented under 10× (**A**) and 20× (**B**) objective magnification.

4. Conclusions

Since α-mangostin is considered a useful agent in anticancer therapy, it seems necessary to try to increase its solubility, efficacy and selectivity by finding a suitable carrier for it. In this study, we showed that non-toxic, third-generation glucoheptoamidated and biotinylated poly(amidoamine) dendrimers can provide an excellent carrier for αM, making its cytotoxic, antiproliferative, and anti-adhesive properties more visible at significantly lower concentrations. Additionally, it gives the opportunity to avoid side effects associated with a too high concentration of αM in therapy. At the same time, the mechanism of action of the described conjugates remains characteristic for αM itself (pro-apoptotic effect and changes in ATP levels related to the mitochondrial pathway). This means that despite the lack of selectivity against neoplastic cells, the studied conjugates may be a valuable tool in the local therapy of glioblastoma and squamous cell carcinoma. It is worth to underline that the dendrimer vehicle at concentrations up to 20 μM showed no anti-proliferative effect against tested cell lines, with a feeble cytotoxicity of the highest concentration seen only with squamous carcinoma cells. The present results indicate that the proposed αM delivery system allows the drug to be more effective in the dendrimer-bound than in the free state against both cultured cancer cells and the model organism, suggesting that this treatment is promising for anticancer as well as anti-nematode chemotherapy.

Supplementary Materials: The following are available online at https://www.mdpi.com/article/10.3390/ijms222312925/s1.

Author Contributions: Conceptualization, J.M., Ł.U. and W.R.; methodology, Ł.U. and S.W.; software, J.M.; validation, J.M., Ł.U. and S.W.; formal analysis, Ł.U. and W.R.; investigation, J.M., Ł.U. and S.W.; resources, Ł.U., S.W. and W.R.; data curation, J.M., Ł.U. and S.W.; writing—original draft preparation, J.M., Ł.U. and S.W.; writing—review and editing, J.M., Ł.U., S.W. and W.R.; visualization, J.M. and Ł.U.; supervision, Ł.U., S.W. and W.R.; project administration, J.M. and Ł.U.; funding acquisition, Ł.U. and W.R. All authors have read and agreed to the published version of the manuscript.

Funding: This research was funded by the National Science Center, Poland, grant no. 2016/21/B/NZ1/00288.

Institutional Review Board Statement: Not applicable.

Informed Consent Statement: Not applicable.

Data Availability Statement: Data supporting the reported results are available on request from the corresponding author.

Acknowledgments: The kind support by Agata Wawrzyniak from the Rzeszow University to enable us to use the confocal microscope to perform measurements is gratefully acknowledged.

Conflicts of Interest: The authors declare no conflict of interest.

References

1. Wang, M.-H.; Zhang, K.-J.; Gu, Q.-L.; Bi, X.-L.; Wang, J.-X. Pharmacology of Mangostins and Their Derivatives: A Comprehensive Review. *Chin. J. Nat. Med.* **2017**, *15*, 81–93. [CrossRef]
2. Akao, Y.; Nakagawa, Y.; Nozawa, Y. Anti-Cancer Effects of Xanthones from Pericarps of Mangosteen. *Int. J. Mol. Sci.* **2008**, *9*, 355–370. [CrossRef] [PubMed]
3. Araújo, J.; Fernandes, C.; Pinto, M.; Tiritan, M.E. Chiral Derivatives of Xanthones with Antimicrobial Activity. *Molecules* **2019**, *24*, 314. [CrossRef] [PubMed]
4. Feng, Z.; Lu, X.; Gan, L.; Zhang, Q.; Lin, L. Xanthones, A Promising Anti-Inflammatory Scaffold: Structure, Activity, and Drug Likeness Analysis. *Molecules* **2020**, *25*, 598. [CrossRef]
5. Markowicz, J.; Uram, Ł.; Sobich, J.; Mangiardi, L.; Maj, P.; Rode, W. Antitumor and Anti-Nematode Activities of α-Mangostin. *Eur. J. Pharmacol.* **2019**, *863*, 172678. [CrossRef]
6. PubChem Alpha-Mangostin. Available online: https://pubchem.ncbi.nlm.nih.gov/compound/5281650 (accessed on 13 August 2021).
7. Patra, J.K.; Das, G.; Fraceto, L.F.; Campos, E.V.R.; Rodriguez-Torres, M.d.P.; Acosta-Torres, L.S.; Diaz-Torres, L.A.; Grillo, R.; Swamy, M.K.; Sharma, S.; et al. Nano Based Drug Delivery Systems: Recent Developments and Future Prospects. *J. Nanobiotechnol.* **2018**, *16*, 71. [CrossRef]
8. Wathoni, N.; Rusdin, A.; Motoyama, K.; Joni, I.M.; Lesmana, R.; Muchtaridi, M. Nanoparticle Drug Delivery Systems for Alpha-Mangostin. *Nanotechnol. Sci. Appl.* **2020**, *13*, 23–36. [CrossRef]
9. Pham, D.T.; Saelim, N.; Tiyaboonchai, W. Alpha Mangostin Loaded Crosslinked Silk Fibroin-Based Nanoparticles for Cancer Chemotherapy. *Colloids Surf. B Biointerfaces* **2019**, *181*, 705–713. [CrossRef]
10. Doan, V.T.H.; Takano, S.; Doan, N.A.T.; Nguyen, P.T.M.; Nguyen, V.A.T.; Pham, H.T.T.; Nakazawa, K.; Fujii, S.; Sakurai, K. Anticancer Efficacy of Cyclodextrin-Based Hyperbranched Polymer Nanoparticles Containing Alpha-Mangostin. *Polym. J.* **2021**, *53*, 481–492. [CrossRef]

11. Samprasit, W.; Opanasopit, P.; Chamsai, B. Mucoadhesive Chitosan and Thiolated Chitosan Nanoparticles Containing Alpha Mangostin for Possible Colon-Targeted Delivery. *Pharm. Dev. Technol.* **2021**, *26*, 362–372. [CrossRef]
12. Chis, A.A.; Dobrea, C.; Morgovan, C.; Arseniu, A.M.; Rus, L.L.; Butuca, A.; Juncan, A.M.; Totan, M.; Vonica-Tincu, A.L.; Cormos, G.; et al. Applications and Limitations of Dendrimers in Biomedicine. *Molecules* **2020**, *25*, 3982. [CrossRef]
13. Kharwade, R.; More, S.; Warokar, A.; Agrawal, P.; Mahajan, N. Starburst Pamam Dendrimers: Synthetic Approaches, Surface Modifications, and Biomedical Applications. *Arab. J. Chem.* **2020**, *13*, 6009–6039. [CrossRef]
14. Patri, A.K.; Kukowska-Latallo, J.F.; Baker, J.R. Targeted Drug Delivery with Dendrimers: Comparison of the Release Kinetics of Covalently Conjugated Drug and Non-Covalent Drug Inclusion Complex. *Adv. Drug Deliv. Rev.* **2005**, *57*, 2203–2214. [CrossRef]
15. Sandoval-Yañez, C.; Castro Rodriguez, C. Dendrimers: Amazing Platforms for Bioactive Molecule Delivery Systems. *Materials* **2020**, *13*, 570. [CrossRef]
16. Zhong, L.; Li, Y.; Xiong, L.; Wang, W.; Wu, M.; Yuan, T.; Yang, W.; Tian, C.; Miao, Z.; Wang, T.; et al. Small Molecules in Targeted Cancer Therapy: Advances, Challenges, and Future Perspectives. *Signal Transduct. Target* **2021**, *6*, 1–48. [CrossRef]
17. Lin, H.-M.; Lin, H.-Y.; Chan, M.-H. Preparation, Characterization, and in Vitro Evaluation of Folate-Modified Mesoporous Bioactive Glass for Targeted Anticancer Drug Carriers. *J. Mater. Chem. B* **2013**, *1*, 6147–6156. [CrossRef]
18. Perumal, D.; Golla, M.; Pillai, K.S.; Raj, G.; PK, A.K.; Varghese, R. Biotin-Decorated NIR-Absorbing Nanosheets for Targeted Photodynamic Cancer Therapy. *Org. Biomol. Chem.* **2021**, *19*, 2804–2810. [CrossRef]
19. Ren, W.X.; Han, J.; Uhm, S.; Jang, Y.J.; Kang, C.; Kim, J.-H.; Kim, J.S. Recent Development of Biotin Conjugation in Biological Imaging, Sensing, and Target Delivery. *Chem. Commun.* **2015**, *51*, 10403–10418. [CrossRef]
20. Uram, Ł.; Szuster, M.; Filipowicz, A.; Zaręba, M.; Wałajtys-Rode, E.; Wołowiec, S. Cellular Uptake of Glucoheptoamidated Poly(Amidoamine) PAMAM G3 Dendrimer with Amide-Conjugated Biotin, a Potential Carrier of Anticancer Drugs. *Bioorg. Med. Chem.* **2017**, *25*, 706–713. [CrossRef]
21. Yang, W.; Cheng, Y.; Xu, T.; Wang, X.; Wen, L. Targeting Cancer Cells with Biotin–Dendrimer Conjugates. *Eur. J. Med. Chem.* **2009**, *44*, 862–868. [CrossRef]
22. Uram, Ł.; Filipowicz, A.; Misiorek, M.; Pieńkowska, N.; Markowicz, J.; Wałajtys-Rode, E.; Wołowiec, S. Biotinylated PAMAM G3 Dendrimer Conjugated with Celecoxib and/or Fmoc-l-Leucine and Its Cytotoxicity for Normal and Cancer Human Cell Lines. *Eur. J. Pharm. Sci.* **2018**, *124*, 1–9. [CrossRef]
23. Tomalia, D.A.; Baker, H.; Dewald, J.; Hall, M.; Kallos, G.; Martin, S.; Roeck, J.; Ryder, J.; Smith, P. A New Class of Polymers: Starburst-Dendritic Macromolecules. *Polym. J.* **1985**, *17*, 117–132. [CrossRef]
24. Kaczorowska, A.; Malinga-Drozd, M.; Kałas, W.; Kopaczyńska, M.; Wołowiec, S.; Borowska, K. Biotin-Containing Third Generation Glucoheptoamidated Polyamidoamine Dendrimer for 5-Aminolevulinic Acid Delivery System. *Int. J. Mol. Sci.* **2021**, *22*, 1982. [CrossRef]
25. Wróbel, K.; Wołowiec, S. Low Generation Polyamidoamine Dendrimers (PAMAM) and Biotin-PAMAM Conjugate—The Detailed Structural Studies by 1H and 13C Nuclear Magnetic Resonance Spectroscopy. *Eur. J. Clin. Exp. Med.* **2020**, *18*, 281–285. [CrossRef]
26. Uram, Ł.; Markowicz, J.; Misiorek, M.; Filipowicz-Rachwał, A.; Wołowiec, S.; Wałajtys-Rode, E. Celecoxib Substituted Biotinylated Poly(Amidoamine) G3 Dendrimer as Potential Treatment for Temozolomide Resistant Glioma Therapy and Anti-Nematode Agent. *Eur. J. Pharm. Sci.* **2020**, *152*, 105439. [CrossRef]
27. Stiernagle, T. *Maintenance of C. elegans*; WormBook: Online, 2006. [CrossRef]
28. Bischof, L.J.; Huffman, D.L.; Aroian, R.V. Assays for Toxicity Studies in C. Elegans with Bt Crystal Proteins. In *C. elegans: Methods and Applications*; da Strange, K., Ed.; Methods in Molecular Biology; Humana Press: Totowa, NJ, USA, 2006; pp. 139–154, ISBN 978-1-59745-151-2.
29. Lewis, J.; Fleming, J. Basic Culture Methods. *Methods Cell Biol.* **1995**, *48*, 3–29.
30. Scanlan, L.; Lund, S.; Coskun, S.; Hanna, S.; Johnson, M.; Sims, C.; Brignoni, K.; Lapasset, P.; Elliott, J.; Nelson, B. Counting Caenorhabditis Elegans: Protocol Optimization and Applications for Population Growth and Toxicity Studies in Liquid Medium. *Sci. Rep.* **2018**, *8*, 904. [CrossRef]
31. Herrera-Aco, D.R.; Medina-Campos, O.N.; Pedraza-Chaverri, J.; Sciutto-Conde, E.; Rosas-Salgado, G.; Fragoso-González, G. Alpha-Mangostin: Anti-Inflammatory and Antioxidant Effects on Established Collagen-Induced Arthritis in DBA/1J Mice. *Food Chem. Toxicol.* **2019**, *124*, 300–315. [CrossRef]
32. Sivaranjani, M.; Leskinen, K.; Aravindraja, C.; Saavalainen, P.; Pandian, S.K.; Skurnik, M.; Ravi, A.V. Deciphering the Antibacterial Mode of Action of Alpha-Mangostin on Staphylococcus Epidermidis RP62A Through an Integrated Transcriptomic and Proteomic Approach. *Front. Microbiol.* **2019**, *10*, 150. [CrossRef]
33. Lee, C.-H.; Ying, T.-H.; Chiou, H.-L.; Hsieh, S.-C.; Wen, S.-H.; Chou, R.-H.; Hsieh, Y.-H. Alpha-Mangostin Induces Apoptosis through Activation of Reactive Oxygen Species and ASK1/P38 Signaling Pathway in Cervical Cancer Cells. *Oncotarget* **2017**, *8*, 47425–47439. [CrossRef] [PubMed]
34. Won, Y.-S.; Lee, J.-H.; Kwon, S.-J.; Kim, J.-Y.; Park, K.-H.; Lee, M.-K.; Seo, K.-I. α-Mangostin-Induced Apoptosis Is Mediated by Estrogen Receptor α in Human Breast Cancer Cells. *Food Chem. Toxicol.* **2014**, *66*, 158–165. [CrossRef] [PubMed]
35. Wang, J.J.; Sanderson, B.J.S.; Zhang, W. Significant Anti-Invasive Activities of α-Mangostin from the Mangosteen Pericarp on Two Human Skin Cancer Cell Lines. *Anticancer Res.* **2012**, *32*, 3805–3816.

36. Tripathi, P.K.; Tripathi, S. 6—Dendrimers for Anticancer Drug Delivery. In *Pharmaceutical Applications of Dendrimers*; Chauhan, A., Kulhari, H., Eds.; Micro and Nano Technologies; Elsevier: Amsterdam, The Netherlands, 2020; pp. 131–150, ISBN 978-0-12-814527-2.
37. Ahmad, M.; Yamin, B.M.; Lazim, A.M. Preliminary study on dispersion of α-Mangostin in Pnipam microgel system. *Malays. J. Anal. Sci.* **2012**, *16*, 256–261.
38. Buravlev, E.V.; Shevchenko, O.G.; Anisimov, A.A.; Suponitsky, K.Y. Novel Mannich Bases of α- and γ-Mangostins: Synthesis and Evaluation of Antioxidant and Membrane-Protective Activity. *Eur. J. Med. Chem.* **2018**, *152*, 10–20. [CrossRef] [PubMed]
39. Czerniecka-Kubicka, A.; Tutka, P.; Pyda, M.; Walczak, M.; Uram, Ł.; Misiorek, M.; Chmiel, E.; Wołowiec, S. Stepwise Glucoheptoamidation of Poly(Amidoamine) Dendrimer G3 to Tune Physicochemical Properties of the Potential Drug Carrier: In Vitro Tests for Cytisine Conjugates. *Pharmaceutics* **2020**, *12*, 473. [CrossRef]
40. Santos, A.; Veiga, F.; Figueiras, A. Dendrimers as Pharmaceutical Excipients: Synthesis, Properties, Toxicity and Biomedical Applications. *Materials* **2020**, *13*, 65. [CrossRef]
41. Luong, D.; Kesharwani, P.; Deshmukh, R.; Mohd Amin, M.C.I.; Gupta, U.; Greish, K.; Iyer, A.K. PEGylated PAMAM Dendrimers: Enhancing Efficacy and Mitigating Toxicity for Effective Anticancer Drug and Gene Delivery. *Acta Biomater.* **2016**, *43*, 14–29. [CrossRef]
42. Miranda-Gonçalves, V.; Honavar, M.; Pinheiro, C.; Martinho, O.; Pires, M.M.; Pinheiro, C.; Cordeiro, M.; Bebiano, G.; Costa, P.; Palmeirim, I.; et al. Monocarboxylate Transporters (MCTs) in Gliomas: Expression and Exploitation as Therapeutic Targets. *Neuro-oncology* **2013**, *15*, 172–188. [CrossRef]
43. Russell-Jones, G.; McTavish, K.; McEwan, J.; Rice, J.; Nowotnik, D. Vitamin-Mediated Targeting as a Potential Mechanism to Increase Drug Uptake by Tumours. *J. Inorg. Biochem.* **2004**, *98*, 1625–1633. [CrossRef]
44. Fröhlich, E. The Role of Surface Charge in Cellular Uptake and Cytotoxicity of Medical Nanoparticles. *Int. J. Nanomed.* **2012**, *7*, 5577–5591. [CrossRef]
45. Seeberg, J.C.; Loibl, M.; Moser, F.; Schwegler, M.; Büttner-Herold, M.; Daniel, C.; Engel, F.B.; Hartmann, A.; Schlötzer-Schrehardt, U.; Goppelt-Struebe, M.; et al. Non-Professional Phagocytosis: A General Feature of Normal Tissue Cells. *Sci. Rep.* **2019**, *9*, 11875. [CrossRef]
46. Swanson, J.A. Shaping Cups into Phagosomes and Macropinosomes. *Nat. Rev. Mol. Cell. Biol.* **2008**, *9*, 639–649. [CrossRef]
47. Matsumoto, K.; Akao, Y.; Yi, H.; Ohguchi, K.; Ito, T.; Tanaka, T.; Kobayashi, E.; Iinuma, M.; Nozawa, Y. Preferential Target Is Mitochondria in α-Mangostin-Induced Apoptosis in Human Leukemia HL60 Cells. *Bioorg. Med. Chem.* **2004**, *12*, 5799–5806. [CrossRef]
48. D'Arcy, M.S. Cell Death: A Review of the Major Forms of Apoptosis, Necrosis and Autophagy. *Cell Biol. Int.* **2019**, *43*, 582–592. [CrossRef]
49. Maj, P.; Mori, M.; Sobich, J.; Markowicz, J.; Uram, Ł.; Zieliński, Z.; Quaglio, D.; Calcaterra, A.; Cau, Y.; Botta, B.; et al. Alvaxanthone, a Thymidylate Synthase Inhibitor with Nematocidal and Tumoricidal Activities. *Molecules* **2020**, *25*, 2894. [CrossRef]
50. Valdés-Rives, S.A.; Casique-Aguirre, D.; Germán-Castelán, L.; Velasco-Velázquez, M.A.; González-Arenas, A. Apoptotic Signaling Pathways in Glioblastoma and Therapeutic Implications. *BioMed Res. Int.* **2017**, *2017*, 7403747. [CrossRef]
51. Furnari, F.B.; Fenton, T.; Bachoo, R.M.; Mukasa, A.; Stommel, J.M.; Stegh, A.; Hahn, W.C.; Ligon, K.L.; Louis, D.N.; Brennan, C.; et al. Malignant Astrocytic Glioma: Genetics, Biology, and Paths to Treatment. *Genes Dev.* **2007**, *21*, 2683–2710. [CrossRef]
52. Chen, Q.; Kang, J.; Fu, C. The Independence of and Associations among Apoptosis, Autophagy, and Necrosis. *Signal Transduct. Target Ther.* **2018**, *3*, 18. [CrossRef]
53. Uram, Ł.; Misiorek, M.; Pichla, M.; Filipowicz-Rachwał, A.; Markowicz, J.; Wołowiec, S.; Wałajtys-Rode, E. The Effect of Biotinylated PAMAM G3 Dendrimers Conjugated with COX-2 Inhibitor (Celecoxib) and PPARγ Agonist (Fmoc-L-Leucine) on Human Normal Fibroblasts, Immortalized Keratinocytes and Glioma Cells in Vitro. *Molecules* **2019**, *24*, 3801. [CrossRef]
54. Mizushina, Y.; Kuriyama, I.; Nakahara, T.; Kawashima, Y.; Yoshida, H. Inhibitory Effects of α-Mangostin on Mammalian DNA Polymerase, Topoisomerase, and Human Cancer Cell Proliferation. *Food Chem. Toxicol.* **2013**, *59*, 793–800. [CrossRef]
55. Kritsanawong, S.; Innajak, S.; Imoto, M.; Watanapokasin, R. Antiproliferative and Apoptosis Induction of α-Mangostin in T47D Breast Cancer Cells. *Int. J. Oncol.* **2016**, *48*, 2155–2165. [CrossRef] [PubMed]
56. Ibrahim, M.Y.; Hashim, N.M.; Mariod, A.A.; Mohan, S.; Abdulla, M.A.; Abdelwahab, S.I.; Arbab, I.A. α-Mangostin from Garcinia Mangostana Linn: An Updated Review of Its Pharmacological Properties. *Arab. J. Chem.* **2016**, *9*, 317–329. [CrossRef]
57. Sasahira, T.; Kirita, T. Hallmarks of Cancer-Related Newly Prognostic Factors of Oral Squamous Cell Carcinoma. *Int. J. Mol. Sci.* **2018**, *19*, 2413. [CrossRef] [PubMed]
58. Gkretsi, V.; Stylianopoulos, T. Cell Adhesion and Matrix Stiffness: Coordinating Cancer Cell Invasion and Metastasis. *Front. Oncol.* **2018**, *8*, 145. [CrossRef]
59. Wu, X.-X.; Yue, G.G.-L.; Dong, J.-R.; Lam, C.W.-K.; Wong, C.-K.; Qiu, M.-H.; Lau, C.B.-S. Actein Inhibits the Proliferation and Adhesion of Human Breast Cancer Cells and Suppresses Migration in Vivo. *Front. Pharmacol.* **2018**, *9*, 1466. [CrossRef]
60. Da Cunha, M.L.V.; Maldaun, M.V.C. Metastasis from Glioblastoma Multiforme: A Meta-Analysis. *Rev. Assoc. Med. Bras.* **2019**, *65*, 424–433. [CrossRef]
61. Hoffman, H.A.; Li, C.H.; Everson, R.G.; Strunck, J.L.; Yong, W.H.; Lu, D.C. Primary Lung Metastasis of Glioblastoma Multiforme with Epidural Spinal Metastasis: Case Report. *J. Clin. Neurosci.* **2017**, *41*, 97–99. [CrossRef]

62. Wu, T.; Xu, H.; Liang, X.; Tang, M. Caenorhabditis Elegans as a Complete Model Organism for Biosafety Assessments of Nanoparticles. *Chemosphere* **2019**, *221*, 708–726. [CrossRef]
63. Zhao, X.; Wan, Q.; Fu, X.; Meng, X.; Ou, X.; Zhong, R.; Zhou, Q.; Liu, M. Toxicity Evaluation of One-Dimensional Nanoparticles Using Caenorhabditis Elegans: A Comparative Study of Halloysite Nanotubes and Chitin Nanocrystals. *ACS Sustain. Chem. Eng.* **2019**, *7*, 18965–18975. [CrossRef]
64. Walczynska, M.; Jakubowski, W.; Wasiak, T.; Kadziola, K.; Bartoszek, N.; Kotarba, S.; Siatkowska, M.; Komorowski, P.; Walkowiak, B. Toxicity of Silver Nanoparticles, Multiwalled Carbon Nanotubes, and Dendrimers Assessed with Multicellular Organism Caenorhabditis Elegans. *Toxicol. Mech. Methods* **2018**, *28*, 432–439. [CrossRef]
65. Meyer, J.N.; Lord, C.A.; Yang, X.Y.; Turner, E.A.; Badireddy, A.R.; Marinakos, S.M.; Chilkoti, A.; Wiesner, M.R.; Auffan, M. Intracellular Uptake and Associated Toxicity of Silver Nanoparticles in Caenorhabditis Elegans. *Aquat. Toxicol.* **2010**, *100*, 140–150. [CrossRef]
66. Hu, C.-C.; Wu, G.-H.; Hua, T.-E.; Wagner, O.I.; Yen, T.-J. Uptake of TiO2 Nanoparticles into C. Elegans Neurons Negatively Affects Axonal Growth and Worm Locomotion Behavior. *ACS Appl. Mater. Interfaces* **2018**, *10*, 8485–8495. [CrossRef]
67. Cagno, S.; Brede, D.A.; Nuyts, G.; Vanmeert, F.; Pacureanu, A.; Tucoulou, R.; Cloetens, P.; Falkenberg, G.; Janssens, K.; Salbu, B.; et al. Combined Computed Nanotomography and Nanoscopic X-Ray Fluorescence Imaging of Cobalt Nanoparticles in Caenorhabditis Elegans. *Anal. Chem.* **2017**, *89*, 11435–11442. [CrossRef]

International Journal of
Molecular Sciences

Article

Manipulating Estrogenic/Anti-Estrogenic Activity of Triphenylethylenes towards Development of Novel Anti-Neoplastic SERMs

Heba E. Elnakib [1,†], Marian M. Ramsis [1,†], Nouran O. Albably [1,†], Merna A. Vector [1], Jan J. Weigand [2], Kai Schwedtmann [2], Jannette Wober [3,*], Oliver Zierau [3], Günter Vollmer [3], Ashraf H. Abadi [1] and Nermin S. Ahmed [1,*]

1 Department of Pharmaceutical Chemistry, Faculty of Pharmacy and Biotechnology, German University in Cairo, Cairo 11835, Egypt; heba.elnakib@guc.edu.eg (H.E.E.); marian.ramsis@guc.edu.eg (M.M.R.); nouran.albably@guc.edu.eg (N.O.A.); mirna.ayad@guc.edu.eg (M.A.V.); ashraf.abadi@guc.edu.eg (A.H.A.)
2 Faculty of Chemistry and Food Chemistry, Institute of Inorganic Molecular Chemistry, Technische Universität Dresden, 01062 Dresden, Germany; jan.weigand@tu-dresden.de (J.J.W.); kai.schwedtmann@tu-dresden.de (K.S.)
3 Faculty of Biology, Institute of Zoology, Technische Universität Dresden, 01062 Dresden, Germany; oliver.zierau@tu-dresden.de (O.Z.); guenter.vollmer@tu-dresden.de (G.V.)
* Correspondence: jannette.wober@tu-dresden.de (J.W.); nermin.salah@guc.edu.eg (N.S.A.)
† The first three authors contributed equally to this work.

Citation: Elnakib, H.E.; Ramsis, M.M.; Albably, N.O.; Vector, M.A.; Weigand, J.J.; Schwedtmann, K.; Wober, J.; Zierau, O.; Vollmer, G.; Abadi, A.H.; et al. Manipulating Estrogenic/Anti-Estrogenic Activity of Triphenylethylenes towards Development of Novel Anti-Neoplastic SERMs. *Int. J. Mol. Sci.* **2021**, *22*, 12575. https://doi.org/10.3390/ijms222212575

Academic Editor: Angela Stefanachi

Received: 10 October 2021
Accepted: 16 November 2021
Published: 22 November 2021

Abstract: Selective estrogen receptor modulators (SERMs) act as estrogen receptor (ERα) agonists or antagonists depending on the target issue. Tamoxifen (TAM) (a non-steroidal triphenylethylene derivative) was the first SERM approved as anti-estrogen for the treatment of metastatic breast cancer. On the hunt for novel SERMs with potential growth inhibitory activity on breast cancer cell lines yet no potential to induce endometrial carcinoma, we designed and synthesized 28 novel TAM analogs. The novel analogs bear a triphenylethylene scaffold. Modifications on rings **A**, **B**, and **C** aim to attenuate estrogenic/anti-estrogenic activities of the novel compounds so they can potentially inhibit breast cancer and provide positive, beneficial estrogenic effects on other tissues with no risk of developing endometrial hyperplasia. Compound **12** (*E/Z*-1-(2-{4-[1-(4-Chloro-phenyl)-2-(4-methoxy-phenyl)-propenyl]-phenoxy}-ethyl)-piperidine) showed an appreciable relative ERα agonistic activity in a yeast estrogen screen (YES) assay. It successfully inhibited the growth of the MCF-7 cell line with GI_{50} = 0.6 μM, and it was approximately three times more potent than TAM. It showed no potential estrogenicity on Ishikawa endometrial adenocarcinoma cell line via assaying alkaline phosphatase (AlkP) activity. Compound **12** was tested in vivo to assess its estrogenic properties in an uterotrophic assay in an ovariectomized rat model. Compared to TAM, it induced less increase in wet uterine wet weight and showed no uterotrophic effect. Compound **12** is a promising candidate for further development due to its inhibition activity on MCF-7 proliferation with moderate AlkP activity and no potential uterotrophic effects. The in vitro estrogenic activity encourages further investigations toward potential beneficial properties in cardiovascular, bone, and brain tissues.

Keywords: tamoxifen; CYP2D6; MCF-7; Ishikawa cells; SERM; TNBC; uterotrophic

1. Introduction

Selective estrogen receptor modulator (SERM) refers to a structurally diverse group of compounds that binds to both estrogen receptor subtypes ERα and/or ERβ despite lacking the estrogen steroid moiety. Whereas estrogens typically exert ER agonist effects, SERMs confer mixed functional ER agonist or antagonist activity depending on the target tissue [1]. An ideal SERM would have ER agonist activity in tissues where mimicking the action of estrogens is desirable (e.g., skeletal, cardiovascular, and central nervous systems), and lack of estrogenicity in tissues where estrogens have been shown to induce cancer initiation and growth (e.g., breast and endometrium) [2]. This definition led to investigations on the clinical profile of an ideal SERM. An ideal SERM prevents bone loss and fractures yet does not stimulate endometrial hyperplasia. It also provides relief of hot flushes and other menopausal symptoms. It should not increase the risk of coronary heart disease, stroke, or deep vein thrombosis. The first-generation triphenylethylene SERM included tamoxifen (TAM) and toremifene. Both SERMs are far from being ideal [3].

TAM (I) (a non-steroidal triphenylethylene derivative) was the first SERM approved as anti-estrogen for the treatment of metastatic breast cancer. It is now widely used as adjuvant chemotherapy for the treatment of hormone-dependent metastatic breast carcinoma in postmenopausal women. Although TAM (I) has been very successful in treating breast cancer, some side effects such as thromboembolic events, vasomotor symptoms, and an increased risk of endometrial hyperplasia are associated with TAM treatment [4].

TAM (I) is regarded as a prodrug that is metabolized to the more active metabolites: 4-OH-TAM (II) and endoxifen (III) Figure 1 [5]. Compared to the parent drug, those metabolites have 100-times more affinity to the ER. This metabolism is mainly mediated via cytochrome P450 (CYP) enzymes, specifically the CYP2D6 and CYP3A4 isoforms. Pharmacogenetics revealed the polymorphic nature of the CYP2D6 enzyme. CYP2D6 poor metabolizers (based on *CYP2D6**4 and *6) were reported to benefit less from TAM compared with extensive metabolizers [6–8].

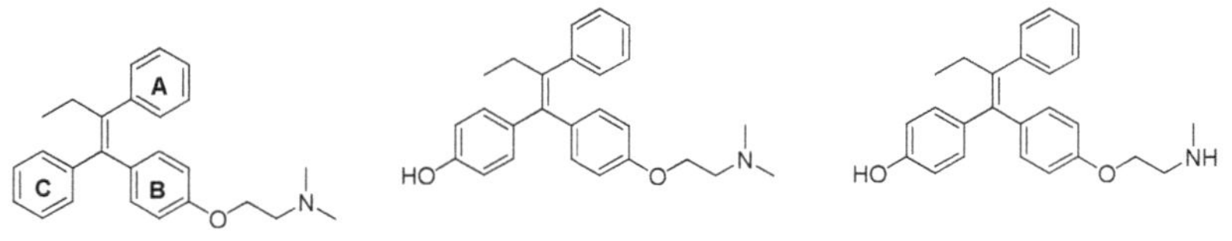

Figure 1. Structures of TAM (**I**), 4-OH-TAM (**II**), and endoxifen (**III**).

The different phenotypes lead to different plasma concentrations of active metabolites among patients of different populations, and hence different clinical outcomes and may lead to drug resistance. Thus, to overcome TAM resistance, TAM is perceived as a clinical target in oncology personalized medicine [9–11].

On the hunt for novel SERMs that possess potential growth inhibitory activity on breast cancer cell lines yet lack the potential to induce endometrial carcinoma, we designed and synthesized 28 novel TAM analogs. The novel analogs bear a triphenylethylene scaffold. Modifications on rings **A**, **B**, and **C** aim to attenuate estrogenic/anti-estrogenic activities of the novel compounds so they can potentially inhibit breast cancer and provide positive estrogenic effects on bones and cardiovascular system without affecting endometrial tissues.

Structural modifications included introducing a chlorine atom at position 4 on ring **C** in all analogs; this ensures the blockage of the site of *para*-hydroxylation; thus, those analogs can bypass *para*-hydroxylation by polymorphic CYP2D6. The effect of this modification on the compounds estrogenic/anti-estrogenic properties is investigated.

The introduction of fluorine into a molecule can productively influence conformation, pK_a, intrinsic potency, membrane permeability, metabolic pathways, and pharmacokinetic properties [12]. Based on these findings, ring **A** was kept unsubstituted or modified to 4-methoxy phenyl, 4-methoxy-3-fluoro phenyl, or 4-fluoro-3-methoxy phenyl. The introduction of a small, highly electronegative fluorine atom on ring **A** can affect the novel analogs stability and lipophilicity. The fluorine atom can further affect the binding affinity either directly or by affecting the polarity of the adjacent methoxy groups.

Previous literature focused on the effect of substitution on position 4 of ring **B** [13]; additionally, recent studies even suggested different substituents on ring **B** to design a homodimeric ER ligand that can act as ER antagonist and SERD (selective estrogen receptor degrader) [14]. In our work, the effect of the length of the alkoxy chain, size, and bulkiness of *N*-substituents and cyclization are thoroughly studied. The novel compounds were depicted in Table 1.

Table 1. Synthetic scheme: Preparation of compounds **1–28**.

a. Zn/$TiCl_4$
b. THF, reflux, dark, 8 h

R1 Cl .HCl
DMF/ K_2CO_3/90°C/24h

1-4 *E/Z*

5-28 *E/Z*

Code	R1	R2	R3
1	—	H	H
2	—	OCH_3	H
3	—	OCH_3	F
4	—	F	OCH_3
5	$-CH_2-N-(CH_3)_2$	H	H
6	$-N-CH_2-CH_2-CH_2-CH_2-$	H	H
7	$-N-CH_2-CH_2-CH_2-CH_2-CH_2-$	H	H
8	$-N-CH_2-CH_2-O-CH_2-CH_2-$	H	H
9	$-N-CH_2-CH_2-CH_2-CH_2-CH_2-CH_2-$	H	H
10	$-CH_2-N-(CH_3)_2$	OCH_3	H
11	$-N-CH_2-CH_2-CH_2-CH_2-$	OCH_3	H
12	$-N-CH_2-CH_2-CH_2-CH_2-CH_2-$	OCH_3	H
13	$-N-CH_2-CH_2-O-CH_2-CH_2-$	OCH_3	H
14	$-N-CH_2-CH_2-CH_2-CH_2-CH_2-CH_2-$	OCH_3	H
15	$-CH_2-N-(CH_3)_2$	OCH_3	F
16	$-N-CH_2-CH_2-CH_2-CH_2-$	OCH_3	F
17	$-N-CH_2-CH_2-CH_2-CH_2-CH_2-$	OCH_3	F
18	$-N-CH_2-CH_2-O-CH_2-CH_2-$	OCH_3	F
19	$-N-CH_2-CH_2-CH_2-CH_2-CH_2-CH_2-$	OCH_3	F
20	$-N-(CH_3)_2$	OCH_3	F
21	$-N-(C_2H_5)_2$	OCH_3	F
22	$-CH_2-N-(CH_3)_2$	F	OCH_3
23	$-N-CH_2-CH_2-CH_2-CH_2-$	F	OCH_3
24	$-N-CH_{2-}CH_2-CH_2-CH_2-CH_2-$	F	OCH_3
25	$-N-CH_2-CH_2-O-CH_2-CH_2-$	F	OCH_3
26	$-N-CH_{2-}CH_2-CH_2-CH_2-CH_2-CH_2-$	F	OCH_3
27	$-N-(CH_3)_2$	F	OCH_3
28	-N-(C2H_5)$_2$	F	OCH_3

All synthesized compounds were tested for their relative activity in β-galactosidase in a yeast estrogen screen (YES) assay. Compounds were tested for their relative estrogenic/ anti-estrogenic activities in comparison to positive and negative controls, respectively. YES assay is a gene reporter assay where the DNA sequence of human ERα is integrated into the yeast genome completed with an expression plasmid carrying estrogen response elements (ERE) in the promoter controlling the expression of the reporter gene *lacZ* (encoding the enzyme β-galactosidase). In the presence of estrogenic compounds, β-galactosidase is synthesized and secreted into the medium, where it converts the chromogenic substrate chlorophenol red-β-D-galactopyranoside (CRPG) from a yellow to a red product, whose absorbance is measured. Agonistic activity is measured directly, whereas antagonistic activity is measured in terms of reduction in color formation in the presence of 0.5 nM/1 nM estradiol (E2) [15–17]. Despite the ability of the YES assay to differentiate between agonists and antagonists, it becomes more and more apparent that compounds exhibit an organ-selective mode of action [18]. Therefore, we decided to test the novel compounds in an organ-specific in vitro model using the human Ishikawa endometrial adenocarcinoma cell line [19,20].

Alkaline phosphatase (AlkP) activity in these human endometrial cancer cells is markedly stimulated by estrogens [9]. In addition and in contrast to yeast assays, which do not mimic human metabolism, the Ishikawa cells, such as normal uterine cells, possess the important capacity to metabolize the compounds, which reflects their true estrogenic activity [21,22].

The anti-proliferative effects of the novel analogs were tested in vitro by the National Cancer Institute (NCI) on a panel of 60 human tumor cell lines at 10 μM. Compounds that elicited mean growth inhibition ≥50% were selected by the NCI for 5-dose testing. The concentration for 50% of maximal inhibition (GI_{50}), total growth inhibition (TGI), and half-maximal lethal concentration (LC_{50}) was measured for each cell line.

Compounds showing appreciable estrogenic activity in the YES assay and that were able to inhibit the growth of MCF-7 cancer cell lines yet with low estrogenic activity on Ishikawa endometrial adenocarcinoma cells might serve as potential ideal SERMS. Compounds **12** and **19** were therefore selected for the in vivo experiments to assess their estrogenic properties in an uterotrophic assay in an ovariectomized rat model. The in vivo uterotrophic rat assay is the gold standard assay to test for the estrogenic effect of compounds; the assay uses adult ovariectomized (OVX) female rats where there is no significant source of endogenous estrogens. Compounds that have estrogenic effects cause uterotrophic response due to the imbibition of water and growth of the uterine cells. Statistically significant uterine weight increases compared to controls provide a positive result [23–29]. Adopting both in vivo and in vitro assays was inevitable due to the limitations of each assay. The cell lines are not properly able to recapitulate the in vivo environment of the uterus within the body. On the other hand, the rat uterotrophic assay merely considers the uterine weight gain as an endpoint of estrogenicity without taking into account all factors that play a role in exerting an estrogenic effect on the organ and body [30].

All our compounds were biologically assayed as *E-Z* mixtures due to synthetic challenges and failure in separating the isomers using available chromatographic techniques. We adopted an in silico model to postulate the isomer with the lowest binding energy. The model also investigates the full agonistic activity of compound **3** despite the lack of an OH group on ring **C**. This group was reported to be essential for ER binding affinity of most synthetic ER ligands.

2. Results and Discussion

2.1. Chemistry Discussion

Compounds (**1–4**) were synthesized using standard McMurry coupling reaction of 4-Chloro-4-hydroxybenzophenone with commercially available ketones using titanium tetrachloride/zinc as a catalyst to yield four condensation products. The condensation products (**1–4**) were then treated with the appropriate base hydrochloride salts in dimethyl formamide (DMF) in the presence of potassium carbonate to form ethers (**5–28**) [31]. The formation of all compounds and their purity were confirmed via UPLC-ESI MS. All compounds were obtained as a mixture of E-Z isomers, as shown from UPLC-UV chromatograms. Some chromatograms showed distinct two peaks of nearly similar area (1:1.1) and having the same molecular ion peak $(M+H)^+$. Attempts to isolate the E-Z isomers using column chromatography as well as preparative HPLC were not successful. ^{1}H-NMR showed peaks integrating for double the number of protons, further confirming the formation of a mixture of E-Z isomers. ^{13}C-NMR further confirmed the formation of isomers since most of the signals were duplicated. Such duplication of signals has been previously reported by Bedford and Richardson [32]. Their masses were confirmed via their molecular ion peaks $[M+H]^+$ and $[M+H+2]^+$ due to the presence of chlorine atoms in all compounds. As previously employed in similar work in the literature, compounds were assayed biologically as E-Z mixtures [33–36].

2.2. Anti-Estrogenic Assays

All compounds lacked significant anti-estrogenic on the ERα except compounds **27** and **28**, which were slightly able to antagonize the β-galactosidase reporter gene activity induced by 1 nM E2 by 11% and 12%, respectively. It seems that the para chlorine substitution at ring **C** has a detrimental effect on the anti-estrogenic activity. This modification has blocked the action of CYP2D6 and therefore prevented the formation of the anti-estrogenic hydroxy metabolite. It is reported that 4-OH-TAM and endoxifen, the active metabolites of TAM, have higher anti-estrogenic potency than the parent drug, TAM [33]. The OH group at position 4 of 4-OH-TAM is presumed to be responsible for its higher anti-estrogenic activity compared to TAM. Additionally, studies have reported that the anti-estrogenic property of SERMs depends on the ability of the cationic nitrogen on the alkylaminoethoxy side chain on ring **B** to neutralize the charge of Asp 351 [37]. Our results showed that the presence of a basic alkylaminoalkoxy group without a phenolic OH on ring **C** or a phenyl ring prone to metabolic hydroxylation could not elicit anti-estrogenic activity regardless of the size and basicity of this group, as shown in compounds **5–26**. Having no tertiary amino group on ring **B** as shown in compounds **3** and **4** or blocking position 4 on ring **C** as shown in compounds **5–26** will mostly abolish the anti-estrogenic action and shift it toward estrogenic activity (Table 2).

Table 2. Relative β-galactosidase activity using YES assay (antagonistic activity).

Code	Anti-Estrogenic Activity *	Code	Anti-Estrogenic Activity *
5	1.34 ± 0.15	**18**	n.d.
6	1.20 ± 0.19	**19**	n.d.
7	1.35 ± 0.16	**20**	n.d.
8	1.11 ± 0.26	**21**	n.d.
9	1.99 ± 0.02	**22**	0.97 ± 0.03
10	3.92 ± 0.58	**23**	1.06 ± 0.05
11	1.55 ± 0.12	**24**	1.19 ± 0.10
12	2.55 ± 0.41	**25**	1.18 ± 0.03
13	n.d. **	**26**	1.05 ± 0.12
14	1.53 ± 0.09	**27**	0.86 ± 0.04
15	n.d.	**28**	0.86 ± 0.07
16	n.d.	**TAM**	0.30 ± 0.08
17	n.d.	**4-OH-TAM**	0.21 ± 0.004

* Relative anti-estrogenic activity is compared to 0.5 nM/1 nM E2 (set as 1), compounds screened at a dose of 1 μM in presence of 0.5 nM/1 nM E2, respectively; compounds were screened in triplicates; ** n.d. = not determined. Compounds were not selected for anti-estrogenic assays due to their high estrogenic activity.

This drives us to the hypothesis that the alkylaminoethoxy side chain on ring **B** is not the only crucial factor for anti-estrogenicity. There are essentially two important features responsible for anti-estrogenic activity. A phenolic OH group is required for high-affinity binding to ER-forming crucial interactions (H-bonds) with Glu 353 and Arg 394 amino acids in the ligand-binding domain (LBD), and the alkylaminoalkoxy bulky group at ring **B** is essential for the ER antagonistic action where it forms a cationic interaction with Asp 351 amino acid of the ER [38].

2.3. Estrogenic Assays

All synthesized compounds were tested for their relative β-galactosidase activity in a yeast estrogen screen (YES) assay at a concentration of 1 μM using DMSO as control (set as 1). The hydroxylated analogs **3** and **4** showed EC_{50} values of 40.1 nM and 258 nM, respectively. E2, the endogenous ligand, showed an EC_{50} = 0.528 nM. The remarkable potency of the two novel analogs can be attributed to the introduction of a chloro group at the para position of ring **C**, the hydroxyl group of ring **B**, and the nature of the substituents on ring **A**.

Compound **3** (EC_{50} = 40 nM) bears a methoxy group at position 4 and a fluoro group at position 3 on ring **A**, and compound **3** showed six-fold more potency than its positional isomer compound **4** (EC_{50} = 258 nM). It seems that a methoxy group at position 4 is essential for agonistic activity.

This could further support the hypothesis that the introduction of a chloro group at ring **C** resulted in an estrogenic property, and the presence of an OH group at ring **B** allows better fitting into the receptor, ensures higher binding affinity, and locking the receptor drug complex into an agonistic conformation.

Replacing the OH group with different alkylaminoalkoxy side chains did not abolish the estrogenic action yet caused a decrease in activity. Comparing compounds (**5**–**9**) bearing a chloro group at ring **C**, unsubstituted ring **A** but different alkylaminoalkoxy side chains, compound **9** with an azepanethoxy side chain at ring **B** induced high relative β-galactosidase activity of 6.74 compared to control; a bulky cyclized side chain on ring **B** seems to improve estrogenic activity.

Compounds (**10**–**14**) bear a methoxy substituent on ring **A**. Both compounds **10** and **13** were the most potent congeners. They bear a dimethylaminopropoxy side chain and a morpholinylethoxy side chain, respectively, on ring **B** (relative β-galactosidase activity = 11.61 and 12.41, respectively). The para methoxy substituent led to an increase in relative estrogenic activity for compounds **10** and **13** compared to their congeners **5** and **8**. A remarkable decrease in relative estrogenic activity was observed for compound **14** compared to its congeners **9**; this may be explained by the fact that the bulky azepanylethoxy group displaced the methoxy substituent of ring **A** outside the binding pocket leading to a possible steric clash.

Compounds (**15**–**21**) bear 3-fluoro 4-methoxy on ring **A**, whereas compounds (**21**–**28**) bear 3-methoxy 4-fluoro substituents on ring **A**. The alkylaminoethoxy side chains on ring **B** were extended to include dimethylaminoethoxy and diethylaminoethoxy side chains. For all compounds (**15**–**21**), the addition of a fluoro group at position **3** enhances the relative estrogenic activity compared to their structural isomers (**22**–**28**) except for compound **18**. The unexpected behavior of compound **18** may be attributed to the less lipophilic character of this compound and lower pK_a value as a result of the morpholinylethoxy substituent on ring **B**. Compounds **15** and **17**, bearing a dimethylaminopropoxy side chain and a piperidinylethoxy side chain, respectively, showed relative estrogenic activities of 7.77 and 7.28, respectively. Compound **17** was the most potent among their series EC_{50} = 252 ± 8 nM. Comparing compound **17** with compound **12**, compound **17** was two-fold more estrogenic at 1 μM, the introduction of a fluoro group at the meta position had a positive impact on estrogenic activity. Compound **19** bearing azepanylethoxy group on ring **B** showed relative estrogenic activities of 3.22 and EC_{50} = 407 ± 86 nM, indicating that estrogenic activity is retained with bulky substituents. Compounds (**22**–**28**) were nearly equipotent. Modifying ring **A** to 3-methoxy 4-fluoro phenyl has resulted in a remarkable decrease in estrogenic activity. It seems that the methoxy substituent at the para position and fluoro substituent at the meta position of ring **A** is the main determinant factors for the higher agonistic action rather than the size or cyclization of substituents on ring **B** (Tables 3 and 4).

Table 3. Relative β-galactosidase activity using YES assay (agonistic activity).

Code	Estrogenic Activity *	Code	Estrogenic Activity *
3	12.83 ± 2.72	**18**	1.76 ± 0.56
4	6.62 ± 1.74	**19**	3.22 ± 1.22
5	2.25 ± 0.92	**20**	2.40 ± 0.64
6	2.88 ± 0.89	**21**	2.40 ± 0.77
7	2.87 ± 0.90	**22**	1.87 ± 0.74
8	1.85 ± 0.11	**23**	1.41 ± 1.01
9	6.74 ± 1.67	**24**	1.19 ± 0.20
10	11.61 ± 0.99	**25**	1.31 ± 0.30
11	1.98 ± 0.12	**26**	1.78 ± 0.64
12	3.65 ± 0.70	**27**	1.15 ± 0.17
13	12.41 ± 0.26	**28**	1.37 ± 0.39
14	1.80 ± 0.09	**TAM**	1.16 ± 0.13
15	7.77 ± 1.9	**4-OH-TAM**	1.46 ± 0.21
16	2.19 ± 0.71	**E2 (10 nM)**	13 ± 2.90
17	7.28 ± 3.10		

* Estrogenic activity is compared to DMSO (set as 1), compounds screened at a dose of 1 μM; compounds were screened in triplicates.

Table 4. EC_{50} values (agonistic activity) of selected compounds.

Code	EC_{50} (nM)
3	40.1 ± 0.5
4	258 ± 80
15	440 ± 10
16	n.c. *
17	252 ± 8
18	n.c.
19	407 ± 86
20	n.c.
21	735 ± 13
E2	0.528 ± 0.051

* n.c. = not calculable because no upper plateau is detectable; compounds were screened in triplicates.

2.4. NCI Growth Inhibition Assays

Compounds were submitted to the Developmental Therapeutics Program (DTP) of the National Cancer Institute (NCI). The program uses a panel of 60 human tumor cell lines representing nine tissue types, including leukemia, non-small cell lung cancer (NSCLC), melanoma, colon cancer, ovarian cancer, CNS cancer, renal cancer, prostate cancer, and breast cancer, to screen for potential new anti-cancer agents. SRB (sulforhodamine B) assay is the preferred high-throughput assay of the National Cancer Institute (NCI) and is the assay used in the NCI's lead compound screening program. Primary screening of synthesized compounds was performed by testing a single high dose of 10 μM in the full NCI-60 panel. The percent growth of treated cells relative to the no-drug control and relative to the time zero number of cells was measured, and a mean graph was provided. The percentage inhibition was then calculated by subtracting the values obtained from 100. In general, all compounds bearing an OH group (**3** and **4**) or a morpholinylethoxy side chain on ring **B** (**8**, **13**, **18**, and **25**) lacked anti-proliferative activity. They had the least percent mean growth inhibition and the lowest percent inhibition on human breast cancer MCF-7 cells. This may be attributed to the partial hydrophilicity of ring **B** (Table 5).

Table 5. Percent mean growth inhibition on 60 NCI tumor cell lines and on MCF-7 cells.

Code	Mean Growth Inhibition (%)	Growth Inhibition on MCF-7 (%) *	Code	Mean Growth Inhibition (%)	Growth Inhibition on MCF-7 (%) *
3	2.34	No inhibition	**17**	69.12	85.01
4	15.41	No inhibition	**18**	4.23	No inhibition
5	15.45	54.99	**19**	60.79	75.88
6	37.82	66.24	**20**	46.97	68.25
7	29	56.39	**21**	31.21	29.69
8	9.4	30.94	**22**	18.62	33.18
9	25.41	24	**23**	46.89	73.52
10	33.44	63.07	**24**	29.19	49.32
11	67.76	86.96	**25**	12.77	2.75
12	55.21	79.65	**26**	28.33	36.51
13	7.86	1.21	**27**	47.17	68.34
14	47.33	63.49	**28**	92.33	>100
15	47.91	86.98	**TAM**	>100	>100
16	77.24	90.04			

* Data obtained from NCI in vitro disease-oriented human tumor cell screen (for details, see the work of [39]), compounds tested at a concentration of 10 μM in triplicates.

Six compounds: **11** (67.76%), **12** (55.21%), **16** (77.24%), **17** (69.12%), **19** (60.79%), **28** (92.33%), showed mean percentage inhibition on all 60 cell lines higher than 50% and were escalated to a dose-dependency assay using five doses on the 60 cell panel. Five of the six compounds share two common features; they bear a para methoxy substituent on ring **A** and bear a cyclic alkylaminoethoxy group on ring **B**.

In the dose-dependency assay, compounds were evaluated against the 60-cell panel at the five doses; 10^{-4} M, 10^{-5} M, 10^{-6} M, 10^{-7} M, and 10^{-8} M. Dose-response curves for each cell line was drawn, and three response parameters are extracted by linear interpolation (GI_{50}, TGI, LC_{50}).

To investigate SERM-like properties of compounds, looking at results from ER-positive cell lines is particularly important. The two most potent compounds on Erα-positive MCF-7 breast cancer cell line were compounds **11** (GI_{50} = 0.89 μM) and **12** (GI_{50} = 0.60 μM). They are almost twice as active as TAM (GI_{50} = 1.58 μM; see Supplementary Materials). Both compounds bear a para methoxy substitution on ring **A** and a cyclic aminoethoxy group on ring **B,** namely a pyrolidine and piperidine, consecutively. The incorporation of the basic nitrogen in a cyclic structure enhances its basicity and significantly improves the anti-proliferative effect of the compounds.

Compounds **16**, **17,** and **19** showed GI_{50} = 2.41, 3.34, and 3.59 μM, respectively. Those compounds bear a para methoxy substituent and a meta fluorine substituent on ring **A**. They exhibited lower anti-proliferative activity on the MCF-7 breast cancer cell line compared to their congeners that lack a fluorine group on meta position, e.g., compounds **11**, **12**, and **14**. This suggests that the presence of a methoxy group increased the electron density on ring **A** resulting in a better anti-proliferative activity, whereas the introduction of an electron-withdrawing group such as fluorine at the meta position lowered the activity. We presumed that introduction of fluorine will increase compounds lipophilicity and therefore improve compounds' cellular uptake and growth inhibition potential. Switching the positions of the methoxy and fluorine substituents in compounds (**22**–**28**) deteriorated the anti-proliferative activity except in compound **28** (GI_{50} = 2.17 μM). This further confirms that fluorine develops essential interactions with specific targets involved in novel compounds' cytotoxic activities.

It is worth mentioning that all six escalated compounds showed more potent anti-proliferative activity than TAM on triple-negative breast cancer (TNBC) cell lines MDA-MDB-231/ATCC and BT-549. Compounds **17**, **19**, and **28** were more potent than TAM on Hs578T, whereas only compound **28** was equipotent to TAM on MDA-MB-468. Since TNBC cell lines do not express ER, this suggests that these novel TAM analogs elicit their anti-proliferative activity via a mechanism that does not involve binding to ER. The six compounds also exhibited mild to high estrogenic activity, but with anti-proliferative activity, this offers an advantage over existing SERM such as TAM (Figure 2).

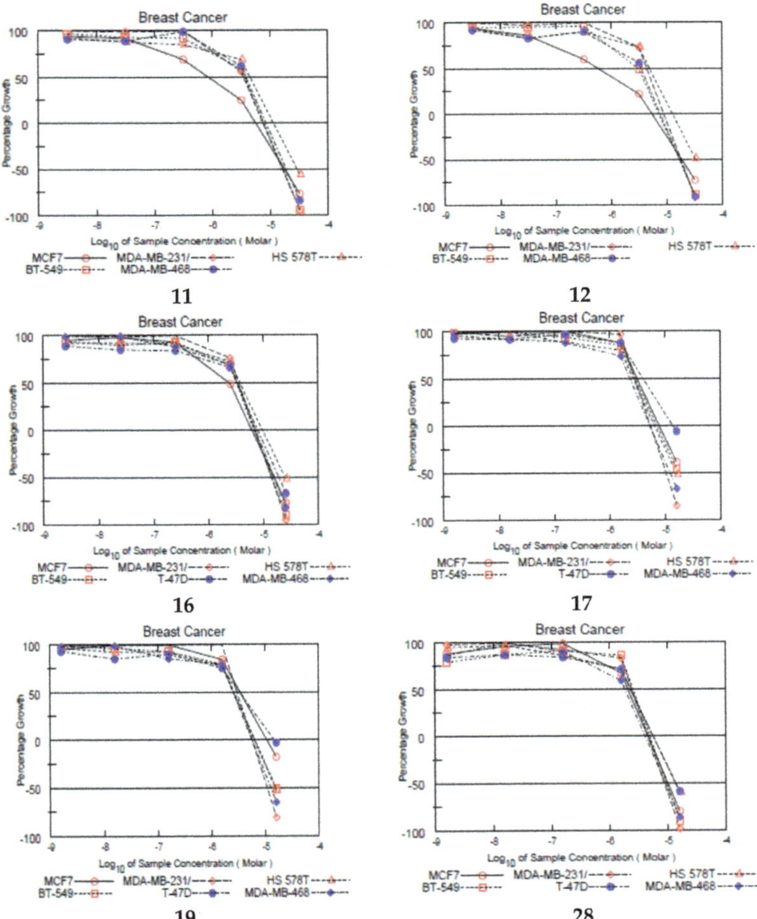

Figure 2. Dose-response curves of selected compounds on breast cancer cell lines. Data obtained from NCI in vitro disease-oriented human tumor cell screen (for details, see the work of [39]) compounds were tested in triplicates.

The ability of the compounds to inhibit the growth of other panels rather than breast cancer was investigated. All six compounds were found to be three times more active than TAM (mean GI_{50} = 6.31 μM) on the colon cancer cell lines with (mean GI_{50} = 1.90 μM). TAM was reported to inhibit the growth of colon cancer cells, yet the mechanism of inhibition is not clear yet, and further studies are warranted before any clinical implications can be postulated (see Supplementary Materials).

Compound **28** (mean GI_{50} = 2.34 μM) was approximately three times as potent as TAM (mean GI_{50} = 6.31 μM) on NSCLC cell lines, and twice as potent as TAM (mean GI_{50} = 5.00 and 5.35 μM) on both renal (mean GI_{50} = 2.40 μM) and prostate (mean GI_{50} = 2.31 μM) cell lines. Compound **28** showed an exceptional broad-spectrum growth inhibition.

The six compounds showed the highest potency on colon cancer cell lines; this might indicate some selectivity toward this particular panel. Further investigations might help understand the reason for this selectivity (Figure 3).

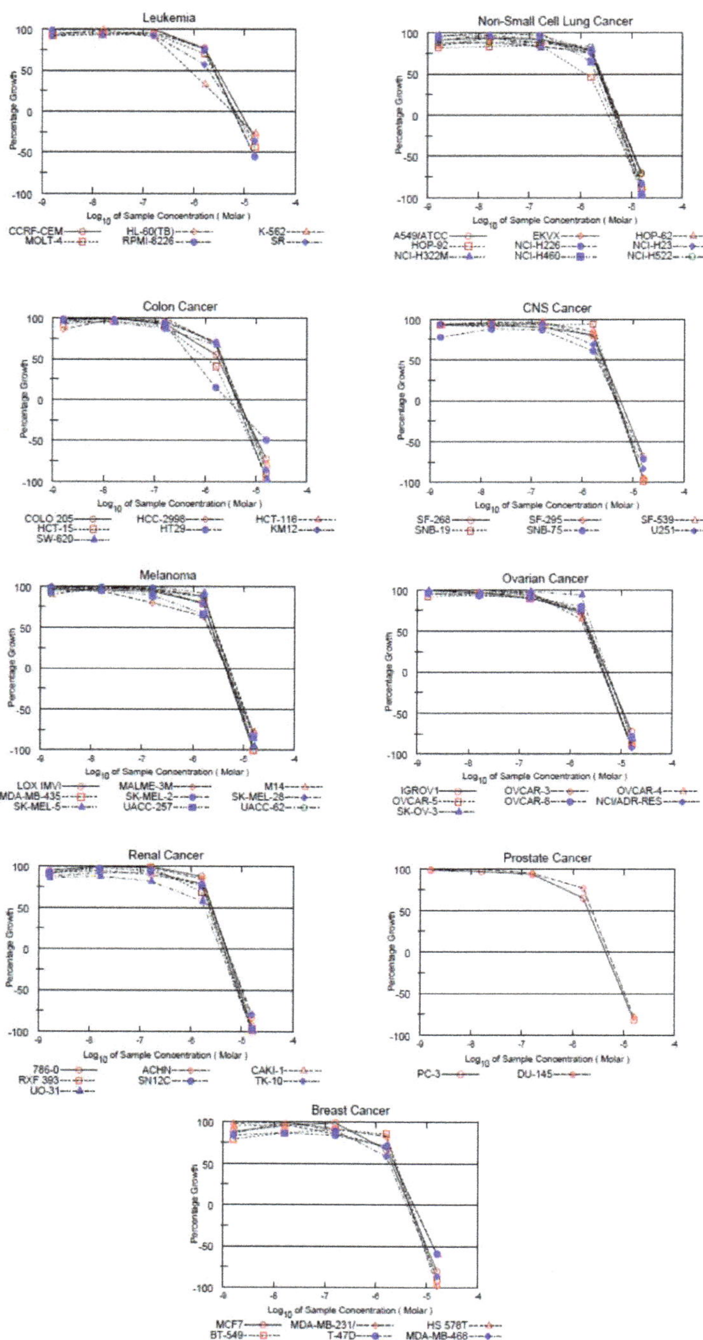

Figure 3. Dose-response curves of compound **28** on different subpanels. Data obtained from NCI in vitro disease-oriented human tumor cell screen (for details, see the work of [39]) compounds were tested in triplicates.

2.5. Alkaline Phosphatase Activity in Ishikawa Cell Line

Because of the potential SERM character of the compounds tested, their estrogenic effects were studied in an endometrial-derived cell culture model, the human endometrial adenocarcinoma cell line Ishikawa. Estrogenic compounds are able to increase the alkaline phosphatase (AlkP) activity mediated by the ERα. All compounds were screened at two concentrations, 0.1 and 1 μM. Its agonistic effect was compared to the vehicle control DMSO (data shown in Supplementary Materials). Estradiol at 10 nM was used as a positive control and TAM and OH-TAM at 1 μM as comparative controls. Most of the compounds showed no significant increase in AlkP activity after a 72 h treatment. Compounds **5**, **11**, **12**, and **19** showed moderate estrogenic activity in YES assay and growth inhibition above 50% on MCF-7 cells at 10 μM; therefore, they were selected for the 5-dose AlkP assay. The four compounds were studied in a concentration range of 1 nM to 10 μM. Compounds **11**, **12**, and **19** were able to increase the AlkP activity in a dose-pendent manner with significant effects at a concentration of 100 nM and 1 μM. No significant effects were observed for compound **5**. The decreased activities at a concentration of 10 μM are caused by a negative influence of the treatment on the cell growth, observed with light microscopy. Compound **12** showed an equipotent activity when compared to TAM and 4-OH-TAM despite its higher relative estrogenic activity in the YES assay (Table 6).

Table 6. Relative alkaline phosphatase activity after an incubation of 72 h in Ishikawa cells.

Code	1 nM	10 nM	100 nM	1 μM	10 μM
E2	n.d. **	6.86 ± 1.60 *	n.d.	n.d.	n.d.
Tam	n.d.	n.d.	n.d.	1.40 ± 0.45	n.d.
OH-Tam	n.d.	n.d.	n.d.	1.47 ± 0.22	n.d.
5	0.93 ± 0.64	0.95 ± 0.37	1.05 ± 0.30	1.08 ± 0.33	0.21 ± 0.16 *
11	1.25 ± 0.61	1.13 ± 0.31	1.75 ± 0.50	2.56 ± 0.83 *	1.36 ± 0.38
12	0.95 ± 0.02	0.96 ± 0.01	1.14 ± 0.01	1.43 ± 0.07 *	0.42 ± 0.13 *
19	1.02 ± 0.13	1.06 ± 0.11	1.75 ± 0.08 *	1.71 ± 0.08 *	0.04 ± 0.04 *

Solvent control (DMSO) was set to 1 * $p < 0.05$ (Tukey test) ** n.d. = not determined.

The observed moderate estrogenic effects of **11**, **12**, and **19** endorse the results obtained by the other in vitro assays reported. Using this Ishikawa cell culture model only gives a hint about possible effects on uterine tissue and needs more investigations.

2.6. Uterotrophic Assay

The most common short-term in vivo assay for estrogenicity/anti-estrogenicity is the uterotrophic assay, suitable for screening ERα agonists and antagonists. The primary endpoint is the uterine wet weight (UWW). An increase in UWW indicates an estrogenic activity of the test compound. Compounds **12** and **19** were screened using the in vivo uterotrophic assay. Both compounds showed less increase in UWW, indicating lower endometrial estrogenic activity and potentially less tendency to induce endometrial carcinoma (Table 7).

Table 7. Relative uterus wet weight of ovariectomized rats.

Code	Mean ± SD g/kg BW
Vehicle	0.61 ± 0.07
E2	3.85 ± 0.71
TAM	1.42 ± 0.30
12	1.23 ± 0.18
19	1.15 ± 0.18

2.7. In Silico Study

The most potent estrogenic compound **3** (EC_{50} = 40.1 nM) bearing an OH group at the para position of ring **B** and 3-fluoro 4-methoxy substituents on ring **A** was selected for the in silico model. Compound **3** was docked into ERα LBD co-crystallized with diethylstilbestrol (DES), a synthetic estrogen with full agonistic activity (PDB: 3ERD) [40]. To validate the docking protocol, the co-crystallized ligand DES was docked into the ERα LBD where all the resultant poses converged to a similar binding mode as that of the experimentally determined position of DES with the best ranking pose having an RMSD value of 1.71 Å.

The crystal structures of ERα bound to DES (PDB code: 3ERD) [9] were downloaded from the PDB database. Only protein molecules were considered where it was optimized using the structure preparation wizard in MOE (version 2009.10) [38] and saved as a mol file. DES was built as E-isomer, whereas compound **3** was built as pure E and Z isomers, minimized using the MMFF94x force field in MOE using a gradient of 0.0001 kcal/(mol Å), and their protonation states at pH 7.0 were generated. A conformational search was adopted for compound **3E** and **3Z** isomers and E-DES. The database obtained was saved as.mdb and used as docking ligands.

Results of the overlay of compounds **3E** and **3Z** on DES showed that the **3E** conformer with the lowest binding energy showed a partial overlay on DES (Figure 4).

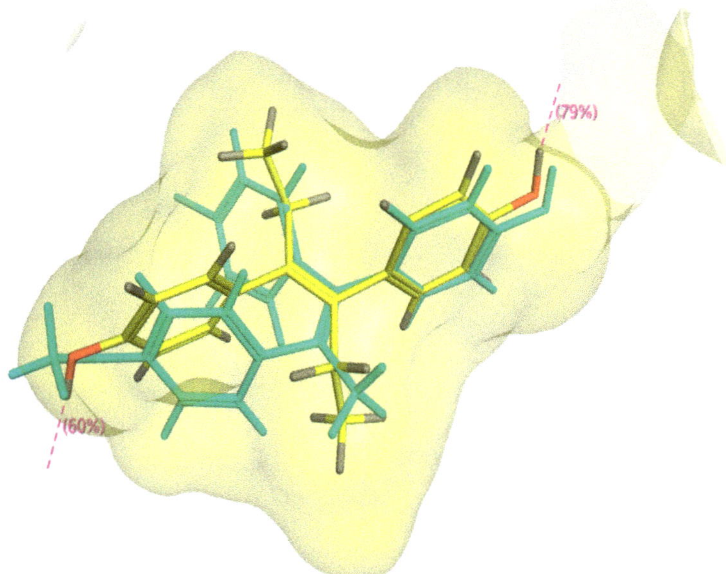

Figure 4. Compound **3*E*** (**cyan**) overlaid with DES (**yellow**) inside ERα LBD.

Compound **3E** retained the two essential interactions with Glu353 and His524, the oxygen of the methoxy group on ring **A** of compound **3E** acted as H-bond acceptor rather than H-bond donor (Figure 5).

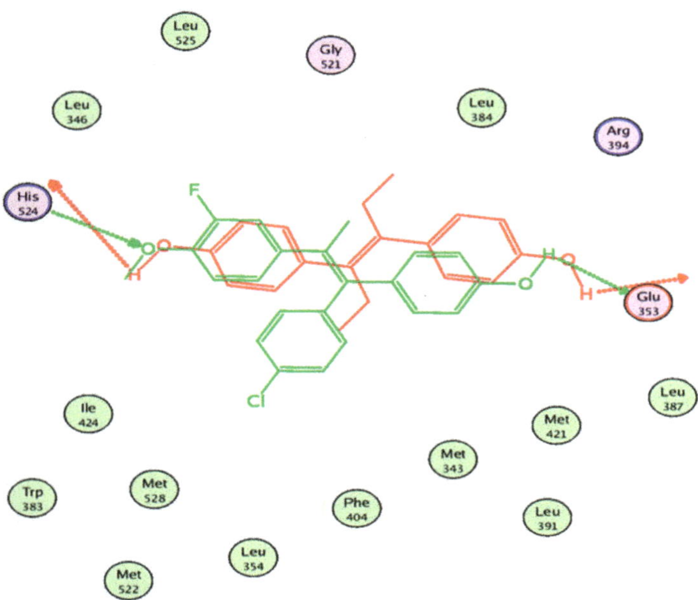

Figure 5. Two-dimensional interactions of DES (**red**) and compound **3E** (**green**) inside ERα LBD.

3. Experimental Section

3.1. Chemistry

All reactions were carried out under nitrogen when an inert atmosphere was needed. Syntheses that required dry and oxygen-free conditions were performed in a Glovebox MB Unilab or using Schlenk techniques under an atmosphere of purified nitrogen or argon, respectively. Dry, oxygen-free solvents (CH_2Cl_2, distilled from CaH_2; THF, distilled from potassium) were employed. All distilled and deuterated solvents were stored over molecular sieves (4 Å). All glassware was oven-dried at 160 °C prior to use. Solvents and reagents were obtained from commercial suppliers and were of pure analytical grade. Purification of compounds was carried out using column chromatography with silica gel 40– 60 μM mesh or using a Biotage® Isolera™ (Uppsala, Sweden) flash purification system using Biotage® KP-Sil SNAP columns. Reaction progress was monitored by TLC using fluorescent pre-coated silica gel plates, and detection of the components was made by short UV light (λ = 254 nm).

^{1}H-NMR spectra were measured on either 400 MHz Bruker or on a Bruker AVANCE III HD Nanobay, 400 MHz UltraSield (^{1}H (400.13 MHz), ^{13}C (100.61 MHz)) or on a Bruker AVANCE III HDX, 500 MHz Ascend (^{1}H (500.13 MHz), ^{13}C (125.75 MHz)) spectrometer. All ^{13}C NMR spectra were exclusively recorded with composite pulse decoupling. Chemical shifts were referenced to δ_{TMS} = 0.00 ppm. (^{1}H, ^{13}C) Chemical shifts (δ) are reported in ppm. Coupling constants (J) are reported in Hz. Multiplicities are abbreviated as: s: singlet; d: doublet; t: triplet; q: quartet; m: multiplet; dd: doublet of doublet; dt: doublet of triplet; brs: broad singlet. Mass spectrometric analysis (UPLC-ESI-MS) was performed using Waters ACQUITY Xevo TQD system, which consisted of an ACQUITY UPLC H-Class system and Xevo™ TQD triple-quadrupole tandem mass spectrometer with an electrospray ionization (ESI) interface (Waters Corp., Milford, MA, USA). Acquity BEH C18 100 × 2.1 mm column (particle size, 1.7 μm) was used to separate analytes (Waters, Dublin, Ireland). The solvent system consisted of water containing 0.1% TFA (A) and 0.1% TFA in acetonitrile (B). UPLC-method: flow rate 200 μL/min. The percentage of B started at an initial of 5% and maintained for 1 min, then increased up to 100% during 10 min, kept at 100% for 2 min, and flushed back to 5% in 3 min. The MS scan was carried out at the following conditions: capillary voltage 3.5 kV, cone voltage 20 V, radio frequency (RF) lens voltage 2.5 V, source temperature 150 °C, and desolvation gas temperature 500 °C. Nitrogen was used as the desolvation and cone gas at a flow rate of 1000 and 20 L/h, respectively. System operation and data acquisition were controlled using Mass Lynx 4.1 software (Waters).

3.1.1. General Procedures for Preparation of Compound **1–4**

Zinc powder (10.11 g, 154 mmol) was suspended in dry THF (100 mL), and the mixture was cooled to 0 °C. $TiCl_4$ (7.5 mL, 70 mmol) was added dropwise under nitrogen/argon. When the addition was complete, the mixture was warmed to room temperature and heated to reflux for 2 h. After cooling down, a solution of 4-Chloro-4-hydroxybenzophenone (2.86 g, 12.3 mmol) and acetophenone/4′-methoxyacetophenone/3′-Fluoro-4′-methoxyacetophen one/4′-Fluoro-3′-methoxyacetophenone (38.4 mmol) in dry THF (100 mL) was added at 0 °C, and the mixture was heated at reflux in the dark for 2.5–7 h. After being cooled to room temperature, the zinc dust was filtered off, and THF was removed under reduced pressure. The residue was dissolved in an aqueous solution containing 30% hydrochloric acid (500 mL) and then extracted with dichloromethane (120 mL × 6). The organic layers were combined and dried over anhydrous Na_2SO_4, concentrated in vacuo, and further purified by silica gel column chromatography or a Biotage® Isolera™ flash purification system using Biotage® KP-Sil SNAP columns (dichloromethane) to yield compounds **1–4** [34].

E/Z-4-[1-(4-Chloro-phenyl)-2-phenylpropenyl]-phenol (**1**)

$C_{21}H_{17}ClO$. Yield: 58%. Orange oil. Purity: 100%. ^{1}H-NMR (400 MHz, $CDCl_3$) δ: 7.33 (d, J = 2.0 Hz, 2H), 7.32 (d, J = 2.0 Hz, 2H), 7.20–7.11 (m, 12H), 6.83 (d, J = 1.5 Hz, 1H), 6.80 (d, J = 1.5 Hz, 1H), 6.74 (d, J = 2.1 Hz, 2H), 6.72 (d, J = 2.1 Hz, 2H), 6.50 (d, J = 2.1 Hz, 2H), 6.49 (d, J = 2.1 Hz, 2H), 2.15 (s, 2H), 2.12 (s, 6H). ^{13}C-NMR (101 MHz, $CDCl_3$) δ: 154.28, 153.60, 143.91, 143.82, 142.12, 141.78, 137.55, 137.51, 136.13, 135.67, 135.51, 135.34, 132.33, 132.18, 132.14, 131.52, 131.39, 129.19, 128.30, 128.03, 127.95, 127.58, 126.33, 126.23, 115.04, 114.45, 23.49, 23.30. MS (ESI): *m/z* = 321.1 $[M+H]^+$ (100%), *m/z* = 323.1 $[M+H+2]^+$ (33%). R_f: 0.42 (100% methylene chloride).

E/Z-4-[1-(4-Chloro-phenyl)-2-(4-methoxy-phenyl)-propenyl]-phenol (**2**)

$C_{22}H_{19}ClO_2$. Yield: 55%. Orange oil. Purity: 95%. ^{1}H-NMR (400 MHz, $CDCl_3$) δ: 7.32 (d, J = 2.0 Hz, 1H), 7.30 (d, J = 1.9 Hz, 1H), 7.18 (d, J = 2.0 Hz, 1H), 7.16 (d, J = 1.8 Hz, 1H), 7.13–6.99 (m, 10H), 6.88 (d, J = 2.2 Hz, 1H), 6.87 (d, J = 2.1 Hz, 1H), 6.83 (d, J = 1.7 Hz, 2H), 6.81 (d, J = 1.8 Hz, 2H), 6.74–6.71 (m, 2H), 6.54 (d, J = 2.2 Hz, 1H), 6.52 (d, J = 2.0 Hz, 1H), 3.76 (d, J = 2.7 Hz, 6H), 2.58 (s, 2H), 2.12 (d, J = 5.0 Hz, 6H). ^{13}C-NMR (101 MHz, $CDCl_3$) δ: 158.33, 157.94, 154.53, 153.80, 142.45, 142.15, 137.09, 137.02, 136.16, 136.07, 135.68, 135.65, 135.47, 135.39, 132.16, 132.12, 131.43, 131.31, 130.75, 130.36, 130.34, 130.15, 128.26, 127.61, 115.06, 114.55, 113.76, 113.72, 113.42, 113.34, 55.49, 55.15, 23.45, 23.24. MS (ESI): *m/z* = 351.1 $[M+H]^+$ (100%) *m/z* = 353.1 $[M+H+2]^+$ (33%). R_f: 0.33 (100% methylene chloride).

E/Z-4-[1-(4-Chlorophenyl)-2-(3-fluoro-4-methoxyphenyl) propenyl]-phenol (**3**)

$C_{22}H_{18}ClFO_2$. Yield: 70%. Orange oil. Purity: 95%. ^{1}H-NMR (500 MHz, $CDCl_3$) (δ 7.33–7.29 (m, 2H), 7.16–7.13 (m, 2H), 7.08–7.05 (m, 2H), 7.04–7.00 (m, 2H), 6.90–6.88 (m, 1H), 6.88–6.85 (m, 1H), 6.83–6.81 (m, 3H), 6.81–6.79 (m, 2H), 6.77–6.76 (m, 1H), 6.76–6.73 (m, 2H), 6.73–6.71 (m, 2H), 6.55–6.51 (m, 2H), 5.30 (s, 2H), 3.84 (d, J = 1.7 Hz, 6H), 2.10 (s, 3H), 2.06 (s, 3H). ^{13}C-NMR (126 MHz, $CDCl_3$) δ 154.47, 153.87, 151.89 (d, J = 245.2 Hz), 151.86 (d, J = 244.9 Hz), 145.94 (t, J = 10.7 Hz), 142.01, 141.70, 137.94, 136.89 (d, J = 6.3 Hz), 136.80 (d, J = 6.3 Hz), 135.45, 135.14, 134.43 (d, J = 1.4 Hz), 133.80 (d, J = 1.4 Hz), 132.42, 132.09, 132.04, 131.71, 131.33, 128.34, 127.78, 125.21 (d, J = 3.3 Hz), 125.13 (d, J = 3.3 Hz), 116.92 (d, J = 18.4 Hz), 116.86 (d, J = 18.3 Hz), 115.08, 114.66, 112.82 (d, J = 8.3 Hz), 112.81 (d, J = 8.3 Hz), 58.59,

56.16, 23.29, 23.09. MS (ESI): m/z = 368.83 $[M+H]^+$, m/z = 370.83 $[M+H+2]^+$. R_f: 0.45 (100% methylene chloride).

E/Z-4-[1-(4-Chlorophenyl)-2-(4-fluoro-3-methoxyphenyl) propenyl]-phenol (**4**)

$C_{22}H_{18}ClFO_2$. Yield: 57%. Orange oil. Purity: 97%. ^{1}H-NMR (400 MHz, $CDCl_3$) δ 7.32 (d, J = 8.3 Hz, 2H), 7.17 (d, J = 8.3 Hz, 2H), 7.08 (d, J = 8.5 Hz, 2H), 7.03 (d, J = 8.4 Hz, 2H), 6.94–6.86 (m, 2H), 6.83 (t, J = 7.8 Hz, 3H), 6.78–6.68 (m, 5H), 6.65 (td, J = 8.2, 1.8 Hz, 2H), 6.54 (d, J = 8.5 Hz, 2H), 3.63 (d, J = 4.4 Hz, 6H), 2.19 (s, 4H), 2.12 (d, J = 13.8 Hz, 2H), 2.10 (s, 3H). ^{13}C-NMR (101 MHz, $CDCl_3$) δ 154.49, 153.91, 151.00 (d, J = 245.6 Hz), 150.93 (d, J = 245.4 Hz), 146.88 (d, J = 10.8 Hz), 146.80 (d, J = 10.8 Hz), 141.88, 141.84, 140.01 (d, J = 4.0 Hz), 139.97 (d, J = 4.0 Hz), 138.01, 138.00, 135.31, 135.25, 135.09, 134.51, 132.47, 132.00, 131.95, 131.76, 131.31, 128.35, 127.77, 121.47 (d, J = 6.6 Hz), 121.36 (d, J = 6.6 Hz), 115.55 (d, J = 18.2 Hz), 115.44 (d, J = 18.3 Hz), 115.19, 115.16, 115.10, 114.64, 56.08, 56.04, 23.22, 23.00. MS (ESI): m/z = 368.83 $[M+H]^+$ (100%), m/z = 370.83 $[M+H+2]^+$. R_f: 0.37 (100% methylene chloride).

3.1.2. General Procedures for Preparation of Compounds **5–28**

A solution of compounds **1–4** (16.28 g, 47 mmol) in DMF (100 mL) was treated with K_2CO_3 (19.5 g, 141 mmol) and heated in an oil bath at 80 °C. The resulting suspension was treated with the appropriate commercially available base hydrochloride salt (51 mmol) portion-wise over a 2 h period and stirred for 16 h. The reaction mixture was cooled to room temperature. K_2CO_3 was filtered off, and DMF was removed under reduced pressure. The final product was purified by silica gel column chromatography or a Biotage® Isolera™ flash purification system using Biotage® KP-Sil SNAP columns (dichloromethane) to yield compounds (**5–28**).

E/Z-(3-{4-[1-(4-Chloro-phenyl)-2-phenyl-propenyl]-phenoxy}-propyl)-dimethyl-amine (**5**)

$C_{26}H_{28}ClNO$. Yield: 48%. Brown oil. Purity: 98%. ^{1}H-NMR (400 MHz, $CDCl_3$) δ: 7.30 (d, J = 8.4 Hz, 2H), 7.19–7.09 (m, 16H), 6.97 (d, J = 8.5 Hz, 2H), 6.86 (d, J = 8.6 Hz, 4H), 6.80–6.73 (m, 2H), 4.06 (t, J = 5.9 Hz, 2H), 3.91 (t, J = 5.9 Hz, 2H), 2.90 (m, 2H), 2.83 (m, 2H), 2.59 (s, 6H), 2.55 (s, 6H), 2.18–2.16 (dd, J = 9.6, 6.1 Hz, 2H), 2.14–2.08 (m, 8H). ^{13}C-NMR (101 MHz, $CDCl_3$) δ: 141.76, 135.78, 132.15, 131.97, 131.37, 131.23, 129.18, 129.16, 128.28, 128.00, 127.94, 127.55, 126.32, 126.22, 114.02, 113.38, 65.30, 56.11, 56.05, 44.18, 44.07, 25.95, 25.78, 23.48, 23.33. MS (ESI): m/z = 406.3 $[M+H]^+$ (100%), m/z = 408.2 $[M+H+2]^+$ (33%). R_f: 0.43 (9:1 methylene chloride: methanol).

E/Z-1-(2-{4-[1-(4-Chloro-phenyl)-2-phenyl-propenyl]-phenoxy}-ethyl)-pyrrolidine (**6**)

$C_{27}H_{28}ClNO$. Yield: 44%. Faint brown oil. Purity: 95.84%. ^{1}H-NMR (400 MHz, $CDCl_3$) δ: 7.33–7.29 (m, J = 2.0 Hz, 2H), 7.20–7.08 (m, 14H), 6.99–6.96 (m, 2H), 6.91–6.87 (m, 2H), 6.81–6.73 (m, 4H), 6.68–6.57 (m, 2H), 4.23 (t, J = 5.6 Hz, 2H), 4.09 (t, J = 5.6 Hz, 2H), 3.09 (t, J = 5.6 Hz, 2H), 3.01 (t, J = 5.5 Hz, 2H), 2.85 (d, J = 18.8 Hz, 8H), 2.13 (s, 3H), 2.10 (s, 3H), 1.96–1.84 (m, 8H). ^{13}C-NMR (101 MHz, $CDCl_3$) δ: 157.17, 156.46, 143.84, 143.79, 142.13, 141.77, 137.52, 137.45, 136.14, 135.84, 135.51, 132.29, 132.18, 131.98, 131.48, 131.41, 131.23, 129.20, 129.17, 128.28, 128.01, 127.97, 127.56, 126.32, 126.25, 114.15, 113.51, 65.95, 65.65, 54.77, 54.69, 54.63, 54.58, 23.50, 23.39, 23.34. MS (ESI): m/z = 418.3 $[M+H]^+$ (100%), m/z = 420.3 $[M+H+2]^+$ (33%). R_f: 0.5 (9:1 methylene chloride: methanol).

E/Z-1-(2-{4-[1-(4-Chloro-phenyl)-2-phenyl-propenyl]-phenoxy}-ethyl)-piperidine (**7**)

$C_{28}H_{30}ClNO$. Yield: 42%. Faint brown oil. Purity: 97.82%. ^{1}H-NMR (400 MHz, $CDCl_3$) δ: 7.32 (d, J = 2.0 Hz, 1H), 7.30 (d, J = 1.6 Hz, 1H), 7.20–7.09 (m, 14H), 6.98 (d, J = 1.9 Hz, 1H), 6.97 (d, J = 2.0 Hz, 1H), 6.89 (d, J = 2.0 Hz, 1H), 6.87 (d, J = 2.0 Hz, 1H), 6.80 (d, J = 2.0 Hz, 1H), 6.79 (d, J = 1.9 Hz, 1H), 6.76 (d, J = 2.0 Hz, 1H), 6.74 (d, J = 2.1 Hz, 1H), 6.56 (d, J = 2.1 Hz, 1H), 6.55 (d, J = 2.0 Hz, 1H), 4.20 (t, J = 5.8 Hz, 2H), 4.05 (t, J = 5.7 Hz, 2H), 2.91 (t, J = 5.7 Hz, 2H), 2.82 (t, J = 5.7 Hz, 2H), 2.72–2.55 (m, 8H), 2.14 (s, 3H), 2.10 (s, 3H), 1.72–1.64

(m, 8H), 1.52–1.44 (m, 4H). ^{13}C-NMR (101 MHz, $CDCl_3$) δ: 157.30, 156.61, 143.87, 143.80, 142.16, 141.79, 137.56, 137.49, 136.09, 135.69, 135.45, 135.34, 132.29, 132.18, 131.95, 131.48, 131.41, 131.21, 129.20, 129.18, 128.28, 128.01, 127.96, 127.56, 126.31, 126.23, 114.13, 113.52, 65.28, 65.03, 57.67, 57.58, 54.87, 54.80, 25.34, 25.23, 23.75, 23.68, 23.50, 23.33. MS (ESI): *m/z* = 432.3 $[M+H]^+$ (100%), *m/z* = 434.3 $[M+H+2]^+$ (33%). R_f: 0.37 (93:7 methylene chloride: methanol).

E/Z-4-(2-{4-[1-(4-Chloro-phenyl)-2-phenyl-propenyl]-phenoxy}-ethyl)-morpholine (**8**)

$C_{27}H_{28}ClNO_2$. Yield: 48%. Orange oil. Purity: 100%. ^{1}H-NMR (400 MHz, $CDCl_3$) δ: 7.32 (d, J = 1.9 Hz, 1H), 7.31 (d, J = 2.0 Hz, 1H), 7.20–7.09 (m, 14H), 6.99 (d, J = 1.9 Hz, 1H), 6.97 (d, J = 2.0 Hz, 1H), 6.90 (d, J = 2.0 Hz, 1H), 6.88 (d, J = 2.0 Hz, 1H), 6.81 (d, J = 2.0 Hz, 1H), 6.79 (d, J = 1.9 Hz, 1H), 6.76 (d, J = 2.0 Hz, 1H), 6.75 (d, J = 2.1 Hz, 1H), 6.57 (d, J = 2.1 Hz, 1H), 6.56 (d, J = 2.0 Hz, 1H), 4.14 (t, J = 5.7 Hz, 2H), 4.00 (t, J = 5.7 Hz, 2H), 3.76 (m, 4H), 3.72 (m, 4H), 2.84 (t, J = 5.7 Hz, 2H), 2.74 (t, J = 5.7 Hz, 2H), 2.62 (m, 4H), 2.55 (m, 4H), 2.14 (s, 3H), 2.11 (s, 3H). ^{13}C-NMR (101 MHz, $CDCl_3$) δ: 157.44, 156.78, 143.90, 143.80, 142.17, 141.80, 137.57, 137.51, 136.09, 135.67, 135.44, 135.28, 132.30, 132.18, 131.93, 131.49, 131.40, 131.20, 129.20, 129.19, 128.29, 128.02, 127.95, 127.56, 126.32, 126.21, 114.14, 113.53, 66.85, 66.80, 65.62, 65.41, 57.66, 57.62, 54.06, 54.02, 23.50, 23.34. MS (ESI): *m/z* = 434.3 $[M+H]^+$ (100%), *m/z* = 436.3 $[M+H+2]^+$ (33%). R_f: 0.68 (95:5 methylene chloride: methanol).

E/Z-1-(2-{4-[1-(4-Chloro-phenyl)-2-phenyl-propenyl]-phenoxy}-ethyl)-azepane (**9**)

$C_{29}H_{32}ClNO$. Yield: 40%. Brown oil. Purity: 95.34%. ^{1}H-NMR (400 MHz, $CDCl_3$) δ: 7.32 (d, J = 1.9 Hz, 1H), 7.30 (d, J = 1.9 Hz, 1H), 7.20–7.08 (m, 14H), 6.99–6.95 (m, 2H), 6.88 (dd, J = 6.7, 4.8 Hz, 2H), 6.80 (d, J = 1.9 Hz, 1H), 6.78 (d, J = 1.9 Hz, 1H), 6.76 (d, J = 2.0 Hz, 1H), 6.74 (d, J = 2.0 Hz, 1H), 6.56 (d, J = 1.9 Hz, 1H), 6.54 (d, J = 2.0 Hz, 1H), 4.21 (t, J = 5.7 Hz, 2H), 4.06 (t, J = 5.7 Hz, 2H), 3.12 (t, J = 5.7 Hz, 2H), 3.04 (t, J = 5.7 Hz, 2H), 3.00–2.89 (m, 8H), 2.14 (s, 3H), 2.10 (s, 3H), 1.80–1.58 (m, 16H). ^{13}C-NMR (101 MHz, $CDCl_3$) δ: 156.76, 156.25, 143.78, 141.76, 136.13, 132.17, 131.97, 131.48, 131.40, 131.23, 129.19, 129.17, 128.29, 128.01, 127.96, 127.56, 126.31, 126.24, 114.15, 113.52, 56.32, 56.20, 55.69, 55.61, 26.96, 26.92, 26.57, 23.50, 23.33. MS (ESI): *m/z* = 446.3 $[M+H]^+$ (100%), *m/z* = 448.3 $[M++H+2]^+$ (33%). R_f: 0.37 (93:7 methylene chloride: methanol).

E/Z-(3-{4-[1-(4-Chloro-phenyl)-2-(4-methoxy-phenyl)-propenyl]-phenoxy}-propyl)-dimethyl-amine (**10**)

$C_{27}H_{30}ClNO_2$. Yield: 48%. Orange oil. Purity: 96.57%. ^{1}H-NMR (400 MHz, $CDCl_3$) δ: 7.30 (d, J = 1.9 Hz, 1H), 7.28 (d, J = 1.9 Hz, 1H), 7.16–6.96 (m, 12H), 6.87–6.68 (m, 8H), 6.56 (d, J = 2.1 Hz, 1H), 6.54 (d, J = 2.0 Hz, 1H), 4.05 (t, J = 6.0 Hz, 2H), 3.92 (t, J = 6.0 Hz, 2H), 3.75 (s, 6H), 2.84–2.73 (m, 4H), 2.51 (d, J = 10.2 Hz, 6H), 2.20–2.11 (m, 4H), 2.10 (s, 6H), 2.07 (s, 6H). ^{13}C-NMR (101 MHz, $CDCl_3$) δ: 157.96, 157.87, 157.32, 156.59, 142.39, 142.08, 136.96, 136.88, 136.05, 135.96, 135.93, 135.65, 135.54, 134.89, 132.21, 131.99, 131.42, 131.29, 131.24, 130.33, 130.31, 128.26, 127.60, 113.99, 113.43, 113.37, 113.31, 65.42, 65.16, 56.21, 56.16, 55.13, 53.44, 44.47, 44.35, 26.28, 26.13, 23.43, 23.28. MS (ESI): *m/z* = 436.3 $[M+H]^+$ (100%), *m/z* = 438.3 $[M+H+2]^+$ (33%). R_f: 0.45 (9:1 methylene chloride: methanol).

E/Z-1-(2-{4-[1-(4-Chloro-phenyl)-2-(4-methoxy-phenyl)-propenyl]-phenoxy}-ethyl)-pyrrolidine (**10**)

$C_{28}H_{30}ClNO_2$. Yield: 40%. Orange oil. Purity: 100%. ^{1}H-NMR (400 MHz, $CDCl_3$) δ: 7.31 (d, J = 1.9 Hz, 1H), 7.29 (d, J = 1.9 Hz, 1H), 7.17–6.96 (m, 12H), 6.89 (d, J = 2.0 Hz, 1H), 6.87 (d, J = 2.0 Hz, 1H), 6.82–6.74 (m, 4H), 6.72 (d, J = 2.1 Hz, 1H), 6.70 (d, J = 2.1 Hz, 1H), 6.59 (d, J = 2.1 Hz, 1H), 6.57 (d, J = 2.0 Hz, 1H), 4.24 (t, J = 5.6 Hz, 2H), 4.12 (t, J = 5.6 Hz, 2H), 3.76 (s, 6H), 3.09 (t, J = 5.5 Hz, 2H), 3.03 (t, J = 5.4 Hz, 2H), 2.88 (s, 8H), 2.11 (s, 3H), 2.07 (s, 3H), 1.95–1.89 (m, 8H). ^{13}C-NMR (101 MHz, $CDCl_3$) δ: 157.96, 157.89, 157.06, 142.06, 136.94, 136.12, 135.99, 135.94, 135.57, 132.22, 131.99, 131.43, 131.30, 131.25, 130.34, 130.31, 128.26, 127.61, 114.13, 113.56, 113.37, 113.33, 55.13, 54.78, 54.71, 54.64, 54.61, 23.44, 23.39,

23.35, 23.27. MS (ESI): *m/z* = 448.3 [M+H]$^+$, *m/z* = 450.2 [M+H+2]$^+$. R_f: 0.5 (9:1 methylene chloride: methanol).

E/Z-1-(2-{4-[1-(4-Chloro-phenyl)-2-(4-methoxy-phenyl)-propenyl]-phenoxy}-ethyl)-piperidine (**12**)

$C_{29}H_{32}ClNO_2$. Yield: 53%. Yellow oil. Purity: 100%. ^{1}H-NMR (400 MHz, $CDCl_3$) δ: 7.31 (d, J = 2.0 Hz, 1H), 7.29 (d, J = 2.1 Hz, 1H), 7.18–6.98 (m, 10H), 6.88 (d, J = 2.2 Hz, 1H), 6.87 (d, J = 2.1 Hz, 1H), 6.83–6.69 (m, 8H), 6.59 (d, J = 2.2 Hz, 1H), 6.57 (d, J = 2.1 Hz, 1H), 4.15 (t, J = 6.0 Hz, 2H), 4.03 (t, J = 6.0 Hz, 2H), 3.76 (s, 6H), 2.84 (t, J = 6.0 Hz, 2H), 2.76 (t, J = 6.0 Hz, 2H), 2.56 (d, J = 18.9 Hz, 8H), 2.11 (s, 3H), 2.08 (s, 2H), 1.64 (tt, J = 11.6, 5.6 Hz, 9H), 1.47 (dd, J = 13.2, 8.4 Hz, 4H). ^{13}C-NMR (101 MHz, $CDCl_3$) δ: 157.98, 157.89, 157.42, 156.73, 142.45, 142.12, 137.06, 136.97, 136.08, 136.00, 135.82, 135.49, 134.79, 132.21, 132.15, 131.91, 131.41, 131.30, 131.17, 130.33, 130.30, 128.25, 127.59, 114.15, 113.61, 113.38, 113.32, 65.65, 65.45, 57.86, 57.81, 55.12, 54.99, 54.95, 25.69, 25.62, 24.01, 23.95, 23.43, 23.25. MS (ESI): *m/z* = 462.3 [M+H]$^+$ (100%), *m/z* = 464.2 [M+H+2]$^+$ (33%). R_f: 0.37 (93:7 methylene chloride: methanol).

E/Z-4-(2-{4-[1-(4-Chloro-phenyl)-2-(4-methoxy-phenyl)-propenyl]-phenoxy}-ethyl)-morpholine (**13**)

$C_{28}H_{30}ClNO_3$. Yield: 45%. Dark orange oil. Purity: 97.45%. ^{1}H-NMR (400 MHz, $CDCl_3$) δ 7.31 (d, J = 1.7 Hz, 1H), 7.29 (d, J = 2.0 Hz, 1H), 7.17–6.97 (m, 10H), 6.89 (d, J = 2.8 Hz, 1H), 6.87 (d, J = 2.0 Hz, 1H), 6.82–6.74 (m, 4H), 6.73–6.68 (m, 4H), 6.59 (d, J = 1.9 Hz, 1H), 6.57 (d, J = 2.0 Hz, 1H), 4.14 (t, J = 5.7 Hz, 2H), 4.01 (t, J = 5.7 Hz, 2H), 3.83–3.69 (m, 14H), 2.83 (t, J = 5.6 Hz, 2H), 2.75 (t, J = 5.6 Hz, 2H), 2.59 (d, J = 20.5 Hz, 8H), 2.11 (s, 3H), 2.08 (s, 3H). ^{13}C-NMR (101 MHz, $CDCl_3$) δ: 157.97, 157.87, 157.36, 156.68, 142.42, 142.10, 136.99, 136.92, 136.07, 135.94, 135.92, 135.59, 135.52, 134.85, 132.22, 132.16, 131.95, 131.43, 131.31, 131.21, 130.34, 130.32, 128.26, 127.61, 114.12, 113.58, 113.37, 113.31, 66.86, 65.61, 65.42, 57.67, 57.65, 55.13, 54.06, 54.04, 53.43, 23.44, 23.28. MS (ESI): *m/z* = 464.3 [M+H]$^+$ (100%), *m/z* = 466.2 [M+H+2]$^+$ (33%). R_f: 0.52 (95:5 methylene chloride: methanol).

E/Z-1-(2-{4-[1-(4-Chloro-phenyl)-2-(4-methoxy-phenyl)-propenyl]-phenoxy}-ethyl)-azepane (**14**)

$C_{30}H_{34}ClNO_2$. Yield: 43%. Orange oil. Purity: 95%. ^{1}H-NMR (400 MHz, $CDCl_3$) δ 7.32–7.25 (m, 2H), 7.17–7.08 (m, 4H), 7.06–6.97 (m, 6H), 6.87 (d, J = 8.7 Hz, 2H), 6.81–6.74 (m, 4H), 6.70 (d, J = 8.7 Hz, 4H), 6.59–6.55 (m, 2H), 4.19 (t, J = 5.6 Hz, 2H), 4.07 (t, J = 5.7 Hz, 2H), 3.76 (s, 6H), 3.09 (t, J = 5.6 Hz, 2H), 3.04–3.00 (m, 2H), 2.97–2.84 (m, 8H), 2.10 (s, 3H), 2.07 (s, 3H), 1.81–1.69 (m, 8H), 1.68–1.60 (m, 8H). ^{13}C-NMR (101 MHz, $CDCl_3$) δ 157.94, 142.40, 142.07, 135.56, 134.91, 132.19, 131.96, 131.51–131.06, 130.31, 128.26, 127.60, 114.16, 113.60, 113.36, 56.25, 55.62, 55.12, 26.97, 23.34. MS (ESI): *m/z* = 476.4 [M+H]$^+$ (100%), *m/z* = 478.4 [M+H+2]$^+$ (33%). R_f: 0.53 (9:1 methylene chloride: methanol).

E/Z-(3-{4-[1-(4-Chlorophenyl)-2-(3-fluoro-4-methoxyphenyl)-propenyl]-phenoxy}-propyl)-dimethyl-amine (**15**)

$C_{27}H_{29}ClFNO_2$. Yield: 74%. Orange oil. Purity: 98.84%. ^{1}H-NMR (400 MHz, $CDCl_3$) δ 7.31–7.28 (m, 2H), 7.15–7.12 (m, 2H), 7.10–7.07 (m, 2H), 7.02–6.99 (m, 2H), 6.88 (s, 2H), 6.85 (d, J = 2.9 Hz, 2H), 6.82–6.79 (m, 3H), 6.77–6.73 (m, 5H), 6.58 (dd, J = 9.1, 2.3 Hz, 2H), 4.02 (t, J = 6.4 Hz, 2H), 3.91 (t, J = 6.4 Hz, 2H), 3.83 (s, 6H), 3.46–3.33 (m, 4H), 2.25 (dd, J = 12.1, 5.9 Hz, 10H), 2.10–2.04 (m, 12H). ^{13}C-NMR: (101 MHz, $CDCl_3$) δ 157.85, 157.27, 145.95 (d, J = 10.5 Hz), 145.92 (d, J = 10.5 Hz), 142.10, 141.75, 138.02, 137.95, 137.99, (d, J = 7.1 Hz), 135.18, 134.82, 134.29 (d, J = 1.4 Hz), 133.60 (d, J = 1.3 Hz), 132.32, 132.07, 131.78, 131.62, 131.30, 128.26, 127.71, 125.18 (d, J = 3.1 Hz), 125.08 (d, J = 3.1 Hz), 116.88 (d, J = 18.1 Hz), 116.83 (d, J = 18.3 Hz), 114.11, 113.66, 112.80 (d, J = 6.0 Hz), 112.78 (d, J = 5.9 Hz), 66.11, 65.93, 56.35 (d, J = 3.8 Hz), 56.10, 45.40, 45.35, 27.49, 27.41, 23.25, 23.06. MS (ESI): *m/z* = 454.30 [M+H]$^+$, *m/z* = 456.30 [M+H+2]$^+$ (100%). R_f: 0.41 (93:7 methylene chloride: methanol).

E/Z-1-(2-{4-[1-(4-Chlorophenyl)-2-(3-fluoro-4-methoxyphenyl)-propenyl]-phenoxy}-ethyl)-pyrrolidine (**16**)

$C_{28}H_{29}ClFNO_2$. Yield: 61%. Orange oil. Purity: 100%. ^{1}H-NMR (500 MHz, $CDCl_3$) δ 7.31–7.27 (m, 2H), 7.15–7.10 (m, 2H), 7.10–7.06 (m, 2H), 7.02–6.98 (m, 2H), 6.88 (d, J = 2.1 Hz, 1H), 6.86 (t, J = 2.6 Hz, 2H), 6.84 (dd, J = 5.9, 3.5 Hz, 1H), 6.81–6.79 (m, 2H), 6.78 (t, J = 4.8 Hz, 2H), 6.76–6.71 (m, 4H), 6.62–6.57 (m, 2H), 4.15 (t, J = 5.9 Hz, 2H), 4.04 (t, J = 5.9 Hz, 2H), 3.82 (s, 6H), 2.65 (d, J = 20.4 Hz, 8H), 2.67–2.64 (m, 8H), 2.10 (s, 3H), 2.07 (s, 3H), 1.90–1.76 (m, 8H). ^{13}C-NMR: (126 MHz, $CDCl_3$) δ 157.64, 157.06, 151.86 (d, J = 245.2 Hz), 151.84 (d, J = 244.9 Hz), 145.93 (t, J = 11.1 Hz), 142.08, 141.74, 138.00, 137.93, 136.87 (d, J = 6.4 Hz), 136.80 (d, J = 6.3 Hz), 135.38, 135.03, 134.36 (d, J = 1.3 Hz), 133.70 (d, J = 1.3 Hz), 132.36, 132.09, 131.80, 131.66, 131.32, 131.09, 128.30, 127.74, 125.20 (d, J = 3.3 Hz), 125.14 (d, J = 3.3 Hz), 116.87 (d, J = 18.4 Hz), 116.85 (d, J = 18.4 Hz), 114.19, 113.76, 112.80 (d, J = 5.9 Hz), 112.79 (d, J = 5.9 Hz), 66.82, 66.63, 56.12, 55.02 (d, J = 5.1 Hz), 54.70, 54.68, 23.48, 23.45, 23.28, 23.08. MS (ESI): *m/z* = 466.40 [M+H]$^+$ (100%), *m/z* = 468.00 [M+H+2]$^+$. R_f: 0.34 (95:5 methylene chloride: methanol).

E/Z-1-(2-{4-[1-(4-Chlorophenyl)-2-(3-fluoro-4-methoxyphenyl)-propenyl]-phenoxy}-ethyl)-piperidine (**17**)

$C_{29}H_{31}ClFNO_2$. Yield: 77%. Orange oil. Purity: 95.88%. ^{1}H-NMR (400 MHz, $CDCl_3$) δ 7.30–7.27 (m, 2H), 7.14–7.11 (m, 2H), 7.08 (d, J = 8.7 Hz, 2H), 7.02–6.98 (m, 2H), 6.88–6.84 (m, 4H), 6.82–6.77 (m, 4H), 6.76–6.73 (m, 4H), 6.60–6.56 (m, 2H), 4.12 (t, J = 6.0 Hz, 2H), 4.00 (t, J = 6.0 Hz, 2H), 3.82 (s, 6H), 2.79 (t, J = 6.0 Hz, 2H), 2.72 (t, J = 6.0 Hz, 2H), 2.58 (m, 8H), 2.07 (s, 3H), 2.05 (s, 3H), 1.60 (dd, J = 11.3, 5.6 Hz, 8H), 1.46–1.40 (m, 4H). ^{13}C-NMR: (101 MHz, $CDCl_3$) δ 157.65, 157.08, 151J = 10.8 Hz.87 (d, J = 245.2 Hz), 151.84 (d, J = 244.9 Hz), 145.97 (d,), 145.88 (d, J = 10.8 Hz), 142.08, 141.74, 138.01, 137.94, 136.88 (d, J = 8.5 Hz), 136.82 (d, J = 8.4 Hz), 135.34, 134.98, 134.35, 133.68, 132.36, 132.08, 131.79, 131.66, 131.31, 131.07, 128.30, 127.74, 125.19 (d, J = 3.4 Hz), 125.11 (d, J = 3.3 Hz), 116.88 (d, J = 18.4 Hz), 116.85 (d, J = 18.3 Hz) 114.20, 113.77, 112.82 (d, J = 5.0 Hz), 112.80 (d, J = 5.1 Hz), 65.72 (d, J = 17.0 Hz), 57.89 (d, J = 4.5 Hz), 56.12, 55.02, 54.99, 25.82, 25.76, 24.11, 24.07, 23.27, 23.08. MS (ESI): *m/z* = 480.01 [M+H]$^+$, *m/z* = 482.01 [M+H+2]$^+$ (100%). R_f: 0.30 (94:6 methylene chloride: methanol).

E/Z-4-(2-{4-[1-(4-Chlorophenyl)-2-(3-fluoro-4-methoxyphenyl)-propenyl]-phenoxy}-ethyl)-morpholine (**18**)

$C_{28}H_{29}ClFNO_3$. Yield: 65%. Brown oil. Purity: 100%. ^{1}H-NMR (400 MHz, $CDCl_3$) δ 7.32–7.25 (m, 3H), 7.16–7.07 (m, 4H), 7.03–6.99 (m, 2H), 6.90–6.73 (m, 11H), 6.62–6.57 (m, 2H), 4.11 (t, J = 5.6 Hz, 2H), 3.99–4.02 (t, J = 5.7 Hz, 2H), 3.84 (s, 6H), 3.77–3.70 (m, 8H), 2.80 (t, J = 5.6 Hz, 2H), 2.73 (t, J = 5.6 Hz, 2H), 2.64–2.50 (m, 8H), 2.09 (s, 3H), 2.06 (s, 3H).^{13}C-NMR: (101 MHz, $CDCl_3$) δ 157.51, 156.93, 151.82 (d, J = 245.2 Hz), 151.79 (d, J = 244.8 Hz), 145.87 (d, J = 10.3 Hz), 142.02, 141.69, 137.91, 137.85, 136.83 (d, J = 6.2 Hz), 136.71 (d, J = 6.2 Hz), 135.47, 135.11, 134.40 (d, J = 1.3 Hz), 133.76 (d, J = 1.3 Hz), 132.34, 132.04, 131.78, 131.63, 131.28, 131.07, 128.27, 127.71, 125.15 (d, J = 3.3 Hz), 125.04 (d, J = 3.4 Hz). 116.88 (d, J = 18.4 Hz), 116.80 (d, J = 18.5 Hz), 114.18, 113.75, 112.80 (d, J = 5.4 Hz), 112.78 (d, J = 5.5 Hz), 66.80, 66.76, 65.65, 65.48, 57.62, 57.58, 57.60 (d, J = 3.3 Hz), 54.03, 54.00, 53.51, 23.22, 23.04. MS (ESI): *m/z* = 482.00 [M+H]$^+$ (100%), *m/z* = 484.00 [M+H+2]$^+$. R_f: 0.58 (95:5 methylene chloride: methanol).

E/Z-1-(2-{4-[1-(4-Chlorophenyl)-2-(3-fluoro-4-methoxyphenyl)-propenyl]-phenoxy}-ethyl)-azepane (**19**)

$C_{30}H_{33}ClFNO_2$. Yield: 54%. Brown oil. Purity: 100%. ^{1}H-NMR (400 MHz, $CDCl_3$) δ 7.30–7.27 (m, 2H), 7.14–7.05 (m, 4H), 7.02–6.98 (m, 2H), 6.88–6.82 (m, 4H), 6.81–6.77 (m, 4H), 6.76–6.71 (m, 4H), 6.61 (dd, J = 6.8, 4.8 Hz, 2H), 4.18 (t, J = 6.0 Hz, 2H), 4.04 (t, J = 6.0 Hz, 2H), 3.83 (s, 6H), 3.03 (t, J = 6.0 Hz, 2H), 2.97 (t, J = 6.0 Hz, 2H), 2.90–2.81 (m, 8H), 2.08 (d, J = 14.2 Hz, 8H), 1.72 (d, J = 4.7 Hz, 6H), 1.65–1.60 (m, 8H). ^{13}C-NMR: (101 MHz, $CDCl_3$) δ 158.79 (d, J =157.7 Hz), 157.56, 151.02, 150.83, 145.97, 144.90, 141.71, 136.39 (d, J =

6.2 Hz), 136.26 (d, J = 6.2 Hz), 135.71, 135.43, 134.38 (d, J = 1.3 Hz), 134.27 (d, J = 1.3 Hz), 132.37, 132.07, 131.81, 131.66, 131.30, 131.09, 128.29, 127.73, 125.18 (d, J = 3.1 Hz), 125.09 (d, J = 3.1 Hz), 116.93 (d, J = 18.4 Hz), 116.74 (d, J = 18.4 Hz), 114.21, 113.76, 112.82 (d, J = 5.7 Hz), 112.79 (d, J = 5.7 Hz), 66.06, 56.37, 56.11, 55.77 (d, J = 5.87 Hz), 53.41, 27.11, 27.01, 26.99, 26.90, 23.26, 23.07. MS (ESI): *m/z* = 494.04 $[M+H]^+$ (100%), *m/z* = 496.04 $[M+H+2]^+$. R_f: 0.34 (93:7 methylene chloride: methanol).

E/Z-(2-{4-[1-(4-Chlorophenyl)-2-(3-fluoro-4-methoxyphenyl)-propenyl]-phenoxy}-ethyl)-dimethyl-amine (**20**)

$C_{26}H_{27}ClFNO_2$. Yield: 55%. Reddish-brown oil. Purity: 98.69%. ^{1}H-NMR (400 MHz, $CDCl_3$) δ 7.32–7.28 (m, 2H), 7.16–7.12 (m, 2H), 7.11–7.07 (m, 2H), 7.03–6.99 (m, 2H), 6.91–6.87 (m, 3H), 6.86 (d, J = 1.9 Hz, 1H), 6.82 (d, J = 1.9 Hz, 1H), 6.79 (dd, J = 5.7, 2.9 Hz, 2H), 6.78–6.74 (m, 4H), 6.72 (dd, J = 8.5, 1.5 Hz, 1H), 6.64–6.59 (m, 2H), 4.10 (t, J = 5.7 Hz, 2H), 3.98 (t, J = 5.7 Hz, 2H), 3.84 (s, 6H), 2.77 (t, J = 5.7 Hz, 2H), 2.70 (t, J = 5.7 Hz, 2H), 2.37 (s, 6H), 2.33 (s, 6H), 2.09 (s, 3H), 2.06 (s, 3H). ^{13}C-NMR: (101 MHz, $CDCl_3$) δ 157.65, 157.07, 151.97 (d, J = 245.1 Hz), 151.94 (d, J = 244.9 Hz), 145.99 (d, J = 9.6 Hz), 142.07, 141.73, 138.01, 137.95, 136.81 (d, J = 6.7 Hz), 136.77 (d, J = 6.7 Hz),135.42, 135.07, 134.37 (d, J = 1.3 Hz), 133.72 (d, J = 1.6 Hz), 132.38, 132.09, 131.79, 131.68, 131.32, 131.09, 128.31, 127.76, 125.20 (d, J = 3.4 Hz), 125.13 (d, J = 3.4 Hz), 116.86 (d, J = 18.4 Hz), 116.98 (d, J = 18.3 Hz), 114.19, 113.75, 112.82 (d, J = 5.1 Hz), 112.80 (d, J = 5.2 Hz), 65.85, 65.64, 58.27, 58.22, 56.14, 45.83, 45.80, 23.29, 23.07. MS (ESI): *m/z* = 440.30 $[M+H]^+$ (100%), *m/z* = 442.30 $[M+2]^+$. R_f: 0.38 (92:8 methylene chloride: methanol).

E/Z-(2-{4-[1-(4-Chlorophenyl)-2-(3-fluoro-4-methoxyphenyl)-propenyl]-phenoxy}-ethyl)-diethyl-amine (**21**)

$C_{28}H_{31}ClFNO_2$. Yield: 51%. Yellow oil. Purity: 98.96%. ^{1}H-NMR (500 MHz, $CDCl_3$) δ 7.30 (d, J = 8.4 Hz, 2H), 7.14 (d, J = 8.4 Hz, 2H), 7.09 (d, J = 8.6 Hz, 2H), 7.02 (d, J = 8.5 Hz, 2H), 6.87 (dd, J = 11.8, 2.6 Hz, 4H), 6.80 (t, J = 8.0 Hz, 3H), 6.78–6.70 (m, 5H), 6.59 (d, J = 8.7 Hz, 2H), 4.10 (d, J = 5.4 Hz, 2H), 4.00 (s, 2H), 3.84 (s, 6H), 2.95 (d, J = 4.8 Hz, 2H), 2.88 (s, 2H), 2.78–2.57 (m, 8H), 2.09 (d, J = 7.5 Hz, 3H), 2.09 (d, J = 7.5 Hz, 3H), 1.14–1.04 (m, 12H). ^{13}C-NMR: (126 MHz, $CDCl_3$) δ 157.58, 156.96, 151.89 (d, J = 245.2 Hz), 151.85 (d, J = 244.8 Hz), 145.94 (t, J = 11.3 Hz), 142.07, 141.73, 137.99, 137.93, 136.89 (d, J = 6.2 Hz), 136.81 (d, J = 6.4 Hz), 135.43, 135.09, 134.38 (d, J = 1.1 Hz), 133.74 (d, J = 1.1 Hz), 132.39, 132.10, 131.84, 131.69, 131.33, 131.12, 128.32, 127.77, 125.21 (d, J = 3.3 Hz), 125.13 (d, J = 3.4 Hz), 116.91 (d, J = 18.4 Hz), 116.86 (d, J = 18.4 Hz), 114.15, 113.71, 112.81 (d, J = 5.9 Hz), 112.79 (d, J = 5.9 Hz), 66.12, 56.14, 53.42, 51.67, 51.54, 47.80, 47.77, 23.30, 23.09, 11.56. MS (ESI): *m/z* = 468.30 $[M+H]^+$ (100%), *m/z* = 470.00 $[M+H+2]^+$. R_f: 0.32 (93:7 methylene chloride: methanol).

E/Z-(3-{4-[1-(4-Chlorophenyl)-2-(4-fluoro-3-methoxyphenyl)-propenyl]-phenoxy}-propyl)-dimethyl-amine (**22**)

$C_{27}H_{29}ClFNO_2$. Yield: 50%. Orange oil. Purity: 97.83%. ^{1}H-NMR (500 MHz, $CDCl_3$) δ 7.31 (d, J = 8.4 Hz, 1H), 7.16 (d, J = 8.3 Hz, 1H), 7.10 (t, J = 5.7 Hz, 3H), 7.02 (d, J = 8.4 Hz, 3H), 6.92–6.85 (m, 5H), 6.81 (t, J = 7.1 Hz, 3H), 6.75 (d, J = 8.7 Hz, 1H), 6.72–6.67 (m, 2H), 6.65 (dd, J = 8.4, 1.9 Hz, 1H), 6.63 (dd, J = 8.3, 1.9 Hz, 1H), 6.59 (d, J = 8.7 Hz, 1H), 4.04 (t, J = 6.3 Hz, 2H), 3.91 (t, J = 6.3 Hz, 2H), 3.62 (d, J = 2.0 Hz, 6H), 2.56 (dt, J = 27.7, 7.4 Hz, 4H), 2.37–2.31 (m, 12H), 2.11 (d, J = 17.4 Hz, 6H), 2.07–1.94 (m, 4H). ^{13}C-NMR: (126 MHz, $CDCl_3$) δ 157.79, 157.21, 150.97 (d, J = 245.6 Hz), 150.89 (d, J = 245.3 Hz), 146.89 (d, J = 10.8 Hz), 146.82 (d, J = 10.7 Hz), 141.93, 141.88, 140.08 (d, J = 4.0 Hz), 139.98 (d, J = 4.0 Hz), 138.09, 138.07, 135.19, 135.11, 135.02, 134.44, 132.44, 132.01, 131.75, 131.73, 131.33, 131.12, 128.34, 127.75, 121.42 (d, J = 6.6 Hz), 115.54 (d, J = 18.2 Hz), 115.43 (d, J = 18.2 Hz), 115.09, 114.13, 113.67, 65.94, 65.80, 56.37, 56.31, 56.03 (d, J = 2.6 Hz), 53.43, 45.17, 45.10, 27.16, 27.04, 23.24, 23.05.MS (ESI): *m/z* = 454.30 $[M+H]^+$ (100%), *m/z* = 456.30 $[M++H+2]^+$. R_f: 0.33 (91:9 methylene chloride: methanol).

E/Z-1-(2-{4-[1-(4-Chlorophenyl)-2-(4-fluoro-3-methoxyphenyl)-propenyl]-phenoxy}-ethyl)-pyrrolidine (**23**)

$C_{28}H_{29}ClFNO_2$. Yield: 71%. Brown oil. Purity: 100%. ^{1}H-NMR (400 MHz, $CDCl_3$) δ 7.31 (d, J = 8.4 Hz, 2H), 7.16 (d, J = 8.4 Hz, 2H), 7.11 (d, J = 8.6 Hz, 2H), 7.02 (d, J = 8.4 Hz, 2H), 6.89 (dt, J = 8.2, 5.5 Hz, 4H), 6.81 (d, J = 8.4 Hz, 2H), 6.76 (d, J = 8.7 Hz, 2H), 6.72–6.67 (m, 3H), 6.64 (t, J = 2.0 Hz, 1H), 6.61 (d, J = 8.6 Hz, 2H), 4.18 (t, J = 5.8 Hz, 2H), 4.05 (t, J = 5.8 Hz, 2H), 3.62 (d, J = 2.8 Hz, 6H), 2.98 (t, J = 5.8 Hz, 2H), 2.90 (t, J = 5.8 Hz, 2H), 2.71 (d, J = 21.8 Hz, 8H), 2.13 (s, 3H), 2.09 (s, 3H), 1.89–1.80 (m, 8H). ^{13}C-NMR: (101 MHz, $CDCl_3$) δ 157.58, 157.01, 152.21, 152.16, 146.90 (d, J = 9.9 Hz), 146.89 (d, J = 9.8 Hz), 141.91, 141.86, 140.03 (d, J = 3.9 Hz), 139.97 (d, J = 4.1 Hz), 138.06, 138.03, 135.34, 135.25, 135.06, 134.47, 132.44, 132.00, 131.73, 131.31, 131.11, 128.33, 127.75, 121.45 (d, J = 6.0 Hz), 115.57 (d, J = 18.2 Hz), 115.39, (d, J = 18.3 Hz), 115.12 (d, J = 2.0 Hz), 115.11, 114.23, 113.79, 66.63, 66.50, 56.03 (d, J = 1.8 Hz), 54.99, 54.91, 54.69, 54.65, 23.48, 23.44, 23.22, 23.03. MS (ESI): *m/z* = 466.30 $[M+H]^+$ (100%), *m/z* = 468.20 $[M+H+2]^+$. R_f: 0.32 (94:6 methylene chloride: methanol).

E/Z-1-(2-{4-[1-(4-Chlorophenyl)-2-(4-fluoro-3-methoxyphenyl)-propenyl]-phenoxy}-ethyl)-piperidine (**24**)

$C_{29}H_{31}ClFNO_2$. Yield: 42%. Orange oil. Purity: 95%. ^{1}H-NMR (400 MHz, $CDCl_3$)) δ 7.33 (d, J = 7.5 Hz, 2H), 7.16 (dd, J = 17.6, 7.6 Hz, 4H), 7.04 (d, J = 7.6 Hz, 2H), 6.91 (t, J = 9.5 Hz, 4H), 6.83 (d, J = 7.4 Hz, 2H), 6.77 (d, J = 7.7 Hz, 2H), 6.70 (d, J = 13.9 Hz, 3H), 6.62 (d, J = 8.6 Hz, 3H), 4.23 (s, 2H), 4.10 (s, 2H), 3.66 (s, 6H), 2.91 (dd, J = 30.7, 19.6 Hz, 4H), 2.63 (s, 6H), 2.13 (d, J = 13.6 Hz, 6H), 1.71 (s, 8H), 1.47 (d, J = 23.0 Hz, 6H). ^{13}C-NMR: (101 MHz, $CDCl_3$) δ 157.79, 157.21, 150.97, 150.89, 146.90 (d, J = 9.2 Hz), 146.81 (d, J = 9.0 Hz), 141.93, 141.88, 140.08 (d, J = 4.0 Hz), 139.98 (d, J = 4.0 Hz), 138.09, 138.07, 135.19, 135.11, 135.02, 134.44, 132.44, 132.01, 131.75, 131.73, 131.33, 131.12, 128.34, 127.75, 121.42 (t, J = 6.6 Hz), 115.56 (d, J = 13.4 Hz), 115.41 (d, J = 13.3 Hz), 115.09,114.33, 114.13, 113.67, 66.22, 57.98, 57.67, 56.04, 54.85 (d, J = 6.4 Hz), 53.40, 26.91, 25.63, 25.29, 23.72, 23.22, 23.04. MS (ESI): *m/z* = 480.01 $[M+H]^+$ (100%), *m/z* = 482.01 $[M+H+2]^+$. R_f: 0.40 (95:5 methylene chloride: methanol).

E/Z-4-(2-{4-[1-(4-Chlorophenyl)-2-(4-fluoro-3-methoxyphenyl)-propenyl]-phenoxy}-ethyl)-morpholine (**25**)

$C_{28}H_{29}ClFNO_3$. Yield: 61%. Orange oil. Purity: 97.07%. ^{1}H-NMR (400 MHz, $CDCl_3$) δ 7.31 (d, J = 8.4 Hz, 2H), 7.14 (dd, J = 15.7, 8.5 Hz, 4H), 7.02 (d, J = 8.4 Hz, 2H), 6.90 (t, J = 9.1 Hz, 4H), 6.81 (d, J = 8.4 Hz, 2H), 6.76 (d, J = 8.7 Hz, 2H), 6.73–6.65 (m, 3H), 6.64 (dd, = 5.4, 3.3 Hz, 1H), 6.61 (t, J = 6.2 Hz, 2H), 4.15 (t, J = 5.5 Hz, 2H), 4.03 (t, J = 5.5 Hz, 2H), 3.80–3.70 (m, 8H), 3.63 (s, 6H), 2.86 (t, J = 5.3 Hz, 2H), 2.77 (t, J = 5.3 Hz, 2H), 2.61 (d, J = 21.5 Hz, 8H), 2.13 (s, 3H), 2.10 (s, 3H). ^{13}C-NMR: (101 MHz, $CDCl_3$) δ 157.67, 157.09, 151.14 (d, J = 245.8 Hz), 151.06 (d, J = 245.2 Hz), 147.06 (d, J = 10.8 Hz), 146.99 (d, J = 10.8 Hz), 142.02, 141.98, 140.19 (d, J = 4.1 Hz), 140.06 (d, J = 4.0 Hz), 138.16, 138.14, 135.58, 135.49, 135.25, 134.70, 132.61, 132.13, 131.90, 131.44, 131.28, 128.49, 127.91, 121.57 (d, J = 5.1 Hz), 121.51 (d, J = 5.2 Hz), 115.69 (d, J = 18.2 Hz), 115.59 (d, J = 18.3 Hz), 115.27 (d, J = 1.4 Hz), 115.22 (d, J = 1.8 Hz), 115.12, 114.37, 113.93, 66.89, 65.70, 65.63, 57.76, 56.19, 54.19, 54.15, 53.56, 23.36, 23.19. MS (ESI): *m/z* = 482.30 $[M+H]^+$ (100%), *m/z* = 484.20 $[M+H+2]^+$. R_f: 0.35 (94:6 methylene chloride: methanol).

E/Z-1-(2-{4-[1-(4-Chlorophenyl)-2-(4-fluoro-3-methoxyphenyl)-propenyl]-phenoxy}-ethyl)-azepane (**26**)

$C_{30}H_{33}ClFNO_2$. Yield: 62%. Orange oil. Purity: 95%. ^{1}H-NMR (400 MHz, $CDCl_3$) δ 7.31 (d, J = 8.4 Hz, 2H), 7.16 (d, J = 8.4 Hz, 2H), 7.11 (d, J = 8.6 Hz, 2H), 7.02 (d, J = 8.5 Hz, 2H), 6.93–6.86 (m, 4H), 6.81 (d, J = 8.4 Hz, 2H), 6.76 (d, J = 8.7 Hz, 2H), 6.73–6.66 (m, 3H), 6.64 (s, 1H), 6.61 (t, J = 6.3 Hz, 2H), 4.13 (t, J = 6.0 Hz, 2H), 4.00 (t, J = 6.0 Hz, 2H), 3.62 (d, J = 3.2 Hz, 6H), 3.01 (t, J = 6.0 Hz, 2H), 2.93 (t, J = 6.0 Hz, 2H), 2.87–2.76 (m, 8H), 2.13 (s, 3H), 2.10 (s, 3H), 1.74–1.65 (m, 8H), 1.64–1.58 (m, 8H). ^{13}C-NMR: (101 MHz, $CDCl_3$)

δ 157.68, 157.10, 150.98 (d, J = 245.8 Hz), 150.91 (d, J = 245.4 Hz), 146.90 (d, J = 10.6 Hz), 146.83 (d, J = 10.8 Hz), 141.93, 141.87, 140.05 (d, J = 4.0 Hz), 139.97 (d, J = 4.0 Hz), 138.08, 138.06, 135.26, 135.17, 135.04, 134.45, 132.44, 132.00, 131.73, 131.31, 131.10, 128.33, 127.75, 121.42 (d, J = 6.9 Hz), 121.31, 115.53 (d, J = 18.4 Hz), 115.43 (d, J = 18.3 Hz), 115.12, 114.24, 113.80,66.16, 66.00, 56.40, 56.30, 56.03 (d, J = 2.3 Hz), 55.83, 55.77, 27.51, 27.47, 27.04, 27.02, 23.23, 23.03. MS (ESI): *m/z* = 494.04 $[M+H]^+$ (100%), *m/z* = 496.04 $[M+2]^+$. R_f: 0.50 (95:5 methylene chloride: methanol).

E/Z-(2-{4-[1-(4-Chlorophenyl)-2-(4-fluoro-3-methoxyphenyl)-propenyl]-phenoxy}-ethyl)-dimethyl-amine (**27**)

$C_{26}H_{27}ClFNO_2$. Yield: 67%. Yellow oil. Purity: 100%. ^{1}H-NMR (400 MHz, $CDCl_3$) δ 7.31 (d, J = 8.4 Hz, 2H), 7.16 (d, J = 8.4 Hz, 2H), 7.11 (d, J = 8.6 Hz, 2H), 7.02 (d, J = 8.5 Hz, 2H), 6.92–6.86 (m, 4H), 6.81 (d, J = 8.5 Hz, 2H), 6.76 (d, J = 8.7 Hz, 2H), 6.73–6.65 (m, 3H), 6.62 (t, J = 7.5 Hz, 3H), 4.11 (t, J = 5.7 Hz, 2H), 3.98 (t, J = 5.7 Hz, 2H), 3.62 (d, J = 4.2 Hz, 6H), 2.78 (t, J = 5.7 Hz, 2H), 2.70 (t, J = 5.7 Hz, 2H), 2.38 (s, 6H), 2.33 (s, 6H), 2.13 (s, 3H), 2.09 (s, 3H). ^{13}C-NMR: (101 MHz, $CDCl_3$) δ 157.14, 156.57, 150.46 (d, J = 245.5 Hz), 150.38 (d, J = 245.3 Hz), 146.37 (d, J = 10.8 Hz), 146.31 (d, J = 10.8 Hz), 141.40, 141.34, 139.51 (d, J = 4.2 Hz), 139.45 (d, J = 4.0 Hz), 137.55, 137.52, 134.77, 134.68, 134.52, 133.93, 131.91, 131.48, 131.19, 130.79, 130.57, 127.81, 127.22, 120.90 (d, J = 6.8 Hz), 120.81 (d, J = 6.7 Hz), 115.04 (d, J = 10.6 Hz), 114.86, 114.81, 114.60, 113.69, 113.25, 65.83, 65.68, 58.25, 58.18, 56.03 (d, J = 3.1 Hz), 56.01, 45.82, 45.76, 23.23, 23.02. MS (ESI): *m/z* = 440.30 $[M+H]^+$ (100%), *m/z* = 442.30 $[M+H+2]^+$. R_f: 0.33 (94:6 methylene chloride: methanol).

E/Z-(2-{4-[1-(4-Chlorophenyl)-2-(4-fluoro-3-methoxyphenyl)-propenyl]-phenoxy}-ethyl)-diethyl-amine (**28**)

$C_{28}H_{31}ClFNO_2$. Yield: 52%. Orange oil. Purity: 97.68%. ^{1}H-NMR ((400 MHz, $CDCl_3$) δ 7.31 (d, J = 7.0 Hz, 2H), 7.14 (dd, J = 16.5, 7.5 Hz, 4H), 7.02 (d, J = 7.2 Hz, 2H), 6.89 (t, J = 8.9 Hz, 4H), 6.81 (d, J = 7.5 Hz, 2H), 6.76 (d, J = 7.4 Hz, 2H), 6.73–6.65 (m, 3H), 6.65–6.57 (m, 3H), 4.17 (s, 2H), 4.04 (s, 2H), 3.63 (s, 6H), 2.97 (d, J = 30.6 Hz, 4H), 2.76 (d, J = 12.1 Hz, 8H), 2.13 (s, 3H), 2.10 (s, 3H), 1.19–1.08 (m, 12H). ^{13}C-NMR: (101 MHz, $CDCl_3$) δ 158.81, 158.37, 151.95, 151.76, 145.82 (d, J = 9.7 Hz), 145.80 (d, J = 9.6 Hz), 142.03, 141.98, 140.95, 140.69, 138.17, 135.25, 134.70, 134.56, 133.47, 132.62, 132.14, 131.92, 131.46, 131.29, 128.49, 127.91, 121.36 (d, J = 6.6 Hz), 117.43 (d, J = 13.6 Hz), 115.26 (d, J = 13.6 Hz), 115.23, 114.34, 113.88, 56.20, 53.56, 53.38, 51.73, 51.59, 47.87, 47.83, 23.38, 23.20, 11.39, 11.31. MS (ESI): *m/z* = 468.30 $[M+H]^+$ (100%), *m/z* = 470.30 $[M+H+2]^+$. R_f: 0.50 (93:7 methylene chloride: methanol).

3.2. *Biology*

3.2.1. Yeast Estrogen Receptor Assay (YES)

The yeast estrogen receptor assay was supplied by Dr. J.P. Sumpter (Brunel University, Uxbridge, UK) and was used to determine the relative transactivation activity of the human ERα as formerly described [15]. Briefly, *Saccharomyces cerevisiae* stably transfected with a human ERα and an estrogen-responsive element fused to the reporter gene *lacZ* encoding for β-galactosidase were treated with the test substances for about 48 h. The β-galactosidase enzymatic activity was measured in a colorimetric assay using a microplate photometer by hydrolysis of the substrate chlorophenol red β-D-galactopyranoside (Roche Diagnostics, Mannheim, Germany), which leads to the formation of chlorophenol red. This can be measured as an increased absorption at 540 nm. All compounds were diluted in DMSO. 17β-estradiol (E2) (Sigma, Deissenhofen, Germany) 10 nM was used as a positive control, and DMSO was used as vehicle control. All compounds, also TAM (TAM) (Biotrend, Cologne, Germany), and 4-hydroxy-TAM (4-OH-TAM), were screened for agonistic and anti-estrogenic activity in a concentration of 1 μM; anti-estrogenic assays were performed in combination with 0.5 nM/1 nM E2 depending on the EC_{50} value in each experimental series. All compounds were tested in technical quadruplicates and biological triplicates. Statistical analysis was performed by analysis of variance (ANOVA) and Tukey's post-hoc test with the significance level of $p < 0.05$. The relative β-galactosidase activity of all compounds is shown in Tables 2–4.

3.2.2. NCI Anti-Cancer Screening

All compounds were subjected to the NCI in vitro disease-oriented human cells screening panel assay. The human tumor cell lines of the cancer-screening panel are grown in RPMI 1640 medium containing 5% fetal bovine serum and 2 mM L-glutamine. For a typical screening experiment, cells are inoculated into 96-well microtiter plates in 100 μL at plating densities ranging from 5000 to 40,000 cells/well depending on the doubling time of individual cell lines. After cell inoculation, the microtiter plates are incubated at 37 °C, 5% CO_2, 95% air, and 100% relative humidity for 24 h prior to the addition of experimental drugs. After 24 h, two plates of each cell line are fixed in situ with TCA to represent a measurement of the cell population for each cell line at the time of drug addition. Experimental drugs are solubilized in dimethyl sulfoxide at 400-fold the desired final maximum test concentration and stored frozen prior to use. At the time of drug addition, an aliquot of frozen concentrate is thawed and diluted to twice the desired final maximum test concentration with complete medium containing 50 μg/mL gentamicin. Additional four, 10-fold, or 1/2 log serial dilutions are made to provide a total of five drug concentrations plus control. Aliquots of 100 μL of these different drug dilutions are added to the appropriate microtiter wells already containing 100 μL of medium, resulting in the required final drug concentrations. Following drug addition, the plates are incubated for an additional 48 h at 37 °C, 5% CO_2, 95% air, and 100% relative humidity. For adherent cells, the assay is terminated by the addition of cold TCA. Cells are fixed in situ by the gentle addition of 50 μL of cold 50% (*w/v*) TCA (final concentration, 10% TCA) and incubated for 60 min at 4 °C. The supernatant is discarded, and the plates are washed five times with tap water and air-dried. Sulforhodamine B (SRB) solution (100 μL) at 0.4% (*w/v*) in 1% acetic acid is added to each well, and plates are incubated for 10 min at room temperature. After staining, unbound dye is removed by washing five times with 1% acetic acid, and the plates are air-dried. Bound stain is subsequently solubilized with a 10 mM trizma base, and the absorbance is read on an automated plate reader at a wavelength of 515 nm. For suspension cells, the methodology is the same except that the assay is terminated by fixing settled cells at the bottom of the wells by gently adding 50 μL of 80% TCA (final concentration, 16% TCA). Compounds are screened at a dose of 10 μM, hits showing mean growth inhibition over 60 cell lines >50% are escalated for 5-dose screening assay. To construct a dose-response curve, about 60 cell lines of nine tumor subpanels were incubated with five concentrations (0.01–100 μM) for each compound. Three response parameters (GI_{50}, TGI, and LC_{50}) were calculated for each cell line. The GI_{50} value corresponds to the compound's concentration causing a 50% decrease in net cell growth, the TGI value is the compound's concentration resulting in total growth inhibition, and the LC_{50} value is the compound's concentration causing a net 50% loss of initial cells at the end of the incubation period (48 h) [41].

3.2.3. Alkaline Phosphatase Activity in Ishikawa Cells

Estrogens stimulate the activity of alkaline phosphatase (AlkP) in Ishikawa cells (human endometrial adenocarcinoma cells; kindly provided by Prof. Masato Nishida, National Hospital Organization, Kasumigaura Medical Center, Japan). This enzyme activity is estimated by using the chromogen substrate (4-nitrophenylphosphate). These cells are very sensitive to estrogens; estradiol already induces the AlkP activity at a concentration of 10^{-12} M [20]. The procedure was modified by Littlefield et al., 1990 [42].

Briefly, cells were cultured in DMEM/F12 medium without phenol red containing 5% dextran-coated charcoal-treated FCS (DCC, BioWest, (Nuaille, France) and insulin–transferrin–selenium A (Invitrogen, Karlsruhe, Germany). Cells were kept in plastic culture flasks at 5% CO_2 and 37 °C and harvested by brief exposure to trypsin (0.05%) EDTA at 37 °C. For experiments, the cells were seeded in 96-well plates at the required density of 11,000 cells per well. Compounds, diluted in DMSO (Carl Roth GmbH, Germany), were tested in a concentration of 1 μM. DMSO was used as a negative control, 10 nM 17β-estradiol as a positive control, respectively. After 72 h incubation, cells were harvested, washed twice with PBS, and incubated at −80 °C for about 30 min to lyse the cells. After thawing, the lysates were resuspended in reaction buffer (274 mM mannitol, 100 mM CAPS, 4 mM $MgCl_2$, pH 10.4) containing 4 mM p-nitrophenylphosphate (NPP). After incubation for 1 h in the dark, AlkP activity was assayed by using the hydrolysis of p-nitrophenylphosphate to p-nitrophenol at pH 10.4 and the spectrometric determination of the kinetic of the product formation at 405 nm. All compounds were tested in technical triplicates and biological triplicates. Statistical analysis was performed by analysis of variance (ANOVA) and Tukey's post-hoc test with the significance level of $p < 0.05$.

3.2.4. Uterotrophic Assay

The most common short-term in vivo assay for (anti)-estrogenicity is the uterine growth test, suitable for screening ERα agonists and antagonists. The primary endpoint is the uterine wet weight (UWW). An increase in UWW indicates an estrogenic activity of the test compound [43]. Sprague Dawley female rats (170–200 g) were obtained from the animal colony of the National Institute of Research (Cairo, Egypt). The rats were housed in a temperature-controlled room (23–24 °C) with a 12 h light:dark cycle and with free access to food and water. They were allowed to acclimatize to the animal house of the German University in Cairo for at least 1 week before initiating the experiments. All efforts were made to minimize animal discomfort and suffering. Animals were ovariectomized. After 14 days of endogenous hormonal decline, the animals were subcutaneously treated for three days with respective compounds. The animals were randomly allocated to treatment and vehicle groups (n = 6). 17β-estradiol were administrated s.c. at a dose of 10 μg/kg/d BW, all test compounds at a dose of 10 mg/kg/d BW daily for a period of three days. Animals were sacrificed by CO_2 inhalation after light anesthesia by inhaling an O_2/CO_2 mixture around 24 h after the third administration. The uterus wet weight was determined.

3.3. *In Silico Study*

A docking experiment was implemented to dock compounds **3** into the active site of estrogen receptor α (ERα) with the program MOE version 2009.10. The Protein Data Bank (PDB) crystal structure of ERα co-crystallized with DES (3ERD) was imported into MOE [40]. All possible hydrogen atoms were added. Atomic charges were assigned using the MMFF94 force field parameters in MOE. The binding pocket was selected and extended 4.5 Å around the pocket. Compound **3E, 3Z,** and DES were built using MOE builder; we run a conformational search to build a database (.mdb) of the most stable conformers of the three compounds. The.mdb file was then docked into the pocket, the poses from the ligand conformation were generated using alpha triangle, the scoring function used was London dG with no refinement. To ensure more accurate docking procedures, DES were redocked to the binding pocket using the same MOE settings as compound **3**. MOE was also used to represent the 2D interactions within ERα LBD.

4. Conclusions

Structural modifications on rings **A**, **B,** and **C** of TAM led to compounds with moderate to high estrogenic activity and potential growth inhibition activity on ER-positive and -negative breast cancer cell lines. Compounds **12** and **19** were tested in vivo in an ovariectomized rat model and are promising candidates for the development of novel SERMs with potent anti-neoplastic activity. This work opens the horizon for further development of triphenylethylenes where the para position of ring **C** bears different substituents; such structural modification can alter both their estrogenic/anti-estrogenic properties and have a prominent effect on their metabolic fate.

Supplementary Materials: The following are available online at https://www.mdpi.com/article/10.3390/ijms222212575/s1.

Author Contributions: Conceptualization, G.V., A.H.A. and N.S.A.; Methodology, H.E.E., M.M.R., N.O.A., M.A.V., O.Z. and N.S.A.; Supervision, J.J.W., K.S., J.W. and N.S.A.; Writing—original draft, J.W. and N.S.A.; Writing—review and editing, J.W., G.V., A.H.A. and N.S.A.; N.S.A. is the PI of the projects that partially financed the work, STDF grants #30298 and STDF grant #5386. N.S.A. is the coordinator of the ERA+ staff mobility grant. All authors have read and agreed to the published version of the manuscript.

Funding: This paper is based upon work supported financially by the Science and Technology Development Funding authority (STDF) under grant no.: 5386 and grant no.: 30298 to Nermin S. Ahmed.

Institutional Review Board Statement: The study was conducted according to the guidelines of the Declaration of Helsinki and approved by the Ethics Committee of German University in Cairo (March 2019). Animal procedures were performed following the approval of the Ethics Committee of the German University in Cairo in association with the recommendations of the National Institutes of Health Guide for Care and Use of Laboratory Animals (publication no. 85-23, revised 1985).

Informed Consent Statement: Not applicable.

Data Availability Statement: Not applicable.

Acknowledgments: The authors are grateful to Susanne Broschk for her technical assistance in the biological screening. The authors are deeply grateful to the authority of the National Cancer Institute, USA, for the antitumor screening. Nouran O. Elbably thank the DAAD (German Academic Exchange Service) for an Erasmus+ stipend (code: **999897729**).

Conflicts of Interest: The authors declared that they have no conflict of interest.

References

1. Riggs, B.L.; Hartmann, L.C. Selective estrogen-receptor modulators–mechanisms of action and application to clinical practice. *N. Engl. J. Med.* **2003**, *348*, 618–629. [CrossRef]
2. Valéra, M.-C.; Fontaine, C.; Dupuis, M.; Noirrit-Esclassan, E.; Vinel, A.; Guillaume, M.; Gourdy, P.; Lenfant, F.; Arnal, J.-F. Towards optimization of estrogen receptor modulation in medicine. *Pharmacol. Ther.* **2018**, *189*, 123–129. [CrossRef] [PubMed]
3. Arnal, J.F.; Lenfant, F.; Metivier, R.; Flouriot, G.; Henrion, D.; Adlanmerini, M.; Katzenellenbogen, J. Membrane and Nuclear Estrogen Receptor Alpha Actions: From Tissue Specificity to Medical Implications. *Physiol. Rev.* **2017**, *97*, 1045–1087. [CrossRef] [PubMed]
4. Maximov, P.Y.; McDaniel, R.E.; Jordan, V.C. (Eds.) *Tamoxifen Goes Forward Alone BT-Tamoxifen: Pioneering Medicine in Breast Cancer*; Springer: Basel, Switzerland, 2013; pp. 31–46.
5. Crewe, H.K.; Notley, L.M.; Wunsch, R.M.; Lennard, M.S.; Gillam, E.M. Metabolism of tamoxifen by recombinant human cytochrome P450 enzymes: Formation of the 4-hydroxy, 4′-hydroxy and N-desmethyl metabolites and isomerization of trans-4-hydroxytamoxifen. *Drug Metab. Dispos.* **2002**, *30*, 869–874. [CrossRef] [PubMed]
6. Goetz, M.P.; Rae, J.M.; Suman, V.J.; Safgren, S.L.; Ames, M.M.; Visscher, D.W.; Reynolds, C.; Couch, F.J.; Lingle, W.L.; Flockhart, D.A.; et al. Pharmacogenetics of Tamoxifen Biotransformation Is Associated With Clinical Outcomes of Efficacy and Hot Flashes. *J. Clin. Oncol.* **2005**, *23*, 9312–9318. [CrossRef]
7. Brauch, H.; Schroth, W.; Goetz, M.P.; Mürdter, T.E.; Winter, S.; Ingle, J.N.; Schwab, M.; Eichelbaum, M. Tamoxifen Use in Postmenopausal Breast Cancer: CYP2D6 Matters. *J. Clin. Oncol.* **2013**, *31*, 176–180. [CrossRef] [PubMed]
8. Schroth, W.; Goetz, M.P.; Hamann, U.; Fasching, P.A.; Schmidt, M.; Winter, S.; Fritz, P.; Simon, W.; Suman, V.J.; Ames, M.M.; et al. Association Between CYP2D6 Polymorphisms and Outcomes among Women with Early Stage Breast Cancer Treated with Tamoxifen. *JAMA* **2009**, *302*, 1429–1436. [CrossRef] [PubMed]
9. Cho, S.-H.; Jeon, J.; Kim, S.I. Personalized Medicine in Breast Cancer: A Systematic Review. *J. Breast Cancer* **2012**, *15*, 265–272. [CrossRef]
10. Dehal, S.S.; Kupfer, D. CYP2D6 catalyzes tamoxifen 4-hydroxylation in human liver. *Cancer Res.* **1997**, *57*, 3402–3406.
11. De Souza, J.A.; Olopade, O.I. CYP2D6 genotyping and tamoxifen: An unfinished story in the quest for personalized medicine. *Semin. Oncol.* **2011**, *38*, 263–273. [CrossRef]
12. Gillis, E.P.; Eastman, K.J.; Hill, M.D.; Donnelly, D.; Meanwell, N. Applications of Fluorine in Medicinal Chemistry. *J. Med. Chem.* **2015**, *58*, 8315–8359. [CrossRef] [PubMed]
13. McCague, R.; Leclercq, G.; Legros, N.; Goodman, J.; Blackburn, G.M.; Jarman, M.; Foster, A.B. Derivatives of tamoxifen. Dependence of antiestrogenicity on the 4-substituent. *J. Med. Chem.* **1989**, *32*, 2527–2533. [CrossRef] [PubMed]

14. Knox, A.; Kalchschmid, C.; Schuster, D.; Gaggia, F.; Manzl, C.; Baecker, D.; Gust, R. Development of bivalent triarylalkene- and cyclofenil-derived dual estrogen receptor antagonists and downregulators. *Eur. J. Med. Chem.* **2020**, *192*, 112191. [CrossRef] [PubMed]
15. Routledge, E.J.; Sumpter, J.P. Estrogenic activity of surfactants and some of their degradation products assessed using a recombinant yeast screen. *Environ. Toxicol. Chem.* **1996**, *15*, 241–248. [CrossRef]
16. Arnold, S.F.; Robinson, M.K.; Notides, A.C.; Guillette, L.J., Jr.; McLachlan, J.A. A yeast estrogen screen for examining the relative exposure of cells to natural and xenoestrogens. *Environ. Health Perspect.* **1996**, *104*, 544–548. [CrossRef] [PubMed]
17. Kretzschmar, G.; Zierau, O.; Wober, J.; Tischer, S.; Metz, P.; Vollmer, G. Prenylation has a compound specific effect on the estrogenicity of naringenin and genistein. *J. Steroid Biochem. Mol. Biol.* **2010**, *118*, 1–6. [CrossRef] [PubMed]
18. Nilsson, S.; Gustafsson, J. Åke Estrogen receptor transcription and transactivation Basic aspects of estrogen action. *Breast Cancer Res.* **2000**, *2*, 360–366. [CrossRef] [PubMed]
19. Nishida, M.; Kasahara, K.; Kaneko, M.; Iwasaki, H.; Hayashi, K. Establishment of a new human endometrial adenocarcinoma cell line, Ishikawa cells, containing estrogen and progesterone receptors. *Nihon Sanka Fujinka Gakkai Zasshi* **1985**, *37*, 1103–1111.
20. Holinka, C.F.; Hata, H.; Kuramoto, H.; Gurpide, E. Effects of steroid hormones and antisteroids on alkaline phosphatase activity in human endometrial cancer cells (Ishikawa line). *Cancer Res.* **1986**, *46*, 2771–2774.
21. Wober, J.; Weißwange, I.; Vollmer, G. Stimulation of alkaline phosphatase activity in Ishikawa cells induced by various phytoestrogens and synthetic estrogens. *J. Steroid Biochem. Mol. Biol.* **2002**, *83*, 227–233. [CrossRef]
22. Hata, H.; Holinka, C.F.; Pahuja, S.L.; Hochberg, R.B.; Kuramoto, H.; Gurpide, E. Estradiol metabolism in ishikawa endometrial cancer cells. *J. Steroid Biochem.* **1987**, *26*, 699–704. [CrossRef]
23. Owens, J.W.; Ashby, J. Critical review and evaluation of the uterotrophic bioassay for the identification of possible estrogen agonists and antagonists: In support of the validation of the OECD uterotrophic protocols for the laboratory rodent. Organisation for Economic Co-oper. *Crit. Rev. Toxicol.* **2002**, *32*, 445–520. [CrossRef] [PubMed]
24. O'Connor, J.C.; Cook, J.C.; Marty, M.S.; Davis, L.G.; Kaplan, A.M.; Carney, E.W. Evaluation of Tier I screening approaches for detecting endocrine-active compounds (EACs). *Crit. Rev. Toxicol.* **2002**, *32*, 521–549. [CrossRef]
25. Yoon, K.; Kwack, S.J.; Kim, H.S.; Lee, B.M. Estrogenic endocrine-disrupting chemicals: Molecular mechanisms of actions on putative human diseases. *J. Toxicol. Environ. Health. Crit. Rev.* **2014**, *17*, 127–174. [CrossRef] [PubMed]
26. Kiyama, R.; Wada-Kiyama, Y. Estrogenic endocrine disruptors: Molecular mechanisms of action. *Environ. Int.* **2015**, *83*, 11–40. [CrossRef]
27. Tyler, C.R.; Jobling, S.; Sumpter, J.P. Endocrine Disruption in Wildlife: A Critical Review of the Evidence. *Crit. Rev. Toxicol.* **1998**, *28*, 319–361. [CrossRef]
28. United States Environmental Protection Agency. *Endocrine Disruptor Screening Program Test Guidelines OPPTS 890.1600: Uterotrophic Assay*; Office of Prevention, Pesticides and Toxic Substances (OPPTS): Washington, DC, USA, 2009; pp. 1–21.
29. United States Environmental Protection Agency. *Uterotrophic Assay OCSPP Guideline 890.1600: Standard Evaluation Procedure (SEP)*; Endocrine Disruptor Screening Program, United States Environmental Protection Agency: Washington, DC, USA, 2011; pp. 1–19.
30. Zierau, O.; Kretzschmar, G.; Möller, F.; Weigt, C.; Vollmer, G. Time dependency of uterine effects of naringenin type phytoestrogens in vivo. *Mol. Cell. Endocrinol.* **2008**, *294*, 92–99. [CrossRef] [PubMed]
31. Ahmed, N.S.; Elghazawy, N.H.; ElHady, A.K.; Engel, M.; Hartmann, R.W.; Abadi, A.H. Design and synthesis of novel tamoxifen analogues that avoid CYP2D6 metabolism. *Eur. J. Med. Chem.* **2016**, *112*, 171–179. [CrossRef] [PubMed]
32. Bedford, G.R.; Richardson, D.N. Preparation and Identification of cis and trans Isomers of a Substituted Triarylethylene. *Nature* **1966**, *212*, 733–734. [CrossRef]
33. Coezy, E.; Borgna, J.L.; Rochefort, H. Tamoxifen and metabolites in MCF7 cells: Correlation between binding to estrogen receptor and inhibition of cell growth. *Cancer Res.* **1982**, *42*, 317–323. [PubMed]
34. Kraft, K.S.; Ruenitz, A.P.C.; Bartlett, M.G. Carboxylic Acid Analogues of Tamoxifen: (Z)-2-[p-(1,2-Diphenyl-1-butenyl)phenoxy]-N,N-dimethylethylamine. Estrogen Receptor Affinity and Estrogen Antagonist Effects in MCF-7 Cells. *J. Med. Chem.* **1999**, *42*, 3126–3133. [CrossRef] [PubMed]
35. Keely, N.O.; Carr, M.; Yassin, B.; Ana, G.; Lloyd, D.G.; Zisterer, D.; Meegan, M.J. Design, Synthesis and Biochemical Evaluation of Novel Selective Estrogen Receptor Ligand Conjugates Incorporating an Endoxifen-Combretastatin Hybrid Scaffold. *Biomedicines* **2016**, *4*, 15. [CrossRef] [PubMed]
36. Lv, W.; Liu, J.; Skaar, T.C.; Flockhart, D.A.; Cushman, M. Design and Synthesis of Norendoxifen Analogues with Dual Aromatase Inhibitory and Estrogen Receptor Modulatory Activities. *J. Med. Chem.* **2015**, *58*, 2623–2648. [CrossRef] [PubMed]
37. Maximov, P.Y.; McDaniel, R.E.; Fernandes, D.J.; Korostyshevskiy, V.R.; Bhatta, P.; Mürdter, T.E.; Jordan, V.C. Simulation with cells in vitro of tamoxifen treatment in premenopausal breast cancer patients with different CYP2D6 genotypes. *Br. J. Pharmacol.* **2014**, *171*, 5624–5635. [CrossRef] [PubMed]
38. Brzozowski, A.M.; Pike, A.C.W.; Dauter, Z.; Hubbard, R.E.; Bonn, T.; Engström, O.; Öhman, L.; Greene, G.L.; Gustafsson, J.A.; Carlquist, M. Molecular basis of agonism and antagonism in the oestrogen receptor. *Nature* **1997**, *389*, 753–758. [CrossRef] [PubMed]
39. Skehan, P.; Storeng, R.; Scudiero, D.; Monks, A.; McMahon, J.; Vistica, D.; Warren, J.T.; Bokesch, H.; Kenney, S.; Boyd, M.R. New Colorimetric Cytotoxicity Assay for Anticancer-Drug Screening. *JNCI J. Natl. Cancer Inst.* **1990**, *82*, 1107–1112. [CrossRef] [PubMed]

40. Shiau, A.K.; Barstad, D.; Loria, P.M.; Cheng, L.; Kushner, P.J.; Agard, D.A.; Greene, G.L. The Structural Basis of Estrogen Receptor/Coactivator Recognition and the Antagonism of This Interaction by Tamoxifen. *Cell* **1998**, *95*, 927–937. [CrossRef]
41. Shoemaker, R.H. The NCI60 human tumour cell line anticancer drug screen. *Nat. Rev. Cancer* **2006**, *6*, 813–823. [CrossRef]
42. Littlefield, B.A.; Gurpide, E.; Markiewicz, L.; McKinley, B.; Hochberg, R.B. A simple and sensitive microtiter plate estrogen bioassay based on stimulation of alkaline phosphatase in Ishikawa cells: Estrogenic action of delta 5 adrenal steroids. *Endocrinology* **1990**, *127*, 2757–2762. [CrossRef]
43. Saarinen, N.M.; Bingham, C.; Lorenzetti, S.; Mortensen, A.; Mäkelä, S.; Penttinen, P.; Zierau, O. Tools to evaluate estrogenic potency of dietary phytoestrogens: A consensus paper from the EU Thematic Network 'Phytohealth' (QLKI-2002-2453). *Genes Nutr.* **2006**, *1*, 143–158. [CrossRef]

International Journal of
Molecular Sciences

Article

Molecular Iodine/Cyclophosphamide Synergism on Chemoresistant Neuroblastoma Models

Winniberg Álvarez-León †, Irasema Mendieta †, Evangelina Delgado-González, Brenda Anguiano and Carmen Aceves *

Instituto de Neurobiología, Universidad Nacional Autónoma de México (UNAM) Juriquilla, Querétaro 76230, Mexico; winnsteph@gmail.com (W.Á.-L.); aliciairasema@hotmail.com (I.M.); edelgado@comunidad.unam.mx (E.D.-G.); anguianoo@unam.mx (B.A.)
* Correspondence: caracev@unam.mx; Tel.: +52-442-238-1067
† W.Á.-L. and I.M. worked together and contributed equally to this paper.

Abstract: Neuroblastoma (Nb), the most common extracranial tumor in children, exhibited remarkable phenotypic diversity and heterogeneous clinical behavior. Tumors with MYCN overexpression have a worse prognosis. MYCN promotes tumor progression by inducing cell proliferation, de-differentiation, and dysregulated mitochondrial metabolism. Cyclophosphamide (CFF) at minimum effective oral doses (metronomic therapy) exerts beneficial actions on chemoresistant cancers. Molecular iodine (I_2) in coadministration with all-trans retinoic acid synergizes apoptosis and cell differentiation in Nb cells. This work analyzes the impact of I_2 and CFF on the viability (culture) and tumor progression (xenografts) of Nb chemoresistant SK-N-BE(2) cells. Results showed that both molecules induce dose-response antiproliferative effects, and I_2 increases the sensibility of Nb cells to CFF, triggering PPARγ expression and acting as a mitocan in mitochondrial metabolism. In vivo oral I_2/metronomic CFF treatments showed significant inhibition in xenograft growth, decreasing proliferation (Survivin) and activating apoptosis signaling (P53, Bax/Bcl-2). In addition, I_2 decreased the expression of master markers of malignancy (MYCN, TrkB), vasculature remodeling, and increased differentiation signaling (PPARγ and TrkA). Furthermore, I_2 supplementation prevented loss of body weight and hemorrhagic cystitis secondary to CFF in nude mice. These results allow us to propose the I_2 supplement in metronomic CFF treatments to increase the effectiveness of chemotherapy and reduce side effects.

Keywords: neuroblastoma; molecular iodine; cyclophosphamide; xenografts; metronomic therapy

Citation: Álvarez-León, W.; Mendieta, I.; Delgado-González, E.; Anguiano, B.; Aceves, C. Molecular Iodine/Cyclophosphamide Synergism on Chemoresistant Neuroblastoma Models. *Int. J. Mol. Sci.* **2021**, *22*, 8936. https://doi.org/10.3390/ijms22168936

Academic Editor: Angela Stefanachi

Received: 13 July 2021
Accepted: 16 August 2021
Published: 19 August 2021

Publisher's Note: MDPI stays neutral with regard to jurisdictional claims in published maps and institutional affiliations.

1. Introduction

Neuroblastoma (Nb) is the most common extracranial tumor in children accounting for 15% of pediatric oncology deaths. The overexpression of the neural MYC gene (MYCN) characterizes the chemoresistant Nb and has a worse prognosis [1]. MYCN induces cell proliferation, inhibits cell differentiation, and maintains the stem-like phenotype [2]. These levels correlate to metastasis and angiogenesis, and MYCN overexpression affects the mitochondria metabolism to support the higher energy demand of chemoresistant Nb cells [1,2]. One of the most widely used drugs in this pathology is cyclophosphamide (CFF), an effective and low-cost chemotherapy. CFF is a prodrug, biotransformed into two metabolites: phosphoramide mustard (amino-[bis(2-chloroethyl)amino] phosphinic acid), which is the active antineoplastic principle, and acrolein (prop-2-enal), associated with several side effects including inflammation and hemorrhagic cystitis [3]. The use of CFF in metronomic therapy has recently been proposed as an effective alternative for chemoresistant cancers [4]. Metronomic therapy uses chronic oral treatments with minimum effective doses that exert their effects by inhibiting angiogenesis, immune modulation of the tumor stroma, and apoptosis of tumor cells. Its long-term and low-dose use reduces side effects and maintains the patient's quality of life [4]. Furthermore, the use of combined therapies that include cellular re-differentiation or immune reactivation messengers are promising approaches that are currently being tested [5,6].

In addition, the antineoplastic effects of molecular iodine (I_2) are well established, triggering apoptotic and redifferentiation mechanisms in several cancer cells, including mammary, ovary, and prostate, among others [7]. These effects are mediated partially by the activation of peroxisome proliferator-activated receptors type gamma (PPARγ) and directly, as a mitocan element, by thiol depletion and disruption of the mitochondrial membrane potential (Mmp), triggering the intrinsic apoptosis [8]. Furthermore, our previous report showed that the I_2 supplement sensitized Nb cells to all-trans retinoic acid (ATRA) in vitro and synergized the antitumor effect of ATRA preventing body-weight loss and diarrhea episodes in nude mice [9]. The present work analyzes the impact of I_2 and metronomic CFF supplements on the viability (culture) and tumor progression (xenografts) of chemoresistant neuroblastoma SK-N-BE(2) cells [10].

2. Results

2.1. Results In Vitro

2.1.1. Viability

Figure 1A shows the effect of molecular iodine (I_2), CFF, and their combination on the viability of the SH-SY5Y and SK-N-BE(2) cell lines at 96 h. Both components generate similar dose-response action, being SH-SY5Y the most sensitive and corroborating that SK-N-BE(2) is a highly resistant cell line. I_2 supplementation showed an IC50 of 205.5 μM in SH-SY5Y cells and 343.7 μM in SK-N-BE(2) cells (Figure S1). In CFF, the IC50 was 0.602 μM in SH-SY5Y and 1.045 μM in SK-N-BE(2) (Figure S2). Specific time-response data are summarized in Figures S1 and S2. All further experiments were analyzed in SK-N-BE(2) cells. The combination of 200 μM I_2 with two concentrations of CFF confirmed that the presence of iodine increases the sensibility of these cells to CFF, improving response by 15% at 1.0 μM CFF and up to 40% at 0.5 μM CFF (Figure 1B). The combination index (CI) for I_2/CFF treatments presented a CI value of 0.00835 at 0.5 μM CFF and 3.39 at 1 μM CFF, indicating synergism at 0.5 μM CFF. To analyze the participation of PPARγ in the I_2 response, we used the agonist rosiglitazone (RGZ 5.0 μM) in the presence or absence of the antagonist GW9662 (1.0 μM) for 96 h. RGZ showed a similar inhibitory effect in viability observed with I_2 (34 vs. 28%). The preincubation (2 h before) with GW9662 canceled the RGZ inhibitor effect but had a partial impact on the I_2 supplement (24%), suggesting that I_2 exerts its actions through other mechanisms besides PPARγ (Figure 1C).

2.1.2. Apoptosis

Figure 2 shows the percentage of apoptosis-positive cells (annexin Cy5) at two different times. At 48 h, 200 μM I_2 and 1.0 μM CFF exhibited similar apoptotic induction (~35% each), and the adjuvant action (I_2 + CFF group) enhanced the apoptosis effect to 40%. At 96 h, apoptosis was maintained primarily in I_2 groups. The apoptotic index Bax/Bcl-2 (RT-PCR), an indicator of caspase pathway activation, indicated that both components exert significant induction, showing a predominant action of I_2 at both times.

2.1.3. Mitochondrial Activity

Figure 3 depicts the fluorescence generated by the MitoTracker entrance after I_2 and CFF treatments. Iodine exerted a rapid and significant increment in mitochondrial permeability at 12 h and maintained until 48 h. In contrast, CFF induced a change in mitochondria metabolism after 48 h. The presence of both components exhibited a synergistic effect only at 48 h.

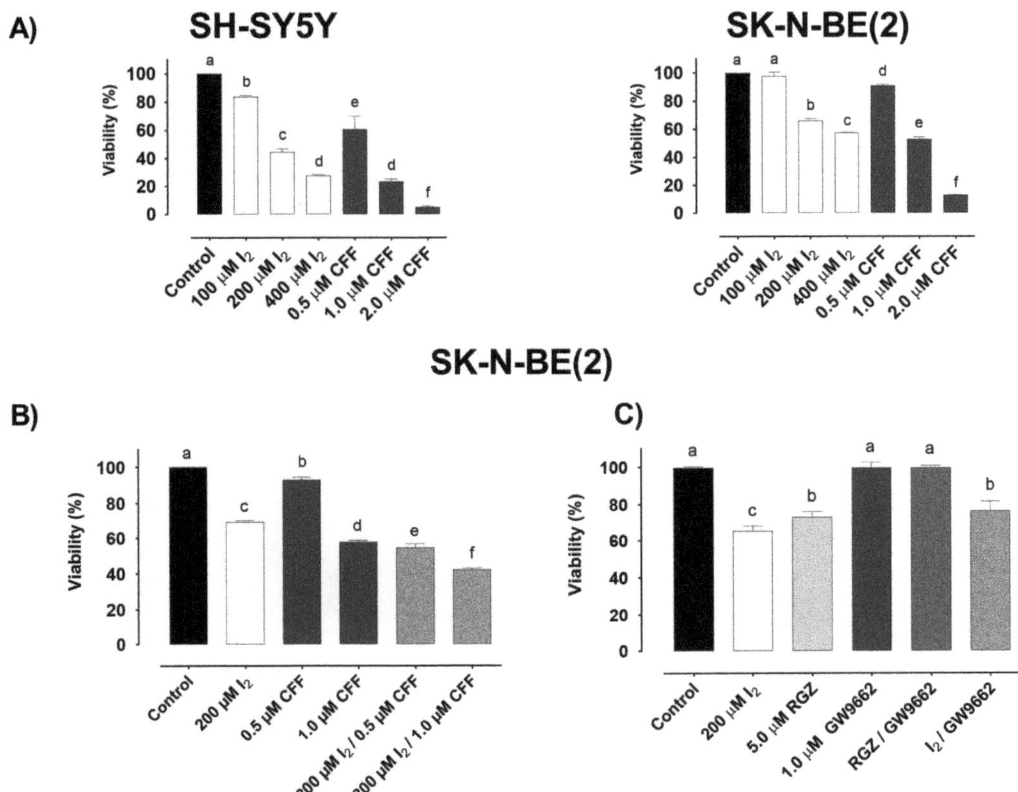

Figure 1. Effect of I_2, CFF and their combination on the viability (exclusion dye trypan blue) of neuroblastoma cell lines. (**A**) Dose response of I_2 and CFF in SH-SY5Y and SK-N-BE(2) cells. (**B**) Synergistic effect (CI < 1.0) of 200 µM I_2 with 0.5 µM CFF (reduction 45%) and 1.0 µM CFF (reduction 25%) concentrations in SK-N-BE(2) cells. (**C**) Participation of PPARγ in the effect of I_2 analyzed with rosiglitazone (RGZ; PPARγ agonist) and GW9662 (PPARγ antagonist) in SK-N-BE(2) cells. All experiments were carried out for 96 h. Data are representative of three independent experiments per triplicate and are expressed as the mean ± SD. Different letters denote statistical differences per group ($p < 0.05$).

2.1.4. Molecular Response

Figure 4 shows the effect of treatments on gene expression. I_2 supplements increased the genes associated with differentiation (PPARγ) and diminished those associated with aggressiveness (MYCN) and resistance (Survivin; SVV). The presence of CFF only reduced SVV expression. In the I_2 + CFF group, only the effects of I_2 remained, suggesting that I_2 is the inductor of differentiation. The multidrug resistance gene MDR1 did not exhibit differences with any treatment.

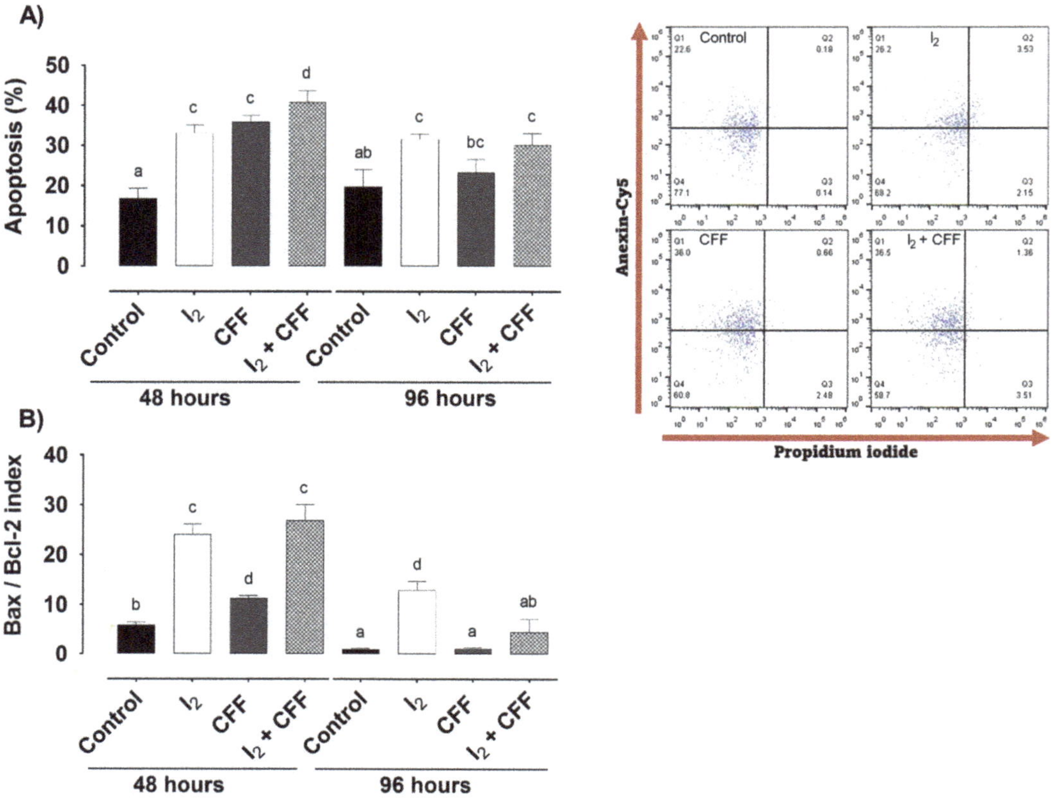

Figure 2. Apoptotic induction by I_2 and CFF supplementation in the neuroblastoma SK-N-BE(2) cells. The cell line was supplemented with 200 μM I_2, 1 μM CFF, and I_2/CFF. (**A**) Apoptotic-positive cell percentage was evaluated with the Attune flow cytometer at 48 and 96 h; representative dot plots at 48 h and quantitative results are shown; (**B**) the apoptotic Bax/Bcl-2 index was assessed using real-time PCR at 48 and 96 h. Figures are representative of three independent experiments per triplicate. Data are expressed as the mean ± SD. Different letters denote statistical differences ($p < 0.05$).

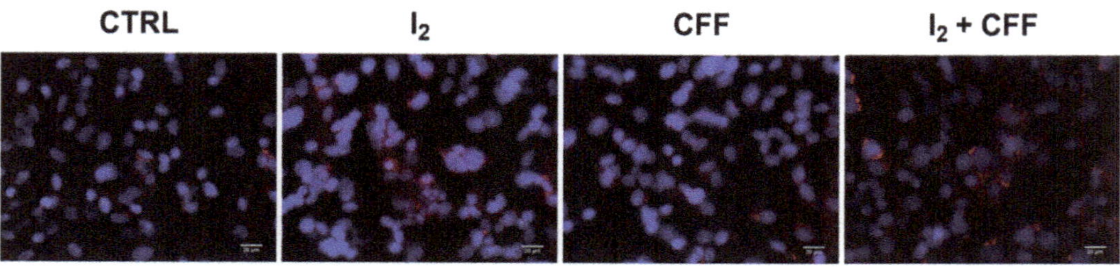

Figure 3. *Cont.*

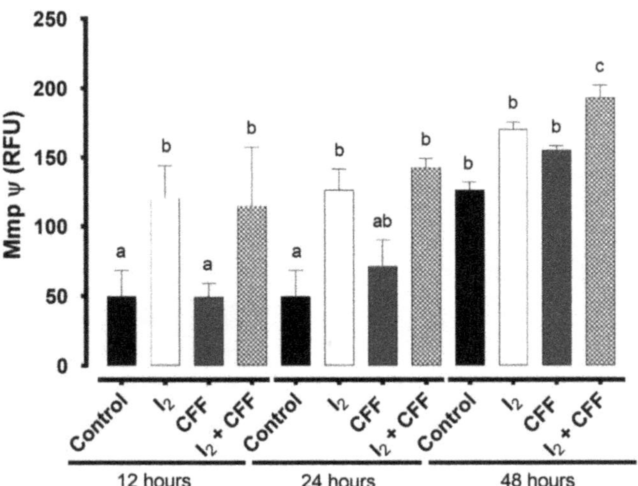

Figure 3. Effect of I_2 and CFF on the mitochondrial functional state in SK-N-BE(2) cells. The cell line was supplemented with 200 μM I_2, 1 μM CFF, and I_2 + CFF, for 12, 24, and 48 h to evaluate the mitochondrial permeability (20 μM, scale bar). Representative micrographs (48 h) and bar graph for all times of the MitoTracker signal (Mmp ψ: mitochondria membrane potential change; RFU: Relative Fluorescence Units). Figures are representative of three independent experiments per triplicate. Data are expressed as the mean ± SD. Different letters denote statistical differences ($p < 0.05$).

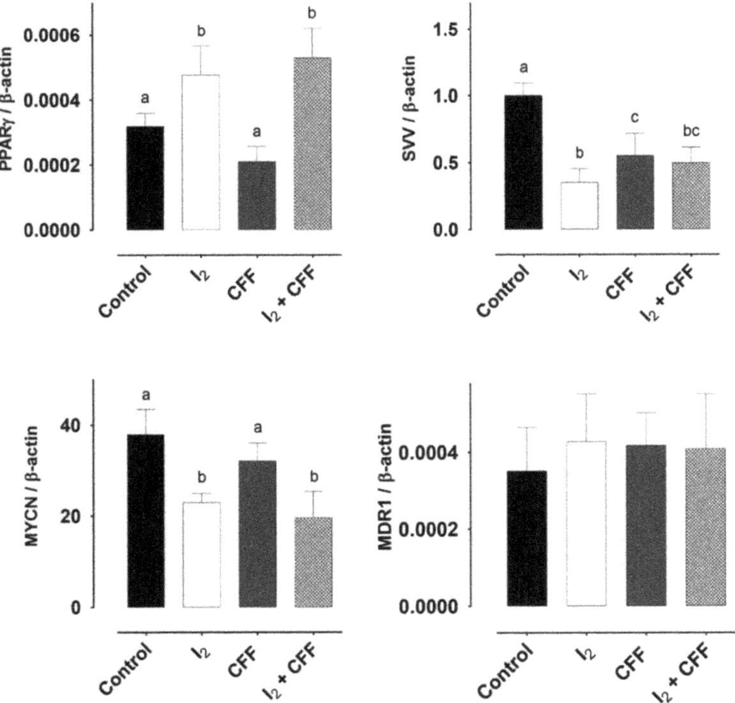

Figure 4. Effect of I_2 and CFF on gene expression in the neuroblastoma SK-N-BE(2) cell line. I_2 (200 μM), CFF (1 μM), and both (I_2 + CFF) were supplemented for 96 h. Aggressive [neural MYC (MYCN)] and

chemoresistant markers (Survivin (SVV), and multidrug resistance mutation 1 gene (MDR1)) were analyzed by RT-qPCR. PPARγ expression and protein content was analyzed by RT-qPCR (PPARγ/β-actin) and Western blot (PPARγ/Actin). Figures are representative of three independent experiments per triplicate. Data are expressed as the mean ± SD. Different letters denote statistical differences ($p < 0.05$).

2.2. Results In Vivo

2.2.1. Tumor Growth

Nude mice with SK-N-BE(2) xenografts were used to analyze the effects of I_2 and CFF under in vivo conditions. Treatments were given when the tumor reached a size of 1.5 cm^3. We evaluated the impact of oral metronomic CFF doses (20 mg/kg/day; 0.06%) with or without the I_2 supplement (8 mg/kg/day; 0.025%) for three weeks. Previous studies from our laboratory found that the presence of xenografts generates a small but consistent reduction in body weight gain (BWG) in all animals and that the I_2 supplement prevents this loss. Figure 5 shows that I_2-supplemented animals exhibited an increase in BWG like that of the control group of xenograft-free (x.f.) animals. The animals with tumors, both the control and the CFF groups, showed a reduction in BWG from the first week. In contrast, the combined group (I_2 + CFF) recovered BWG in the last week of treatment, indicating a beneficial effect of I_2.

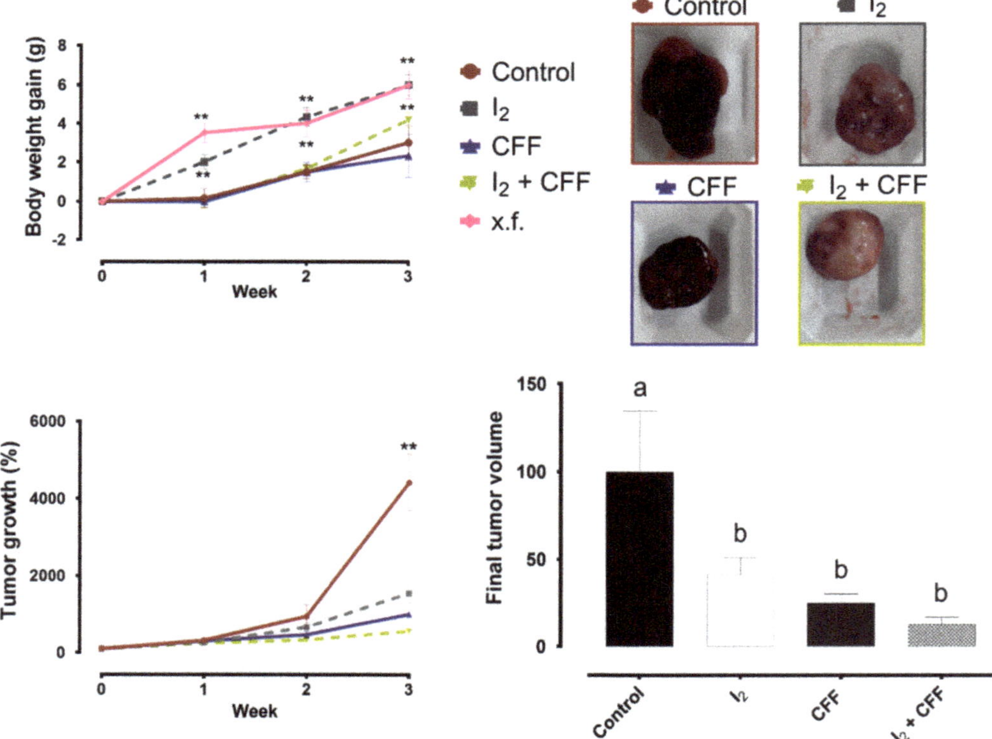

Figure 5. Effect of I_2 and CFF on nude mice and xenografts. Nude mice with SK-N-BE(2) xenografts were supplemented with I_2 (0.025%) and metronomic CFF (0.060%) in drinking water for three weeks. The line graphs show the body weight gain, % tumor growth and final tumor volume compared to the control group. The pictures are representative of the tumor bleeding appearance. Data are expressed as the mean ± SD ($n = 4$). Different letters and ** denote statistical differences for the control group ($p < 0.01$).

Concerning tumor progression, after the third week, all treatments significantly inhibited tumor growth. The tumor volume decreased 44.72% with the I_2 supplement and 68.11% with CFF. The coadministration of both components (I_2 + CFF) showed a considerable, yet not statistically significant, tumor inhibition, decreasing the tumor size by 78.78%.

A characteristic of these types of tumors is their aberrant vascularization and abundant bleeding, and so they are known as blue tumors. As shown in Figure 5, I_2 supplementation considerably decreased bleeding patterns regardless of tumor size. However, the CFF group alone showed a decrease in tumor size but did not change bleeding appearance.

2.2.2. Histopathology

Figure 6 shows the analysis of vascularization (H&E stain and CD34 immunohistology and quantification) and collagen fibrosis content (Masson's stain and quantification). Control and CFF xenograft micrography showed aberrant vascular patterns and abundant extravascular erythrocytes (H&E stain). In contrast, treatments with I_2 alone and combined (I_2 + CFF) showed a consistent reduction in the vasculature, minor invasion of tumor cells in vascular structures, and fewer extravascular erythrocytes. The quantification of mean vascular density (MVD) by CD34 immunohistochemistry showed that I_2 and CFF decrease vasculature area in comparison with control group. A significant increase in positive collagen fibers (blue stain) is observed in I_2 and CFF groups suggesting a hypoxic microenvironment, with tumoral cell death fibrosis substitution. Following this hypoxic status, I_2-treated tumors exhibited an elevated expression of Hypoxia-inducible factor (HIF1) but not vascular endothelial growth factor (VEGF), which corroborates that the I_2 treatments do not induce an increase in vascularization.

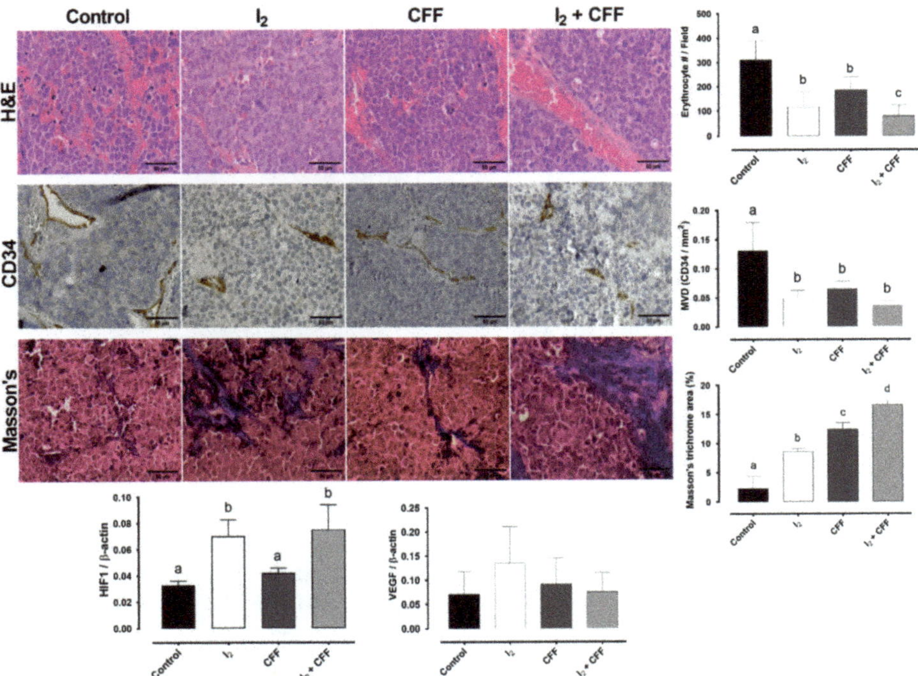

Figure 6. Effect of I_2 and CFF on the histopathology of SK-N-BE(2) xenografts. H&E stain (40×). Immunohistochemistry of endothelial protein CD34 and quantification of mean vascular density (area CD34+/field). Masson's trichrome stain and percent of positive fibrosis area. Epithelial (red) and collagen fibers (blue) (40×) (50 μM, bar graph). Hypoxia-inducible factor (HIF1) and vascular endothelial growth factor (VEGF) expression (RT-qPCR). Data are expressed as the mean ± SD (n = 4). Different letters denote statistical differences ($p < 0.05$).

2.2.3. Molecular Response

Similar to the in vitro results, xenografts from the control animal expressed an elevated amount of MYCN (Figure 7). The I_2 supplement modified the xenografts' aggressivity pattern, increasing the expression of differentiation promoters (PPARγ and TrkA) and decreasing those related to resistance (MYCN and TrkB). Interestingly, the higher TrkB expression promoted by the CFF supplement was suppressed by the presence of I_2 (I_2 + CFF group). In addition, both components (I_2 and CFF) induced the expression of p53 and increased the apoptotic index Bax/Bcl-2.

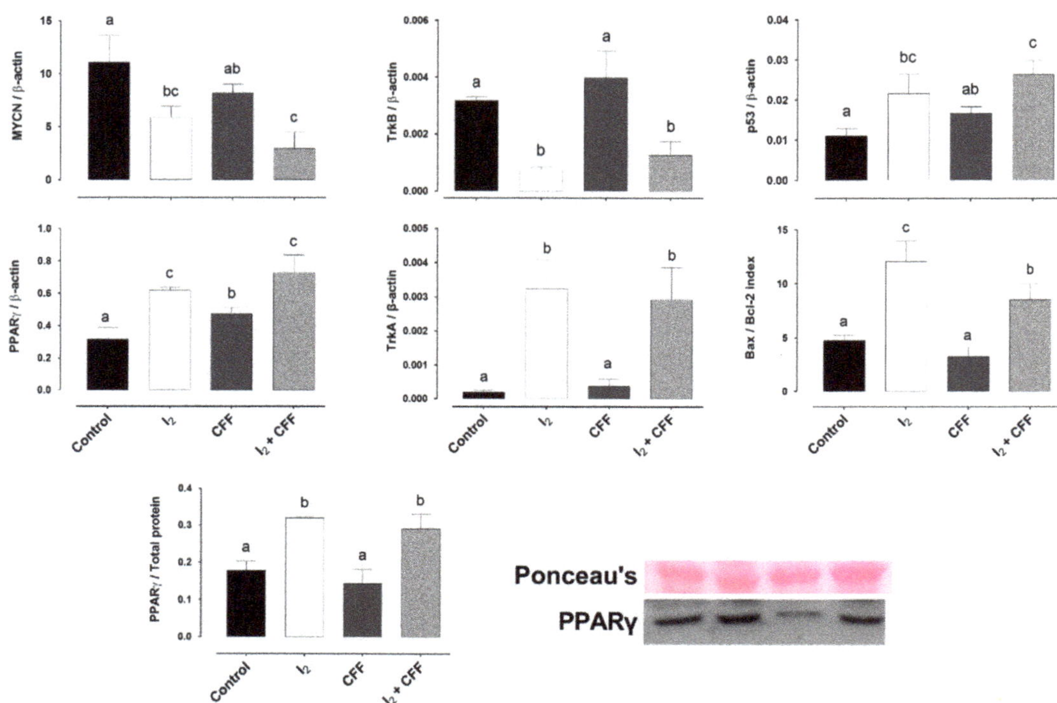

Figure 7. Effect of I_2 and CFF on gene expression in neuroblastoma xenografts. Nude mice with SK-N-BE(2) xenografts were supplemented with I_2 (0.025%) or the metronomic dose of CFF (0.060%) administered in drinking water for three weeks. Genes related to neuronal differentiation (PPARγ and TrkA), aggressiveness (MYCN and TrkB), and apoptosis induction (p53 and Bax/Bcl-2 index) were analyzed by RT-qPCR. PPARγ content was analyzed by Western blot. Data are expressed as the mean ± SD (n = 4). Different letters denote statistical differences ($p < 0.05$).

2.2.4. Preventive Effect of I_2 in Bladder Damage

The more frequent side effect of CFF treatment is hemorrhagic cystitis, evidenced by hypervascularity, edema, inflammation, and bleeding. We examined the bladder morphology at the end of the experiment and evaluated the vasculature (blood vases/field) through expression of CD34 (immunohistochemistry) and histopathology (H&E) (Figure 8). No significant differences in the vasculature were found in the treatments. However, the stains with H&E revealed clear signs of edema in the lamina propria and an increased thickness in the urothelium in the CFF group (arrows). I_2 and I_2 + CFF group bladders did not show these alterations, indicating the I_2 alone did not cause any irritation and that, in the presence of CFF, this halogen exerts a significant preventive action.

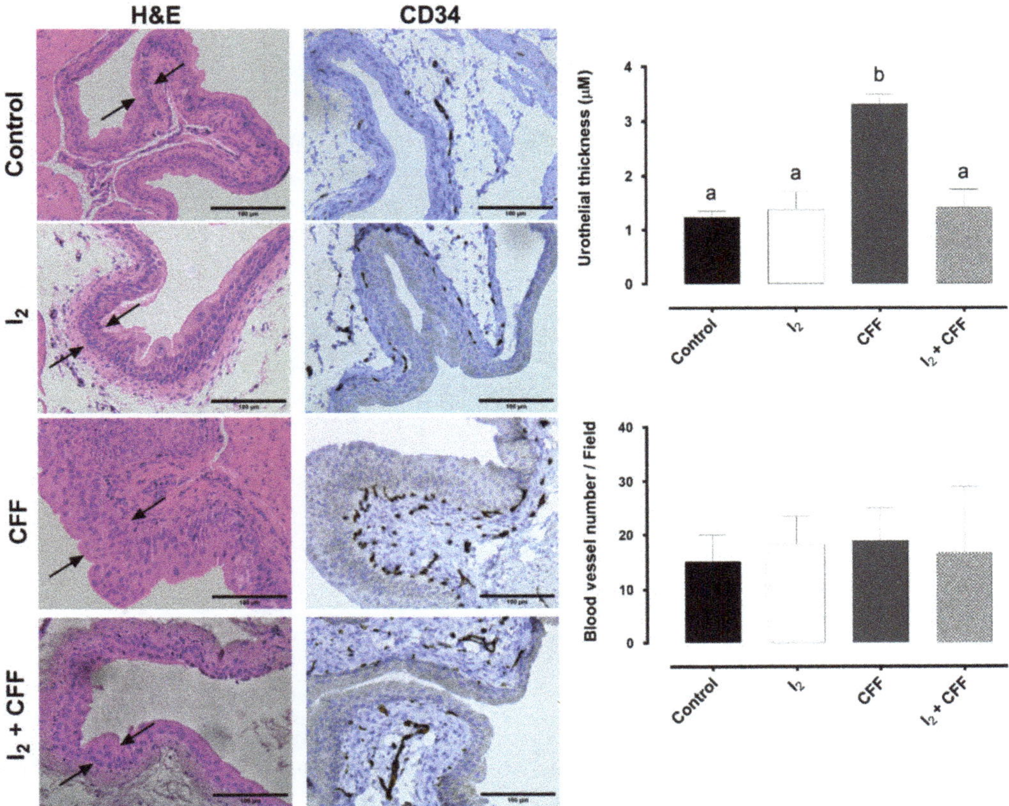

Figure 8. Effect of I_2 and CFF on the bladder morphology of the chemoresistant neuroblastoma model. The figure shows the urothelial thickness indicated by arrows in H&E representative micrographs (100 µM, bar graph) and the vasculature quantification (blood vessel number/field) of CD34 immunohistochemical staining (representative micrographs and bar graph). Three sections per tumor sample were analyzed. Data are expressed as the mean ± SD (n = 4). Different letters denote statistical differences ($p < 0.05$).

3. Discussion

Previously our group showed that I_2 in coadministration with ATRA synergizes apoptosis and cell differentiation in Nb cells [9]. The present work analyzes the impact of I_2 and CFF in viability and tumor progression of Nb chemoresistant SK-N-BE(2) cells. Our results corroborate that these cells have a high resistance to various antineoplastic components since both I_2 and CFF exert attenuated effects up to 40% compared to the more sensitive SH-SY5Y cells [10]. However, an important finding is that the I_2 supplement, which does not generate any side effects, increased the sensitivity to CFF by 25 to 40%. It is well established that the primary mechanism of CFF is the induction of p53 apoptosis via DNA adducts formation [11]. Conversely, I_2 actions are more complex since this halogen could act directly on mitochondria by inducing an apoptosis cascade [12], or indirectly by activating PPARγ and triggering redifferentiation or apoptotic signaling [7,13]. Our results showed that both components (I_2 and CFF) increased apoptosis (exposure of annexin-Cy5 and low expression of SVV), but only I_2 groups modified master differentiation genes, decreasing the expression of MYCN, and significantly inducing PPARγ. Iodine in neoplastic cells binds with arachidonic acid and generates an iodolipid called 6-iodolactone (6-IL) [14,15]. This iodolipid is a specific activator of PPARγ [16,17]. These receptors are expressed in Nb, and the use of agonists impairs proliferation and induces differentiation of these cells [9,18]. PPARγ agonists reduce levels of MYCN by inhibiting critical molecules within the PI3K/AKT/mTOR signaling pathway increasing GSK-3β activity as well as MYCN phosphorylation and its proteasome degradation [19–21]. We corroborated that PPARγ reduces the viability of these SK-N-BE(2) Nb cells since the supplement with 1 uM RGZ decreased its proliferation. Moreover, we found that the I_2 effect was partially canceled with the agonist GW9662, suggesting that the antineoplastic effects of I_2 include the activation of PPARγ, but other mechanisms also contribute.

Mitochondria metabolism is considered a hallmark of cancer, showing a crucial contributor in the process as metabolic reprogramming, generation of reactive oxygen species (ROS) and production of metabolites that enhance oncogenesis [22]. Recently, several therapeutic approaches for these processes identified the "mitocans," a category of drug that targets the mitochondria of cancer cells [8]. Many natural agents can target mitochondria and exert anticancer activities with minimal or no side effects. In the light of this process, I_2 seems to be a mitocan since, in cancer cells, I_2 depleted thiol generation and disrupted the Mmp, inducing significant increases in its permeability and triggering apoptosis [12,23]. The substantial rise in MitoTracker signaling observed in I_2 groups starting at 12 h corroborated these direct mitochondrial effects and explains, in part, the increase in CFF sensibility. Previous data from our laboratory showed that this increase in Mmp was accompanied by a decrease in SVV content in both intramitochondrial and cytoplasmic compartments [9]. SVV is an apoptosis-inhibiting factor (IAP) [24]. SVV overexpression in Nb cells makes them resistant to ATRA, protecting them against agents that damage DNA, and stabilizes the mitochondrial membrane by decreasing apoptosis induction [20,25].

The xenograft model was efficient and reproducible since we obtained between 95 and 100% implantation. Moreover, we corroborated that this is an aggressive Nb type generating fast-growing tumors and hypervascularization characteristics (bluish). The results showed that both the metronomic CFF and the I_2 supplement effectively decreased tumor growth (68.11% and 44.72%, respectively), especially when both components were co-administered (78.78%). In addition, the I_2 supplement was accompanied by decreased aberrant vascularization and bleeding, associated with a significant increase in the expression of HIF1, which is the first signaling secondary to lack of oxygen. No change in the expression of VEGF, the inducer of new vessels, was observed. This pattern can be interpreted as a modification in the vascular cytoarchitecture rather than a process of angiogenesis, but the possible mechanisms involved in this I_2 effect have not yet been elucidated. However, conventional mechanisms of I_2 could participate. Angiogenesis is an active process that involves a significant increase in ROS generation, promoting the angiogenic switch from quiescent to active endothelial cells [25,26]. It is possible that the mitocan effect of iodine neutralized ROS [8]. An alternative way might be the significant decrease in MYCN expression observed with I_2 treatments. MYCN induces angiogenesis by its direct action in VEGF amplification [2].

At molecular level, the decrease in tumor size was accompanied by an increase in the apoptosis markers p53 and Bax/Bcl-2 index in all treated groups. This apoptotic induction also agrees with the finding that the tumors supplemented with both components had a higher proportion of type 1A collagen fibers (Masson's trichrome staining), indicating the replacement of epithelial cancer cells with fibrous tissue in response to rapid induction of apoptosis.

In addition, one important finding is the differential gene response between tumors treated with I_2 vs. CFF. Our results agreed with the in vitro results, that only I_2 supplements exert an evident modulation in the differentiation master genes increasing PPARγ and TrkA, decreasing the basal expression of MYCN and TrkB, and canceling the increase of TrkB secondary to CFF treatment. MYCN amplification in Nb is typically associated with epigenetic abnormalities to impair apoptosis, followed by the overexpression of the anti-apoptotic proteins Bcl-2, SVV, and TrkB [1,19,20]. TrkB stimulates cell survival and angiogenesis and activates the survival pathway PI3K/AKT, contributing to increased drug resistance [26,27]. Therefore, the rise in TrkB in the presence of CFF could be interpreted as a response of tumor cells to the drug's toxic effect. Studies analyzing this hypothesis are needed; however, the prevalence of I_2 action indicates an antitumor benefit.

Finally, we explored the possible role of the I_2 supplement in the side effects prevention associated with xenograft signalization and the bladder injury secondary to CFF administration. Previous studies in our group had detected that xenografts and tumor growth generate stress in the mouse, evidenced in the loss of BWG [9]. This effect has been described in preclinical and clinical studies and is known as cachexia [28]. Cachexia is accompanied by loss of adipocytes and muscle tissue with chronic inflammation and increases in proinflammatory factors such as TNFα and IL-6 [28,29]. The prevention of weight loss in the I_2-supplemented groups might be due to two conditions: first, the antineoplastic effect of I_2 that prevents tumor growth and decreases the tumor mass signaling; and second, a direct impact on chronic inflammation processes due to I_2 antioxidant action [9,30]. This effect also appears to be exercised in the prevention of bladder injury. It is well known that the hepatic biotransformation of CFF produces, in addition to phosphoramide mustard (an antineoplastic metabolite), acrolein, which causes hemorrhagic cystitis [3]. We did not expect severe effects of acrolein at the metronomic dose used; however, the histological evaluation showed a thickening of the urothelium and the lamina propria of the bladder, which indicates moderate hemorrhagic cystitis. The I_2 supplement in coadministration with CFF prevented these alterations. This protective mechanism might be due to its antioxidant action. Previously, it has been shown that the chemical form of I_2 has an in vitro reducing capacity (FRAP test) ten times greater than ascorbic acid and 60 times greater than potassium iodide [31]. In vivo studies showed that the iodine supplement decreases the oxidative potential in the serum of rodents and patients [30]. The administration of other antioxidants, such as ascorbic acid, retinol, and resveratrol, improves oxidative stress by reducing ROS levels in bladder tissues generated by CFF treatment [32,33]. An unexplored alternative is that the I_2 could be binding directly to acrolein, decreasing its irritating action, and preventing its contact or entry into the urothelium of the bladder. This alternative is based on the acrolein structure that contains double bonds capable of being iodinated [34].

4. Materials and Methods

4.1. Chemicals and Reagents

The Cyclophosphamide for in vivo assays and the CFF active metabolite 4-Hydroperoxycyclophosphamide for in vitro assays were obtained by Cryofarma (Jalisco, Mexico) and Toronto Research Chemicals (Toronto, Ontario, CA, USA) respectively, and we denominate both with the same abbreviation (CFF). Rosiglitazone (RGZ; PPARγ-specific agonist, by Cayman Chemical, Los Angeles, CA, USA), GW9662 (PPARγ-specific antagonist, by Corning, Bedford, MA, USA) and Matrigel (basement membrane matrix, Corning, Bedford, MA, USA). Sublimed iodine was obtained from Macron-Avantor (Center Valley, PA, USA). The concentration of iodine solutions was verified by sodium thiosulfate titration. All other chemicals were of the highest purity grade available.

4.2. Cell Culture

The Nb cell lines SH-SY5Y (CRL-2266) and SK-N-BE(2) (CRL-2271) were obtained from the company American Type Culture Collection (ATCC, Manassas, VA, USA). All the experiments were performed with passages 1–5 and recently tested and authenticated by STR profiling (BIMODI Invoice number 190320-029). The conditions for cell culture were Dulbecco's Modified Eagle's Medium (DMEM) supplemented with fetal bovine serum (FBS, 10%) and penicillin/streptomycin (2%) by Invitrogen (Carlsbad, CA, USA) in a humidified chamber with 5% CO_2 atmosphere and 95% air at 37 °C.

4.3. Cell Viability

A total of 50,000 cells/well were seeded onto 12-well plates. After 24 h, different concentrations of CFF (0.5, 1, and 2 μM), I_2 (100, 200 and 400 μM) and I_2 + CFF (200 + 0.5 or 200 + 1.0, respectively) were added for 0, 24, 48, 72, and 96 h. Control groups were followed at the same times using deionized water as treatment (vehicle of I_2 and CFF). In the GW9662/RGZ or I_2-treated groups, GW9662 (1 μM) was administered 2 h before RGZ (5 μM) or I_2 (200 μM) treatment.

After treatment, cells were detached and mixed with the exclusion dye trypan blue (0.04%) to count the cells using a hemocytometer in light microscopy; viability was reported as fold change against control. All experiments were carried out in three independent experiments per triplicate. To measure the extent of interaction between I_2 and CFF, data were analyzed by CompuSyn software 1.0 (ComboSyn, Inc., Paramus, NJ, USA) based on the combination index (CI) of the multiple drug effect equation of Chou-Talalay [35].

4.4. Apoptosis

Apoptosis was evaluated by flow cytometry using the apoptosis kit (ABCAM No. 14190, Cambridge, UK) with the Attune NXT flow cytometer (BRVY). Briefly, the pellet of cells was resuspended in PBS, and the monoclonal antibody for annexin and propidium iodide were added, according to manufacturer's instructions. The mixture was incubated for 30 min at room temperature and protected from light. After incubation, the cells were washed twice with PBS and resuspended in 500 μL PBS. Data analysis of 10,000 events was performed using FlowJo v10 (Trial version) software.

4.5. Gene Expression

TrkA, TrkB, PPARγ, SVV, MDR-1, MYCN, P53, Bax, Bcl-2, VEGF, HIF1, and β-actin were analyzed by RT-qPCR from SK-N-BE(2) cell cultures and xenografts after the corresponding treatments. Briefly, total RNA was obtained using Trizol reagent (Life Technologies, Inc., Carlsbad, CA, USA). RNA (2 μg) was reverse transcribed (RT) using oligodeoxythymidine (Invitrogen, Waltham, MA, USA). Real-time PCR was performed on the Rotor-Gene 3000 sequence detector system (Corbett Research, Mortlake, NSW, Australia) using SYBR Green as a DNA amplification marker (gene-specific primers are listed in Table 1). Relative mRNA levels were normalized to the mRNA level of β-actin.

Table 1. Oligonucleotide sequences.

Gen	Reference	Sense	Antisense	bp	Ta (°C)
TrkA	NM_002529.3	CATCGTGAAGAGTGGTCTCCG	GAGAGAGACTCCAGAGCGTTGAA	102	60
TrkB	NM_001007097.3	TCGTGGCATTTCCGAGATTGG	TCGTCAGTTTGTTTCGGGTAAA	231	60
PPARγ	NM_001354666.3	TCTCTCCGTAATGGAAGACC	GCATTATGAGACATCCCCAC	474	62
SVV	NM_001168.3	TTCTCAAGGACCACCGCATC	CCAAGTCTGGCTCGTTCTCA	126	60
MDR1	NM_001348945.2	GAGAGATCCTCACCAAGCGG	ATCATTGGCGAGCCTGGTAC	122	60
MYCN	NM_001293228.2	ACCCTGAGCGATTCAGATGAT	GTGGTGACAGCCTTGGTGTT	113	62
P53	NM_001126118.2	CCATGAGCGCTGCTCAGATA	GGGCACCACCACACTATGTC	124	60
Bax	NM_138764.5	AAGCTGAGCGAGTGTCTCAAGCGC	TCCCGCCACAAAGATGGTCACG	327	60
Bcl-2	NM_000633.3	CTCGTCGCTACCGTCGTGACTTCG	CAGATGCCGGTTCAGGTACTCAGTC	242	60
VEGF	NM_001025366.3	CTCGATTGGATGGCAGTAGCT	AGGAGGAGGGCAGAATCATCA	76	60
HIF1	NM_001530.4	TTGATGGGATATGAGCCAGA	TGTCCTGTGGTGACTTGTCC	128	60
β-actin	NM_001101.5	CCATCATGAAGTGTGACGTTG	ACAGAGTACTTGCGCTCAGGA	175	60

4.6. Mitochondrial Membrane Potential

After 12, 24, and 48-h treatments (I_2, CFF, or I_2 + CFF), the cells were PBS-washed and labeled with 200 nM MitoTracker Red CM-H2Xros (Thermo Fisher; Waltham, MA, USA) for 45 min. Then, cells were fixed for 10 min with ethanol, PBS-washed, and mounted with anti-FADE and DAPI. Micrographs were taken with an epifluorescence microscope (Axio Imager, Carl Zeiss, Jena, Germany). The software Image J 1.8 (National Institutes of Health, Bethesda, MD, USA) was used to quantify the relative fluorescence units (RFUs) and determine the mitochondrial functional state.

4.7. Tumoral Implantation and Progression

Xenografts with SK-N-BE(2) cells were generated using 5×10^6 cells/injection of Nb cells in 6–7-week-old male immunodeficient athymic nude mice (Foxn1 nu/nu, Harlan Mexico, Ciudad de Mexico, Mexico) as previously described [36]. Mice were housed in barrier conditions under a 12-h light/dark cycle with food and water supplied ad libitum. All the procedures followed the Animal Care and Use Program of the National Institutes of Health (NIH; Bethesda, MD, USA) and were approved by the Ethics Committee of the Instituto de Neurobiología (ethical approval number 035).

When palpable tumors reached a volume of 1 cm^3, animals were randomly assigned to each group ($n = 4$). I_2 (8 mg/kg/day; 0.025%), CFF (20 mg/Kg/day; 0.060%), or a mixture of both. The treatments were supplied in drinking water ad libitum. The control group received only water. Animals were sacrificed after anesthesia with a ketamine/xylazine mixture (30 mg/Kg and 6 mg/Kg from Pisa Agropecuaria, Hgo., Mexico, and Cheminova CDMX, Mexico, respectively). The bladder and a tumor section were fixed in 10% formalin for at least 24 h and processed for immunohistochemistry. The remaining tumors were frozen in dry ice for RNA analysis.

4.8. Immunohistochemistry

Tumor sections and bladders were stained with hematoxylin-eosin (H&E) and Masson's trichrome techniques for histopathological analysis. In addition to the vasculature analysis, the endothelial protein antibody CD34 (ab182981; 1:2500, Abcam, Cambridge, UK) was used to detect endothelial-positive cells (Vector Labs, Burlingame, CA, USA). Sections were counterstained by hematoxylin. The Mean Vascular Density (CD34/mm^2) or vascular number per field were quantified by randomly analyzing three fields from three different sections of each tumor and bladder, using the software ImageJ version 1.8 (National Institutes of Health, Bethesda, MD, USA).

4.9. Western Blot

Western blot (WB) analysis of PPARγ proteins for tumor tissue was performed with the chemiluminescence technique [9]. Briefly, 50 μg of protein per lane were separated by electrophoresis in 10% acrylamide gel, proteins were later transferred to a nitrocellulose membrane (Bio-Rad, Hercules, CA, USA). The unspecified reaction was blocked overnight with PBS containing 5% skimmed milk powder. The membranes were treated with polyclonal antibodies (Santa Cruz Biotechnology, Los Angeles, CA, USA) against anti-PPARγ (ab209350, 54 kDa, 1: 1000, Abcam, Cambridge, MA, USA). As a secondary antibody, goat anti-rabbit (Thermo scientific 656120, 1: 10,000, Invitrogen, Waltham, MA, USA) was used. Proteins were visualized using chemiluminescent detection (ECL, Amersham Biosciences, Buckinghamshire, UK). The blots were visualized and pictured with Image LabTM (Biorad), and the densitometry analysis was performed with Image ImageJ V1.53e; PPAR levels were normalized to total protein of Ponceau red staining signal [37].

4.10. Statistical Analysis

Data for in vitro experiments are the media of three independent tests in triplicate. In vivo, four animals per group were used. Tissue analysis for PCR is the average of four samples, and three sections of each tumor were used for immunohistochemistry. Statistical analysis was performed by one-way ANOVA followed by Tukey's test for analysis between groups. Values with $p < 0.05$ were considered statistically significant.

5. Conclusions

Molecular iodine exerts antiproliferative and differentiation effects in Nb cell lines, increasing their sensitivity to CFF. Molecular mechanisms include decreased expression of master regulators related to malignancy (MYCN, TrkB), remodeling of the vasculature, and increased differentiation signaling (PPARγ and TrkA). Furthermore, I_2 supplementation prevents loss of body weight and hemorrhagic cystitis secondary to CFF in nude mice. These results allow us to propose the I_2 supplement in metronomic CFF treatments to increase the effectiveness of chemotherapy and reduce side effects.

Supplementary Materials: The following are available online at https://www.mdpi.com/article/10.3390/ijms22168936/s1.

Author Contributions: W.Á.-L. and I.M. contributed equally to this paper. W.Á.-L. and I.M. performed the SK-N-BE(2) in vitro and in vivo assays and prepared the original draft. E.D.-G. supervised and analyzed the real-time polymerase chain reactions (RT-qPCR). B.A. provided coordination, supervised the statistical analysis, and corrected the draft. C.A. conceived the study, coordinated the research, and corrected the manuscript. All authors have read and agreed to the published version of the manuscript.

Funding: This research was partially funded by PAPIIT-UNAM, grant numbers 203919, 205920; DGAPA postdoctoral fellowship award for Irasema Mendieta and CONACYT fellowship for Winniberg Álvarez-León.

Institutional Review Board Statement: The study was conducted according to the Animal Care and Use Program of the National Institutes of Health (NIH; Bethesda, MD, USA) and approved by the Research Ethics Committee of the Instituto de Neurobiología (ethical permit number 035).

Informed Consent Statement: Not applicable.

Data Availability Statement: The data presented in this study are available on request from the corresponding author.

Acknowledgments: The authors are grateful to Laura Ines Garcia, Elsa Nydia Hernández Ríos, Ericka de los Rios, Martín García Servín, Alejandra Castilla, and Antonieta Carbajo from INB-UNAM; Alejandra Castillo Carbajal, Carina Uribe Díaz and Rafael Palacios de la Lama from LIIGH-UNAM for technical assistance; Francisco Javier Valles and Rafael Silva for bibliographic assistance; Nuri Aranda and Sofia Gutierrez Ramirez for academic support; Alberto Lara, Omar Gonzalez, Ramon Martinez, and Maria Eugenia Rosas Alatorre for computer assistance; and Jessica Gonzalez Norris for proofreading.

Conflicts of Interest: The authors declare no conflict of interest.

References

1. Otte, J.; Dyberg, C.; Pepich, A.; Johnsen, J.I. MYCN Function in Neuroblastoma Development. *Front. Oncol.* **2020**, *10*, 1–12. [CrossRef]
2. Montemurro, L.; Raieli, S.; Angelucci, S.; Bartolucci, D.; Amadesi, C.; Lampis, S.; Scardovi, A.L.; Venturelli, L.; Nieddu, G.; Cerisoli, L.; et al. A novel MYCN-specific antigene oligonucleotide deregulates mitochondria and inhibits tumor growth in MYCN-amplified neuroblastoma. *Cancer Res.* **2019**, *79*, 6166–6177. [CrossRef]
3. Emadi, A.; Jones, R.J.; Brodsky, R.A. Cyclophosphamide and cancer: Golden anniversary. *Nat. Rev. Clin. Oncol.* **2009**, *6*, 638–647. [CrossRef] [PubMed]
4. Morscher, R.J.; Aminzadeh-Gohari, S.; Hauser-Kronberger, C.; Feichtinger, R.G.; Sperl, W.; Kofler, B. Combination of metronomic cyclophosphamide and dietary intervention inhibits neuroblastoma growth in a CD1-nu mouse model. *Oncotarget* **2016**, *7*, 17060–17073. [CrossRef]

5. Fathpour, G.; Jafari, E.; Hashemi, A.; Dadgar, H.; Shahriari, M.; Zareifar, S.; Jenabzade, A.R.J.; Vali, R.; Ahmadzadehfar, H.; Assadi, M. Feasibility and Therapeutic Potential of Combined Peptide Receptor Radionuclide Therapy With Intensive Chemotherapy for Pediatric Patients With Relapsed or Refractory Metastatic Neuroblastoma. *Clin. Nucl. Med.* **2021**, *46*, 540–548. [CrossRef] [PubMed]
6. Tuyaerts, S.; van Nuffel, A.M.T.; Naert, E.; van Dam, P.A.; Vuylsteke, P.; de Caluwé, A.; Aspeslagh, S.; Dirix, P.; Lippens, L.; de Jaeghere, E.; et al. PRIMMO study protocol: A phase II study combining PD-1 blockade, radiation and immunomodulation to tackle cervical and uterine cancer. *BMC Cancer* **2019**, *19*, 506. [CrossRef] [PubMed]
7. Aceves, C.; Mendieta, I.; Anguiano, B.; Delgado-González, E. Molecular iodine has extrathyroidal effects as an antioxidant, differentiator, and immunomodulator. *Int. J. Mol. Sci.* **2021**, *22*, 1228. [CrossRef]
8. Mani, S.; Swargiary, G.; Singh, K.K. Natural agents targeting mitochondria in cancer. *Int. J. Mol. Sci.* **2020**, *2*, 6992. [CrossRef]
9. Mendieta, I.; Rodríguez-Gómez, G.; Rueda-Zarazúa, B.; Rodríguez-Castelán, J.; Álvarez-León, W.; Delgado-González, E.; Anguiano, B.; Vázquez-Martínez, O.; Díaz-Muñoz, M.; Aceves, C. Molecular iodine synergized and sensitized neuroblastoma cells to the antineoplastic effect of ATRA. *Endocr. Relat. Cancer* **2020**, *27*, 699–710. [CrossRef]
10. Coulter, D.W.; McGuire, T.R.; Sharp, J.G.; McIntyre, E.M.; Dong, Y.; Wang, X.; Gray, S.; Alexander, G.R.; Chatuverdi, N.K.; Joshi, S.S.; et al. Treatment of a chemoresistant neuroblastoma cell line with the antimalarial ozonide OZ513. *BMC Cancer* **2016**, *16*, 867. [CrossRef]
11. Chesler, L.; Goldenberg, D.D.; Collins, R.; Grimmer, M.; Kim, G.E.; Tihan, T.; Nguyen, K.; Yakovenko, S.; Matthay, K.K.; Weiss, W.A. Chemotherapy-Induced Apoptosis in a Transgenic Model of Neuroblastoma Proceeds Through p53 Induction. *Neoplasia* **2008**, *10*, 1268–1274. [CrossRef] [PubMed]
12. Shrivastava, A.; Tiwari, M.; Sinha, R.A.; Kumar, A.; Balapure, A.K.; Bajpai, V.K.; Sharma, R.; Mitra, K.; Tandon, A.; Godbole, M.M. Molecular iodine induces caspase-independent apoptosis in human breast carcinoma cells involving the mitochondria-mediated pathway. *J. Biol. Chem.* **2006**, *281*, 19762–19771. [CrossRef] [PubMed]
13. Yousefnia, S.; Momenzadeh, S.; Forootan, F.S.; Ghaedi, K.; Esfahani, M.H.N. The influence of peroxisome proliferator-activated receptor γ (PPARγ) ligands on cancer cell tumorigenicity. *Gene* **2018**, *649*, 14–22. [CrossRef]
14. Arroyo-Helguera, O.; Rojas, E.; Delgado, G.; Aceves, C. Signaling pathways involved in the antiproliferative effect of molecular iodine in normal and tumoral breast cells: Evidence that 6-iodolactone mediates apoptotic effects. *Endocr. Relat. Cancer* **2008**, *15*, 1003–1011. [CrossRef]
15. Anguiano, B.; Aceves, C. Iodine in Mammary and Prostate Pathologies. *Curr. Chem. Biol.* **2011**, *5*, 177–182. [CrossRef]
16. Nuñez-Anita, R.E.; Arroyo-Helguera, O.; Cajero-Juárez, M.; López-Bojorquez, L.; Aceves, C. A complex between 6-iodolactone and the peroxisome proliferator-activated receptor type gamma may mediate the antineoplastic effect of iodine in mammary cancer. *Prostaglandins Other Lipid Mediat.* **2009**, *89*, 34–42. [CrossRef]
17. Nava-Villalba, M.; Nuñez-Anita, R.E.; Bontempo, A.; Aceves, C. Activation of peroxisome proliferator-activated receptor gamma is crucial for antitumoral effects of 6-iodolactone. *Mol. Cancer.* **2015**, *14*, 1–11. [CrossRef] [PubMed]
18. Vella, S.; Conaldi, P.G.; Florio, T.; Pagano, A. PPAR Gamma in Neuroblastoma: The Translational Perspectives of Hypoglycemic Drugs. *PPAR Research.* **2016**, *2016*, 3038164. [CrossRef] [PubMed]
19. Kushner, B.H.; Cheung, N.K.V.; Modak, S.; Becher, O.J.; Basu, E.M.; Roberts, S.S.; Kramer, K.; Dunkel, I.J. A phase I/Ib trial targeting the Pi3k/Akt pathway using perifosine: Long-term progression-free survival of patients with resistant neuroblastoma. *Int. J. Cancer.* **2017**, *140*, 480–484. [CrossRef]
20. Islam, A.; Kageyama, H.; Takada, N.; Kawamato, T.; Takayasu, H.; Isogai, E.; Ohira, M.; Hashizume, K.; Kobayashi, H.; Kaneko, Y.; et al. High expression of Survivin, mapped to 17q25, is significantly associated with poor prognostic factors and promotes cell survival in human neuroblastoma. *Oncogene* **2000**, *19*, 617–623. [CrossRef]
21. Nakagawara, A.; Ohira, M. Comprehensive genomics linking between neural development and cancer: Neuroblastoma as a model. *Cancer Lett.* **2004**, *204*, 213–224. [CrossRef]
22. Fukai, T.; Ushio-Fukai, M. Cross-Talk between NADPH Oxidase and Mitochondria: Role in ROS Signaling and Angiogenesis. *Cells* **2020**, *9*, 1849. [CrossRef] [PubMed]
23. Rösner, H.; Torremante, P.; Möller, W.; Gärtner, R. Antiproliferative/cytotoxic activity of molecular iodine and iodolactones in various human carcinoma cell lines. No interfering with EGF-signaling, but evidence for apoptosis. *Exp. Clin. Endocrinol. Diabetes* **2010**, *118*, 410–419. [CrossRef] [PubMed]
24. Chen, X.; Duan, N.; Zhang, C.; Zhang, W. Survivin and tumorigenesis: Molecular mechanisms and therapeutic strategies. *J. Cancer* **2016**, *7*, 314–323. [CrossRef]
25. Ausserlechner, M.J.; Hagenbuchner, J. Mitochondrial survivin–an Achilles' heel in cancer chemoresistance. *Mol. Cell. Oncol.* **2016**, *3*, 1–3. [CrossRef]
26. Brodeur, G.M.; Minturn, J.E.; Ho, R.; Simpson, A.M.; Iyer, R.; Varela, C.R.; Light, J.E.; Kolla, V.; Evans, A.E. Trk receptor expression and inhibition in neuroblastomas. *Clin. Cancer Res.* **2009**, *15*, 3244–3250. [CrossRef]
27. Li, Z.; Zhang, Y.; Tong, Y.; Tong, J.; Thiele, C.J. Trk inhibitor attenuates the BDNF/TrkB-induced protection of neuroblastoma cells from etoposide in vitro and in vivo. *Cancer Biol. Ther.* **2015**, *16*, 477–483. [CrossRef]
28. Pin, F.; Barreto, R.; Kitase, Y.; Mitra, S.; Erne, C.E.; Novinger, L.J.; Zimmers, T.A.; Couch, M.E.; Bonewald, L.F.; Bonetto, A. Growth of ovarian cancer xenografts causes loss of muscle and bone mass: A new model for the study of cancer cachexia. *J. Cachexia Sarcopenia Muscle* **2018**, *9*, 685–700. [CrossRef] [PubMed]

29. Chambaut-Guérin, A.-M.; Martinez, M.-C.; Hamimi, C.; Gauthereau, X.; Nunez, J. Tumor Necrosis Factor Receptors in Neuroblastoma SKNBE Cells and Their Regulation by Retinoic Acid. *J. Neurochem.* **1995**, *65*, 537–544. [CrossRef]
30. Winkler, R. Iodine—A Potential Antioxidant and the Role of Iodine/Iodide in Health and Disease. *Nat. Sci.* **2015**, *7*, 548–557. [CrossRef]
31. Alfaro, Y.; Delgado, G.; Cárabez, A.; Anguiano, B.; Aceves, C. Iodine and doxorubicin, a good combination for mammary cancer treatment: Antineoplastic adjuvancy, chemoresistance inhibition, and cardioprotection. *Mol. Cancer* **2013**, *12*, 1–11. [CrossRef] [PubMed]
32. Song, J.; Liu, L.; Li, L.; Liu, J.; Song, E.; Song, Y. Protective effects of lipoic acid and mesna on cyclophosphamide-induced haemorrhagic cystitis in mice. *Cell Biochem. Funct.* **2014**, *32*, 125–132. [CrossRef] [PubMed]
33. Keles, I.; Bozkurt, M.F.; Cemek, M.; Karalar, M.; Hazini, A.; Alpdagtas, S.; Keles, H.; Yildiz, T.; Ceylan, C.; Buyukokuroglu, M.E. Prevention of cyclophosphamide-induced hemorrhagic cystitis by resveratrol: A comparative experimental study with mesna. *Int. Urol. Nephrol.* **2014**, *46*, 833. [CrossRef] [PubMed]
34. Monach, P.A. Incidence and prevention of bladder toxicity from cyclophosphamide in the treatment of rheumatic diseases: A data-driven review. *Arthritis Rheum.* **2010**, *62*, 165. [CrossRef]
35. Chou, T.C. Drug combination studies and their synergy quantification using the chou-talalay method. *Cancer Res.* **2010**, *70*, 440–446. [CrossRef] [PubMed]
36. Mendieta, I.; Nuñez-Anita, R.E.; Nava-Villalba, M.; Zambrano-Estrada, X.; Delgado-González, E.; Anguiano, B.; Aceves, C. Molecular iodine exerts antineoplastic effects by diminishing proliferation and invasive potential and activating the immune response in mammary cancer xenografts. *BMC Cancer* **2019**, *19*, 261. [CrossRef] [PubMed]
37. Sander, H.; Wallace, S.; Plouse, R.; Tiwari, S.; Gomes, A.V. Ponceau S waste: Ponceau S staining for total protein normalization. *Anal. Biochem.* **2019**, *575*, 44–53. [CrossRef]

International Journal of
Molecular Sciences

Review

Strategies to Improve the Antitumor Effect of $\gamma\delta$ T Cell Immunotherapy for Clinical Application

Masatsugu Miyashita [1,2,*], Teruki Shimizu [1], Eishi Ashihara [3] and Osamu Ukimura [1]

1 Department of Urology, Kyoto Prefectural University of Medicine, Kyoto 602-8566, Japan; shimizu1@koto.kpu-m.ac.jp (T.S.); ukimura@koto.kpu-m.ac.jp (O.U.)
2 Department of Urology, Japanese Red Cross Kyoto Daini Hospital, Kyoto 602-8026, Japan
3 Department of Clinical and Translational Physiology, Kyoto Pharmaceutical University, Kyoto 607-8414, Japan; ash@mb.kyoto-phu.ac.jp
* Correspondence: miyash@koto.kpu-m.ac.jp; Tel.: +81-752515595

Abstract: Human $\gamma\delta$ T cells show potent cytotoxicity against various types of cancer cells in a major histocompatibility complex unrestricted manner. Phosphoantigens and nitrogen-containing bisphosphonates (N-bis) stimulate $\gamma\delta$ T cells via interaction between the $\gamma\delta$ T cell receptor (TCR) and butyrophilin subfamily 3 member A1 (BTN3A1) expressed on target cells. $\gamma\delta$ T cell immunotherapy is classified as either in vivo or ex vivo according to the method of activation. Immunotherapy with activated $\gamma\delta$ T cells is well tolerated; however, the clinical benefits are unsatisfactory. Therefore, the antitumor effects need to be increased. Administration of $\gamma\delta$ T cells into local cavities might improve antitumor effects by increasing the effector-to-target cell ratio. Some anticancer and molecularly targeted agents increase the cytotoxicity of $\gamma\delta$ T cells via mechanisms involving natural killer group 2 member D (NKG2D)-mediated recognition of target cells. Both the tumor microenvironment and cancer stem cells exert immunosuppressive effects via mechanisms that include inhibitory immune checkpoint molecules. Therefore, co-immunotherapy with $\gamma\delta$ T cells plus immune checkpoint inhibitors is a strategy that may improve cytotoxicity. The use of a bispecific antibody and chimeric antigen receptor might be effective to overcome current therapeutic limitations. Such strategies should be tested in a clinical research setting.

Keywords: $\gamma\delta$ T cells; immunotherapy; tumor resistance; combination therapy; tumor microenvironment; immune checkpoint inhibitor

Citation: Miyashita, M.; Shimizu, T.; Ashihara, E.; Ukimura, O. Strategies to Improve the Antitumor Effect of $\gamma\delta$ T Cell Immunotherapy for Clinical Application. *Int. J. Mol. Sci.* **2021**, 22, 8910. https://doi.org/10.3390/ijms22168910

Academic Editor: Angela Stefanachi

Received: 14 July 2021
Accepted: 17 August 2021
Published: 18 August 2021

1. Introduction

Cancer is one of the most serious and potentially fatal diseases in humans. According to statistical reports, there were an estimated 18.1 million new cancer cases and 9.6 million cancer-related deaths worldwide in 2018 [1]. Surgery, chemotherapy, and radiotherapy are the three pillars of antitumor therapy. Surgery and radiotherapy are curative for localized cancers; however, most cancer-related deaths are due to metastasis, which requires systemic therapy. Chemotherapy is the first-line systemic therapy against metastatic cancers; however, many cancers become resistant, which leads to treatment failure. Recently, immunotherapy, now regarded as the fourth pillar of antitumor therapy, has been used for systemic antitumor therapy.

T cell-based immunotherapy is an effective cancer treatment strategy. T cells are divided into two major subpopulations based on surface expression of $\alpha\beta$ and $\gamma\delta$ T cell receptors (TCRs). $\alpha\beta$ T cells recognize peptide antigens in the context of non-self; for example, antigens expressed by cancer cells. $\alpha\beta$ T cells are effector cells that operate within the adaptive arm of the immune system; these cells exert cytotoxicity in a major histocompatibility complex (MHC)-restricted manner. However, due to loss of MHC molecules, tumor cells are often resistant to attack by $\alpha\beta$ T cells [2]. By contrast, $\gamma\delta$ T cells are effectors that operate within the innate arm of the immune system; these cells act in an

MHC-unrestricted manner, making them interesting mediators of cancer immunotherapy. Human γδ T cells were first identified in the mid-1980s [3–5]. They are abundant in the intestine and skin and play a role in defense against microbial infections in an MHC-unrestricted manner [6]. Recent studies show that γδ T cells exert potent cytotoxic effects against various types of cancer cell [7–12]. Their activation induces release of cytotoxic molecules such as perforin and granzymes. Activated γδ T cells also secrete cytokines such as interferon-γ (IFN-γ) and tumor necrosis factor-α (TNF-α). These cytotoxic molecules and cytokines induce cancer cell apoptosis. However, γδ T cells comprise only a small percentage of circulating lymphocytes and require stimulation to exert antitumor effects. In this review, we will outline the methods used to stimulate γδ T cells and improve their antitumor effects. We also discuss strategies for clinical application.

2. Phosphoantigens and Nitrogen-Containing Bisphosphonates Stimulate γδ T Cells

Human peripheral blood γδ T cells, which predominantly express the Vδ2 chain paired with the Vγ9 chain, are activated upon recognition of phosphoantigens (PAgs) such as (E)-4-hydroxy-3-methylbut-2-enyl pyrophosphate (HMBPP), which is synthesized in bacteria via isoprenoid biosynthesis [13], and isopentenyl pyrophosphate (IPP), which is produced in eukaryotic cells via the mevalonate pathway [14]. Activation of γδ T cells by PAgs was first reported in the 1990s [15,16]; however, it is unclear how the γδ TCR recognizes PAgs. Butyrophilin subfamily 3 member A1 (BTN3A1) molecules, which are isoforms of the BTN3A (also termed CD277) subfamily, play an indispensable role in activation of γδ T cells by PAgs [17]. BTN3A1, which is expressed ubiquitously on the surface of cells, comprises two immunoglobulin-like extracellular domains and an intracellular B30.2 domain. The precise mechanism by which γδ T cells recognize BTN3A1 is not completely clear, but several studies demonstrate that binding of PAgs directly to a positively-charged pocket in the intracellular B30.2 domain of BTN3A1 recruits the cytoskeletal adaptor protein periplakin and the GTPase RhoB, which increases membrane mobility and induces a conformational change in BTN3A1; the altered conformation is recognized by the γδ TCR [18,19]. Recent studies show that BTN2A1, which binds directly to the TCRs via germline-encoded regions of Vγ9, is also essential to BTN3A-mediated γδ T cell cytotoxicity and BTN2A1 expression at the plasma membrane of cancer cells correlated with γδ T cell cytotoxicity [20,21]. BTN2A1 interacts with BTN3A1, leading to enhance plasma membrane export, and BTN2A1/BTN3A1 interaction is enhanced by PAgs. Anti-BTN2A monoclonal antibodies (mAbs) inhibit BTN2A1 biding to the γδ TCR and modulate γδ T cell killing of cancer cells [21]. These studies demonstrate the potential of butyrophilin subfamily cooperation pathway as a therapeutic target in γδ T cell activation.

In general, the concentrations of PAgs is not high enough to stimulate γδ T cells under physiological conditions; however, tumor cells show upregulated production of PAgs due to metabolic reprogramming, which increases mevalonate pathway activity [22,23]. Moreover, PAgs concentrations can be increased pharmacologically. Nitrogen-containing bisphosphonates (N-bis) such as pamidronate (Pam) and zoledronate acid (ZOL), which are used to treat hypercalcemia or bone metastases of cancer, inhibit the enzyme farnesyl diphosphate (FPP) synthase, which is the rate determining enzyme in the mevalonate pathway [24]. As a result, the concentration of IPP (derived from the upstream FPP synthase metabolite) increases, thereby activating γδ T cells (Figure 1).

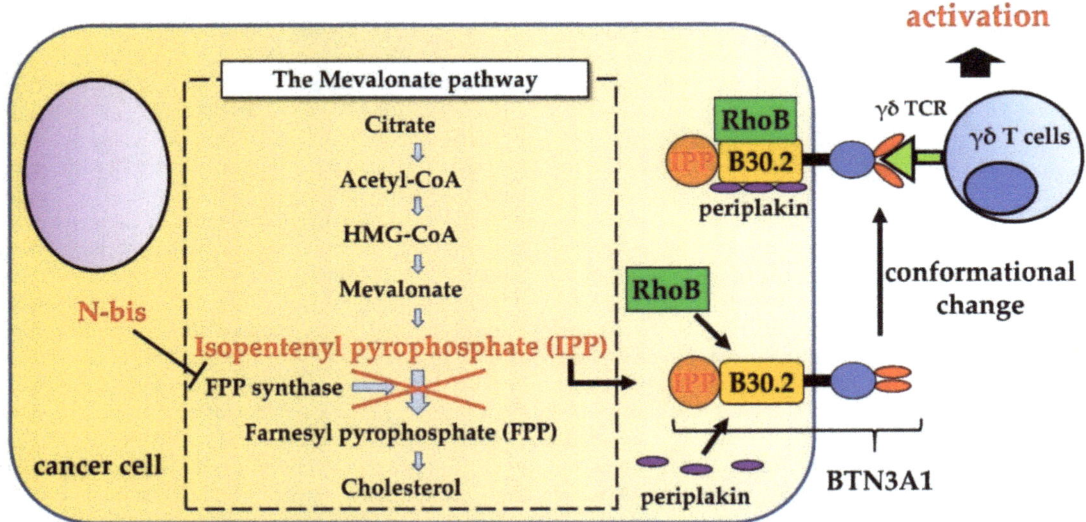

Figure 1. Mechanism of γδ T cell activation by N-bis. N-bis inhibits FPP synthase in the mevalonate pathway and induces accumulation of IPP. Binding of IPP to the intracellular B30.2 domain of BTN3A1 recruits the cytoskeletal adaptor protein periplakin and the GTPase RhoB, which increases membrane mobility and induces a conformational change in BTN3A1, which is then recognized by the γδ TCR.

γδ T cell-based immunotherapy is classified according to the method used to activate and expand the cells [25]. The first method involves in vivo activation by systemic administration of PAgs or N-bis, along with exogenous interleukin (IL)-2 [26–32] (Table 1). Dieli et al. conducted a phase I clinical trial involving patients with metastatic hormone-refractory prostate cancer. The aim was to examine the antitumor effect of single or combined administration of ZOL and IL-2. Nine patients were enrolled in each arm. Six of the nine patients received combined administration of ZOL and IL-2, but only two of nine patients received single administration of ZOL, and showed a significant long-term shift in peripheral blood γδ T cells toward an activated state in which they produced IFN-γ and perforin; also, the number of activated γδ T cells showed a significant correlation with favorable clinical outcomes [26]. This indicates the importance of the administration of IL-2 to maintain peripheral γδ T cells. Wilhelm et al. reported a pilot study of patients with low-grade non-Hodgkin lymphoma and multiple myeloma; this study involved in vivo activation of γδ T cells by combined administration of Pam and IL-2. The results showed that γδ T cell activation/proliferation and response to treatment were disappointing, with only one of ten patients that received an intravenous infusion of IL-2 on Day 3 through Day 8 achieving stable disease. On the other hand, the next nine patients selected had shown positive in vitro proliferation of γδ T cells in response to Pam/IL-2; when these patients received an intravenous infusion of IL-2 on Day 1 through Day 6, five showed in vivo activation/proliferation of γδ T cells, and three showed a partial response [27]. Therefore, if patients are to have any chance of a clinical responses, they must show positive in vitro proliferation of γδ T cells in response to stimulation with Pam, and IL-2 must be administered immediately after in vivo Pam stimulation. Lang et al. reported a pilot trial of in vivo γδ T cell activation in 12 patients with metastatic renal cell carcinoma (RCC); they used different doses of ZOL in combination with low-dose IL-2. Two patients experienced a prolonged period of stable disease; however, no objective clinical responses were observed [28]. The most common adverse events associated with in vivo-activated γδ T cell immunotherapy are the same as those reported for IL-2 monotherapy; they include fever, fatigue, elevation of liver transaminase, and eosinophilia. These adverse events are usually grade 1 or 2, meaning that in vivo-activated therapy is well tolerated. However, the clinical benefits appear to be mild to moderate [25]. This problem could be related to anergy and exhaustion of activation-induced $\gamma\delta$ T cells. The mechanisms underlying this anergy and exhaustion remain unclear. The second category of $\gamma\delta$ T cell immunotherapy involves ex vivo expansion of $\gamma\delta$ T cells by PAgs or N-bis, followed by administration of the cultured $\gamma\delta$ T cells to the patient (i.e., adoptive immunotherapy) [33–42] (Table 1). The mechanism by which N-bis expands $\gamma\delta$ T cells from peripheral blood is as follows: treatment of peripheral blood mononuclear cells with N-bis leads to accumulation of IPP in monocytes because these cells take up N-bis efficiently; the monocytes that accumulate IPP become antigen-presenting cells and stimulate $\gamma\delta$ T cells in the peripheral blood [43,44]. Kobayashi et al. conducted a pilot study of adoptive immunotherapy in patients with advanced RCC using autologous $\gamma\delta$ T cells stimulated by PAg (namely, 2-methyl-3-butenyl-1-pyrophosphate (2M3B1-PP)). Seven patients were enrolled and all received an intravenous infusion of recombinant human IL-2 plus autologous $\gamma\delta$ T cells expanded from their own peripheral blood nuclear cells. All patients had IL-2-related adverse events, which were graded as 1 or 2. The antitumor effects in five patients were evaluated by comparing the tumor-doubling time, assessed by computed tomography (CT), between pre- and post-treatment. Three of the five showed a prolonged tumor-doubling time; however, the other two patients showed a shorter tumor-doubling time. One died within 2 months of $\gamma\delta$ T cell administration, and the other showed a shorter tumor-doubling time for liver metastases [33]. In this study, no patient received systemic ZOL. ZOL treatment is important for the antitumor effects of $\gamma\delta$ T cells because it inhibits FPP synthase, leading to accumulation of IPP in cancer cells and specific antitumor cytolysis by $\gamma\delta$ T cells in a TCR-dependent manner. Kobayashi et al. also conducted a phase I/II study of adoptive $\gamma\delta$ T cell immunotherapy in combination with ZOL and IL-2. Enrolled patients had advanced RCC. Eleven patients were enrolled and all received 4 mg ZOL intravenously, followed by administration of autologous $\gamma\delta$ T cells starting 2 h after completion of ZOL infusion. Patients then received low-dose recombinant human IL-2 on Day 0 through Day 4. Clinical responses were examined by CT and evaluated using the Response Evaluation Criteria in Solid Tumors. One patient exhibited a complete response, five patients had stable disease (SD), and five had progressive disease (PD) [34]. Nicol et al. reported a clinical study of autologous $\gamma\delta$ T cell immunotherapy for various types of metastatic solid tumors (i.e., melanoma, breast cancer, cervical cancer, ovarian cancer, colon cancer, cholangiocarcinoma, and duodenal cancer). Eighteen patients were enrolled. Three of the 14 evaluable patients showed a SD and 11 had PD. Interestingly, this study also examined the migratory pattern of intravenously-infused ex vivo-expanded $\gamma\delta$ T cells labeled with radioactive ^{111}indium oxine (^{111}In) in three patients (two melanoma patients, one colon cancer patient). In all three, labeled $\gamma\delta$ T cells migrated rapidly to the lungs and remained there for 4 to 7 h. Cell numbers (estimated by measurement of γ-ray radioactivity in the lungs) decreased slowly, corresponding with gradual migration into the liver and spleen. After 24 h, almost all cells were located in the liver and spleen and virtually no activity remained in the lungs. Moreover, assessment of the number of peripheral blood $\gamma\delta$ T cells at multiple time points during the 48 h after $\gamma\delta$ T cell infusion showed no substantial change compared with pre-infusion levels. These data indicate that few of the $\gamma\delta$ T cells remained in the bloodstream. However, in one melanoma patient of the three patients, the ^{111}In-labeled $\gamma\delta$ T cells appeared to have migrated to the metastatic mass on the left adrenal gland by 1 h after infusion. Maximal activity was seen at the metastatic tumor site at 4 h, and the tracer remained detectable for 48 h [35]. Adoptive immunotherapy using ex vivo-expanded $\gamma\delta$ T cells is also safe and well tolerated; however, expanding $\gamma\delta$ T cells from some cancer patients is difficult. The reasons for this are unclear. Moreover, favorable clinical outcomes require higher effector ($\gamma\delta$ T cells)-to-target cell (cancer cells) ratios (E/T ratio) at the tumor site. Although potent cytotoxic activity against various cancer cells has been confirmed in vitro, there is much room for improvement.

Table 1. γδ T cell-based clinical trials.

Author	Year	Tumor	Interventions	Phase	Ref. or Clinical Trials. Gov Identifier
Wilhelm et al.	2003	MM, NHL	Pam + IL-2 (in vivo)	Pilot study	[27]
Kobayashi et al.	2006	RCC	Ex-vivo γδ T cell + IL-2	Pilot study	[33]
Kobayashi et al.	2007	RCC	Ex-vivo γδ T cell + ZOL + IL-2	I/II	[34]
Dieli et al.	2007	Prostate cancer	ZOL/ZOL + IL-2 (in vivo)	I	[26]
Bennouna et al.	2008	RCC	BrHPP + IL-2 (in vivo)	I	[31]
Abe et al.	2009	MM	Ex-vivo γδ T cell + ZOL + IL-2	Pilot study	[36]
Meraviglia et al.	2010	Breast cancer	ZOL + IL-2 (in vivo)	I	[29]
Bennouna et al.	2010	Solid cancer	BrHPP + IL-2 (in vivo)	I	[30]
Nakajima et al.	2010	NSCLC	Ex-vivo γδ T cell + ZOL + IL-2	I	[37]
Lang et al.	2011	RCC	ZOL + IL-2 (in vivo)	Pilot study	[28]
Nicol et al.	2011	Solid cancer	Ex-vivo γδ T cell + ZOL	I	[35]
Sakamoto et al.	2011	NSCLC	Ex-vivo γδ T cell + ZOL + IL-2	I	[39]
Noguchi et al.	2011	Solid cancer	Ex-vivo γδ T cell	Pilot study	[40]
Kanzmann et al.	2012	RCC, MM, AML	ZOL + IL-2 (in vivo)	I/II	[32]
Izumi et al.	2013	Colorectal cancer	Ex-vivo γδ T cell	Pilot study	[41]
Wada et al.	2014	Gastric cancer	Ex-vivo γδ T cell + ZOL (intraperitoneal injection)	Pilot study	[38]
Kakimi et al.	2014	NSCLC	Ex-vivo γδ T cell	I	[42]
Ghigo et al.	2020	Solid cancerHematopoietic/ Lymphoid cancer	ICT01 (anti-BTN3A mAbs)/ICT01 plus pembrolizumab	I	NCT04243499

MM: multiple myeloma; NHL: non-Hodgkin's lymphoma; RCC: renal cell carcinoma; NSCLC: non-small-cell lung cancer; AML: acute myeloid leukemia; Pam: pamidronate; IL-2: interleukin-2; ZOL: zoledronate acid; BrHPP: bromohydrin pyrophosphate; BTN3A: Butyrophilin subfamily 3 member A; mAbs: monoclonal antibodies.

3. Administration of γδ T Cells into a Local Cavity Improves the E/T Ratio to Achieve a Maximum Cytotoxic Effect

The E/T ratio at the tumor site is an important factor that determines cytotoxicity. Administration of effector cells into a local cavity might improve the E/T ratio at the tumor site, making it more likely that γδ T cells make direct contact with cancer cells. Several studies describe administration of γδ T cells into a local cavity, such as the intraperitoneal cavity, enucleated cavity, or intravesical cavity. Wada et al. reported injection of ex vivo-expanded γδ T cells following ZOL administration into the intraperitoneal cavity of seven patients with symptomatic malignant ascites secondary to gastric adenocarcinoma. Two of the seven dropped out of the study after a single injection due to disease progression. In one patient, the bloody ascites became clear and reduced in volume. In another patient, the ascites almost disappeared. The most commonly observed treatment-related adverse events were fever and ZOL-induced hypocalcemia. These events were reversible, and none of the patients experienced abdominal pain or any toxicity related to the intraperitoneal injection of γδ T cells [38]. Nichole et al. reported intracranial infusion of ex vivo-expanded γδ T cells from healthy volunteers into athymic nude mice bearing xenografts of the human glioblastoma (GBM) cell line, U251. Intracranial infusion of γδ T cells led to regression of GBM tumors and improved survival [45]. Intravesical administration of drugs (mitomycin C, adriamycin, or Bacillus Calmette-Guerin) is the standard treatment for bladder cancers. Yuasa et al. implanted a human bladder cancer cell line (UMUC3 cells transfected with the luciferase gene (UMUC3-luc)) into the murine bladder cavity and then administered ex vivo-expanded γδ T cells from healthy volunteers along with 5 μM ZOL by the transurethral and intravesical routes on Day 4 through 8 after cancer cell transplantation [46]. In our previous study, we used an in vivo orthotopic xenograft model to test a protocol based on weekly bladder instillation of γδ T cells, as this is a clinically acceptable schedule [47]. The results of these studies showed that intravesical administration of ex vivo-expanded γδ T cells combined with ZOL inhibits the growth of bladder cancers and prolongs survival significantly. Administration of ex vivo-expanded γδ T cells into a local cavity, rather than systemically, is one strategy that improves the antitumor effects of γδ T cells for clinical application.

4. Other Interactions between γδ T Cells and Cancer Cells

γδ T cells recognize not only PAgs via the γδ TCR, but also stress-associated antigens via the natural killer (NK) group 2 member D (NKG2D) receptor; as for natural killer cells, this method of recognition is MHC unrestricted [48–53]. In 1999, Bauer et al. reported that MHC class I chain-related molecule A (MICA) is a functional ligand that stimulates the NKG2D receptor [49]. In addition to MICA, the MICB and UL16-binding proteins 1–4 (ULBP 1–4) in human NKG2D ligands, as well as interactions between these ligands and the NKG2D receptor, are important for cancer cell recognition and γδ T cell-mediated cytotoxicity [51–53]. Anticancer agents inhibit immune function in cancer patients, mainly through bone marrow suppression [54]. However, recent studies show that some agents amplify the cytotoxic effects of immune cells against cancer cells [55]. Anticancer agents induce the DNA damage response, which in turn upregulates expression of NKG2D ligands [56]. Todaro et al. reported that low concentrations of anticancer agents 5-fluorouracyl and doxorubicin sensitize colon cancer-initiating stem cells to γδ T cell-mediated cytotoxicity via NKG2D receptor:ligand interactions [57]. Lamb et al. showed that temozolomide (TMZ), the main chemotherapeutic agent used to treat GBM, increases expression of NKG2D ligands on TMZ-resistant glioma cells, making them more susceptible to recognition and lysis by γδ T cells [58]. In our previous study, we showed that pretreatment of an orthotopic xenograft model with low-dose gemcitabine upregulates expression of MICA/B in bladder cancer cells and increases the cytotoxic effects of γδ T cells plus ZOL [47]. Molecularly targeted agents also could affect NKG2D ligands. Huang et al. reported that tyrosine kinase inhibitors, sorafenib and sunitinib, markedly increased NK cells cytotoxicity against multidrug-resistant nasopharyngeal carcinoma cells in association with up-regulation of NKG2D ligands, MICA, MICB, and ULBP1-3 [59]. Inhibition of epidermal growth factor receptor (EGFR) pathway also leads to induction of NKG2D ligands. Kim et al. reported that EGFR inhibitors, gefitinib and erlotinib enhanced the susceptibility to NK cell mediated lysis of lung cancer cells by induction of ULBP1 by inhibition of protein kinase C (PKC) pathway [60]. In the γδ T cells field, Story et al. reported that proteasome inhibitor bortezomib significantly increased expression of ULBP 2/5/6 in both acute myeloid leukemia (AML) and T-cell acute lymphoblastic leukemia (T-ALL) cells, and enhanced ex vivo expanded γδ T cell-mediated killing of these cells [61]. Histone deacetylase (HDAC) inhibitors, which are epigenetic agents, are also candidates for combined therapy with γδ T cells. Skov et al. reported that HDAC inhibitors upregulate NKG2D ligands on the surface of several cancer cells [62].

Expression of Fas ligand (FasL) and TNF-related apoptosis-inducing ligand (TRAIL) is upregulated in activated γδ T cells [63]. FasL interacts with CD95, also called Fas or APO-1, which was the first death receptor within the apoptotic chain to be molecularly characterized [64]. CD95 is expressed by various human cancer cells; ligation of CD95 by FasL activates the caspase cascade, which initiates cancer cell apoptosis. TRAIL interacts with five receptors (TRAIL-Rs): death receptor 4 (DR4), DR5, decoy receptor 1 (DcR1), DcR2, and osteoprotegerin [65–69]. Death receptors DR4 and DR5 contain a cytoplasmic region known as the death domain, which enables these receptors to initiate cytotoxic signals when engaged by TRAIL [70]. For these reasons, upregulation of CD95 or death receptors DR4 or DR5 in cancer cells might enhance γδ T cell-mediated cytotoxicity. Several anticancer agents upregulate CD95 or death receptors in cancer cells, thereby sensitizing cancer cells to apoptosis mediated by FasL and TRAIL. Shankar et al. report that paclitaxel, vincristine, vinblastine, camptothecin, etoposide, and doxorubicin upregulate DR4 and DR5 in prostate cancer cells, leading to augmentation of TRAIL-induced apoptosis via caspase activation [71]. Mattarollo et al. reported that etoposide, cisplatin, and doxorubicin upregulate CD95 and DR5 in various cancer cells, and that ex vivo-expanded NK cells kill sensitized targets via FasL- and TRAIL-mediated mechanisms [72]. Indeed, they showed that pretreatment of target cells with anticancer agents increased cytotoxicity to 60–70% (compared with the 5–30% observed when either chemotherapy or NK cells were used alone).

Thus, combination therapy with γδ T cells plus anticancer agents, molecularly targeted agents, and epigenetic agents are a promising strategy to improve the antitumor effects of γδ T cells for clinical application (Figure 2).

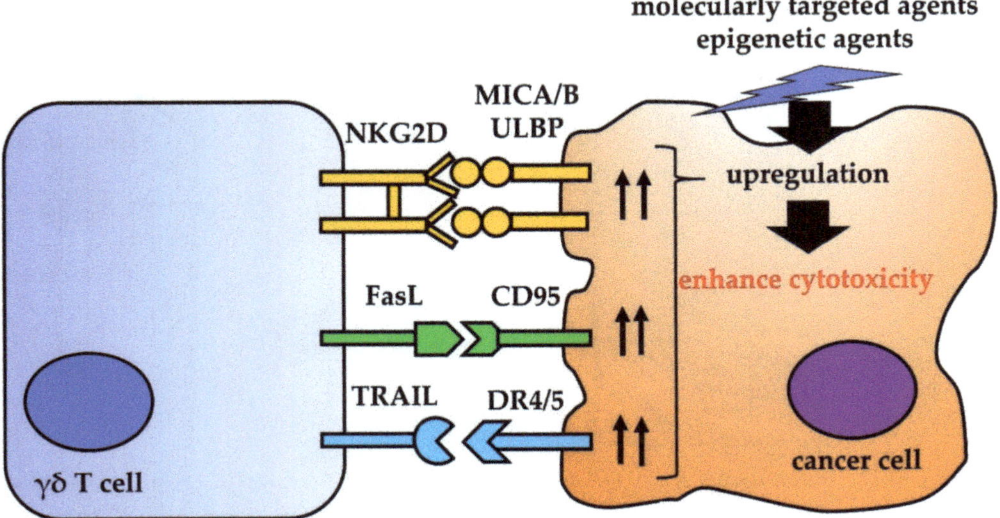

Figure 2. Interaction between γδ T cells and cancer cells. Anticancer agents, molecularly targeted agents, and epigenetic agents upregulate ligands that activate γδ T cells, thereby increasing cytotoxicity.

5. The Tumor Microenvironment (TME) Limits the Cytotoxicity of γδ T Cells by Promoting Their Regulatory Functions, by Secreting Immunosuppressive Cytokines, and by Inhibiting Immune Checkpoint Molecules

Several studies demonstrate the plasticity of γδ T cells. After activation by PAgs, γδ T cells promote a Th1 immune response by secreting TNF-α and IFN-γ; however, γδ T cells can be polarized into cells with properties similar to those of Th2 cells, Th17 cells, or regulatory T cells (Tregs) [73–76]. Unlike monolayer 2D models and mouse models injected with tumor cells, an actual tumor comprises not only cancer cells but also an extracellular matrix (ECM), stromal cells (such as fibroblasts and mesenchymal stromal cells), vascular networks, and immune cells such as T and B cells, NK cells, and tumor-associated macrophages (TAM). This is the TME. The TME plays a significant role in the subsequent evolution of malignancy [77]. For example, the TME harbors various cytokines, chemokines, growth factors, inflammatory mediators, and matrix remodeling enzymes to facilitate crosstalk between TME-constituting cells [78]; this environment can promote polarization of γδ T cells into Th17-or Treg-like cells that produce IL-17 and transforming growth factor (TGF)-β, which favor cancer cell proliferation [79,80]. IL-17-producing γδ T cells induce angiogenesis and support cancer progression [81,82]. TGF-β secreted by Treg cells can negatively regulate γδ T cells [83]. Moreover, the TME harbors various immunosuppressive cells (Figure 3).

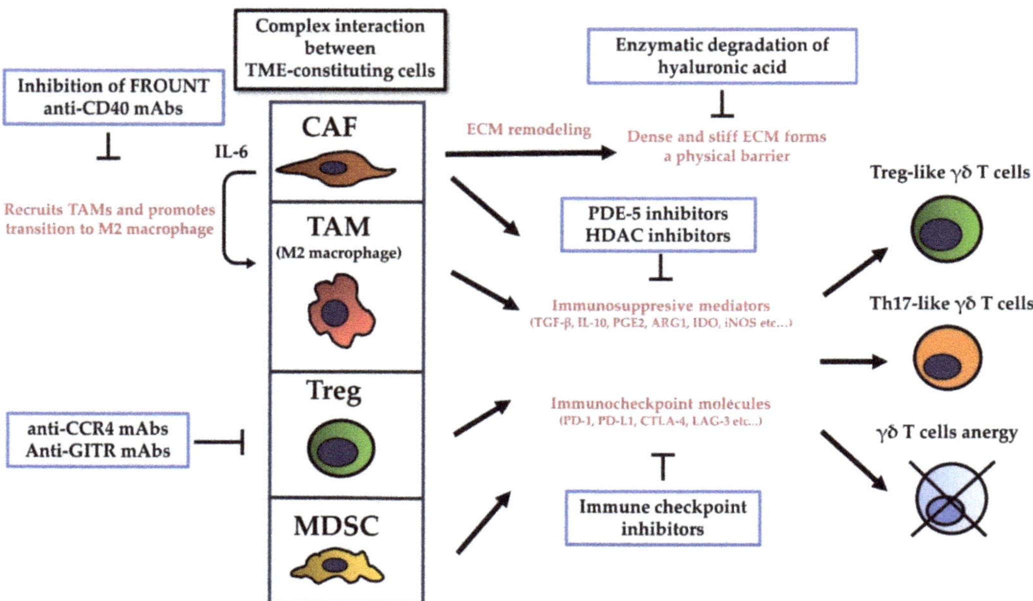

Figure 3. Cells in the TME induce polarization and anergy of γδ T cells in various ways (with red font) and some potential strategies to overcome negative effect from the TME are conceivable (in the blue boxes).

Cancer-associated fibroblasts (CAFs), which are recruited to the tumor stroma by growth factors secreted by cancer cells, are key components that maintain an immunosuppressive TME. CAFs produce matrix-crosslinking enzymes and mediate ECM remodeling, resulting in a dense and stiff ECM [84]. The dense and stiff ECM compresses intratumoral blood and lymphatic vessels to increase interstitial tissue pressure, which induces hypoxia and impedes delivery of anticancer agents. The dense and stiff ECM also forms a physical barrier that prevents immune cells from infiltrating the cancer [85]. Provenzano et al. reported that hyaluronic acid (HA) is the primary determinant of the ECM barrier. They showed that enzymatic degradation of HA reduces interstitial tissue pressure to facilitate tumor penetration by gemcitabine, leading to improved antitumor effects in preclinical pancreatic ductal adenocarcinoma transgenic mouse models [86]. HA targeting might permit efficient delivery of γδ T cells to the tumor, thereby improving the E/T ratio on the tumor site. CAFs produce various immunosuppressive cytokines and factors such as IL-6, TGF-β, and prostaglandin E2 (PGE2) [87,88]. IL-6 recruits TAMs and promotes their transition to an immunosuppressive phenotype (i.e., M2 macrophages). CAFs can also inhibit activation of cytotoxic T cells and NK cells directly by expressing inhibitory immune checkpoint molecules such as programmed death-ligand (PD-L)1 and PDL-2 [89].

Myeloid-derived suppressor cells (MDSCs) also play a crucial role in maintaining an immunosuppressive TME. They are converted from immature myeloid cells in the bone marrow by inflammatory mediators released by cancer and immune cells and are recruited to the tumor site through interaction between C-C motif receptors (CCR) and their respective chemokines, such as C-C motif chemokine ligand. They produce different immunosuppressive mediators such as arginase-1 (ARG1), indoleamine 2,3 dioxygenase (IDO), and nitric oxide synthase (iNOS), all of which induce T cell anergy via different pathways [90]. Sacchi et al. reported that MDSCs inhibit IFN-γ production by PAgs-activated γδ T cells and suppress their cytotoxic activity [91]. Several strategies to target MDSCs have been investigated. Blocking migration of MDSCs is one strategy for targeting this cell type. CCR5 plays a key role in migration of MDSCs. The interaction between CCR5 and its ligand CCL5 supports tumor growth and invasion, and migration of MDSCs to the tumor site; tumor growth and invasiveness are suppressed by targeting the CCR5-CCL5 interaction [92–94]. Inhibiting MDSCs-producing immunosuppressive mediators is another strategy for targeting MDSCs. Serafini et al. reported that sildenafil and tadalafil, both of which are inhibitors of phosphodiesterase-5 (PDE-5), increase antitumor cytotoxic T lymphocyte activity and act synergistically with adoptive vaccine-primed $CD8^+$ T cell therapy to delay tumor outgrowth in preclinical mouse models by downregulating ARG1 and iNOS activity [95]. Entinostat, a class I HDAC inhibitor, is another candidate agent that neutralizes MDSCs-producing immunosuppressive mediators. Orillion et al. reported that entinostat reduced the expression of ARG1, iNOS, and COX2 by MDSCs, and that the combination of entinostat plus anti-PD-1 antibodies increased survival and delayed tumor growth significantly in several preclinical mouse models [96]. Combination of $\gamma\delta$ T cell immunotherapy with PDE-5 inhibitors and HDAC inhibitors is a good strategy for overcoming the immunosuppressive effects of MDSCs.

Tregs, which suppress aberrant immune responses against self-antigens, promote immune evasion of the TME. Infiltration of tumor tissue by a large number of Tregs is often associated with a poor prognosis. They not only exert immunosuppressive activity indirectly by releasing soluble inhibitory molecules such as TGF-β and IL-10, but also directly by inhibiting effector T cells via immune checkpoint receptor cytotoxic T lymphocyte antigen-4 (CTLA-4) and lymphocyte activation gene-3 (LAG-3) [97,98]. Molecules that are relatively specific for Tregs are good candidates for targeting Tregs in combination with $\gamma\delta$ T cell immunotherapy. Several studies suggest that an anti-CTLA-4 monoclonal antibody (mAb) predominantly targets Treg cells and strengthens antitumor immune responses [99–101]. Moreover, the clinical efficacy of ipilimumab, a mAb specific for CTLA-4, correlates with a reduction in Treg numbers in tumor tissue [102,103]. CCR4 is expressed predominantly by effector Tregs, which are the most abundant cell type among $FOXP3^+$ T cells in tumor tissue; in addition, CCR4 ligands produced by cancer cells or by infiltrating macrophages appear to be involved in migration and infiltration of Tregs into various tumor tissues [104,105]. Sugiyama et al. reported that anti-CCR4 mAb treatment selectively depleted effector Tregs and efficiently induced tumor antigen-specific $CD4^+$ and $CD8^+$ T cells both in vitro and in vivo [106]. Glucocorticoid-induced TNF receptor-related protein (GITR) is another molecule expressed by Tregs. Ko et al. reported that administration of an agonistic anti-GITR mAb affects tumor-infiltrating Tregs and evokes a potent antitumor immune response, which can eradicate established mouse tumors without eliciting overt autoimmune disease [107].

TAMs also play a pivotal role in the TME by behaving as M2 macrophages; these cells secrete anti-inflammatory factors such as IL-10, TGF-β, and vascular endothelial growth factor (VGEF)-A [108]. These inhibitory cytokines cause cancer cells to become refractory to immunotherapy. Therefore, therapeutic strategies to target TAMs might be effective. Inhibiting differentiation of systemic monocytes once they enter tumor tissue is one strategy to target TAMs. Interaction between CCR2 on monocytes with its ligand (CCL2) induces migration of monocytes from the circulation to the tumor tissue and promotes tumor proliferation. The cytoplasmic protein, FROUNT, binds directly to activated CCR2 and facilitates monocyte infiltration. Inhibition of FROUNT decreased the number of TAMs in an osteosarcoma mouse model [109,110]. Reprogramming of TAMs, i.e., transdifferentiating M2 macrophages to M1 macrophages, is an alternative strategy to target TAMs for cancer immunotherapy. First, M1 macrophages are induced by IFN-γ, and then combined treatment with IL-2 and anti-CD40 induces a switch from an M2 to an M1 phenotype [111]. Moreover, a recent study shows that PD-1 expressed by TAMs inhibits antitumor immunity [112]. Therefore, anti-PD/PD-L1 therapies are expected to have a direct effect on TAMs.

Among these TME-targeting therapies, therapeutic antibodies specific for inhibitory immune checkpoint molecules are an attractive strategy for overcoming the immunosuppressive effects of the TME; this is because various inhibitory immune checkpoint molecules are associated with immunosuppression by various TME-constituting cells. Therapeutic antibodies specific for PD-1, PD-L1, and CTLA-4, namely immune checkpoint inhibitors, have had a huge impact on cancer immunotherapy over the past decade [113–116]. The combination of adoptive γδ T cells plus immune checkpoint inhibitors is a hopeful strategy for improving their cytotoxicity because PAgs-stimulated γδ T cells express PD-1 [117] and Rossi et al. reported that blockade of PD-1 can boost antitumor effect of γδ T cells against follicular lymphoma [118]. However, we recently reported that PD-1 blockade did not increase the cytotoxicity of γδ T cell against PD-L1 ^{high} solid cancer cells and PD-L1 knockdown did not increase the cytotoxicity [119]. The augmentation effect of blockade of PD-1/PD-L1 axis is still controversial. Further studies should investigate how other inhibitory immune checkpoint molecules such as CTLA-4, IDO, and LAG-3, mediate their immunosuppressive effects against γδ T cells, and how these immunosuppressive effects can be circumvented.

6. Cancer Stem Cells (CSCs) Could Mediate Resistance to γδ T Cell Immunotherapy

According to the American Association for Cancer Research (AACR), CSCs are defined as cells within a tumor that possess the capacity to self-renew and to cause the heterogeneous lineages of cancer cells that comprises the tumor [120]. CSCs are a rare cell population within the tumor, but they are spared after conventional therapy because they are resistant and have the capacity to self-renew, ultimately causing tumor relapse and metastasis. Recent studies indicate that CSCs in various solid tumors play an important role in tumor resistance to conventional chemotherapy and radiotherapy [121–123]. Therefore, unsatisfactory clinical responses reported by past clinical trials of γδ T cell immunotherapy against various advanced and recurrent cancers might be due to the presence of CSCs. Moreover, CSCs can modulate immune cell activity by interacting with the TME. Jinushi reported that chemoresistant CSCs promote M2 macrophage differentiation through interferon-regulatory factor-5 (IRF5)- and macrophage-colony stimulating factor (M-CSF)-dependent mechanisms [124]. Schatton et al. reported that malignant melanoma CSCs possess the capacity to inhibit IL-2-dependent T cell activation and support induction of Tregs [125]. In addition, CSCs secrete several immunosuppressive cytokines into the TME, including TGF-β, IL-10, IL-4, and IL-13 [126,127]. CSCs also express high levels of immune checkpoint molecules, which enable them to evade to immune system [128]. Few studies have investigated the relationship between CSCs and γδ T cells. Previously, we generated prostate cancer spheres and used them to examine the cytotoxicity of *ex vivo*-expanded γδ T cells against sphere-derived prostate cancer cells. Sphere-derived prostate cancer cells were resistant to ex vivo-expanded γδ T cells; in addition, their stem cell markers, including CD133, NANOG, SOX2, and OCT4, were upregulating compared with those of parental cells [129]. These results suggest that ex vivo-expanded γδ T cells will not be effective against CSCs. Further research is needed to clarify the mechanisms underlying the resistance of CSCs to human γδ T cells.

7. Novel Forms of γδ T Cell Therapy Overcome Current Therapeutic Limitations

Recently, several strategies have been developed to improve the antitumor effect of γδ T cell immunotherapy. The use of a bispecific antibody, which is typically equipped with a first specificity for an antigen expressed by cancer cells and a second specificity for an activating molecule on effector cells [130], improved the cytotoxicity significantly. Hoh et al. reported that EpCAM/CD3 bispecific antibody enhanced γδ T cell -mediated lysis of hepatoblastoma and paediatric hepatocellular carcinoma cells in spheroid culture models [131]. Oberg et al. reported that ex-vivo expanded γδ T cell administration with the HER2/Vγ9 bispecific antibody significantly reduced the growth of pancreatic cancer and colon cancer in preclinical models [132,133]. They also reported that tribody [$(HER2)_2$xCD16], which comprises two HER2-specific single chain fragment variables fused to a fragment antigen biding directed to the CD16 antigen expressed on γδ T cells and NK cells, enhanced γδ T cells and NK cells-mediated lysis of HER2-expressing tumor cells, such as pancreatic ductal adenocarcinoma, breast cancer, and autologous primary ovarian tumors [134]. Bispecific antibodies may be promising strategy to overcome current therapeutic limitations. Chimeric antigen receptor-transduced γδ T cells (CAR-γδ T cells) is another novel strategy to overcome current therapeutic limitations. Chimeric antigen receptors (CARs) are usually derived from single-chain variable fragments (scFvs) of antibodies specific for tumor antigens and transduced using viral vectors. Unlike TCRs, which have narrow range of affinities, CARs typically have a much higher and broader range of affinities [135], thus enabling the CAR-γδ T cells to recognize tumor epitopes independently on their TCR. Deniger et al. reported that polyclonal γδ T cells with CD19-specific CAR-γδ T cells enhanced killing of $CD19^+$ tumor cells compared with CAR^{neg} γδ T cells in vitro, and CD19-specific CAR-γδ T cells reduced $CD19^+$ leukemia xenografts in mice [136]. CAR-T cell immunotherapy has an off-target effect problem. Fisher et al. designed GD2-specific CAR-γδ T cells in order to limit the toxic effects on normal cells. GD2 is abundantly expressed on the surface of neuroblastoma cells and on several other cancer cell types. In this study, γδ T cells recognized the tumor antigen, then the monoclonal antibody against GD2 recognized GD2 and activated the downstream signal domain to exert antitumor effects. Consequently, GD2-expressing neuroblastoma cells which engaged γδ TCR were efficiently lysed, whereas cells that expressed GD2 equivalently bud did not engage γδ TCR were untouched [137]. Currently, several clinical studies have been ongoing (Table 2). CAR-γδ T cells are expected to be a new type of γδ T cell immunotherapy in the future.

Table 2. CAR-γδ T cell-based clinical trials.

Clinical Trials. Gov Identifier	Interventions	Cancers	Phase
NCT02656147	Anti-CD19-CAR-γδ T cell	Leukemia and lymphoma	I
NCT04107142	NKG2DL-targeting CAR-γδ T cell	Solid cancer	I
NCT04702841	CAR-γδ T cell	Relapsed and refractory CD7 positive T cell-derived malignant tumor	I
NCT04796441	CAR-γδ T cell	AML	Not Applicable

CAR: chimeric antigen receptor; NKG2DL: natural killer group 2 member D ligand; AML: acute myeloid leukemia.

8. Conclusions

In this review, we have discussed different ways of activating γδ T cells, along with various strategies aimed at improving their antitumor effects during clinical application. γδ T cell-based immunotherapy is very attractive because these cells show cytotoxic effects against various cancer types, both in vitro and in mouse models. However, clinical trials have reported limited clinical benefit. In vivo activation of γδ T cells by systemic administration of PAgs or N-bis, along with exogenous interleukin (IL)-2, is well tolerated; however, the clinical benefits appear to be mild to moderate, likely due to anergy and exhaustion of activation-induced γδ T cells. However, adoptive immunotherapy using ex vivo-expanded γδ T cells could be achieved by repeated administration of activated γδ T cells, although it is difficult to acquire adequate numbers of activated γδ T cells from some patients. Further research into the mechanisms underlying this problem is needed. Another problem with adoptive immunotherapy conferred by ex vivo-expanded γδ T cells is that systematic intravenous administration of these cells does not achieve a high E/T ratio at the target tumor site. Administration of ex vivo-expanded γδ T cells into a local cavity resolves this problem and is a promising approach to making the most out of their cytotoxic potential. Moreover, pretreatment with anticancer agents, molecularly targeted agents, and epigenetic agents sensitizes cancer cells to γδ T cells by upregulating expression of several stress-induced ligands. Immunosuppression of γδ T cells by the TME and CSCs is less clear-cut, and might operate via multiple mechanisms; however, they affect the immune system via common inhibitory immune checkpoint molecules. Therefore, co-immunotherapy with γδ T cells plus immune checkpoint inhibitors is one strategy that may improve cytotoxicity. Bispecific antibodies and CAR-γδ T cells are novel strategies which are expected to overcome current therapeutic limitations. Further basic studies of the immunosuppressive effects of the TME and CSCs on γδ T cells, along with clinical studies examining administration into local cavities, combination therapy with anticancer agents, molecularly targeted agents, epigenetic agents, and bispecific antibodies, and CAR-γδ T cell immunotherapy are needed to ensure successful clinical application of γδ T cell-based immunotherapy.

Author Contributions: M.M. and T.S. wrote the manuscript and drew the Figures. E.A. and O.U. reviewed the manuscript and finalized it for publication. All authors have read and agreed to the published version of the manuscript.

Funding: This work was supported by Scholarship donation to the Department of Urology, Kyoto Prefectural University of Medicine.

Institutional Review Board Statement: Not applicable.

Informed Consent Statement: Not applicable.

Data Availability Statement: Not applicable.

Conflicts of Interest: The authors declare no conflict of interest.

References

1. Bray, F.; Ferlay, J.; Soerjomataram, I.; Siegel, R.L.; Torre, L.A.; Jemal, A. Global cancer statistics 2018: GLOBOCAN estimates of incidence and mortality worldwide for 36 cancers in 185 countries. *CA Cancer J. Clin.* **2018**, *68*, 394–424. [CrossRef]
2. Aptsiauri, N.; Cabrera, T.; Garcia-Lora, A.; Lopez-Nevot, M.A.; Ruiz-Cabello, F.; Garrido, F. MHC class I antigens and immune surveillance in transformed cells. *Int. Rev. Cytol.* **2007**, *256*, 139–189. [CrossRef]
3. Saito, H.; Kranz, D.M.; Takagaki, Y.; Hayday, A.C.; Eisen, H.N.; Tonegawa, S. Complete primary structure of a heterodimeric T-cell receptor deduced from cDNA sequences. *Nature* **1984**, *309*, 757–762. [CrossRef] [PubMed]
4. Brenner, M.B.; McLean, J.; Dialynas, D.P.; Strominger, J.L.; Smith, J.A.; Owen, F.L.; Seidman, J.G.; Ip, S.; Rosen, F.; Krangel, M.S. Identification of a putative second T-cell receptor. *Nature* **1986**, *322*, 145–149. [CrossRef] [PubMed]
5. Chien, Y.H.; Iwashima, M.; Kaplan, K.B.; Elliott, J.F.; Davis, M.M. A new T-cell receptor gene located within the alpha locus and expressed early in T-cell differentiation. *Nature* **1987**, *327*, 677–682. [CrossRef]
6. Morita, C.T.; Beckman, E.M.; Bukowski, J.F.; Tanaka, Y.; Band, H.; Bloom, B.R.; Golan, D.E.; Brenner, M.B. Direct presentation of nonpeptide prenyl pyrophosphate antigens to human γδ T cells. *Immunity* **1995**, *3*, 495–507. [CrossRef]
7. Liu, Z.; Guo, B.L.; Gehrs, B.C.; Nan, L.; Lopez, R.D. Ex vivo expanded human Vλ9Vδ 2+ λδ-T cells mediate innate antitumor activity against human prostate cancer cells in vitro. *J. Urol.* **2005**, *173*, 1552–1556. [CrossRef]
8. Corvaisier, M.; Moreau-Aubry, A.; Diez, E.; Bennouna, J.; Mosnier, J.F.; Scotet, E.; Bonneville, M.; Jotereau, F. Vγ9Vδ2 T cell response to colon carcinoma cells. *J. Immunol.* **2005**, *175*, 5481–5488. [CrossRef]
9. Viey, E.; Fromont, G.; Escudier, B.; Morel, Y.; Da Rocha, S.; Chouaib, S.; Caignard, A. Phosphostim-activated γδ T cells kill autologous metastatic renal cell carcinoma. *J. Immunol.* **2005**, *174*, 1338–1347. [CrossRef] [PubMed]
10. Uchida, R.; Ashihara, E.; Sato, K.; Kimura, S.; Kuroda, J.; Takeuchi, M.; Kawata, E.; Taniguchi, K.; Okamoto, M.; Shimura, K.; et al. γδ T cells kill myeloma cells by sensing mevalonate metabolites and ICAM-1 molecules on cell surface. *Biochem. Biophys. Res. Commun.* **2007**, *354*, 613–618. [CrossRef]
11. Toutirais, O.; Cabillic, F.; Le Friec, G.; Salot, S.; Loyer, P.; Le Gallo, M.; Desille, M.; de La Pintière, C.T.; Daniel, P.; Bouet, F.; et al. DNAX accessory molecule-1 (CD226) promotes human hepatocellular carcinoma cell lysis by Vγ9Vδ2 T cells. *Eur. J. Immunol.* **2009**, *39*, 1361–1368. [CrossRef]
12. Ashihara, E.; Munaka, T.; Kimura, S.; Nakagawa, S.; Nakagawa, Y.; Kanai, M.; Hirai, H.; Abe, H.; Miida, T.; Yamato, S.; et al. Isopentenyl pyrophosphate secreted from Zoledronate-stimulated myeloma cells, activates the chemotaxis of γδT cells. *Biochem. Biophys. Res. Commun.* **2015**, *463*, 650–655. [CrossRef]
13. Hintz, M.; Reichenberg, A.; Altincicek, B.; Bahr, U.; Gschwind, R.M.; Kollas, A.K.; Beck, E.; Wiesner, J.; Eberl, M.; Jomaa, H. Identification of (E)-4-hydroxy-3-methyl-but-2-enyl pyrophosphate as a major activator for human γδ T cells in Escherichia coli. *FEBS Lett.* **2001**, *509*, 317–322. [CrossRef]
14. Gober, H.J.; Kistowska, M.; Angman, L.; Jenö, P.; Mori, L.; De Libero, G. Human T cell receptor γδ cells recognize endogenous mevalonate metabolites in tumor cells. *J. Exp. Med.* **2003**, *197*, 163–168. [CrossRef] [PubMed]
15. Pfeffer, K.; Schoel, B.; Gulle, H.; Kaufmann, S.H.; Wagner, H. Primary responses of human T cells to mycobacteria: A frequent set of γ/δ T cells are stimulated by protease-resistant ligands. *Eur. J. Immunol.* **1990**, *20*, 1175–1179. [CrossRef]
16. Constant, P.; Davodeau, F.; Peyrat, M.A.; Poquet, Y.; Puzo, G.; Bonneville, M.; Fournié, J.J. Stimulation of human gamma delta T cells by nonpeptidic mycobacterial ligands. *Science* **1994**, *264*, 267–270. [CrossRef] [PubMed]
17. Harly, C.; Peigné, C.M.; Scotet, E. Molecules and Mechanisms Implicated in the Peculiar Antigenic Activation Process of Human Vγ9Vδ2 T Cells. *Front. Immunol.* **2015**, *5*, 657. [CrossRef]

18. Rhodes, D.A.; Chen, H.C.; Price, A.J.; Keeble, A.H.; Davey, M.S.; James, L.C.; Eberl, M.; Trowsdale, J. Activation of human γδ T cells by cytosolic interactions of BTN3A1 with soluble phosphoantigens and the cytoskeletal adaptor periplakin. *J. Immunol.* **2015**, *194*, 2390–2398. [CrossRef] [PubMed]
19. Boutin, L.; Scotet, E. Towards Deciphering the Hidden Mechanisms That Contribute to the Antigenic Activation Process of Human Vγ9Vδ2 T Cells. *Front. Immunol.* **2018**, *9*, 828. [CrossRef] [PubMed]
20. Karunakaran, M.M.; Willcox, C.R.; Salim, M.; Paletta, D.; Fichtner, A.S.; Noll, A.; Starick, L.; Nöhren, A.; Begley, C.R.; Berwick, K.A.; et al. Butyrophilin-2A1 Directly Binds Germline-Encoded Regions of the Vγ9Vδ2 TCR and Is Essential for Phosphoantigen Sensing. *Immunity* **2020**, *52*, 487–498.e6. [CrossRef]
21. Cano, C.E.; Pasero, C.; De Gassart, A.; Kerneur, C.; Gabriac, M.; Fullana, M.; Granarolo, E.; Hoet, R.; Scotet, E.; Rafia, C.; et al. BTN2A1, an immune checkpoint targeting Vγ9Vδ2 T cell cytotoxicity against malignant cells. *Cell Rep.* **2021**, *36*, 109359. [CrossRef]
22. Clendening, J.W.; Pandyra, A.; Boutros, P.C.; El Ghamrasni, S.; Khosravi, F.; Trentin, G.A.; Martirosyan, A.; Hakem, A.; Hakem, R.; Jurisica, I.; et al. Dysregulation of the mevalonate pathway promotes transformation. *Proc. Natl. Acad. Sci. USA* **2010**, *107*, 15051–15056. [CrossRef] [PubMed]
23. Gruenbacher, G.; Thurnher, M. Mevalonate Metabolism in Cancer Stemness and Trained Immunity. *Front. Oncol.* **2018**, *8*, 394. [CrossRef]
24. Kavanagh, K.L.; Guo, K.; Dunford, J.E.; Wu, X.; Knapp, S.; Ebetino, F.H.; Rogers, M.J.; Russell, R.G.; Oppermann, U. The molecular mechanism of nitrogen-containing bisphosphonates as antiosteoporosis drugs. *Proc. Natl. Acad. Sci. USA* **2006**, *103*, 7829–7834. [CrossRef]
25. Kobayashi, H.; Tanaka, Y. γδ T Cell Immunotherapy-A Review. *Pharmaceuticals* **2015**, *8*, 40–61. [CrossRef] [PubMed]
26. Dieli, F.; Vermijlen, D.; Fulfaro, F.; Caccamo, N.; Meraviglia, S.; Cicero, G.; Roberts, A.; Buccheri, S.; D'Asaro, M.; Gebbia, N.; et al. Targeting human γδ T cells with zoledronate and interleukin-2 for immunotherapy of hormone-refractory prostate cancer. *Cancer Res.* **2007**, *67*, 7450–7457. [CrossRef]
27. Wilhelm, M.; Kunzmann, V.; Eckstein, S.; Reimer, P.; Weissinger, F.; Ruediger, T.; Tony, H.P. γδ T cells for immune therapy of patients with lymphoid malignancies. *Blood* **2003**, *102*, 200–206. [CrossRef]
28. Lang, J.M.; Kaikobad, M.R.; Wallace, M.; Staab, M.J.; Horvath, D.L.; Wilding, G.; Liu, G.; Eickhoff, J.C.; McNeel, D.G.; Malkovsky, M. Pilot trial of interleukin-2 and zoledronic acid to augment γδ T cells as treatment for patients with refractory renal cell carcinoma. *Cancer Immunol. Immunother.* **2011**, *60*, 1447–1460. [CrossRef]
29. Meraviglia, S.; Eberl, M.; Vermijlen, D.; Todaro, M.; Buccheri, S.; Cicero, G.; La Mendola, C.; Guggino, G.; D'Asaro, M.; Orlando, V.; et al. In vivo manipulation of Vγ9Vδ2 T cells with zoledronate and low-dose interleukin-2 for immunotherapy of advanced breast cancer patients. *Clin. Exp. Immunol.* **2010**, *161*, 290–297. [CrossRef]
30. Bennouna, J.; Levy, V.; Sicard, H.; Senellart, H.; Audrain, M.; Hiret, S.; Rolland, F.; Bruzzoni-Giovanelli, H.; Rimbert, M.; Galéa, C.; et al. Phase I study of bromohydrin pyrophosphate (BrHPP, IPH 1101), a Vγ9Vδ2 T lymphocyte agonist in patients with solid tumors. *Cancer Immunol. Immunother.* **2010**, *59*, 1521–1530. [CrossRef] [PubMed]
31. Bennouna, J.; Bompas, E.; Neidhardt, E.M.; Rolland, F.; Philip, I.; Galéa, C.; Salot, S.; Saiagh, S.; Audrain, M.; Rimbert, M.; et al. Phase-I study of Innacell γδ, an autologous cell-therapy product highly enriched in γ9δ2 T lymphocytes, in combination with IL-2, in patients with metastatic renal cell carcinoma. *Cancer Immunol. Immunother.* **2008**, *57*, 1599–1609. [CrossRef] [PubMed]
32. Kunzmann, V.; Smetak, M.; Kimmel, B.; Weigang-Koehler, K.; Goebeler, M.; Birkmann, J.; Becker, J.; Schmidt-Wolf, I.G.; Einsele, H.; Wilhelm, M. Tumor-promoting versus tumor-antagonizing roles of γδ T cells in cancer immunotherapy: Results from a prospective phase I/II trial. *J. Immunother.* **2012**, *35*, 205–213. [CrossRef] [PubMed]
33. Kobayashi, H.; Tanaka, Y.; Yagi, J.; Osaka, Y.; Nakazawa, H.; Uchiyama, T.; Minato, N.; Toma, H. Safety profile and anti-tumor effects of adoptive immunotherapy using gamma-delta T cells against advanced renal cell carcinoma: A pilot study. *Cancer Immunol. Immunother.* **2007**, *56*, 469–476. [CrossRef] [PubMed]
34. Kobayashi, H.; Tanaka, Y.; Yagi, J.; Minato, N.; Tanabe, K. Phase I/II study of adoptive transfer of γδ T cells in combination with zoledronic acid and IL-2 to patients with advanced renal cell carcinoma. *Cancer Immunol. Immunother.* **2011**, *60*, 1075–1084. [CrossRef]
35. Nicol, A.J.; Tokuyama, H.; Mattarollo, S.R.; Hagi, T.; Suzuki, K.; Yokokawa, K.; Nieda, M. Clinical evaluation of autologous gamma delta T cell-based immunotherapy for metastatic solid tumours. *Br. J. Cancer* **2011**, *105*, 778–786. [CrossRef]
36. Abe, Y.; Muto, M.; Nieda, M.; Nakagawa, Y.; Nicol, A.; Kaneko, T.; Goto, S.; Yokokawa, K.; Suzuki, K. Clinical and immunological evaluation of zoledronate-activated Vγ9γδ T-cell-based immunotherapy for patients with multiple myeloma. *Exp. Hematol.* **2009**, *37*, 956–968. [CrossRef]
37. Nakajima, J.; Murakawa, T.; Fukami, T.; Goto, S.; Kaneko, T.; Yoshida, Y.; Takamoto, S.; Kakimi, K. A phase I study of adoptive immunotherapy for recurrent non-small-cell lung cancer patients with autologous γδ T cells. *Eur. J. Cardiothorac. Surg.* **2010**, *37*, 1191–1197. [CrossRef]
38. Wada, I.; Matsushita, H.; Noji, S.; Mori, K.; Yamashita, H.; Nomura, S.; Shimizu, N.; Seto, Y.; Kakimi, K. Intraperitoneal injection of in vitro expanded Vγ9Vδ2 T cells together with zoledronate for the treatment of malignant ascites due to gastric cancer. *Cancer Med.* **2014**, *3*, 362–375. [CrossRef]

39. Sakamoto, M.; Nakajima, J.; Murakawa, T.; Fukami, T.; Yoshida, Y.; Murayama, T.; Takamoto, S.; Matsushita, H.; Kakimi, K. Adoptive immunotherapy for advanced non-small cell lung cancer using zoledronate-expanded γδTcells: A phase I clinical study. *J. Immunother.* **2011**, *34*, 202–211. [CrossRef]
40. Noguchi, A.; Kaneko, T.; Kamigaki, T.; Fujimoto, K.; Ozawa, M.; Saito, M.; Ariyoshi, N.; Goto, S. Zoledronate-activated Vγ9γδ T cell-based immunotherapy is feasible and restores the impairment of γδ T cells in patients with solid tumors. *Cytotherapy* **2011**, *13*, 92–97. [CrossRef]
41. Izumi, T.; Kondo, M.; Takahashi, T.; Fujieda, N.; Kondo, A.; Tamura, N.; Murakawa, T.; Nakajima, J.; Matsushita, H.; Kakimi, K. Ex vivo characterization of γδ T-cell repertoire in patients after adoptive transfer of Vγ9Vδ2 T cells expressing the interleukin-2 receptor β-chain and the common γ-chain. *Cytotherapy* **2013**, *15*, 481–491. [CrossRef]
42. Yoshida, Y.; Nakajima, J.; Wada, H.; Kakimi, K. γδ T-cell immunotherapy for lung cancer. *Surg. Today* **2011**, *41*, 606–611. [CrossRef] [PubMed]
43. Roelofs, A.J.; Jauhiainen, M.; Mönkkönen, H.; Rogers, M.J.; Mönkkönen, J.; Thompson, K. Peripheral blood monocytes are responsible for γδ T cell activation induced by zoledronic acid through accumulation of IPP/DMAPP. *Br. J. Haematol.* **2009**, *144*, 245–250. [CrossRef] [PubMed]
44. Dieli, F.; Gebbia, N.; Poccia, F.; Caccamo, N.; Montesano, C.; Fulfaro, F.; Arcara, C.; Valerio, M.R.; Meraviglia, S.; Di Sano, C.; et al. Induction of γδ T-lymphocyte effector functions by bisphosphonate zoledronic acid in cancer patients in vivo. *Blood* **2003**, *102*, 2310–2311. [CrossRef]
45. Bryant, N.L.; Gillespie, G.Y.; Lopez, R.D.; Markert, J.M.; Cloud, G.A.; Langford, C.P.; Arnouk, H.; Su, Y.; Haines, H.L.; Suarez-Cuervo, C.; et al. Preclinical evaluation of ex vivo expanded/activated γδ T cells for immunotherapy of glioblastoma multiforme. *J. Neurooncol.* **2011**, *101*, 179–188. [CrossRef]
46. Yuasa, T.; Sato, K.; Ashihara, E.; Takeuchi, M.; Maita, S.; Tsuchiya, N.; Habuchi, T.; Maekawa, T.; Kimura, S. Intravesical administration of γδ T cells successfully prevents the growth of bladder cancer in the murine model. *Cancer Immunol. Immunother.* **2009**, *58*, 493–502. [CrossRef]
47. Shimizu, T.; Tomogane, M.; Miyashita, M.; Ukimura, O.; Ashihara, E. Low dose gemcitabine increases the cytotoxicity of human Vγ9Vδ2 T cells in bladder cancer cells in vitro and in an orthotopic xenograft model. *Oncoimmunology* **2018**, *7*, e1424671. [CrossRef] [PubMed]
48. Rincon-Orozco, B.; Kunzmann, V.; Wrobel, P.; Kabelitz, D.; Steinle, A.; Herrmann, T. Activation of Vγ9Vδ2 T cells by NKG2D. *J. Immunol.* **2005**, *175*, 2144–2151. [CrossRef]
49. Bauer, S.; Groh, V.; Wu, J.; Steinle, A.; Phillips, J.H.; Lanier, L.L.; Spies, T. Activation of NK cells and T cells by NKG2D, a receptor for stress-inducible MICA. *Science* **1999**, *285*, 727–729. [CrossRef]
50. Das, H.; Groh, V.; Kuijl, C.; Sugita, M.; Morita, C.T.; Spies, T.; Bukowski, J.F. MICA engagement by human Vγ2Vδ2 T cells enhances their antigen-dependent effector function. *Immunity* **2001**, *15*, 83–93. [CrossRef]
51. Groh, V.; Steinle, A.; Bauer, S.; Spies, T. Recognition of stress-induced MHC molecules by intestinal epithelial γδ T cells. *Science* **1998**, *279*, 1737–1740. [CrossRef] [PubMed]
52. Cosman, D.; Müllberg, J.; Sutherland, C.L.; Chin, W.; Armitage, R.; Fanslow, W.; Kubin, M.; Chalupny, N.J. ULBPs, novel MHC class I-related molecules, bind to CMV glycoprotein UL16 and stimulate NK cytotoxicity through the NKG2D receptor. *Immunity* **2001**, *14*, 123–133. [CrossRef]
53. Chalupny, N.J.; Sutherland, C.L.; Lawrence, W.A.; Rein-Weston, A.; Cosman, D. ULBP4 is a novel ligand for human NKG2D. *Biochem. Biophys. Commun.* **2003**, *305*, 129–135. [CrossRef]
54. Chen, G.; Emens, L.A. Chemoimmunotherapy: Reengineering tumor immunity. *Cancer Immunol. Immunother.* **2013**, *62*, 203–216. [CrossRef]
55. Green, D.R.; Ferguson, T.; Zitvogel, L.; Kroemer, G. Immunogenic and tolerogenic cell death. *Nat. Rev. Immunol.* **2009**, *9*, 353–363. [CrossRef]
56. Gasser, S.; Raulet, D. The DNA damage response, immunity and cancer. *Semin. Cancer Biol.* **2006**, *16*, 344–347. [CrossRef]
57. Todaro, M.; Orlando, V.; Cicero, G.; Caccamo, N.; Meraviglia, S.; Stassi, G.; Dieli, F. Chemotherapy sensitizes colon cancer initiating cells to Vγ9Vδ2 T cell-mediated cytotoxicity. *PLoS ONE* **2013**, *8*, e65145. [CrossRef] [PubMed]
58. Lamb, L.S., Jr.; Bowersock, J.; Dasgupta, A.; Gillespie, G.Y.; Su, Y.; Johnson, A.; Spencer, H.T. Engineered drug resistant γδ T cells kill glioblastoma cell lines during a chemotherapy challenge: A strategy for combining chemo- and immunotherapy. *PLoS ONE* **2013**, *8*, e51805. [CrossRef] [PubMed]
59. Huang, Y.; Wang, Y.; Li, Y.; Guo, K.; He, Y. Role of sorafenib and sunitinib in the induction of expressions of NKG2D ligands in nasopharyngeal carcinoma with high expression of ABCG2. *J. Cancer Res. Clin. Oncol.* **2011**, *137*, 829–837. [CrossRef] [PubMed]
60. Kim, H.; Kim, S.H.; Kim, M.J.; Kim, S.J.; Park, S.J.; Chung, J.S.; Bae, J.H.; Kang, C.D. EGFR inhibitors enhanced the susceptibility to NK cell-mediated lysis of lung cancer cells. *J. Immunother.* **2011**, *34*, 372–381. [CrossRef] [PubMed]
61. Story, J.Y.; Zoine, J.T.; Burnham, R.E.; Hamilton, J.; Spencer, H.T.; Doering, C.B.; Raikar, S.S. Bortezomib enhances cytotoxicity of ex vivo-expanded gamma delta T cells against acute myeloid leukemia and T-cell acute lymphoblastic leukemia. *Cytotherapy* **2021**, *23*, 12–24. [CrossRef]
62. Skov, S.; Pedersen, M.T.; Andresen, L.; Straten, P.T.; Woetmann, A.; Odum, N. Cancer cells become susceptible to natural killer cell killing after exposure to histone deacetylase inhibitors due to glycogen synthase kinase-3-dependent expression of MHC class I-related chain A and B. *Cancer Res.* **2005**, *65*, 11136–11145. [CrossRef] [PubMed]

63. Braza, M.S.; Klein, B. Anti-tumour immunotherapy with Vγ9Vδ2 T lymphocytes: From the bench to the bedside. *Br. J. Haematol.* **2013**, *160*, 123–132. [CrossRef]
64. Nagata, S. Early work on the function of CD95, an interview with Shige Nagata. *Cell Death Differ.* **2004**, *11* (Suppl. S1), S23–S27. [CrossRef]
65. Pan, G.; O'Rourke, K.; Chinnaiyan, A.M.; Gentz, R.; Ebner, R.; Ni, J.; Dixit, V.M. The receptor for the cytotoxic ligand TRAIL. *Science* **1997**, *276*, 111–113. [CrossRef]
66. Wu, G.S.; Burns, T.F.; McDonald, E.R., 3rd; Jiang, W.; Meng, R.; Krantz, I.D.; Kao, G.; Gan, D.D.; Zhou, J.Y.; Muschel, R.; et al. KILLER/DR5 is a DNA damage-inducible p53-regulated death receptor gene. *Nat. Genet.* **1997**, *17*, 141–143. [CrossRef] [PubMed]
67. Pan, G.; Ni, J.; Wei, Y.F.; Yu, G.; Gentz, R.; Dixit, V.M. An antagonist decoy receptor and a death domain-containing receptor for TRAIL. *Science* **1997**, *277*, 815–818. [CrossRef]
68. Degli-Esposti, M.A.; Dougall, W.C.; Smolak, P.J.; Waugh, J.Y.; Smith, C.A.; Goodwin, R.G. The novel receptor TRAIL-R4 induces NF-kappaB and protects against TRAIL-mediated apoptosis, yet retains an incomplete death domain. *Immunity* **1997**, *7*, 813–820. [CrossRef]
69. Emery, J.G.; McDonnell, P.; Burke, M.B.; Deen, K.C.; Lyn, S.; Silverman, C.; Dul, E.; Appelbaum, E.R.; Eichman, C.; DiPrinzio, R.; et al. Osteoprotegerin is a receptor for the cytotoxic ligand TRAIL. *J. Biol. Chem.* **1998**, *273*, 14363–14367. [CrossRef] [PubMed]
70. Wang, S. TRAIL: A sword for killing tumors. *Curr. Med. Chem.* **2010**, *17*, 3309–3317. [CrossRef]
71. Shankar, S.; Chen, X.; Srivastava, R.K. Effects of sequential treatments with chemotherapeutic drugs followed by TRAIL on prostate cancer in vitro and in vivo. *Prostate* **2005**, *62*, 165–186. [CrossRef] [PubMed]
72. Mattarollo, S.R.; Kenna, T.; Nieda, M.; Nicol, A.J. Chemotherapy pretreatment sensitizes solid tumor-derived cell lines to Vα24+ NKT cell-mediated cytotoxicity. *Int. J. Cancer* **2006**, *119*, 1630–1637. [CrossRef]
73. Wesch, D.; Glatzel, A.; Kabelitz, D. Differentiation of resting human peripheral blood gamma delta T cells toward Th1- or Th2-phenotype. *Cell. Immunol.* **2001**, *212*, 110–117. [CrossRef]
74. Vermijlen, D.; Ellis, P.; Langford, C.; Klein, A.; Engel, R.; Willimann, K.; Jomaa, H.; Hayday, A.C.; Eberl, M. Distinct cytokine-driven responses of activated blood γδ T cells: Insights into unconventional T cell pleiotropy. *J. Immunol.* **2007**, *178*, 4304–4314. [CrossRef]
75. Bansal, R.R.; Mackay, C.R.; Moser, B.; Eberl, M. IL-21 enhances the potential of human γδ T cells to provide B-cell help. *Eur. J. Immunol.* **2012**, *42*, 110–119. [CrossRef]
76. Casetti, R.; Agrati, C.; Wallace, M.; Sacchi, A.; Martini, F.; Martino, A.; Rinaldi, A.; Malkovsky, M. Cutting edge: TGF-β1 and IL-15 Induce FOXP3+ γδ regulatory T cells in the presence of antigen stimulation. *J. Immunol.* **2009**, *183*, 3574–3577. [CrossRef] [PubMed]
77. Roma-Rodrigues, C.; Mendes, R.; Baptista, P.V.; Fernandes, A.R. Targeting Tumor Microenvironment for Cancer Therapy. *Int. J. Mol. Sci.* **2019**, *20*, 840. [CrossRef] [PubMed]
78. Baghban, R.; Roshangar, L.; Jahanban-Esfahlan, R.; Seidi, K.; Ebrahimi-Kalan, A.; Jaymand, M.; Kolahian, S.; Javaheri, T.; Zare, P. Tumor microenvironment complexity and therapeutic implications at a glance. *Cell Commun. Signal.* **2020**, *18*, 59. [CrossRef] [PubMed]
79. Lo Presti, E.; Toia, F.; Oieni, S.; Buccheri, S.; Turdo, A.; Mangiapane, L.R.; Campisi, G.; Caputo, V.; Todaro, M.; Stassi, G.; et al. Squamous Cell Tumors Recruit γδ T Cells Producing either IL17 or IFNγ Depending on the Tumor Stage. *Cancer Immunol. Res.* **2017**, *5*, 397–407. [CrossRef] [PubMed]
80. Caccamo, N.; La Mendola, C.; Orlando, V.; Meraviglia, S.; Todaro, M.; Stassi, G.; Sireci, G.; Fournié, J.J.; Dieli, F. Differentiation, phenotype, and function of interleukin-17-producing human Vγ9Vδ2 T cells. *Blood* **2011**, *118*, 129–138. [CrossRef] [PubMed]
81. Patil, R.S.; Shah, S.U.; Shrikhande, S.V.; Goel, M.; Dikshit, R.P.; Chiplunkar, S.V. IL17 producing γδT cells induce angiogenesis and are associated with poor survival in gallbladder cancer patients. *Int. J. Cancer* **2016**, *139*, 869–881. [CrossRef]
82. Wakita, D.; Sumida, K.; Iwakura, Y.; Nishikawa, H.; Ohkuri, T.; Chamoto, K.; Kitamura, H.; Nishimura, T. Tumor-infiltrating IL-17-producing γδ T cells support the progression of tumor by promoting angiogenesis. *Eur. J. Immunol.* **2010**, *40*, 1927–1937. [CrossRef]
83. Yi, Y.; He, H.W.; Wang, J.X.; Cai, X.Y.; Li, Y.W.; Zhou, J.; Cheng, Y.F.; Jin, J.J.; Fan, J.; Qiu, S.J. The functional impairment of HCC-infiltrating γδ T cells, partially mediated by regulatory T cells in a TGFβ- and IL-10-dependent manner. *J. Hepatol.* **2013**, *58*, 977–983. [CrossRef]
84. Piersma, B.; Hayward, M.K.; Weaver, V.M. Fibrosis and cancer: A strained relationship. *Biochim. Biophys. Acta Rev. Cancer* **2020**, *1873*, 188356. [CrossRef]
85. Salmon, H.; Franciszkiewicz, K.; Damotte, D.; Dieu-Nosjean, M.C.; Validire, P.; Trautmann, A.; Mami-Chouaib, F.; Donnadieu, E. Matrix architecture defines the preferential localization and migration of T cells into the stroma of human lung tumors. *J. Clin. Invest.* **2012**, *122*, 899–910. [CrossRef] [PubMed]
86. Provenzano, P.P.; Cuevas, C.; Chang, A.E.; Goel, V.K.; Von Hoff, D.D.; Hingorani, S.R. Enzymatic targeting of the stroma ablates physical barriers to treatment of pancreatic ductal adenocarcinoma. *Cancer Cell* **2012**, *21*, 418–429. [CrossRef]
87. Zhang, J.; Liu, J. Tumor stroma as targets for cancer therapy. *Pharmacol. Ther.* **2013**, *137*, 200–215. [CrossRef]
88. Li, T.; Yang, Y.; Hua, X.; Wang, G.; Liu, W.; Jia, C.; Tai, Y.; Zhang, Q.; Chen, G. Hepatocellular carcinoma-associated fibroblasts trigger NK cell dysfunction via PGE2 and IDO. *Cancer Lett.* **2012**, *318*, 154–161. [CrossRef]

89. Nazareth, M.R.; Broderick, L.; Simpson-Abelson, M.R.; Kelleher, R.J., Jr.; Yokota, S.J.; Bankert, R.B. Characterization of human lung tumor-associated fibroblasts and their ability to modulate the activation of tumor-associated T cells. *J. Immunol.* **2007**, *178*, 5552–5562. [CrossRef] [PubMed]
90. Fleming, V.; Hu, X.; Weber, R.; Nagibin, V.; Groth, C.; Altevogt, P.; Utikal, J.; Umansky, V. Targeting Myeloid-Derived Suppressor Cells to Bypass Tumor-Induced Immunosuppression. *Front. Immunol.* **2018**, *9*, 398. [CrossRef] [PubMed]
91. Sacchi, A.; Tumino, N.; Sabatini, A.; Cimini, E.; Casetti, R.; Bordoni, V.; Grassi, G.; Agrati, C. Myeloid-Derived Suppressor Cells Specifically Suppress IFN-γ Production and Antitumor Cytotoxic Activity of Vδ2 T Cells. *Front. Immunol.* **2018**, *9*, 1271. [CrossRef]
92. Blattner, C.; Fleming, V.; Weber, R.; Himmelhan, B.; Altevogt, P.; Gebhardt, C.; Schulze, T.J.; Razon, H.; Hawila, E.; Wildbaum, G.; et al. CCR5+ Myeloid-Derived Suppressor Cells Are Enriched and Activated in Melanoma Lesions. *Cancer Res.* **2018**, *78*, 157–167. [CrossRef]
93. Tan, M.C.; Goedegebuure, P.S.; Belt, B.A.; Flaherty, B.; Sankpal, N.; Gillanders, W.E.; Eberlein, T.J.; Hsieh, C.S.; Linehan, D.C. Disruption of CCR5-dependent homing of regulatory T cells inhibits tumor growth in a murine model of pancreatic cancer. *J. Immunol.* **2009**, *182*, 1746–1755. [CrossRef] [PubMed]
94. Velasco-Velázquez, M.; Jiao, X.; De La Fuente, M.; Pestell, T.G.; Ertel, A.; Lisanti, M.P.; Pestell, R.G. CCR5 antagonist blocks metastasis of basal breast cancer cells. *Cancer Res.* **2012**, *72*, 3839–3850. [CrossRef] [PubMed]
95. Serafini, P.; Meckel, K.; Kelso, M.; Noonan, K.; Califano, J.; Koch, W.; Dolcetti, L.; Bronte, V.; Borrello, I. Phosphodiesterase-5 inhibition augments endogenous antitumor immunity by reducing myeloid-derived suppressor cell function. *J. Exp. Med.* **2006**, *203*, 2691–2702. [CrossRef] [PubMed]
96. Orillion, A.; Hashimoto, A.; Damayanti, N.; Shen, L.; Adelaiye-Ogala, R.; Arisa, S.; Chintala, S.; Ordentlich, P.; Kao, C.; Elzey, B.; et al. Entinostat Neutralizes Myeloid-Derived Suppressor Cells and Enhances the Antitumor Effect of PD-1 Inhibition in Murine Models of Lung and Renal Cell Carcinoma. *Clin. Cancer Res.* **2017**, *23*, 5187–5201. [CrossRef] [PubMed]
97. Tanaka, A.; Sakaguchi, S. Regulatory T cells in cancer immunotherapy. *Cell Res.* **2017**, *27*, 109–118. [CrossRef] [PubMed]
98. Fujio, K.; Yamamoto, K.; Okamura, T. Overview of LAG-3-Expressing, IL-10-Producing Regulatory T Cells. *Curr. Top. Microbiol. Immunol.* **2017**, *410*, 29–45. [CrossRef]
99. Bulliard, Y.; Jolicoeur, R.; Windman, M.; Rue, S.M.; Ettenberg, S.; Knee, D.A.; Wilson, N.S.; Dranoff, G.; Brogdon, J.L. Activating Fc γ receptors contribute to the antitumor activities of immunoregulatory receptor-targeting antibodies. *J. Exp. Med.* **2013**, *210*, 1685–1693. [CrossRef] [PubMed]
100. Simpson, T.R.; Li, F.; Montalvo-Ortiz, W.; Sepulveda, M.A.; Bergerhoff, K.; Arce, F.; Roddie, C.; Henry, J.Y.; Yagita, H.; Wolchok, J.D.; et al. Fc-dependent depletion of tumor-infiltrating regulatory T cells co-defines the efficacy of anti-CTLA-4 therapy against melanoma. *J. Exp. Med.* **2013**, *210*, 1695–1710. [CrossRef]
101. Selby, M.J.; Engelhardt, J.J.; Quigley, M.; Henning, K.A.; Chen, T.; Srinivasan, M.; Korman, A.J. Anti-CTLA-4 antibodies of IgG2a isotype enhance antitumor activity through reduction of intratumoral regulatory T cells. *Cancer Immunol. Res.* **2013**, *1*, 32–42. [CrossRef] [PubMed]
102. Liakou, C.I.; Kamat, A.; Tang, D.N.; Chen, H.; Sun, J.; Troncoso, P.; Logothetis, C.; Sharma, P. CTLA-4 blockade increases IFNgamma-producing CD4+ICOShi cells to shift the ratio of effector to regulatory T cells in cancer patients. *Proc. Natl. Acad. Sci. USA* **2008**, *105*, 14987–14992. [CrossRef]
103. Hodi, F.S.; Butler, M.; Oble, D.A.; Seiden, M.V.; Haluska, F.G.; Kruse, A.; Macrae, S.; Nelson, M.; Canning, C.; Lowy, I.; et al. Immunologic and clinical effects of antibody blockade of cytotoxic T lymphocyte-associated antigen 4 in previously vaccinated cancer patients. *Proc. Natl. Acad. Sci. USA* **2008**, *105*, 3005–3010. [CrossRef] [PubMed]
104. Curiel, T.J.; Coukos, G.; Zou, L.; Alvarez, X.; Cheng, P.; Mottram, P.; Evdemon-Hogan, M.; Conejo-Garcia, J.R.; Zhang, L.; Burow, M.; et al. Specific recruitment of regulatory T cells in ovarian carcinoma fosters immune privilege and predicts reduced survival. *Nat. Med.* **2004**, *10*, 942–949. [CrossRef] [PubMed]
105. Faget, J.; Biota, C.; Bachelot, T.; Gobert, M.; Treilleux, I.; Goutagny, N.; Durand, I.; Léon-Goddard, S.; Blay, J.Y.; Caux, C.; et al. Early detection of tumor cells by innate immune cells leads to T(reg) recruitment through CCL22 production by tumor cells. *Cancer Res.* **2011**, *71*, 6143–6152. [CrossRef]
106. Sugiyama, D.; Nishikawa, H.; Maeda, Y.; Nishioka, M.; Tanemura, A.; Katayama, I.; Ezoe, S.; Kanakura, Y.; Sato, E.; Fukumori, Y.; et al. Anti-CCR4 mAb selectively depletes effector-type FoxP3+CD4+ regulatory T cells, evoking antitumor immune responses in humans. *Proc. Natl. Acad. Sci. USA* **2013**, *110*, 17945–17950. [CrossRef]
107. Ko, K.; Yamazaki, S.; Nakamura, K.; Nishioka, T.; Hirota, K.; Yamaguchi, T.; Shimizu, J.; Nomura, T.; Chiba, T.; Sakaguchi, S. Treatment of advanced tumors with agonistic anti-GITR mAb and its effects on tumor-infiltrating Foxp3+CD25+CD4+ regulatory T cells. *J. Exp. Med.* **2005**, *202*, 885–891. [CrossRef] [PubMed]
108. Tamura, R.; Tanaka, T.; Yamamoto, Y.; Akasaki, Y.; Sasaki, H. Dual role of macrophage in tumor immunity. *Immunotherapy* **2018**, *10*, 899–909. [CrossRef]
109. Terashima, Y.; Onai, N.; Murai, M.; Enomoto, M.; Poonpiriya, V.; Hamada, T.; Motomura, K.; Suwa, M.; Ezaki, T.; Haga, T.; et al. Pivotal function for cytoplasmic protein FROUNT in CCR2-mediated monocyte chemotaxis. *Nat. Immunol.* **2005**, *6*, 827–835. [CrossRef]
110. Toda, E.; Terashima, Y.; Sato, T.; Hirose, K.; Kanegasaki, S.; Matsushima, K. FROUNT is a common regulator of CCR2 and CCR5 signaling to control directional migration. *J. Immunol.* **2009**, *183*, 6387–6394. [CrossRef]

111. Weiss, J.M.; Back, T.C.; Scarzello, A.J.; Subleski, J.J.; Hall, V.L.; Stauffer, J.K.; Chen, X.; Micic, D.; Alderson, K.; Murphy, W.J.; et al. Successful immunotherapy with IL-2/anti-CD40 induces the chemokine-mediated mitigation of an immunosuppressive tumor microenvironment. *Proc. Natl. Acad. Sci. USA* **2009**, *106*, 19455–19460. [CrossRef]
112. Gordon, S.R.; Maute, R.L.; Dulken, B.W.; Hutter, G.; George, B.M.; McCracken, M.N.; Gupta, R.; Tsai, J.M.; Sinha, R.; Corey, D.; et al. PD-1 expression by tumour-associated macrophages inhibits phagocytosis and tumour immunity. *Nature* **2017**, *545*, 495–499. [CrossRef] [PubMed]
113. Pardoll, D.M. The blockade of immune checkpoints in cancer immunotherapy. *Nat. Rev. Cancer* **2012**, *12*, 252–264. [CrossRef] [PubMed]
114. Wolchok, J.D.; Chiarion-Sileni, V.; Gonzalez, R.; Rutkowski, P.; Grob, J.J.; Cowey, C.L.; Lao, C.D.; Wagstaff, J.; Schadendorf, D.; Ferrucci, P.F.; et al. Overall Survival with Combined Nivolumab and Ipilimumab in Advanced Melanoma. *N. Engl. J. Med.* **2017**, *377*, 1345–1356. [CrossRef]
115. Hamid, O.; Robert, C.; Daud, A.; Hodi, F.S.; Hwu, W.J.; Kefford, R.; Wolchok, J.D.; Hersey, P.; Joseph, R.W.; Weber, J.S.; et al. Safety and tumor responses with lambrolizumab (anti-PD-1) in melanoma. *N. Engl. J. Med.* **2013**, *369*, 134–144. [CrossRef]
116. Flippot, R.; Escudier, B.; Albiges, L. Immune Checkpoint Inhibitors: Toward New Paradigms in Renal Cell Carcinoma. *Drugs* **2018**, *78*, 1443–1457. [CrossRef]
117. Iwasaki, M.; Tanaka, Y.; Kobayashi, H.; Murata-Hirai, K.; Miyabe, H.; Sugie, T.; Toi, M.; Minato, N. Expression and function of PD-1 in human $\gamma\delta$ T cells that recognize phosphoantigens. *Eur. J. Immunol.* **2011**, *41*, 345–355. [CrossRef] [PubMed]
118. Rossi, C.; Gravelle, P.; Decaup, E.; Bordenave, J.; Poupot, M.; Tosolini, M.; Franchini, D.M.; Laurent, C.; Morin, R.; Lagarde, J.M.; et al. Boosting $\gamma\delta$ T cell-mediated antibody-dependent cellular cytotoxicity by PD-1 blockade in follicular lymphoma. *Oncoimmunology* **2018**, *8*, 1554175. [CrossRef]
119. Tomogane, M.; Sano, Y.; Shimizu, D.; Shimizu, T.; Miyashita, M.; Toda, Y.; Hosogi, S.; Tanaka, Y.; Kimura, S.; Ashihara, E. Human Vγ9Vδ2 T cells exert anti-tumor activity independently of PD-L1 expression in tumor cells. *Biochem. Biophys. Res. Commun.* **2021**, *573*, 132–139. [CrossRef]
120. Clarke, M.F.; Dick, J.E.; Dirks, P.B.; Eaves, C.J.; Jamieson, C.H.; Jones, D.L.; Visvader, J.; Weissman, I.L.; Wahl, G.M. Cancer stem cells–perspectives on current status and future directions: AACR Workshop on cancer stem cells. *Cancer Res.* **2006**, *66*, 9339–9344. [CrossRef]
121. Ojo, D.; Lin, X.; Wong, N.; Gu, Y.; Tang, D. Prostate Cancer Stem-like Cells Contribute to the Development of Castration-Resistant Prostate Cancer. *Cancers* **2015**, *7*, 2290–2308. [CrossRef] [PubMed]
122. Duru, N.; Fan, M.; Candas, D.; Menaa, C.; Liu, H.C.; Nantajit, D.; Wen, Y.; Xiao, K.; Eldridge, A.; Chromy, B.A.; et al. HER2-associated radioresistance of breast cancer stem cells isolated from HER2-negative breast cancer cells. *Clin. Cancer Res.* **2012**, *18*, 6634–6647. [CrossRef] [PubMed]
123. Ong, C.W.; Kim, L.G.; Kong, H.H.; Low, L.Y.; Iacopetta, B.; Soong, R.; Salto-Tellez, M. CD133 expression predicts for non-response to chemotherapy in colorectal cancer. *Mod. Pathol.* **2010**, *23*, 450–457. [CrossRef]
124. Jinushi, M. Role of cancer stem cell-associated inflammation in creating pro-inflammatory tumorigenic microenvironments. *Oncoimmunology* **2014**, *3*, e28862. [CrossRef]
125. Schatton, T.; Schütte, U.; Frank, N.Y.; Zhan, Q.; Hoerning, A.; Robles, S.C.; Zhou, J.; Hodi, F.S.; Spagnoli, G.C.; Murphy, G.F.; et al. Modulation of T-cell activation by malignant melanoma initiating cells. *Cancer Res.* **2010**, *70*, 697–708. [CrossRef] [PubMed]
126. Clara, J.A.; Monge, C.; Yang, Y.; Takebe, N. Targeting signalling pathways and the immune microenvironment of cancer stem cells—A clinical update. *Nat. Rev. Clin. Oncol.* **2020**, *17*, 204–232. [CrossRef]
127. Di Tomaso, T.; Mazzoleni, S.; Wang, E.; Sovena, G.; Clavenna, D.; Franzin, A.; Mortini, P.; Ferrone, S.; Doglioni, C.; Marincola, F.M.; et al. Immunobiological characterization of cancer stem cells isolated from glioblastoma patients. *Clin. Cancer Res.* **2010**, *16*, 800–813. [CrossRef]
128. Lee, Y.; Shin, J.H.; Longmire, M.; Wang, H.; Kohrt, H.E.; Chang, H.Y.; Sunwoo, J.B. CD44+ Cells in Head and Neck Squamous Cell Carcinoma Suppress T-Cell-Mediated Immunity by Selective Constitutive and Inducible Expression of PD-L1. *Clin. Cancer Res.* **2016**, *22*, 3571–3581. [CrossRef]
129. Miyashita, M.; Tomogane, M.; Nakamura, Y.; Shimizu, T.; Fujihara, A.; Ukimura, O.; Ashihara, E. Sphere-derived Prostate Cancer Stem Cells Are Resistant to $\gamma\delta$ T Cell Cytotoxicity. *Anticancer Res.* **2020**, *40*, 5481–5487. [CrossRef] [PubMed]
130. Weidle, U.H.; Kontermann, R.E.; Brinkmann, U. Tumor-antigen-binding bispecific antibodies for cancer treatment. *Semin. Oncol.* **2014**, *41*, 653–660. [CrossRef]
131. Hoh, A.; Dewerth, A.; Vogt, F.; Wenz, J.; Baeuerle, P.A.; Warmann, S.W.; Fuchs, J.; Armeanu-Ebinger, S. The activity of $\gamma\delta$ T cells against paediatric liver tumour cells and spheroids in cell culture. *Liver Int.* **2013**, *33*, 127–136. [CrossRef] [PubMed]
132. Oberg, H.H.; Peipp, M.; Kellner, C.; Sebens, S.; Krause, S.; Petrick, D.; Adam-Klages, S.; Röcken, C.; Becker, T.; Vogel, I.; et al. Novel bispecific antibodies increase $\gamma\delta$ T-cell cytotoxicity against pancreatic cancer cells. *Cancer Res.* **2014**, *74*, 1349–1360. [CrossRef] [PubMed]
133. Oberg, H.H.; Kellner, C.; Gonnermann, D.; Peipp, M.; Peters, C.; Sebens, S.; Kabelitz, D.; Wesch, D. $\gamma\delta$ T cell activation by bispecific antibodies. *Cell. Immunol.* **2015**, *296*, 41–49. [CrossRef]
134. Oberg, H.H.; Kellner, C.; Gonnermann, D.; Sebens, S.; Bauerschlag, D.; Gramatzki, M.; Kabelitz, D.; Peipp, M.; Wesch, D. Tribody [$(HER2)_2$xCD16] Is More Effective Than Trastuzumab in Enhancing $\gamma\delta$ T Cell and Natural Killer Cell Cytotoxicity Against HER2-Expressing Cancer Cells. *Front. Immunol.* **2018**, *9*, 814. [CrossRef] [PubMed]

135. Maus, M.V.; Grupp, S.A.; Porter, D.L.; June, C.H. Antibody-modified T cells: CARs take the front seat for hematologic malignancies. *Blood* **2014**, *123*, 2625–2635. [CrossRef] [PubMed]
136. Deniger, D.C.; Switzer, K.; Mi, T.; Maiti, S.; Hurton, L.; Singh, H.; Huls, H.; Olivares, S.; Lee, D.A.; Champlin, R.E.; et al. Bispecific T-cells expressing polyclonal repertoire of endogenous $\gamma\delta$ T-cell receptors and introduced CD19-specific chimeric antigen receptor. *Mol. Ther.* **2013**, *21*, 638–647. [CrossRef]
137. Fisher, J.; Abramowski, P.; Wisidagamage Don, N.D.; Flutter, B.; Capsomidis, A.; Cheung, G.W.; Gustafsson, K.; Anderson, J. Avoidance of On-Target Off-Tumor Activation Using a Co-stimulation-Only Chimeric Antigen Receptor. *Mol. Ther.* **2017**, *25*, 1234–1247. [CrossRef]

International Journal of
Molecular Sciences

Article

Platelet Microparticles Decrease Daunorubicin-Induced DNA Damage and Modulate Intrinsic Apoptosis in THP-1 Cells

Daniel Cacic [1,*], Oddmund Nordgård [1], Peter Meyer [1] and Tor Hervig [2,3]

[1] Department of Hematology and Oncology, Stavanger University Hospital, 4068 Stavanger, Norway; oddmund.nordgard@sus.no (O.N.); peter.albert.meyer@sus.no (P.M.)
[2] Department of Clinical Science, University of Bergen, 5021 Bergen, Norway; tor.audun.hervig@helse-fonna.no
[3] Laboratory of Immunology and Transfusion Medicine, Haugesund Hospital, 5528 Haugesund, Norway
* Correspondence: daniel.limi.cacic@sus.no

Abstract: Platelets can modulate cancer through budding of platelet microparticles (PMPs) that can transfer a plethora of bioactive molecules to cancer cells upon internalization. In acute myelogenous leukemia (AML) this can induce chemoresistance, partially through a decrease in cell activity. Here we investigated if the internalization of PMPs protected the monocytic AML cell line, THP-1, from apoptosis by decreasing the initial cellular damage inflicted by treatment with daunorubicin, or via direct modulation of the apoptotic response. We examined whether PMPs could protect against apoptosis after treatment with a selection of inducers, primarily associated with either the intrinsic or the extrinsic apoptotic pathway, and protection was restricted to the agents targeting intrinsic apoptosis. Furthermore, levels of daunorubicin-induced DNA damage, assessed by measuring gH2AX, were reduced in both 2N and 4N cells after PMP co-incubation. Measuring different BCL2-family proteins before and after treatment with daunorubicin revealed that PMPs downregulated the pro-apoptotic PUMA protein. Thus, our findings indicated that PMPs may protect AML cells against apoptosis by reducing DNA damage both dependent and independent of cell cycle phase, and via direct modulation of the intrinsic apoptotic pathway by downregulating PUMA. These findings further support the clinical relevance of platelets and PMPs in AML.

Keywords: acute myelogenous leukemia; platelets; microparticles; apoptosis

Citation: Cacic, D.; Nordgård, O.; Meyer, P.; Hervig, T. Platelet Microparticles Decrease Daunorubicin-Induced DNA Damage and Modulate Intrinsic Apoptosis in THP-1 Cells. *Int. J. Mol. Sci.* **2021**, *22*, 7264. https://doi.org/10.3390/ijms22147264

Academic Editor: Angela Stefanachi

Received: 30 May 2021
Accepted: 2 July 2021
Published: 6 July 2021

1. Introduction

Platelets were originally discovered in the late 19th century as a key player in hemostasis [1]. It is now clear, however, that they serve a broader role in both health and disease [2–7]. Platelets contain many different biologically active molecules, which include proteins [8,9], regulatory microRNAs [10,11], and long RNA sequences, such as ribosomal RNAs and protein-coding transcripts inherited from parental megakaryocytes [12,13]. The long RNA sequences are prone to time-dependent decay [12,14], and correlation with the proteome is weak [13], suggesting only a limited protein synthesis capacity, which may be confined to reticulated platelets [12].

Bioactive substances can be secreted from platelets as paracrine or endocrine factors that are able to modify various cancers [15–17]. These bioactive molecules can also be transferred via platelet microparticles (PMPs), which in turn have been shown to be internalized by many different cancer cells, altering crucial functions of the cells, namely invasiveness, proliferation, and viability [18–20]. The pro-tumoral properties of platelets are further supported by retrospective and observational data showing an association between platelet inhibition and decreased risk for development of cancer, and increased cancer-specific survival [21–24]. However, the mechanism underlying this potential effect remains unknown, and the data from the few prospective studies that have been done are less convincing [25–27].

Acute myelogenous leukemia (AML) is a bone marrow disease affecting hematopoietic stem and progenitor cells [28,29]. The genomic landscape of AML has been thoroughly analyzed and the first study performing whole-genome sequencing in AML was already published in 2008 [30]. AML usually has a lower frequency of somatic mutations than most other cancers [31,32], with a median of 13 different coding mutations per case [33,34]. Despite the low mutational burden, there are large variations in transcriptomic and proteomic signatures in AML cells, compared with healthy hematopoietic progenitors, and even between different subclones [35–37]. Despite increased knowledge in the genomics of AML, treatment strategies have essentially remained unchanged for several decades, with a few exceptions [38]. Curative treatment, which is restricted to younger patients, consists of intensive chemotherapy with consolidating hematopoietic stem cell treatment for high-risk cases. Despite this, median survival is just 11 months when including all age groups [39], underscoring the need for a more profound understanding of progression of the disease and development of treatment resistance, in addition to established genomic mechanisms.

Targeting apoptosis in cancer is a novel treatment strategy that is finally maturing into clinical use. Apoptosis can be divided into two separate pathways, which are interlinked with common feedback mechanisms [40]; the death receptor initiated extrinsic pathway (FAS/CD95, TNFR1, TRAIL-R1, TRAIL-R2, DR3, and DR6), and the intrinsic, or mitochondrial, pathway. Dysregulation of the latter has proven to be an important feature in cancer biology [41]. The regulatory and anti-apoptotic proteins in the BCL2-family are also known to be upregulated in hematological malignancies [42,43]. Hence, numerous drugs that target major apoptotic regulators, such as BCL2 or MCL1, are currently either under development, or have just been approved, to treat a variety of hematological malignancies, including AML [41].

The intrinsic apoptotic pathway is initiated by several factors, including DNA damage or cellular stress, which is accompanied by upregulation of the pro-apoptotic BH3-only proteins (including BAD, BID, NOXA, HRK, BMF, PUMA, BIM), which then activate the effector proteins BAK and BAX directly or through inhibition of anti-apoptotic regulator proteins [44,45]. Upon activation, the predominantly mitochondrial outer membrane (MOM)-bound BAK, and predominantly cytosolic BAX protein, oligomerize in the MOM, leading to cytochrome c leakage from the mitochondria [46,47]. Cytochrome c then forms an apoptosome with apoptotic protease activating factor-1 (APAF1), which recruits pro-caspase 9, both activating and regulating its function [48,49]. Caspase-9 activates caspase-3, where the intrinsic and extrinsic pathways converge. Caspase-3 has multiple substrates [50], including a caspase-dependent DNase, which leads to DNA degradation upon activation by caspase-3 [51].

Our group has previously shown that co-incubation of the monocytic AML cell line, THP-1, or primary AML samples, with platelet microparticles, protects against daunorubicin (DNR)-induced apoptosis and cell death, at least partially via a decrease in cell activity [52]. We also found that miR-125a and miR-125b levels were elevated in THP-1 cells after PMP co-incubation. These microRNAs have been associated with chemotherapy resistance [53,54]. However, whether the PMP-associated increase in resistance to DNR is caused solely by protection against DNR-induced cell damage, or a modulation of the intrinsic apoptotic pathway regulators, remains unknown. Thus, we sought to further examine the anti-apoptotic effects of PMPs in the monocytic AML cell line THP-1.

2. Results

2.1. PMPs Offered Protection from Apoptosis Induced by Multiple Agents

We have previously demonstrated that PMPs increase resistance to DNR-induced apoptosis and cell death [52]. To investigate whether co-incubation with PMPs provided a general anti-apoptotic effect, we compared apoptosis and cell death after treatment with several agents associated with inducing apoptosis, primarily through intrinsic (alantolactone, staurosporine, MG 132), or extrinsic (piceatannol, TRAIL) apoptosis. Co-incubation of PMPs with THP-1 cells decreased the relative frequency of dead and apoptotic cells induced by alantolactone, staurosporine, and MG 132, but not piceatannol (Figure 1). In our experiments 50 ng/mL TRAIL was not sufficient to induce apoptosis in THP-1 cells, but it slightly potentiated the apoptotic effect of piceatannol. Surprisingly, PMP co-incubation increased the relative frequency of dead and apoptotic cells in the case of the combination of piceatannol and TRAIL ($p = 0.003$). However, as this effect was marginal (mean difference of 1.89 percentage points; SD 0.43), it could be biologically insignificant. From these analyses, we suggest that PMPs may provide general protection from apoptosis, but seemingly only against agents that primarily activate the intrinsic apoptotic pathway.

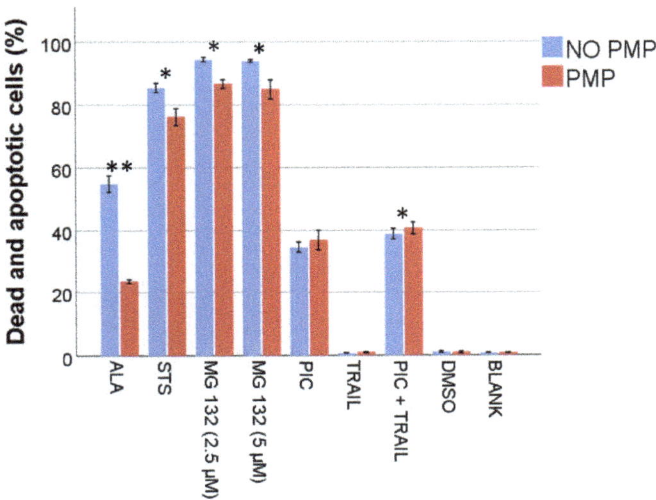

Figure 1. Apoptosis inhibition by platelet microparticles (PMPs). THP-1 cells with or without PMP co-incubation for 24 h were treated with an apoptosis-inducing molecule at a concentration and an incubation time as described in Table S1 ($n = 4$). Relative frequency of dead and apoptotic cells were analyzed by flow cytometry, and gated out in a single gate (annexin V vs. propidium iodide). Data were compared using the paired-sample t-test for data pairs. * $p < 0.05$, ** $p < 0.001$. ALA, alantolactone. STS, staurosporine. Pic, piceatannol. TRAIL, tumor necrosis factor-related apoptosis-inducing ligand.

2.2. PMPs Reduced Both Caspase-8 and Caspase-9 Activation Induced by DNR

The cytotoxic effect of DNR is commonly associated with an increase in DNA damage, i.e., an intrinsic stimulus. However, it is also suggested to activate the extrinsic apoptotic pathway [55]. To evaluate activation of intrinsic and extrinsic apoptosis after DNR-treatment, we measured levels of active caspase-8 and caspase-9 by flow cytometry, and gated the cells in "lo", "mid", and "hi" populations. In the case of caspase-8, it was not possible to accurately discriminate between the "mid" and "hi" populations, and consequently these populations were gated as one. Our analyses indicated that both caspases were highly activated after DNR-treatment, but this was partially inhibited by PMP co-incubation (Figure 2A, Figure S3). In addition, fluorescence of the respective caspases were decreased for all subpopulations in PMP co-incubated cells (Figure 2B, Figure S3). The relative decrease in frequency of caspase-8 or caspase-9 "mid/hi" cells associated with PMP co-incubation were equal (Figure 2C; $p = 0.756$). These findings indicated that activation of caspase-8 is important in DNR-induced apoptosis, and is most likely inhibited by PMPs via an upstream mechanism common with caspase-9 activation.

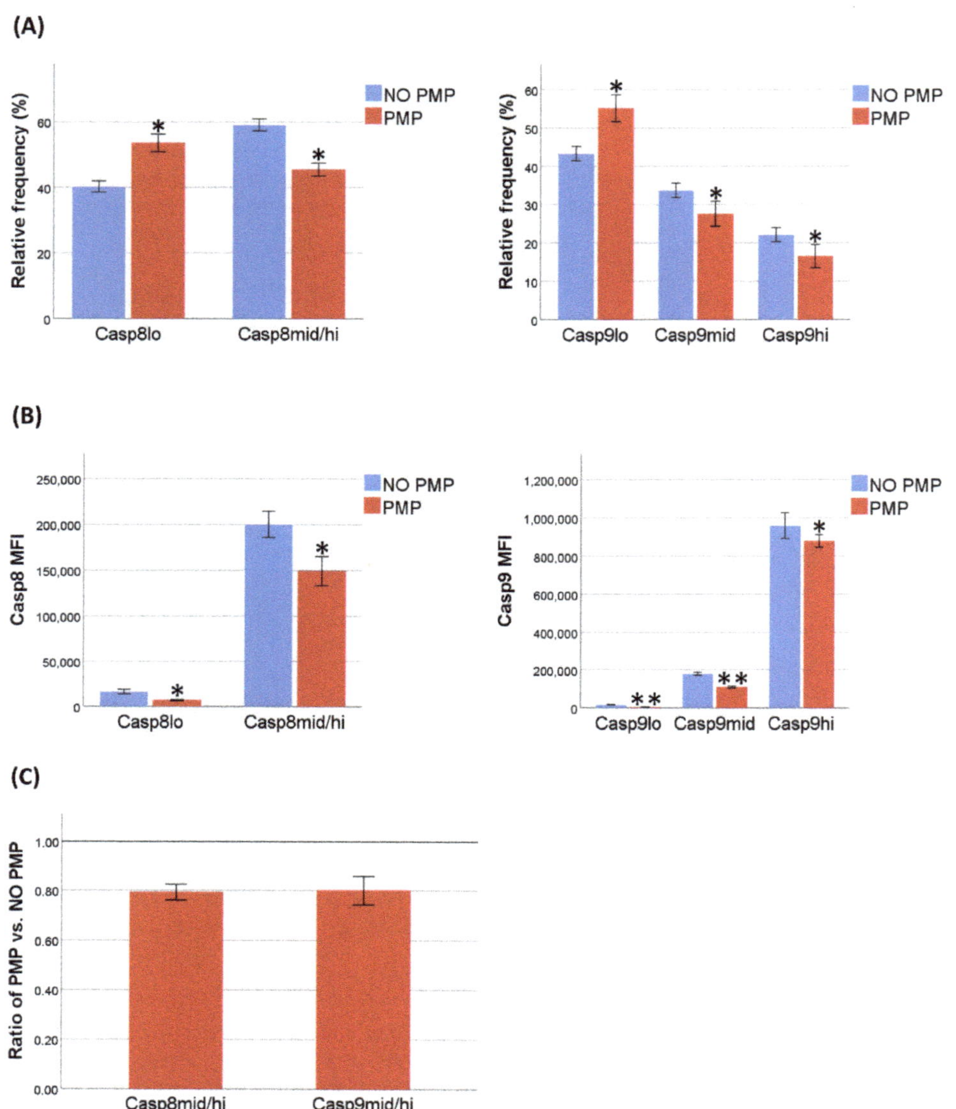

Figure 2. Caspase-8 and caspase-9 activation after daunorubicin (DNR)-treatment. (**A**) Active caspase-8 and caspase-9 were analyzed by flow cytometry after 24 h, with or without platelet microparticle (PMP) co-incubation, and an additional 24 h with DNR-treatment at 0.5 μM ($n = 4$). Events were gated as either "lo", "mid", or "hi", according to fluorescence intensity. Data are presented as the relative frequencies of the populations. (**B**) Background corrected mean fluorescence intensity (MFI) of the respective caspases for different populations. (**C**) Ratio of the relative frequencies of the subpopulations "mid" and "hi" combined for respective caspases between THP-1 cells, with versus without PMP co-incubation. Data were compared using the paired-sample t-test for data pairs. * $p < 0.05$. ** $p < 0.001$. Casp8, caspase-8. Casp9, caspase-9.

2.3. PMP Co-Incubation Downregulated Pro-Apoptotic PUMA Protein

To further investigate if PMPs could independently affect intrinsic apoptosis, we analyzed levels of BCL2-family proteins with and without PMP co-incubation, and both with and without DNR. Gating strategy is summarized in Figure 3. For the cell population only visible with DNR-treatment (P2), levels of BAK, BCL2, MCL1, and PUMA were relatively less increased with PMP co-incubation (Figure 4), when compared to non-DNR-treated THP-1 cells (P1), and the decrease seen in BMF levels was relatively less. We also identified the P1 population in DNR-treated cells, and antibody fluorescence intensity was more or less unaffected, except for BMF, which had a somewhat higher level than the P1 population in non-DNR-treated cells. The relative change in fluorescence intensity accompanying PMP co-incubation was as anticipated, and followed the expected trend of protection from DNR-induced cell damage with PMP co-incubation. For example, the fluorescence intensity of BAK increased with DNR in both groups, but the increase was less with PMP co-incubation than without. However, in the case of PUMA we found a reduced signal intensity with PMP co-incubation in all cell populations, independent of DNR. Thus, we suggest that PMP co-incubation may protect THP-1 cells against DNR-induced cell death, at least partially through downregulation of the pro-apoptotic PUMA protein.

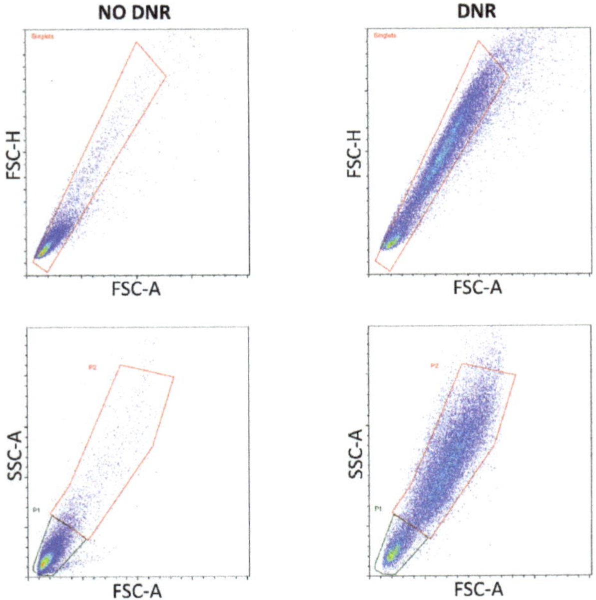

Figure 3. Gating strategy for intracellular flow cytometry of BCL2-family proteins. Doublets were first discriminated in FSC-A vs. FSC-H plots. P1 represents the population of daunorubicin (DNR)-treated cells gated in FSC-A vs. SSC-A plots that aligned well with non-DNR-treated cells. P2 represents a population generated by DNR-treatment and with increased light scatter.

2.4. Inhibitors of Caspase-9 and BAX Protected Against DNR-Induced Cell Death, but Less so with PMP Co-Incubation

As PMPs can decrease PUMA protein levels, DNR-induced apoptosis in cells co-incubated with PMPs may be less driven by the intrinsic apoptotic pathway. We investigated whether the protective effect of two inhibitors of intrinsic apoptosis, iMAC1 (BAX) and Q-LEHD-Oph (caspase-9), was affected by PMP co-incubation prior to DNR-treatment. We found a lower relative reduction in the relative frequency of dead and apoptotic cells, both for iMAC1 and Q-LEHD-Oph, with PMP co-incubation, which may indicate that caspase-9 activation was a weaker driver in apoptosis (Figure 5A). In addition, inhibiting the activity of caspase-9 or BAX with Q-LEHD-Oph and iMAC1 only yielded a reduction in levels of active caspase-9 in the "NO PMP" setting (Figure 5B). Thus, inhibitors of the intrinsic apoptotic pathway were less effective when THP-1 cells were co-incubated with PMPs, suggesting a direct modulation of this pathway.

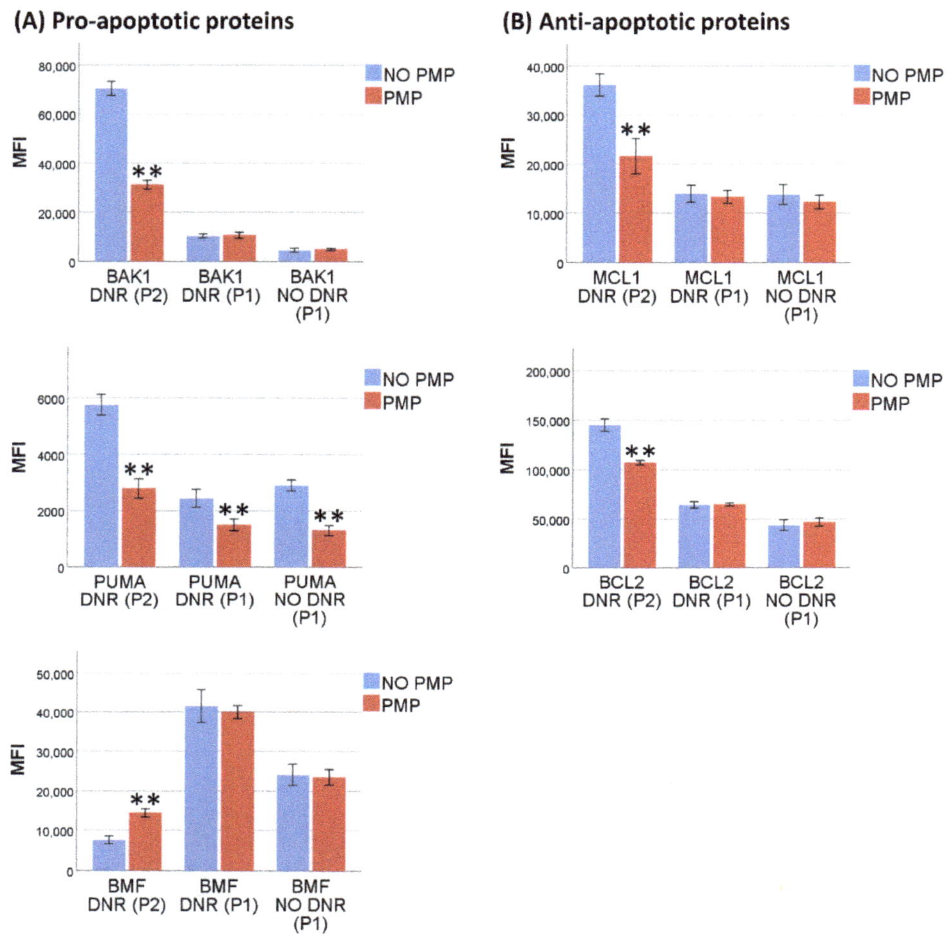

Figure 4. Levels of BCL2-family proteins before and after daunorubicin treatment. (**A**) THP-1 cells were co-incubated with or without platelet microparticles (PMPs) for 24 h before treatment with or without 0.5 μM daunorubicin (DNR) for an additional 24 h. Cells were then analyzed by intracellular flow cytometry. Data were collected as mean fluorescence intensity (MFI) levels corrected for a "no primary antibody" sample of pro-apoptotic BCL2-family proteins ($n = 5$). P1 represents the population determined by viable DNR untreated cells, but also visible in DNR-treated cells, with minimal changes of protein expression. P2 represents the population generated by DNR-treatment. (**B**) MFI levels corrected for a "no primary antibody" sample of anti-apoptotic BCL2-family proteins ($n = 5$). Data were compared using the paired-sample t-test for data pairs. ** $p < 0.001$.

2.5. PMP Co-Incubation Reduced DNA Damage After DNR-Treatment Independently of Cell Cycle Phase

To evaluate the anti-apoptotic effect of PMP co-incubation, we indirectly analyzed double-stranded DNA-breaks (DSBs) through measurement of phosphorylated histone H2AX, or gH2AX, after four h of DNR-treatment. As the process of apoptosis increases DSBs, we first investigated if apoptosis was induced within this time frame. We found that after four h apoptosis was still at the baseline level (Figure 6A). As expected, fluorescence of gH2AX was increased after DNR-treatment in 4N cells compared to 2N cells for both groups, and the relative frequency of 2N cells was increased with PMP co-incubation (Figure 6B,C). Additionally, the fluorescence of gH2AX was decreased, both for 4N cells, and more surprisingly, for 2N cells with co-incubation of PMPs, compared to the "NO PMP" setting (Figure 6B). These findings indicated that PMP co-incubation protected THP-1 cells against DNR-induced apoptosis by decreasing the amount of DNA damage produced by DNR-treatment, both dependently and independently of cell cycle inhibition.

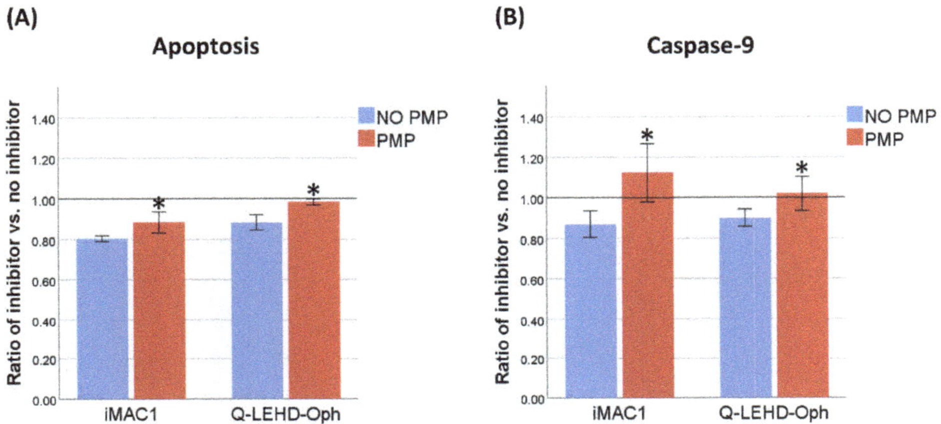

Figure 5. Effects of BAX and caspase-9 inhibitors on apoptosis and caspase-9 activation. (**A**) THP-1 cells were incubated with or without platelet microparticles (PMPs) for 23 h, and then with or without iMAC1 (BAX inhibitor), or Q-LEHD-Oph (caspase-9 inhibitor), for 1 h before treatment with 0.5 μM daunorubicin (DNR). Ratio of relative frequency of dead and apoptotic cells with or without inhibitor was calculated after 24 h ($n = 3$). (**B**) Ratio of relative frequency of caspase-$9^{mid/hi}$ cells after 24 h of DNR-treatment with or without inhibitor ($n = 3$). Data were compared using the paired-sample t-test for data pairs. * $p < 0.05$.

3. Discussion

Platelets are now recognized as an important contributor in cancer biology through several mechanisms involving immune evasion, metastasis, and development of cancer microenvironments [15,56–60]. We have previously shown that platelet microparticles increase resistance to DNR in acute myelogenous leukemia cells as a result of decreasing cell activity [52]. Here we provide evidence that this effect is multifactorial. We showed that PMPs protected equally against caspase-8 and caspase-9 activation in DNR-induced apoptosis, and that PMPs decreased DNR-induced DNA damage, not just by inhibiting cell cycle progression. The PMPs also directly modulated intrinsic apoptosis via the downregulation of the pro-apoptotic PUMA protein.

The anti-apoptotic effect of PMPs was evident with alantolactone, staurosporine, and MG 132, all primarily associated with activation of the intrinsic apoptotic pathway in THP-1 cells [61–63]. On the other hand, PMPs did not protect THP-1 cells against piceatannol or a combination of TRAIL + piceatannol, which are known to activate death receptor 5 and the extrinsic apoptotic pathway [64]. This indicates that PMPs may have broader anti-apoptotic properties, albeit restricted to intrinsic apoptosis. However, it yields no insight into the distinct mechanisms, which could be a common upstream effect on the intrinsic apoptotic pathway, e.g., cell cycle inhibition decreasing induced cellular stress or DNA damage. Interestingly, MG 132 has been shown to induce apoptosis in THP-1 cells arrested in either G1 or G2/M phases, but not when macrophage differentiation is induced [63]. Previously we have shown that PMPs inhibit cell cycle progression, and stimulate differentiation towards macrophages [52]. Thus, the latter could represent an anti-apoptotic mechanism independent of cell cycle inhibition by PMPs.

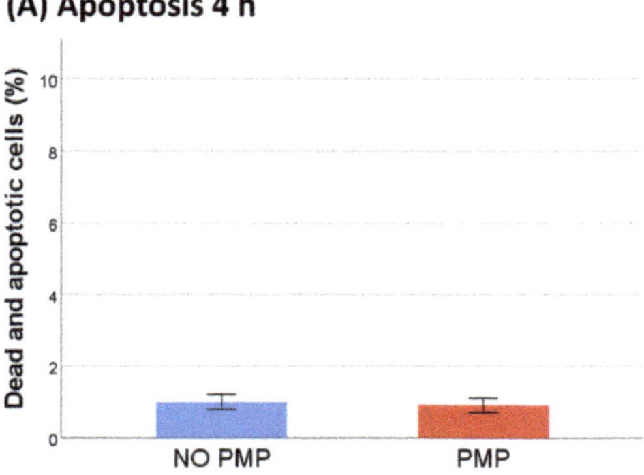

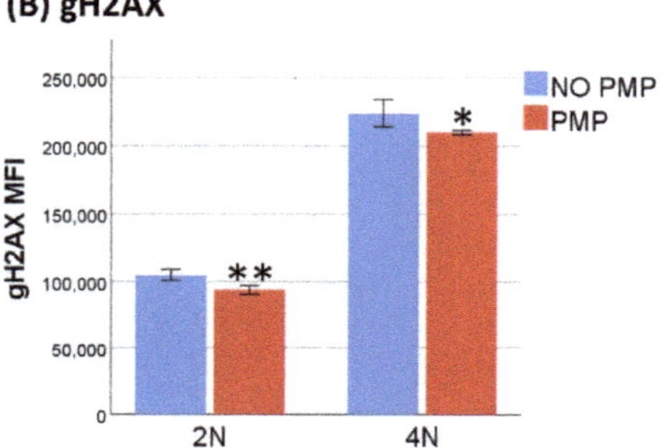

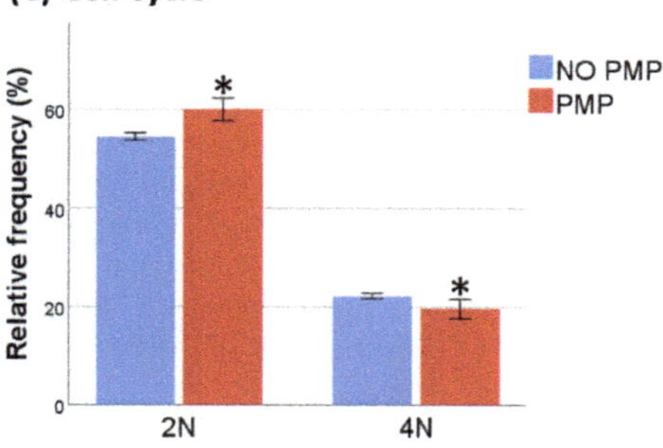

Figure 6. DNA damage after daunorubicin (DNR) treatment. (**A**) THP-1 cells were co-incubated with or without platelet microparticles (PMPs) for 24 h and treated with 0.5 μM DNR for 4 h before analysis with flow cytometry. Relative frequency of dead and apoptotic cells (n = 3). (**B**) Mean fluorescence intensity (MFI) of gH2AX corrected for both an unstained sample and background corrected MFI of representative experiment without DNR (n = 4). (**C**) Relative frequency of 2N (G1) and 4N (G2/M) cells (n = 4). Data were compared using the paired-sample t-test for data pairs. * $p < 0.05$.

Apoptosis induction by DNR is generally believed to be a result of inhibition of topoisomerase II (Top2) enzyme activity, leading to a rise in DNA-Top2 cleavage complexes [65]. The dependence of intact p53 protein for apoptosis induction by doxorubicin, a related Top2 poison, suggests a strong reliance on the activation of the intrinsic apoptotic pathway for this class of chemotherapeutics [66]. However, there is evidence that DNR-treatment upregulates death receptors and activate caspase-8 in multiple leukemic cell lines, thereby inducing extrinsic apoptosis [55]. In our experiments, caspase-8 and caspase-9 activation was equally inhibited by PMPs, suggesting that PMPs interfere with a common activation mechanism. However, this does not completely rule out a skew in upstream initiation of these pathways, as the levels of active caspase-8 and caspase-9 are also regulated by the downstream caspase-3 and caspase-7 as important feedback mechanisms [40].

We showed that not only was the relative frequency of caspase-9 positive cells lower with PMP co-incubation, but the potency of caspase-9 and BAX inhibitors was also reduced. Both these findings suggest a weaker drive from the intrinsic apoptotic pathway in PMP co-incubated cells, but also correlated with a reduction in the ultimate function of these molecules, which is the inhibition of caspase-9 activation. iMAC1 inhibits conformational activation of BAX, and maybe BAK, without competing with BH3 only proteins [67,68]. The anti-apoptotic effect of iMAC1 is also known to decrease with higher levels of BAX [69], but this should increase chemosensitivity [70], which is the opposite of the effects associated with PMPs. We suggest a common mechanism to explain the reduction in potency of both inhibitors. LEHD (leu-glu-his-asp)-sequence based peptides block the catalytic activity of caspase-9 [71]. iMAC1 will also lead to a decrease in caspase-9 activity by inhibiting mitochondrial outer membrane permeabilization [72]. Thus, both inhibitors ultimately lead to a decrease in the activation of caspase-3, which is not only essential for apoptosis induction, but also for caspase-9 activation in a feedback loop [40]. Thus, if the intrinsic apoptotic pathway is inhibited by PMP co-incubation, the relative contribution of this pathway to caspase-3 activation is reduced compared to the extrinsic pathway, which is also activated by DNR. This should lead to a relative reduction in efficiency of apoptosis inhibition through the intrinsic apoptotic pathway, as extrinsic apoptosis is presumably unaffected by both PMPs and the inhibitors. However, one important caveat for this conclusion is the selectivity of the caspase-inhibitor, which, at least in the older generation inhibitors, is proven to be poor [71]. There are some indications that the second generation inhibitor Q-LEHD-Oph also inhibits caspase-8, but this has not been analyzed in a cell-free system and it was less extensive then the caspase-9 inhibition [73]. Furthermore, our conclusion is supported by results involving two independent inhibitors of separate stages in the intrinsic apoptotic pathway.

The inhibitory effect of PMPs on cell cycle progression is a possible mechanism for increased DNR-resistance, since Top2 poisons are believed to be most effective in proliferative cells [65]. We have previously provided evidence for this, showing that serum starvation of THP-1 cells significantly reduces DNR-induced apoptosis [52]. By measuring gH2AX we showed that PMP co-incubation decreased the level of DNA damage after DNR-treatment. Phosphorylation of histone H2AX is induced by double-stranded DNA-breaks as a DNA damage response [74]. Thus, the level of gH2AX is a widely used proxy for DSBs in biological research [75–77]. As expected, gH2AX levels increased more in dividing 4N cells (G2/M), compared with non-dividing 2N cells (G1) across both groups. As PMPs inhibit cell cycle progression, this would necessarily decrease the level of DNA damage. However, we identified a relative decrease in the signal intensity of gH2AX with PMP co-incubation for both cell phases, suggesting a de facto protective mechanism against the effects of DNR, independent of cell cycle inhibition. Somewhat surprisingly, gH2AX levels were also lower with PMP co-incubation in cells in the G1 cell phase. Previously we have found a decrease in mitochondrial membrane potential associated with PMP co-incubation [52], which may decrease the level of reactive oxygen species (ROS). Significant DNA damage in cells in G1 cell phase is also found in doxorubicin-treated U2OS osteosarcoma cells [78]. This probably has a different etiology compared to the mechanism in the G2/M cell phase and may be explained by an increase in ROS [79,80].

An important question regarding the anti-apoptotic effect of PMPs is if they can directly modulate the apoptotic response. We measured anti-apoptotic (BCL2 and MCL1) and pro-apoptotic (BAK, BMF, and PUMA) BCL2-familiy proteins, both in response to DNR and at baseline, in a "NO DNR" setting. The decreased levels of PUMA associated with PMP co-incubation probably represent a de facto downregulation, as it was present in all cell populations both with and without DNR. This could be due to increased levels of microRNAs, miR-125a and miR-125b, which are transferred by PMPs [52], and proven to downregulate the protein at the translational level, inducing chemoresistance [53,54]. Furthermore, the "readiness" for activation of intrinsic apoptosis in AML cells has clinical relevance as it is a predictor of outcome with conventional treatment [81]. The other proteins analyzed were also altered, but not in the viable, non-DNR-treated cells, and always in sync with an expected decrease in apoptosis and cell damage associated with PMP co-incubation. Thus, it cannot be stated that these proteins were directly downregulated as a result of PMP-internalization. These differences could be a result of an altered regulation of BCL2-family proteins caused by downregulation of other proteins, such as PUMA. However, they may also stem from a shift in ratio of apoptotic to dead cells, which we did not discriminate. Surprisingly, DNR increased the fluorescence intensity for both the anti-apoptotic proteins tested (BCL2 and MCL1); one would expect a decrease in the level of anti-apoptotic proteins when apoptosis is induced. However, this pattern has been observed for some anti-apoptotic proteins in select leukemic cell lines and is presumably transitory [82].

PUMA is regulated by several factors, including different transcription factors and proteins like forkhead box O (FOXO) and p53 family members [83]. However, these mechanisms may be shared with other pro-apoptotic BCL2-familiy proteins [84], and therefore do not coincide with our observations of isolated PUMA downregulation. PUMA is also post-translationally regulated by phosphorylation and proteosomal degradation, which is proven to be induced by interleukin-3 and HER2 [85,86], but none of these proteins are considered to be a part of the platelet granule or releasate proteome [8,9]. In addition, there are other microRNAs that are present in PMPs, like miR-221 and miR-222 [11], which also are known to downregulate PUMA [87]. However, this has not been investigated in THP-1 or other acute myelogenous leukemia cell lines.

The evidence provided here supplements our previously published work that PMPs have anti-apoptotic properties in acute myelogenous leukemia. This effect could stem partially from inhibition of cell cycle progression and cell activity, making the cells less susceptible to damage induced by chemotherapy. In addition, we showed that PMPs may modulate the intrinsic apoptotic pathway through downregulation of PUMA, as a mechanism independent of cell cycle inhibition. The mechanistic findings from this study were derived solely from one cell line and need to be confirmed in primary AML cells. Nonetheless, translational research with PMPs in AML is warranted, as the indications for platelet inhibition to decrease PMP production, and thus potentially increase chemosensitivity, are further supported.

4. Materials and Methods

4.1. Cell Line

The THP-1 cell line was purchased from the American Type Culture Collection (Manassas, VA, USA) and cultured in Iscove's Modified Dulbecco's Medium (IMDM; Thermo Fisher Scientific, Waltham, MA, USA) + 10% FBS (Sigma Aldrich, St. Louis, MO, USA). Culture medium was partially replaced approximately every second day to keep the total cell concentration in the range of 2–6 $\times$ 10^5 per mL, and cells were only used in experiments once the exponential growth phase was reached. Cultures were kept for less than three months.

4.2. Platelet Concentrate

Platelet concentrates pooled from four donors were produced using the automated Tacsi system (Terumo BCT, Lakewood, CO, USA) and were provided by the Department of Immunology and Transfusion Medicine, Stavanger University Hospital (Stavanger, Norway). The platelet concentrations were 0.94–1.06 $\times$ 10^9 per mL. Leukocytes were removed by filtration to a residual level of <1.00 $\times$ 10^6. The storage medium for the platelets was approximately 35% plasma and 65% additive solution (PAS-III; Baxter, Lake Zurich, IL, USA). Written consent was obtained from all donors.

4.3. Platelet Releasate

Platelet releasate was produced by adding human thrombin (Sigma Aldrich) at a final concentration of 1 U/mL to the platelet concentrates in 50 mL tubes, and incubating for one hour in a 37 °C water bath, as described in [52]. The releasates were mixed by gentle shaking every 5 min. To separate the releasate from the clot, the tubes were centrifuged for 10 min at 900 g and the supernatant was transferred to new 50 mL tubes. The samples were stored at −80 °C. Fibrin clots that appeared after thawing were removed using a 10 mL serological pipette.

4.4. Platelet Microparticle Production

Platelet microparticles were isolated as previously described [52]. Briefly, platelet releasate was centrifuged at 15,000 g for 90 min at room temperature, and the supernatant carefully poured off. The PMPs were then resuspended in IMDM + 10% FBS and transferred to cell culture, thoroughly mixing with the cells by pipetting. The final concentration of PMPs was 1.5×10^7 per mL culture medium in all experiments. The wells were mixed again by pipetting 2 h after the PMPs were added to the cell cultures.

4.5. Platelet Microparticle Quantitation

One mL of platelet releasate, washed with 9 mL of Dulbecco's phosphate-buffered saline (Sigma Aldrich), was centrifuged as described above and the supernatant carefully poured off. The platelet microparticles were resuspended in 400 µL of 0.22 µm filtered Annexin V Binding Buffer (Miltenyi Biotec, Bergisch Gladbach, Germany), and 200 µL of the solution was transferred to a second tube. The solution was then stained with 20 µL of Annexin V FITC (Milteny Biotec), and 2 µL of anti-CD61 APC (clone Y2/51; Miltenyi Biotec), or 22 µL of 0.22 µm filtered Annexin V Binding Buffer for a negative control and incubated for 15 min at room temperature. Finally, 278 µL of 0.22 µm filtered Annexin V Binding Buffer and 50 µL CountBright beads (Thermo Fisher Scientific) were added before analysis. Microparticle gates were set using Megamix-PLUS FSC beads (bead size range: 0.3 to 0.9 µm; BioCytex, Marseille, France), according to our previous report [52]. At least 2500 bead events were collected. This, and all other flow cytometric analyses, were performed on a CytoFLEX flow cytometer (Beckman Coulter, Brea, CA, USA) using CytExpert ver. 2.4 acquisition and analysis software (Beckman Coulter). An example of gating strategy for PMP quantitation can be found in Figure S1 (see Supplementary Materials).

4.6. Apoptosis Assay

Approximately 5×10^5 cells per mL THP-1 cells were cultured with or without PMPs for 24 h. The cells were then treated with an apoptosis inductor at a concentration and time interval as indicated in Table S1. Cell viability was analyzed with the Annexin V FITC Kit (Miltenyi Biotec), strictly following the manufacturer's instructions. Dead and apoptotic cells were analyzed using flow cytometry and gated out in a single gate using a dot plot of FITC-A versus PerCP Cy 5.5-A after doublet discrimination. At least 20,000 gated cells were collected. An example of gating strategy can found in Figure S2.

4.7. Apoptosis Inhibition

For select experiments, after the initial 23 h of incubation with or without PMPs, the THP-1 cells were pretreated with either 20 µM of the caspase-9 inhibitor, Q-LEHD-Oph (Abcam, Cambridge, UK), or 10 µM of the BAX inhibitor, iMAC1 (Sigma Aldrich), and incubated for one hour before adding DNR, as described in the previous section.

4.8. Caspase Activity

Caspase-8 and caspase-9 activity in THP-1 cells after 24 h of DNR-treatment was measured using the CaspGLOW Fluorescein Active Staining Kit for the respective caspases (Thermo Fisher Scientific). Approximately 5×10^5 cells in 0.3 mL IMDM + 10% FBS were incubated with 1 μL of either FITC-IETD-FMK or FITC-LEHD-FMK for 30 min in a CO_2 incubator before washing twice with the supplied wash medium, and analyzing with flow cytometry. Both untreated and treated, but not stained, THP-1 cells were used as controls to determine low, medium, and high caspase populations. At least 20,000 gated cells were collected.

4.9. gH2AX

Measurement of gH2AX by flow cytometry was performed according to Darzynkiewicz et al. [75]. After PMP co-incubation, cells were fixed with 1% methanol free formaldehyde for 15 min on ice, then fixed and permeabilized in 70% ethanol, and stored overnight. Fixed and permeabilized cells were stained with FITC conjugated anti-phospho-Histone H2A.X (Ser139) antibody (clone JBW301; 1 μg/100 μL; Sigma Aldrich), and propidium iodide solution (5 μg/mL; Thermo Fisher Scientific) containing DNase free RNASE A/T1 cocktail (25 U/1000 U per mL; Thermo Fisher Scientific).

4.10. BCL2-Family Proteins

THP-1 cells were cultured for 24 h with or without PMPs, and an additional 24 h with or without treatment with DNR, before analysis of intracellular proteins using the published protocol by Ludwig et al. [88]. Briefly, cells were fixed and permeabilized using the eBioscience Foxp3/Transcription Factor Staining Buffer Set (Thermo Fisher Scientific). Cells were then incubated with unconjugated antibodies and labeled with the proper conjugated secondary antibodies. A list of the antibodies used, dilutions, and incubation times, can be found in Table S2. A "no primary antibody" sample was used to subtract the background signal. At least 25,000 gated cells were collected.

4.11. Statistical Analyses

Statistical analyses were performed using the IBM SPSS 26 software (IBM Corp, Armonk, NY, USA). All figures show mean values with 95% confidence intervals. A comparison of means was performed using tests for paired data, or one-sample tests, when appropriate. The data were checked for normality using PP plots, the Shapiro–Wilks test, and the Kolmogorov–Smirnov test. A p value < 0.05 was considered significant. "n" denotes technical replicates.

Supplementary Materials: The following are available online at https://www.mdpi.com/article/10.3390/ijms22147264/s1.

Author Contributions: Conceptualization, D.C.; methodology, D.C.; software, D.C.; validation, D.C.; formal analysis, D.C.; investigation, D.C.; resources, D.C. and T.H.; writing—original draft preparation, D.C.; writing—review and editing, D.C., O.N., P.M. and T.H.; visualization, D.C.; supervision, O.N., P.M., T.H.; project administration, D.C. and T.H.; funding acquisition, D.C. and T.H. All authors have read and agreed to the published version of the manuscript.

Funding: This work was supported by Western Norway Regional Health Authorities and Bergen Stem Cell Consortium.

Institutional Review Board Statement: The study was conducted according to the guidelines of the Declaration of Helsinki, and approved by the regional ethics committee, Regional Etisk Komite Vest (ref 8144) on 16.05.2017.

Informed Consent Statement: Informed consent was obtained from all subjects involved in the study.

Data Availability Statement: The data presented in this study are available on reasonable request from the corresponding author.

Acknowledgments: We would like to thank the Department of Immunology and Transfusion Medicine, Stavanger University Hospital for providing the platelet concentrates used in this study. Graphical abstracted is created with BiroRender.com (accessed on 19 May 2021).

Conflicts of Interest: The authors declared no potential conflict of interest.

References

1. De Gaetano, G. Historical Overview of the Role of Platelets in Hemostasis and Thrombosis. *Haematology* **2001**, *86*, 349–356.
2. Varon, D.; Shai, E. Platelets and Their Microparticles as Key Players in Pathophysiological Responses. *J. Thromb. Haemost.* **2015**, *13*, S40–S46. [CrossRef] [PubMed]
3. Walsh, T.G.; Metharom, P.; Berndt, M.C. The Functional Role of Platelets in the Regulation of Angiogenesis. *Platelets* **2014**, *26*, 199–211. [CrossRef] [PubMed]
4. Gawaz, M.P.; Vogel, S. Platelets in Tissue Repair: Control of Apoptosis and Interactions with Regenerative Cells. *Blood* **2013**, *122*, 2550–2554. [CrossRef]
5. Kurokawa, T.; Zheng, Y.; Ohkohchi, N. Novel Functions of Platelets in the Liver. *J. Gastroenterol. Hepatol.* **2016**, *31*, 745–751. [CrossRef]
6. Dovizio, M.; Bruno, A.; Contursi, A.; Grande, R.; Patrignani, P. Platelets and Extracellular Vesicles in Cancer: Diagnostic and Therapeutic Implications. *Cancer Metastasis Rev.* **2018**, *37*, 455–467. [CrossRef] [PubMed]
7. Xu, X.R.; Yousef, G.M.; Ni, H. Cancer and Platelet Crosstalk: Opportunities and Challenges for Aspirin and Other Anti-Platelet Agents. *Blood* **2018**, *131*, 1777–1789. [CrossRef] [PubMed]
8. Parsons, M.E.M.; Szklanna, P.B.; Guerrero, J.A.; Wynne, K.; Dervin, F.; O'Connell, K.; Allen, S.; Egan, K.; Bennett, C.; McGuigan, C.; et al. Platelet Releasate Proteome Profiling Reveals a Core Set of Proteins with Low Variance Between Healthy Adults. *Proteomics* **2018**, E1800219. [CrossRef]
9. Zufferey, A.; Schvartz, D.; Nolli, S.; Reny, J.-L.; Sanchez, J.-C.; Fontana, P. Characterization of the Platelet Granule Proteome: Evidence of the Presence of MHC1 in Alpha-Granules. *J. Proteom.* **2014**, *101*, 130–140. [CrossRef] [PubMed]
10. Pienimaeki-Roemer, A.; Konovalova, T.; Musri, M.; Sigruener, A.; Boettcher, A.; Meister, G.; Schmitz, G. Transcriptomic Profiling of Platelet Senescence and Platelet Extracellular Vesicles. *Transfusion* **2017**, *57*, 144–156. [CrossRef]
11. Ambrose, A.; Alsahli, M.A.; Kurmani, S.A.; Goodall, A. Comparison of the Release of MicroRNAs and Extracellular Vesicles from Platelets in Response to Different Agonists. *Platelets* **2018**, *29*, 446–454. [CrossRef] [PubMed]
12. Mills, E.W.; Green, R.; Ingolia, N.T. Slowed Decay of MRNAs Enhances Platelet Specific Translation. *Blood* **2017**, *129*, e38–e48. [CrossRef] [PubMed]
13. Londin, E.R.; Hatzimichael, E.; Loher, P.; Edelstein, L.; Shaw, C.; Delgrosso, K.; Fortina, P.; Bray, P.F.; McKenzie, S.E.; Rigoutsos, I. The Human Platelet: Strong Transcriptome Correlations Among Individuals Associate Weakly with the Platelet Pro-Teome. *Biol. Direct.* **2014**, *9*, 3. [CrossRef] [PubMed]
14. Angénieux, C.; Maître, B.; Eckly, A.; Lanza, F.; Gachet, C.; De La Salle, H. Time-Dependent Decay of MRNA and Ribosomal RNA During Platelet Aging and Its Correlation with Translation Activity. *PLoS ONE* **2016**, *11*, e0148064. [CrossRef]
15. Takemoto, A.; Okitaka, M.; Takagi, S.; Takami, M.; Sato, S.; Nishio, M.; Okumura, S.; Fujita, N. A Critical Role of Platelet TGF-β Release in Podoplanin-Mediated Tumour Invasion and Metastasis. *Sci. Rep.* **2017**, *7*, 42186. [CrossRef]
16. Guo, Y.; Cui, W.; Pei, Y.; Xu, D. Platelets Promote Invasion and Induce Epithelial to Mesenchymal Transition in Ovarian Cancer Cells by TGF-β Signaling Pathway. *Gynecol. Oncol.* **2019**, *153*, 639–650. [CrossRef]
17. Pucci, F.; Rickelt, S.; Newton, A.P.; Garris, C.; Nunes, E.; Evavold, C.; Pfirschke, C.; Engblom, C.; Mino-Kenudson, M.; Hynes, R.O.; et al. PF4 Promotes Platelet Production and Lung Cancer Growth. *Cell Rep.* **2016**, *17*, 1764–1772. [CrossRef]
18. Liang, H.; Yan, X.; Pan, Y.; Wang, Y.; Wang, N.; Li, L.; Liu, Y.; Chen-Yu, Z.; Zhang, C.-Y.; Gu, H.; et al. MicroRNA-223 Delivered by Platelet-Derived Microvesicles Promotes Lung Cancer Cell Invasion via Targeting Tumor Suppressor EPB41L3. *Mol. Cancer* **2015**, *14*, 58. [CrossRef]
19. Tang, M.; Jiang, L.; Lin, Y.; Wu, X.; Wang, K.; He, Q.; Wang, X.; Li, W. Platelet Microparticle-Mediated Transfer of MiR-939 to Epithelial Ovarian Cancer Cells Promotes Epithelial to Mesenchymal Transition. *Oncotarget* **2017**, *8*, 97464–97475. [CrossRef]
20. Michael, J.V.; Wurtzel, J.G.T.; Mao, G.F.; Rao, A.K.; Kolpakov, M.A.; Sabri, A.; Hoffman, N.E.; Rajan, S.; Tomar, D.; Madesh, M.; et al. Platelet Microparticles Infiltrating Solid Tumors Transfer MiRNAs That Suppress Tumor Growth. *Blood* **2017**, *130*, 567–580. [CrossRef]
21. Chan, A.T.; Ogino, S.; Fuchs, C.S. Aspirin Use and Survival After Diagnosis of Colorectal Cancer. *JAMA* **2009**, *302*, 649–658. [CrossRef]
22. Bains, S.J.; Mahic, M.; Myklebust, T.Å.; Småstuen, M.C.; Yaqub, S.; Dørum, L.M.; Bjørnbeth, B.A.; Møller, B.; Brudvik, K.W.; Taskén, K. Aspirin as Secondary Prevention in Patients with Colorectal Cancer: An Unselected Population-Based Study. *J. Clin. Oncol.* **2016**, *34*, 2501–2508. [CrossRef]
23. Cao, Y.; Nishihara, R.; Wu, K.; Wang, M.; Ogino, S.; Willett, W.C.; Spiegelman, D.; Fuchs, C.S.; Giovannucci, E.L.; Chan, A.T. Population-Wide Impact of Long-Term Use of Aspirin and the Risk for Cancer. *JAMA Oncol.* **2016**, *2*, 762–769. [CrossRef]
24. Qiao, Y.; Yang, T.; Gan, Y.; Li, W.; Wang, C.; Gong, Y.; Lu, Z. Associations Between Aspirin Use and the Risk of Cancers: A Meta-Analysis of Observational Studies. *BMC Cancer* **2018**, *18*, 288. [CrossRef] [PubMed]

25. McNeil, J.J.; Gibbs, P.; Orchard, S.G.; Lockery, J.E.; Bernstein, W.B.; Cao, Y.; Ford, L.; Haydon, A.; Kirpach, B.; Macrae, F.; et al. Effect of aspirin on cancer incidence and mortality in older adults. *J. Natl. Cancer Inst.* **2020**. [CrossRef] [PubMed]
26. Hurwitz, L.M.; Pinsky, P.F.; Huang, W.-Y.; Freedman, N.D.; Trabert, B. Aspirin Use and Ovarian Cancer Risk Using Extended Follow-up of the PLCO Cancer Screening Trial. *Gynecol. Oncol.* **2020**, *159*, 522–526. [CrossRef] [PubMed]
27. Burn, J.; Sheth, H.; Elliott, F.; Reed, L.; Macrae, F.; Mecklin, J.-P.; Möslein, G.; McRonald, F.E.; Bertario, L.; Evans, D.G.; et al. Cancer prevention with aspirin in hereditary colorectal cancer (Lynch syndrome), 10-year follow-up and registry-based 20-year data in the CAPP2 study: A double-blind, randomised, placebo-controlled trial. *Lancet* **2020**, *395*, 1855–1863. [CrossRef]
28. Sarry, J.-E.; Murphy, K.; Perry, R.; Sanchez, P.V.; Secreto, A.; Keefer, C.; Swider, C.R.; Strzelecki, A.-C.; Cavelier, C.; Récher, C.; et al. Human Acute Myelogenous Leukemia Stem Cells Are Rare and Heterogeneous When Assayed in NOD/SCID/IL2Rγc-Deficient Mice. *J. Clin. Investig.* **2011**, *121*, 384–395. [CrossRef]
29. Krivtsov, A.; Twomey, D.; Feng, Z.; Stubbs, M.C.; Wang, Y.; Faber, J.; Levine, J.E.; Wang, J.; Hahn, W.C.; Gilliland, D.G.; et al. Transformation from Committed Progenitor to Leukaemia Stem Cell Initiated by MLL–AF9. *Nat. Cell Biol.* **2006**, *442*, 818–822. [CrossRef]
30. Ley, T.J.; Mardis, E.R.; Ding, L.; Fulton, B.; McLellan, M.D.; Chen, K.; Dooling, D.; Dunford-Shore, B.H.; McGrath, S.; Hickenbotham, M.; et al. DNA Sequencing of a Cytogenetically Normal Acute Myeloid Leukaemia Genome. *Nat. Cell Biol.* **2008**, *456*, 66–72. [CrossRef]
31. Lawrence, M.S.; Stojanov, P.; Mermel, C.; Robinson, J.T.; Garraway, L.A.; Golub, T.R.; Meyerson, M.; Gabriel, S.B.; Lander, E.S.; Getz, G. Discovery and Saturation Analysis of Cancer Genes across 21 Tumour Types. *Nat. Cell Biol.* **2014**, *505*, 495–501. [CrossRef] [PubMed]
32. Lawrence, M.S.; Stojanov, P.; Polak, P.; Kryukov, G.V.; Cibulskis, K.; Sivachenko, A.; Carter, S.L.; Stewart, C.; Mermel, C.H.; Roberts, S.A.; et al. Mutational heterogeneity in cancer and the search for new cancer-associated genes. *Nat. Cell Biol.* **2013**, *499*, 214–218. [CrossRef]
33. Tyner, J.W.; Tognon, C.E.; Bottomly, D.; Wilmot, B.; Kurtz, S.E.; Savage, S.L.; Long, N.; Schultz, A.R.; Traer, E.; Abel, M.; et al. Functional Genomic Landscape of Acute Myeloid Leukaemia. *Nat. Cell Biol.* **2018**, *562*, 526–531. [CrossRef] [PubMed]
34. The Cancers Genome Atlas Research Network. Genomic and Epigenomic Landscapes of Adult De Novo Acute Myeloid Leukemia. *N. Engl. J. Med.* **2013**, *368*, 2059–2074. [CrossRef]
35. De Boer, B.; Prick, J.; Pruis, M.G.; Keane, P.; Imperato, M.R.; Jaques, J.; Brouwers-Vos, A.Z.; Hogeling, S.M.; Woolthuis, C.M.; Nijk, M.T.; et al. Prospective Isolation and Characterization of Genetically and Functionally Distinct AML Subclones. *Cancer Cell* **2018**, *34*, 674–689.e8. [CrossRef]
36. Van Galen, P.; Hovestadt, V.; Wadsworth, M.H.; Huges, T.K.; Griffin, G.K.; Battaglia, S.; Verga, J.A.; Stephansky, J.; Pastika, T.T.; Lombardy Story, J.; et al. Single-Cell RNA-Seq Reveals AML Hierarchies Relevant to Disease Progression and Immunity. *Cell* **2019**, *176*, 1265–1281.e24. [CrossRef]
37. Glass, J.L.; Hassane, D.; Wouters, B.J.; Kunimoto, H.; Avellino, R.; Garrett-Bakelman, F.E.; Guryanova, O.A.; Bowman, R.; Redlich, S.; Intlekofer, A.M.; et al. Epigenetic Identity in AML Depends on Disruption of Nonpromoter Regulatory Elements and Is Affected by An-Tagonistic Effects of Mutations in Epigenetic Modifiers. *Cancer Discov.* **2017**, *7*, 868–883. [CrossRef]
38. Kantarjian, H.M.; Kadia, T.M.; Di Nardo, C.D.; Welch, M.A.; Ravandi, F. Acute Myeloid Leukemia: Treatment and Research Outlook for 2021 and the MD Anderson Approach. *Cancer* **2021**, *127*, 1186–1207. [CrossRef] [PubMed]
39. Sasaki, K.; Ravandi, F.; Kadia, T.M.; DiNardo, C.D.; Short, N.J.; Borthakur, G.; Jabbour, E.; Kantarjian, H.M. De Novo Acute Myeloid Leukemia: A population-based Study of Outcome in the United States Based on the Surveillance, Epidemiology, and End Results (SEER) Database, 1980 to 2017. *Cancer* **2021**, *127*, 2049–2061. [CrossRef]
40. McComb, S.; Chan, P.K.; Guinot, A.; Hartmannsdottir, H.; Jenni, S.; Dobay, M.P.; Bourquin, J.-P.; Bornhauser, B.C. Efficient Apoptosis Requires Feedback Amplification of Upstream Apoptotic Signals by Effector Caspase-3 or -7. *Sci. Adv.* **2019**, *5*, eaau9433. [CrossRef] [PubMed]
41. Ngoi, N.Y.L.; Choong, C.; Lee, J.; Bellot, G.; Wong, A.L.A.; Goh, B.C.; Pervaiz, S. Targeting Mitochondrial Apoptosis to Overcome Treatment Resistance in Cancer. *Cancers* **2020**, *12*, 574. [CrossRef] [PubMed]
42. Zhou, J.-D.; Zhang, T.-J.; Xu, Z.-J.; Gu, Y.; Ma, J.-C.; Li, X.-X.; Guo, H.; Wen, X.-M.; Zhang, W.; Yang, L.; et al. BCL2 Overexpression: Clinical Implication and Biological Insights in Acute Myeloid Leukemia. *Diagn. Pathol.* **2019**, *14*, 1–10. [CrossRef] [PubMed]
43. Li, X.-X.; Zhou, J.-D.; Wen, X.-M.; Zhang, T.-J.; Wu, D.-H.; Deng, Z.-Q.; Zhang, Z.-H.; Lian, X.-Y.; He, P.-F.; Yao, X.-Y.; et al. Increased *MCL-1* Expression Predicts Poor Prognosis and Disease Recurrence in Acute Myeloid Leukemia. *OncoTargets Ther.* **2019**, *12*, 3295–3304. [CrossRef] [PubMed]
44. Singh, R.; Letai, A.; Sarosiek, K. Regulation of Apoptosis in Health and Disease: The Balancing Act of BCL-2 Family Proteins. *Nat. Rev. Mol. Cell Biol.* **2019**, *20*, 175–193. [CrossRef] [PubMed]
45. Letai, A.; Bassik, M.C.; Walensky, L.D.; Sorcinelli, M.D.; Weiler, S.; Korsmeyer, S.J. Distinct BH3 Domains Either Sensitize or Activate Mitochondrial Apoptosis, Serving as Prototype Cancer Therapeutics. *Cancer Cell* **2002**, *2*, 183–192. [CrossRef]
46. Subburaj, Y.D.; Cosentino, K.; Axmann, M.; Villalmanzo, E.P.; Hermann, E.; Bleicken, S.; Spatz, J.P.; García-Sáez, A.J. Bax Monomers Form Dimer Units in the Membrane That Further Self-Assemble into Multiple Oligomeric Species. *Nat. Commun.* **2015**, *6*, 8042. [CrossRef]
47. Todt, F.; Cakir, Z.; Reichenbach, F.; Emschermann, F.; Lauterwasser, J.; Kaiser, A.; Ichim, G.; Tait, S.; Frank, R.; Langer, H.F.; et al. Differential Retrotranslocation of Mitochondrial Bax and Bak. *EMBO J.* **2015**, *34*, 67–80. [CrossRef]

48. Fullstone, G.; Bauer, T.L.; Gutta, C.; Salvucci, M.; Prehn, J.H.M.; Rehm, M. The Apoptosome Molecular Timer Synergises with XIAP to Suppress Apoptosis Execution and Contributes to Prog-Nosticating Survival in Colorectal Cancer. *Cell Death Differ.* **2020**, *27*, 2828–2842. [CrossRef] [PubMed]
49. Wu, C.-C.; Lee, S.; Malladi, S.; Chen, M.-D.; Mastrandrea, N.J.; Zhang, Z.; Bratton, S.B. The Apaf-1 Apoptosome Induces Formation of Caspase-9 Homo- and Heterodimers with Distinct Activities. *Nat. Commun.* **2016**, *7*, 13565. [CrossRef]
50. Agard, N.J.; Mahrus, S.; Trinidad, J.C.; Lynn, A.; Burlingame, A.L.; Wells, J.A. Global Kinetic Analysis of Proteolysis via Quantitative Targeted Proteomics. *Proc. Natl. Acad. Sci. USA* **2012**, *109*, 1913–1918. [CrossRef]
51. Enari, M.; Sakahira, H.; Yokoyama, H.; Okawa, K.; Iwamatsu, A.; Nagata, S. A Caspase-Activated DNase That Degrades DNA During Apoptosis, and Its Inhibitor ICAD. *Nat. Cell Biol.* **1998**, *391*, 43–50. [CrossRef]
52. Cacic, D.; Reikvam, H.; Nordgård, O.; Meyer, P.; Hervig, T. Platelet Microparticles Protect Acute Myelogenous Leukemia Cells Against Daunorubicin-Induced Apoptosis. *Cancers* **2021**, *13*, 1870. [CrossRef] [PubMed]
53. Bai, H.; Zhou, L.; Wang, C.; Xu, X.; Jiang, J.; Qin, Y.; Wang, X.; Zhao, C.; Shao, S. Involvement of MiR-125a in Resistance to Daunorubicin by Inhibiting Apoptosis in Leukemia Cell Lines. *Tumor Biol.* **2017**, *39*, 1010428317695964. [CrossRef] [PubMed]
54. Zhou, L.; Bai, H.; Wang, C.; Wei, D.; Qin, Y.; Xu, X. MicroRNA-125b Promotes Leukemia Cell Resistance to Daunorubicin by Inhibiting Apoptosis. *Mol. Med. Rep.* **2014**, *9*, 1909–1916. [CrossRef]
55. Al-Aamri, H.M.; Irving, H.R.; Bradley, C.; Meehan-Andrews, T. Intrinsic and Extrinsic Apoptosis Responses in Leukaemia Cells Following Daunorubicin Treatment. *BMC Cancer* **2021**, *21*. [CrossRef] [PubMed]
56. Nieswandt, B.; Hafner, M.; Echtenacher, B.; Männel, D.N. Lysis of Tumor Cells by Natural Killer Cells in Mice Is Impeded by Platelets. *Cancer Res.* **1999**, *59*, 1295–1300.
57. Kopp, H.G.; Placke, T.; Salih, H.R. Platelet-Derived Transforming Growth Factor-Beta down-Regulates NKG2D Thereby Inhibiting Natural Killer Cell Antitumor Reactivity. *Cancer Res.* **2009**, *69*, 7775–7783. [CrossRef]
58. Yu, M.; Bardia, A.; Wittner, B.S.; Stott, S.L.; Smas, M.E.; Ting, D.T.; Isakoff, S.J.; Ciciliano, J.C.; Wells, M.N.; Shah, A.M.; et al. Circulating Breast Tumor Cells Exhibit Dynamic Changes in Epithelial and Mesenchymal Composition. *Science* **2013**, *339*, 580–584. [CrossRef]
59. Palacios-Acedo, A.L.; Mège, D.; Crescence, L.; Dignat-George, F.; Dubois, C.; Panicot-Dubois, L. Platelets, Thrombo-Inflammation, and Cancer: Collaborating with the Enemy. *Front. Immunol.* **2019**, *10*, 1805. [CrossRef]
60. Yan, M.; Jurasz, P. The Role of Platelets in the Tumor Microenvironment: From Solid Tumors to Leukemia. *Biochim. Biophys. Acta (BBA) Bioenerg.* **2016**, *1863*, 392–400. [CrossRef]
61. Ahmad, B.; Gamallat, Y.; Su, P.; Husain, A.; Rehman, A.U.; Zaky, M.Y.; Bakheet, A.M.H.; Tahir, N.; Xin, Y.; Liang, W. Alantolactone Induces Apoptosis in THP-1 Cells through STAT3, Survivin Inhibition, and Intrinsic Apoptosis Pathway. *Chem. Biol. Drug Des.* **2021**, *97*, 266–272. [CrossRef] [PubMed]
62. Gupta, S.; Prasad, G.V.R.K.; Mukhopadhaya, A. Vibrio Cholerae Porin OmpU Induces Caspase-Independent Programmed Cell Death Upon Translocation to the Host Cell Mitochondria. *J. Biol. Chem.* **2015**, *290*, 31051–31068. [CrossRef] [PubMed]
63. Chen, C.; Lin, H.; Karanes, C.; Pettit, G.R.; Chen, B.D. Human THP-1 Monocytic Leukemic Cells Induced to Undergo Monocytic Differentiation by Bryostatin 1 Are Refractory to Proteasome Inhibitor-Induced Apoptosis. *Cancer Res.* **2000**, *60*, 4377–4385. [PubMed]
64. Kang, C.-H.; Moon, D.-O.; Choi, Y.H.; Choi, I.-W.; Moon, S.-K.; Kim, W.-J.; Kim, G.-Y. Piceatannol Enhances TRAIL-Induced Apoptosis in Human Leukemia THP-1 Cells through Sp1- and ERK-Dependent DR5 up-Regulation. *Toxicol. In Vitro* **2011**, *25*, 605–612. [CrossRef]
65. Marinello, J.; Delcuratolo, M.; Capranico, G. Anthracyclines as Topoisomerase II Poisons: From Early Studies to New Perspec-Tives. *Int. J. Mol. Sci.* **2018**, *19*, 3480. [CrossRef]
66. Demir, S.; Boldrin, E.; Sun, Q.; Hampp, S.; Tausch, E.; Eckert, C.; Ebinger, M.; Handgretinger, R.; Kronnie, G.T.; Wiesmüller, L.; et al. Therapeutic Targeting of Mutant p53 in Pediatric Acute Lymphoblastic Leukemia. *Haematology* **2019**, *105*, 170–181. [CrossRef] [PubMed]
67. Garner, T.P.; Amgalan, D.; Reyna, D.E.; Li, S.; Kitsis, R.N.; Gavathiotis, E. Small-Molecule Allosteric Inhibitors of BAX. *Nat. Chem. Biol.* **2019**, *15*, 322–330. [CrossRef] [PubMed]
68. Peixoto, P.M.; Teijido, O.; Mirzalieva, O.; Dejean, L.M.; Pavlov, E.V.; Antonsson, B.; Kinnally, K.W. MAC Inhibitors Antagonize the Pro-Apoptotic Effects of TBid and Disassemble Bax/Bak Oligomers. *J. Bioenerg. Biomembr.* **2017**, *49*, 65–74. [CrossRef]
69. Amgalan, D.; Garner, T.P.; Pekson, R.; Jia, X.F.; Yanamandala, M.; Paulino, V.; Liang, F.G.; Corbalan, J.J.; Lee, J.; Chen, Y.; et al. A Small-Molecule Allosteric Inhibitor of BAX Protects Against Doxorubicin-Induced Cardiomyopathy. *Nat. Rev. Cancer* **2020**, *1*, 315–328. [CrossRef]
70. Sarosiek, K.; Fraser, C.; Muthalagu, N.; Bhola, P.D.; Chang, W.; McBrayer, S.K.; Cantlon, A.; Fisch, S.; Golomb-Mello, G.; Ryan, J.A.; et al. Developmental Regulation of Mitochondrial Apoptosis by C-Myc Governs Age- and Tissue-Specific Sensitivity to Cancer Therapeutics. *Cancer Cell* **2017**, *31*, 142–156. [CrossRef] [PubMed]
71. Pereira, N.A.; Song, Z. Some Commonly Used Caspase Substrates and Inhibitors Lack the Specificity Required to Monitor Individual Caspase Activity. *Biochem. Biophys. Res. Commun.* **2008**, *377*, 873–877. [CrossRef] [PubMed]
72. Peixoto, P.M.; Ryu, S.; Bombrun, A.; Antonsson, B.; Kinnally, K.W. MAC Inhibitors Suppress Mitochondrial Apoptosis. *Biochem. J.* **2009**, *423*, 381–387. [CrossRef] [PubMed]

73. Zhao, Z.; Liu, J.; Deng, Y.; Huang, W.; Ren, C.; Call, D.R.; Hu, C. The Vibrio Alginolyticus T3SS Effectors, Val1686 and Val1680, Induce Cell Rounding, Apoptosis and Lysis of Fish Epithelial Cells. *Virulence* **2018**, *9*, 318–330. [CrossRef]
74. Collins, P.L.; Purman, C.; Porter, S.I.; Nganga, V.; Saini, A.; Hayer, K.E.; Gurewitz, G.L.; Sleckman, B.P.; Bednarski, J.J.; Bassing, C.H.; et al. DNA Double-Strand Breaks Induce H2Ax Phosphorylation Domains in a Contact-Dependent Manner. *Nat. Commun.* **2020**, *11*, 1–9. [CrossRef]
75. Darzynkiewicz, Z.; Halicka, D.H.; Tanaka, T. Cytometric Assessment of DNA Damage Induced by DNA Topoisomerase Inhibitors. *Breast Cancer* **2009**, *582*, 145–153. [CrossRef]
76. Johansson, P.; Fasth, A.; Ek, T.; Hammarsten, O. Validation of a Flow Cytometry-Based Detection of γ-H2AX, to Measure DNA Damage for Clinical Applications. *Cytom. Part B Clin. Cytom.* **2017**, *92*, 534–540. [CrossRef]
77. Lee, Y.; Wang, Q.; Shuryak, I.; Brenner, D.J.; Turner, H.C. Development of a High-Throughput γ-H2AX Assay Based on Imaging Flow Cytometry. *Radiat. Oncol.* **2019**, *14*, 1–10. [CrossRef]
78. Tentner, A.R.; Lee, M.; Ostheimer, G.J.; Samson, L.D.; A Lauffenburger, D.; Yaffe, M.B. Combined Experimental and Computational Analysis of DNA Damage Signaling Reveals context-dependent Roles for Erk in Apoptosis and G1/S Arrest After Genotoxic Stress. *Mol. Syst. Biol.* **2012**, *8*, 568. [CrossRef] [PubMed]
79. Al-Aamri, H.M.; Ku, H.; Irving, H.R.; Tucci, J.; Meehan-Andrews, T.; Bradley, C. Time Dependent Response of Daunorubicin on Cytotoxicity, Cell Cycle and DNA Repair in Acute Lymphoblastic Leukaemia. *BMC Cancer* **2019**, *19*, 179. [CrossRef]
80. Doroshow, J.H. Mechanisms of Anthracycline-Enhanced Reactive Oxygen Metabolism in Tumor Cells. *Oxidative Med. Cell. Longev.* **2019**, *2019*, 9474823. [CrossRef] [PubMed]
81. Vo, T.-T.; Ryan, J.; Carrasco, R.; Neuberg, D.; Rossi, D.J.; Stone, R.M.; DeAngelo, D.J.; Frattini, M.G.; Letai, A. Relative Mitochondrial Priming of Myeloblasts and Normal HSCs Determines Chemotherapeutic Success in AML. *Cell* **2012**, *151*, 344–355. [CrossRef] [PubMed]
82. López-Royuela, N.; Pérez-Galán, P.; Galán-Malo, P.; Yuste, V.J.; Anel, A.; Susin, S.A.; Naval, J.; Marzo, I. Different Contribution of BH3-Only Proteins and Caspases to Doxorubicin-Induced Apoptosis in p53-Deficient Leukemia Cells. *Biochem. Pharmacol.* **2010**, *79*, 1746–1758. [CrossRef] [PubMed]
83. Li, M. The Role of P53 up-Regulated Modulator of Apoptosis (PUMA) in Ovarian Development, Cardiovascular and Neurodegenerative Diseases. *Apoptosis* **2021**, *26*, 235–247. [CrossRef] [PubMed]
84. Mandal, M.; Crusio, K.M.; Meng, F.; Liu, S.; Kinsella, M.; Clark, M.R.; Takeuchi, O.; Aifantis, I. Regulation of Lymphocyte Progenitor Survival by the Proapoptotic Activities of Bim and Bid. *Proc. Natl. Acad. Sci. USA* **2008**, *105*, 20840–20845. [CrossRef]
85. Sandow, J.J.; Jabbour, A.; Condina, M.R.; Daunt, C.P.; Stomski, F.C.; Green, B.D.; Riffkin, C.D.; Hoffmann, P.; Guthridge, M.; Silke, J.; et al. Cytokine Receptor Signaling Activates an IKK-Dependent Phosphorylation of PUMA to Prevent Cell Death. *Cell Death Differ.* **2011**, *19*, 633–641. [CrossRef]
86. Carpenter, R.L.; Han, W.; Paw, I.; Lo, H.-W. HER2 Phosphorylates and Destabilizes Pro-Apoptotic PUMA, Leading to Antagonized Apoptosis in Cancer Cells. *PLoS ONE* **2013**, *8*, e78836. [CrossRef]
87. Zhang, C.-Z.; Zhang, J.-X.; Zhang, A.-L.; Shi, Z.-D.; Han, L.; Jia, Z.-F.; Yang, W.-D.; Wang, G.-X.; Jiang, T.; You, Y.-P.; et al. MiR-221 and MiR-222 Target PUMA to Induce Cell Survival in Glioblastoma. *Mol. Cancer* **2010**, *9*, 229. [CrossRef] [PubMed]
88. Ludwig, L.M.; Maxcy, K.L.; LaBelle, J.L. Flow Cytometry-Based Detection and Analysis of BCL-2 Family Proteins and Mito-Chondrial Outer Membrane Permeabilization (MOMP). *Methods Mol. Biol.* **2019**, *1877*, 77–91.

International Journal of
Molecular Sciences

Article

Cytotoxic Activity against A549 Human Lung Cancer Cells and ADMET Analysis of New Pyrazole Derivatives

Agnieszka Czylkowska [1,*], Małgorzata Szczesio [1], Anita Raducka [1], Bartłomiej Rogalewicz [1], Paweł Kręcisz [2], Kamila Czarnecka [2,3], Paweł Szymański [2,3], Monika Pitucha [4] and Tomasz Pawlak [5]

1 Institute of General and Ecological Chemistry, Faculty of Chemistry, Lodz University of Technology, 90-924 Lodz, Poland; malgorzata.szczesio@p.lodz.pl (M.S.); anita.raducka@edu.p.lodz.pl (A.R.); 211150@edu.p.lodz.pl (B.R.)

2 Department of Pharmaceutical Chemistry, Drug Analyses and Radiopharmacy, Faculty of Pharmacy, Medical University of Lodz, 90-151 Lodz, Poland; pawel.krecisz@umed.lodz.pl (P.K.); kamila.czarnecka@umed.lodz.pl (K.C.); pawel.szymanski@umed.lodz.pl (P.S.)

3 Department of Radiobiology and Radiation Protection, Military Institute of Hygiene and Epidemiology, 01-163 Warsaw, Poland

4 Independent Radiopharmacy Unit, Faculty of Pharmacy, Medical University of Lublin, 20-093 Lublin, Poland; monika.pitucha@umlub.pl

5 Centre of Molecular and Macromolecular Studies, Polish Academy of Sciences, 90-363 Lodz, Poland; tpawlak@cbmm.lodz.pl

* Correspondence: agnieszka.czylkowska@p.lodz.pl

Abstract: Two new pyrazole derivatives, namely compound **1** and compound **2,** have been synthesized, and their biological activity has been evaluated. Monocrystals of the obtained compounds were thoroughly investigated using single-crystal X-ray diffraction analysis, FTIR spectroscopy, and NMR spectroscopy. The results gathered from all three techniques are in good agreement, provide complete information about the structures of **1** and **2**, and confirm their high purity. Thermal properties were studied using thermogravimetric analysis; both **1** and **2** are stable at room temperature. In order to better characterize **1** and **2**, some physicochemical and biological properties have been evaluated using ADMET analysis. The cytotoxic activity of both compounds was determined using the MTT assay on the A549 cell line in comparison with etoposide. It was determined that compound **2** was effective in the inhibition of human lung adenocarcinoma cell growth and may be a promising compound for the treatment of lung cancer.

Keywords: cytotoxic activity; pyrazole derivatives; MTT assay; ADMET analysis; single-crystal diffraction; FTIR spectroscopy; NMR spectroscopy thermogravimetric analysis

Citation: Czylkowska, A.; Szczesio, M.; Raducka, A.; Rogalewicz, B.; Kręcisz, P.; Czarnecka, K.; Szymański, P.; Pitucha, M.; Pawlak, T. Cytotoxic Activity against A549 Human Lung Cancer Cells and ADMET Analysis of New Pyrazole Derivatives. *Int. J. Mol. Sci.* **2021**, *22*, 6692. https://doi.org/10.3390/ijms22136692

Academic Editor: Angela Stefanachi

Received: 27 May 2021
Accepted: 17 June 2021
Published: 22 June 2021

Publisher's Note: MDPI stays neutral with regard to jurisdictional claims in published maps and institutional affiliations.

1. Introduction

There are many different types of compounds, both natural and synthetic, with potential medical properties. Substances, such as imidazoles, oxadiazoles, pyrroles, and many of their derivatives, are well studied and described in the literature [1–5]. The main goal of medicinal chemistry is to synthesize compounds with promising activity and therapeutic agents that exhibit lower toxicity. The search for new, pharmacologically active chemical compounds is related to the modification of existing molecules. A group of compounds that is potentially interesting, due to its structure and biological activity, is those substances containing a pyrazole ring. Pyrazoles exist in many compounds that are used as pharmaceuticals and other active compounds [6–8]. Diseases caused by microbial infection are a serious menace to the health of human beings and often are connected with some other diseases whenever the body system becomes debilitated. The growing incidence of microorganisms that resist antimicrobials is a constant concern for the scientific community. Pyrazoles have always been considered as a scaffold-of-choice in designing novel therapeutic agents because of their anti-inflammatory, anti-tumor, antihyperglycemic, antipyretic, analgesic, antibacterial, and antifungal properties [9–16]. In this work, we present a synthesis, structural characterization, and investigation of the biological, spectroscopic, and thermal properties of two new pyrazole derivatives. The compounds were also tested for cellular survival through MTT assays on the A549 cell line, which investigated their biological properties.

2. Results and Discussion

2.1. Synthesis

The title compounds **1** and **2** have been synthesized in the reaction of 1-cyanophenyl acetic acid hydrazide with isocyanates (butyl and 2,4-dichlorophenyl), according to Scheme 1.

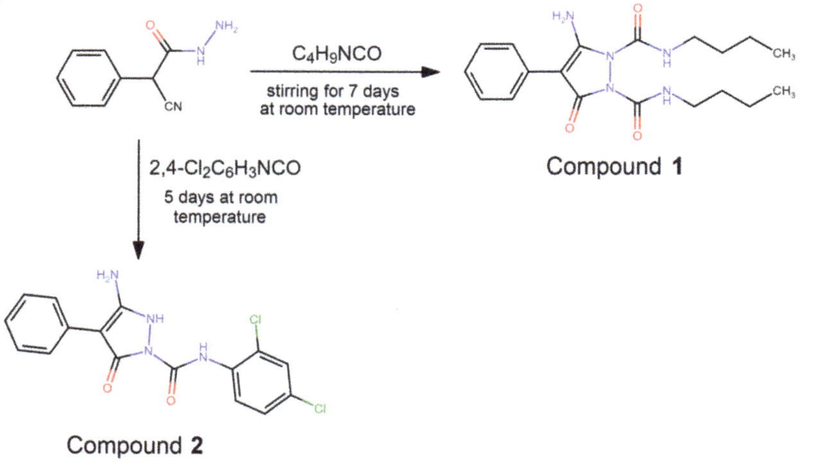

C_4H_9NCO
stirring for 7 days at room temperature

Compound **1**

$2,4\text{-}Cl_2C_6H_3NCO$
5 days at room temperature

Compound **2**

Scheme 1. Synthesis route of the obtained pyrazole derivatives.

Compound **1**: 3-amino-N^1,N^2-dibutyl-5-oxo-4-phenyl-pyrazole-1,2-dicarboxamide.

To synthesize compound **1**, 1.75 g of 1-cyanophenylacetic acid hydrazide (0.01 mol) was dissolved while hot and in 15 mL of anhydrous acetonitrile. Then, 1.98 g (0.02 mol) of butyl isocyanate was added to the warm mixture. Stirring was allowed for 7 days at room temperature. The precipitate was then filtered off and dried. The reaction yield was 67%, and the melting point was 131.1 °C.

Compound **2**: 3-amino-*N*-(2,4-dichlorophenyl)-5-oxo-4-phenyl-2H-pyrrole-1-carboxamide.

To synthesize compound **2**, 1.75 g of 1-cyanophenylacetic acid hydrazide (0.01 mol) was dissolved in 15 mL of anhydrous acetonitrile. Then, 1.88 g (0.01 mol) of 2,4-dichlorophenyl isocyanate was added. The mixture was kept at room temperature for 5 days. The precipitate was then filtered off and dried. The reaction yield was 81%, and the melting point was 188.9 °C.

2.2. Crystal Structure Analysis

Both structures crystallized in the $P2_1/c$ space group in a monoclinic crystal system. The crystal data and structure refinement details are summarized in Table 1. Compound **1**, crystallized from CH_3OH, has one molecule in the asymmetric *unit* (Figure 1a). One of the butyl fragments shows a disorder. The single crystal of compound **2** crystallized from DMSO. In this case, the molecules of solvents are built into the structure; this involves two DMSO molecules, including one disordered and one water molecule (Figure 1b). Tautomerism was described in an earlier publication from 2014 containing an analogous compound [17]. Compound **2** adopts a keto tautomeric form in a crystalline state. There are typical carbonyl bonds in the structure (C2-O3 = 1.221(4) Å, C9-O10 = 1.242(4) Å). Compound **1** also contains only typical carbonyl groups.

Table 1. Crystal and structure refinement data for **1** and **2** compounds.

Compound	1	2
Chemical formula	$C_{19}H_{27}N_5O_3$	$C_{16}H_{12}Cl_2N_4O_2{\cdot}2(C_2H_6OS){\cdot}H_2O$
M_r	373.45	537.47
Crystal system, space group	Monoclinic, $P2_1/c$	Monoclinic, $P2_1/c$
Temperature (K)	100	100
a, b, c (Å)	11.9591 (2), 16.2222 (2), 11.6359 (2)	7.0863 (2), 25.7334 (8), 13.7460 (4)
β (°)	117.695 (2)	97.800 (3)
V (Å^3)	1998.78 (6)	2483.45 (13)
Z	4	4
Radiation type	Cu $K\alpha$	Mo $K\alpha$
No. of measured, independent and observed $[I > 2\sigma(I)]$ reflections	24,501, 4143, 3729	37,018, 7334, 5953
R_{int}	0.039	0.078
$(\sin \theta/\lambda)_{max}$ (Å^{-1})	0.635	0.736
$R[F^2 > 2\sigma(F^2)]$, $wR(F^2)$, S	0.044, 0.119, 1.03	0.079, 0.203, 1.15
No. of reflections	4143	7334
No. of parameters	266	337

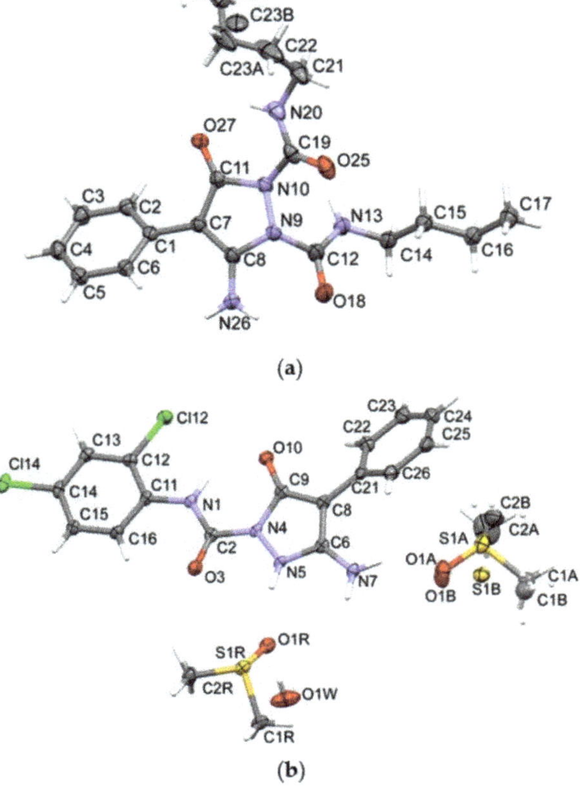

Figure 1. The molecular structure and atom-numbering schemes for: compound **1**—figure (**a**), and compound **2**— figure (**b**), with displacement ellipsoids drawn at the 50% probability level.

The conformation of the molecules is stabilized by intra-molecular hydrogen bonds (S(6)), according to the theory of Bernstein [18]. These are N-H ... O bonds (N26—H26B··· O18 and N20—H20··· O27) (Table 2). The packing of the molecules of the structure **1** is layered (Figure 2). These layers are stabilized by chain hydrogen bonds N26—H26A ··· O25 (C (7)—graph-set) and additionally form dimers N13—H13··· O27 ($R_2^2(14)$).

Table 2. Hydrogen-bond geometry (Å, °) for **1**.

D—H···*A*	*D*—H	H···*A*	*D*···*A*	*D*—H···*A*
N26—H26*A*···O25 [i]	0.86	1.99	2.8055 (16)	158
N26—H26*B*···O18	0.86	2.11	2.7088 (17)	127
N20—H20···O27	0.86	2.00	2.6844 (15)	136
N13—H13···O27 [ii]	0.86	2.00	2.8210 (15)	160

Symmetry codes: (i) $x, -y+3/2, z-1/2$; (ii) $-x+1, -y+1, -z+1$.

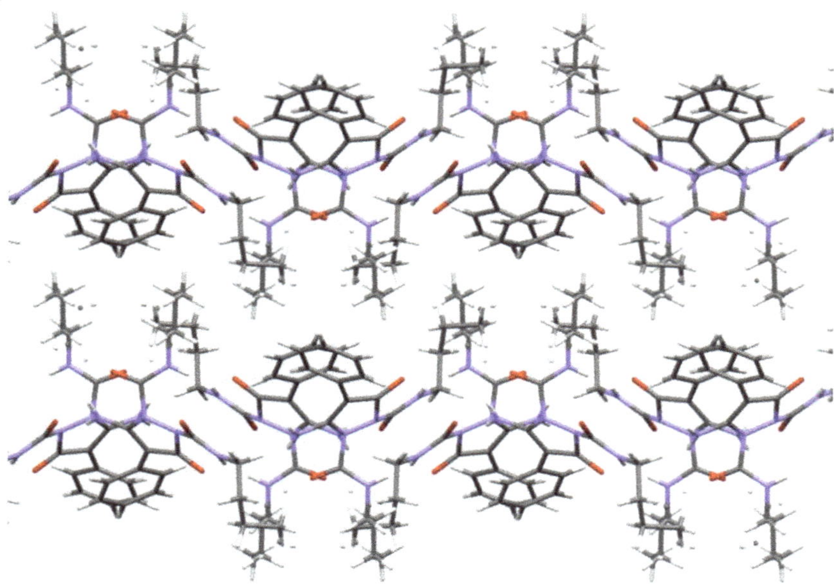

Figure 2. The crystal packing of **1**, viewed along the c axis.

The molecules in structure **2** create gaps into which the disordered solvent molecules are fit (Figure 3). The DMSO molecules form hydrogen bonds of the N-H ... O(DMSO) type. A water molecule stabilizes the structure with three hydrogen bonds with two **2** molecules and one DMSO molecule (Table 3).

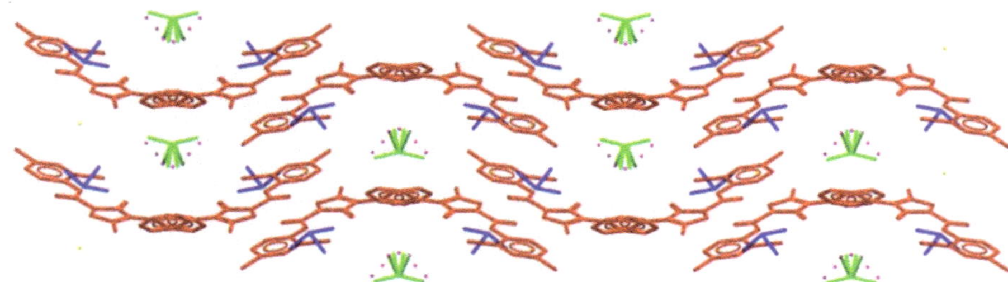

Figure 3. The crystal packing of **2**, viewed along the c axis.

Table 3. Hydrogen-bond geometry (Å, °) for **2**.

D—H···*A*	*D*—H	H···*A*	*D*···*A*	*D*—H···*A*
N1—H1···O10	0.86	1.97	2.689 (3)	140 S(6)
N5—H5···S1*R*	0.86	2.66	3.459 (3)	155
N5—H5···O1*R*	0.86	2.26	2.799 (4)	121
N7—H7*A*···O1*W* i	0.86	2.01	2.835 (4)	160
N7—H7*B*···O1Aa	0.86	2.06	2.81 (4)	146
N7—H7*B*···O1Bb	0.86	2.17	2.86 (3)	137
O1*W*—H1*W*1···O1*R*	0.81 (2)	2.02 (3)	2.817 (4)	166 (7)
O1*W*—H1*W*2···O3 ii	0.81 (2)	2.08 (3)	2.867 (3)	163 (6)

Symmetry codes: (i) −x + 2, −y + 1, −z + 2 (ii) x + 1, y, z.

2.3. NMR Studies

It is well known that monocrystals selected for single-crystal X-ray analysis do not always represent the bulk material. Therefore, we additionally performed a validation analysis of the studied materials using the solution NMR technique (Figures S1 and S2 are in the Supplementary Materials). For both compounds **1** and **2**, we recorded series of NMR spectra as ^{1}H, ^{13}C, ^{1}H-^{1}H COSY, ^{1}H-^{13}C HSQC, and ^{1}H-^{13}C HMBC, and made full assignments of ^{1}H as well as ^{13}C signals. The following NMR parameters were noted for studied compounds: compound **1**, ^{1}H (DMSO-d_6) δ [ppm]: 0.87 (3 × H17, 3 × H24), 1.30 (2 × H16, 2 × H23), 1.44 (2 × H15, 2 × H22), 3.10–3.20 (2 × H14, 2 × H21), 7.20 (H4), 7.28 (H26), 7.37 (H3, H5), 7.49 (H2, H6), 8.10–8.26 (H13, H20) and ^{13}C (DMSO-d_6) δ [ppm]: 13.5–13.6 (C17, C24), 19.2–19.3 (C16–C23), 30.7–31.2 (C15, C22), 39.2–40.1 (C14, C21), 86.0 (C7), 125.6 (C4), 127.3 (C3, C5), 128.2 (C2, C6), 130.2 (C1), 150.5–152.7 (C12, C19), 156.0 (C8), 165.6 (C11); compound **2**, ^{1}H (DMSO-d_6) δ [ppm]: 6.90 (2 × H7), 7.16 (H24), 7.34–7.37 (H23, H25), 7.46 (H15), 7.56 (H22, H26), 7.70 (H13), 8.31 (H16), 10.81 (H5), 11.73 (H1) and ^{13}C (DMSO-d_6) δ [ppm]: 85.9 (C8), 121.6 (C16), 122.9 (C21), 124.9 (C24), 126.6 (C22, C26), 127.1 (C12), 147.1 (C2), 128.0 (C15), 128.2 (C23, C25), 128.8 (C13), 131.4 (C14), 134.3 (C11), 156.3 (C6), 163.2 (C9). Apart from the listed resonances, the residual DMSO, as well as the H_2O and HOD signals, were clearly visible on the spectra. The results obtained herein confirmed that both the **1** and **2** samples are homogenous, pure, and stable in DMSO solution. The single-crystal structures presented in the previous section are fully consistent with the NMR results.

2.4. FTIR Spectra

The FTIR spectra of both compounds confirm their molecular structures. Both spectra (Figures 4 and 5) show several bands of various intensities and shapes that can be ascribed to ν(NH) in the range of 3500–3100 cm^{-1}—with the sharpest peaks at 3398, 3328, and 3297 cm^{-1} for **1** and 3459, 3308, and 3184 cm^{-1} for **2**, respectively. Due to the presence of a butyl group in the **1** molecule, we can observe characteristic bands in the range of 3090–2850 cm^{-1} that correspond to alkane ν(CH) modes, with peaks at 3081, 3060, 3028, 2954, 2931, and 2869 cm^{-1}. In both spectra, we can observe sharp bands in the range of 1730–1680 cm^{-1} that result from ν(C=O) modes—in the **1** spectrum at 1707 cm^{-1} and in the **2** spectrum at 1710 and 1694 cm^{-1}. In both spectra, there are several bands present in the 1630–1500 cm^{-1} region that can be ascribed to ν(CN), ν(C=C), and δ(NH) modes. Sharp bands in the **1** spectrum at 1461 and 1440 cm^{-1} most likely correspond to the δ(CH) methyl group modes. When moving to the lower wavenumbers, we can observe bands for both compounds in ranges of 1280–1110 cm^{-1} and 790–690 cm^{-1} that correspond to β(CH) and γ(CH) modes, respectively. Several sharp bands that result from ν(NN) vibrations can also be observed in the **1** spectrum (1050 and 1016 cm^{-1}) and in the **2** spectrum (1055, 1019 cm^{-1}). The bands that are present in the 870–790 cm^{-1} range in the **2** spectrum, but are absent in the **1** spectrum, can most likely be assigned to ν(CCl).

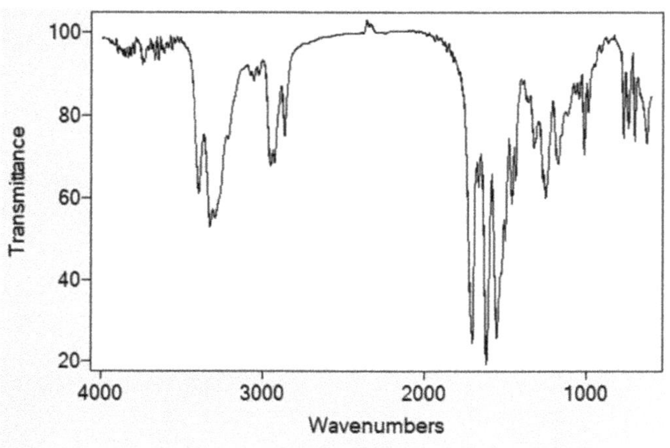

Figure 4. FTIR spectrum of compound **1**.

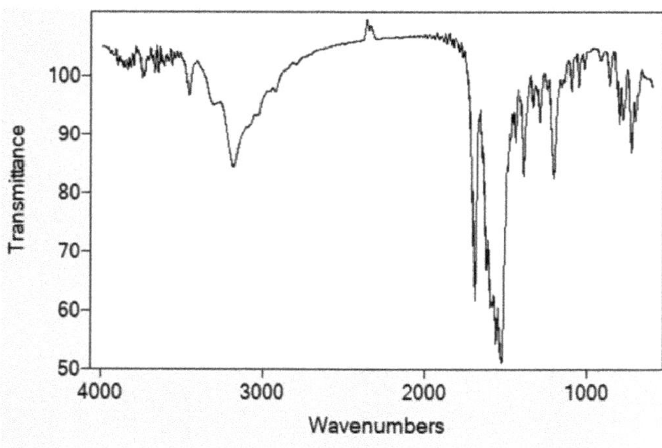

Figure 5. FTIR spectrum of compound **2**.

2.5. Thermogravimetric Studies in Air

The thermal decomposition of **1** is shown in Figure 6. This compound is thermally stable up to 125 °C. In the first stage of thermolysis, one of the aliphatic chains is destroyed. In the temperature range of 125–175, the exothermic effect on the DTA curve is observed (175 °C). When the temperature rises, further destruction of **1** takes place. On the TG curve, there are three mass losses of 8.0% (calc. 7.78%), 19.0% (calc. 19.32%), and 40.0% (calc. 39.68%) within the temperature ranges of 125–175 °C, 175–250 °C, and 250–525 °C, respectively. The final step of decomposition is the burning of organic residues with corresponding exothermic effects on the DTA curve at 680 °C. In Figure 7, the TG, DTG, and DTA curves of **2** are shown. Compound **2** starts to decompose at 175 °C. The first step of pyrolysis is the destruction of the benzene ring. This process is accompanied by an exothermic peak on the DTA curve at 225 °C. The thermolysis of **2** is also a multi-stage and overlapping process. In the temperature range of 175–240 °C, the experimental mass loss is 21.0% and it is calculated at 21.23%. The next step is connected with the 52.0% (calc. 52.06%) loss of mass and occurs between 240 and 300 °C; on the DTA there are peaks at 270 °C, and when the temperature rises above 900 °C, the process stops.

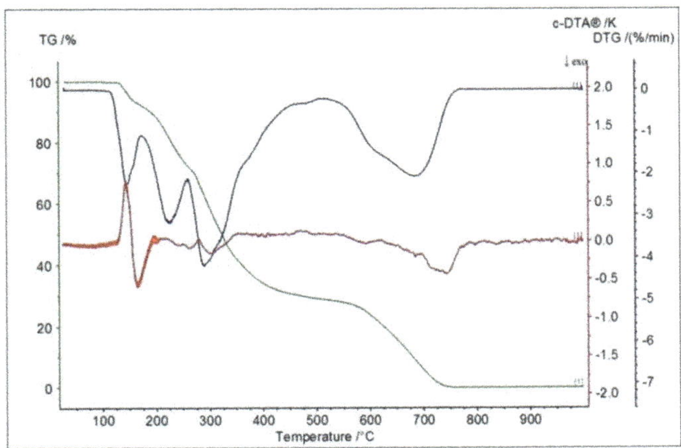

Figure 6. Thermal decomposition of compound **1**.

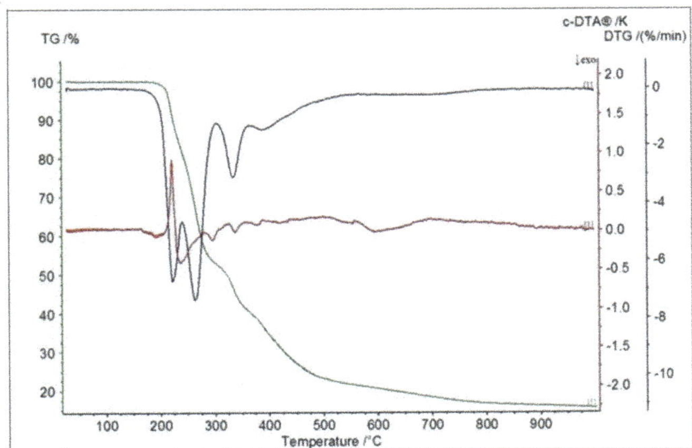

Figure 7. Thermal decomposition of compound **2**.

2.6. *Biological Assays*

In A549 cells, the 50% effective concentration (EC_{50}) for compounds **1** and **2** was found to be 613.22 and 220.20 μM, respectively. The values of the effective concentration after the treatment of the compounds are given in Table 4. It was observed that all synthesized molecules were very active; compound **1** showed much less toxicity than **2**. These results revealed that **2** showed the highest cytotoxicity and the most significant decrease in cell viability relative to the A549 lung cancer cell line. This is very interesting, as both compounds are active in between the activities of the etoposide.

Table 4. Cytotoxicity activity at the EC_{50}.

Compound	Cytotoxicity Activity EC_{50} [μM]
1	613.22 +/−23.56
2	220.20+/−22.47
etoposide	451.47+/−18.27 *

Results are presented as the means ± SD; EC_{50}, 50% inhibition of the cell viability. Statistical significance was assessed using a one-way ANOVA analysis. * $p < 0.01$ was considered significantly different between cancer and non-cancer cell lines.

2.7. ADMET Analysis

The pharmacokinetic profile of compound **1** is very promising. ACD/Percepta software indicated optimal human plasma protein binding (74.49%); 4.6 L/kg of distribution volume, which means good distribution to all parts of the human body; and 91.7% of single 50 mg dose bio-availability per os. Compound **2** exhibited more problematic distribution properties: 96.81% human plasma protein binding, 0.34 L/kg of distribution volume, and 36.9% of single 50 mg dose bio-availability per os. Moreover, the prediction results (ACD/Percepta, admetSAR 2.0) indicate the possibility of blood–brain barrier penetration for both structures. The physicochemical profiles of compounds **1** and **2** indicate that both structures are good candidates for drug agents, as they both show the fulfillment of the Lipinski rule [19], the Ghose rule [20], the Egan rule [21], and the Muegge rule [22]. The basic physicochemical properties of compounds **1** and **2** were gathered in Table 5. The analysis results indicate that compounds can affect the pharmacokinetics of other drugs because of their effects on cytochromes P450 isoenzymes. Compound **1** showed inhibition properties for CYP2C19, CYP2C9, CYP3A4, and compound **2** showed inhibition properties for CYP1A2. Both compounds have very promising physicochemical properties for oral bio-availability (Figure 8). Both compounds showed a very low probability of positive AMES test results and hERG inhibition test results. A ProTox II analysis classified both compounds to toxicity class 4 (predicted LD_{50} 1000 mg/kg). Moreover, both compounds had very promising results of the detailed prediction of the toxicity profile—compound **1** showed a 0.6 probability of carcinogenicity (1 positive test result out of 17 different predictions), while compound **2** showed a 0.52 probability of carcinogenicity and a 0.64 probability of hepatotoxicity (2 positive test results out of 17 different predictions).

Table 5. Basic physicochemical properties of the two compounds.

Compound	Molecular Weight [g/mol]	LogP	pKa (Acid)	pKa (Base)	TPSA [Å^2]	Molar Refractivity [m^3/mol]
1	373.45	2.83	11.70	3.90	111.15	106.29
2	363.20	3.17	9.60	3.48	92.91	95.59

Log p value is an average of 5 prediction algorithms (iLOGP, XLOGP, WLOGP, MLOGP, SILICOS-IT); TPSA—topological polar surface area.

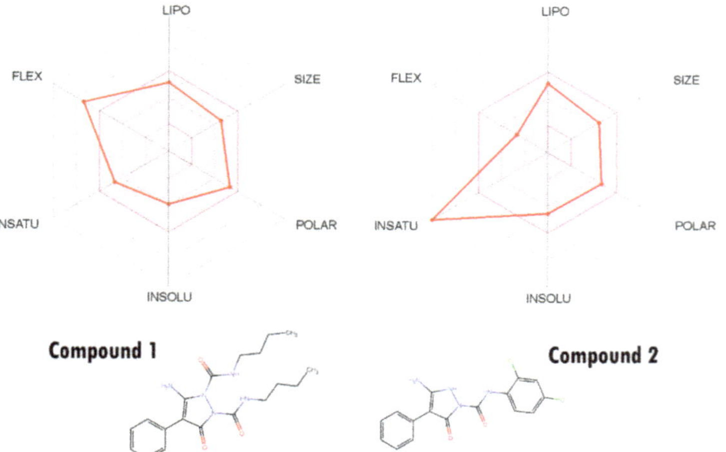

Figure 8. Oral bio-availability graph generated using the SwissADME service. The red–coloured zone is physicochemically suitable for oral bio-availability. LIPO—lipophility (−0.7 < XlogP3 < +5.0); SIZE—molecular weight (150 g/mol < MW < 500 g/mol); POLAR—polarity (20 Å^2 < TPSA < 130 Å^2); INSOLU—insolubility (0 < logS < 6); INSATU—insaturation (0.25 < fraction Csp3 < 1); FLEX—flexibility (0 < num. of rotatable bonds < 9).

3. Materials and Methods

3.1. Chemistry

All of the chemicals used for the synthesis were purchased from Sigma-Aldrich, AlfaAesar, and POCH, and were used without further purification. The FTIR spectra were recorded with an IRTracer-100 Schimadzu Spectrometer (4000–600 cm^{-1}), with an accuracy of recording 1 cm^{-1} using KBr pellets. The thermolysis of the compounds in the air atmosphere was studied using TG-DTG-DTA techniques in the temperature range of 25–1000 °C at a heating rate of 10 °C min^{-1}; TG, DTG, and DTA curves were recorded on a Netzsch TG 209 apparatus under air atmosphere (v = 20 mL × min^{-1}) using ceramic crucibles. Ceramic crucibles were also used as a reference material. All NMR experiments were run at 298 K on a 500 MHz Bruker Avance III spectrometer, which was equipped with ^{1}H with a ^{13}C BB probehead (^{1}H-detected experiment) and operating at 500.13 and 125.76 MHz for ^{1}H and ^{13}C nuclei, respectively. The samples were prepared in DMSO-d_6 (99.8% + D) from Armar Chemicals. The chemical shifts in ^{1}H and ^{13}C were referenced to the methyl groups of DMSO (2.50 and 39.5 ppm, respectively). The ^{13}C NMR data were assigned by using the standard 2D ^{1}H-^{13}C NMR correlation techniques, gradient-selected heteronuclear single-quantum correlation (gs-HSQC) [23], and gradient-selected heteronuclear multiple-bond correlation (gs-HMBC) [24,25].

3.2. Crystal Structure Determination

X-ray data were collected at 100 K on an XtaLAB Synergy, Dualflex, Pilatus 300K diffractometer apparatus (Rigaku Corporation, Tokyo, Japan) equipped with a PhotonJet microfocus X-ray tube apparatus (Rigaku Corporation, Tokyo, Japan). Data reduction was performed using CrysAlisPro (Agilent Technologies UK Ltd., Yarnton, UK) [26]. The structure was refined in ShelXL [27]. Molecular plots and packing diagrams were drawn using Mercury [28]. Molecular geometry parameters were computed using PLATON and publCIF [29,30]. The crystallographic information files for the crystal structures are available under the deposition numbers: 2075915 and 2075918.

3.3. ADMET Analysis

Compound **1** and compound **2** were analyzed using ACDLabs Percepta software version 14.0.0 (Advanced Chemistry Development, Inc., Metropolitan, Toronto, ON, Canada), SwissADME service (Swiss Institute of Bioinformatics, Lausanne, Switzerland, 2021) [31], admetSAR 2.0 service (admetSAR 2019) [32], and ProTOX II service [33] to obtain the computational pharmacokinetic and toxicologic profiles of the tested compounds.

3.4. Biological Assays

To evaluate the active metabolic cells, the MTT (3-(4,5-dimethylthiazol-2-yl))-2,5 diphenyltetrazoliumbromide) assay was used [34]. In this method, EC_{50} (the effective concentration of the tested drug, where a 50% growth reduction is observed in cell growth compared to the untreated control) was used. An MTT assay was performed to test the in vitro cytotoxicity against the A549 cells, which were from a human lung adenocarcinoma that was obtained from the European Collection of Cell Cultures (ECACC, Salisbury, UK). The cells were cultured in Dulbecco's Modified Eagle's Medium (PAN-Biotech, Aidenbach, Germany), 100 units of penicillin/mL (Sigma Aldrich, St. Louis, MO, USA), 100 μg of streptomycin/mL (Sigma Aldrich, St. Louis, MO, USA), 2 mM L-glutamine (Sigma Aldrich, St. Louis, MO, USA), 10% Fetal Bovine Serum (FBS) (Sigma Aldrich, St. Louis, MO, USA), and MTT (3-(4,5-dimethylthiazol-2-yl))-2,5 diphenyltetrazoliumbromide) (Sigma Aldrich, St. Louis, MO, USA). To complete the analyses of the new compounds, the cells were cultured overnight at 37 °C with 5% CO_2 in a standard 96-well flat-bottomed plate containing 10^4 cells/well. The following day, the medium was replaced by 100 μL of **1**, **2**, and etoposide added in varying concentrations to the wells. After 24 h of incubation, 50 μL MTT was added to each well for the last 2 h. The final absorbance was measured in analytical wavelengths (570 nm for blue-violet insoluble formazan) using a microplate reader

(Synergy H1, BioTek, Winooski, VT, USA). The viability and cell cycle analysis results were presented as mean ± standard deviation. All assays were performed in triplicate, and the results were obtained in three independent experiments [35].

4. Conclusions

In summary, etoposide is one of the most commonly used anticancer agents. For many years, it has been the standard therapy for small cell lung cancer, leukemia, lymphoma, germ-cell tumors, and neuroblastoma [36]. Our present findings have shown that both derivatives were very effective in the inhibition of human lung adenocarcinoma cell growth in comparison with etoposide. In conclusion, our present findings showed that the new 3-amino-N-(2,4-dichlorophenyl)-5-oxo-4-phenyl-2,5-dihydro-1H-pyrazole-1-carboxamide **2** (EC_{50} = 220.20+/−22.47 μM,) was much more effective in the inhibition of human lung adenocarcinoma cell growth in comparison to compound **1** with 2,4-dichlorophenyl moiety. These results suggest that compound **2** may be a promising molecule for the treatment of lung cancer. In addition, our studies gained new knowledge about pyrazole derivatives.

Supplementary Materials: The following are available online at https://www.mdpi.com/article/10.3390/ijms22136692/s1. Figure S1. Solution-state 1H NMR (DMSO-d6) of **1** (a) and **2** (b). Figure S2. Solution-state 13C NMR (DMSO-d6) of **1** (a) and **2** (b).

Author Contributions: Conceptualization, A.C. and M.S.; methodology, A.C., M.S., K.C., P.S. and M.P.; software, M.S., B.R. and T.P.; formal analysis, A.C. and M.S.; investigation, A.R., B.R., P.K., K.C. and T.P.; data curation, M.S., B.R. and T.P.; writing—original draft preparation, A.C., M.S., A.R. and B.R.; writing—review and editing, A.C., M.S. and B.R.; supervision, A.C., M.S. and P.S.; project administration, A.C. All authors have read and agreed to the published version of the manuscript.

Funding: This research received no external funding.

Institutional Review Board Statement: Not applicable.

Informed Consent Statement: Not applicable.

Data Availability Statement: The conducted research was presented for the first time in this publication.

Conflicts of Interest: The authors declare no conflict of interest.

References

1. Pitucha, M.; Rzymowska, J. Anticancer screening and structure activity relationship study of some semicarbazides and 1,2,4-Triazolin-5-ones. *Lett. Drug Des. Discov.* **2012**, *9*, 568–572. [CrossRef]
2. Kumar, V.; Kaur, K.; Gupta, G.K.; Sharma, A.K. Pyrazole containing natural products: Synthetic preview and biological significance. *Eur. J. Med. Chem.* **2013**, *69*, 735–753. [CrossRef] [PubMed]
3. Thawabteh, A.; Juma, S.; Bader, M.; Karaman, D.; Scrano, L.; Bufo, S.A.; Karaman, R. The biological activity of natural alkaloids against herbivores, cancerous cells and pathogens. *Toxins* **2019**, *11*, 656. [CrossRef] [PubMed]
4. Dai, H.; Xiao, Y.S.; Li, Z.; Xu, X.Y.; Qian, X.H. The thiazoylmethoxy modification on pyrazole oximes: Synthesis and insecticidal biological evaluation beyond acaricidal activity. *Chin. Chem. Lett.* **2014**, *25*, 1014–1016. [CrossRef]
5. Saundane, A.R.; Verma, V.A.; Katkar, V.T. Synthesis and antimicrobial and antioxidant activities of some new 5-(2-Methyl-1 H -indol-3-yl)-1,3,4-oxadiazol-2-amine derivatives. *J. Chem.* **2013**. [CrossRef]
6. El-Borai, M.A.; Rizk, H.F.; Sadek, M.R.; El-Keiy, M.M. An eco-friendly synthesis of heterocyclic moieties condensed with pyrazole system under green conditions and their biological activity. *Green Sustain. Chem.* **2016**, *6*, 88–100. [CrossRef]
7. Yang, B.; Lu, Y.; Chen, C.J.; Cui, J.P.; Cai, M.S. The synthesis of 5-amino-4-arylazo-3-methyl-1H-pyrazoles and 5-aryl-3-methylpyrazole[3,4-e] [1–4] tetrazines. *Dyes Pigment.* **2009**, *83*, 144–147. [CrossRef]
8. Sener, N.; Sener, I.; Yavuz, S.; Karci, F. Synthesis, absorption properties and biological evaluation of some novel disazo dyes derived from pyrazole derivatives. *Asian J. Chem.* **2015**, *27*, 3003–3012. [CrossRef]
9. Abdel-Mohsen, S.A.; El-Emary, T.I. New pyrazolo [3,4-b] pyridines: Synthesis and antimicrobial Activity. *Der Pharma Chem.* **2018**, *10*, 44–51.
10. El-Dean, A.M.K.; Zaki, R.M.; Geies, A.A.; Radwan, S.M.; Tolba, M.S. Synthesis and antimicrobial activity of new heterocyclic compounds containing thieno[3,2-c]coumarin and pyrazolo[4,3-c]coumarin frameworks. *Russ. J. Bioorgan. Chem.* **2013**, *39*, 553–564. [CrossRef]
11. Skehan, P.; Storeng, R.; Scudiero, D.; Monks, A.; Mcmahon, J.; Vistica, D.; Warren, J.T.; Bokesch, H.; Kenney, S.; Boyd, M.R. New colorimetric cytotoxicity assay for anticancer-drug screening. *J. Natl. Cancer Inst.* **1990**, *82*, 1107–1112. [CrossRef]

12. Brand-Williams, W.; Cuvelier, M.E.; Berset, C. Use of a free radical method to evaluate antioxidant activity. *LWT Food Sci. Technol.* **1995**, *28*, 25–30. [CrossRef]
13. Liu, X.H.; Qiao, L.; Zhai, Z.W.; Cai, P.P.; Cantrell, C.L.; Tan, C.X.; Weng, J.Q.; Han, L.; Wu, H.K. Novel 4-pyrazole carboxamide derivatives containing flexible chain motif: Design, synthesis and antifungal activity. *Pest. Manag. Sci.* **2019**, *75*, 2892–2900. [CrossRef]
14. Yan, Z.; Liu, A.; Huang, M.; Liu, M.; Pei, H.; Huang, L.; Yi, H.; Liu, W.; Hu, A. Design, synthesis, DFT study and antifungal activity of the derivatives of pyrazolecarboxamide containing thiazole or oxazole ring. *Eur. J. Med. Chem.* **2018**, *149*, 170–181. [CrossRef]
15. Ren, Z.L.; Liu, H.; Jiao, D.; Hu, H.T.; Wang, W.; Gong, J.X.; Wang, A.L.; Cao, H.Q.; Lv, X.H. Design, synthesis, and antifungal activity of novel cinnamon–pyrazole carboxamide derivatives. *Drug Dev. Res.* **2018**, *79*, 307–312. [CrossRef]
16. Isyaku, Y.; Uzairu, A.; Uba, S. In silico studies of novel pyrazole-furan and pyrazole-pyrrole carboxamide as fungicides against Sclerotinia sclerotiorum. *Beni Suef Univ. J. Basic Appl. Sci.* **2020**, *9*. [CrossRef]
17. Kaczor, A.A.; Wrobel, T.; Karczmarzyk, Z.; Wysocki, W.; Fruzinski, A.; Brodacka, M.; Matosiuk, D.; Pitucha, M. Structural studies on N-(1-naphthyl)-3-amino-5-oxo-4-phenyl-1hpyrazole-1-carboxamide with antibacterial activity. *Lett. Organ. Chem.* **2014**, *11*, 40–44. [CrossRef]
18. Bernstein, J.; Davis, R.E.; Shimoni, L.; Chang, N.-L. Patterns in hydrogen bonding: Functionality and graph set analysis in crystals. *Angew. Chem. Int. Ed. Engl.* **1995**, *34*, 1555–1573. [CrossRef]
19. Lipinski, C.A.; Lombardo, F.; Dominy, B.W.; Feeney, P.J. Experimental and computational approaches to estimate solubility and permeability in drug discovery and development settings. *Adv. Drug Deliv. Rev.* **2001**, *46*, 3–26. [CrossRef]
20. Ghose, A.K.; Viswanadhan, V.N.; Wendoloski, J.J. A knowledge-based approach in designing combinatorial or medicinal chemistry libraries for drug discovery. 1. A qualitative and quantitative characterization of known drug databases. *J. Comb. Chem.* **1999**, *1*, 55–68. [CrossRef]
21. Egan, W.J.; Merz, K.M.; Baldwin, J.J. Prediction of drug absorption using multivariate statistics. *J. Med. Chem.* **2000**, *43*, 3867–3877. [CrossRef]
22. Muegge, I.; Heald, S.L.; Brittelli, D. Simple selection criteria for drug-like chemical matter. *J. Med. Chem.* **2001**, *44*, 1841–1846. [CrossRef]
23. Bodenhausen, G.; Ruben, D.J. Natural abundance nitrogen-15 NMR by enhanced heteronuclear spectroscopy. *Chem. Phys. Lett.* **1980**, *69*, 185–189. [CrossRef]
24. Bax, A.; Summers, M.F. Proton and carbon-13 assignments from sensitivity-enhanced detection of heteronuclear multiple-bond connectivity by 2D multiple quantum NMR. *J. Am. Chem. Soc.* **1986**, *108*, 2093–2094. [CrossRef]
25. Willker, W.; Leibfritz, D.; Kerssebaum, R.; Bermel, W. Gradient selection in inverse heteronuclear correlation spectroscopy. *W. Magn. Reson. Chem.* **1993**, *31*, 287–292. [CrossRef]
26. Agilent Technologies UK. *CrysAlisPro 1.171.39.33c*; Rigaku Oxford Diffraction; Agilent Technologies UK Ltd.: Yarnton, UK, 2017.
27. Sheldrick, G.M. SHELXT—Integrated space-group and crystal-structure determination. *Acta Crystallogr. Sect. C Struct. Chem.* **2015**, *71*, 3–8. [CrossRef]
28. Macrae, C.F.; Bruno, I.J.; Chisholm, J.A.; Edgington, P.R.; Mccabe, P.; Pidcock, E.; Rodriguez–Monge, L.; Taylor, R.; Van De Streek, J.; Wood, P.A. Mercury: Visualization and analysis of crystal structures. *J. Appl. Cryst.* **2008**, *41*, 466–470. [CrossRef]
29. Spek, A.L. Structure validation in chemical crystallography. *Acta Crystallogr. Sect. D Biol. Crystallogr.* **2009**, *65*, 148–155. [CrossRef]
30. Westrip, S.P. publCIF: Software for editing, validating and formatting crystallographic information files. *J. Appl. Cryst.* **2010**, *43*, 920–925. [CrossRef]
31. Daina, A.; Michielin, O.; Zoete, V. SwissADME: A free web tool to evaluate pharmacokinetics, drug-likeness and medicinal chemistry friendliness of small molecules. *Sci. Rep.* **2017**, *7*. [CrossRef]
32. Yang, H.B.; Lou, C.F.; Sun, L.X.; Li, J.; Cai, Y.C.; Wang, Z.; Li, W.H.; Liu, G.X.; Tang, Y. admetSAR 2.0: Web-service for prediction and optimization of chemical ADMET properties. *Bioinformatics* **2019**, *35*, 1067–1069. [CrossRef] [PubMed]
33. Banerjee, P.; Eckert, A.O.; Schrey, A.K.; Preissner, R. ProTox-II: A webserver for the prediction of toxicity of chemicals. *Nucleic Acids Res.* **2018**, *46*, W257–W263. [CrossRef] [PubMed]
34. Plumb, J.A. Cell sensitivity assays: The MTT assay. In *Cancer Cell Culture: Methods and Protocols*; Langdon, S.P., Ed.; Humana Press: Totowa, NJ, USA, 2004; pp. 165–169.
35. Girek, M.; Kłosiński, K.; Grobelski, B.; Pizzimenti, S.; Cucci, M.A.; Daga, M.; Barrera, G.; Pasieka, Z.; Czarnecka, K.; Szymański, P. Novel tetrahydroacridine derivatives with iodobenzoic moieties induce G0/G1 cell cycle arrest and apoptosis in A549 non-small lung cancer and HT-29 colorectal cancer cells. *Mol. Cell. Biochem.* **2019**, *460*, 123–150. [CrossRef] [PubMed]
36. Mascaux, C.; Paesmans, M.; Berghmans, T.; Branle, F.; Lafitte, J.J.; Lemaitre, F.; Meert, A.P.; Vermylen, P.; Sculier, J.P. European Lung Cancer Working Party (ELCWP). A systematic review of the role of etoposide and cisplatin in the chemotherapy of small cell lung cancer with methodology assessment and meta-analysis. *Lung Cancer* **2000**, *30*, 23–36. [CrossRef]

International Journal of
Molecular Sciences

Article

New Succinimides with Potent Anticancer Activity: Synthesis, Activation of Stress Signaling Pathways and Characterization of Apoptosis in Leukemia and Cervical Cancer Cells

Marcin Cieślak [1,*], Mariola Napiórkowska [2,*], Julia Kaźmierczak-Barańska [1], Karolina Królewska-Golińska [1], Anna Hawrył [3], Iwona Wybrańska [4] and Barbara Nawrot [1]

1 Centre of Molecular and Macromolecular Studies, Department of Bioorganic Chemistry, Polish Academy of Sciences, 112 Sienkiewicza, 90-363 Lodz, Poland; juliakazmierczak@tlen.pl (J.K.-B.); kkrolews@cbmm.lodz.pl (K.K.-G.); bnawrot@cbmm.lodz.pl (B.N.)
2 Department of Biochemistry, Faculty of Medicine, Medical University of Warsaw, 1 Banacha, 02-097 Warsaw, Poland
3 Department of Inorganic Chemistry, Faculty of Pharmacy, Medical University of Lublin, Chodźki 4a, 20-093 Lublin, Poland; anna.hawryl@umlub.pl
4 Clinical Biochemistry, Department of Genetics and Nutrigenomics, Faculty of Medicine, Jagiellonian University Medical College, 15a Kopernika, 31-501 Kraków, Poland; mbwybran@cyf-kr.edu.pl
* Correspondence: marcin@cbmm.lodz.pl (M.C.); mariola.krawiecka@wum.edu.pl (M.N.); Tel.: +48-42-680-32-30 (M.C.); +48-22-572-06-39 (M.N.)

Citation: Cieślak, M.; Napiórkowska, M.; Kaźmierczak-Barańska, J.; Królewska-Golińska, K.; Hawrył, A.; Wybrańska, I.; Nawrot, B. New Succinimides with Potent Anticancer Activity: Synthesis, Activation of Stress Signaling Pathways and Characterization of Apoptosis in Leukemia and Cervical Cancer Cells. *Int. J. Mol. Sci.* **2021**, *22*, 4318. https://doi.org/10.3390/ijms22094318

Academic Editor: Angela Stefanachi

Received: 4 March 2021
Accepted: 19 April 2021
Published: 21 April 2021

Abstract: Based on previously identified dicarboximides with significant anticancer and immunomodulatory activities, a series of 26 new derivatives were designed and synthesized by the Diels–Alder reaction between appropriate diene and maleimide or hydroxymaleimide moieties. The resulting imides were functionalized with alkanolamine or alkylamine side chains and subsequently converted to their hydrochlorides. The structures of the obtained compounds were confirmed by 1H and 13C NMR and by ESI MS spectral analysis. Their cytotoxicity was evaluated in human leukemia (K562, MOLT4), cervical cancer (HeLa), and normal endothelial cells (HUVEC). The majority of derivatives exhibited high to moderate cytotoxicity and induced apoptosis in K562 cells. Microarray gene profiling demonstrated upregulation of proapoptotic genes involved in receptor-mediated and mitochondrial cell death pathways as well as antiapoptotic genes involved in NF-kB signaling. Selected dicarboximides activated JNK and p38 kinases in leukemia cells, suggesting that MAPKs may be involved in the regulation of apoptosis. The tested dicarboximides bind to DNA as assessed by a plasmid DNA cleavage protection assay. The selected dicarboximides offer new scaffolds for further development as anticancer drugs.

Keywords: dicarboximides; chemical synthesis; cytotoxicity; apoptosis; kinases; anticancer; gene profiling; SAR

1. Introduction

Cancer diseases are an important problem in the clinical medicine and pharmacology of the 21st century. Annually, more than eight million people die from cancer worldwide. In Europe alone, three million new cases are diagnosed and 1.7 million people die from cancer. In Poland, cancer kills nearly 110,000 people, of which about 100,000 die from malignant tumors. It is estimated that the ongoing increase in the incidence of cancer will make it the main cause of death in Poland and worldwide in the coming decades [1,2]. Despite recent progress in cancer treatment (for example development of immune checkpoint inhibitors), chemotherapy (either alone or in combination with radiotherapy or biologic drugs) still remains the first-line treatment in the majority of cancers. Therefore, new cytotoxic low molecular weight compounds are actively sought and developed because they may offer better pharmacokinetic properties and/or less toxic profiles compared to already applied chemotherapeutics (for example platinum compounds or taxanes).

Dicarboximides comprise a large group of compounds and can be divided into four categories: succinimides, maleimides, glutarimides, and phthalimides. Succinimides and phthalimides offer a wide range of chemical and biological applications. They are also very important from a therapeutic perspective, as they are used as drugs (for example, anticonvulsant–ethosuximide) or tested as drug candidates. These derivatives show antifungal, antibacterial, antidepressant, or analgesic activities. Several reports describe the antitumor activity of succinimides and phthalimides. Pyridinyl- and quinolinyl-derivatives of succinimide demonstrated antitumor activity in leukemia cells by inhibition of DNA synthesis [3]. Trifluoromethylated mono- and bicyclic succinimides were found to inhibit the growth of myeloma, non-small cell lung cancer, and renal cancer cells [4]. N-benzylcantharidinamide promoted the degradation of matrix metalloproteinase-9 (MMP-9) mRNA in an antigen R-dependent manner in human hepatocellular carcinoma Hep3B cells. This resulted in decreased expression of MMP-9 and reduced invasion of the highly metastatic Hep3B cells [5]. Benzothiazole derivatives of phthalimide showed significant cytotoxicity and induced apoptosis in human hepatoma (SKHep1), Burkitt's B-cell lymphoma (CA46), and chronic myelogenous leukemia (K562) cells [6]. Dual-acting thiazolidinone derivates of phthalimide with antiproliferative and immunosuppressive activity were also identified. These compounds inhibited the proliferation of cancer cells and blocked the secretion of pro-inflammatory cytokines, for example, TNF-α, IL-2, or Il-6 [7]. Probably the best-known phthalimide-based drug is thalidomide (N-α-phthalimidoglutarimide). This is a synthetic derivative of phthalimide N-substituted with glutarimide. Since 1957, thalidomide has been used as an anti-emetic agent, but in 1961 it was withdrawn from the market because of severe teratogenicity. However, further studies demonstrated the potent anti-angiogenic and anti-inflammatory activity of thalidomide (and its derivatives—lenalidomide and pomalidomide), which warranted its return as an essential medicine. Currently, thalidomide is used as an anti-cancer drug (for multiple myeloma), for the treatment of myelodysplastic syndrome and inflammatory disorders, like erythema nodosum leprosum [8]. We have previously identified several lead dicarboximides (in particular **Ic**, **Ie**, and **II**, Figure 1) structurally related to succinimides and phthalimides, which exhibited high cytotoxicity and selectivity toward leukemia cells.

We also found that they targeted the ABC50 protein, showed immunomodulatory activity, and induced apoptosis in leukemia cells [9–11]. In particular, compound **Ic** inhibited endonuclease-catalyzed digestion of DNA, suggesting its interaction with nucleic acids [9] and activated caspase 8 and 9, which are markers of receptor-mediated and mitochondrial apoptotic pathways, respectively [11]. However, the detailed mechanism of apoptosis and the involvement of mitogen-activated protein kinases (MAPKs) were not investigated. These anticancer and immunomodulatory activities of dicarboximides indicate their importance as potential drug candidates and significance for further development.

The main goal of this study was to obtain new derivatives of dicarboximides with decreased lipophilicity and improved anticancer activity as compared to the lead compounds (i.e., **Ia**–**Ie** and **II**). To achieve this aim, we designed, synthesized, and examined the biological activity of new derivatives of dicarboximides. These new derivatives showed potent cytotoxic and proapoptotic activity in cancer cells, and DNA intercalating properties. Microarray analysis of gene expression revealed that dicarboximides upregulated genes involved in receptor-mediated and mitochondrial apoptotic pathways. Interestingly, antiapoptotic genes associated with NF-kB signaling were also upregulated. They also activated stress-induced MAPK signaling pathways in cancer cells.

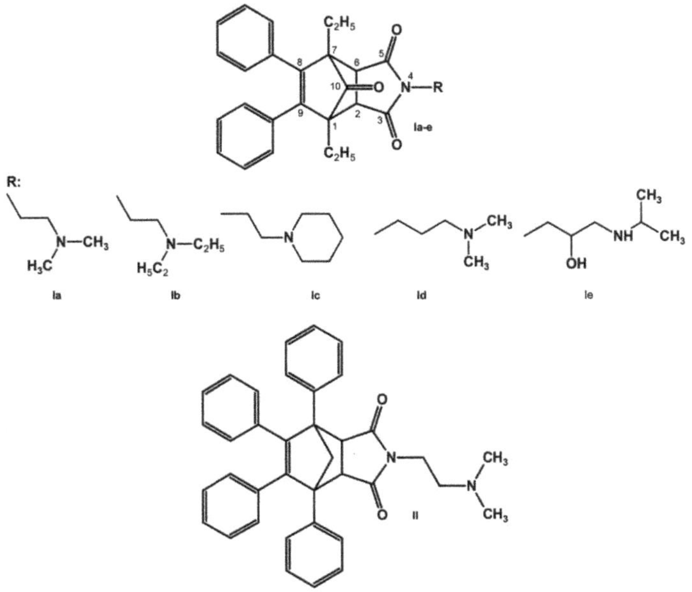

Figure 1. Structures of lead compounds **Ia–e** and **II**.

2. Results

2.1. Synthesis

The starting imides **1–3** were obtained in Diels–Alder reactions between appropriate dienes and maleimide, while the compounds **4** and **5** were obtained in the reaction with N-(hydroxyethyl)-maleimide (Scheme 1). The epoxy-derivatives were obtained by condensation of starting imides with epichlorohydrin, and the products were condensed with appropriate amines to give alkanolamine derivatives (Scheme 1). In the reaction between imide **1** and 1,2-dichloroethane, the chloroalkyl derivative was obtained and condensed with an appropriate amine (Scheme 1). All obtained amino-derivatives were converted to their hydrochlorides by using methanol saturated with hydrochloride. The lead compounds with the best activity (**Ic**, **Ie**, and **II**) described previously [9] were converted to their acetates by using boiling acetic acid. The structures of new compounds were established by spectroscopic methods (1H NMR, 13C NMR, and HRMS). In this case, ethyl groups in positions 1 and 7 were replaced by methyl groups.

The main goal of the synthetic part was the synthesis of new derivatives of dicarboximides with decreased lipophilicity in comparison to the lead compounds, i.e., **Ia–Ie** and **II**. To reduce the lipophilicity of the new dicarboximides, a hydroxyalkyl chain connected with an aliphatic amine was introduced to the imides (Scheme 1). At the same time, the imide scaffold in derivatives from group **2** and **3** was the same as in lead compounds **Ia–Ie** and **II**, respectively (Figure 1). The structure of the imide was changed in derivatives **1b–1g**. The lipophilicity of new derivatives was calculated using the Osiris Property Explorer program. For some compounds, lipophilicity was determined experimentally by a reversed-phase chromatography (detailed description is given in Materials and Methods and in the Supplementary Data, Table S2). Table 1 summarizes the calculated (*clogP*) and experimental ($logk_w$) parameters of the lipophilicity of new derivatives.

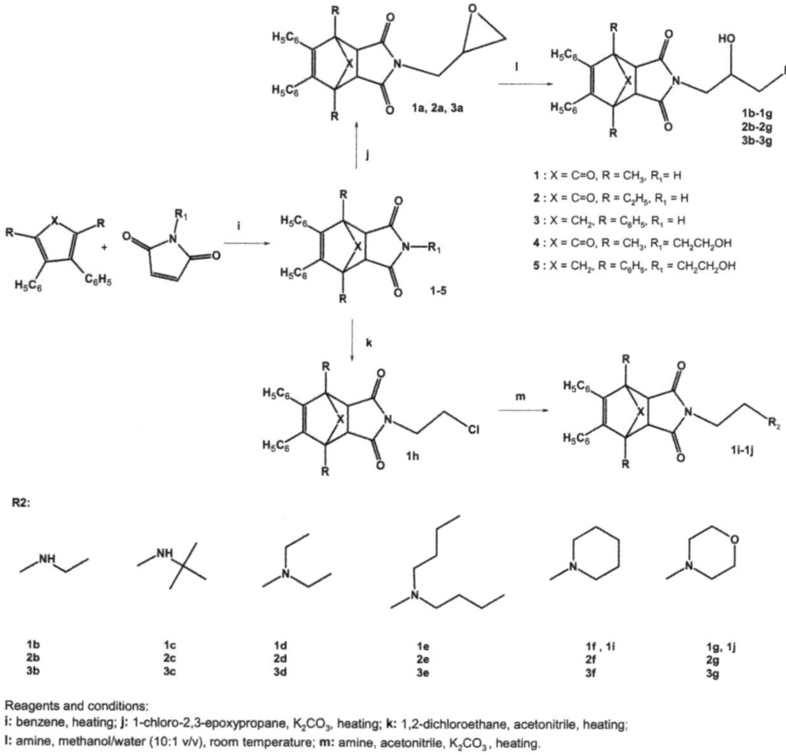
1a, 2a, 3a
1b-1g
2b-2g
3b-3g
1 : X = C=O, R = CH3, R1 = H
2 : X = C=O, R = C2H5, R1 = H
3 : X = CH2, R = C6H5, R1 = H
4 : X = C=O, R = CH3, R1 = CH2CH2OH
5 : X = CH2, R = C6H5, R1 = CH2CH2OH
1-5
1h
1i-1j
R2:
1b 2b 3b
1c 2c 3c
1d 2d 3d
1e 2e 3e
1f, 1i 2f 3f
1g, 1j 2g 3g
Reagents and conditions:
i: benzene, heating; j: 1-chloro-2,3-epoxypropane, K2CO3, heating; k: 1,2-dichloroethane, acetonitrile, heating;
l: amine, methanol/water (10:1 v/v), room temperature; m: amine, acetonitrile, K2CO3, heating.

Scheme 1. Synthesis of starting imides, N–alkanolamine and N–alkylamino derivatives of dicarboximides.

In some cases, the new derivatives had lower lipophilicity in comparison to the lead compounds. For example, derivatives of imide **1** had a median *clogP* and $logk_w$ of 3.19 and 4.85, respectively, while corresponding values for lead dicarboximides **Ia–Ie** were 4.27 and 5.05. This decrease in lipophilicity can be attributed to the replacement of ethyl groups by methyl groups in the imide scaffold and the presence of the hydroxyalkyl chain. The introduction of a hydroxyalkyl chain in derivatives from the group of imide **2** resulted in a slight decrease in lipophilicity (median *clogP* and $logk_w$ of 4.1 and 5.03, respectively) compared to parent compounds **Ia–Ie**. However, pronounced exceptions were alkanoldibutylamine derivatives **1e** and **2e**, which showed significantly increased lipophilicity. The modifications introduced to imide **3** derivatives did not significantly affect their lipophilicity (median *clogP* and $logk_w$ of 5.69 and 6.28, respectively) compared to the lead dicarboximide **II** (*clogP* 5.45 and $logk_w$ 6.34).

2.2. MTT Cytotoxicity Studies

The cytotoxicity of new derivatives of dicarboximides was tested in HeLa (cervix carcinoma cell line), K562 (myelogenous leukemia cell line), MOLT-4 (Human T-Lymphoid Cell Line), and HUVEC (normal endothelial) cells. First, the viability of cells was determined after 48 h incubation with the screened compounds used at the concentration of 100 μM (data not shown). For compounds that reduced the viability of cancer cells by more than 50%, the IC_{50} values were determined (Table 2). Cells exposed to 1% DMSO (as vehicle) served as the control with 100% survival. Cells treated with 1 μM staurosporine served as the internal control of the cytotoxicity experiments (the viability of cancer cells was reduced to 44–49%, data not shown).

Table 1. Calculated (*clogP*) and experimental ($logk_w$) parameters of lipophilicity for synthesized compounds. nd—not determined.

Imide Group	Compound	*clogP* (Calculated)	Median *clogP*	$logk_w$ (Experimental)	Median $logk_w$
2	**Ia**	3.82	4.27	4.94	5.05
	Ib	4.63		5.09	
	Ic	4.95		5.02	
	Id	4.27		5.05	
	Ie	4.06		5.17	
1	**1b**	2.48	3.19	4.78	4.85
	1c	3.22		4.91	
	1d	3.15		4.87	
	1e	4.97		nd	
	1f	3.47		nd	
	1g	2.31		4.83	
	4	2.69		nd	
	1h	3.84		nd	
	1j	2.88		nd	
	1i	4.04		nd	
2	**2b**	3.39	4.1	5.08	5.03
	2c	4.13		4.74	
	2d	4.06		5.10	
	2e	5.88		nd	
	2f	4.38		nd	
	2g	3.22		4.98	
3	**II**	5.45		6.34	
3	**3b**	5.02	5.69	6.31	6.28
	3c	5.76		6.06	
	3d	5.69		6.28	
	3e	7.51		nd	
	3f	6.01		nd	
	3g	4.85		6.28	
	5	5.23		nd	

Test compounds showed significant cytotoxicity toward all tested cancer cell lines. Derivatives of imide **2** and **3** (i.e., **Ic*CH_3COOH, Ie*CH_3COOH, 2b–2g, 3c–3g, 5**), possessing the same imide skeleton as the lead compounds, displayed the best activity with IC_{50} in the range of 1.9–26 µM. In the group of derivatives of imide **1**, the highest activity was demonstrated by **1e** against all tested cancer cell lines (IC_{50} of 3.2, 5.8, and 8 µM, respectively). It is noteworthy that derivatives **1b**, **1h**, and **1i** were the most toxic to MOLT-4 cells (IC_{50} of 7, 20, and 15 µM, respectively), whereas compound **1f** was cytotoxic to K562 (IC_{50} of 18 µM). Other derivatives of this class showed moderate activity (IC_{50} in the range of 28–105 µM). Similar cytotoxicity as for cancer cells was observed in normal endothelial cells, with IC_{50} values in the range of 0.3–64 µM.

2.3. Activation of Apoptosis and Expression Profile of Apoptotic Genes

Six derivatives of dicarboximides, i.e., **3c**, **3d**, **3f**, **2c**, **2d**, and **2f**, with the highest cytotoxicity against K562 leukemia cells ($IC_{50} < 6$ µM) and with the same alkanolamine fragment in the structure were further evaluated for the ability to induce apoptosis in cancer cells. Caspases 3 and 7 (caspase 3/7) are markers of apoptosis, and their activity was upregulated during this process. K562 cells were treated with 1% DMSO (negative control), 1 µM staurosporine (a strong inducer of apoptosis, a positive control), or a test compound at the concentration of 5xIC_{50}. After incubation for 18 h, K562 cells were lysed and the activity of caspase 3/7 was measured using a profluorescent peptide substrate. Staurosporine strongly activated caspase 3/7 in K562 cells, while DMSO had no effect (Figure 2). We also observed a significant increase in the activity of caspase 3/7 in the presence of all tested dicarboximides. Particularly, derivatives **2c**, **2d**, and **2f** were potent activators of apoptosis, as they induced an over eight-fold increase in caspase activity compared to control. Incubation of cells with compounds **3c**, **3d**, and **3f** activated caspase 3/7 to a lesser extent. Altogether, these data suggest that the cytotoxicity of new derivatives of dicarboximides is most likely due to the induction of apoptosis in the chronic myelogenous leukemia (CML) K562 cells.

Table 2. The IC_{50} values (μM) for dicarboximides after 48 h incubation with a given cell line. Average values and standard deviations (in parenthesis) from three experiments are shown. Compounds were used as HCl salts except for **Ic** *, **Ie** *, and **II** * which were used as CH_3COOH salts.

Compound	HeLa	K562	MOLT-4	HUVEC
1b	30 (7.5)	32 (8.3)	7 (1.8)	20 (2.4)
1c	26 (5.2)	28 (6.7)	28 (5)	23 (3.2)
1d	59 (8.9)	39 (9.4)	37 (4.4)	22 (3.3)
1e	3.2 (0.4)	5.8 (1.0)	8 (0.9)	30 (4.2)
1f	55 (17.6)	18 (3.8)	40 (10.8)	21 (1.6)
1g	>100	>100	>100	nd
4	61 (14.8)	105 (15.2)	71 (14.8)	56 (4.8)
1h	nd	nd	20 (4.7)	64 (7)
1j	>100	>100	>100	nd
1i	52 (19.2)	30 (4.1)	15 (1.2)	23 (1.8)
Ic *	18 (3.1)	20 (3.4)	15 (3.2)	15 (6.3)
Ie *	10 (2.6)	8 (1.7)	15 (2.3)	3.5 (0.2)
2b	26 (5.6)	8 (1.5)	14 (0.9)	3.4 (0.3)
2c	3.2 (0.8)	4 (0.5)	2.5 (0.1)	3.3 (0.2)
2d	18 (4.9)	5.8 (1.2)	14 (2.6)	3.6 (0.2)
2e	19 (4.5)	4.1 (0.5)	3.1 (0.2)	1.2 (0.06)
2f	5.1 (0.7)	3.9 (0.5)	3.4 (0.3)	3.1 (0.3)
2g	23 (3.6)	20 (1.4)	18 (2.5)	17 (1.6)
II *	20 (2.4)	16 (4.5)	18 (5)	4 (0.5)
3b	>100	>100	>100	nd
3c	2 (0.4)	2 (0.6)	2.6 (0.3)	0.4 (0.03)
3d	1.9 (0.6)	3.2 (0.4)	2.2 (0.3)	0.3 (0.02)
3e	40 (8.4)	6 (1.4)	5.4 (0.5)	7 (0.6)
3f	3 (0.9)	3 (0.6)	3 (1)	0.4 (0.02)
3g	120 (16)	31 (3.6)	5.2 (0.9)	17 (1.8)
5	8 (1)	6.2 (1)	2.1 (0.1)	0.7 (0.08)

* denotes compounds as acetate salts; nd—not determined.

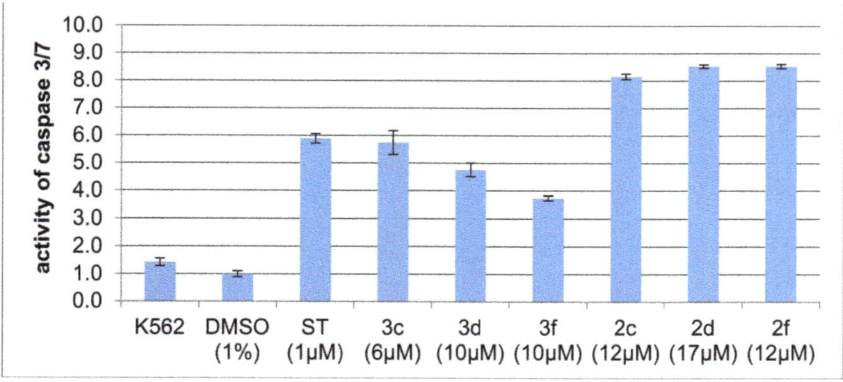

Figure 2. Activity of caspase 3/7 in K562 cells treated with the test dicarboximides or staurosporine for 18 h. Abbreviations: K562—untreated cells; DMSO—K562 cells treated with 1% DMSO; ST—staurosporine. The activity of caspase 3/7 in cells exposed to 1% DMSO was normalized to 1.0. Mean values ± SD are shown.

We also profiled the expression of 93 genes associated with apoptosis in leukemia cells treated with the lead dicarboximide **Ic**. We showed previously that compound **Ic** induced apoptosis in CML K562 cells and acute lymphoblastic leukemia (ALL) cells MOLT-4, and that both receptor-mediated and mitochondrial apoptotic pathways may be involved [9,11]. K562 leukemia cells were incubated for 18 h with 1% DMSO (control–comparator), staurosporine (1 μM, positive control), or compound **Ic** (10 μM). Subsequently, the total RNA was isolated and the expression of 93 mRNAs associated with apoptosis was determined using DNA microarrays. The expression profile of apoptotic genes upon treatment with **Ic** was similar to staurosporine, a known proapoptotic agent (Figure 3 and Supplementary Materials Table S1). The expression of 12 genes was significantly upregulated in response to **Ic** (Figure 3).

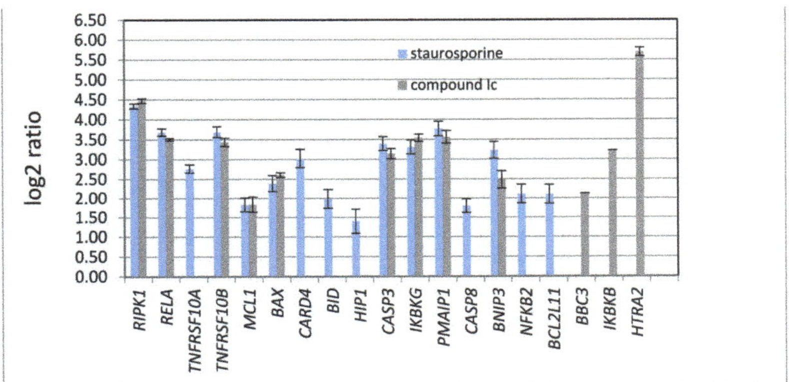

Figure 3. Human apoptosis array analysis of gene expression in K562 exposed to compound **Ic** at 10 μM ($5 \times IC_{50}$) for 18 h. Staurosporine (1 μM) served as a positive control. Only apoptosis-related genes with statistically significant changes in expression are shown ($p < 0.05$, *t*-Student). The fold of change was expressed as log2 ratio of gene expression in a test sample vs control sample (cells + DMSO).

Eight of these genes encode proteins with proapoptotic activity. For example, TNFRSF10B (a membrane receptor from the TNF family) and RIPK1 (a Ser-Thr kinase, which is a key regulator of TNF-mediated apoptosis) activate the receptor pathway of apoptosis. PMAIP1 (Phorbol-12-Myristate-13-Acetate-Induced Protein 1), BAX (BCL2 Associated X, a Bcl-2 family member), BBC3 (BCL2 Binding Component 3, a Bcl-2 family member), and BNIP3 (BCL2 Interacting Protein 3) induce apoptosis via the mitochondrial pathway. HTRA2 (HtrA Serine Peptidase 2) promotes apoptosis by binding and inhibiting IAPs (inhibitor of apoptosis proteins), while CASP3 (caspase 3) is a major executioner protease during cell apoptosis. Only one of the upregulated genes encodes a protein that inhibits apoptosis, i.e., MCL1 (Myeloid Cell Leukemia Sequence 1), which belongs to the Bcl-2 family. Interestingly, compound **Ic** also increased the expression of genes that encode anti-apoptotic proteins involved in the activation of the transcription factor NF-kB. Specifically, IKBKB (Inhibitor Of Nuclear Factor Kappa B Kinase Subunit Beta) and IKBKG (Inhibitor Of Nuclear Factor Kappa B Kinase Subunit Gamma) are protein kinases that activate NF-kB in cells, and RelA is a subunit of NF-kB heterodimer.

2.4. *Effect of Dicarboximides on MAP Kinase Signaling Pathways*

In this study, we have demonstrated that new derivatives of dicarboximides are cytotoxic and induce apoptosis in leukemia cells. Similar effects were observed for previously studied lead compounds **Ic** and **II** [9,11]; therefore, we investigated whether mitogen-activated protein kinases (MAPKs) were involved in mediating dicarboximides-induced cytotoxic effects and induction of apoptosis. MAPKs include extracellular signal-regulated kinases (ERK1/2), p38 kinase, and c-Jun N-terminal kinase (JNK), which regulate cellular responses to multiple stimuli like growth factors, cytokines, stress signals, or cytotoxic drugs. In cells exposed to stress conditions (for example toxic compounds), p38 kinase and JNK kinase are often activated, leading to cell cycle arrest and apoptosis. To investigate the effect of dicarboximides on MAPKs, leukemia cells were incubated with lead compounds **Ic** and **II**, and subsequently, cell lysates were immunoblotted with specific antibodies recognizing activated (i.e., phosphorylated) forms of MAP kinases. As positive controls, we used cells treated with staurosporine (activates ERK1/2) or anisomycin (activator of JNK and p38 kinases). As expected, the incubation of K562 cells with compounds **Ic** or **II** had no effect on the ERK pathway, which is usually activated by growth factors and promotes cell proliferation (Figure 4a). However, dicarboximides **Ic** and **II** activated JNK and p38 kinases. This is evidenced by increased phosphorylation of JNK and p38 kinase (Figure 4b,c, lanes 3 and 4) compared to control (lane 1). These data suggest that in leukemia cells dicarboximides activate stress signaling pathways regulated by JNK and p38 kinases, which subsequently leads to apoptotic death.

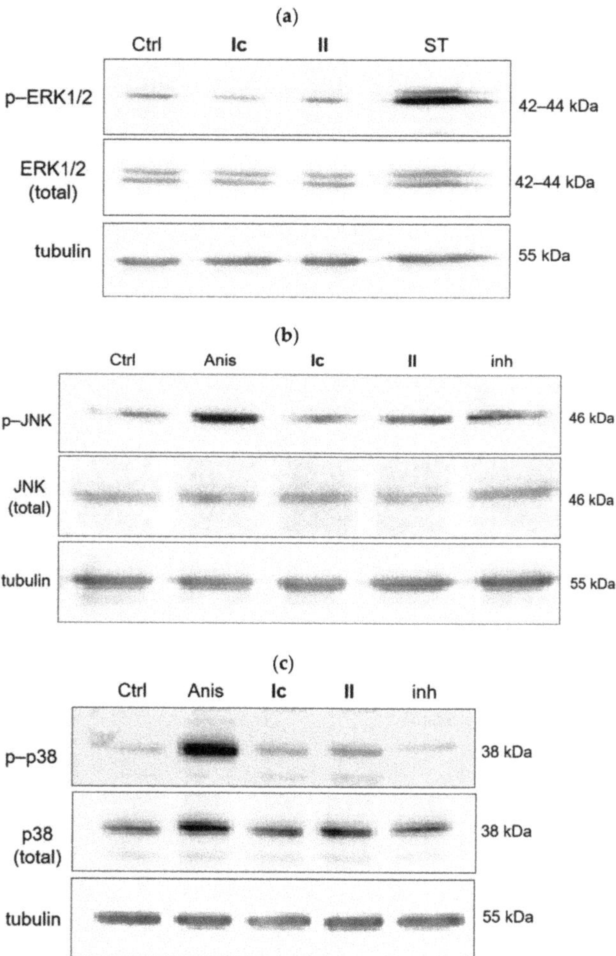

Figure 4. Effect of dicarboximides **Ic** and **II** on activation of ERK1/2 (**a**), JNK (**b**), and p38 (**c**) kinases in K562 leukemia cells. Ctrl—control cells treated with 1% DMSO; Anisomycin 20 μM (positive control, activator of JNK and p38 kinases); **Ic** (6 μM); **II** (3 μM); ST—Staurosporine 1 μM (positive control, activator of ERK1/2); inh—JNK inhibitor 20 μM or p38 inhibitor 10 μM.

2.5. Interaction with DNA

A DNA-cleavage protection assay has been successfully used for the identification of compounds that bind DNA either covalently or noncovalently. Such binding may result in DNA being less susceptible to enzymatic cleavage [12]. Using this assay, we showed previously that compound **Ic** may interact with DNA [9,10]. It was also reported that naphtalenoimides or indolomaleimides intercalate into DNA [13,14]. With the aid of circular dichroism (CD) spectroscopy, we demonstrated the direct binding of **Ic** to DNA (Supplementary Materials Figure S1). In CD studies, a synthetic 23-nt duplex DNA or bovine thymus DNA was incubated with compound **Ic** or daunorubicin (reference compound). CD spectra indicated that compound **Ic** induced conformational changes in DNA compared to the control sample (Supplementary Materials Figure S1). This observation suggests that the cytotoxicity of dicarboximides may result from binding to and damaging the double-stranded DNA. Using a DNA-cleavage protection assay, we investigated whether selected compounds **2c**, **2d**, and **2f** and **3c**, **3d**, and **3f** interact with a plasmid DNA, resulting in increased resistance to endonucleolytic cleavage. Plasmid DNA (pcDNA3.1HisC) containing a single *BamHI* cleavage site was incubated with test compounds and subsequently digested with *BamH1* restriction endonuclease. Daunorubicin (a strong DNA-intercalating agent) was used as a reference compound. Non-digested plasmid DNA existed predominantly in the circular form; however, the superhelical and linear forms were also detected (Figure 5, lane 1). Digestion with the *BamH1* enzyme converted the plasmid into a linear form (Figure 5, lane 2; no circular DNA), while daunorubicin (at 100 μM) prevented linearization of DNA (Figure 5, lane 3; circular DNA present). In the presence of test dicarboximides, the linear plasmid DNA was predominant; however, the circular form was also detected (Figure 5 lane 4–9). This indicates that test dicarboximides interact with DNA. Interestingly, this interaction seems to be stronger for **3c**, **3d**, and **3f** (as more circular DNA is observed in lanes 4–6) than for the remaining test compounds (lanes 7–9). This observation can be explained by the higher lipophilicity of **3c**, **3d**, and **3f** compared to **2c**, **2d**, and **2f**, which may facilitate their more efficient binding to DNA.

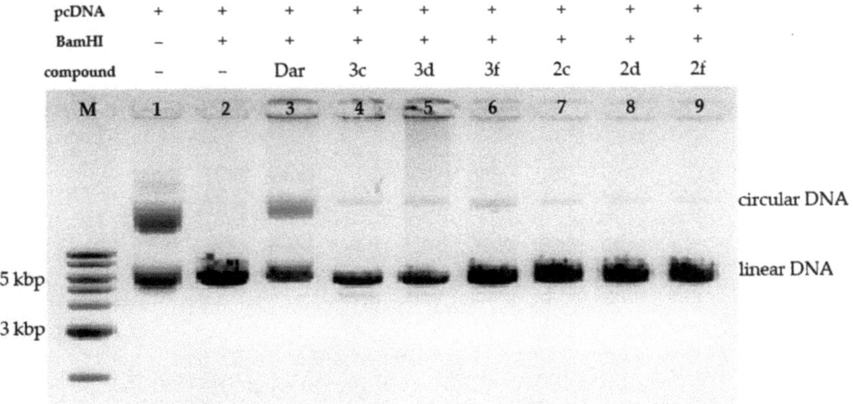

Figure 5. Effect of dicarboximides on digestion of pcDNA3.1HisC (total length 5.5 kbp) with *BamH1* endonuclease observed by agarose gel electrophoresis. M—marker DNA; Dar—daunorubicin.

3. Discussion

In our previous studies, we identified several derivatives of dicarboximides (**Ia–Ie**, **II**) that efficiently killed cancer cells and were not toxic to normal endothelial cells [9–11]. Compounds **Ic**, **Ie**, and **II** were particularly promising, as they showed specificity to cancer cells only and induced apoptosis. However, they are lipophilic (Table 1) and have limited water solubility. Based on their structures, we have designed and synthesized a library of

26 new derivatives of dicarboximides with lower lipophilicity, which is a desired property for potential new drug candidates. This was achieved mainly by the replacement of ethyl groups in the imide scaffold by methyl groups at positions 1 and 7 as well as by the introduction of a side chain with the hydroxyalkyl group substituted with alkylamino groups (Scheme 1). This resulted in significantly lower lipophilicity of derivatives of imide **1** (median *clogP* 3.19), and imide **2** (median *clogP* of 4.1) compared to parent compounds **Ia–Ie** (median *clogP* 4.27). However, chemical modifications introduced to derivatives of imide **3** rather increased their lipophilicity (median *clogP* 5.69) when compared to the parent dicarboximide **II** (*clogP* 5.45) (Table 1). Lipophilicity is the most important property of new drug candidates. According to Lipinski's rule, *clogP* > 5 predicts poor absorption and/or permeability of a drug [15]. In addition, high lipophilicity may increase the unspecific cytotoxicity [16].

The cytotoxic properties of all investigated analogs were tested in human cells of cancer and normal origin. Fourteen compounds (**1b**, **1e**, **2b–2f**, **3c–3g**, **5**, **Ie***) displayed significant cytotoxicity toward cancer cell lines (IC_{50} < 10 μM); however, they were less toxic than the lead compounds and exhibited cytotoxicity in normal endothelial cells (Table 2). Comparison of IC_{50} values indicates that compounds with lower lipophilicity are less cytotoxic. This is particularly evident for derivatives of dicarboximid **1**, which show the lowest cytotoxicity.

Preliminary studies on derivatives **3c**, **3d**, **3f**, **2c**, **2d**, and **2f** showed that these compounds activated caspase 3 and 7 in K562 cells (Figure 2). This suggests that the cytotoxicity of the new derivatives is associated with the induction of apoptosis in CML cells. Interestingly, proapoptotic activity of dicarboximides seems to depend on lipophilicity. Derivatives of imide **2** (median *clogP* 4.1) were significantly more potent than derivatives of imide **3** (median *clogP* 5.69). Apoptosis may proceed through two major pathways: receptor-mediated (extrinsic) and mitochondrial (intrinsic) [17]. The effect of dicarboximides on apoptotic pathways was investigated by microarray-based gene expression profiling. We found that several pro-apoptotic genes involved in the receptor and mitochondrial apoptotic pathways were upregulated in K562 cells treated with **Ic**. The expression profile induced by **Ic** was similar to staurosporine, a natural compound that activates mainly the mitochondrial pathway of apoptosis [18]. Specifically, compound **Ic** upregulated the expression of TNFRSF10B and *RIPK1*, which promote apoptosis via the receptor (extrinsic) pathway. TNFRSF10B belongs to a TNF family of transmembrane receptors. Upon binding the TRAIL ligand (TNF-related apoptosis-inducing ligand), TNFRSF10B is activated leading to caspase 8-mediated apoptosis [19]. Interestingly, in response to cytotoxic agents cancer cells overexpress TNFRSF10B, and this seems to be sufficient to trigger ligand-independent apoptosis [20]. RIPK1 (Receptor Interacting Serine/Threonine Kinase 1) is involved in signaling events downstream of TNFR1 that (depending on the cellular context) may lead to apoptosis through recruitment of FADD and subsequent activation of caspase 8 [21] or promote necroptosis through phosphorylation of RIPK3 and MLKL [22]. Compound **Ic** also increased the expression of protein genes involved in the mitochondrial (intrinsic) pathway of apoptosis: *PMAIP1* (Phorbol-12-Myristate-13-Acetate-Induced Protein 1), *BAX* (BCL2 Associated X, a Bcl-2 family member), *BBC3* (BCL2 Binding Component 3, a Bcl-2 family member), and *BNIP3* (BCL2 Interacting Protein 3). BAX is a member of the Bcl-2 family of proteins and a key activator of the intrinsic apoptotic pathway. In response to cellular stress, BAX is activated and oligomerizes, forming pores in the mitochondrial outer membrane. This enables cytochrome c and/or SMAC/Diablo proteins to leak into the cytoplasm and activate caspase 3 [23]. BBC3 is a pro-apoptotic protein that belongs to a class of BH3-only proteins. BBC3 binds other anti-apoptotic Bcl2 proteins (for example antiapoptotic MCL1) using its BH3 domains and thus promotes apoptosis. Interestingly, the expression of BBC3 is induced by DNA damaging agents and regulated by p53 [24]. This could explain the increased expression of BBC3 in K562 cells treated with **Ic**, as we demonstrated previously that this compound interacts with DNA and thus may be considered a DNA-damaging agent [9]. Altogether, these results suggest that dicarboximides activate both receptor-mediated and mitochondrial pathways of apoptosis by the upregulation of several genes involved in both pathways.

Interestingly, compound **Ic** also increased the expression of anti-apoptotic genes encoding IKBKB/IKKβ (Inhibitor Of Nuclear Factor Kappa B Kinase Subunit Beta), IKBKG/IKKγ (Inhibitor Of Nuclear Factor Kappa B Kinase Subunit Gamma), and RelA, which serves as a subunit of NF-κB heterodimer. These proteins are involved in the activation and function of transcription factor NF-κB, which regulates the expression of several genes involved in inflammation, immune responses, cell survival, and tumorigenesis and inhibits apoptosis. Serine kinase IKKβ and regulatory protein Ikkγ are key components of the IKK complex, which is essential for the activation of NF-κB transcription factors. The anti-apoptotic activity of IKKβ and IKKγ has been confirmed in studies using knockout mice. IKKβ-deficient mice died during embryonic development due to the uncontrolled apoptosis of liver cells [25]. A similar phenotype, i.e., liver apoptosis was observed in the case of Ikkγ-deficient male embryos, leading to their death on day (E)12 [26].

The cytotoxic effects of dicarboximides may also be mediated by MAPKs, a family of Ser-Thr kinases that consists of ERK1/2, p38, and JNK kinase. We demonstrated that in cells exposed to lead dicarboximides **Ic** and **II**, p38 and JNK kinases are activated while there was no effect on ERK1/2 (Figure 4). ERK1/2 are activated mainly by growth factors (for example PDGF or EGF) and positively control cell proliferation by promoting G1- to S-phase transition [27]. On the other hand, p38 and JNK kinases are activated by stress conditions, for example, oxidative stress, ionizing radiation, cytotoxic/genotoxic agents, DNA damage, inflammatory cytokines, or deprivation of growth factors. In response to stress conditions, JNK transactivates the c-Jun transcription factor or phosphorylates proteins associated with apoptosis, leading to activation of both mitochondrial and receptor-mediated pathways of apoptosis [28]. Similarly, activation of p38 leads to inhibition of cell cycle progression and induction of apoptosis [29,30]. Our results suggest that p38 and JNK can contribute to the activation of apoptosis induced by **Ic** and **II**.

We also examined whether DNA may be targeted by test derivatives. The obtained results showed that dicarboximides inhibited the digestion of plasmid DNA with *BamHI* endonuclease, albeit to a lower extent than daunorubicin (Figure 5). These data suggest that the test dicarboximides may intercalate to DNA and possibly cause DNA damage. This property of dicarboximides (including compound **Ic**) could explain the increased expression of the BBC3 gene in K562 cells exposed to **Ic**, as discussed above.

The SAR analysis indicated that the volume of substituents at positions 1 and 7 of the imide system is an important factor in determining the cytotoxicity of dicarboximides (Figure 1). This conclusion is supported by comparison of the cytotoxic and apoptotic activities in three sets of compounds: **1c**, **2c**, and **3c**; **1d**, **2d**, and **3d**; or **1f**, **2f**, and **3f**. The derivatives containing ethyl or phenyl groups at positions 1 and 7 showed lower IC_{50} towards HeLa, K456, and MOLT-1 cells than compounds with methyl groups. The type of alkylamine substituent in the side chain is also important for inducing cytotoxicity and apoptosis-signaling pathways. Dicarboximides with alkylamine substituent bearing a protonated nitrogen atom at physiological pH (pKa of amines is > 9.0, [31]) were the most active. For example, *tert*-butyl derivatives (**2c** and **3c**), piperidynyl derivatives (**2f** and **3f**), dibutyl derivatives (**1e**, **2e**, and **3e**), and diethylamine derivative (**3d**) showed the highest cytotoxicity against cancer cells tested. It is worth noting that all of the above-mentioned compounds have values of *clogP* > 4. On the other hand, compounds containing the morpholine system i.e., **1g**, **2g**, and **3g** (with the lowest *clogP* values among other derivatives within the corresponding group of imides **1**, **2**, and **3**, respectively) exhibited the lowest cytotoxicity towards tested cell lines. This result suggests that the presence of an acceptor oxygen atom at the morpholine ring, but not the cyclic structure of the substituent, is responsible for undesirable interactions that are not offered by the piperidine ring of **1f**, **2f**, and **3f**. The favorable role of the protonated nitrogen atom of the alkylamino group is further supported by comparison of compounds of scaffold **1**, which are less or not cytotoxic when the imide function is substituted with hydroxyalkyl (**4**) or chloroethyl substituent (**1h**), respectively, but exhibit higher cytotoxicity when **1h** is substituted with a *tert*-butyl amine residue (**1i**). However, the presence of the protonated alkylamino group is not an obligatory feature to provide high cytotoxicity, since, e.g., compound **5**, containing a hydroxyalkyl side chain, exhibited significant cytotoxicity. The activity of the lead compounds is also influenced by the type of salts, i.e., acetate vs. hydrochloride. The acetate salts showed lower cytotoxicity and lacked selectivity against the studied cell lines. It can be hypothesized that the change of polarization of the molecule by the acetate anion and the decrease in the hydrophilicity could affect the activity of these derivatives.

4. Materials and Methods

4.1. Chemistry

All chemicals used in the reaction were supplied from Sigma-Aldrich (Saint Louis, MO, USA) or POCH (Polskie Odczynniki Chemiczne, Gliwice, Poland). Melting points were determined by the capillary method using the Electrothermal 9100 apparatus and were uncorrected. The nuclear magnetic resonance spectra were recorded in DMSO-d_6 or $CDCl_3$ on VMNRS300 spectrometer (Bruker, Billerica MA, USA) operating at 300 MHz (^{1}HNMR) and 75 MHz (^{13}CNMR). Chemical shifts (δ) are expressed in parts per million relative to tetramethylsilane (TMS) used as the internal reference. Coupling constants (*J*) values are given in hertz (Hz), and spin multiples are given as s (singlet), d (doublet), t (triplet), and m (multiplet). Mass spectral ESI (electrospray ionization) measurements were carried out on a MicrOTOF II Bruker instrument with a TOF detector. The spectra were obtained in the positive ion mode. Flash chromatography was performed on Merck Kieselgel (Darmstadt, Germany) 0.05–0.2 mm reinst (70–325 mesh ASTM) silica gel using chloroform or chloroform–methanol as an eluent system at the v/v ratio of 50:0.2 or 50:0.5. The progress of the reactions described in the experimental section was monitored by TLC on silica gel (plates with fluorescent indicator 254 nm, layer thickness 0.2 mm, Kieselgel G. Merck), using chloroform–methanol as an eluent system at the v/v ratio of 9.8:0.2 or 9.5:05.

4.1.1. Synthesis of Imides **1–5** and **1h**

The starting imides **1–3** were prepared by the methods described earlier [10,32].

4.1.2. Synthesis of 1,7-Dimethyl-8,9-diphenyl-4-azatricyclo[5.2.1.0^{2,6}]dec-8-ene-3,5,10-trione (**1**)

A mixture of 2,5-dimethyl-3,4-diphenylcyclopenta-2,4-dien-1-one (0.02 mol) and maleimide (0.022 mol) was heated for 6 h in 20 mL of benzene. During the synthesis, the solution became red, and a white solid precipitated from it. When the reaction was completed, the solvent was evaporated under reduced pressure, and the solid residue was crystallized from benzene.

M.W. = 357.4018; $C_{23}H_{19}NO_3$; Yield: 77%, white powder, m.p. 220.6–221.1 °C (from benzen); ^{1}HNMR (300 MHz, DMSO-d_6+ TMS, δ/ppm): 1.39 (6H, *s*, -CH_3); 3.38 (2H, *s*, C2-H, C6-H); 6.98 (4H, *m*, Ar-H); 7.19 (6H, *m*, Ar-H); 11.66 (1H, *s*, NH); Anal. Calc for $C_{23}H_{19}N_3O_3$: 77.28% C; 5.32% H, 3.92% N; found 77.40% C; 5.30% H; 3.99% N; ^{13}CNMR: δ 11.7, 48.5, 55.5, 127.4, 127.9, 129.4, 133.2, 141.3, 177.5, 199.5; HRMS (m/z): calculated value for [M + Na^+] 380.1257; found 380.1259.

4.1.3. Synthesis of 1,7-Diethyl-8,9-tetraphenyl-4-azatricyclo[5.2.1.0^{2,6}]dec-8-ene-3,5,10-trione (**2**)

A mixture of 2,5-diethyl-3,4-diphenylcyclopenta-2,4-dien-1-one (0.01 mol) and maleimide (0.012 mol) in benzene (15 mL) was refluxed for 14 h. When the reaction was completed, the solvent was evaporated under reduced pressure, and the solid residue was crystallized from benzene.

M.W. = 385.455; $C_{25}H_{23}NO_3$; Yield: 92%; white powder, m.p. 206.8–209 °C (from benzen); ^{1}HNMR (300 MHz, DMSO-d_6+ TMS, δ/ppm): 0.87 (6H, *t*, *J* = 7.5 Hz, -CH_3); 2.05 (m, 4H, -CH_2-), 3.58 (2H, *s*, C2-H, C6-H); 6.97 (4H, *m*, Ar-H); 7.18 (6H, *m*, Ar-H); 11.70 (1H, *s*, NH);

Anal. Calc for $C_{25}H_{23}NO_3$: 77.93% C; 5.98% H, 3.63% N; found 77.94% C; 5.97% H; 3.62% N; ^{13}CNMR: δ 9.1, 19.2, 43.9, 55.3, 60.4, 128.0, 129.4, 133.1, 142.4, 171.6, 197.8; HRMS (*m*/*z*): calculated value for [M + Na$^+$] 408.1570; found 408.1570.

4.1.4. Synthesis of 1,7,8,9-Tetraphenyl-4-azatricyclo[5.2.1.0^{2,6}]dec-8-ene-3,5-dione (**3**)

A mixture of 1,2,3,4-tetraphenylcyclopenta-1,3-diene (0.02 mol) and maleimide (0.022 mol) was heated for 6 h in 20 mL of benzene. When the reaction was completed, the solvent was evaporated under reduced pressure, and the solid residue was crystallized from benzene.

M.W. = 467.5571; $C_{33}H_{25}NO_2$; Yield: 81%, white powder, m.p. 246.5–246.9 °C (from benzen); ^{1}HNMR (300 MHz, DMSO-d$_6$+ TMS, δ/ppm): 2.23 (1H, *d*, *J* = 8.4, C10-H), 3.23 (1H, *d*, *J* = 8.7, C10-H), 4.20 (2H, *s*, C2-H, C6-H); 6.60 (4H, *m*, Ar-H); 6.88 (6H, *m*, Ar-H); 7.24 (6H, *m*, Ar-H); 7.74 (4H, *d*, *J* = 7.2, Ar-H); 11.44 (1H, *s*, NH); Anal. Calc for $C_{33}H_{25}NO_2$: 84.79% C; 5.35% H, 3.00% N; found 84.85% C; 5.39% H; 3.08% N; ^{13}CNMR: δ 52.9, 62.9, 65.1, 126.4, 126.6, 127.1, 127.7, 129.1, 129.6, 134.5, 139.8, 145.0, 178.0; HRMS (*m*/*z*): calculated value for [M + Na$^+$] 490.1778; found 490.1777.

4.1.5. Synthesis of N-Hydroxyethylimides **4** and **5**

A mixture of an appropriate diene:

- 2,5-dimethyl-3,4-diphenylcyclopenta-2,4-dien-1-one (0.02 mol) for compound **4**
- 1,2,3,4-tetraphenylcyclopenta-1,3-diene (0.02 mol) for compound **5** and N-hydroxymaleimide (0.022 mol) was heated for 6 h in 20 mL of benzene. When the reaction was completed, the solvent was evaporated under reduced pressure, and the solid residue was crystallized from benzene.

***4-(2-hydroxyethyl)-1,7-dimethyl-8,9-diphenyl-4-azatricyclo[5.2.1.0^{2,6}]dec-8-ene-3,5,10-trione* (4)**

M.W. = 401.4544; $C_{25}H_{23}NO_4$; Yield: 84%; white powder, m.p. 186–187 °C (from MeOH/(Et)$_2$O); ^{1}HNMR (300 MHz, DMSO-d$_6$+ TMS, δ/ppm): 1.57 (6H, *s*, -CH$_3$), 2.16 (1H, *s*, -OH), 3.25 (2H, *m*, C2-H, C6-H), 3.73 (4H, *m*, -CH$_2$-), 6.93 (4H, *m*, Ar-H), 7.26 (6H, *m*, Ar-H); ^{13}CNMR: δ 12.2, 47.4, 47.9, 56.5, 60.1, 127.7, 128.1, 129.4, 132.9, 141.5, 175.8, 176.1, 199.3; HRMS (*m*/*z*): calculated value for [M + Na$^+$] 424.1519; found 424.1519.

***4-(2-hydroxyethyl)-1,7,8,9-tetraphenyl-4-azatricyclo[5.2.1.02,6]dec-8-ene-3,5-dione* (5)**

M.W. = 511.6096; $C_{35}H_{29}NO$; Yield: 61%, white powder, m.p. 197–198 °C (from benzen); ^{1}HNMR (300 MHz, DMSO-d$_6$+ TMS, δ/ppm): 2.12 (1H, *br.s*, -OH), 2.32 (1H, *d*, *J* = 4.5, C6-H), 3.17 (1H, *d*, *J* = 4.5, C2-H), 3.74 (2H, *m*, C1′-H, C2′-H); 4.17 (2H, *s*, C10-H), 6.55 (4H, *m*, Ar-H); 6.89 (6H, *m*, Ar-H); 7.32 (6H, *m*, Ar-H); 7.70 (4H, *m*, Ar-H); ^{13}CNMR: δ 42.2, 52.2, 60.5, 63.6, 65.9, 126.8, 127.1, 127.4, 128.1, 129.0, 129.8, 133.8, 139.2, 145.0, 177.1; HRMS (*m*/*z*): calculated value for [M + Na$^+$] 543.5988; found 534.2045.

4.1.6. Synthesis of 4-(2-Chloroethyl)-1,7-dimethyl-8,9-diphenyl-4-azatricyclo[5.2.1.0^{2,6}] dec-8-ene-3,5,10-trione (**1h**)

To the solution of imide **a** (0.01 mol) in acetonitrile (30 mL), 1,2-dichloroethane (0.02 mol) was added. The reaction was heated for 72 h. When the reaction was completed, the solvent was evaporated under reduced pressure, and the residue was purified by column chromatography on silica gel (eluent: chloroform).

***4-(2-chloroethyl)-1,7-dimethyl-8,9-diphenyl-4-azatricyclo[5.2.1.0^{2,6}]dec-8-ene-3,5,10-trione* (1h)**

M.W. = 419.9000; $C_{25}H_{22}ClNO_3$; Yield: 74%; white powder, m.p. 190–191 °C (from hexane); ^{1}HNMR (300 MHz, CDCl$_3$, δ/ppm): 1.58 (6H, -CH$_3$), 3.28 (2H, *m*, C2-H, C6-H), 3.64 (2H, *t*, *J* = 6.4 Hz, C1′-H,), 3.86 (2H, *t*, *J* = 6 Hz, C2′-H), 6.91 (4H, *m*, Ar-H), 7.25 (6H, *m*, Ar-H); ^{13}CNMR: δ 12.2, 39.5, 40.4, 47.9, 56.5, 127.7, 128,1, 129.4, 132.9, 141.5, 175.1, 199.2; HRMS (*m*/*z*): calculated value for [M + Na$^+$] 442.8893; found 442.1175.

4.1.7. General Conditions of Synthesis of New N-Alkylamino Derivatives of 1,7-Dimethyl-8,9-diphenyl-4-azatricyclo[5.2.1.0^{2,6}]dec-8-ene-3,5,10-trione (**1i–1j**)

The 4-(2-chloroethyl)-1,7-dimethyl-8,9-diphenyl-4-azatricyclo[5.2.1.02,6]dec-8-ene-3,5,10-trione was dissolved in acetonitrile (30 mL), and then anhydrous K_2CO_3 (0.01 mol) and an appropriate amine (0.01 mol) were added. The reaction was carried out in reflux, respectively, for 24–48 h in the boiling temperature of the solvent. The reaction was monitored by TLC. When the reaction was completed, the solvent was evaporated, and the residue was purified by column chromatography on silica gel (eluent: chloroform or chloroform/methanol 50:0.2 *v*/*v*).

***4-[2-(morpholin-4-yl)ethy]-1,7-dimethyl-8,9-diphenyl-4-azatricyclo[5.2.1.0^{2,6}]dec-8-ene-3,5,10-trione* (1j)**

M.W. = 470.5595 (free amine); $C_{29}H_{30}N_2O_4$*HCl; Yield: 70%; white powder, m.p. 227–228 °C (from EtOH/(Et)$_2$O); ^{1}HNMR (300 MHz, DMSO-d$_6$+ TMS, δ/ppm): 1.41 (6H, *s*, -CH_3), 3.09 (2H, *m*, -CH_2-morph), 3.30 (2H, *m*, -CH_2-morph), 3.39 (3H, *m*, C2-H, C6-H), 3.86 (4H, *m*, C1′-H, -CH_2-morph), 3.91 (2H, *m*, C2′-H), 6.87 (4H, *m*, Ar-H), 7.18 (6H, *m*, Ar-H), 11.57 (1H, *br.s*, NH$^+$); ^{13}CNMR: δ 11.8, 32.5, 48.1, 50.9, 52.3, 55.6, 62.8, 127.5, 128,1, 129.2, 133.1, 141.4, 176.3, 199.2; HRMS (*m*/*z*): calculated value for [M + H$^+$] 471.2278; found 471.2276.

***4-[2-(piperidin-1-yl)ethy]-1,7-dimethyl-8,9-diphenyl-4-azatricyclo[5.2.1.0^{2,6}]dec-8-ene-3,5,10-trione* (1i)**

M.W. = 468.5866 (free amine); $C_{30}H_{32}N_2O_3$*HCl; Yield: 64%; white powder, m.p. 214–216 °C (from EtOH/(Et)$_2$O); ^{1}HNMR (300 MHz, $CDCl_3$, δ/ppm): 1.42 (6H, -CH_3), 1.76 (4H, *m*, -CH_2-piper), 2.85 (2H, *m*, C1′-H), 3.16 (2H, *m*, C2′-H), 3.36 (1H, *m*, C6-H), 3.49 (1H, *m*, C2-H), 3.54 (4H, *m*, -CH_2-piper), 3.82 (2H, *m*, -CH_2-piper), 6.86 (4H, *m*, Ar-H), 7.19 (6H, *m*, Ar-H), 10.37 (1H, *br.s*, NH$^+$); ^{13}CNMR: δ 11.8, 15.1, 21.2, 22.0, 32.9, 48.0, 51.9, 55.6, 127.5, 128,1, 129.2, 133.0, 141.4, 176.1, 199.1; HRMS (*m*/*z*): calculated value for [M + H$^+$] 469.2486; found 469.2485.

4.1.8. General Conditions of the Synthesis of New Alkanoloamine Derivatives of Dicarboximides (**1b–1g**, **2b–2g**, **3b–3g**)

The new alkanolamine derivatives were obtained in two-step reactions according to the method described previously [10]. Step **I**: the mixture of an appropriate imide:

1: 1,7-dimethyl-8,9-diphenyl-4-azatricyclo[5.2.1.0^{2,6}]dec-8-ene-3,5,10-trione
2: 1,7-diethyl-8,9-diphenyl-4-azatricyclo[5.2.1.0^{2,6}]dec-8-ene-3,5,10-trione
3: 1,7,8,9-tetraphenyl-4-azatricyclo[5.2.1.0^{2,6}]dec-8-ene-3,5-dione

(0.01 mol) with 1-chloro-2,3-epoxypropane (50–60 mL), in the presence of an anhydrous K_2CO_3 (0.01 mol) was stirred on a magnetic stirrer under reflux with a tube with $CaCl_2$. The reaction was carried out. at room temperature for 15 h. When the reaction was completed (TLC control), the inorganic precipitate was filtered off and the filtrate was concentrated. The obtained oily product was purified by column chromatography on silica gel (eluent: chloroform and chloroform/methanol 50:0.2). Step **II**: to the solution of an appropriate 4-(oxirane-2-ylmethyl)-4-imide (**1a**, **2a**, **3a**) (0.001 mol) in a mixture of methanol/water (10:1 *v*/*v*), the appropriate amine (0.001 mol) was added. Reactions were carried out at room temperature on a magnetic stirrer under reflux for 15–20 h. When the reaction was completed (TLC control), the excess of the solvent was evaporated and the crude product was purified by column chromatography on silica gel (eluent: chloroform or chloroform/methanol (50:0.2, 50:0.5)).

***4-[3-(ethylamino)-2-hydroxypropyl]-1,7-dimethyl-8,9-diphenyl-4-azatricyclo [5.2.1.0^{2,6}] dec-8-ene-3,5,10-trione* (1b)**

M.W. = 458.5488 (free amine); $C_{28}H_{30}N_2O_4$*HCl; Yield: 55%; white powder, m.p. 198–201 °C (from MeOH/(Et)$_2$O); ^{1}HNMR (300 MHz, DMSO-d$_6$+ TMS, δ/ppm): 1.18 (3H, *t*,

J = 7.2 Hz, -CH_3), 1.42 (6H, *s*, -CH_3), 2.78 (1H, *m*, C6-H), 2.80 (2H, *m*, -CH_2-) 3.05 (1H, *m*, C2-H), 3.37(2H, *m*, C3′-H), 3.51 (2H, *m*, C1′-H), 4.10 (1H, *s*, C2′-H), 5.88 (1H, *s*, -OH), 6.87 (4H, *m*, Ar-H), 7.18 (6H, *m*, Ar-H), 8.65 (1H, *br.s*, -NH_2^+), 9.09 (1H, *br.s*, -NH_2^+); ^{13}CNMR: δ 10.7, 11.8, 42.3, 42.7, 47.4, 47.6, 49.3, 55.7, 63.4, 127.5, 128,1, 129.3, 133.2, 141.3, 175.8, 176.2, 199.4; HRMS (m/z): calculated value for [M + H^+] 459.2284; found 459.2284.

***4-[3-(tert-butylamino)-2-hydroxypropyl]-1,7-dimethyl-8,9-diphenyl-4-azatricyclo [5.2.1.0^{2,6}] dec-8-ene-3,5,10-trione* (1c)**

M.W. = 486.6019 (free amine); $C_{30}H_{34}N_2O_4$*HCl; Yield: 57%; white powder, m.p. 248–250 °C (from MeOH/$(Et)_2O$); ^{1}HNMR (300 MHz, DMSO-d_6+ TMS, δ/ppm): 1.26 (9H, *s*, -$C(CH_3)_3$), 1.43 (6H, *d*, J = 5.1 Hz, -CH_3), 2.50 (1H, *m*, C6-H), 3.16 (1H, *m*, C2-H), 3.52 (4H, *m*, C3′-H, C1′-H), 4.14 (1H, *m*, C2′-H), 5.83 (1H, *s*, -OH), 6.88 (4H, *m*, Ar-H), 7.18 (6H, *m*, Ar-H), 8.39 (1H, *br.s*, -NH_2^+), 9.04 (1H, *br.s*, -NH_2^+); ^{13}CNMR: δ 11.8, 24.9, 42.8, 44.5, 47.3, 47.6, 48.5, 55.6, 56.2, 63.7, 127.4, 128.0, 129.3, 133.2, 141.3, 175.7, 176.3, 199.4; HRMS (m/z): calculated value for [M + H^+] 487.2598; found 487.2597.

***4-[3-(diethylamino)-2-hydroxypropyl]-1,7-dimethyl-8,9-diphenyl-4-azatricyclo[5.2.1.0^{2,6}] dec-8-ene-3,5,10-trione* (1d)**

M.W. = 486.6019 (free amine); $C_{30}H_{34}N_2O_4$*HCl; Yield: 72%; white powder, m.p. 177–179 °C (from MeOH/$(Et)_2O$); ^{1}HNMR (300 MHz, DMSO-d_6+ TMS, δ/ppm): 1.18 (6H, *t*, J = 7.2 Hz, -CH_3), 1.42 (6H, *d*, J = 4.2 Hz, -CH_3), 3.16 (6H, *m*, C6-H, C2-H, C3′-H, C1′-H), 3.52 (2H, *m*, -CH_2-), 3.73 (2H, m, -CH2-), 4.19 (1H, *s*, -OH), 6.87 (4H, *m*, Ar-H), 7.17 (6H, *m*, Ar-H), 9.87 (1H, *br.s*, -NH^+); ^{13}CNMR: δ 8.3, 11.8, 42.8, 46.7, 47.4, 47.6, 54.0, 55.7, 62.5, 127.5, 128.0, 129.3, 133.1, 141.3, 175.8, 176.2, 199.3; HRMS (m/z): calculated value for [M + H^+] 487.2598; found 487.2597.

***4-[3-(dibutylamino)-2-hydroxypropyl]-1,7-dimethyl-8,9-diphenyl-4-azatricyclo [5.2.1.0^{2,6}] dec-8-ene-3,5,10-trione* (1e)**

M.W. = 542.7082 (free amine); $C_{34}H_{42}N_2O_4$*HCl; Yield: 74%; white powder, m.p. 192–196 °C (from MeOH/$(Et)_2O$); ^{1}HNMR (300 MHz, DMSO-d_6+ TMS, δ/ppm): 0.89 (6H, *t*, J = 7.35 Hz, -CH_3), 1.27 (4H, *m*, -CH_2-), 1.42 (6H, *d*, J = 4.2 Hz, -CH_3), 1.60 (4H, *m*, -CH_2-), 3.05 (4H, *m*, -CH_2-), 3.20 (1H, *m*, C6-H), 3.31 (2H, *m*, C3′-H), 3.51 (3H, *m*, C1′-H, C2′-H), 4.19 (1H, *m*, C2′-H) 5.94 (1H, *s*, -OH), 6.88 (4H, *m*, Ar-H), 7.17 (6H, *m*, Ar-H), 9.81 (1H, *br.s*, -NH^+); ^{13}CNMR: δ 11.8, 13.4, 19.3, 24.5, 47.4, 47.6, 52.2, 52.8, 55.1, 55.7, 62.6, 127.5, 128.0, 129.3, 133.1, 141.3, 175.8, 176.2, 199.2; HRMS (m/z): calculated value for [M + H^+] 543.3224; found 543.3223.

***4-[2-hydroxy-3-(piperidin-1-yl)propyl]-1,7-dimethyl-8,9-diphenyl-4-azatricyclo[5.2.1.0^{2,6}] dec-8-ene-3,5,10-trione* (1f)**

M.W. = 498.6126 (free amine); $C_{31}H_{34}N_2O_4$*HCl; Yield: 45%; white powder, m.p. 170–171 °C (from MeOH/$(Et)_2O$); ^{1}HNMR (300 MHz, DMSO-d_6+ TMS, δ/ppm): 1.42 (6H, *m*, -CH_3), 1.74 (4H, *m*, -CH_2-pip), 2.50 (3H, m, C6-H, C3′-H), 2.88 (2H, *m*, -C1′-H), 3.32 (7H, *m*, C2-H, -CH_2-pip), 4.42 (1H, m, C2′-H) 5.97 (1H, *s*, -OH), 6.88 (4H, *m*, Ar-H), 7.18 (6H, *m*, Ar-H), 9.97 (1H, *br.s*, -NH^+); ^{13}CNMR: δ 11.8, 21.2, 22.1, 42.7, 47.6, 52.1, 53.4, 55.6, 58.7, 62.3, 127.5, 128.0, 129.3, 133.1, 141.4, 175.8, 176.1, 199.3; HRMS (m/z): calculated value for [M + H^+] 499.2598; found 499.2597.

***4-[-2-hydroxy-3-(morpholin-4-yl)propyl]-1,7-dimethyl-8,9-diphenyl-4-azatricyclo [5.2.1.0^{2,6}] dec-8-ene-3,5,10-trione* (1g)**

M.W. = 500.5854 (free amine); $C_{30}H_{32}N_2O_5$*HCl; Yield: 54%; white powder, m.p. 210–212 °C (from MeOH/$(Et)_2O$); ^{1}HNMR (300 MHz, DMSO-d_6+ TMS, δ/ppm): 1.42 (6H, *m*, -CH_3), 3.07 (2H, *m*, C2-H, C6-H), 3.38 (6H, *m*, -CH_2-morpholin), 3.52 (2H, *m*, -CH_2-morpholin), 3.88 (4H, *m*, C3′-H, C1′-H), 4.30 (1H, *m*, C2′-H), 6.01 (1H, *s*, -OH), 6.87 (4H, *m*, Ar-H), 7.18 (6H, *m*, Ar-H), 10.07 (1H, *br.s*, -NH^+); ^{13}CNMR: δ 11.8, 42.6, 47.5, 51.0, 52.5, 55.6,

58.7, 62.2, 62.9, 127.5, 128.1, 129.3, 133.1, 141.4, 175.8, 176.1, 199.3; HRMS (m/z): calculated value for [M + H$^+$] 501.2384; found 501.2379.

***4-[3-(ethylamino)-2-hydroxypropyl]-1,7-diethyl-8,9-diphenyl-4-azatricyclo [5.2.1.0^{2,6}]dec-8-ene-3,5,10-trione* (2b)**

M.W. = 486.6019 (free amine); $C_{30}H_{34}N_2O_4$*HCl; Yield: 55%; white powder, m.p. 197–200 °C (from MeOH/(Et)$_2$O); ^{1}HNMR (300 MHz, DMSO-d$_6$+ TMS, δ/ppm): 0.89 (6H, *m*, -CH$_3$), 1.18 (3H, *t*, *J* = 7.2 Hz, -CH$_3$), 1.87 (2H, *m*, -CH$_2$-), 2.11 (2H, *m*, -CH$_2$-), 2.80 (1H, *m*, C6-H), 2.88 (2H, *m*, -CH$_2$-) 3.06 (1H, *m*, C2-H), 3.41(2H, *m*, C3′-H), 3.54 (2H, *m*, C1′-H), 4.13 (1H, *s*, C2′-H), 5.89 (1H, *br.s*, -OH), 6.87 (4H, *m*, Ar-H), 7.16 (6H, *m*, Ar-H), 8.67 (1H, *br.s*, -NH$_2^+$), 9.10 (1H, *br.s*, -NH$_2^+$); ^{13}CNMR: δ 9.0, 10.7, 18.9, 39.5, 42.2, 42.7, 43.6, 43.7, 49.2, 59.2, 63.3, 127.4, 128.0, 129.2, 133.5, 141.7, 141.8, 176.0, 176.4, 199.0; HRMS (m/z): calculated value for [M + H$^+$] 487.6093; found 487.2591.

***4-[3-(tert-butylamino)-2-hydroxypropyl]-1,7-diethyl-8,9-diphenyl-4-azatricyclo [5.2.1.0^{2,6}] dec-8-ene-3,5,10-trione* (2c)**

M.W. = 514.6551 (free amine); $C_{32}H_{38}N_2O_4$*HCl; Yield: 67%; white powder, m.p. 231–235 °C (from MeOH/(Et)$_2$O); ^{1}HNMR (300 MHz, DMSO-d$_6$+ TMS, δ/ppm): 0.88 (6H, *d*, *J* = 5.1 Hz, -CH$_3$), 1.26 (9H, *m*, -C(CH$_3$)$_3$), 1.87 (2H, *m*, -CH$_2$-), 2.09 (2H, m, -CH$_2$-), 2.78 (1H, *m*, C6-H), 3.08 (1H, *m*, C2-H), 3.40 (2H, *m*, C3′-H0, 3.69 (2H, *m*, C1′-H), 4.14 (1H, *m*, C2′-H), 5.80 (1H, *s*, -OH), 6.88 (4H, *m*, Ar-H), 7.16 (6H, *m*, Ar-H), 8.36 (1H, *br.s*, -NH$_2^+$), 8.86 (1H, *br.s*, -NH$_2^+$); ^{13}CNMR: δ 9.0, 18.8, 24.9, 42.9, 43.6, 44.3, 56.2, 59.2, 59.3, 63.6, 127.4, 128.0, 129.1, 133.5, 133.5, 141.7, 141.8, 175.9, 176.5, 198.9; HRMS (m/z): calculated value for [M + H$^+$] 515.6625; found 515.2904.

***4-[3-(diethylamino)-2-hydroxypropyl]-1,7-diethyl-8,9-diphenyl-4-azatricyclo [5.2.1.0^{2,6}] dec-8-ene-3,5,10-trione* (2d)**

M.W. = 514.6551 (free amine); $C_{30}H_{34}N_2O_4$*HCl; Yield: 62%; white powder, m.p. 177–180 °C (from MeOH/(Et)$_2$O); ^{1}HNMR (300 MHz, DMSO-d$_6$+ TMS, δ/ppm): 0.88 (6H, *m*, -CH$_3$), 1.19 (6H, *t*, *J* =·7.1 Hz, -CH$_3$), 2.06 (2H, *m*, -CH$_2$-), 2.08 (2H, *m*, -CH$_2$-),3.16 (6H, *m*, C6-H, C2-H, C3′-H, C1′-H), 3.51 (4H, *m*, -CH$_2$-), 4.20 (1H, m, C2′-H), 5.98 (1H, *br. s*, -OH), 6.88 (4H, *m*, Ar-H), 7.16 (6H, *m*, Ar-H), 9.87 (1H, *br.s*, -NH$^+$); ^{13}CNMR: δ 9.3, 9.0, 18.8, 42 8, 43.7, 46.7, 47.4, 48.5, 54.0, 59.2, 62.5, 127.4, 128.0, 129.1, 133.5, 141.7, 176.0, 176.4, 199.0; HRMS (m/z): calculated value for [M + H$^+$] 515.6625; found 515.2904.

***4-[3-(dibutylamino)-2-hydroxypropyl]-1,7-diethyl-8,9-diphenyl-4-azatricyclo[5.2.1.0^{2,6}] dec-8-ene-3,5,10-trione* (2e)**

M.W. = 570.7614 (free amine); $C_{36}H_{46}N_2O_4$*HCl; Yield: 64%; white powder, m.p. 92–110 °C (from MeOH/(Et)$_2$O); ^{1}H NMR (300 MHz, DMSO-d$_6$+ TMS, δ/ppm): 0.88 (12H, *m*, -CH$_3$), 1.29 (4H, *m*, -CH$_2$-), 1.62 (4H, *m*, -CH$_2$-), 1.87 (2H, *m*, -CH$_2$-), 2.07 (2H, *m*, -CH$_2$-), 3.06 (4H, *m*, -CH$_2$-), 3.19 (2H, *m*, C3-H), 3.52 (2H, *m*, C1′-H), 3.70 (2H, *m*, C2-H, C6-H), 4.23 (1H, *m*, C2′-H), 6.88 (4H, *m*, Ar-H), 7.16 (6H, *m*, Ar-H), 9.99 (1H, *br.s*, -NH$^+$); ^{13}CNMR: δ 9.0, 13.5, 18.9, 19.4, 24.5, 42.8, 43.6, 52.3, 52.7, 55.2, 59.2, 62.6, 127.4, 127.9, 129.1, 133.5, 141.7, 176.0, 176.4, 198.9; HRMS (m/z): calculated value for [M + H$^+$] 571.7688; found 571.3530.

***4-[2-hydroxy-3-(piperidin-1-yl)propyl]-1,7-diethyl-8,9-diphenyl-4-azatricyclo[5.2.1.0^{2,6}] dec-8-ene-3,5,10-trione* (2f)**

M.W. = 526.6658 (free amine); $C_{33}H_{38}N_2O_4$*HCl; Yield: 55%; white powder, m.p. 190–195 °C (from MeOH/(Et)$_2$O); ^{1}HNMR (300 MHz, DMSO-d$_6$+ TMS, δ/ppm): 0.89 (6H, *m*, -CH$_3$), 1.36 (2H, *m*, -CH$_2$-), 1.74 (4H, *m*, -CH$_2$-, -CH$_2$-pip), 1.87 (2H, *m*, -CH$_2$-pip), 2.07 (2H, *m*, -CH$_2$-pip), 2.90 (4H, *m*, C3′-H, -CH$_2$-pip), 3.16 (2H, *m*, -CH$_2$-pip), 3.46 (1H, m, C6-H), 3.49 (2H, C1′-H, C2-H), 4.26 (1H, m, C2′-H), 6.87 (4H, *m*, Ar-H), 7.16 (6H, *m*, Ar-H), 10.08 (1H, *br.s*, -NH$^+$); ^{13}CNMR: δ 9.0, 18.8, 21.2, 22.1, 42.7, 43.7, 52.1, 53.3, 58.8, 59.2, 62.3, 127.4, 128.0, 129.1, 133.5, 141.7, 176.0, 176.3, 199.01; HRMS (m/z): calculated value for [M + H$^+$] 527.6732; found 527.2904.

***4-[-2-hydroxy-3-(morpholin-4-yl)propyl]-1,7-diethyl-8,9-diphenyl-4-azatricyclo[5.2.1.0^{2,6}] dec-8-ene-3,5,10-trione* (2g)**

M.W. = 528.6386 (free amine); $C_{32}H_{36}N_2O_5$*HCl; Yield: 54%; white powder, m.p. 226–228 °C (from MeOH/(Et)$_2$O); ^{1}HNMR (300 MHz, DMSO-d$_6$+ TMS, δ/ppm): 0.89 (6H, *m*, -CH$_3$), 1.87 (2H, *m*, -CH$_2$-), 2.06 (2H, *m*, -CH$_2$-), 3.1 (3H, *m*, C3-H, C6-H), 3.36 (3H, *m*, -CH$_2$-morpholin, C2-H), 3.47 (4H, *m*, -CH$_2$-morpholin), 3.89 (4H, *m*, -CH$_2$-morpholin, C1′-H), 4.30 (1H, *m*, C2′-H), 6.01 (1H, *br.s*, -OH), 6.87 (4H, *m*, Ar-H), 7.68 (6H, *m*, Ar-H), 10.57 (1H, *br.s*, -NH$^+$); ^{13}CNMR: δ 9.0, 18.8, 39.5, 43.7, 51.0, 52.5, 58.8, 59.2, 62.1, 62.9, 127.4, 128.0, 129.1, 133.5, 141.7, 176.0, 176.3, 199.0; HRMS (*m/z*): calculated value for [M + H$^+$] 529.6460; found 529.2697.

***4-[3-(ethylamino)-2-hydroxypropyl]-1,7,8,9-tetraphenyl-4-azatricyclo[5.2.1.0^{2,6}]dec-8-ene-3,5-dione* (3b)**

M.W. = 568.7040 (free amine); $C_{38}H_{36}N_2O_3$*HCl; Yield: 45%; white powder, m.p. 254–255 °C (from MeOH/(Et)$_2$O); ^{1}HNMR (300 MHz, DMSO-d$_6$+ TMS, δ/ppm): 1.19 (3H, *t*, *J* = 7.2 Hz, -CH$_3$), 2.34 (1H, *m*, C10-H), 2.78 (1H, *m*, C10-H), 2.91 (2H, *m*, -CH$_2$-) 3.05 (1H, *m*, C2-H), 3.16 (1H, m, C6-H), 3.30 (2H, *m*, C3′-H), 3.46 (1H, *s*, C2′-H), 4.29 (2H, *m*, C1′-H), 5.85 (1H, *s*, -OH), 6.51 (4H, *m*, Ar-H), 6.88 (6H, *m*, Ar-H), 7.32 (6H, *m*, Ar-H), 7.76 (4H, *m*, Ar-H), 8.60 (1H, *br.s*, -NH$_2^+$), 8.94 (1H, *br.s*, -NH$_2^+$); ^{13}CNMR: δ 10.7, 42.2, 49.4, 51.9, 63.0, 63.4, 65.1, 126.5, 126.8, 127.2, 127.8, 129.2, 129.5, 134.4, 139.7, 145.1, 176.6, 176.9; HRMS (*m/z*): calculated value for [M + H$^+$] 569.7119; found 569.2798.

***4-[3-(tert-butylamino)-2-hydroxypropyl]-1,7,8,9-tetraphenyl-4-azatricyclo[5.2.1.0^{2,6}]dec-8-ene-3,5-dione* (3c)**

M.W. = 596.7575 (free amine); $C_{40}H_{40}N_2O_3$*HCl; Yield: 60%; white powder, m.p. 177–179 °C (from MeOH/(Et)$_2$O); ^{1}HNMR (300 MHz, DMSO-d$_6$+ TMS, δ/ppm): 1.28 (9H, *s*, -C(CH$_3$)$_3$), 2.36 (1H, *d*, *J* = 8.7 Hz C6-H), 2.76 (1H, *m*, C10-H), 3.06 (1H, *m*, C10-H), 3.27 (1H, *m*, C2-H), 3.35 (4H, *m*, C3′-H, C1′-H), 3.52 (1H, *m*, C2′-H), 4.21 (1H, *s*, -OH), 6.51 (4H, *m*, Ar-H), 6.88 (6H, *m*, Ar-H), 7.31 (6H, *m*, Ar-H), 7.76 (4H, *m*, Ar-H), 8.41 (1H, *br.s*, -NH$_2^+$), 9.06 (1H, *br.s*, -NH$_2^+$); ^{13}CNMR: δ 24.9, 42.7, 44.6, 51.7, 52.0, 56.2, 63.0, 63.7, 65.1, 126.4, 126.5, 127.2, 127.8, 129.1, 129.5, 134.4, 139.7, 145.0, 145.1, 176.5, 177.0; HRMS (*m/z*): calculated value for [M + H$^+$] 597.7651; found 597.3117.

***4-[3-(diethylamino)-2-hydroxypropyl]-1,7,8,9-tetraphenyl-4-azatricyclo[5.2.1.0^{2,6}]dec-8-ene-3,5-dione* (3d)**

M.W. = 596.7575 (free amine); $C_{40}H_{40}N_2O_3$*HCl; Yield: 62%; white powder, m.p. 240–247 °C (from MeOH/(Et)$_2$O); ^{1}HNMR (300 MHz, DMSO-d$_6$+ TMS, δ/ppm): 1.20 (6H, *t*, *J* = 7.2 Hz, -CH$_3$), 2.33 (1H, *d*, *J* = 8.4 Hz, C6-H), 3.16 (6H, *m*, -CH$_2$-, -CH$_2$- C3′-H, C1′-H), 3.38 (1H, *m*, C2-H), 3.50 (1H, *m*, C2′-H), 4.26 (2H, *m*, C10-H), 5.98 (1H, *s*, -OH), 6.52 (4H, *m*, Ar-H), 6.88 (6H, *m*, Ar-H), 7.31 (6H, *m*, Ar-H), 7.77 (4H, *m*, Ar-H), 9.93 (1H, *br.s*, -NH$^+$); ^{13}CNMR: δ 8.4, 42.7, 46.7, 47.4, 51.9, 52.0, 54.2, 62.5, 63.0, 65.1, 126.5, 126.8, 127.2, 127.8, 129.2, 129.5, 134.4, 139.7, 145.0, 145.1, 176.6, 177.0; HRMS (*m/z*): calculated value for [M + H$^+$] 597.7651; found 597.3117.

***4-[3-(dibutylamino)-2-hydroxypropyl]-1,7,8,9-tetraphenyl-4-azatricyclo[5.2.1.0^{2,6}]dec-8-ene-3,5-dione* (3e)**

M.W. = 652.8635 (free amine); $C_{44}H_{48}N_2O_3$*HCl; Yield: 64%; white powder, m.p. 237–238 °C (from MeOH/(Et)$_2$O); ^{1}HNMR (300 MHz, DMSO-d$_6$+ TMS, δ/ppm): 0.90 (6H, *t*, *J* = 7.2 Hz, -CH$_3$), 1.29 (4H, *m*, -CH$_2$-), 1.62 (4H, *m*, -CH$_2$-), 2.37 (1H, *m*, C6-H), 3.16 (4, *m*, -CH$_2$-), 3.34 (8H, *m*, C2-H, C1′-H, C2′-H, C3′-H, C10-H), 5.94 (1H, *s*, -OH), 6.51 (4H, *m*, Ar-H), 6.88 (6H, *m*, Ar-H), 7.31 (6H, *m*, Ar-H), 7.74 (4H, *m*, Ar-H), 9.71 (1H, *br.s*, -NH$^+$); ^{13}CNMR: δ 13.5, 19.4, 24.6, 48.5, 51.8, 52.0, 52.3, 52.7, 55.4, 62.6, 63.0, 65.1, 126.5, 126.8, 127.3, 127.8, 129.2, 129.5, 134.4, 139.7, 145.0, 145.1, 176.6, 177.0; HRMS (*m/z*): calculated value for [M + H$^+$] 653.8714; found 653.3743.

***4-[2-hydroxy-3-(piperidin-1-yl)propyl]-1,7,8,9-tetraphenyl-4-azatricyclo[5.2.1.0^{2,6}]dec-8-ene-3,5-dione* (3f)**

M.W. = 608.7679 (free amine); $C_{41}H_{40}N_2O_3$*HCl; Yield: 54%; white powder, m.p. 268–270 °C (from MeOH/(Et)$_2$O); ^{1}HNMR (300 MHz, DMSO-d$_6$+ TMS, δ/ppm): 1.35 (1H, *m*, C2-H), 1.75 (5H, *m*, -CH$_2$-pip., C6-H), 2.36 (1H, *m*, C10-H), 2.88 (2H, m, C1′-H), 2.99 (1H, *m*, C2′-H), 3.40 (9H, *m*, -CH$_2$-pip, C3′-H, C10-H), 5.98 (1H, *s*, -OH), 6.51 (6H, *m*, Ar-H), 6.88 (4H, *m*, Ar-H), 7.31 (6H, *m*, Ar-H), 7.74 (4H, *m*, Ar-H), 9.93 (1H, *br.s*, -NH$^+$); ^{13}CNMR: δ 21.22, 22.1, 42.6, 51.9, 53.4, 59.0, 62.28, 63.0, 65.1, 126.5, 126.7, 127.2, 127.8, 129.1, 129.5, 134.3, 139.7, 144.9, 145.1, 176.6, 176.9; HRMS (*m*/*z*): calculated value for [M + H$^+$] 609.7758; found 609.3117.

***4-[-2-hydroxy-3-(morpholin-4-yl)propyl]-1,7,8,9-tetraphenyl-4-azatricyclo[5.2.1.0^{2,6}]dec-8-ene-3,5-dione* (3g)**

M.W. = 610.7407 (free amine); $C_{40}H_{38}N_2O_3$*HCl; Yield: 64%; white powder, m.p. 276–277 °C (from MeOH/(Et)$_2$O); ^{1}HNMR (300 MHz, DMSO-d$_6$+ TMS, δ/ppm): 2.34 (1H, *m*, C6-H), 3.07 (3H, *m*, -CH$_2$-morpholin, C2-H), 3.29 (5H, *m*, -CH$_2$-morpholin, C10-H), 3.46 (3H, *m*, C3′-H, C2′-H), 3.73 (3H, *m*, C1′-H, C10-H), 3.92 (2H, *m*, -CH$_2$-morpholin), 6.48 (4H, *m*, Ar-H), 6.88 (6H, *m*, Ar-H), 7.31 (6H, *m*, Ar-H), 7.74 (4H, *m*, Ar-H), 10.45 (1H, *br.s*, -NH$^+$); ^{13}CNMR: δ 21.22.0, 50.7, 51.9, 52.7, 58.9, 61.9, 63.0, 65.0, 126.5, 126.8, 127.2, 127.8, 129.1, 129.5, 145.1, 176.6, 176.9; HRMS (*m*/*z*): calculated value for [M + H$^+$] 611.7486; found 611.2904.

4.1.9. Preparation of Acetate Salt of Derivatives **Ic**, **Ie**, **II**

Alkylamine derivatives were obtained according to the methods published previously [9,10]. Next, the appropriate derivative (0.01 mol) was dissolved in acetic acid (60%) and heated to boiling. After cooling, the excess acid was evaporated to give a crystalline precipitate.

Acetate of 4-[2-(piperidin-1-yl)ethyl]-1,7-diethyl-8,9-diphenyl-4-azatricyclo[5.2.1.0^{2,6}]-dec-8-ene-3,5,10-trione (Ic*CH$_3$COOH)

M.W. = 496.2720 (free amine); $C_{32}H_{36}N_2O_3$*CH$_3$COOH; Yield: 64%; white powder, m.p. 85–89 °C (from CH$_3$COOH); ^{1}HNMR (300 MHz, CDCl$_3$, δ/ppm): 0.99 (6H, *t*, -CH$_3$, *J* = 7.5 Hz), 1.46 (2H, *m*, -CH$_2$-), 1.62 (4H, *m*, -(CH$_2$)$_2$-), 2.02 (2H, *m*, -CH$_2$-piper), 2.19 (2H, m, -CH$_2$-), 2.67 (4H, m, -CH$_2$-piper), 3.48 (2H, *m*, C2-H, C6-H), 3.52 (2H, *m*, -CH$_2$-piper), 3.74 (2H, *m*, -CH$_2$-piper), 6.90 (4H, *m*, Ar-H), 7.14 (6H, *m*, Ar-H); ^{13}CNMR: δ 9.0, 18.9, 23.7, 25.2, 36.2, 43.5, 53.8, 55.2, 59.3, 127.4, 127.9, 129.1, 133.5, 141.5, 176.0, 199.8; HRMS (*m*/*z*): calculated value for [M + H$^+$] 497.2799; found 498.2799.

Acetate of 4-[2-hydroxy-3-(propan-2-ylamino)propyl]-1,7-diethyl-8,9-diphenyl-4-azatricyclo[5.2.1.0^{2,6}]dec-8-ene-3,5,10-trione (Ie * CH$_3$COOH)

M.W. = 500.2669 (free amine); $C_{31}H_{36}N_2O_4$*CH$_3$COOH; Yield: 77%; white powder, m.p. 156–158 °C (from CH$_3$COOH); ^{1}HNMR (300 MHz, CDCl$_3$, δ/ppm): 0.89 (6H, *t*, -CH$_3$, *J* = 7.5 Hz), 1.24 (6H, *d*, -CH$_3$, *J* = 6.3 Hz), 2.00 (2H, *m*, -CH2-), 2.19 (2H, *m*, -CH$_2$-), 2.80 (1H, *m*, C1′-H), 2.89 (1H, *m*, C1′-H), 3.11 (1H, *m*, -CH-), 3.57 (3H, *m*, C2′-H, C3′-H), 3.65 (2H, *m*, C2-H, C6-H), 4.17 (1H, *m*, -OH), 6.87 (4H, *m*, Ar-H), 7.17 (6H, *m*, Ar-H), ^{13}CNMR: δ 9.0, 18.9, 21.7, 43.5, 48.4, 50.0, 59.2, 65.7, 127.3, 127.9, 129.2, 133.5, 141.7, 172.7, 176.2, 198.9; HRMS (*m*/*z*): calculated value for [M + H$^+$] 501.2748; found 501.2748.

Acetate of 4-[2-(dimethylamino)ethyl]-1,7,8,9-tetraphenyl-4-azatricyclo[5.2.1.0^{2,6}]dec-8-ene-3,5-dione (II*CH$_3$COOH)

M.W. = 538.2614 (free amine); $C_{37}H_{34}N_2O_2$*CH$_3$COOH, Yield: 82%; white powder, m.p. 203–205 °C (from CH$_3$COOH); ^{1}HNMR (300 MHz, CDCl$_3$, δ/ppm): 2.32 (6H, *m*, -CH3), 2.35 (1H, *m*, C10-H), 2.65 (2H, *t*, *J* = 6.3, C1′-H), 3.17 (1H, *m*, C10-H), 3.64 (2H, *t*, *J* = 6.3, C2′-H), 4.18 (2H, *s*, C2-H, C6-H), 6.55 (4H, *m*, Ar-H), 6.92 (6H, *m*, Ar-H), 7.27 (6H, *m*, Ar-H), 7.72 (4H, *m*, Ar-H); ^{13}CNMR: δ 21.0, 36.5, 44.8, 51.8, 55.7, 63.0, 65.3, 126.4, 126.7, 127.1, 127.7,

129.1, 129.5, 134.3, 139.8, 144.9, 176.6; HRMS (m/z): calculated value for [M + H^+] 539.2693; found 539.2693.

4.1.10. Determination of Lipophilicity by Reversed-Phase Chromatography

The reversed-phase high-performance liquid chromatography (RP-HPLC) analysis was performed on a Merck–Hitachi LaChrom Elite System (Merck, Darmstadt, Germany) with diode array detector L-2455, thermostat L-2300, pump L-2130, and autosampler L-2200 and with the Waters XTerra MS RP-18 (3.5 µm, 150 × 4.6 mm) chromatographic column (Waters, Milford, MA, USA) as the stationary phase. The mobile phase consisted of methanol, water, and 0.1% (v/v) formic acid and was degassed by the use of the built-in membrane degasser. The methanol gradient grade for HPLC was purchased from Merck (Merck, Darmstadt, Germany) and formic acid from Polish Reagents (Polish Reagents, Gliwice, Poland); double-distilled water was used. Then, 5 µL of methanolic solutions (0.1%; m/v) of selected samples were applied to the chromatographic column by the use of an autosampler Hitachi L-2200 (LaChrom Elite, Hitachi-Merck, Darmstadt, Germany). The analysis was performed with the flow rate of 0.7 mL min^{-1} in isocratic mode using various concentrations of organic modifier (methanol) in binary polar mobile phases; percentages of methanol in water were 45–70% (%, v/v) and changed by 5% per step. Chromatograms were detected at 254 nm and the temperature of the column was 25 °C. All experiments were repeated in triplicate, and the final results were taken to be the arithmetic means. Dead time was measured by the use of uracil (Calbiochem. Merck, Darmstadt, Germany). Statistical and regression analyses were performed using Statistica (ver. 13.3 for Windows). The chromatographic lipophilicity parameters ($logk_W$) for selected samples were obtained by the extrapolation of the retention parameter $logk$ to pure water, according to Equation (1):

$$logk = logk_w - S{\cdot}\phi; \tag{1}$$

where $logk_W$ is the value of the retention factor of a substance in pure water, S is the slope of the regression curve, and ϕ is the concentration of the organic modifier [33]. The values of $logk$ were calculated based on the raw HPLC data using the formula (2):

$$logk = \log\left(\frac{t_R - t_0}{t_0}\right); \tag{2}$$

where t_R is the retention time and t_0 is the dead retention time (determined for uracil).

4.2. *Anticancer Activity*

4.2.1. Cells and Cytotoxicity Assay

Human umbilical vein endothelial cells (Life Technologies, Carlsbad, CA, USA) were cultured (according to the manufacturer's instructions) in Medium 200 containing low serum growth supplement; 1×10^4 cells were seeded on each well on a 96-well plate (Nunc, Roskilde, Denmark). The HeLa (human cervix carcinoma), K562, and MOLT-4 (leukemia) cells were cultured in RPMI 1640 medium supplemented with antibiotics and 10% fetal calf serum in a 5% CO_2–95% air atmosphere; 7×10^3 cells were seeded on each well on a 96-well plate (Nunc). Twenty-four hours later, cells were exposed to the test compounds for additional 48 hours. Stock solutions of test compounds were freshly prepared in DMSO. The final concentrations of the compounds tested in the cell cultures were 2×10^{-1}, 1×10^{-1}, 5×10^{-2}, 1×10^{-2}, 1×10^{-3}, and 1×10^{-4} mM. The concentration of DMSO in the cell culture medium was 1%. The values of IC_{50} (the concentration of test compound required to reduce the cell survival fraction to 50% of the control) were calculated from dose–response curves and used as a measure of cellular sensitivity to a given treatment. The cytotoxicity of all compounds was determined by the MTT [3-(4,5-dimethylthiazol-2-yl)-2,5-diphenyltetrazolium bromide; Sigma, St. Louis, MO, USA] assay. Briefly, after 24 or 48 h of incubation with drugs, the cells were treated with the MTT reagent, and incubation was continued for 2 h. MTT-formazan crystals were dissolved in 20% SDS and 50% DMF at pH 4.7, and absorbance was read at 570 and 650 nm on a microplate reader FLUOstar Omega (BMG Labtech, Offenburg, Germany). As a control (100% viability), we used cells grown in the presence of the vehicle (1% DMSO) only.

4.2.2. Analysis of Cell Apoptosis by Caspase-3/7 Activity Assay

First, 20×10^3 K562 cells were seeded on each well of a 96-well plate in RPMI 1640 medium supplemented with 10% fetal calf serum and antibiotics. Cells were grown for 24 h at 37 °C and 5% CO_2. Test dicarboximides were dissolved in DMSO and added to cell culture to the final concentration of $5 \times IC_{50}$. Cells treated with 1% DMSO served as a negative control, while cells incubated with staurosporine (Sigma, St. Louis, MO, USA) were used as a positive control. Cells were exposed to test compounds for 18 h at 37 °C and 5% CO_2. Subsequently, the activity of caspase 3 and 7 was measured by Apo-ONE® Homogeneous Caspase-3/7 Assay (Promega, Madison, WI, USA) according to the manufacturer's instructions. Briefly, the cells were lysed and incubated for 1.5 h with a profluorescent substrate for caspase 3 and 7. Next, the fluorescence was read at an excitation wavelength of 485 nm and emission of 520 nm with a FLUOStar Omega microplate reader (BMG-Labtech, Offenburg, Germany).

4.2.3. Gene Expression Analysis with DNA Microarray

The expression of apoptotic genes was studied using predesigned 384-well microfluidic cards (TLDA TaqMan®human apoptosis array, Life Technologies, Applied Biosystems, Foster City, CA, USA). The RT-PCR experiments were executed twice, each on three separate TLDA plates. K562 tumor cells were seeded on a six-well plate (Nunc) in an amount of 3×10^6 cells/well and were exposed to the test compounds at a concentration of $5 \times IC_{50}$ for another 18 h. The control cells were exposed to 1% DMSO or 1 μM staurosporine. The total RNA pool was isolated from the cell lysates using TriPure Isolation Reagent (Roche, Basel, Switzerland) according to the manufacturer's instruction. The RNA quality was checked with a NanoDrop ND-1000 spectrophotometer (Thermo Fisher Scientific, Waltham, MA, USA). Two micrograms of total RNA were reverse transcribed using a High Capacity cDNA Reverse Transcription kit (Applied Biosystems, Life Technologies). The 384-well TLDA (the qRT-PCR based TaqMan low-density array) cards were configured into four identical 96-gene sets consisting of 93 human genes and three endogenous controls: 18S, ACTB, and GAPDH. The primers (TaqMan Gene Expression Assay, Applied Biosystems) were selected to produce amplicons not longer than 100 bp (a mean of 74 bp). Those 93 genes were categorized into multiple classes including intrinsic, extrinsic, regulatory, and execution apoptosis traits. A reaction mixture with a cDNA template (100 ng) and an equal volume of TaqMan® universal master mix (Applied Biosystems) was immediately loaded into each line of the TLDA microfluidic card. Each card was spun twice (each time for 1 min) at 1200 rpm, then was sealed and loaded into an ABI 7900HT fast real-time PCR system (Applied Biosystems, Waltham, MA, USA), operating for 2 min at 50 °C, 10 min at 94.5 °C, 30 s at 97 °C, and 1 min at 59.7 °C (40 cycles). The results were statistically analyzed using the Student's *t*-test, assuming a level of significance value of $p < 0.05$. The MetaCoreTM software (Thomson Reuters, from GeneGo) was used to perform the pathway analysis of the differentially expressed genes.

4.2.4. MAPKs Immunoblotting

First, 2×10^6 K562 cells per well were seeded on a six-well plate in the complete medium (RPMI 1640 + 10% FBS). Then, 20 μM JNK kinase inhibitor SP600125 or 10 μM p38 inhibitor SB203580 (Cell Signaling Technology, Danvers, MA, USA) was added for 24 h. Alternatively, K562 cells were exposed to 1% DMSO (control sample), test compound **Ic** (6 μM), **II** (3 μM), or staurosporine (1 μM) for 18 h. Cells exposed to staurosporine served as a positive control in ERK1/2 experiments. K562 cells treated with 20 μM anisomycin (BioShop, Burlington, ON, Canada) for 30 min served as a positive control in JNK and p38 experiments. After incubation, cells were centrifuged (600 rpm, 6 min, 24 °C), washed once with PBS, and lysed in the buffer containing 20 mM Tris (pH 7.5), 150 mM NaCl, and 1% Triton X-100, containing protease (complete protease inhibitor cocktail, Roche) and phosphatase inhibitors (PhosSTOP phosphatase inhibitor cocktail, Roche). Cells were lysed for 10 min on ice and centrifuged (12,000 rpm, 10 min, 4 °C), and supernatants were collected. Total protein concentration in cell lysates was measured with DC Protein Assay (Bio-Rad, Hercules, CA, USA). Cell lysates containing 25 μg of total protein were mixed with Laemli buffer supplemented with SDS and β-Mercaptoethanol, denatured (10 min, 95 °C), and subjected to SDS-PAGE electrophoresis (4% stacking and 10% resolving gels). Resolved proteins were electrotransferred (semi-dry, 90 min, 1.5 mA/cm^2) to nitrocellulose membranes (0.45 μm, ThermoScientific). Membranes were blocked for 2 h in TBST buffer (20 mM Tris-HCl pH 7.5, 0.9% NaCl, 0.1% Tween 20) containing 5% BSA (BioShop) and probed with primary antibodies (overnight at 4 °C) diluted in TBST buffer with 0.5% BSA. Tubulin served as a loading control for IB experiments. Anti-phospho ERK1/2 (rabbit polyclonal, dilution 1:1000) and anti-ERK1/2 (rabbit polyclonal, dilution 1:1000) were from Cell Signaling Technology. Anti-phospho JNK (mouse polyclonal, dilution 1:250) and anti-JNK (mouse polyclonal, dilution 1:250) were from Santa Cruz Biotechnology (Dallas, TX, USA). Anti-phospho p38 (mouse polyclonal, dilution 1:250) and anti-p38 (mouse polyclonal, dilution 1:250) were from BD Transduction Laboratories (San Diego, CA, USA). Anti-α-tubulin (mouse monoclonal IgG, 1:2000) was from Sigma. After extensive washing with the TBST buffer, membranes were probed with secondary goat anti-rabbit (1:5000 in TBST buffer) or goat anti-mouse HRP-conjugated IgG (1:5000 in TBST buffer, from Santa Cruz Biotechnology) for 45 min at room temperature. Membranes were washed in the TBST buffer, and the chemiluminescent signal was developed with ECL Plus Western Blotting Substrate (Thermo Scientific, Waltham, MA, USA) and visualized in a Syngene G:Box detection system.

4.2.5. Digestion of Plasmid DNA with BamHI Restriction Nuclease

First, 0.5 μg of plasmid DNA (pcDNAHisC, total length 5.5 kbp) containing a unique *BamHI* restriction site was dissolved in a 1× *BamHI* reaction buffer and incubated overnight at 37 °C with the test compounds or daunorubicin. Daunorubicin, a strong intercalating agent, was used as a positive control. The concentration of the test compounds and daunorubicin samples was 100 μM. In the next step, the reaction mixtures were digested with *BamHI* restriction endonuclease (2 U/μL) for 3 h at 37 °C. The total reaction volume was 10 μL. Products of the reaction were subjected to the 1% agarose gel electrophoresis in TBE buffer. The gel was stained with ethidium bromide, and DNA fragments were visualized under a UV lamp (GBox, Syngene).

Supplementary Materials: Supplementary Materials are available online at https://www.mdpi.com/article/10.3390/ijms22094318/s1.

Author Contributions: Conceptualization: M.C., M.N., and B.N.; methodology: M.C. and M.N.; validation: M.C., M.N., and I.W.; investigation: M.C., M.N., J.K.-B., K.K.-G., I.W., and A.H.; resources: M.C., M.N., B.N., and I.W.; data curation: M.C., M.N., and I.W.; writing—original draft preparation: M.C. and M.N.; writing—review and editing: M.C., M.N., and B.N.; visualization: M.C. and M.N.; supervision: M.C., M.N., and B.N.; project administration: M.C., M.N., and B.N.; funding acquisition: M.N. and B.N. All authors have read and agreed to the published version of the manuscript.

Funding: This research was funded by the National Science Centre in Poland, project number UMO-2014/15/B/NZ7/00966 (for B.N. and M.N.), by statutory funds of the Centre of Molecular and Macromolecular Studies, Polish Academy of Sciences, and by Warsaw Medical University.

Institutional Review Board Statement: Not applicable.

Informed Consent Statement: Not applicable.

Data Availability Statement: Not applicable.

Acknowledgments: The cytotoxicity, activation of caspases, and DNA interaction studies were performed in the Screening Laboratory at the Department of Bioorganic Chemistry, Centre of Molecular and Macromolecular Studies of the Polish Academy of Sciences (CMMS PAS). The cost of this publication was covered in part by the statutory funds of CMMS PAS. The authors wish to thank Markus Düchler for revising the manuscript.

Conflicts of Interest: The authors declare no conflict of interest.

References

1. Didkowska, J.; Wojciechowska, U.; Zatoński, W. Prediction of Cancer Incidence and Mortality in Poland Up to the Year 2025. Warszawa 2009, ISSN 0867-8251. Available online: http://onkologia.org.pl/wp-content/uploads/Prognozy_2025.pdf (accessed on 20 April 2021).
2. Bray, F.; Ferlay, J.; Soerjomataram, I.; Siegel, R.L.; Torre, L.A.; Jemal, A. Global cancer statistics 2018: GLOBOCAN estimates of incidence and mortality worldwide for 36 cancers in 185 countries. *CA Cancer J. Clin.* **2018**, *68*, 394–424. [CrossRef] [PubMed]
3. Hall, I.H.; Wong, O.T.; Scovill, J.P. The cytotoxicity of N-pyridinyl and N-quinolinyl substituted derivatives of phthalimide and succinimide. *Biomed. Pharmacother.* **1995**, *49*, 251–258. [CrossRef]
4. Luzina, E.L.; Popov, A.V. Synthesis and anticancer activity evaluation of 3,4-mono- and bicyclosubstituted *N*-(het)aryl trifluoromethyl succinimides. *J. Fluor. Chem.* **2014**, *168*, 121–127. [CrossRef] [PubMed]
5. Lee, J.-Y.; Chung, T.-W.; Choi, H.-J.; Lee, C.H.; Eun, J.S.; Han, Y.T.; Choi, J.-Y.; Kim, S.-Y.; Han, C.-W.; Jeong, H.-S.; et al. A novel cantharidin analog *N*-Benzylcantharidinamide reduces the expression of MMP-9 and invasive potentials of Hep3B via inhibiting cytosolic translocation of HuR. *Biochem. Biophys. Res. Commun.* **2014**, *447*, 371–377. [CrossRef]
6. Kok, S.H.L.; Gambari, R.; Chui, C.H.; Yuen, M.C.W.; Lin, E.; Wong, R.S.M.; Lau, F.Y.; Cheng, G.Y.M.; Lam, W.S.; Chan, A.S.C. Synthesis and anti-cancer activity of benzothiazole containing phthalimide on human carcinoma cell lines. *Bioorg. Med. Chem.* **2008**, *16*, 3626–3631. [CrossRef]
7. Cardoso, M.V.D.O.; Moreira, D.R.M.; Filho, G.B.O.; Cavalcanti, S.M.T.; Coelho, L.C.D.; Espíndola, J.W.P.; Gonzalez, L.R.; Rabello, M.M.; Hernandes, M.Z.; Ferreira, P.M.P.; et al. Design, synthesis and structure–activity relationship of phthalimides endowed with dual antiproliferative and immunomodulatory activities. *Eur. J. Med. Chem.* **2015**, *96*, 491–503. [CrossRef]
8. Millrine, D.; Kishimoto, T. A Brighter Side to Thalidomide: Its Potential Use in Immunological Disorders. *Trends Mol. Med.* **2017**, *23*, 348–361. [CrossRef]
9. Kuran, B.; Napiórkowska, M.; Kossakowski, J.; Cieślak, M.; Kaźmierczak-Barańska, J.; Królewska, K.; Nawrot, B. New, Substituted Derivatives of Dicarboximides and their Cytotoxic Properties. *Anti-Cancer Agents Med. Chem.* **2016**, *16*, 852–864. [CrossRef]
10. Kuran, B.; Krawiecka, M.; Kossakowski, J.; Cieslak, M.; Kazmierczak-Baranska, J.; Królewska, K.; Nawrot, B. Dicarboximides Derivatives for Use in the Treatment of Cancer. European Patent Application EP13176421, 13 July 2013.
11. Cieślak, M.; Kaźmierczak-Barańska, J.; Królewska-Golińska, K.; Napiórkowska, M.; Stukan, I.; Wojda, U.; Nawrot, B. New Thalidomide-Resembling Dicarboximides Target ABC50 Protein and Show Antileukemic and Immunomodulatory Activities. *Biomolecules* **2019**, *9*, 446. [CrossRef]
12. Budzisz, E.; Krajewska, U.; Rozalski, M.; Szulawska, A.; Czyz, M.; Nawrot, B. Biological evaluation of novel Pt(II) and Pd(II) complexes with pyrazole-containing ligands. *Eur. J. Pharmacol.* **2004**, *502*, 59–65. [CrossRef]
13. Xu, G.; He, Q.; Yang, B.; Hu, Y. Synthesis and Antitumor Activity of Novel 4-Chloro-3-Arylmaleimide Derivatives. *Lett. Drug Des. Discov.* **2009**, *6*, 51–55. [CrossRef]
14. Hsiang, Y.H.; Jiang, J.B.; Liu, L.F. Topoisomerase II-mediated DNA cleavage by amonafide and its structural analogs. *Mol. Pharmacol.* **1989**, *36*, 2550774.
15. Lipinski, C.A.; Lombardo, F.; Dominy, B.W.; Feeney, P.J. Experimental and computational approaches to estimate solubility and permeability in drug discovery and development settings. *Adv. Drug Deliv. Rev.* **2001**, *46*. [CrossRef]
16. Leeson, P.D.; Springthorpe, B. The influence of drug-like concepts on decision-making in medicinal chemistry. *Nat. Rev. Drug Discov.* **2007**, *6*, 881–890. [CrossRef]
17. Jan, R.; Chaudhry, G.-E.-S. Understanding Apoptosis and Apoptotic Pathways Targeted Cancer Therapeutics. *Adv. Pharm. Bull.* **2019**, *9*, 205–218. [CrossRef]
18. Manns, J.; Daubrawa, M.; Driessen, S.; Paasch, F.; Hoffmann, N.; Löffler, A.; Lauber, K.; Dieterle, A.; Alers, S.; Iftner, T.; et al. Triggering of a novel intrinsic apoptosis pathway by the kinase inhibitor staurosporine: Activation of caspase-9 in the absence of Apaf. *FASEB J.* **2011**, *25*, 3250–3261. [CrossRef]
19. Kelley, S.K.; Ashkenazi, A. Targeting death receptors in cancer with Apo2L/TRAIL. *Curr. Opin. Pharmacol.* **2004**, *4*, 333–339. [CrossRef]
20. Su, L.; Liu, G.; Hao, X.; Zhong, N.; Zhong, D.; Liu, X.; Singhal, S. Death Receptor 5 and cellular FLICE-inhibitory protein regulate pemetrexed-induced apoptosis in human lung cancer cells. *Eur. J. Cancer* **2011**, *47*, 2471–2478. [CrossRef]
21. Amin, P.; Florez, M.; Najafov, A.; Pan, H.; Geng, J.; Ofengeim, D.; Dziedzic, S.A.; Wang, H.; Barrett, V.J.; Ito, Y.; et al. Regulation of a distinct activated RIPK1 intermediate bridging complex I and complex II in TNFα-mediated apoptosis. *Proc. Natl. Acad. Sci. USA* **2018**, *115*, E5944–E5953. [CrossRef]
22. Petrie, E.J.; Czabotar, P.E.; Murphy, J.M. The Structural Basis of Necroptotic Cell Death Signaling. *Trends Biochem. Sci.* **2019**, *44*, 53–63. [CrossRef]

23. Westphal, D.; Dewson, G.; Czabotar, P.E.; Kluck, R.M. Molecular biology of Bax and Bak activation and action. *Biochim. Biophys. Acta (BBA) Mol. Cell Res.* **2011**, *1813*, 521–531. [CrossRef]
24. Han, J.-W.; Flemington, C.; Houghton, A.B.; Gu, Z.; Zambetti, G.P.; Lutz, R.J.; Zhu, L.; Chittenden, T. Expression of bbc3, a pro-apoptotic BH3-only gene, is regulated by diverse cell death and survival signals. *Proc. Natl. Acad. Sci. USA* **2001**, *98*, 11318–11323. [CrossRef]
25. Li, Z.-W.; Chu, W.; Hu, Y.; Delhase, M.; Deerinck, T.J.; Ellisman, M.H.; Johnson, R.S.; Karin, M. The IKKβ Subunit of IκB Kinase (IKK) is Essential for Nuclear Factor κB Activation and Prevention of Apoptosis. *J. Exp. Med.* **1999**, *189*, 1839–1845. [CrossRef]
26. Makris, C.; Godfrey, V.L.; Krähn-Senftleben, G.; Takahashi, T.; Roberts, J.L.; Schwarz, T.; Feng, L.; Johnson, R.S.; Karin, M. Female Mice Heterozygous for IKKγ/NEMO Deficiencies Develop a Dermatopathy Similar to the Human X-Linked Disorder Incontinentia Pigmenti. *Mol. Cell* **2000**, *5*, 969–979. [CrossRef]
27. Meloche, S.; Pouysségur, J. The ERK1/2 mitogen-activated protein kinase pathway as a master regulator of the G1- to S-phase transition. *Oncogene* **2007**, *26*, 3227–3239. [CrossRef]
28. Dhanasekaran, D.N.; Reddy, E.P. JNK signaling in apoptosis. *Oncogene* **2008**, *27*, 6245–6251. [CrossRef]
29. Thornton, T.M.; Rincon, M. Non-Classical P38 Map Kinase Functions: Cell Cycle Checkpoints and Survival. *Int. J. Biol. Sci.* **2009**, *5*, 44–52. [CrossRef]
30. Cuenda, A.; Rousseau, S. p38 MAP-Kinases pathway regulation, function and role in human diseases. *Biochim. Biophys. Acta (BBA) Mol. Cell Res.* **2007**, *1773*, 1358–1375. [CrossRef]
31. Eckert, F.; Klamt, A. Accurate prediction of basicity in aqueous solution with COSMO-RS. *J. Comput. Chem.* **2005**, *27*, 11–19. [CrossRef]
32. Kuran, B.; Krawiecka, M.; Kossakowski, J.; Koronkiewicz, M.; Chilmonczyk, Z. Proapoptotic activity of heterocyclic compounds containing succinimide moiety in the promyelocytic leukemia cell line HL-60. *Acta Pol. Pharm. Drug Res.* **2013**, *70*, 459–468.
33. Snyder, L.R.; Dolan, J.W.; Gant, J.R. Gradient Elution in High performance Liquid Chromatography: II. Practical Application to Reversed-phase Systems. *J. Chromatogr. A* **1979**, *165*, 31–58. [CrossRef]

International Journal of *Molecular Sciences*

Review

Profiling Colorectal Cancer in the Landscape Personalized Testing—Advantages of Liquid Biopsy

Donatella Verbanac [1,*], Andrea Čeri [1], Iva Hlapčić [1], Mehdi Shakibaei [2], Aranka Brockmueller [2], Božo Krušlin [3,4], Neven Ljubičić [3,5,6], Neven Baršić [3,5], Dijana Detel [7], Lara Batičić [7], Lada Rumora [1], Anita Somborac-Bačura [1], Mario Štefanović [1,8], Ivana Ćelap [8], Alma Demirović [4], Roberta Petlevski [1], József Petrik [1], Marija Grdić Rajković [1], Andrea Hulina-Tomašković [1], Ivana Rako [9], Luciano Saso [10] and Karmela Barišić [1]

1 Department of Medical Biochemistry and Hematology, Faculty of Pharmacy and Biochemistry, University of Zagreb, Ante Kovačića 1, 10000 Zagreb, Croatia; andrea.ceri@pharma.unizg.hr (A.Č.); iva.hlapcic@pharma.unizg.hr (I.H.); lada.rumora@pharma.unizg.hr (L.R.); asomborac@pharma.unizg.hr (A.S.-B.); mario.stefanovic@kbcsm.hr (M.Š.); roberta.petlevski@pharma.unizg.hr (R.P.); jozsef.petrik@pharma.unizg.hr (J.P.); mgrdic@pharma.unizg.hr (M.G.R.); andrea.hulina@pharma.unizg.hr (A.H.-T.); karmela.barisic@pharma.unizg.hr (K.B.)
2 Musculoskeletal Research Group and Tumour Biology, Faculty of Medicine, Institute of Anatomy, Ludwig-Maximilian-University Munich, Pettenkoferstrasse 11, D-80336 Munich, Germany; mehdi.shakibaei@med.uni-muenchen.de (M.S.); Aranka.Brockmueller@med.uni-muenchen.de (A.B.)
3 School of Medicine, University of Zagreb, Šalata 3, 10000 Zagreb, Croatia; bozo.kruslin@mef.hr (B.K.); neven.ljubicic@kbcsm.hr (N.L.); neven.barsic@gmail.com (N.B.)
4 Department of Pathology and Cytology "Ljudevit Jurak", University Hospital Centre "Sestre milosrdnice", University of Zagreb, Vinogradska 29, 10000 Zagreb, Croatia; alma.demirovic@kbcsm.hr
5 Department of Internal Medicine, University Hospital Centre "Sestre milosrdnice", Division of Gastroenterology and Hepatology, University of Zagreb, Vinogradska 29, 10000 Zagreb, Croatia
6 School of Dental Medicine, Gundulićeva 5, 10000 Zagreb, Croatia
7 Department of Medical Chemistry, Biochemistry and Clinical Chemistry, Faculty of Medicine, Braće Branchetta 20/1, 51000 Rijeka, Croatia; dijana.detel@medri.uniri.hr (D.D.); lara.baticic@medri.uniri.hr (L.B.)
8 Department of Clinical Chemistry, University Hospital Centre "Sestre milosrdnice", University of Zagreb, Vinogradska 29, 10000 Zagreb, Croatia; ivana.celap@gmail.com
9 Department of Laboratory Diagnostics, University Hospital Centre Zagreb, University of Zagreb, Kišpatićeva 12, 10000 Zagreb, Croatia; ivana.rako@kbc-zagreb.hr
10 Department of Physiology and Pharmacology "Vittorio Erspamer", Sapienza University of Rome, Piazzale Aldo Moro 5, 00185 Roma, Italy; luciano.saso@uniroma1.it
* Correspondence: donatella.verbanac@pharma.unizg.hr

Citation: Verbanac, D.; Čeri, A.; Hlapčić, I.; Shakibaei, M.; Brockmueller, A.; Krušlin, B.; Ljubičić, N.; Baršić, N.; Detel, D.; Batičić, L.; et al. Profiling Colorectal Cancer in the Landscape Personalized Testing—Advantages of Liquid Biopsy. *Int. J. Mol. Sci.* **2021**, 22, 4327. https://doi.org/10.3390/ijms22094327

Academic Editor: Angela Stefanachi

Received: 26 March 2021
Accepted: 18 April 2021
Published: 21 April 2021

Abstract: Drug-specific therapeutic approaches for colorectal cancer (CRC) have contributed to significant improvements in patient health. Nevertheless, there is still a great need to improve the personalization of treatments based on genetic and epigenetic tumor profiles to maximize the quality and efficacy while limiting cytotoxicity. Currently, CEA and CA 19-9 are the only validated blood biomarkers in clinical practice. For this reason, laboratories are trying to identify new specific prognostics and, more importantly, predictive biomarkers for CRC patient profiling. Thus, the unique landscape of personalized biomarker data should have a clinical impact on CRC treatment strategies and molecular genetic screening tests should become the standard method for diagnosing CRC. This review concentrates on recent molecular testing in CRC and discusses the potential modifications in CRC assay methodology with the upcoming clinical application of novel genomic approaches. While mechanisms for analyzing circulating tumor DNA have been proven too inaccurate, detecting and analyzing circulating tumor cells and protein analysis of exosomes represent more promising options. Blood liquid biopsy offers good prospects for the future if the results align with pathologists' tissue analyses. Overall, early detection, accurate diagnosis and treatment monitoring for CRC with specific markers and targeted molecular testing may benefit many patients.

Keywords: biomarkers; colorectal cancer; early detection examination; liquid biopsy; personalized medicine; tumor treatment; exosomes; ctDNA; CTC

1. Introduction

Colorectal cancer (CRC) represents the second most common cause of cancer-related death globally [1], with annual incidence approaching two million cases worldwide [2] (Figure 1). Moreover, CRC incidence is rising in low-income and middle-income countries [1]. The disease results from the accumulation of multiple genetic and epigenetic modifications that lead to the transformation of colonic epithelial cells into invasive and aggressive adenocarcinomas [3,4]. The lack of and inadequate response to numerous mono-target therapies in cancer treatments emphasizes that personalized diagnostic and therapeutic approaches are necessary for effective strategies that target not only tumor cells, but more importantly, the multicellular tumor microenvironment for improved patient outcomes. Nevertheless, one of the most important keys to successful treatment of this malignant tumor and patient survival is not only the early diagnosis of the disease but also controlling tumor dissemination and progression [5]. For example, the 5-year survival rate for patients with early diagnosis is approximately 90%. In contrast, the survival rate for patients with regional lymph node metastasis is around 70%, and for those with distant metastases it is only 13% [6,7].

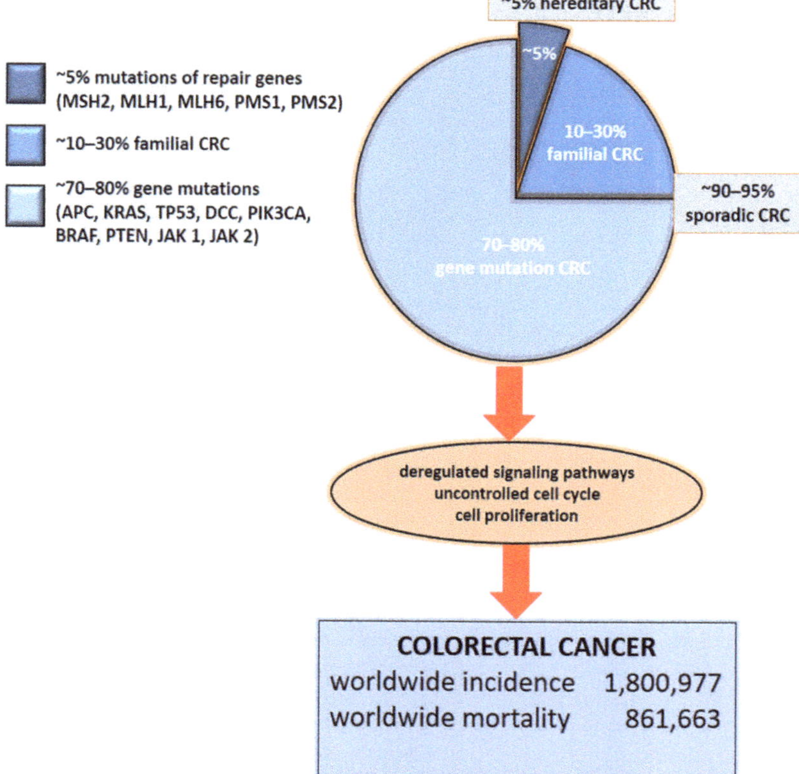

Figure 1. Molecular basis of CRC. Colorectal cancer is based on gene mutations, familial or hereditary CRC. The indication of total numbers refers to Global Cancer Statistics 2018 [3]. For worldwide incidence and mortality, colorectal cancer cases from 185 countries in 2018 were totaled.

Other important keys to improving CRC therapies are enhancements in surgical modalities and adjuvant chemotherapy, which has increased the cure rates in early-stage disease. Still, unfortunately, a significant proportion of patients will develop recurrence or advanced illness. Nevertheless, the efficacy of chemotherapy for recurrence and advanced CRC has improved significantly over the last decade. Previously, the historical drug 5-fluorouracil was the only chemotherapeutic agent used. With the addition of other chemotherapeutic agents such as capecitabine, irinotecan, oxaliplatin, bevacizumab, cetuximab, panitumumab, vemurafenib, and dabrafenib, the median survival of patients with oligometastatic CRC has improved significantly from less than one year to the current standard of nearly two years [8]. However, many side effects of systemic therapy, such as toxicity, may cause fatal complications and significantly affect the patients' quality of life. In parallel, a plethora of biologically active compounds are tested in vitro and in vivo and promising hits/leads compounds that may be used in the development as adjunct to the therapy are continuously identified [9,10]. An overview of existing CRC-targeted agents and their underlying mechanisms, as well as a discussion of their limitations and future trends, has been published recently [11]. Still there is an urgent need for crucial biomarkers to select optimal drugs individually or in combination for an individual patient. The application of personalized therapy based on DNA testing could help clinicians provide the most effective chemotherapy agents and dose modifications for each patient. Yet, some of the current findings are controversial, and the evidence is conflicting [12]. The current trend is to achieve successful personalized therapeutic approaches based on monitoring of disease-specific biomarker(s). However, the data in this respect is scarce and studies which include the personalized testing vs treatment are needed. The aim of our ongoing translational research is to contribute to this unmet medical need.

2. Etiology of Colorectal Cancer

The etiology of CRC is extensively described in the literature but is still not known in detail [13]. In this review, we describe specifically the CRC-related genes and pathways, while knowing that they often overlap with other solid tumors, such breast and prostate cancer. In approximately 70–90% of patients, CRC develops sporadically due to point mutations of the *APC*, *KRAS*, *TP53*, and *DCC* genes.

In approximately 1–5% of cases, it is a consequence of a hereditary polypoid and non-polypoid syndrome and 10–30% of patients have a familial CRC [14] (Figure 1). It is important to note that 1–2% of CRC has been associated with chronic inflammatory conditions such as ulcerative colitis and Crohn's disease. The risk increases with the longer duration of ongoing inflammation [15] caused also by the dysbiosis in the gut [16] and inappropriate nutrition patterns and deteriorating life-style conditions [17]. Chromosomal instability (CIN), as an essential molecular pathway of malignant transformation, mainly affects genes such as *APC*, *KRAS*, *PIK3CA*, and *TP53* [18]. In addition, the adenoma–carcinoma sequence offers potential for screening and surveillance; e.g., connexin 43 expression in colonic adenomas is linked with high-grade dysplasia and colonic mucosa surrounding adenomas [19]. *APC* mutations lead to nuclear beta-catenin translocations and the transcription of genes participating in carcinogenesis and invasion processes. *KRAS* and *PIK3CA* mutations lead to continuous activation of mitogen-activated protein kinase (MAPK) pathways, which, in turn, increases cell proliferation, whereas *TP53* mutations lead to the loss of the p53 function and uncontrolled cell cycle [18]. Finally, epigenetic instability (CIMP) is associated with hypermethylation of the oncogenes promotor and loss of expression of the corresponding proteins [20].

3. The impact of Genetic Alterations on Disease Outcome

The most common mutations, chromosomal alterations, and translocations affect critical wingless-related integration sites (WNT), MAPK/phosphatidylinositol 3-kinase (PI3K), and transforming growth factor β (TGF-β) signaling pathways and intracellular protein functions such as p53, as well as cell cycle regulation [21]. The WNT pathway, which is a critical mediator of tissue homeostasis and repair, is frequently co-opted during tumor development. Almost all colorectal cancers demonstrate hyper-activation of the WNT pathway, which is considered to be the initiating and driving event in many cases [22]. *APC* gene mutations represent the most significant genetic change associated with the WNT signaling pathway, regulating stem cell differentiation and cell growth. Nevertheless, they do not represent a good predictor of the disease progression due to their high CRC frequency and the number of various mutations identified within the gene [23]. Increased β-catenin expression associated with the WNT signaling pathway has also been recognized as a non-reliable marker for disease prognosis. In contrast, overexpression of the *c-MYC* gene triggered by the activation of WNT signaling pathway represents a good predictor of metastasis and disease progression [24,25]. *KRAS*, *BRAF*, and *PIK3CA* mutations are common and are associated with the MAPK/PI3K signaling pathways.

Furthermore, mutations of the *KRAS* gene in exon 2, codon 13 are associated with poor prognosis and a low survival rate, while mutations in exon 2 and codon 12 are associated with tumor progression and metastasis [26,27]. Recently, AMG 510, the first *KRAS* G12C inhibitor, after promising preclinical results, has entered into the clinical development [28]. This represent fascinating efforts that could overcome the perception that *KRAS* is in principle "undruggable" as a therapeutic target and may contribute to the development of effective drugs for targeting traditionally difficult signaling pathways in the clinical setting [29]. *BRAF* gene mutations are associated with poor prognosis and survival [30–32].

The associations between disease outcome or survival and *PIK3CA* mutations have not yet been established. Still, evidence supports that these mutations, combined with the *KRAS* gene mutations, are associated with poor outcomes [33]. Furthermore, CRC patients with multiple *PIK3CA* mutations, e.g., a combination of mutations in exons 9 and 20, have a poorer prognosis than patients with only one of these mutations [34]. Protein phosphatase and tensin homolog (PTEN) adversely affects the PI3K signaling pathway and CRC in which the loss of the *PTEN* gene has been associated with a poor prognosis [35]. In CRC patients, changes in the TGF-β signaling pathway are associated with CIN [36]. Chromosome 18q is bearing the tumor suppressor genes *SMAD2* and *SMAD4* and their encoded proteins are functionally associated with apoptosis and cell cycle regulation [37,38]. Likewise, they play a role in tumor cell migration by regulating the activity of proteins such as matrix metallopeptidase 9 (MMP9) [39]. A significant association between the loss of chromosome 18q and poor prognosis and survival has not been found [37,38]. In CRC, the loss of the 17q-*TP53* gene, which encodes a tumor suppressor protein p53 that regulates the cell cycle, is quite common. Without it, cells proliferate uncontrollably and tumor progresses [40]. Janus kinases, JAK1 and JAK2, are associated with cytokine receptors [41,42], and cytokine binding leads to their activation and phosphorylation. Afterwards, Janus kinases phosphorylate signal transducer and activator of transcription (STAT) proteins, leading to their translocation to the nucleus and transcription of their target genes [41,42]. There is evidence of JAK1 and JAK2 gene mutations that inhibit the function of the corresponding JAK1 and JAK2 proteins. The JAK1 frameshift mutations (positions 142/143, 430/431, and 860/861) have been described as the consequence of insertion/deletion of one nucleotide [42]. The V617E mutation leads to the JAK2 loss-of-function mutation [41]. These mutations have been found in tumors with high MSI resulting from dysfunctional DNA repair during replication, known as mismatched repair [43]. Indeed, they were associated with tumor resistance to treatment targeting the programmed cell death protein 1 (PD-1) [43]. Determination of molecular changes at the DNA level, particularly derived from tumor-specific liquid components such as circulating tumor DNA (ctDNA), exosomes and circulating tumor cells (CTC), can improve prediction of disease development and help in adjustment of therapy for each patient individually, as part of personalized health care.

4. Liquid Biopsy

Much has been learned about the molecular background of CRC development and progression, which may help to tailor therapy for each patient and improve their survival prognosis. However, advances in early CRC diagnosis have not been made as far as expected, and still rely on biomarkers from readily available biological materials (Table 1). Currently, there are only two validated protein-based blood biomarkers used in routine clinical practice: carcinoembryonic antigen (CEA) and carbohydrate antigen 19-9 (CA 19-9). CEA is an embryo-specific glycoprotein that can also be found in CRC. In clinical practice, it is used to monitor the tumor's progression after its diagnosis [44]. However, it shows an insufficient sensitivity and specificity since it is hereditarily determined and in the case of recessive homozygote, the levels of CA 19-9 would not be increased (approximately in 15% of individuals) [45–47]. During the search for new biomarkers that would replace the old ones, a promising non-invasive, and a repeatable procedure called "liquid biopsy" was developed for different body fluids (blood, saliva and urine). Peripheral blood liquid biopsy is used for diagnostic screening, as well as for determining a response to therapy and evaluating the outcome of the disease [48]. Peripheral blood can contain CTC, ctDNA and exosomes (vesicular structures, which contain proteins and RNA molecules, that may be released into circulation by different cells, including tumor cells) [49]. This could make it possible to determine the molecular profile of the disease, the degree of affected tissue, and the response to therapy in a non-invasive way. The founder and establisher of new principles and methods of healing, Leroy Hood, has relentlessly emphasized that in the new era of personalized approaches undertaken while assessing different conditions and diseases, "the blood becomes a window through which we observe what is happening in the body" [50]. The same idea was accepted and maintained by other biomedical disciplines, from genetics to personalized nutrition [51]. Future molecular profiling, ideally assessed and monitored by liquid biopsy, might personalize decision-making even more in CRC patients' adjuvant scenery [52,53].

However, liquid biopsy results need to be combined and evaluated with the tissue's pathological findings before final validation of the proposed approach. The existing testing landscape presents additional challenges in the application of liquid biopsy in clinical practice, and consideration needs to be given to how the pathologist should be involved in interpreting liquid biopsy data in the context of the patient's cancer diagnosis and stage assessment [54].

Table 1. Biomarker usage related to CRC.

Biomarker	Signification	Structure	Experience/Implication	Reference
CEA	carcinoembryonic antigen	glycoprotein	Validated blood biomarker in clinical practice. Not recommended as sole CRC screening test. Preoperative CEA > 5 mg/mL may correlate with poorer CRC prognosis. Used as postoperative serum testing and monitoring during active CRC treatment every 3 months.Diagnostic sensitivity 54.5%; specificity 98.4%.	Locker et al.,2006 [44] Wu et al., 2020 [55]
CA 19-9	carbohydrate antigen	glycoprotein	Validated blood biomarker in clinical practice. Not recommended as sole screening or monitoring CRC marker. Used as supplementary progress monitoring during pancreatic cancer treatment every 1–3 months. Individual values for each patient. Diagnostic sensitivity 64.4%; specificity 96.8%.	Locker et al., 2006 [44] Wu et al., 2020 [55]

Table 1. *Cont.*

Biomarker	Signification	Structure	Experience/Implication	Reference
CTC	circulating tumor cells	tumor cells	Epithelial marker in the peripheral blood via automatic detection system. Detected in different cancer types. 1-10 CTCs per ml blood were found in patients with metastases but rarely in healthy people. Poor prognosis for CRC patients with ≥ 5 CTC per 7.5 ml blood. Diagnostic sensitivity 62.7%; specificity 82.0%. Multivariate analysis of the disease-free survival data of examined patient group showed that a CTC count ≥ 5 was an independent prognostic factor of distant metastasis (Hazard ratio = 7.5, 95% CI: 1.6 to 34.7, $p = 0.01$).	Dominguez-Vigil et al., 2018 [49] Tsai et al., 2016 [56]
ctDNA	circulating tumor DNA	small DNA fragments released by tumor cells	Tumor mutation search in the peripheral blood, plasma and serum. Patients with 100 g tumor burden released 3.3% of ctDNA into circulation. In CRC, ctDNA is more sensitive than CEA. KRAS mutations were detected with 87.2% sensitivity and a 99.2% specificity.	Osumi et al., 2020 [57] Said et al., 2020 [58] Dominguez-Vigil et al., 2018 [49]
exosomes	nanovesicles	vesicular structures released by different cell types, including tumors	Tumor miRNA molecules in biological fluid like blood and urine. Associated with several types of CRC. Each tumor is characterized by specific protein profile. Positive correlation from miRNA exosomes and proteins with the stage of tumor progression.	Dominguez-Vigil et al., 2018 [49] Wang et al., 2016 [59]

5. Circulating Tumor DNA (ctDNA)

Tumor cells release small DNA fragments through various mechanisms, including apoptosis, necrosis and active secretion from tumor cells [60]. These single- or double-stranded DNA fragments in the circulation may contain cancer-related gene mutations such as point mutations, copy number variations, chromosomal rearrangements, and DNA methylation. ctDNA reflects the genetic and epigenetic properties of the genomic DNA in tumor cells [61]. Therefore, identical variants present in the genomic DNA of tumor cells, such as mutations in *KRAS*, *NRAS*, *BRAF*, and other genes, can be identified in ctDNA. Furthermore, due to a relatively short half-life and therefore a rapid turnover in the circulation, ctDNA is considered a real-time biomarker of mutation dynamics and tumor burden [57]. The concentration of ctDNA is not related to a specific type of tumor, its size, or stage of progression, although there are reports of higher ctDNA levels in patients with advanced disease and distant metastases [58]. ctDNA can represent between 0.01% and 90% of total cell-free circulating DNA (cfDNA) [61,62]. The concordance between cfDNA within liquid biopsy and genomic DNA within tumor tissue biopsy is still under debate. Kang et al. compared somatic mutations of the 10 genes between cfDNA and genomic DNA from CRC metastatic tumor tissues and observed an overall 93% concordance rate between the two types of samples [63].

On the other hand, there is evidence that some types of tumor, like gliomas or sarcomas, are not good shedders of ctDNA, although the reason is still unclear [64]. Numerous methods for the analysis of free, non-cellular, circulating DNA in the diagnosis of tumors have been used (Figure 2) [65,66]. However, in clinical practice, their value has not been fully known and accepted yet. Additional studies and data are needed to evaluate the potential and significance of ctDNA further and move it into the clinical mainstream.

Currently, several clinical trials are ongoing [67]. Our goal is to point out the importance and provide additional justification to include such procedures in clinical analyses. We hope that in the near future, a faster, non-invasive, more timely, less costly diagnosis of CRC and other malignant tumors will be possible.

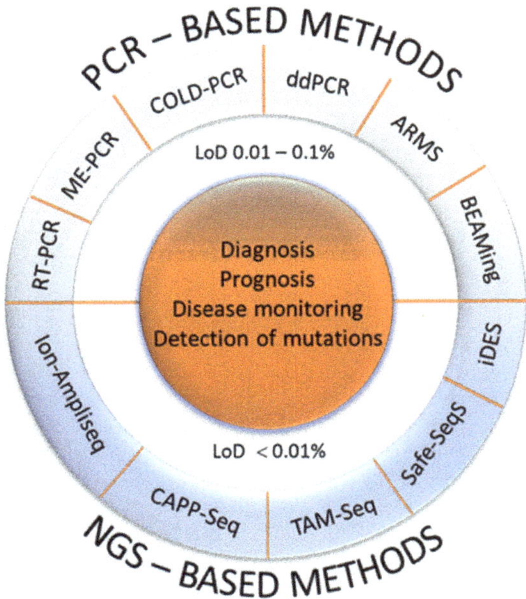

Figure 2. Current methods for identification and analysis of ctDNA. Based on the technologies which are used for ctDNA analysis, PCR- and NGS-based methods emerged. In general, PCR-based methods are cost-effective, rapid and no specific informatic skills are needed but the main disadvantage is that they can detect a limited number of known mutations. On the other hand, NGS is based on the analysis of several millions of short DNA sequences in parallel, followed by either sequence alignments to a reference genome or a de novo sequence assembly. Therefore, the NGS-based methods are expensive and time-consuming, but can detect a large number of mutations. Furthermore, according to the methodological approaches and analytical sensitivity, there are also two strategies for ctDNA analysis: (i) targeted methods with high resolution such as ARMS, ddPCR, BEAMing which in most cases determine only a single or a few mutations with a high analytical sensitivity and (ii) more comprehensive or untargeted, genome-wide approaches, which require a certain amount of tumor DNA in the circulation, typically 5–10%, in order to achieve informative results. To date, there is no consensus regarding methods that could eventually find a practical application since the clinical and practical application depends on individual situation and the goal of the ctDNA analysis [65,66]. (PCR, polymerase chain reaction, RT-PCR, real-time PCR; ME-PCR, mutant-enriched PCR; COLD-PCR, co-amplification at lower denaturation temperature PCR assays; ddPCR, droplet-based digital PCR; ARMS, amplification refractory mutation system; BEAMing, beads-emulsion-amplification-and-magnetics; NGS, next generation sequencing; CAPP-Seq, cancer personalized profiling by deep sequencing; TAM-Seq, tagged-amplicon deep sequencing; Safe-SeqS, Safe-Sequencing System; iDES, integrated digital error suppression; LoD, limit of detection).

6. Exosomes

Exosomes, together with apoptotic bodies and microvesicles, belong to the group of extracellular vesicles. They are considered as nanovesicles (because of their diameter between 30 and 120 nm), composed of a phospholipid bilayer, originating from multi-vesicular bodies generated during the endocytic cycle [68,69]. There is evidence of their potential role in various biological events, such as in intercellular communication [70], cell signaling [71], tissue regeneration [72], immune response [73], cancer development [74], and metastasis [75]. They have a unique capability to transfer different contents including DNA, RNA and proteins. Exosomes may contain different heat shock proteins (Hsps) (Hsc70, Hsp70, Hsp60, and Hsp90) [76,77], which mediate protein distribution in intraluminal vesicles (exosome precursors) and inclusion of cytoskeleton proteins such as actin, tubulin and cofilin [78]. Exosomes also express proteins from the dipeptidyl-peptidase IV (DPP IV) and MMP9 families, involved in the extracellular matrix remodeling, representing the reason why exosomes are associated with tumor invasion and metastasis [59]. They are usually enriched with lipid rafts containing cholesterol, sphingolipids, ceramides, and glycerophospholipids with long-chain saturated fatty acids [70].

Evidence shows that exosomes originating from the human colon carcinoma cell line LIM1215 contain A33 antigens and epithelial cell adhesion molecules (EpCAM), also known as cluster of differentiation 326 (CD326), molecules specific for colonic epithelial cells [79]. It is worth mentioning that the A33 antigen is a glycoprotein highly expressed in CRC [80], while EpCAM expression is increased in most CRCs. EpCAM is significantly associated with uncontrolled cell proliferation and CRC invasion, and metastasis [81].

CRC-generated exosomes contain intracellular CRC proteins [82]. Some of them, such as cadherins, CEA, and TGF-β, may be used for early detection of CRC [82,83]. Recent protein analysis of exosomes isolated from the blood of CRC patients and the blood of healthy volunteers showed that the levels of proteins involved in the remodeling of the extracellular matrix, intercellular communication, and cell signaling, increased vascular permeability, and tumor-promoting inflammation (α-1 antitrypsin (SERPINA1), α -2 antiplasmin (SERPINF2), and MMP9), are increased in CRC patients [84]. In contrast, the level of proteins involved in immune evasion, complement binding, cell adhesion, and tumor growth (integrin-linked protein kinase (ILK), calpain small subunit 1 (CAPNS1), and neuroblastoma RAS (NRAS)) were decreased, although many of them are known to show higher expression in tumor tissue [84]. The proteome profile of exosomes generated from human metastatic colon cancer cells SW620 differs significantly from the proteome profile of exosomes found in non-metastatic primary CRC [85]. Exosomes derived from metastatic CRC contained ample amounts of metastatic factors, signaling molecules, lipid rafts and their associated elements [85].

In general, exosomes contain different RNA molecules, including mRNA, microRNA, long non-coding RNA (lncRNA), and circular RNA (circRNA). Elevated levels of exosomal microRNA (namely, miR-17-92a, miR-19a, miR-210, miR193a) were found in invasive metastatic tumors and are associated with poor prognosis [86].

7. Circulating Tumor Cells (CTC)

Circulating tumor cells (CTC) are a very rare subset of cells found in the blood of patients with solid tumors. One milliliter of tumor patients' peripheral blood may contain approximately ten CTCs [87]. The study on patients with breast cancer showed interconnection between the CTC count and survival rate [88]. A similar finding was reported for patients suffering from CRC [89]. In patients with solid tumors, metastasis is the primary cause of death, and CTC quite likely acts as a seed for metastases [90]. For the early diagnosis, recurrence and response to therapy, it may be essential to determine tumor cells or cells with epithelial markers in CRC patients' peripheral blood [91]. Current research shows that CTC counts are associated with overall and progression-free survival in patients with various metastasizing cancers. CTC count is also considered a reliable indicator of CRC treatment response [87,92–94].

Nowadays, many innovative methods are available to detect and analyze CTC, including CTC microchips, filtration devices, molecular analytical methods, CellSearch™ system, flow cytometry methods and automated microscopy [93,94].

Finally, researchers and clinicians can use CTC to identify gene mutations and changes in the signaling pathways, and to monitor malignancy development and response to therapy. The immense benefit of using these approaches is to ensure patients' safety and reduce manipulative efforts involved in standard diagnostic procedures, which require significantly more resources and time. Therefore, further research and clinical trials are needed to clarify relevant questions and to highlight important clinical aspects of these strategies.

8. Conclusions

Taking into consideration and summarizing the molecular disturbances specific to CRC (Figure 1), the present review addresses the usefulness of liquid biopsy in diagnosis of the disease, the choice of efficient treatment, the monitoring of the response to treatment, the progression of the disease, and the detection of recurrence at different molecular levels. Studying various biomarkers (Table 1) and evaluating their potential is crucial to achieving the best therapeutic approaches for patients. Likewise, we presented the contribution of liquid biopsy in providing biological samples (ctDNA, exosomes, CTC) containing potent biomarkers (genes, RNA, proteins), that can give an accurate description and profile of CRC, as well as being used to diagnose and predict disease progression and outcome. We believe that with such a comprehensive approach, one should be able to identify biomarkers useful for CRC diagnosis and predict recurrence and potential for metastasis, monitor the response to treatment and predict outcomes, at least suggestively. A combination of different molecular markers will likely be necessary to make CRC treatment approaches more specific.

Author Contributions: Conceptualization, K.B. and D.V.; Writing—Original Draft Preparation, K.B., M.Š., L.B., D.D., N.L., A.Č., I.R., R.P., A.D., I.Ć., J.P., A.S.-B., L.R., I.H., B.K., N.B., A.H.-T., M.G.R., D.V., A.B., M.S., L.S.; Writing—Review and Editing, L.S., D.V., A.Č., I.H., L.R., A.S.-B., D.D., M.Š., N.L., B.K.; Visualization, A.Č., D.D., M.Š., A.B.; Supervision, D.V., K.B.; Project Administration, K.B. All authors have read and agreed to the published version of the manuscript.

Funding: This research was funded by the Croatian Science Foundation, grant number HRZZ IP-2019-04-4624 (project "CRCMolProfil—Genetic, protein and RNA profiling of colorectal cancer using liquid biopsy").

Conflicts of Interest: The authors declare no conflict of interest. The funders had no role in the design of the study; in the collection, analyses, or interpretation of data; in the writing of the manuscript, or in the decision to publish the results.

Abbreviations

APC	adenomatous polyposis coli protein gene
BRAF	serine/threonine-protein kinase B-Raf gene
CA	carbohydrate antigen
CAPNS1	calpain small subunit 1
CD326	cluster of differentiation 326
c-MYC	cellular MYC gene
CEA	carcinoembryonic antigen
cfDNA	cell-free circulating deoxyribonucleic acid
CIMP	epigenetic instability
CIN	chromosomal instability
circRNA	circular ribonucleic acid
CRC	colorectal cancer
CTC	circulating tumor cells
ctDNA	circulating tumor deoxyribonucleic acid
DCC	netrin-1 receptor gene
DPP IV	dipeptidyl-peptidase IV
EpCAM	epithelial cell adhesion molecule
Hsp	heat shock protein
ILK	integrin-linked protein kinase
JAK	Janus kinase

KRAS	c-K-ras protein gene
lncRNA	long non-coding ribonucleic acid
MAPK	mitogen-activated protein kinase
MMP9	matrix metallopeptidase 9
MLH1	mutL homolog 1 gene
MLH6	mutL homolog 6 gene
MSH2	mutS homolog 2 gene
MSI	microsatellite instability
NRAS	neuroblastoma RAS
NRAS	neuroblastoma RAS gene
PD-1	programmed cell death protein 1
PIK3CA	phosphatidylinositol-4,5-bisphosphate 3-kinase catalytic subunit alpha gene
PI3K	phosphatidylinositol 3-kinase
PMS1	PMS1 protein homolog 1 gene
PMS2	PMS1 protein homolog 2 gene
PTEN	phosphatase and tensin homolog
PTEN	phosphatase and tensin homolog gene
SERPINA1	α-1 antitrypsin
SERPINAF1	α-2 antiplasmin
SMAD2	mothers against decapentaplegic homolog 2 gene
SMAD4	mothers against decapentaplegic homolog 4 gene
STAT	signal transducer and activator of transcription
TGF-β	transforming growth factor β
TP53	tumor protein p53 gene
WNT	wingless-related integration site

References

1. Knight, S.R.; Shaw, A.C.; Pius, R.; Drake, T.M.; Norman, L.; Ademuyiwa, O.A.; Adisa, O.A.; Aguilera, M.L.; Al-Saqqa, S.W.; Al-Slaibi, I.; et al. Global variation in postoperative mortality and complications after cancer surgery: A multicentre, prospective cohort study in 82 countries. *Lancet* **2021**, *397*, 387–397. [CrossRef]
2. Xu, P.; Zhu, Y.; Sun, B.; Xiao, Z. Colorectal cancer characterization and therapeutic target prediction based on microRNA expression profile. *Sci. Rep.* **2016**, *6*, 20616. [CrossRef]
3. Bray, F.; Ferlay, J.; Soerjomataram, I.; Siegel, R.L.; Torre, L.A.; Jemal, A. Global cancer statistics 2018: GLOBOCAN estimates of incidence and mortality worldwide for 36 cancers in 185 countries. *CA Cancer J. Clin.* **2018**, *68*, 394–424. [CrossRef] [PubMed]
4. Tripathi, S.; Belkacemi, L.; Cheung, M.S.; Bose, R.N. Correlation between Gene Variants, Signaling Pathways, and Efficacy of Chemotherapy Drugs against Colon Cancers. *Cancer Inform.* **2016**, *15*, 1–13. [CrossRef] [PubMed]
5. Rodriguez-Casanova, A.; Costa-Fraga, N.; Bao-Caamano, A.; López-López, R.; Muinelo-Romay, L.; Diaz-Lagares, A. Epigenetic Landscape of Liquid Biopsy in Colorectal Cancer. *Front. Cell Dev. Biol.* **2021**, *9*, 622459. [CrossRef]
6. Miyamoto, Y.; Hiyoshi, Y.; Tokunaga, R.; Akiyama, T.; Daitoku, N.; Sakamoto, Y.; Yoshida, N.; Baba, H. Postoperative complications are associated with poor survival outcome after curative resection for colorectal cancer: A propensity-score analysis. *J. Surg. Oncol.* **2020**, *122*, 344–349. [CrossRef] [PubMed]
7. Howlader, N.; Noone, A.M.; Krapcho, M.; Miller, D.; Bishop, K.; Altekruse, S.F.; Kosary, C.L.; Yu, M.; Ruhl, J.; Tatalovich, Z.; et al. *SEER Cancer Statistics Review*; National Cancer Institute: Bethesda, MD, USA; pp. 1975–2013, Based on November 2015 SEER data submission, posted to the SEER web site, April 2016.
8. Nakamura, M.; Kageyama, S.-I.; Seki, M.; Suzuki, A.; Okumura, M.; Hojo, H.; Motegi, A.; Akimoto, T. Liquid Biopsy Cell-free DNA Biomarkers in Patients with Oligometastatic Colorectal Cancer Treated by Ablative Radiotherapy. *Anticancer Res.* **2021**, *41*, 829–834. [CrossRef] [PubMed]
9. Najmanova, I.; Vopršalová, M.; Saso, L.; Mladěnka, P. The pharmacokinetics of flavanones. *Crit. Rev. Food Sci. Nutr.* **2020**, *60*, 3155–3171. [CrossRef]
10. Buhrmann, C.; Shayan, P.; Brockmueller, A.; Shakibaei, M. Resveratrol Suppresses Cross-Talk between Colorectal Cancer Cells and Stromal Cells in Multicellular Tumor Microenvironment: A Bridge between In Vitro and In Vivo Tumor Microenvironment Study. *Molecules* **2020**, *25*, 4292. [CrossRef]
11. Xie, Y.-H.; Chen, Y.-X.; Fang, J.-Y. Comprehensive review of targeted therapy for colorectal cancer. *Signal Transduct. Target. Ther.* **2020**, *5*, 1–30. [CrossRef]
12. Ab Mutalib, N.-S.; Yusof, N.F.M.; Abdul, S.-N.; Jamal, R. Pharmacogenomics DNA Biomarkers in Colorectal Cancer: Current Update. *Front. Pharmacol.* **2017**, *8*, 736. [CrossRef] [PubMed]
13. Etiology of CRC—Colorectal Cancer Research. Available online: https://www.colorectalcancer.dk/en/our-research/four-areas/crc-etiology/ (accessed on 1 February 2021).

14. Mármol, I.; Sánchez-De-Diego, C.; Dieste, A.P.; Cerrada, E.; Yoldi, M.J.R. Colorectal Carcinoma: A General Overview and Future Perspectives in Colorectal Cancer. *Int. J. Mol. Sci.* **2017**, *18*, 197. [CrossRef] [PubMed]
15. Ahmed, M. Colon Cancer: A Clinician's Perspective in 2019. *Gastroenterol. Res.* **2020**, *13*, 1–10. [CrossRef]
16. Matijašić, M.; Meštrović, T.; Perić, M.; Paljetak, H. Čipčić; Panek, M.; Bender, D.V.; Kelečić, D.L.; Krznarić, Željko; Verbanac, D. Modulating Composition and Metabolic Activity of the Gut Microbiota in IBD Patients. *Int. J. Mol. Sci.* **2016**, *17*, 578. [CrossRef]
17. Verbanac, D.; Maleš, Ž.; Barišić, K. Nutrition—Acts and myths. *Acta Pharm.* **2019**, *69*, 497–510. [CrossRef] [PubMed]
18. Pino, M.S.; Chung, D.C. The Chromosomal Instability Pathway in Colon Cancer. *Gastroenterology* **2010**, *138*, 2059–2072. [CrossRef] [PubMed]
19. Bišćanin, A.; Ljubičić, N.; Boban, M.; Baličević, D.; Pavić, I.; Bišćanin, M.M.; Budimir, I.; Dorosulic, Z.; Duvnjak, M. CX43 Expression in Colonic Adenomas and Surrounding Mucosa Is a Marker of Malignant Potential. *Anticancer Res.* **2016**, *36*, 5437–5442. [CrossRef]
20. Lao, V.V.; Grady, W.M. Epigenetics and colorectal cancer. *Nat. Rev. Gastroenterol. Hepatol.* **2011**, *8*, 686–700. [CrossRef]
21. Herzig, D.O.; Tsikitis, V.L. Molecular markers for colon diagnosis, prognosis and targeted therapy. *J. Surg. Oncol.* **2014**, *111*, 96–102. [CrossRef]
22. Schatoff, E.M.; Leach, B.I.; Dow, L.E. WNT Signaling and Colorectal Cancer. *Curr. Color. Cancer Rep.* **2017**, *13*, 101–110. [CrossRef]
23. Brocardo, M.; Henderson, B.R. APC shuttling to the membrane, nucleus and beyond. *Trends Cell Biol.* **2008**, *18*, 587–596. [CrossRef]
24. Rennoll, G.Y.S. Regulation ofMYCgene expression by aberrant Wnt/β-catenin signaling in colorectal cancer. *World J. Biol. Chem.* **2015**, *6*, 290–300. [CrossRef] [PubMed]
25. Toon, C.W.; Chou, A.; Clarkson, A.; DeSilva, K.; Houang, M.; Chan, J.C.Y.; Sioson, L.L.; Jankova, L.; Gill, A.J. Immunohistochemistry for Myc Predicts Survival in Colorectal Cancer. *PLoS ONE* **2014**, *9*, e87456. [CrossRef]
26. Chen, J.; Guo, F.; Shi, X.; Zhang, L.; Zhang, A.; Jin, H.; He, Y. BRAF V600E mutation and KRAS codon 13 mutations predict poor survival in Chinese colorectal cancer patients. *BMC Cancer* **2014**, *14*, 802. [CrossRef]
27. Li, W.; Qiu, T.; Zhi, W.; Shi, S.; Zou, S.; Ling, Y.; Shan, L.; Ying, J.; Lu, N. Colorectal carcinomas with KRAS codon 12 mutation are associated with more advanced tumor stages. *BMC Cancer* **2015**, *15*, 1–9. [CrossRef] [PubMed]
28. Canon, J.; Rex, K.; Saiki, A.Y.; Mohr, C.; Cooke, K.; Bagal, D.; Gaida, K.; Holt, T.; Knutson, C.G.; Koppada, N.; et al. The clinical KRAS(G12C) inhibitor AMG 510 drives anti-tumour immunity. *Nature* **2019**, *575*, 217–223. [CrossRef] [PubMed]
29. Porru, M.; Pompili, L.; Caruso, C.; Biroccio, A.; Leonetti, C. Targeting KRAS in metastatic colorectal cancer: Current strategies and emerging opportunities. *J. Exp. Clin. Cancer Res.* **2018**, *37*, 1–10. [CrossRef] [PubMed]
30. Ogino, S.; Nosho, K.; Kirkner, G.J.; Kawasaki, T.; Meyerhardt, J.A.; Loda, M.; Giovannucci, E.L.; Fuchs, C.S. CpG island methylator phenotype, microsatellite instability, BRAF mutation and clinical outcome in colon cancer. *Gut* **2008**, *58*, 90–96. [CrossRef]
31. Goldstein, J.; Tran, B.; Ensor, J.; Gibbs, P.; Wong, H.L.; Wong, S.F.; Vilar, E.; Tie, J.; Broaddus, R.; Kopetz, S.; et al. Multicenter retrospective analysis of metastatic colorectal cancer (CRC) with high-level microsatellite instability (MSI-H). *Ann. Oncol.* **2014**, *25*, 1032–1038. [CrossRef]
32. Kadowaki, S.; Kakuta, M.; Takahashi, S.; Takahashi, A.; Arai, Y.; Nishimura, Y.; Yatsuoka, T.; Ooki, A.; Yamaguchi, K.; Matsuo, K.; et al. Prognostic value ofKRASandBRAFmutations in curatively resected colorectal cancer. *World J. Gastroenterol.* **2015**, *21*, 1275–1283. [CrossRef]
33. Yaeger, R.; Cercek, A.; O'Reilly, E.M.; Reidy, D.L.; Kemeny, N.; Wolinsky, T.; Capanu, M.; Gollub, M.J.; Rosen, N.; Berger, M.F.; et al. Pilot Trial of Combined BRAF and EGFR Inhibition in BRAF-Mutant Metastatic Colorectal Cancer Patients. *Clin. Cancer Res.* **2015**, *21*, 1313–1320. [CrossRef]
34. Liao, X.; Morikawa, T.; Lochhead, P.; Imamura, Y.; Kuchiba, A.; Yamauchi, M.; Nosho, K.; Qian, Z.R.; Nishihara, R.; Meyerhardt, J.A.; et al. Prognostic Role of PIK3CA Mutation in Colorectal Cancer: Cohort Study and Literature Review. *Clin. Cancer Res.* **2012**, *18*, 2257–2268. [CrossRef]
35. Atreya, C.E.; Sangale, Z.; Xu, N.; Matli, M.R.; Tikishvili, E.; Welbourn, W.; Stone, S.; Shokat, K.M.; Warren, R.S. PTEN expression is consistent in colorectal cancer primaries and metastases and associates with patient survival. *Cancer Med.* **2013**, *2*, 496–506. [CrossRef] [PubMed]
36. Sarli, L.; Bottarelli, L.; Bader, G.; Iusco, D.; Pizzi, S.; Costi, R.; D'Adda, T.; Bertolani, M.; Roncoroni, L.; Bordi, C. Association Between Recurrence of Sporadic Colorectal Cancer, High Level of Microsatellite Instability, and Loss of Heterozygosity at Chromosome 18q. *Dis. Colon Rectum* **2004**, *47*, 1467–1482. [CrossRef] [PubMed]
37. Popat, S.; Houlston, R.S. A systematic review and meta-analysis of the relationship between chromosome 18q genotype, DCC status and colorectal cancer prognosis. *Eur. J. Cancer* **2005**, *41*, 2060–2070. [CrossRef]
38. Popat, S.; Zhao, D.; Chen, Z.; Pan, H.; Shao, Y.; Chandler, I.; Houlston, R.S. Relationship between chromosome 18q status and colorectal cancer prognosis: A prospective, blinded analysis of 280 patients. *Anticancer Res.* **2007**, *27*, 627–633.
39. Said, A.H.; Raufman, J.-P.; Xie, G. The Role of Matrix Metalloproteinases in Colorectal Cancer. *Cancers* **2014**, *6*, 366–375. [CrossRef] [PubMed]
40. Munro, A.J.; Lain, S.; Lane, D.P. P53 abnormalities and outcomes in colorectal cancer: A systematic review. *Br. J. Cancer* **2005**, *92*, 434–444. [CrossRef]
41. Lay, M.; Mariappan, R.; Gotlib, J.; Dietz, L.; Sebastian, S.; Schrijver, I.; Zehnder, J.L. Detection of the JAK2 V617F Mutation by LightCycler PCR and Probe Dissociation Analysis. *J. Mol. Diagn.* **2006**, *8*, 330–334. [CrossRef]

42. Albacker, L.A.; Wu, J.; Smith, P.; Warmuth, M.; Stephens, P.J.; Zhu, P.; Yu, L.; Chmielecki, J. Loss of function JAK1 mutations occur at high frequency in cancers with microsatellite instability and are suggestive of immune evasion. *PLoS ONE* **2017**, *12*, e0176181. [CrossRef]
43. Shin, D.S.; Zaretsky, J.M.; Escuin-Ordinas, H.; Garcia-Diaz, A.; Hu-Lieskovan, S.; Kalbasi, A.; Grasso, C.S.; Hugo, W.; Sandoval, S.; Torrejon, D.Y.; et al. Primary Resistance to PD-1 Blockade Mediated by JAK1/2 Mutations. *Cancer Discov.* **2017**, *7*, 188–201. [CrossRef] [PubMed]
44. Locker, G.Y.; Hamilton, S.; Harris, J.; Jessup, J.M.; Kemeny, N.; Macdonald, J.S.; Somerfield, M.R.; Hayes, D.F.; Bast, R.C. ASCO 2006 Update of Recommendations for the Use of Tumor Markers in Gastrointestinal Cancer. *J. Clin. Oncol.* **2006**, *24*, 5313–5327. [CrossRef]
45. Lee, W.-S.; Baek, J.-H.; Kim, K.K.; Park, Y.H. The prognostic significant of percentage drop in serum CEA post curative resection for colon cancer. *Surg. Oncol.* **2012**, *21*, 45–51. [CrossRef] [PubMed]
46. Su, B.-B. Role of serum carcinoembryonic antigen in the detection of colorectal cancer before and after surgical resection. *World J. Gastroenterol.* **2012**, *18*, 2121–2126. [CrossRef] [PubMed]
47. Tümay, V.; Guner, O.S. The utility and prognostic value of CA 19-9 and CEA serum markers in the long-term follow up of patients with colorectal cancer. A single-center experience over 13 years. *Ann. Ital. Chir.* **2020**, *91*, 494–503. [PubMed]
48. Jung, A.; Kirchner, T. Liquid Biopsy in Tumor Genetic Diagnosis. *Dtsch. Aerzteblatt Online* **2018**, *115*, 169–174. [CrossRef] [PubMed]
49. Domínguez-Vigil, I.G.; Moreno-Martínez, A.K.; Wang, J.Y.; Roehrl, M.H.A.; Barrera-Saldaña, H.A. The dawn of the liquid biopsy in the fight against cancer. *Oncotarget* **2017**, *9*, 2912–2922. [CrossRef]
50. Hood, L. Systems Biology and P4 Medicine: Past, Present, and Future. *Rambam Maimonides Med. J.* **2013**, *4*, e0012. [CrossRef]
51. O'Sullivan, A.; Henrick, B.; Dixon, B.; Barile, D.; Zivkovic, A.; Smilowitz, J.; Lemay, D.; Martin, W.; German, J.B.; Schaefer, S.E. 21st century toolkit for optimizing population health through precision nutrition. *Crit. Rev. Food Sci. Nutr.* **2018**, *58*, 3004–3015. [CrossRef]
52. Taghizadeh, H.; Prager, G.W. Personalized Adjuvant Treatment of Colon Cancer. *Visc. Med.* **2020**, *36*, 397–406. [CrossRef]
53. Chakrabarti, S.; Xie, H.; Urrutia, R.; Mahipal, A. The Promise of Circulating Tumor DNA (ctDNA) in the Management of Early-Stage Colon Cancer: A Critical Review. *Cancers* **2020**, *12*, 2808. [CrossRef]
54. Sholl, L.M.; Oxnard, G.R.; Paweletz, C.P. Traditional Diagnostics versus Disruptive Technology: The Role of the Pathologist in the Era of Liquid Biopsy. *Cancer Res.* **2020**, *80*, 3197–3199. [CrossRef]
55. Wu, T.; Mo, Y.; Wu, C. Prognostic values of CEA, CA19-9, and CA72-4 in patients with stages I-III colorectal cancer. *Int. J. Clin. Exp. Pathol.* **2020**, *13*, 1608–1614.
56. Tsai, W.-S.; Chen, J.-S.; Shao, H.-J.; Wu, J.-C.; Lai, J.-M.; Lu, S.-H.; Hung, T.-F.; Chiu, Y.-C.; You, J.-F.; Hsieh, P.-S.; et al. Circulating Tumor Cell Count Correlates with Colorectal Neoplasm Progression and Is a Prognostic Marker for Distant Metastasis in Non-Metastatic Patients. *Sci. Rep.* **2016**, *6*, 24517. [CrossRef]
57. Osumi, H.; Shinozaki, E.; Yamaguchi, K. Circulating Tumor DNA as a Novel Biomarker Optimizing Chemotherapy for Colorectal Cancer. *Cancers* **2020**, *12*, 1566. [CrossRef] [PubMed]
58. Said, R.; Guibert, N.; Oxnard, G.R.; Tsimberidou, A.M. Circulating tumor DNA analysis in the era of precision oncology. *Oncotarget* **2020**, *11*, 188–211. [CrossRef] [PubMed]
59. Wang, Z.; Chen, J.-Q.; Liu, J.-L.; Tian, L. Exosomes in tumor microenvironment: Novel transporters and biomarkers. *J. Transl. Med.* **2016**, *14*, 1–9. [CrossRef]
60. Han, X.; Wang, J.; Sun, Y. Circulating Tumor DNA as Biomarkers for Cancer Detection. *Genom. Proteom. Bioinform.* **2017**, *15*, 59–72. [CrossRef] [PubMed]
61. Mouliere, F.; El Messaoudi, S.; Pang, D.; Dritschilo, A.; Thierry, A.R. Multi-marker analysis of circulating cell-free DNA toward personalized medicine for colorectal cancer. *Mol. Oncol.* **2014**, *8*, 927–941. [CrossRef] [PubMed]
62. Schøler, L.V.; Reinert, T.; Ørntoft, M.-B.W.; Kassentoft, C.G.; Árnadóttir, S.S.; Vang, S.; Nordentoft, I.; Knudsen, M.; Lamy, P.; Andreasen, D.; et al. Clinical Implications of Monitoring Circulating Tumor DNA in Patients with Colorectal Cancer. *Clin. Cancer Res.* **2017**, *23*, 5437–5445. [CrossRef] [PubMed]
63. Kang, J.-K.; Heo, S.; Kim, H.-P.; Song, S.-H.; Yun, H.; Han, S.-W.; Kang, G.H.; Bang, D.; Kim, T.-Y. Liquid biopsy-based tumor profiling for metastatic colorectal cancer patients with ultra-deep targeted sequencing. *PLoS ONE* **2020**, *15*, e0232754. [CrossRef] [PubMed]
64. Demoret, B.; Gregg, J.; Liebner, D.A.; Tinoco, G.; Lenobel, S.; Chen, J.L. Prospective Evaluation of the Concordance of Commercial Circulating Tumor DNA Alterations with Tumor-Based Sequencing across Multiple Soft Tissue Sarcoma Subtypes. *Cancers* **2019**, *11*, 1829. [CrossRef] [PubMed]
65. Perakis, S.; Auer, M.; Belic, J.; Heitzer, E. Advances in Circulating Tumor DNA Analysis. *Adv. Clin. Chem.* **2017**, *80*, 73–153. [PubMed]
66. Zhao, H.; Chen, K.-Z.; Hui, B.-G.; Zhang, K.; Yang, F.; Wang, J. Role of circulating tumor DNA in the management of early-stage lung cancer. *Thorac. Cancer* **2018**, *9*, 509–515. [CrossRef]
67. Rothwell, D.G.; Ayub, M.; Cook, N.; Thistlethwaite, F.; Carter, L.; Dean, E.; Smith, N.; Villa, S.; Dransfield, J.; Clipson, A.; et al. Utility of ctDNA to support patient selection for early phase clinical trials: The TARGET study. *Nat. Med.* **2019**, *25*, 738–743. [CrossRef]

68. El Andaloussi, S.; Mäger, I.; Breakefield, X.O.; Wood, M.J.A. Extracellular vesicles: Biology and emerging therapeutic opportunities. *Nat. Rev. Drug Discov.* **2013**, *12*, 347–357. [CrossRef] [PubMed]
69. Bobrie, A.; Colombo, M.; Krumeich, S.; Raposo, G.; Théry, C. Diverse subpopulations of vesicles secreted by different intracellular mechanisms are present in exosome preparations obtained by differential ultracentrifugation. *J. Extracell. Vesicles* **2012**, *1*. [CrossRef] [PubMed]
70. Record, M.; Carayon, K.; Poirot, M.; Silvente-Poirot, S. Exosomes as new vesicular lipid transporters involved in cell–cell communication and various pathophysiologies. *Biochim. Biophys. Acta (BBA) Mol. Cell Biol. Lipids* **2014**, *1841*, 108–120. [CrossRef] [PubMed]
71. Xiao, H.; Lässer, C.; Shelke, G.V.; Wang, J.; Rådinger, M.; Lunavat, T.R.; Malmhäll, C.; Lin, L.H.; Li, J.; Li, L.; et al. Mast cell exosomes promote lung adenocarcinoma cell proliferation—Role of KIT-stem cell factor signaling. *Cell Commun. Signal.* **2014**, *12*, 1–10. [CrossRef]
72. Borges, F.T.; Melo, S.A.; Özdemir, B.C.; Kato, N.; Revuelta, I.; Miller, C.A.; Ii, V.H.G.; LeBleu, V.S.; Kalluri, R. TGF-β1–Containing Exosomes from Injured Epithelial Cells Activate Fibroblasts to Initiate Tissue Regenerative Responses and Fibrosis. *J. Am. Soc. Nephrol.* **2012**, *24*, 385–392. [CrossRef]
73. Liu, Q.; Rojas-Canales, D.M.; DiVito, S.J.; Shufesky, W.J.; Stolz, D.B.; Erdos, G.; Sullivan, M.L.; Gibson, G.A.; Watkins, S.C.; Larregina, A.T.; et al. Donor dendritic cell–derived exosomes promote allograft-targeting immune response. *J. Clin. Investig.* **2016**, *126*, 2805–2820. [CrossRef] [PubMed]
74. Greening, D.W.; Gopal, S.K.; Mathias, R.A.; Liu, L.; Sheng, J.; Zhu, H.-J.; Simpson, R.J. Emerging roles of exosomes during epithelial–mesenchymal transition and cancer progression. *Semin Cell Dev. Biol.* **2015**, *40*, 60–71. [CrossRef] [PubMed]
75. Hoshino, A.; Costa-Silva, B.; Shen, T.-L.; Rodrigues, G.; Hashimoto, A.; Mark, M.T.; Molina, H.; Kohsaka, S.; Di Giannatale, A.; Ceder, S.; et al. Tumour exosome integrins determine organotropic metastasis. *Nature* **2015**, *527*, 329–335. [CrossRef]
76. Lv, L.-H.; Wan, Y.-L.; Lin, Y.; Zhang, W.; Yang, M.; Li, G.-L.; Lin, H.-M.; Shang, C.-Z.; Chen, Y.-J.; Min, J. Anticancer Drugs Cause Release of Exosomes with Heat Shock Proteins from Human Hepatocellular Carcinoma Cells That Elicit Effective Natural Killer Cell Antitumor Responses in Vitro. *J. Biol. Chem.* **2012**, *287*, 15874–15885. [CrossRef]
77. Kastelowitz, N.; Yin, H. Exosomes and Microvesicles: Identification and Targeting By Particle Size and Lipid Chemical Probes. *ChemBioChem* **2014**, *15*, 923–928. [CrossRef]
78. Kowal, J.; Tkach, M.; Théry, C. Biogenesis and secretion of exosomes. *Curr. Opin. Cell Biol.* **2014**, *29*, 116–125. [CrossRef] [PubMed]
79. Tauro, B.J.; Greening, D.W.; Mathias, R.A.; Mathivanan, S.; Ji, H.; Simpson, R.J. Two Distinct Populations of Exosomes Are Released from LIM1863 Colon Carcinoma Cell-derived Organoids. *Mol. Cell. Proteom.* **2013**, *12*, 587–598. [CrossRef]
80. Welt, S.; Ritter, G.; Williams, C.; Cohen, L.S.; John, M.; Jungbluth, A.; Richards, A.E.; Old, L.J.; Kemeny, E.N. Phase I study of anticolon cancer humanized antibody A33. *Clin. Cancer Res.* **2003**, *9*, 1338–1346.
81. Liu, D.; Sun, J.; Zhu, J.; Zhou, H.; Zhang, X.; Zhang, Y. Expression and clinical significance of colorectal cancer stem cell marker EpCAMhigh/CD44+ in colorectal cancer. *Oncol. Lett.* **2014**, *7*, 1544–1548. [CrossRef] [PubMed]
82. Bernhard, O.K.; Greening, D.W.; Barnes, T.W.; Ji, H.; Simpson, R.J. Detection of cadherin-17 in human colon cancer LIM1215 cell secretome and tumour xenograft-derived interstitial fluid and plasma. *Biochim. Biophys. Acta (BBA) Proteins Proteom.* **2013**, *1834*, 2372–2379. [CrossRef] [PubMed]
83. Jurčić, P.; Radulović, P.; Balja, M.P.; Milošević, M.; Krušlin, B. E-cadherin and NEDD9 expression in primary colorectal cancer, metastatic lymph nodes and liver metastases. *Oncol. Lett.* **2019**, *17*, 2881–2889. [CrossRef]
84. Chen, Y.; Xie, Y.; Xu, L.; Zhan, S.; Xiao, Y.; Gao, Y.; Wu, B.; Ge, W. Protein content and functional characteristics of serum-purified exosomes from patients with colorectal cancer revealed by quantitative proteomics. *Int. J. Cancer* **2017**, *140*, 900–913. [CrossRef]
85. Ji, H.; Greening, D.W.; Barnes, T.W.; Lim, J.W.; Tauro, B.J.; Rai, A.; Xu, R.; Adda, C.; Mathivanan, S.; Zhao, W.; et al. Proteome profiling of exosomes derived from human primary and metastatic colorectal cancer cells reveal differential expression of key metastatic factors and signal transduction components. *Proteomics* **2013**, *13*, 1672–1686. [CrossRef] [PubMed]
86. Ogata-Kawata, H.; Izumiya, M.; Kurioka, D.; Honma, Y.; Yamada, Y.; Furuta, K.; Gunji, T.; Ohta, H.; Okamoto, H.; Sonoda, H.; et al. Circulating Exosomal microRNAs as Biomarkers of Colon Cancer. *PLoS ONE* **2014**, *9*, e92921. [CrossRef] [PubMed]
87. Lianidou, E.S.; Strati, A.; Markou, A. Circulating tumor cells as promising novel biomarkers in solid cancers. *Crit. Rev. Clin. Lab. Sci.* **2014**, *51*, 160–171. [CrossRef] [PubMed]
88. Cristofanilli, M.; Budd, G.T.; Ellis, M.J.; Stopeck, A.; Matera, J.; Miller, M.C.; Reuben, J.M.; Doyle, G.V.; Allard, W.J.; Terstappen, L.W.; et al. Circulating Tumor Cells, Disease Progression, and Survival in Metastatic Breast Cancer. *N. Engl. J. Med.* **2004**, *351*, 781–791. [CrossRef]
89. Miller, M.C.; Doyle, G.V.; Terstappen, L.W.M.M. Significance of Circulating Tumor Cells Detected by the CellSearch System in Patients with Metastatic Breast Colorectal and Prostate Cancer. *J. Oncol.* **2009**, *2010*, 1–8. [CrossRef]
90. Yang, C.; Xia, B.-R.; Jin, W.-L.; Lou, G. Circulating tumor cells in precision oncology: Clinical applications in liquid biopsy and 3D organoid model. *Cancer Cell Int.* **2019**, *19*, 1–13. [CrossRef] [PubMed]
91. Mueller, M.M.; Fusenig, N.E. Friends or foes—bipolar effects of the tumour stroma in cancer. *Nat. Rev. Cancer* **2004**, *4*, 839–849. [CrossRef]
92. Hardingham, J.E.; Grover, P.; Winter, M.; Hewett, P.J.; Price, T.J.; Thierry, B. Detection and Clinical Significance of Circulating Tumor Cells in Colorectal Cancer—20 Years of Progress. *Mol. Med.* **2015**, *21*, 25–31. [CrossRef]

93. Alix-Panabières, C.; Pantel, K. Circulating Tumor Cells: Liquid Biopsy of Cancer. *Clin. Chem.* **2013**, *59*, 110–118. [CrossRef] [PubMed]
94. Eliasova, P.; Pinkas, M.; Kolostova, K.; Gurlich, R.; Bobek, V. Circulating tumor cells in different stages of colorectal cancer. *Folia Histochem. Cytobiol.* **2017**, *55*, 1–5. [CrossRef] [PubMed]

International Journal of
Molecular Sciences

Article

A Novel Synthetic Compound (E)-5-((4-oxo-4H-chromen-3-yl)methyleneamino)-1-phenyl-1H-pyrazole-4-carbonitrile Inhibits TNFα-Induced *MMP9* Expression via EGR-1 Downregulation in MDA-MB-231 Human Breast Cancer Cells

Munki Jeong [1], Euitaek Jung [1], Young Han Lee [1,2], Jeong Kon Seo [3], Seunghyun Ahn [4], Dongsoo Koh [4], Yoongho Lim [2,5] and Soon Young Shin [1,2,*]

1 Department of Biological Sciences, Sanghuh College of Lifesciences, Koknkuk University, Seoul 05029, Korea; Jmk4919@naver.com (M.J.); mylife4sci@naver.com (E.J.); yhlee58@konkuk.ac.kr (Y.H.L.)
2 Cancer and Metabolism Institute, Konkuk University, Seoul 05029, Korea; yoongho@konkuk.ac.kr
3 UNIST Central Research Facilities, Ulsan National Institute of Science and Technology, Ulsan 44919, Korea; jkseo6998@unist.ac.kr
4 Department of Applied Chemistry, Dongduk Women's University, Seoul 02748, Korea; mistahn321@naver.com (S.A.); dskoh@dongduk.ac.kr (D.K.)
5 Division of Bioscience and Biotechnology, BMIC, Konkuk University, Seoul 05029, Korea
* Correspondence: shinsy@konkuk.ac.kr; Tel.: +82-2-2030-7946

Received: 13 July 2020; Accepted: 15 July 2020; Published: 18 July 2020

Abstract: Breast cancer is a common malignancy among women worldwide. Gelatinases such as matrix metallopeptidase 2 (MMP2) and MMP9 play crucial roles in cancer cell migration, invasion, and metastasis. To develop a novel platform compound, we synthesized a flavonoid derivative, (E)-5-((4-oxo-4H-chromen-3-yl)methyleneamino)-1-phenyl-1H-pyrazole-4-carbonitrile (named DK4023) and characterized its inhibitory effects on the motility and *MMP2* and *MMP9* expression of highly metastatic MDA-MB-231 breast cancer cells. We found that DK4023 inhibited tumor necrosis factor alpha (TNFα)-induced motility and F-actin formation of MDA-MB-231 cells. DK4023 also suppressed the TNFα-induced mRNA expression of *MMP9* through the downregulation of the TNFα-extracellular signal-regulated kinase (ERK)/early growth response 1 (EGR-1) signaling axis. These results suggest that DK4023 could serve as a potential platform compound for the development of novel chemopreventive/chemotherapeutic agents against invasive breast cancer.

Keywords: EGR-1; flavonoid; (E)-5-((4-oxo-4H-chromen-3-yl)methyleneamino)-1-phenyl-1H-pyrazole-4-carbonitrile; MDA-MB-231; MMP9; TNFα

---

1. Introduction

Breast cancer is a common malignancy among women worldwide [1]. Metastasis, the spread of primary tumor cells to different organs through the blood or lymphatic vessels, is a typical hallmark of all malignant tumors and is responsible for high mortality among patients with cancer. Breast cancer commonly spreads to the regional lymph nodes, bone, liver, lungs, and brain [2]. Thus, the control over the process of metastasis at the early stages is imperative for the successful prevention and treatment of cancer, including breast cancer.

Tumor cell invasion involves their migration and penetration into neighboring tissues. Lymphovascular invasion is, in general, the first stage of carcinogenic events that initiate tumor metastasis [3]. The initial step of invasion is characterized by the breakdown of the basement membrane, a sheet-like structure of the extracellular matrix secreted by the epithelium [4]. Type IV collagen is a major component of the basement membrane, and gelatinases such as matrix metallopeptidase 9 (MMP9, also known as gelatinase B and 92 kDa type IV collagenase) and MMP2 (also known as gelatinase A and 72 kDa type IV collagenase) are responsible for the breakdown of type IV collagen [5]. MMP2 and MMP9 expression is known to be highly upregulated in almost every metastatic cancer cell type [5,6], and elevated MMP9 level is associated with the high potential of metastasis in several human carcinomas, including breast cancer [6–8]. MMP9 expression has been observed in invasive mammary carcinomas, but not in carcinomas in situ or hyperplastic mammary glands [9]. Experimental metastasis studies have also shown that the downregulation of MMP9 expression in cancer cells by ribozyme could reduce tumor foci in the lungs of mice [10], consistent with the results observed in *Mmp2*- [10] or *Mmp9*-deficient mice [11]. These observations suggest that a therapeutic strategy to manipulate MMP2 or MMP9 expression could be potentially advantageous for the development of anti-metastatic agents.

Flavonoids have a common feature of C6-C3-C6 skeleton. One of the flavonoids, flavone is composed of 4H-chromen-4-one for C6-C3- and benzene ring for the -C6. Compounds carrying the 1-phenyl-1H-pyrazole moiety have been reported to exhibit antitumor activities [12] and induce apoptosis [13]. In addition, compounds with the 4H-chromen-4-one moiety are known to demonstrate BRAF kinase inhibitory activity [14], cytotoxicity against the breast cancer cell line MCF-7 [15], and inhibitory effects on the p53-mouse double minute 2 (MDM2) pathway [16]. Therefore, we synthesized a novel compound combining 1-phenyl-1H-pyrazole and 4H-chromen-4-one moieties to obtain (E)-5-((4-oxo-4H-chromen-3-yl)methyleneamino)-1-phenyl-1H-pyrazole-4-carbonitrile (termed as DK4023). When C2- and C3-position of flavonoids are substituted with benzene ring, they are classified as flavone and isoflavone, respectively. Since the C3-position is substituted with 5-(methyleneamino)-1-phenyl-1H-pyrazole-4-carbonitrile, DK4023 can be classified as isoflavone derivative. Among isoflavones, genistein and daidzein are known to inhibit TNFα-induced migration and invasion of human breast cancer cells by preventing the inhibition of NF-κB [17]. TNFα is a central inflammatory cytokine in the tumor microenvironment that promotes migration, invasion, and metastasis [18–20]. In the current study, we examined the inhibitory effects of DK4023 on TNFα-induced motility and invasive capability of the highly metastatic MDA-MB-231 breast cancer cells. In addition, we determined the inhibitory effect of DK4023 on TNFα-induced *MMP2* and *MMP9* mRNA expression.

2. Results and Discussion

2.1. Chemical Synthesis and Cytotoxicity of DK4023 against MDA-MB-231 Cells

The synthesis of DK4023 was started from phenylhydrazine (**I**) and 2-(ethoxymethylidene) propanedinitrile (**II**). The resulting 5-amino-1-phenyl-1H-pyrazole-4-carbonitrile (**III**) was reacted with 4-oxo-4H-chromene-3-carbaldehyde (**IV**) to yield (E)-5-((4-oxo-4H-chromen-3-yl)methyleneamino)-1-phenyl-1H-pyrazole-4-carbonitrile (**V**; named DK4023) (Scheme 1).

To determine the cytotoxicity of DK4023, we treated MDA-MB-231 human breast cancer cells with DK4023 (0, 10, 25, 50, and 100 μM) for 24 h. Cellular cytotoxicity was determined using the water-soluble 2-(2-methoxy-4-nitrophenyl)-3-(4-nitrophenyl)-5-(2,4-disulfophenyl)-2*H*-tetrazolium, monosodium salt (WST-8). As no significant cytotoxicity was observed at up to 50 μM DK4023 concentration (Figure 1), we used 25 and 50 μM concentrations for subsequent experiments.

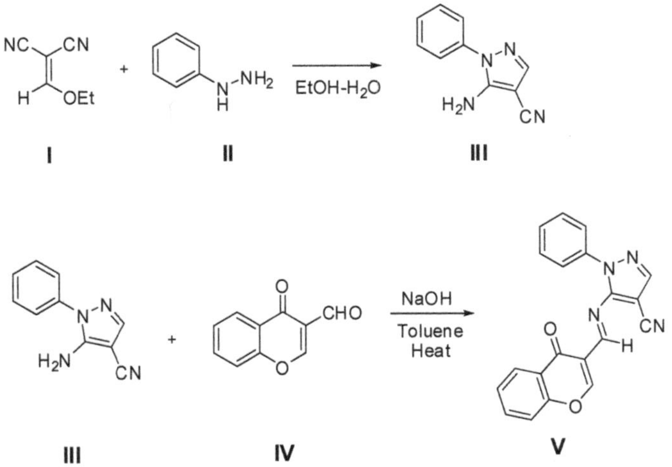

Scheme 1. Synthesis of (E)-5-((4-oxo-4H-chromen-3-yl)methyleneamino)-1-phenyl-1H-pyrazole-4-carbonitrile (DK4023).

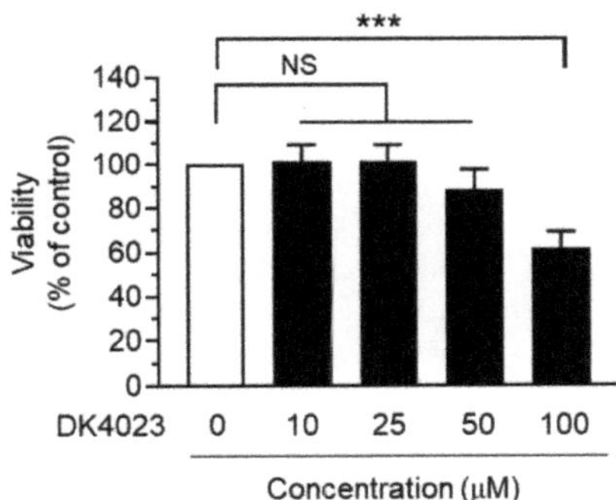

Figure 1. Cytotoxicity of DK4023 against MDA-MB-231 cells. Values are expressed as mean ± SD. *** $p < 0.001$, ^{NS} not significant ($p > 0.200$) by Sidak's multiple comparison test ($n = 3$).

2.2. *Effect of DK4023 on the TNFα-Induced Migration of MDA-MB-231 Cells*

The tumor mass is composed of tissue-resident fibroblasts, peripherally recruited immune cells, and endothelial cells of surrounding blood vessels, as well as cancer cell populations. The local environment around the tumor mass includes various growth factors and cytokines, which are collectively referred to as a tumor microenvironment [21]. It has been well characterized that the inflammatory tumor microenvironment is closely associated with tumor development and progression [22,23]. TNFα is a major proinflammatory cytokine that is released from many cell types, including cancer cells, immune cells, and fibroblasts, in the tumor microenvironment [20]. It has been demonstrated that TNFα increases the expression of other cytokines and chemokines, including IL-1, IL-6, CCL2, CXCL8, and CXCL12 [24], induces epithelial-to-mesenchymal transition (EMT), through the activation of NF-κB and AP1, and facilitates the invasion and metastasis of breast cancer cells [21,25].

A crucial feature of invasive and metastatic breast cancer cells is the increase in their motility. To evaluate whether DK4023 could modulate the motility of metastatic MDA-MB-231 cells, we used an in vitro scratch-wound healing assay and measured the thickness of the scratched area. After scratching a confluent monolayer, cells were treated with TNFα (10 ng/mL) or TNFα (10 ng/mL) plus DK4023 (25 and 50 μM) (Figure 2a). At 12 h post-scratching, the scratched area decreased following TNFα treatment as compared that observed after vehicle treatment. In contrast, the TNFα-induced closure of the scratched area was significantly suppressed in the presence of DK4023 (Figure 2b). As DK4023 did not exhibit cytotoxicity at concentrations around 50 μM (Figure 1), its inhibitory effect on the motility of MDA-MB-231 cells was not related to its cytotoxicity.

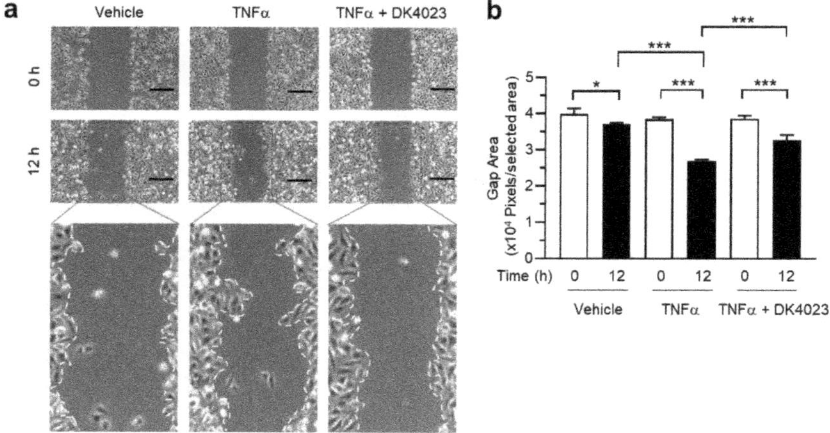

Figure 2. Effect of DK4023 on the migration of MDA-MB-231 cells. (**a**) Scratch wound-healing assay. DK4023 was pretreated for 30 min before addition of TNFα. Migration images were captured after 12 h of TNFα treatment. Box areas were enlarged underneath the image. Arrow indicates the elongated spindle shape of leader cells at the leading edge. Vehicle, phosphate-buffered saline. Scale bars, 500 μm. (**b**) A plot of the scratched gap area was expressed as the number of pixels in the selected boxes. Values were expressed as mean ± SD. * $p = 0.021$, *** $p < 0.001$ by Sidak's multiple comparison test ($n = 3$).

2.3. Effect of DK4023 on the Actin Reorganization of MDA-MB-231 Cells

As shown in Figure 3, DK4023 reduced the TNFα-induced branched structures of the leader cells at the edge (arrow). Monomeric globular actin (G-actin) is polymerized into filamentous actin (F-actin), which is known to build up higher-ordered structures such as stress fibers, lamellipodia, and filopodia during cell movement [26]. As the dynamic rearrangement of the actin cytoskeleton plays a crucial role in cell migration [27], we assessed whether DK4023 affects actin cytoskeletal rearrangement. We used rhodamine-conjugated phalloidin to stain F-actin and found that TNFα treatment stimulated cytoskeletal rearrangement, as evident from the formation of F-actin-rich protrusions that appeared like lamellipodia (arrows) at the cell periphery (Figure 3). After the treatment of cells with DK4023, the TNFα-induced F-actin-rich protrusions were substantially reduced. These data suggest that DK4023 prevents dynamic F-actin polymerization, resulting in the inhibition of cell motility.

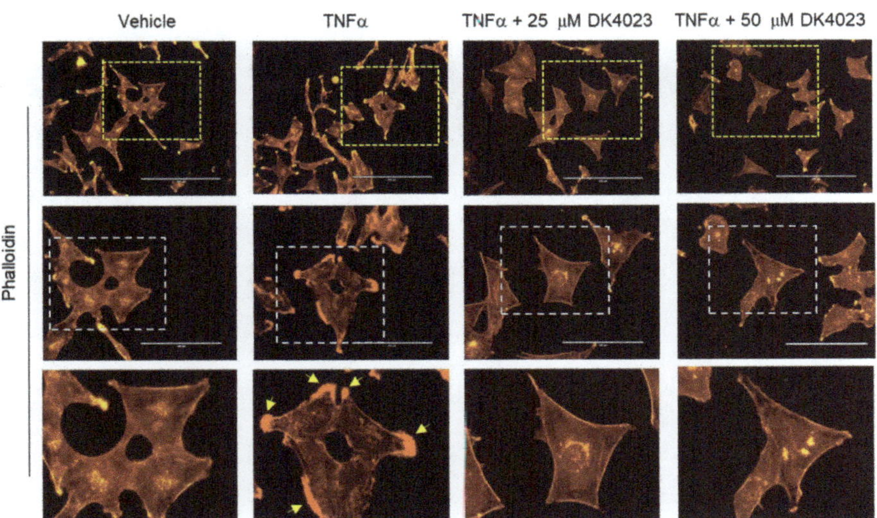

Figure 3. Effect of DK4023 on the actin rearrangement of MDA-MB-231 cells. Polymerized actin (F-actin) was stained with rhodamine-conjugated phalloidin. Box area was enlarged underneath the image. Arrows indicate lamellipodia. Bar size: top panels, 200 μm; middle panels, 100 μm.

2.4. Effect of DK4023 on the Expression of MMP9

As gelatinases play a critical role in cell migration, invasion, and metastasis, we investigated whether DK4023 modulates gelatinase activation. Gelatinase activity was determined using gelatin-gel zymography. DK4023 reduced the density of TNFα-induced gelatinolytic white band at 92 kDa MMP9, but not that of the band at 72 kDa MMP2 (Figure 4a). To investigate whether DK4023-induced suppression of MMP9 activity was associated with *MMP9* downregulation, we determined the effect of DK4023 on the expression of *MMP9* mRNA using conventional reverse-transcription polymerase chain reaction (RT-PCR). We found that DK4023 dose-dependently inhibited TNFα-induced mRNA expression of *MMP9* but not *MMP2*, while glyceraldehyde 3-phosphate dehydrogenase (*GAPDH*) mRNA levels remained unchanged (Figure 4b). As TNFα induced MMP9 mRNA expression and proteolytic activity of MMP9 more efficiently than MMP2, we focused on the inhibitory effect of DK4023 on MMP9 expression in substantial experiments. To further evaluate the inhibitory effect of DK4023 on *MMP9* expression at the transcriptional level, we measured *MMP9* promoter activity using the pMMP9(−925/+13)-Luc luciferase-based promoter-reporter plasmid [28]. TNFα stimulated *MMP9* promoter activity by 3.47 ± 0.351-fold, but this effect was significantly suppressed by 2.13 ± 0.351- and 1.47 ± 0.208-fold in the presence of 25 and 50 μM DK4023, respectively (Figure 4c). Thus, DK4023 inhibits *MMP9* mRNA expression.

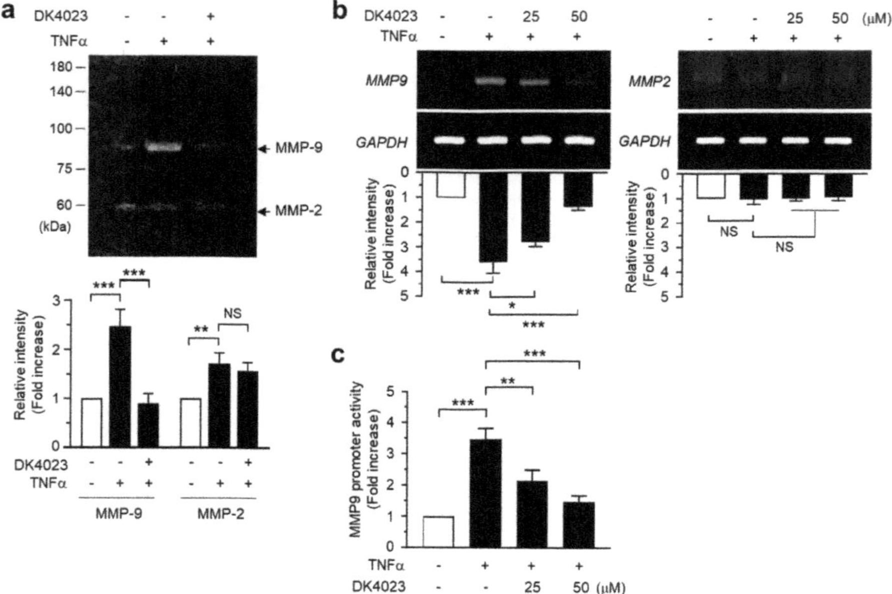

Figure 4. Effect of DK4023 on TNFα-induced MMP9 activation. Effect of DK4023 on TNFα-induced gelatinolytic activity determined by gelatin-gel zymography (**a**), *MMP9 and MMP2* mRNA expression examined by RT-PCR (**b**), and luciferase-based *MMP9* promoter-reporter activity (**c**). The full-length gel images are presented in Supplementary Figure S1. Relative band intensities were measured by ImageJ. Values are expressed as mean ± SD. * $p = 0.016$, ** $p < 0.01$, ***$p < 0.001$, NS, not significant ($p > 0.5$) by Sidak's multiple comparison test ($n = 3$).

2.5. Effect of DK4023 on the TNFα-Induced Invasion of MD-MB-231 Cells

MMP9 plays a critical role in the invasion of tumor cells through the degradation of extracellular matrix proteins [5]. To investigate the effect of DK4023 on the invasive capability of MDA-MB-231 cells, we used a matrigel-based three-dimensional (3-D) spheroid culture model that closely resembles the situation inside the in vivo tumor mass. Up to 1 day, the cells within the spheroid were noninvasive and remained as cell aggregates. (Figure 5, top panels). Following 5 days of culture, invasive cells could be detected in vehicle-treated spheroid (Figure 5, left lane). Following treatment with 10 ng/mL TNFα, the sizes of spheroids were bigger than vehicle-treated groups, and the cells efficiently invaded the surrounding matrix as spindle-like protrusions (Figure 5, middle lane). However, in the presence of DK4023, TNFα-induced invasive capability was substantially attenuated (Figure 5, right lane). These results suggest that DK4023 inhibits TNFα-induced tumor cell invasion through the downregulation of MMP9 expression.

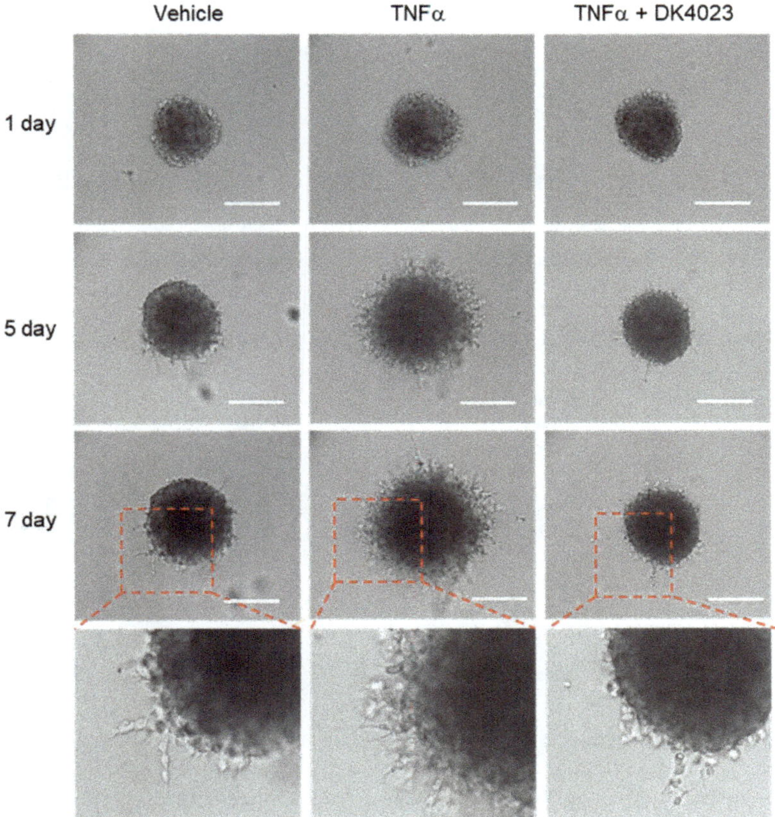

Figure 5. Effect of DK4023 on the invasion of MDA-MB-231 cells. Schematic representation of three-dimensional spheroid formation and protrusion of MDA-MB-231 cells (left). MDA-MB-231 cell spheroids in the extracellular matrix were treated with vehicle (DMSO) or 10 ng/mL TNFα for 7 days in the presence or absence of 25 μM DK4023. The morphology of the invasive protrusions was captured with an EVOS FL Auto Cell Imaging System. Scale bars, 400 μm. The boxed areas are enlarged.

2.6. Effect of DK4023 on the TNFα-Induced Activation of Nuclear Factor-Kappa B (NF-κB) Pathway

NF-κB is a well-known transcription factor involved in the regulation of TNFα-induced *MMP9* expression [29–32]. In unstimulated cells, RelA/p65 and p50, which are the most abundant forms of NF-κB heterodimers, are inhibited by IκB in the cytoplasm. Upon exposure of cells to TNFα, IκB kinase (IKK) gets activated and phosphorylates IκB, subsequently resulting in the degradation of IκB through a ubiquitin-mediated proteolytic pathway [33]. We observed that TNFα enhanced the phosphorylation of IKK and RelA/p65 in a time-dependent manner, while IκB phosphorylation level increased within 10 min, then rapidly decreased, and slowly recovered thereafter in MDA-MB-231 cells (Figure 6a). To determine whether DK4023 modulates the NF-κB pathway, we pretreated MDA-MB-231 cells with DK4023 and measured the TNFα-induced phosphorylation status of three key proteins, IKK, IκB, and RelA/p65. DK4023 failed to significantly decrease the TNFα-induced phosphorylation of IKKα/β or RelA/p65 but could significantly increase the phosphorylation of IκB (Figure 6b). These data suggest that DK4023 may affect the NF-κB pathway through the stabilization of IκB, which could be necessary but not sufficient to downregulate *MMP9* expression.

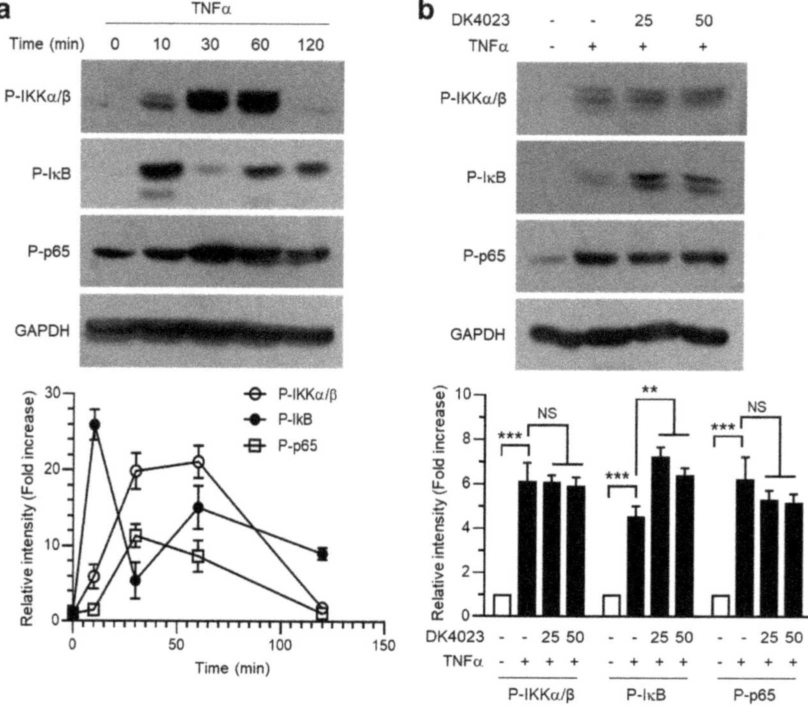

Figure 6. Effect of DK4023 on the TNFα-induced activation of NF-κB pathway. (**a**) The TNFα-induced phosphorylation of IKKα/β on Ser-180/181, IκB on Ser-32, and RelA/p65 on Ser-536 were determined by immunoblot analysis. (**b**) The effect of DK4023 on the TNFα-induced phosphorylation of NF-κB pathway proteins was determined by immunoblotting. Relative band intensities were measured by ImageJ. Values are expressed as mean ± SD. *** $p < 0.001$, ** $p < 0.01$, ^{NS} not significant ($p > 0.05$) by Sidak's multiple comparison test ($n = 3$). The full-length blot images are presented in Supplementary Figure S2.

2.7. Effect of DK4023 on the TNFα-Induced Expression of Early Growth Response-1 (EGR-1)

EGR-1 is a Cys_2His_2-type zinc-finger transcription factor induced by mitogenic stimulation and DNA damage signals [34]. EGR-1 is known to directly bind to the EGR-1-binding element within the *MMP9* gene promoter and transactivate the *MMP9* promoter activity upon TNFα stimulation in HeLa cervical cancer cells [28]. In MDA-MB-231 cells, TNFα increased the level of EGR-1 protein in a time-dependent manner (Figure 7a). However, DK4023 decreased the TNFα-induced expression of EGR-1 in a dose-dependent manner (Figure 7b). To evaluate the role of EGR-1 in DK4023-induced *MMP9* suppression, we used the wild-type *MMP9* promoter construct and a site-directed mutant construct obtained by disrupting the EGR-1-binding site (EBS) at *MMP9* promoter-reporter. TNFα stimulated *MMP9* promoter activation in the wild-type construct, but its effect was lost following the disruption of EBS (Figure 7c). DK4023 treatment inhibited the TNFα-induced *MMP9* promoter activation in the wild-type construct. These data suggest that EGR-1 is critical for mediating the TNFα-induced activation of *MMP9* promoter and that EGR-1 downregulation by DK4023 is associated with the suppression of TNFα-induced *MMP9* promoter activation.

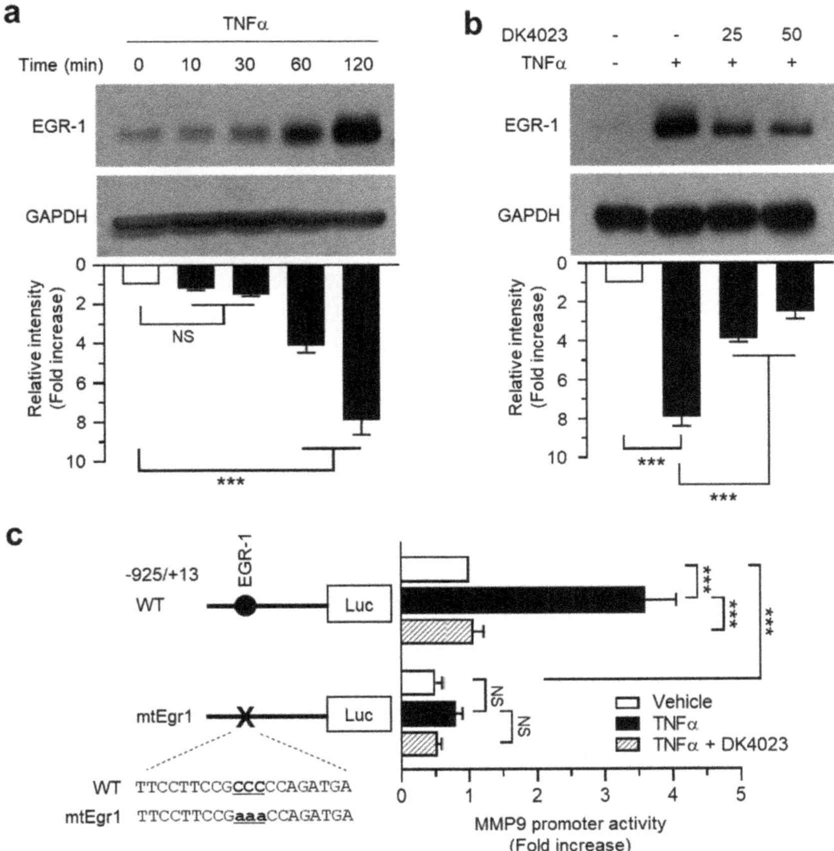

Figure 7. Effect of DK4023 on the TNFα-induced expression of EGR-1. (**a**) The TNFα-induced accumulation of EGR-1 was determined by immunoblotting. (**b**) Inhibitory effect of DK4023 on TNFα-induced EGR-1 accumulation was determined by immunoblotting. Relative band intensities were measured by ImageJ. The full-length blot images are presented in Supplementary Figure S3. (**c**) The effect of the mutation of the EGR-1-binding element in *MMP9* promoter-reporter construct, pMMP9-Luc(−925/+13), on the TNFα-induced activation of *MMP9* promoter was determined by luciferase reporter activity. WT, wild-type; mtEgr1, mutation of the EGR-1-binding sequence. Values were expressed as mean ± SD. *** $p < 0.001$, ^{NS} not significant ($p > 0.05$) by Dunnett's (**a**,**b**) or Sidak's multiple comparison test (**c**).

2.8. Effect of DK4023 on the TNFα-Induced Activation of Mitogen-Activated Protein Kinase (MAPK) Pathway

MAPK signaling pathway plays an essential role in mediating the TNFα-induced expression of EGR-1 in MDA-MB-231 cells [35]. We investigated if DK4023 modulates MAPK pathways by evaluating its effects on the TNFα-induced phosphorylation of three key MAPKs, extracellular signal-regulated kinase 1/2 (ERK1/2), c-Jun N-terminal kinase 1/2 (JNK1/2), and p38 kinase, in MDA-MB-231 cells using phospho-specific antibodies. We found that the phosphorylation of ERK1/2 on Thr-201/Tyr-204, JNK1/2 on Thr-183/Tyr-185, and p38 kinase on Thr-180/Tyr-182 increased within 15 min of TNFα treatment as compared to that observed under unstimulated condition (Figure 8a). Under these conditions, pretreatment of MDA-MB-231 cells with DK4023 significantly reduced ($p < 0.01$, $n = 3$) the levels of TNFα-induced ERK1/2, whereas did not affect or slightly increased the phosphorylation of JNK1/2 and p38 kinase (Figure 8b). These data suggest that DK4023 reduces TNFα-induced *MMP9* expression by selectively inhibiting the ERK1/2 and EGR-1 signaling axis.

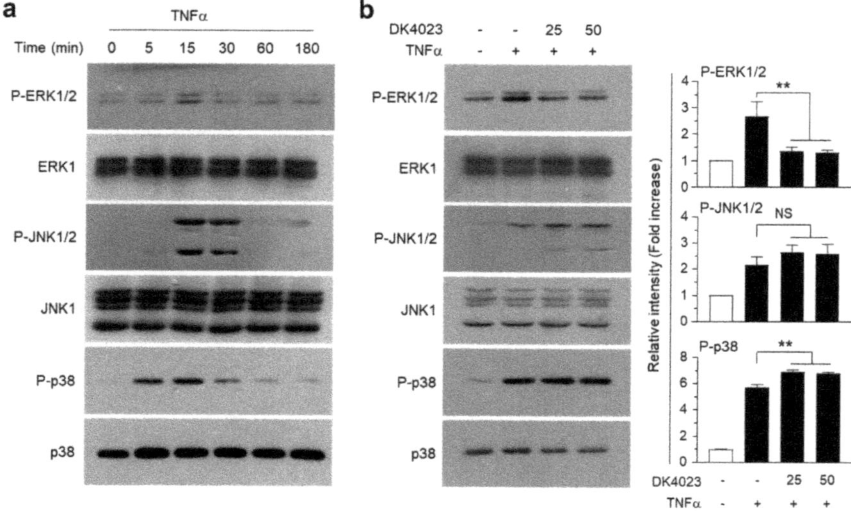

Figure 8. Effect of DK4023 on the TNFα-induced activation of MAPK pathways. (**a**) The TNFα-induced phosphorylation of ERK1/2 on Thr-201/Tyr-204, JNK1/2 on Thr-183/Tyr-185, and p38 on Thr-180/Tyr-182 were determined by immunoblotting. (**b**) The effect of DK4023 on the TNFα-induced phosphorylation of MAPKs was determined by immunoblotting. Relative band intensities were measured by ImageJ. Values were expressed as mean ± SD. ** $p < 0.01$, ^{NS} not significant ($p > 0.05$) by Dunnett's multiple comparison test ($n = 3$). The full-length blot images are presented in Supplementary Figure S4.

3. Materials and Methods

3.1. Cells and Chemicals

MDA-MB-231 human breast cancer cells were obtained from the American Type Culture Collection (ATCC, Rockville, MD, USA) and cultured in Dulbecco's modified Eagle's medium (Corning Cellgro, Manassas, VA, USA) supplemented with 10% (v/v) heat-inactivated fetal bovine serum (Corning Cellgro). TNFα was purchased from Sigma-Aldrich (Saint Louis, MO, USA), and Pierce BCA Protein Assay Reagent was supplied by Thermo Scientific (Rockford, IL, USA). The Firefly and *Renilla* Dual-Glo™ Luciferase Assay System was procured from Promega (Madison, WI, USA), and nitrocellulose membrane, from Bio-Rad Laboratories (Hercules, CA, USA).

3.2. Chemical Synthesis

Commercially available phenylhydrazine (**I**, 1.10 g, 10 mmol) and 2-(ethoxymethylidene) propanedinitrile (**II**, 10 mmol, 1.22 g) were dissolved in 20 mL of an ethanol-water (1:3) solution, and the resulting mixture was stirred at 25 °C for 20 h to obtain an orange color solid. The precipitate was filtered and washed with cold ethanol. The resulting solid of **III**, 5-amino-1-phenyl-1H-pyrazole-4-carbonitrile, was dried and used for the next reaction without any further purification. In brief, 5 mL of toluene solvent was mixed with equimolar amounts of 5-amino-1-phenyl-1H-pyrazole-4-carbonitrile (**III**, 60 mg, 0.3 mmol) and 4-oxo-4H-chromene-3-carbaldehyde (**IV**, 51 mg, 0.3 mmol), and the mixture was treated with a catalytic amount of sodium hydroxide (NaOH). The reaction mixture was stirred at 75 °C for 5 h. After reaction completion, the mixture was cooled to room temperature and the solvent was evaporated under the vacuum. The residues were purified with medium-pressure chromatography to obtain an analytically pure compound of **V** (named DK4023). The structure of DK4023 was verified using nuclear magnetic resonance (NMR) spectroscopy and high-resolution electrospray ionization mass spectrometry (HR-ESI-MS) as described previously [36]. Its spectroscopic data are summarized as follows: IR (cm^{-1}) 3112.55(N-H), 2216.77(-CN), 1657.62(conj. C=O); ^{1}H NMR (500 MHz, DMSO-d6) δ 9.23 (d, 1H, J = 0.5 Hz), 9.04 (m, 1H), 8.32 (s, 1H), 8.17 (dd, 1H, J = 8.0, 1.5 Hz), 7.90 (m, 1H), 7.76 (dd, 1H, J = 8.5, 0.7 Hz), 7.71 (m, 2H), 7.60 (m, 1H), 7.55 (m, 2H), 7.46 (m, 1H); ^{13}C NMR (125 MHz, DMSO-d6) δ 174.64 161.00, 160.03, 155.65, 152.23, 142.90, 137.64, 135.18, 129.09, 128.13, 126.75, 125.35, 124.05, 123.64, 118.88, 118.77, 113.86, 81.23, 48.56.; HR-ESI-MS (m/z): Calcd. for $C_{20}H_{12}N_4O_2$ $[M+H]^+$: 341.1039; found 341.1036. The raw spectra, including IR, ^{1}H-NMR, ^{13}C-NMR, and HR-ESI-Mass spectrum, are provided as a Supplementary Figure S5.

3.3. Cytotoxicity Assay

MDA-MB-231 cells cultured in 96-well plates (2×10^3 cells/well) were treated with different concentrations (10–100 μM) of DK4023 for 24 h. Cell viability was measured by the Cell Counting Kit-8 (Dojindo Molecular Technologies, Gaithersburg, MD) assay using the water-soluble tetrazolium salt WST-8, as previously described [37]. The absorbance was read at 540 nm using a microplate reader (Molecular Devices Corp., Menlo Park, CA).

3.4. Cell Migration Assay

Cell motility was examined using an in vitro scratch-wound healing assay, as previously described [19]. In brief, MDA-MB-231 cells were cultured in six-well plates and allowed to reach confluency. A scratch wound was made with a pipette tip on the confluent cell layer, which was then pretreated with either vehicle (DMSO) or 25 μM DK4023. After 30 min, cells were treated with either vehicle (PBS) or 10 ng/mL TNFα. Gap closure was imaged at 12 h post-scratching using an EVOS FL Auto Cell Imaging System (Life Technologies, Carlsbad, CA, USA) and the area of the closed gap was quantified using ImageJ version 1.52a (http://imagej.nih.gov/ij/; Center for Information Technology, National Institute of Health, Bethesda, MA, USA).

3.5. Actin Rearrangement

MDA-MB-231 cells were pretreated with either vehicle (DMSO) or DK4023 (25 or 50 μM). After 30 min, cells were treated with either vehicle (PBS) or TNFα (10 ng/mL) for 24 h, and then fixed in 4% paraformaldehyde and permeabilized in 0.3% Triton X-100. Actin polymerization was examined using the rhodamine/phalloidin-based F-Actin Visualization Biochem Kit (Cytoskeleton, Inc.; Denver, CO, USA) [38]. Images were captured using EVOS FL Auto Cell Imaging System.

3.6. In-Gel Gelatinase Activity Assay

Gelatin zymography was carried out using the Novex Zymogram gel system (Novex, San Diego, CA), as previously described [39]. In brief, MDA-MB-231 cells (3×10^5) were cultured in a serum-free medium in 24-well plates for 24 h and then pretreated with either vehicle (DMSO) or DK4023 (25 or 50 μM). After 30 min, cells were treated with either vehicle (PBS) or TNFα (10 ng/mL). After 24 h, the culture medium was collected following the removal of cell debris and mixed with a non-reducing sample buffer (62.5 mM Tris-HCl [pH 6.8], 4% SDS, 25% glycerol, and 0.01% bromophenol blue). Equal amounts of each sample were electrophoresed using Novex pre-cast gels (10% acrylamide-0.1% gelatin). After electrophoresis, gels were equilibrated by incubation first in Novex Zymogram renaturation buffer for 30 min and then overnight in Zymogram developing buffer at 37 °C. Destaining was performed in methanol/acetic acid/distilled water (25:7:68, by volume) until gelatinolytic white bands appeared on the blue background. Relative gelatinolytic activities of MMP2 and MMP9 were determined by quantifying the white band intensities using ImageJ version 1.52a.

3.7. RT-PCR

The experimental procedures for total RNA isolation, cDNA synthesis, and PCR conditions are described elsewhere [39]. Briefly, cDNA was synthesized using 1 μg of total RNA and subjected to PCR using the following gene-specific primers: *MMP9* (forward) 5′-agattccaaacctttgag-3′ and (reverse) 5′-ggccttggaagatgaatg-3′; and *GAPDH* (forward) 5′-acccactcctccacctttg-3′ and (reverse) 5′-ctcttgtgctcttgctggg-3′. PCR conditions were as follows: denaturation, 94 °C for 30 s; annealing, 55 °C for 30 s; elongation, 72 °C for 1 min. The amplified products were subjected to electrophoresis on a 1% agarose gel. The relative *MMP9* mRNA level was quantitated after normalizing to *GAPDH* mRNA level using ImageJ version 1.52a.

3.8. MMP9 Promoter-Reporter Assay

The generation of the *MMP9* promoter construct pMMP9-Luc(−925/+13) and the mutated construct for EGR-1 binding site, pMMP9-Luc(−925/+13)mtEgr1, was described elsewhere [28]. MDA-MB-231 cells cultured on 12-well plates were transfected with 0.1 μg of MMP9 promoter-reporter constructs along with 25 ng of the pRL-null plasmid encoding *Renilla* luciferase to monitor transfection efficiency. At 48 h post-transfection, cells were pretreated with either vehicle (DMSO) or DK4023 (25 or 50 μM). After 30 min, cells were treated with either vehicle (PBS) or TNFα (10 ng/mL). After 8 h, cells were collected and the firefly luciferase activity was measured using a Centro LB960 dual luminometer (Berthold Technologies) as previously described [28].

3.9. 3-D Spheroid Culture and Invasion Assay

The 3-D invasion assay was carried out using the Cultrex 3-D Spheroid Cell Invasion Assay kit (Trevigen, Inc., Gaithersburg, MD, USA), as previously described [39]. Briefly, after forming spheroids of MDA-MB-231 cells, spheroids were embedded in matrigel-based extracellular matrix components and treated with or without 10 ng/mL TNFα in the presence or absence of 25 μM DK4023 for 7 days. Invasive protrusions into the surrounding extracellular matrix were visualized with an EVOS® FL Auto Cell Imaging System (Life Technologies).

3.10. Immunoblot Analysis

Immunoblotting was performed as previously described [28]. In brief, MDA-MB-231 cells were lysed in a buffer comprising 20 mM HEPES (pH 7.2), 1% Triton X-100, 10% glycerol, 150 mM sodium hydroxide (NaCl), 10 μg/mL leupeptin, and 1 mM phenylmethylsulfonyl fluoride and electrophoresed on 10% SDS polyacrylamide gel electrophoresis (PAGE) gels. The separated protein bands were transferred onto nitrocellulose membranes, which were incubated with appropriate primary and horseradish peroxidase-conjugated secondary antibodies. The blots were then developed using an Amersham ECL Western Blotting Detection Kit (GE Healthcare Life Science, Chicago, IL, USA).

3.11. Statistical Analysis

Statistical significance was analyzed using GraphPad PRISM software version 8.2.1 (GraphPad Software Inc., La Jolla, CA, USA). A value of $p < 0.05$ was considered significant.

4. Conclusions

The present study identified a novel synthetic isoflavone derivative, DK4023, as an anti-invasive agent against metastatic breast cancer cells. Natural isoflavones, such as genistein and daidzein, are known to inhibit TNFα-induced migration and invasion of human breast cancer cells by preventing the inhibition of NF-κB [17]. This study demonstrated that DK4023 inhibited TNFα-induced migration and lamellipodium formation of highly metastatic MDA-MB-231 cells. Our findings also show that DK4023 suppressed TNFα-induced expression of *MMP9* mRNA through the downregulation of the ERK1/2-mediated expression of EGR-1, independently of NF-κB. More importantly, DK4023 substantially attenuated the invasive capability of MDA-MB-231 breast cancer cells. Although additional in vivo studies are warranted to validate the clinical efficacy of DK4023 in a metastasis animal model, we propose that DK4023 can serve as a promising agent to target the TNFα-ERK1/2 MAPK-EGR-1-MMP9 signaling pathway for the development of a chemopreventive or chemotherapeutic adjuvant, which can be used in combination with conventional chemotherapy, radiotherapy, or immunotherapy against metastatic cancers, particularly breast cancer.

Supplementary Materials: The following are available online at http://www.mdpi.com/1422-0067/21/14/5080/s1.

Author Contributions: M.J. and E.J. performed experiments and analyzed the data; S.A., D.K., J.K.S. and Y.L. synthesized chemicals. Y.H.L., Y.L., and S.Y.S. conceived and designed the experiments; S.Y.S. wrote the paper; Y.H.L. and Y.L. edited the manuscript; Y.L. supervised the project. All authors have read and agreed to the published version of the manuscript.

Funding: This study was supported by the National Research Foundation of Korea (NRF), funded by the Korea government (MSIT), under Grant number NRF-2019R1A2C1002677.

Acknowledgments: The paper was supported by the KU Research Professor Program of Konkuk University.

Conflicts of Interest: These authors contributed equally to this work.

Abbreviations

DK4023	(E)-5-((4-oxo-4H-chromen-3-yl)methyleneamino)-1-phenyl-1H-pyrazole-4-carbonitrile
EBS	EGR-1-binding site
EGR-1	Early growth response-1
ERK	Extracellular signal-regulated kinase
GAPDH	Glyceraldehyde 3-phosphate dehydrogenase
IκB	Inhibitor of nuclear factor-κB
IKK	IκB kinase
JNK	c-Jun N-terminal kinase
MAPK	Mitogen-activated protein kinase
MMP	Matrix metallopeptidase
NF-κB	Nuclear factor kappa B
RelA	v-rel Avian reticuloendotheliosis viral oncogene homolog A
RT-PCR	Reverse-transcription polymerase chain reaction
TNFα	Tumor necrosis factor alpha

References

1. Jemal, A.; Bray, F.; Center, M.M.; Ferlay, J.; Ward, E.; Forman, D. Global cancer statistics. *CA Cancer J. Clin.* **2011**, *61*, 69–90. [CrossRef] [PubMed]
2. Lacroix, M. Significance, detection and markers of disseminated breast cancer cells. *Endocr. Relat. Cancer* **2006**, *13*, 1033–1067. [CrossRef] [PubMed]
3. Kurozumi, S.; Joseph, C.; Sonbul, S.; Alsaeed, S.; Kariri, Y.; Aljohani, A.; Raafat, S.; Alsaleem, M.; Ogden, A.; Johnston, S.J.; et al. A key genomic subtype associated with lymphovascular invasion in invasive breast cancer. *Br. J. Cancer* **2019**, *120*, 1129–1136. [CrossRef] [PubMed]
4. Gatseva, A.; Sin, Y.Y.; Brezzo, G.; Van Agtmael, T. Basement membrane collagens and disease mechanisms. *Essays Biochem.* **2019**, *63*, 297–312.
5. Egeblad, M.; Werb, Z. New functions for the matrix metalloproteinases in cancer progression. *Nat. Rev. Cancer* **2002**, *2*, 161–174. [CrossRef]
6. Morini, M.; Mottolese, M.; Ferrari, N.; Ghiorzo, F.; Buglioni, S.; Mortarini, R.; Noonan, D.M.; Natali, P.G.; Albini, A. The alpha 3 beta 1 integrin is associated with mammary carcinoma cell metastasis, invasion, and gelatinase B (MMP-9) activity. *Int. J. Cancer* **2000**, *87*, 336–342. [CrossRef]
7. Hidalgo, M.; Eckhardt, S.G. Development of matrix metalloproteinase inhibitors in cancer therapy. *J. Natl. Cancer Inst.* **2001**, *93*, 178–193. [CrossRef]

8. Jones, J.L.; Shaw, J.A.; Pringle, J.H.; Walker, R.A. Primary breast myoepithelial cells exert an invasion-suppressor effect on breast cancer cells via paracrine down-regulation of MMP expression in fibroblasts and tumour cells. *J. Pathol.* **2003**, *201*, 562–572. [CrossRef]
9. Kupferman, M.E.; Fini, M.E.; Muller, W.J.; Weber, R.; Cheng, Y.; Muschel, R.J. Matrix metalloproteinase 9 promoter activity is induced coincident with invasion during tumor progression. *Am. J. Pathol.* **2000**, *157*, 1777–1783. [CrossRef]
10. Itoh, T.; Tanioka, M.; Yoshida, H.; Yoshioka, T.; Nishimoto, H.; Itohara, S. Reduced angiogenesis and tumor progression in gelatinase A-deficient mice. *Cancer Res.* **1998**, *58*, 1048–1051.
11. Itoh, T.; Tanioka, M.; Matsuda, H.; Nishimoto, H.; Yoshioka, T.; Suzuki, R.; Uehira, M. Experimental metastasis is suppressed in MMP-9-deficient mice. *Clin. Exp. Metastasis* **1999**, *17*, 177–181. [CrossRef] [PubMed]
12. Insuasty, B.; Tigreros, A.; Orozco, F.; Quiroga, J.; Abonia, R.; Nogueras, M.; Sanchez, A.; Cobo, J. Synthesis of novel pyrazolic analogues of chalcones and their 3-aryl-4-(3-aryl-4,5-dihydro-1H-pyrazol-5-yl)-1-phenyl-1H-pyrazole derivatives as potential antitumor agents. *Bioorg. Med. Chem.* **2010**, *18*, 4965–4974. [CrossRef]
13. Mellini, P.; Marrocco, B.; Borovika, D.; Polletta, L.; Carnevale, I.; Saladini, S.; Stazi, G.; Zwergel, C.; Trapencieris, P.; Ferretti, E.; et al. Pyrazole-based inhibitors of enhancer of zeste homologue 2 induce apoptosis and autophagy in cancer cells. *Philos. Trans. R. Soc. Lond. B Biol. Sci.* **2018**, *373*.
14. Lu, X.; Dong, G.; Zheng, Y.; Zhang, C.; Qiu, Y.; Lua, T.; Zhou, X. Synthesis and Anticancer Study of Novel 4H-Chromen Derivatives. *Anticancer Agents Med. Chem.* **2017**, *17*, 1070–1083. [CrossRef]
15. Singh, A.K.; Saxena, G.; Sahabjada, S.; Arshad, M. Synthesis, characterization and biological evaluation of ruthenium flavanol complexes against breast cancer. *Spectrochim. Acta A Mol. Biomol. Spectrosc.* **2017**, *180*, 97–104. [CrossRef] [PubMed]
16. Bhatia, R.K.; Singh, L.; Garg, R.; Kaur, M.; Yadav, M.; Madan, J.; Kancherla, S.; Pissurlenkar, R.R.S.; Coutinho, E.C. Novel p-Functionalized Chromen-4-on-3-yl Chalcones Bearing Astonishing Boronic Acid Moiety as MDM2 Inhibitor: Synthesis, Cytotoxic Evaluation and Simulation Studies. *Med. Chem.* **2020**, *16*, 212–228. [CrossRef] [PubMed]
17. Valachovicova, T.; Slivova, V.; Bergman, H.; Shuherk, J.; Sliva, D. Soy isoflavones suppress invasiveness of breast cancer cells by the inhibition of NF-kappaB/AP-1-dependent and -independent pathways. *Int. J. Oncol.* **2004**, *25*, 1389–1395. [PubMed]
18. Balkwill, F. TNF-alpha in promotion and progression of cancer. *Cancer Metastasis Rev.* **2006**, *25*, 409–416. [CrossRef]
19. Sethi, G.; Sung, B.; Aggarwal, B.B. TNF: A master switch for inflammation to cancer. *Front. Biosci.* **2008**, *13*, 5094–5107. [CrossRef]
20. Balkwill, F. Tumour necrosis factor and cancer. *Nat. Rev. Cancer* **2009**, *9*, 361–371. [CrossRef]
21. Lim, B.; Woodward, W.A.; Wang, X.; Reuben, J.M.; Ueno, N.T. Inflammatory breast cancer biology: The tumour microenvironment is key. *Nat. Rev. Cancer* **2018**, *18*, 485–499. [CrossRef] [PubMed]
22. Ham, M.; Moon, A. Inflammatory and microenvironmental factors involved in breast cancer progression. *Arch. Pharm. Res.* **2013**, *36*, 1419–1431. [CrossRef] [PubMed]
23. Mantovani, A.; Allavena, P.; Sica, A.; Balkwill, F. Cancer-related inflammation. *Nature* **2008**, *454*, 436–444. [CrossRef]
24. Krones-Herzig, A.; Adamson, E.; Mercola, D. Early growth response 1 protein, an upstream gatekeeper of the p53 tumor suppressor, controls replicative senescence. *Proc. Natl. Acad. Sci. USA* **2003**, *100*, 3233–3238. [CrossRef]
25. Ham, B.; Fernandez, M.C.; D'Costa, Z.; Brodt, P. The diverse roles of the TNF axis in cancer progression and metastasis. *Trends Cancer Res.* **2016**, *11*, 1–27. [PubMed]
26. Merino, F.; Pospich, S.; Raunser, S. Towards a structural understanding of the remodeling of the actin cytoskeleton. *Semin. Cell Dev. Biol.* **2020**, *102*, 51–64. [CrossRef] [PubMed]
27. Schaks, M.; Giannone, G.; Rottner, K. Actin dynamics in cell migration. *Essays Biochem.* **2019**, *63*, 483–495.
28. Shin, S.Y.; Kim, J.H.; Baker, A.; Lim, Y.; Lee, Y.H. Transcription factor Egr-1 is essential for maximal matrix metalloproteinase-9 transcription by tumor necrosis factor alpha. *Mol. Cancer Res.* **2010**, *8*, 507–519. [CrossRef]

29. Esteve, P.O.; Chicoine, E.; Robledo, O.; Aoudjit, F.; Descoteaux, A.; Potworowski, E.F.; St-Pierre, Y. Protein kinase C-zeta regulates transcription of the matrix metalloproteinase-9 gene induced by IL-1 and TNF-alpha in glioma cells via NF-kappa B. *J. Biol Chem.* **2002**, *277*, 35150–35155. [CrossRef]
30. Huang, L.; Lin, H.; Chen, Q.; Yu, L.; Bai, D. MPPa-PDT suppresses breast tumor migration/invasion by inhibiting Akt-NF-kappaB-dependent MMP-9 expression via ROS. *BMC Cancer* **2019**, *19*, 1159. [CrossRef]
31. Wang, J.; Li, S.; Li, X.; Li, B.; Li, Y.; Xia, K.; Yang, Y.; Aman, S.; Wang, M.; Wu, H. Circadian protein BMAL1 promotes breast cancer cell invasion and metastasis by up-regulating matrix metalloproteinase9 expression. *Cancer Cell Int.* **2019**, *19*, 182. [CrossRef] [PubMed]
32. Yang, H.L.; Thiyagarajan, V.; Shen, P.C.; Mathew, D.C.; Lin, K.Y.; Liao, J.W.; Hseu, Y.C. Anti-EMT properties of CoQ0 attributed to PI3K/AKT/NFKB/MMP-9 signaling pathway through ROS-mediated apoptosis. *J. Exp. Clin. Cancer Res.* **2019**, *38*, 186. [CrossRef] [PubMed]
33. Taniguchi, K.; Karin, M. NF-kappaB, inflammation, immunity and cancer: Coming of age. *Nat. Rev. Immunol.* **2018**, *18*, 309–324. [CrossRef] [PubMed]
34. Gashler, A.; Sukhatme, V.P. Early growth response protein 1 (Egr-1): Prototype of a zinc-finger family of transcription factors. *Prog. Nucleic Acid Res. Mol. Biol.* **1995**, *50*, 191–224.
35. Shin, S.Y.; Lee, J.M.; Lim, Y.; Lee, Y.H. Transcriptional regulation of the growth-regulated oncogene alpha gene by early growth response protein-1 in response to tumor necrosis factor alpha stimulation. *Biochim. Biophys. Acta* **2013**, *1829*, 1066–1074. [CrossRef]
36. Koh, D.; Jung, Y.; Ahn, S.; Mok, K.H.; Shin, S.Y.; Lim, Y. Synthesis and structure elucidation of polyphenols containing the N′-methyleneformohydrazide scaffold as aurora kinase inhibitors. *Magn. Reson. Chem.* **2017**, *55*, 864–876. [CrossRef]
37. Shin, S.Y.; Lee, J.; Park, J.; Lee, Y.; Ahn, S.; Lee, J.H.; Koh, D.; Lee, Y.H.; Lim, Y. Design, synthesis, and biological activities of 1-aryl-(3-(2-styryl)phenyl)prop-2-en-1-ones. *Bioorg. Chem.* **2019**, *83*, 438–449. [CrossRef]
38. Lee, D.Y.; Lee, D.H.; Jung, J.Y.; Koh, D.; Kim, G.S.; Ahn, Y.S.; Lee, Y.H.; Lim, Y.; Shin, S.Y. A synthetic chalcone derivative, 2-hydroxy-3′,5,5′-trimethoxychalcone (DK-139), suppresses the TNFalpha-induced invasive capability of MDA-MB-231 human breast cancer cells by inhibiting NF-kappaB-mediated GROalpha expression. *Bioorg. Med. Chem. Lett.* **2016**, *26*, 203–208. [CrossRef]
39. Shin, S.Y.; Kim, C.G.; Jung, Y.J.; Lim, Y.; Lee, Y.H. The UPR inducer DPP23 inhibits the metastatic potential of MDA-MB-231 human breast cancer cells by targeting the Akt-IKK-NF-kappaB-MMP-9 axis. *Sci. Rep.* **2016**, *6*, 34134. [CrossRef]

International Journal of
Molecular Sciences

Article

Biotin-Containing Third Generation Glucoheptoamidated Polyamidoamine Dendrimer for 5-Aminolevulinic Acid Delivery System

Aleksandra Kaczorowska [1], Małgorzata Malinga-Drozd [2], Wojciech Kałas [3], Marta Kopaczyńska [1], Stanisław Wołowiec [2,*] and Katarzyna Borowska [4]

1 Department of Biomedical Engineering, Faculty of Fundamental Problems of Technology, Wrocław University of Science and Technology, 27 Wybrzeże Wyspiańskiego Str., 50-370 Wrocław, Poland; aleksandra.kaczorowska@pwr.edu.pl (A.K.); marta.kopaczynska@pwr.edu.pl (M.K.)
2 Medical College, University of Rzeszów, Warzywna 1a, 35-310 Rzeszów, Poland; malgorzatamalinga@wp.pl
3 Department of Experimental Oncology, Ludwik Hirszfeld Institute of Immunology and Experimental Therapy, Polish Academy of Sciences, Rudolfa Weigla 12 Str., 53-114 Wrocław, Poland; wojciech.kalas@hirszfeld.pl
4 Department of Histology and Embryology with Experimental Cytology Unit, Medical University of Lublin, 11 Radziwiłowska Str., 20–080 Lublin, Poland; k_borowska@wp.pl
* Correspondence: swolowiec@ur.edu.pl; Tel.: +48-604-505-241

Citation: Kaczorowska, A.; Malinga-Drozd, M.; Kałas, W.; Kopaczyńska, M.; Wołowiec, S.; Borowska, K. Biotin-Containing Third Generation Glucoheptoamidated Polyamidoamine Dendrimer for 5-Aminolevulinic Acid Delivery System. *Int. J. Mol. Sci.* **2021**, *22*, 1982. https://doi.org/10.3390/ijms22041982

Academic Editor: Angela Stefanachi

Received: 13 January 2021
Accepted: 15 February 2021
Published: 17 February 2021

Publisher's Note: MDPI stays neutral with regard to jurisdictional claims in published maps and institutional affiliations.

Abstract: Polyamidoamine PAMAM dendrimer generation 3 (G3) was modified by attachment of biotin via amide bond and glucoheptoamidated by addition of α-D-glucoheptono-1,4-lacton to obtain a series of conjugates with a variable number of biotin residues. The composition of conjugates was determined by detailed 1-D and 2-D NMR spectroscopy to reveal the number of biotin residues, which were 1, 2, 4, 6, or 8, while the number of glucoheptoamide residues substituted most of the remaining primary amine groups of PAMAM G3. The conjugates were then used as host molecules to encapsulate the 5-aminolevulinic acid. The solubility of 5-aminolevulinic acid increased twice in the presence of the 5-mM guest in water. The interaction between host and guest was accompanied by deprotonation of the carboxylic group of 5-aminolevulinic acid and proton transfer into internal ternary nitrogen atoms of the guest as evidenced by a characteristic chemical shift of resonances in the ^{1}H NMR spectrum of associates. The guest molecules were most likely encapsulated inside inner shell voids of the host. The number of guest molecules depended on the number of biotin residues of the host, which was 15 for non-biotin-containing glucoheptoamidated G3 down to 6 for glucoheptoamidated G3 with 8 biotin residues on the host surface. The encapsulates were not cytotoxic against Caco-2 cells up to 200-μM concentration in the dark. All encapsulates were able to deliver 5-aminolevulinic acid to cells but aqueous encapsulates were more active in this regard. Simultaneously, the reactive oxygen species were detected by staining with H2DCFDA in Caco-2 cells incubated with encapsulates. The amount of PpIX was sufficient for induction of reactive oxygen species upon 30-s illumination with a 655-nm laser beam.

Keywords: photosensitizer delivery system; PAMAM dendrimer; photodynamic therapy; cytotoxicity; phototoxicity; colorectal adenocarcinoma

1. Introduction

Photodynamic therapy (PDT) is a non-invasive and an effective procedure that has been clinically approved for treating a number of diseases, including cancer. PDT is widely used in dermatology in the treatment of actinic keratoses [1], Bowen's disease [2,3], and cutaneous microbial infections, for example, acne, onychomycosis, and verrucae. [4]. With its range of indications continually expanding, PDT has also demonstrated potential as a treatment for dermatological malignancies such as squamous cell carcinoma [5] and superficial basal cell carcinomas [6]. In addition, PDT has been applied in treatment with other types of human cancers such as cervical cancer [7], breast cancer [8], glioma [9], prostate cancer [10], and colorectal cancer (CRC) [11,12]. CRC is one of the most common diagnosed cancers and one of the leading causes of death worldwide [13]. Due to great metastatic potential and high invasiveness, both radical and selective methods for CRC treatment are required [14]. The method that meets these criteria and can be used in personalized therapy of CRC is PDT [15]. One of the most common agents used in PDT of colorectal with satisfactory results cancer is 5-aminolevulinic acid (ALA), a precursor of Protoporphyrin IX (PpIX) [14,16].

PDT offers the advantages of minimal invasiveness, better cosmetic outcomes, and minimal functional disturbances. PDT is usually well tolerated and can be applied repeatedly at the same site [17]. The efficacy of PDT depends on the level of reactive oxygen species (ROS), such as singlet oxygen generated by photosensitizers (for example ALA) upon specific laser irradiation to induce tumor cell apoptosis and/or necrosis [18,19]. ALA has been shown to be effective at inducing PpIX after topical, oral, and intravenous applications in vivo [20]. Although ALA is widely used in PDT, cellular uptake of ALA is limited by its solubility and ability to penetrate biological barriers [21]. The efficiency of PDT is then far from satisfactory as optimal tissue accumulation and localization of ALA remains a clinical problem. Although it was shown that ABCG2 is an important efflux transporter of PpIX, especially in glioma cells [22] limiting intracellular PpIX concentration, the mechanism of PpIX accumulation is far more complex and involves the ALA influx and biosynthesis rate [23]. But in the case of Caco-2, the role of ABCG2 in transport of its other known substrates in cells is unclear [24], underlining a wide spectrum of influx/efflux mechanisms.

In order to avoid protein-assisted ALA influx barrier, we used dendrimeric drug carriers which showed an appreciable efficacy for ALA delivery [25–27]. Thus, we modified polyamidoamine (PAMAM) G3 and synthesized encapsulates with ALA. The cationic character of host PAMAM was eradicated by primary amine groups substitution with glucoheptoamide substituents. In addition, biotin was covalently attached into amine groups in order to facilitate selective binding of dendrimer into cell membrane of Caco-2, which is one of many cancer cells with overexpression of biotin receptors [28]. It has been previously shown that biotin-attached glucoheptoamidated PAMAM G3 dendrimer accumulated four times more effectively in fibroblasts (BJ), squamous cell carcinoma (SCC-15), and glioblastoma (U-118MG) cells than non-biotinylated analogues, in a time- and concentration-dependent manner [29]. Biotin-attached dendrimers were also less toxic than non-biotinylated analogues within 10–50 μM concentrations for all cell lines. We adopted these molecules as the host of ALA guest and tested the encapsulates for photocytotoxicity.

2. Results and Discussion

2.1. Chemistry

2.1.1. Synthesis and Characterization of Modified PAMAM G3 Dendrimers

PAMAM G3 dendrimer was modified by addition of *α-D*-glucoheptono-1,4-lactone (GHL) and biotin (Figure 1). Both substituents were amide-bonded to terminal (surface) primary amine groups of G3 in order to eradicate the cationic character of PAMAM G3 in neutral aqueous solution.

The conjugates were characterized by NMR spectroscopy. The ^{1}H NMR spectra of conjugates allowed us to determine the average stoichiometry of the conjugates by integration of PAMAM resonances which covered the 2.2–3.4 ppm region, versus glucoheptoamide (gh) –C$\underline{H}$ resonances which were multiplets within the 3.4–4.2 ppm region, and versus biotin resonances, some of which were well behind the G3 and gh resonances envelope. The multiplets of 3b, 4b, and 5b (total of six protons) were in the region 1.0–1.7 ppm and 8b and 9b quartets at 4.34 and 4.25 ppm, respectively (Figure 1; for atom numbering see also Scheme 1). Considering that all PAMAM CH_2(b) signals (triplets) were overlapped and located at 2.3 ppm, this resonance of intensity [120H] was used as an internal intensity reference. Other PAMAM G3 resonances were assigned as before [29]. Two sets of –C$\underline{H}$ resonances were observed for gh substituents, which were combined by scalar coupling in 2-D ^{1}H-^{1}H correlations spectroscopy (COSY), ^{1}H-^{13}C heteronuclear single quantum correlation (HSQC), and heteronuclear multiple bond correlation (HMBC) experiments (Figures A1 and A2 (Appendix A), respectively). Two doublets of 2g protons were observed at 4.10 and 3.96 ppm with a 3.63:1 intensity ratio. Such a considerable chemical shift difference only at 2g resonance suggested that gh substituents are bound by trans- and cis-amide bonds into terminal amine groups of G3. Considering the intensity of 2g and 2g′ doublets the average 26 gh substituents were trans- and 6 gh substituents are cis-amide-bonded in $G3^{32gh}$. Moreover, two triplet resonances from the G3 terminal ethylene group, namely d_3' and c_3' (Figure 1) had a double intensity of the 2g′ resonance. The patterns of major and minor resonances from gh occurred reproducibly in all compounds obtained here, i.e., biotin-containing conjugates $G3^{1B31gh}$, $G3^{2B27gh}$, $G3^{4B24gh}$, $G3^{6B21gh}$, and $G3^{8B17gh}$, as well as in various generation glucoheptoamidated PAMAM glycodendrimers, $G2^{16gh}$, $G4^{64gh}$, and $G5^{128gh}$.

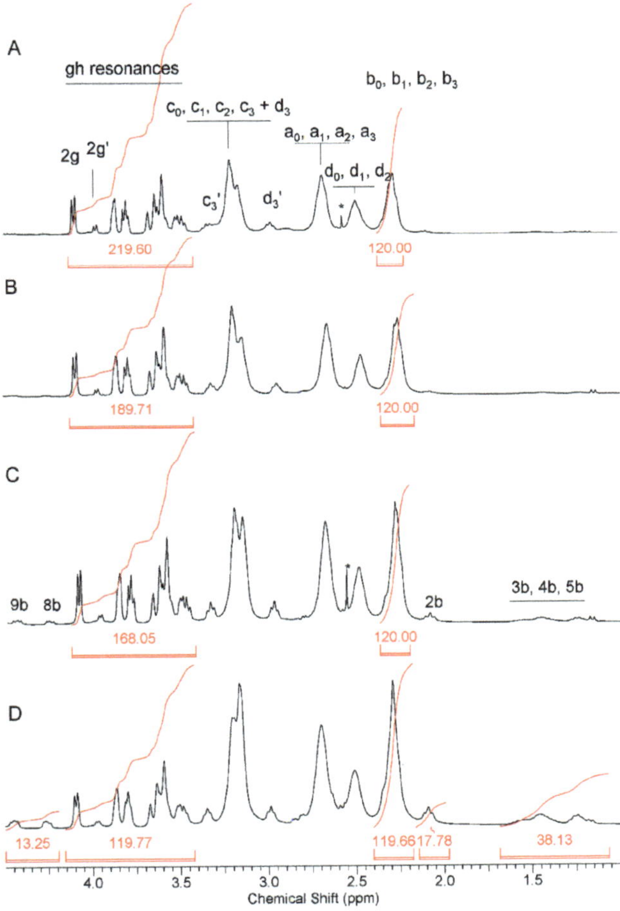

Figure 1. The ^{1}H NMR spectra of $G3^{1B31gh}$ (**A**), $G3^{2B27gh}$ (**B**), $G3^{4B24gh}$ (**C**), and $G3^{8B17gh}$ (**D**) in deuterium oxide. The broad signal for all CH_2(b) resonances of the PAMAM G3 core was used as internal intensity reference [120H]. The residual DMSO resonance is labeled with asterisk.

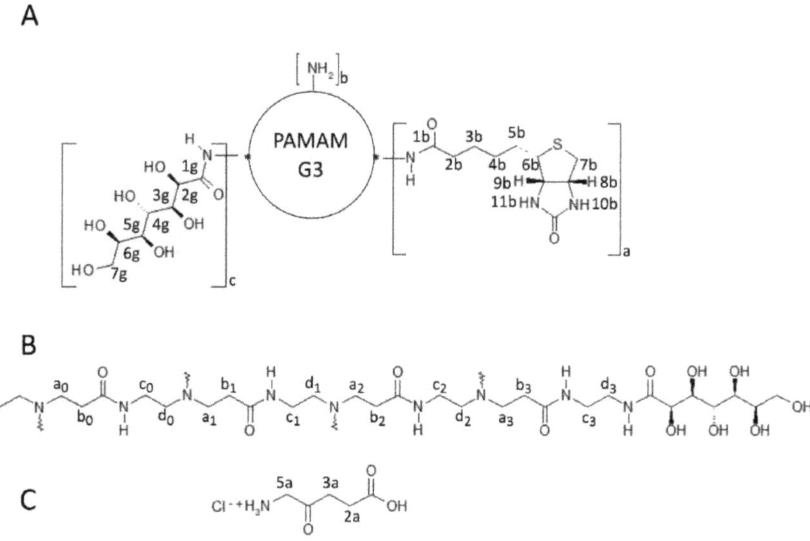

Scheme 1. General formula of polyamidoamine (PAMAM) generation three (G3) conjugates used herein with atom numbering of glucoheptoamide and biotin substituents (**A**) and one arm of PAMAM G3 starting from nitrogen atom of ethylenediamine core (**B**). The following conjugates were synthesized: $G3^{1B31gh}$ (a = 1, b = 0, c = 31); $G3^{2B27gh}$ (a = 2, b = 3, c = 27); $G3^{4B24gh}$ (a = 4, b = 4, c = 24); $G3^{6B21gh}$ (a = 6, b = 5, c = 21); $G3^{8B17gh}$ (a = 8, b = 7, c = 17). Atom numbering for 5-aminolevulinic acid (in protonated form) is given in (**C**).

The average stoichiometry of $G3^{1B31gh}$, $G3^{2B27gh}$, $G3^{4B24gh}$, $G3^{6B21gh}$, and $G3^{8B17gh}$ was determined by integration of gh –C<u>H</u> resonances versus internal reference of the PAMAM G3 $C\underline{H}_2$(b) signal at 2.30 ppm corresponding to [120H]. Thus, in the series of $G3^{1B31gh}$, $G3^{2B27gh}$, $G3^{4B24gh}$, $G3^{6B21gh}$, and $G3^{8B17gh}$, the intensities gh –C<u>H</u> resonances were: 220, 190, 168, 147, and 120, which after dividing the intensity by 7 (number of –C<u>H</u> protons in every gh substituent) gave 31, 27, 24, 21, and 17 gh substituents per conjugate molecule, respectively (Figure 1).

2.1.2. Interaction of 5-Aminolevulinic Acid with $G3^{Bgh}$ Conjugates; Stability of ALA@$G3^{gh}$ Encapsulates

ALA in aqueous solution occurs in the zwitterion form in pH between 5 and 7, according to determined $pK_a^{(COOH)} = 3.90$ and $pK_a^{(NH3)} = 8.05$ [30]. When this compound was added into the 5.1 mM solution of $G3^{32gh}$, all the ^{1}H and ^{13}C resonances in NMR spectra of ALA shifted remarkably and remained unchanged until 68 mM concentration. The ^{1}H NMR spectra of the solutions are presented in Figure 2 (titration experiment). Thus, the singlet $C\underline{H}_2$(5), and $C\underline{H}_2$(3) and $C\underline{H}_2$(2) resonances (5a, 3a, and 2a in Figure 2A) centered at 4.00, 2.77, and 2.58 for ALAxHCl shifted upfield to 3.96, 2.65, and 2.36 ppm, respectively, in the presence of $G3^{32gh}$ (Figure 2B–G). The largest was shift $\delta 2a = -0.22$ ppm compared to $\Delta 3a = -0.12$ ppm and negligible $\Delta 5a = -0.04$ ppm. This suggests the involvement of carboxylic group of ALA in interaction with $G3^{32gh}$ host molecule. Similar chemical shift od ^{13}C resonances were observed, with $\Delta C2 = -2.7$ ppm, $\Delta C3 = -1.4$ ppm, $\Delta C5 = 0.0$ ppm, $\Delta C4 = 1.0$ ppm, and $\Delta C1 = 2.4$ ppm (Figures A3 and A4).

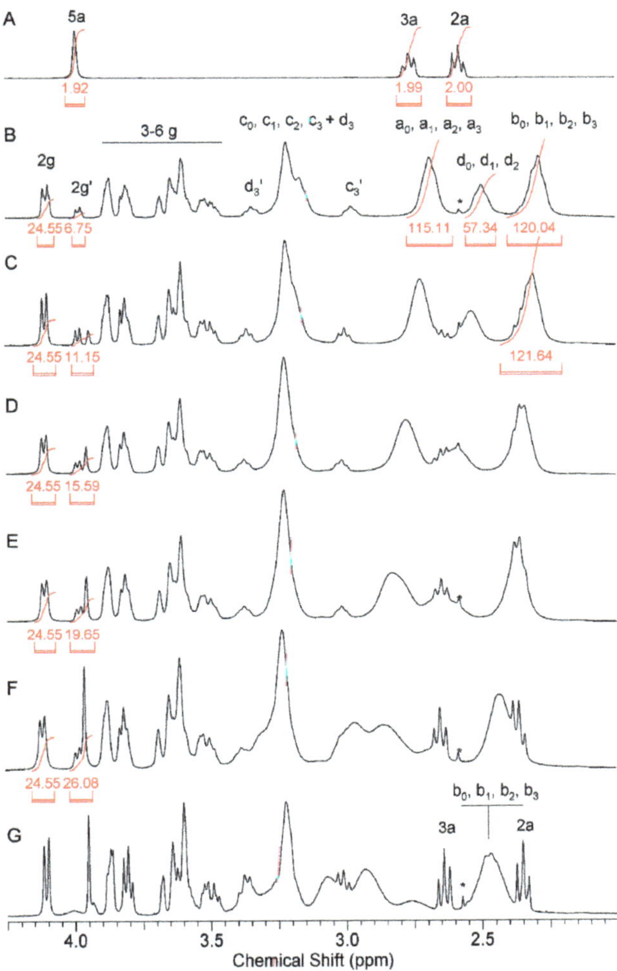

Figure 2. The ^{1}H NMR spectra of (**A**) 5-aminolevulinic acid (ALA) in D_2O; (**B**) G3^{32gh} (5.1 mM in D_2O); (**C–F**) G3^{32gh} (5.1 mM in D_2O) containing 2.2 (**C**), 4.4 (**D**), 6.5 (**E**), 9.7 (**F**), and 15 (**G**) equivalents of ALA. Final concentration of ALA in case of solution F was 76 mM, which is twice compared to ALA concentration in saturated aqueous solution. The residual DMSO resonance is labeled with asterisk.

The largest chemical shifts for carbon and proton nuclei next to carboxylic group in the ^{1}H and ^{13}C NMR spectra accompanying the interaction of ALA with G3^{32gh} indicated clearly that the carboxylic group underwent deprotonation.

Moreover, the ^{1}H NMR spectra of the G3 core changed regioselectively upon interaction with ALA with G3^{32gh}. Namely, the –CH_2- proton resonances neighboring ternary nitrogen atoms shifted downfield gradually upon addition of ALA. The common resonance of d_0, d_1, and d_2 as well as a_0, a_1, a_2, and a_3 (all methylene groups of internal shells) shifted from 2.49 and 2.65 (Figure 2B) into eventually 2.93 and 3.07 ppm (Figure 2G), respectively. In addition, the resonances of b protons of the PAMAM G3 core shifted downfield from 2.27 into eventually 2.46 ppm, while resonances of c protons and all gh C$\underline{H}$ protons remained unaltered after the addition of ALA into solution containing G3^{32gh} in deuterium oxide.

Similar NMR spectral patterns of solutions containing G3^{1B31gh}, G3^{2B27gh}, G3^{4B24gh}, G3^{6B21gh}, and G3^{8B17gh} conjugates and ALA were observed. The ^{1}H NMR spectra of the

$G3^{4B24gh}$ solution titrated with ALA are presented in Figure 3. The downfield shifts of PAMAM G3 core methylene proton resonances a, d, and b were consistent with protonation of ternary nitrogen atoms involved in dendrimer branching (see Scheme 1), while gh and B resonances remained unaltered upon addition of ALA. In addition, the methylene resonances d_3' and c_3' (outer sphere of G3 with cis-glucoheptoamidated amine groups) did not shift upon addition of ALA (Figure 3). This is presumably because the neighboring ternary nitrogen atoms are not involved in protonation due to steric hindrance of cis-gh groups which can be folded inside the outer sphere dendrimer cavity. In fact, four kinds of cavities in synthesized G3 conjugates could be expected based upon the formula of G3 conjugates, i.e., the number of them corresponding to branching nitrogen atoms: 2 in zero sphere, 4 in first sphere, 8 in second sphere, and finally 16 in third (outer-)sphere, some of the latter occupied by cis-gh substituents, which reduces the number of outer-sphere voids into 12. In total, the number of voids would then be 28, while the number of ternary nitrogen atoms, which are proton acceptors is 30. Thus, $G3^{32gh}$ could accept 30 proton cations from 15 HClxALA donors and encapsulate 15 anionic ALA carboxylates. In fact, we found such final stoichiometry for the ALA:$G3^{32gh}$ solution. In the case of the $G3^{1B31gh}$, $G3^{2B27gh}$, $G3^{4B24gh}$, $G3^{6B21gh}$, and $G3^{8B17gh}$ conjugates, the voids' availability was lower due to increasing steric hindrance imposed by B substituents, which not only enable total substitution of the remaining primary amine groups of G^B conjugates with GHL, but also reduce the ability of a conjugate to encapsulate ALA. Thus, the overall capacity of ALA encapsulation estimated by NMR spectral monitoring was 12, 10, 8, 7, and 6 equivalents of ALA per one equivalent of $G3^{1B31gh}$, $G3^{2B27gh}$, $G3^{4B24gh}$, $G3^{6B21gh}$, and $G3^{8B17gh}$ conjugates, respectively. The numbers corresponded to 5-mM solutions of conjugates in water and can be even higher, although the encapsulates became water insoluble in the case of 7-mM solutions of the conjugates. Nonetheless, the encapsulates containing 6 equivalents of ALA were soluble in dimethylsulfoxide and water, and stable as it was evidenced by exhaustive dialysis with water. Therefore, we prepared the solutions of all conjugates in dimethylsulfoxide at a 6-mM dendrimeric host and 36-mM ALA guest and initially used them as stock solutions for biological tests (6ALA@$G3^{Bgh}$). However, the stock solutions in DMSO were finally replaced by aqueous stock solutions (2.5 mM host $G3^{2B27gh}$ or $G3^{6B21gh}$, containing 10 mM ALA; 5ALA@$G3^{Bgh}$ encapsulates) for phototoxicity studies in order to avoid interference from water–DMSO mixtures on metabolic behavior of the colorectal adenocarcinoma model line, i.e., Caco-2 cells. For simplicity the encapsulates would be further abbreviated as A@D^n (where n—the number of biotin moieties in the conjugate, D—glucoheptoamidated PAMAM G3 dendrimer).

The encapsulation of ALA by conjugates was accompanied by an increase of molecular size of conjugates, comparable to those observed for protonation. Thus, the number-average size determined by dynamic light scattering (DLS) of $G3^{32gh}$ in water was 1.8 (±0.2) nm, while the molecule expanded into 4.0 (±0.2) nm at pH 5 [31]. The molecules of $G3^{32gh}$, $G3^{2B27gh}$, and $G3^{6B21gh}$ expanded from 2.0 (±0.2) nm in water into 5.0, 5.3, and 5.4 (±0.3) nm upon addition of 5 equivalents of ALAxHCl and further for solution containing 16 equivalents of ALA. This was attributed to both protonation of tertiary amine groups and encapsulation of ALA.

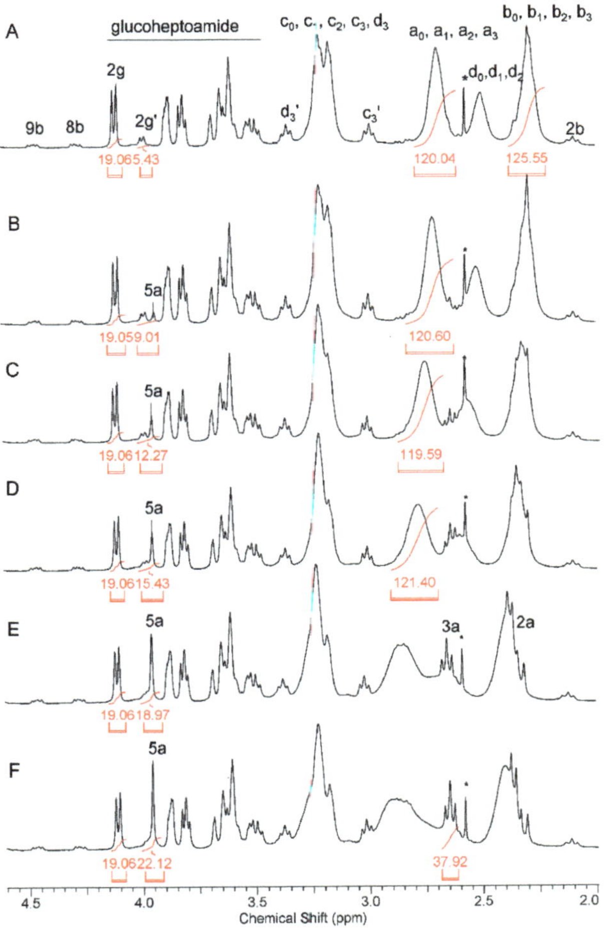

Figure 3. The ^{1}H NMR spectra of (**A**) $G3^{4B24gh}$ in D_2O; (**B–F**) $G3^{4B24gh}$ (6.2 mM in D_2O) containing 1.8 (**B**), 3.4 (**C**), 5.0 (**D**), 8.0 (**E**), and 10 (**F**) equivalents of ALA. Final concentration of ALA in case of solution E was 62 mM, which is 1.7 higher if compared to ALA concentration in saturated aqueous solution. The residual DMSO resonance is labeled with asterisk.

2.2. Biological Studies

2.2.1. Cytotoxicity of Glucoheptoamidated PAMAM-Biotin Conjugates and ALA Encapsulates

To assess the dark cytotoxicity of aqueous and DMSO solutions of PAMAM-biotin conjugates and its ALA@$G3^{gh}$ encapsulates, the MTS cell viability assay was performed. The concentration of dendrimers was normalized to the concentration of ALA in acid-containing conjugates which was equal 180, 540 µM, and 1.08 mM, respectively, in solutions in H_2O/DMSO (A@D^1, A@D^2, A@D^4, and A@D^8) and water (A@D^2 and A@D^6). The viability of Caco-2 cells treated with dendrimer conjugates at concentrations of 30, 90, and 180 µM in H_2O/DMSO is shown in Figure 4. Results of the experiments conducted with the use of aqueous solutions of dendrimers at concentrations of 45, 135, and 270 µM were shown in Figure 5. All the conjugates exhibited acceptable cytotoxicity to Caco-2 colorectal adenocarcinoma cells and did not affect the morphology of cells (Figure 5C,D). Dendrimer conjugates containing ALA were not more cytotoxic, indicating low dark cytotoxicity.

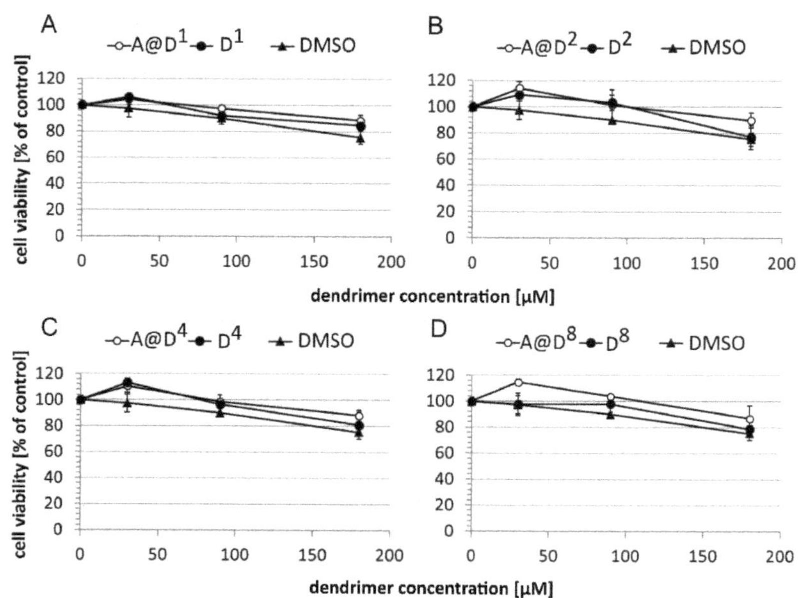

Figure 4. Viability of Caco-2 cells after treatment with G3 PAMAM dendrimer conjugates: (**A**) A@D^1, D^1; (**B**) A@D^2, D^2; (**C**) A@D^4, D^4, and (**D**) A@D^8, D^8 at concentrations of 30, 90, and 180 µM. The concentrations of DMSO were 0.5%, 1.5%, and 3%, respectively. Data are presented as means ± SD normalized to untreated control.

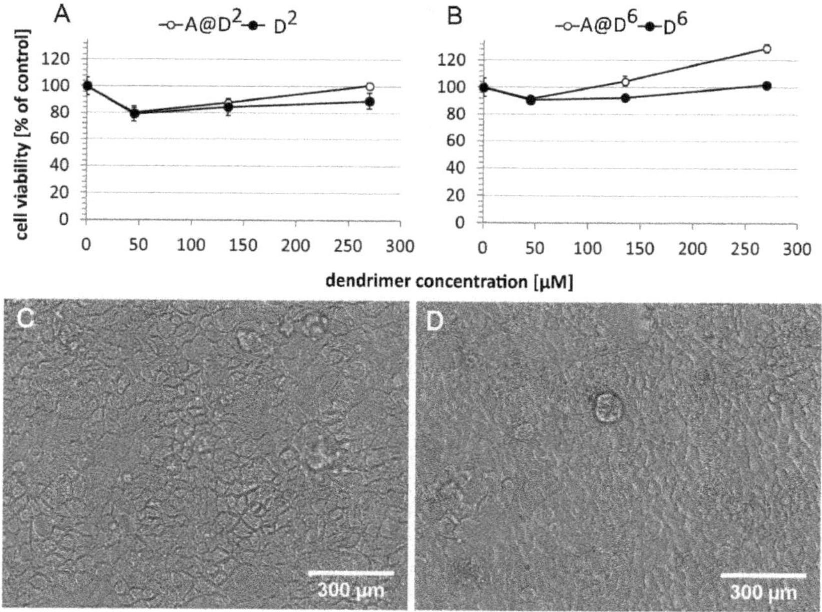

Figure 5. Viability of Caco-2 cells after treatment with aqueous solutions of encapsulates and dendrimer conjugates: (**A**) A@D^2, D^2 and (**B**) A@D^6, D^6 at concentrations of 45, 135, and 270 µM. Data are presented as means ± SD normalized to untreated control. Images of Caco-2 cells: (**C**) Control and (**D**) incubated with D^2 at concentration of 270 µM.

2.2.2. Accumulation of Protoporphyrin IX in Caco-2 Cells

To assess the accumulation of Protoporphyrin IX, Caco-2 cells were treated with 180 μM of dendrimers in H_2O/DMSO (prepared by dilution of stock solutions in DMSO; 6 D^n and 36 mM ALA, where n = 1B, 2B, 4B, or 8B). After 24-h incubation, fluorescence at 605 nm was assayed using cytofluorimeter. This wavelength corresponded to the peak of PpIX fluorescence [32]. As expected, the highest fluorescence intensity was observed in cells treated with conjugates containing Protoporphyrin IX precursor–ALA (encapsulates A@D^1, A@D^2, A@D^4, and A@D^8). The greatest shift of the mean fluorescence channel in relation to host dendrimers was observed for A@D^1 and A@D^2 encapsulates (Figure 6), which indicates the highest accumulation of Protoporphyrin IX in cells.

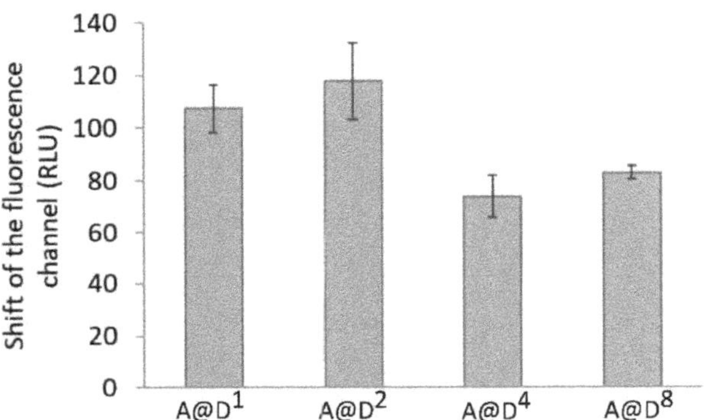

Figure 6. Shift of the mean fluorescence channel of A@D^1, A@D^2, A@D^4, and A@D^8 encapsulates in relation to conjugates without ALA (D^1, D^2, D^4, and D^8, respectively). Fluorescence detection wavelength was 605 nm. Data are presented as means ± SD.

In another experiment, cells were treated with 180 μM of encapsulates in H_2O/DMSO or with 270 μM of encapsulates in water (Figure 7) (the concentration of ALA was 1.08 mM in case of both solvents). After 24-h incubation, a fluorescence at 711 nm was assayed using flow cytofluorimeter. The 710 nm wavelength corresponded to another peak of PpIX fluorescence [33].

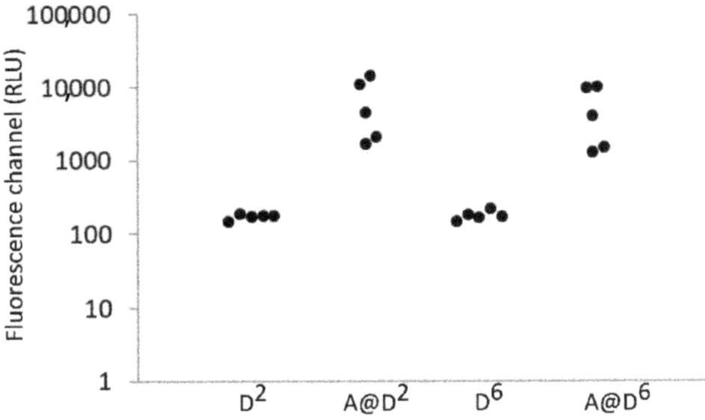

Figure 7. Mean fluorescence channel of D^2, A@D^2, D^6, and A@D^6 dendrimers upon excitation with 405 nm laser. Fluorescence detection wavelength was 711 nm.

Fluorescence detection at this wavelength provided the greatest values of the mean fluorescence channel of PpIX synthesised from ALA contained in aqueous solutions of dendrimers. Using this method of fluorescence detection, the shift of the mean fluorescence channel for conjugates dissolved in H_2O/DMSO was about 10 times smaller than in the case of detection at 605 nm. For aqueous solutions of encapsulates, the significant increase of mean fluorescence channel was observed for both A@D^2 and A@D^6 (Figure 7). This indicates that 24-h incubation of cells with ALA-containing conjugates is a sufficient time for Protoporphyrin IX synthesis. Short incubations resulted in an undetectable increase of mean fluorescence channel.

Interestingly, we found that an increased number of biotin residues in the host conjugate did not improve the intracellular level of PpIX in comparable conditions. In fact, in the case of host $G3^{4B24gh}$ and $G3^{8B17gh}$, a decrease of the PpIX level was noticed. This inverse effect was probably due to an association of a higher biotin-substituted conjugate. Indeed, we observed considerably higher values of volume-average molecular size of $G3^{4B24gh}$ and $G3^{8B17gh}$ (>6.5 nm in both cases) in comparison with the number-average diameter (5.0 nm both). Considerably, a larger volume-average in relation to the number-average molecular size evidenced the association of dendrimers and was previously found for cytisine-$G3^{gh}$ conjugates [31]. Another reason for such an effect can be related to the elevated efflux of high biotin-substituted conjugates, especially in the presence of DMSO.

2.2.3. Intracellular Reactive Oxygen Species Level

In order to study the phototoxicity of ALA encapsulates, we examined the level of reactive oxygen species (ROS) induced upon the illumination of treated cells. Caco-2 cells were incubated with conjugates for 24 h and then illuminated with a 655-nm laser beam, corresponding to the PDT window of Protoporphyrin IX. Illumination was performed for (A) 30s (E = 137 mW/cm^2, H = 4.1 J/cm^2) and (B) 60s (E = 137 mW/cm^2, H = 8.2 J/cm^2). With the applied illumination parameters and ALA concentration, we did not observe the phototoxic effect of conjugates dissolved in DMSO and water upon 1-min illumination. The phototoxic effect, manifested by the ROS production in cells illuminated for 30 s, was observed for A@D^2 and A@D^6 conjugates in aqueous solution, not in the presence of traces of DMSO (Table 1).

Table 1. Shift of the mean fluorescence channel of A@D^1, A@D^2, A@D^4, A@D^8 (aqueous/DMSO solution), and A@D^2 and A@D^6 (in aqueous solution) dendrimers in relation to conjugates without ALA (D^1, D^2, D^4, D^8, and D^2 and D^6, respectively). The parameters of cell illumination were: (A) λ = 655 nm, E = 137 mW/cm^2, H = 4.1 J/cm^2, and (B) λ = 655 nm, E = 137 mW/cm^2, H = 8.2 J/cm^2. Excitation wavelength was 488 nm; fluorescence was detected at 530 nm.

	Shift of the Fluorescence Channel of H2DCFDA (RLU)					
	H_2O/DMSO Solution		**in H_2O (DMSO Excluded)**			
Experiment	A@D^2	A@D^6	A@D^1	A@D^2	A@D^4	A@D^8
A (30 s illumination)	263.28	443.29				
B (60 s illumination)	−206.1	44.49	−529.43	−363.42	−514.6	−1084.1

3. Materials and Methods

3.1. Reagents and Methods

All the chemicals used in synthesis of the PAMAM G3 dendrimer and its conjugates as well as 5-aminolevulinic acid hydrochloride were purchased from Merck (KGaA, Darmstadt, Germany).

The 1-D ^1H and ^{13}C NMR spectra as well as 2-D ^1H-^1H correlation spectroscopy (COSY), ^1H-^{13}C heteronuclear single quantum correlation (HSQC), and the heteronuclear multiple bond correlation spectra (HMBC), were recorded in deuterated water using Bruker 300 MHz (Rheinstetten, Germany) and worked up with TopSpin 3,5 software at the College of Natural Sciences, University of Rzeszów.

3.2. Chemical Syntheses

3.2.1. PAMAM G3 Substituted with Biotin

PAMAM G3 dendrimer was obtained at the 5 millimolar scale according to the procedure published by Tomalia et al. [34], and stored as 20.1 mM solution in methanol for further use. Then G3 was substituted with 1, 2, 4, and 8 equivalents of biotin by stepwise addition of solid *N*-hydroxysuccinimide ester of biotin (NHS-B) into 0.414 g of G3 (60.3 μM) dissolved in 3 mL dimethylsulfoxide (DMSO) with vigorous stirring. The mixture was left at an ambient temperature for 12 h, transferred into dialytic tube (nitrocellulose, MW_{cutoff}—3.5 kD), and dialyzed against water for 3 days (5 times 3 dm^3). Water was evaporated under reduced pressure and products were identified by ^1H NMR spectroscopy as G3 substituted with average 1, 2, 4, 6, and 8 equivalents of biotin per one PAMAM G3 molecule: $G3^{1B}$, $G3^{2B}$, $G3^{4B}$, $G3^{6B}$, and $G3^{8B}$, respectively. The isolated yield was above 80% in every case.

3.2.2. PAMAM G3 Substituted with Biotin and Glucoheptoamide

The biotin-substituted G3 PAMAM dendrimers were further converted by blocking amine groups in reaction of *ca* 20 μM of $G3^{1B}$, $G3^{2B}$, $G3^{4B}$, $G3^{6B}$, and $G3^{8B}$ with 20% excess of α-*D*-glucoheptono-1,4-lactone (GHL) in relation to terminal amine group of G3. In a typical procedure to the 120 mg of $G3^{2B}$ (16.3 μM), 2 mL DMSO solid GHL was added stepwise (135 mg, 648 μM) with magnetic stirring until it dissolved. The mixture was kept at 50 °C for 6 h and then dialyzed against water for 2 days. Afterwards, water was removed under reduced pressure and solid products were isolated at >80% yield.

The stoichiometry of obtained conjugates was examined by the ^1H NMR spectroscopy (Figure 2) and it was noticed that not all amine groups were converted into glucoheptoamide derivative. The following conjugates were obtained: $G3^{1B31gh}$, $G3^{2B27gh}$, $G3^{4B24gh}$, $G3^{6B20gh}$, and $G3^{8B17gh}$. Thus, in these conjugates, average 0, 3, 4, 5, and 7 amine groups were left unsubstituted, respectively. We also prepared biotin-free conjugate $G3^{32gh}$, with all primary amine groups converted into glucoheptoamide as described elsewhere [31,35].

The molecular size of conjugates was determined by the DLS method as before [31]. The number-average diameters of $G3^{32gh}$, $G3^{2B27gh}$, and $G3^{6B21gh}$ at pH 5 (0.05 M acetate buffer) were 4.0 (±0.2), 4.2 (±0.3), and 4.4 (±0.3) nm, while the volume-average diameters were 4.6 (±0.2), 5.0 (±0.3), and 5.0 (±0.3) nm, respectively. The size of all conjugates in water were within 1.8–2.0 (±0.3) nm (number-average).

3.2.3. Encapsulation of ALA in the $G3^{Bgh}$ Conjugates

Interaction between the $G3^{32gh}$ conjugate and ALA in aqueous solution was monitored by the ^1H NMR spectroscopy. Thus, solid ALA was added into an NMR tube containing 5.1 mM $G3^{32gh}$, and NMR spectra were recorded after ALA was dissolved. Addition of ALA was continued until final portion of ALA remained undissolved. The spectrum of the final solution in equilibrium with the precipitate was taken after one day equilibration. The precipitate was separated from the mixture and identified as pure ALA. The final concentration of ALA in the presence of 5.1 $G3^{32gh}$ was 76 mM, which was *ca* twice higher in comparison to the concentration of ALA in a saturated aqueous solution (D_2O). The ^1H NMR spectra of starting compounds and mixtures of $G3^{32gh}$ and ALA are presented in Figure 3.

Similar experiments were performed in the case of all $G3^{1B31gh}$, $G3^{2B27gh}$, $G3^{4B24gh}$, and $G3^{8B17gh}$ conjugates and ALA. The results are illustrated by series of ^1H NMR spectra of $G3^{4B24gh}$ and ALA in D_2O in Figure 4. In the case of this experiment, the concentration of $G3^{4B24gh}$ conjugate was 6.2 mM, while the final concentration of ALA was *ca* 50 mM in equilibrium with the solid. The solid was isolated and identified as a $G3^{4B24gh}$: ALA 1:8 complex. Similar results were also obtained for other conjugates containing a variable amount of biotin.

Finally, the 6-mM solutions of $G3^{Bgh}$ conjugates and 36 mM ALA in DMSO were prepared and used as stock solution for biological tests. The ^1H NMR spectra of all solutions were examined after one month storage at room temperature and showed no traces of converted ALA.

In order to determine the stability of ALA@G3^{4B24gh} and other encapsulates, the 5-mM aqueous solution of 8ALA@G3^{4B24gh} encapsulate (10 mL) was dialyzed in a cellulose bag (MW_{cutoff} = 3.5 kDa) against 0.1 M phosphate buffer pH 7.2 (3 dm^3) four times for 4 h and at every step. The ^{1}H NMR spectrum of remaining encapsulate was recorded. The final composition of dialyzed encapsulate was determined as 6ALA@G3^{4B24gh}.

The hydrodynamic diameters determined by DLS in aqueous solutions containing G3^{32gh}, G3^{2B27gh}, and G3^{6B21gh}, and 6 equivalents of ALA were 5.0, 5.3, and 5.4 (±0.3) nm, which increased slightly upon addition of further 10 equivalents of ALA into: 5.2, 5.4, and 5.5 (±0.3) nm, respectively.

3.3. Cell Culture Conditions and Materials

Colorectal adenocarcinoma cells (Caco-2) were cultured in eagle medium (Ludwik Hirszfeld Institute of Immunology and Experimental Therapy) supplemented with 10% fetal bovine serum (FBS, Gibco, Thermo Fisher Scientific, Waltham, MA, USA) and 2 mM L-glutamine (Sigma–Aldrich, Saint Louis, MO, USA). Cells were cultured on Petri dishes (Sarstedt AG and Co. KG, Nümbrecht, Germany), flasks, and 24- or 96-well plates (Corning Life Sciences, Kennebunk, ME, USA) at 37 °C in a humidified atmosphere of 5% CO_2.

3.4. Toxicity Studies

Caco-2 cells were seeded on a 96-well plate (Corning Life Sciences, Kennebunk, ME, USA). Cells were treated with ALA encapsulates at concentrations of 30, 90, and 180 μM in H_2O/DMSO and with aqueous solutions of dendrimers at concentrations of 45, 135, and 270 μM (Table 2). After 24-h incubation at 37 °C, the viability of cells was evaluated with an MTS assay, and 15 μL of CellTiter 96®AQueous One Solution Reagent (Promega, Madison, WI, USA) was added to each well. After 2 h at 37 °C in a humidified, 5% CO_2 atmosphere, the absorbance at 490 nm was recorded using the Wallac 1420 Victor 2 plate reader (Perkin Elmer, Waltham, MA, USA). Cells incubated with H_2O/DMSO (Sigma–Aldrich, Saint Louis, MO, USA) at the same dilutions as the dendrimer served as a solvent control.

Table 2. Concentrations of dendrimers, ALA, and DMSO in samples used in toxicity studies.

Compound	Dendrimer Concentration	ALA Concentration	DMSO Concentration
D^1, D^2, D^4, D^8	30 μM	0	0.5%
	90 μM	0	1.5%
	180 μM	0	3%
A@D^1, A@D^2, A@D^4, A@D^8	30 μM	180 μM	0.5%
	90 μM	540 μM	1.5%
	180 μM	1.08 mM	3%
D^2, D^6	45 μM	0	–
	135 μM	0	
	270 μM	0	
A@D^2, A@D^6	45 μM	180 μM	
	135 μM	540 μM	
	270 μM	1.08 mM	

3.5. Accumulation of Protoporphyrin IX in Caco-2 Cells

Caco-2 cells were seeded on a 24-well plate (Corning Life Sciences, Kennebunk, ME, USA). Cells were treated with A@D^n conjugates at a concentration of 180 μM (solutions in H_2O/DMSO) and at a concentration of 270 μM (aqueous solutions). After 24-h incubation at 37 °C in a humidified, 5% CO_2 atmosphere, cells were harvested and washed with PBS (phosphate-buffered saline solution, Ludwik Hirszfeld Institute of Immunology and

Experimental Therapy). Cells were analyzed using LSRFortessa flow cytometer (Becton Dickinson, Franklin Lakes, NJ, USA). Excitation and emission wavelengths were chosen to detect the presence of Protoporphyrin IX in cells: a 405-nm laser was used as a fluorescence excitation source, and the fluorescence was measured at 605 and 710 nm. For data analysis, Flowing Software 2.5.1 was used (Turku University, Turku, Finland).

3.6. Intracellular Reactive Oxygen Species Level

Reactive oxygen species (ROS) induction was assessed by measuring the 488-/530-nm fluorescence of H2DCFDA (6-carboxy-20, 70-dichlorodihydrofluorescein diacetate, di(acetoxymethy ester), Molecular Probes, Thermo Fisher Scientific, Inc., Waltham, MA, USA. In this regard, Caco-2 cells were seeded on a 24-well plate (Corning Life Sciences, Kennebunk, ME, USA). Cells were treated with A@D^n encapsulates at concentration of 180 μM (solutions in H_2O/DMSO) and at a concentration of 270 μM (aqueous solutions). After 24-h incubation at 37 °C in a humidified, 5% CO_2 atmosphere, culture medium was harvested. Cells were stained with 10 μM of H2DCFDA for 30 min in PBS with 10% FBS and washed twice with warm PBS with 2.5% FBS. Next, cells were illuminated ((A) λ = 655 nm, E = 137 mW/cm^2, H = 4.1 J/cm^2 and (B) λ = 655 nm, E = 137 mW/cm^2, H = 8.2 J/cm^2). Then cells were harvested, washed in PBS, and analyzed using LSRFortessa flow cytometer (Becton Dickinson, Franklin Lakes, NJ, US). For data analysis, Flowing Software 2.5.1 was used (Turku University, Turku, Finland).

4. Conclusions

Nanosized dendrimers are currently tested as drug delivery systems in many laboratories. The current state of nanotechnology applications in colorectal cancer has been recently reviewed, including clinical trials and status [11]. The chitosan nanoparticles loaded with ALA were already demonstrated as a photosensitizer delivery system (PSDS) and convenient tool for fluorescent endoscopic detection of colorectal cancer cells [12]. The chitosan was further equipped with folic acid for targeting cell membrane and enhancing nanoparticle endocytosis via the folate receptor [36]. Generally, a dendrimer-based PSDS can be constructed either as encapsulates of PS in dendrimer [37–40] or conjugates of the dendrimer with the PS molecule attached covalently [31]. In both strategies, the dendrimeric carrier can be designed to enter the cell; in the latter, the active molecule of PS needs to be cleaved from the conjugate [41]. Another important factor of PSDS is to equip the carrier with a cancer cell membrane-targeting molecule, which is commonly folate [37] or biotin [29]. We used the glucoheptoamidated PAMAM G3, equipped with variable equivalents of biotin, to address the encapsulated ALA to colorectal adenocarcinoma cells in vitro using the Caco-2 line. From previous studies, we know that the G3^{Bgh} carriers enter the cells of normal fibroblasts (BJ), squamous cell carcinoma (SCC-15), and glioblastoma (U-118 MG) within 24 h in a time- and concentration-dependent manner and are 3–4 times less cytotoxic than the non-biotinylated carrier. All cell lines survived a 50-μM concentration of G3^{Bgh} as well as keratinocytes (HaCat) [29,42].

In this paper, we found that the third generation polyamidoamine dendrimer, with amide-linked biotin and glucoheptoamide substituents (G3^{Bgh}), is a highly effective host to encapsulate 12-6 molecules of 5-aminolevulinic acid (ALA) in water. Aminolevulinic acid was deprotonated and hydrogen cation was transferred from its carboxylic group into internal ternary nitrogen atoms of the host. The encapsulated guest molecules were bound by ionic interaction and were released slowly in neutral pH. The payload of encapsulates depended on the number of biotin residues in the conjugate and equaled 12 for one biotin-containing conjugate.

The conjugates containing 2 to 6 biotin residues were able to encapsulate at least 7 guest molecules of ALA. When aqueous solutions were applied to colorectal adenocarcinoma cells (Caco-2 line) at a 1-mM concentration of ALA for 24 h, the rapid increase of Protoporphyrin IX was observed. The Protoporphyrin IX-induced cells produced single oxygen upon 30-s irradiation with a 655-nm laser pulse. Both phenomena accompanied the regular response and photocytotoxicity pattern for photodynamic anticancer therapy, including colorectal cancer [11,12].

The encapsulates of ALA in G3^{Bgh} offered the possibility to use this PSDS in local treatment if the concentration of deposited encapsulates was higher. In order to estimate the effectiveness of the designed PSDS, the PK profiles were needed from in vivo studies on model animals. In local PS delivery, the concentration of G3^{Bgh} could be higher than 0.3 mM; thus, the ALA concentration could be at least 3 mM. Another issue to optimize in vivo is the number of biotins in G3^{Bgh}; here we did not find the difference in Protoporphyrin IX level between the 2 and 6 biotin-containing host. The elaborated PSDS is currently being tested for treatment of the skin impairments encountered in the introduction [43].

Author Contributions: Conceptualization, K.B., M.K., and S.W.; methodology, M.K., W.K., and S.W.; formal analysis, M.K. and S.W.; investigation, A.K., M.M.-D., W.K., M.K., and S.W.; writing—original draft preparation, A.K., M.K., and S.W.; writing—review and editing, A.K., M.K., K.B., and S.W.; supervision, K.B., M.K., and S.W. All authors have read and agreed to the published version of the manuscript.

Funding: This work was partially supported by statutory funds of the Department of Biomedical Engineering, Wrocław University of Science and Technology, statutory funds of the Medical College, University of Rzeszów, and statutory program no. 3/2020 of the Institute of Immunology and Experimental Therapy, PAS, Wrocław.

Institutional Review Board Statement: Not applicable.

Informed Consent Statement: Not applicable.

Data Availability Statement: The data presented in this study are available on request from the corresponding author. The data are not publicly available due to data are collected and stored in 4 different laboratories, encountered in affiliations of all authors. The corresponding author is responsible for transferring the data.

Acknowledgments: Not applicable.

Conflicts of Interest: The authors declare no conflict of interest.

Appendix A

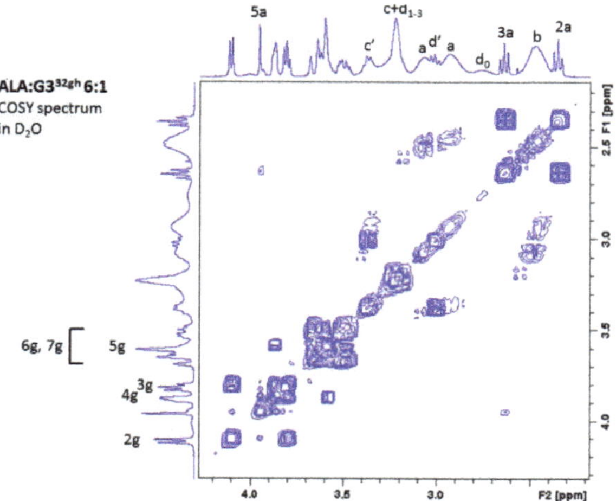

Figure A1. ^{1}H-^{1}H- COSY spectrum of G3^{32gh} in D$_2$O.

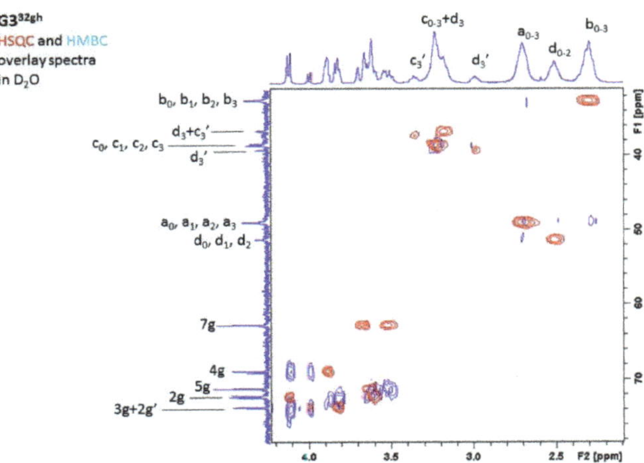

Figure A2. ^{1}H-^{13}C HSQC (red cross-peaks) and HMBC (blue cross-peaks) spectra of G3^{32gh} in D_2O.

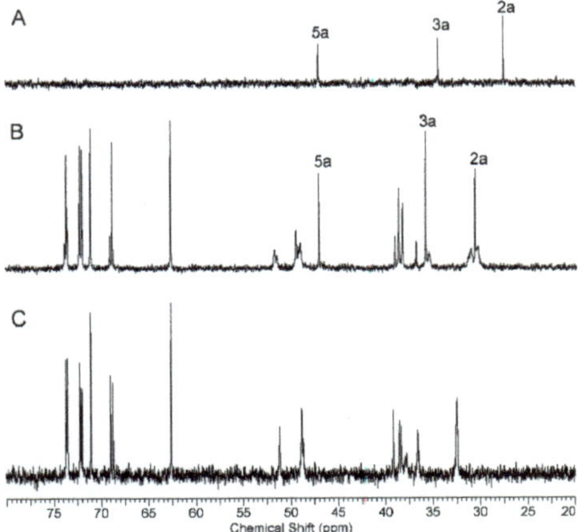

Figure A3. Aliphatic region of ^{13}C NMR spectra of ALA (**A**), G3^{32gh} an d 8 equivalents of ALA (**B**), and G3^{32gh} (**C**) in D_2O.

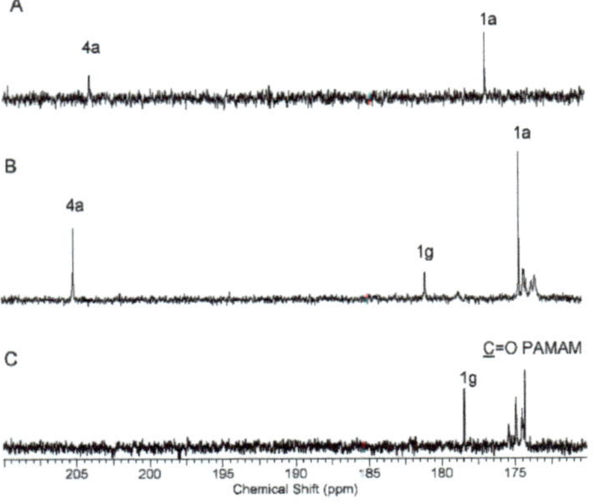

Figure A4. Downfield region of ^{13}C NMR spectra of ALA (**A**), G3^{32gh} and 8 equivalents of ALA (**B**), and G3^{32gh} (**C**) in D_2O.

References

1. Nguyen, M.; Sandhu, S.S.; Sivamani, R.K. Clinical utility of daylight photodynamic therapy in the treatment of actinic keratosis—A review of the literature. *Clin. Cosmet. Investig. Dermatol.* **2019**, *12*, 427–435. [CrossRef]
2. Hasegawa, T.; Suga, Y.; Mizuno, Y.; Haruna, K.; Ogawa, H.; Ikeda, S. Efficacy of photodynamic therapy with topical 5-aminolevulinic acid using intense pulsed light for Bowen's disease. *J. Dermatol.* **2010**, *37*, 623–628. [CrossRef]
3. Zaar, O.; Fougelberg, J.; Hermansson, A.; Gillstedt, M.; Wennberg-Larko, A.M.; Paoli, J. Effectiveness of photodynamic therapy in Bowen's disease: A retrospective observational study in 423 lesions. *J. Eur. Acad. Dermatol. Venereol.* **2017**, *31*, 1289–1294. [CrossRef]
4. Dai, T.; Huang, Y.; Hamblin, M. Photodynamic therapy for localized infections—State of the art. *Photodiagn. Photodyn. Ther.* **2009**, *6*, 170–188. [CrossRef]
5. Wang, H.; Li, J.; Lv, T.; Tu, Q.; Huang, Z.; Wang, X. Therapeutic and immune effects of 5-aminolevulinic acid photodynamic therapy on UVB-induced squamous cell carcinomas in hairless mice. *Exp. Dermatol.* **2013**, *22*, 362–363. [CrossRef]
6. de Albuquerque, I.O.; Nunes, J.; Longo, J.P.; Muehlmann, L.A.; de Azevedo, R.B. Photodynamic therapy in superficial basal cell carcinoma treatment. *Photodiagn. Photodyn. Ther.* **2019**, *27*, 428–432. [CrossRef]
7. Chizenga, E.P.; Chandran, R.; Abrahamse, H. Photodynamic therapy of cervical cancer by eradication of cervical cancer cells and cervical cancer stem cells. *Oncotarget* **2019**, *10*, 4380–4396. [CrossRef]
8. Monroe, J.D.; Belekov, E.; Er, A.O.; Smith, M.E. Anti-cancer photodynamic therapy properties of sulphur-doped graphene quantum dot and methylene blue preparations in MCF-7 breast cancer cell culture. *Photochem. Photobiol.* **2019**, *95*, 1473–1481. [CrossRef]
9. Jia, Y.; Chen, L.; Chi, D.; Cong, D.; Zhou, P.; Jin, J.; Ji, H.; Liang, B.; Gao, S.; Hu, S. Photodynamic therapy combined with temozolomide inhibits C6 glioma migration and invasion and promotes mitochondrial-associated apoptosis by inhibiting sodium-hydrogen exchanger isoform 1. *Photodiagn. Photodyn. Ther.* **2019**, *26*, 405–412. [CrossRef]
10. Wang, Q.; Zhang, X.; Sun, Y.; Wang, L.; Ding, L.; Zhu, W.-H.; Di, W.; Duan, Y.-R. Gold-caged copolymer nanoparticles as multimodal synergistic photodynamic/photothermal/ chemotherapy platform against lethality androgen-resistant prostate cancer. *Biomaterials* **2019**, *212*, 73–86. [CrossRef]
11. Carvalho, M.R.; Reis, R.L.; Oliverira, J.M. Dendrimer nanoparticles for colorectal cancer applications. *J. Mater. Chem. B* **2020**, *8*, 1128–1138. [CrossRef] [PubMed]
12. Yang, S.-J.; Shieh, M.-J.; Lin, F.-H.; Lu, P.-J.; Peng, C.-L.; Wei, M.-F.; Jao, C.-J.; Lai, P.-S.; Young, T.H. Colorectal cancer cell detection by 5-aminolaevulinic acid-loaded chitosan nano-particles. *Cancer Lett.* **2009**, *273*, 210–220. [CrossRef] [PubMed]
13. Bray, F.; Ferlay, J.; Soerjomataram, I.; Siegel, R.L.; Torre, L.A.; Jemal, A. Global cancer statistics 2018: GLOBOCAN estimates of incidence and mortality worldwide for 36 cancers in 185 countries. *CA Cancer J. Clin.* **2018**, *68*, 394–424. [CrossRef] [PubMed]
14. Kawczyk-Krupka, A.; Bugaj, A.M.; Latos, W.; Zaremba, K.; Wawrzyniec, K.; Sieroń, A. Photodynamic therapy in colorectal cancer treatment: The state of the art in clinical trials. *Photodiagn. Photodyn. Ther.* **2015**, *12*. [CrossRef] [PubMed]
15. Kawczyk-Krupka, A.; Czuba, Z.P.; Kwiatek, B.; Kwiatek, S.; Krupka, M.; Sieroń, K. The effect of ALA-PDT under normoxia and cobalt chloride (CoCl2)-induced hypoxia on adhesion molecules (ICAM-1, VCAM-1) secretion by colorectal cancer cells. *Photodiagn. Photodyn. Ther.* **2017**, *19*, 103–115. [CrossRef] [PubMed]
16. Simelane, N.W.N.; Kruger, C.A.; Abrahamse, H. Photodynamic diagnosis and photodynamic therapy of colorectal cancer in vitro and in vivo. *RSC Adv.* **2020**, *10*, 41560. [CrossRef]
17. Shi, L.; Wang, X.; Zhao, F.; Luan, H.; Tu, Q.; Huang, Z.; Wang, H.; Wang, H.W. In vitro evaluation of 5-aminolevulinic acid (ALA) loaded PLGA nanoparticles. *Int. J. Nanomed.* **2013**, *8*, 2669–2676. [CrossRef]
18. Wang, Z.; Gai, S.; Wang, C.; Yang, G.; Zhong, C.; Dai, Y.; He, F.; Yang, D.; Yang, P. Self-assembled zinc phthalocyanine nanoparticles as excellent photothermal/photodynamic synergistic agent for antitumor treatment. *Chem. Eng. J.* **2019**, *361*, 117–128. [CrossRef]
19. Sun, W.; Zhou, Z.; Pratx, G.; Chen, X.; Chen, H. Nanoscintillator-mediated x-ray induced photodynamic therapy for deep-seated tumors: From concept to biomedical applications. *Theranostics* **2020**, *10*, 1296–1318. [CrossRef]
20. Namikawa, T.; Yatabe, T.; Inoue, K.; Shuin, T.; Hanazaki, K. Clinical applications of 5-aminolevulinic acid-mediated fluorescence for gastric cancer. *World J. Gastroenterol.* **2015**, *21*, 8769–8775. [CrossRef]
21. Fotinos, N.; Campo, M.A.; Popowycz, F.; Gurny, R.; Lange, N. 5-Aminolevulinic acid derivatives in photomedicine: Characteristics, application and perspectives. *Photochem. Photobiol.* **2006**, *82*, 994–1015. [CrossRef] [PubMed]
22. Müller, P.; Abdel Gaber, S.A.; Zimmermann, W.; Wittig, R.; Stepp, H. ABCG2 influence on the efficiency of photodynamic therapy in glioblastoma cells. *J. Photochem. Photobiol. B Biol.* **2020**, *210*, 111963. [CrossRef]
23. Nakanishi, T.; Ogawa, T.; Yanagihara, C.; Tamai, I. Kinetic Evaluation of Determinant Factors for Cellular Accumulation of Protoporphyrin IX Induced by External 5-Aminolevulinic Acid for Photodynamic Cancer Therapy. *J. Pharm. Sci.* **2015**, *104*, 3092–3100. [CrossRef] [PubMed]
24. Saito, K.; Fujiwara, T.; Ota, U.; Hatta, S.; Ichikawa, S.; Kobayashi, M.; Okitsu, Y.; Fukuhara, N.; Onishi, Y.; Ishizuka, M.; et al. Dynamics of absorption, metabolism, and excretion of 5-aminolevulinic acid in human intestinal Caco-2 cells. *Biochem. Biophys. Rep.* **2017**, *11*, 105–111. [CrossRef] [PubMed]

25. Sourdon, A.; Gary-Bobo, M.; Maynadier, M.; Garcia, M.; Majoral, J.-P.; Caminade, A.-M.; Mongin, O.; Blanchard-Desce, M. Dendrimeric Nanoparticles for Two-Photon Photodynamic Therapy and Imaging: Synthesis, Photophysical Properties, Innocuousness in Daylight and Cytotoxicity under Two-Photon Irradiation in the NIR. *Chem. Eur. J.* **2019**, *25*, 3637–3649. [CrossRef]
26. Pandey, P.K.; Maheshwari, R.; Raval, N.; Gondaliya, P.; Kalia, K.; Tekade, R.K. Nanogold-core multifunctional dendrimer for pulsatile chemo-, photothermal- and photodynamic- therapy of rheumatoid arthritis. *J. Colloid Interface Sci.* **2019**, *15*, 61–77. [CrossRef]
27. Zhou, T.; Battah, S.; Mazzacuva, F.; Hider, R.C.; Dobbin, P.; MacRobert, A.J. Design of Bifunctional Dendritic 5-Aminolevulinic Acid and Hydroxypyridinone Conjugates for Photodynamic Therapy. *Bioconjug. Chem.* **2018**, *29*, 3411–3428. [CrossRef] [PubMed]
28. Pfister, A.B.; Wood, R.C.; Salas, P.J.I.; Zea, D.L.; Ramsauer, V.P. Early Response to ErbB2 Over-Expression in Polarized Caco-2 Cells Involves Partial Segregation from ErbB3 by Relocalization to the Apical Surface and Initiation of Survival Signaling. *J. Cell Biochem.* **2010**, *111*, 643–652. [CrossRef]
29. Uram, Ł.; Szuster, M.; Filipowicz, A.; Zaręba, M.; Wałajtys-Rode, E.; Wołowiec, S. Cellular uptake of glucoheptoamidated poly(amidoamine) PAMAM G3 dendrimer with amide-conjugated biotin, a potential carrier of anticancer drugs. *Bioorg. Med. Chem.* **2017**, *25*, 706–713. [CrossRef] [PubMed]
30. Elfsson, B.; Wallin, I.; Eksborg, S.; Rudaeusa, K.; Ros, A.M.; Ehrsson, H. Stability of 5-aminolevulinic acid in aqueous solution. *Eur. J. Pharm. Sci.* **1998**, *7*, 87–91. [CrossRef]
31. Czerniecka-Kubicka, A.; Tutka, P.; Pyda, M.; Walczak, M.; Uram, Ł.; Misiorek, M.; Chmiel, E.; Wołowiec, S. Stepwise glucoheptoamidation of poly(amidoamine) dendrimer G3 to tune physicochemical properties of the potential drug carrier; in vitro tests for cytisine conjugates. *Pharmaceutics* **2020**, *12*, 473. [CrossRef]
32. Lin, Y.-H.; Chang, H.-M.; Chang, F.-P.; Shen, C.-R.; Liu, C.-L.; Mao, W.-Y.; Lin, C.-C.; Lee, H.-S.; Shen, C.-N. Protoporphyrin IX accumulation disrupts mitochondrial dynamics and function in ABCG2-deficient hepatocytes. *FEBS Lett.* **2013**, *587*, 3202–3209. [CrossRef]
33. Wirth, D.J.; Sibai, M.; Wilson, B.C.; Roberts, D.W.; Paulsen, K. First experience with spatial frequency domain imaging and red-light excitation of protoporphyrin IX fluorescence during tumor resection. *Biomed. Opt. Express* **2020**, *11*, 4306–4315. [CrossRef]
34. Tomalia, D.; Baker, H.; Dewald, J.; Hall, M.; Kallos, G.; Martin, S.; Roeck, J.; Ryder, J.; Smith, P. A new Class of Polymers: Starburst-Dendritic Macromolecules. *Polym. J.* **1985**, *17*, 117–132. [CrossRef]
35. Czarnik-Kwaśniak, J.; Kwaśniak, K.; Tutaj, K.; Filiks, I.; Uram, Ł.; Stompor, M.; Wołowiec, S. Glucoheptoamidated polyamidoamine PAMAM G3 dendrimer as a vehicle for succinate linked doxorubicin; enhanced toxicity of DOX against grade IV glioblastoma U-118 MG cells. *J. Drug Deliv. Sci. Technol.* **2020**, *55*, 101424. [CrossRef]
36. Yang, S.-J.; Lin, F.-H.; Tsai, K.-C.; Wei, M.-F.; Tsai, H.-M.; Wong, J.-M.; Shieh, M.-J. Folic Acid-Conjugated Chitosan Nanoparticles Enhanced Protoporphyrin IX Accumulation in Colorectal Cancer Cells. *Bioconjug. Chem.* **2010**, *21*, 679–689. [CrossRef]
37. Avci, P.; Erdem, S.S.; Hamblin, M.R. Photodynamic Therapy: One Step Ahead with Self-Assembled Nanoparticles. *J. Biomed. Nanotechnol.* **2014**, *10*, 1937–1952. [CrossRef] [PubMed]
38. Borowska, K.; Laskowska, B.; Magoń, A.; Myśliwiec, B.; Pyda, M.; Wołowiec, S. PAMAM dendrimers as solubilizers and hosts for 8-methoxypsoralene enabling transdermal diffusion of the guest. *Int. J. Pharm.* **2010**, *398*, 185–189. [CrossRef]
39. Borowska, K.; Wolowiec, S.; Rubaj, A.; Głowniak, K.; Sieniawska, E.; Radej, S. Effect of polyamidoamine dendrimer G3 and G4 on skin permeation of 8-methoxypsoralene—In vivo study. *Int. J. Pharm.* **2012**, *426*, 280–283. [CrossRef]
40. Borowska, K.; Wolowiec, S.; Głowniak, K.; Sieniawska, E.; Radej, S. Transdermal delivery of 8-methoxypsoralene mediated by polyamidoamine dendrimer G2.5 and G3.5—In vitro and in vivo study. *Int. J. Pharm.* **2012**, *436*, 764–770. [CrossRef]
41. Battah, S.; Balaratnam, S.; Casas, A.; O'Neill, S.; Edwards, C.; Batlle, A.; Dobbin, P.; MacRobert, A.J. Macromolecular delivery of 5-aminolaevulinic acid for photodynamic therapy using dendrimer conjugates. *Mol. Cancer Ther.* **2007**, *6*, 876–885. [CrossRef] [PubMed]
42. Szuster, M.; Uram, Ł.; Filipowicz-Rachwał, A.; Wołowiec, S.; Wałajtys-Rode, E. Evaluation of the localization and biological effects of PAMAM G3 dendrimer-biotin/pyridoxal conjugate as HaCaT keratinocyte targeted nanocarrier. *Acta Biochim. Polon.* **2019**, *66*, 1–10. [CrossRef] [PubMed]
43. Borowska, K.; Wołowiec, S. *Pol. Patent* P.432139. 2020. Available online: https://uprp.gov.pl/sites/default/files/bup/2020/Wynalazki%20i%20wzory%20użytkowe/06_12-14/12/bup12_2020.pdf (accessed on 17 February 2021).

International Journal of
Molecular Sciences

Article

Cell-Penetrating Doxorubicin Released from Elastin-Like Polypeptide Kills Doxorubicin-Resistant Cancer Cells in In Vitro Study

Jung Su Ryu [1], Felix Kratz [2] and Drazen Raucher [1,*]

[1] Department of Cell and Molecular Biology, University of Mississippi Medical Center, Jackson, MS 39213, USA; jryu@umc.edu
[2] CytRx Corporation, 79104 Freiburg, Germany; fkratz@cytrx.com
* Correspondence: draucher@umc.edu; Tel.: +1-601-984-1510

Abstract: Elastin-like polypeptides (ELPs) undergo a characteristic phase transition in response to ambient temperature. Therefore, it has been be used as a thermosensitive vector for the delivery of chemotherapy agents since it can be used to target hyperthermic tumors. This novel strategy introduces unprecedented options for treating cancer with fewer concerns about side effects. In this study, the ELP system was further modified with an enzyme-cleavable linker in order to release drugs within tumors. This system consists of an ELP, a matrix metalloproteinase (MMP) substrate, a cell-penetrating peptide (CPP), and a 6-maleimidocaproyl amide derivative of doxorubicin (Dox). This strategy shows up to a 4-fold increase in cell penetration and results in more death in breast cancer cells compared to ELP-Dox. Even in doxorubicin-resistant cells (NCI/ADR and MES-SA/Dx5), ELP-released cell-penetrating doxorubicin demonstrated better membrane penetration, leading to at least twice the killing of resistant cells compared to ELP-Dox and free Dox. MMP-digested CPP-Dox showed better membrane penetration and induced more cancer cell death in vitro. This CPP-complexed Dox released from the ELP killed even Dox-resistant cells more efficiently than both free doxorubicin and non-cleaved ELP-CPP-Dox.

Keywords: drug delivery; tumor targeting; elastin-like polypeptide; cell penetrating peptide; matrix metalloproteinase; doxorubicin resistance

Citation: Ryu, J.S.; Kratz, F.; Raucher, D. Cell-Penetrating Doxorubicin Released from Elastin-Like Polypeptide Kills Doxorubicin-Resistant Cancer Cells in In Vitro Study. *Int. J. Mol. Sci.* **2021**, *22*, 1126. https://doi.org/10.3390/ijms22031126

Academic Editor: Angela Stefanachi
Received: 23 December 2020
Accepted: 20 January 2021
Published: 23 January 2021

Publisher's Note: MDPI stays neutral with regard to jurisdictional claims in published maps and institutional affiliations.

1. Introduction

It is acknowledged that current conventional chemotherapy is mostly comprised of cytotoxic drugs, which have a strong anticancer efficacy but cause collateral damage to non-tumor tissues. These unwanted side effects are usually a dose-limiting factor for chemotherapy and are a main reason for the unsatisfactory prognosis of the therapy. Many efforts have been made to resolve these problems, usually by attempting to raise the therapeutic index of the chemotherapy.

Curing cancer is certainly one of the greatest challenges of our time, and to confront it our knowledge of cancer has grown greatly over the last decades. In recent years, there has been a surge of new technologies for cancer treatment such as molecular targeted therapies (i.e., anti-tyrosine kinase and anti-HER2) [1], immunotherapies such as cancer vaccines or anti-PD1 [2,3], sophisticated radiation therapy [4], and advanced tumor-targeting technologies such as nanoparticles and antibody–drug conjugates. These technologies could make a big difference in the way we treat cancer, bringing us closer to being able to "cure" this disease. In particular, the nanosized drug delivery technologies have been significantly improved, and many of them are currently being used to solubilize the drugs, bypass immune surveillance, sensitize current therapies, and target tumor tissues [5,6]. A tumor-targeting technology delivers drugs specifically to tumor tissue so that the concentration of drugs in the tissue will increase compared to the concentration in normal tissue. This allows more opportunity for a drug to express its activity on tumor cells, resulting in the selective death of cancer cells with tolerable side effects.

An elastin-like polypeptide (ELP) is a thermo-responsive bio-polymeric carrier for targeted drug delivery. ELPs derived from tropoelastin consist of repeats of pentapeptides (mainly comprised of valine, proline, and glycine) [7,8]. The repeats of these hydrophobic amino acids permit an ELP to have a unique re-arrangement of molecules in response to the surrounding temperature, which is a thermoresponsive phase transition. At low temperature, an ELP remains a monomer and is soluble in solution; however, it co-acervates and precipitates in solution when the ambient temperature rises above its phase-transition temperature [9–11]. This co-acervation can also be reversed by decreasing the temperature of the ELP solution. Thus, this reversible phase transition of an ELP is mainly controlled by temperature, and the ELP is highlighted as a controllable carrier for the delivery of anticancer drugs in active tumor-targeting strategies.

Additionally, and ELP exploits an "enhanced permeability and retention (EPR) effect" and can progressively accumulate in tumor tissue due to the abnormal histological structures of the tumor. These unique properties turn the ELP itself into a drug carrier that can exploit both the EPR effect and tumor targeting using the hyperthermic technique [12]. Furthermore, the ELP has been modified by the addition of cell-penetrating peptides (CPPs) to allow enhanced cellular uptake, improved penetration of physiological barriers like the blood–brain barrier, and preferential intracellular distribution such as in the cytoplasm or the nucleus [13,14]. Many previous researchers have verified the potential of this polymer [15–18], and animal studies have demonstrated that ELPs are able to deliver a sufficient amount of a drug to the tumor area to produce significant tumor reduction efficacy in combination with the use of local hyperthermia.

In this study, we further modified an ELP drug delivery system to release drugs in response to an enzyme that is abundant in tumor tissue (Figure 1). The suggested system is composed of an ELP; a matrix metalloproteinase (MMP) substrate, mmpL; a cell-penetrating peptide, CPP; and a 6-maleimidocaproyl amide derivative of doxorubicin, Dox (Figure 1A). We report the potential use of this strategy, an MMP-responsive ELP drug delivery system releasing CPP-Dox, to overcome Dox resistance by investigating the cellular uptake and anti-proliferation properties of the proposed system.

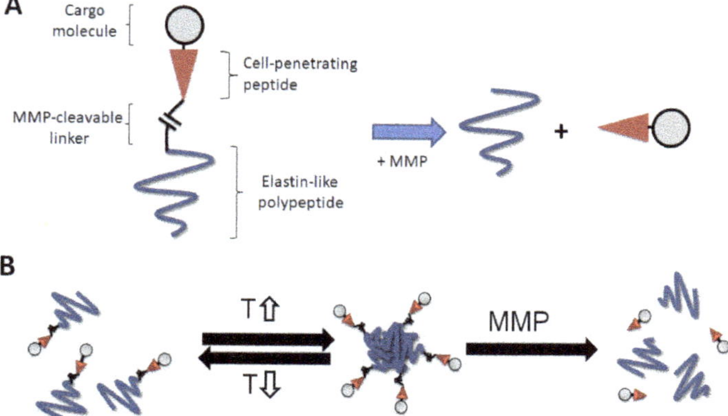

Figure 1. Elastin-like peptide (ELP) drug delivery system. (**A**) The proposed ELP system consists of an elastin-like polypeptide, matrix metalloproteinase (MMP)-cleavable linker, cell-penetrating peptide (CPP), and cargo molecules. The constructs are digested by the MMP, releasing the CPP cargo molecule. (**B**) The hypothetical model proposed in this system. The ELP constructs can form aggregates and release the CPP cargo molecule in hyperthermic tumors (T, temperature).

2. Results

2.1. Incubation of ELP-mmpL-CPP with MMP-2 Produces Cleaved CPP

Figure 2A depicts how ELP-mmpL-CPP would be cleaved by MMP, producing ELP (60 kDa) and cleaved CPP (Tat peptide, 3 kDa), while the other construct, ELP-CPP-Dox (63 kDa), would not be digested by MMP. This hypothesis was verified by the following experiments. MMP-2 was used for the digestion in this experiment since MMP-2 is involved in the degradation of extracellular matrices in tumors and is overexpressed in most tumors compared with normal tissues [19]. After incubation of each rhodamine (rho)-labeled construct (ELP-mmpL-CPP-rho and ELP-CPP-rho) with MMP-2, the reactant was run on SDS-PAGE and analyzed by both silver staining and fluorescence scanning (Figure 2B). Silver-stained gels revealed that MMP-digested ELP-mmpL-CPP-rho produced two bands (lane②in the left panel); the upper one for ELP (60 kDa) and the other for cleaved CPP-rho (3 kDa), while ELP-CPP-rho digestion produced only one band (lane①in the left panel) which represents undigested ELP-CPP-rho. However, when the gel was scanned for fluorescence, each reactant showed only one band. Since rhodamine was conjugated to the C-terminal of the CPP (Tat peptide), MMP digestion produced one single fluorescent band (CPP-rho, 3 kDa) without ELP (lane②in the right panel), while a band of undigested ELP-CPP-rho fluoresced at around 63 kDa, as with the silver-stained gel (lane①in the right panel).

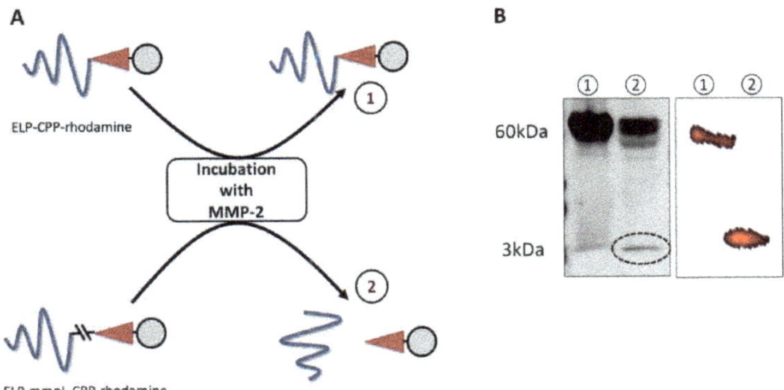

Figure 2. MMP-2 digestion of ELP-mmpL-CPP. (**A**) ELP-mmpL-CPP-rhodamine and ELP-CPP-rhodamine were incubated with MMP-2 for 4 h in $ZnCl_2$ buffer (pH 7). (**B**) MMP-2 incubation of the constructs produced ELP-CPP-Dox (63 kDa, upper band) and CPP-rhodamine (3 kDa, lower band). Left panel: silver-stained gel; Right panel: fluorescence-scanned gel. Dox: 6-maleimidocaproyl amide derivative of doxorubicin; mmpL: MMP substrate.

2.2. MMP-2 Digestion Increases the Cellular Uptake of CPP -Rhodamine in Breast Cancer Cells

MMP digestion will produce CPP-rhodamine (rho), which is smaller than the whole construct, ELP-mmpL-CPP-rho. This smaller size would facilitate its cellular uptake. Cells treated with MMP-digested ELP-mmpL-CPP-rho and ELP-CPP-rho, respectively, were analyzed for uptake ability via flow cytometry. In Figure 3A, cells treated with cleaved CPP-rho (from ELP-mmpL-CPP-rho) showed up to five times higher uptake rates than the ELP-CPP-rho-treated group in three cancer cell lines. This improved cellular uptake was also evident in observation with a fluorescence microscope (Figure 3B).

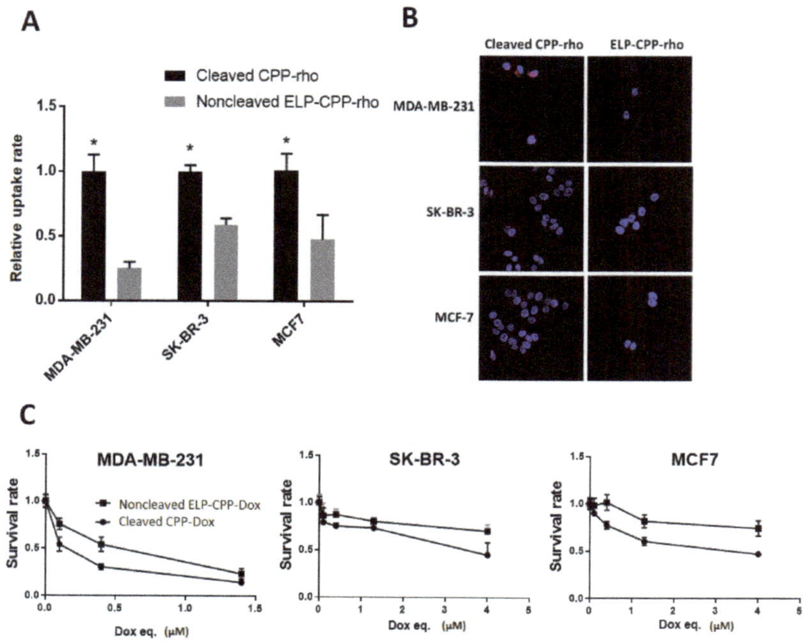

Figure 3. Cellular uptake rate of cleaved CPP-rhodamine in breast cancer cells. (**A**) Each cellular uptake rate was measured through flow cytometry (* $p < 0.05$). (**B**) Merged image of Dox (red) and DAPI (blue). (**C**) Cytotoxicity of cleaved CPP-Dox on breast cancer cells. Cells were treated with ELP-CPP-Dox and ELP-mmpL-CPP-Dox, both digested by MMP incubation.

2.3. *Cleaved CPP-Dox Kills Breast Cancer Cells More Efficiently than Non-Cleaved ELP-CPP*

Rhodamine was replaced by doxorubicin to investigate whether improved uptake of cleaved CPP would contribute to cytotoxicity. Figure 3C compares the cytotoxicities of MMP-2-digested ELP-mmpL-CPP-Dox and ELP-CPP-Dox against three cancer cell lines. Improved cytotoxicity was observed in MMP-2-digested ELP-mmpL-CPP-Dox-treated cells than those treated with ELP-CPP-Dox. These results suggest that the MMP digestion of ELP-mmpL-CPP-Dox results in increased uptake of cargo molecules and facilitated the death of cancer cells by cleaved CPP-Dox.

2.4. *Cleaved CPP-Dox Deposits in and Kills Dox-Resistant Cancer Cells*

To investigate whether cleaved CPP-Dox is able to penetrate and kill even Dox-resistant cancer cells, comparisons of cytotoxicities and uptake rates of MMP-cleaved CPP-Dox were made between Dox-resistant cells (NCI/ADR, MES-SA/Dx5) and Dox-sensitive cells (MCF7, MES-SA).

Figure 4A shows the validated Dox resistance in NCI/ADR and MES-SA/Dx5, and cleaved CPP-Dox from ELP-mmpL-CPP-Dox showed more cell killing than ELP-CPP-Dox at 4 μM Dox equivalence. Confocal microscopic images of NCI/ADR cells show that cleaved CPP-Dox from ELP-mmpL-CPP-Dox was taken up by NCI/ADR more than the other constructs (i.e., free Dox and ELP-CPP-Dox; Figure 4B). This was also confirmed by flow cytometry (Figure 4C). The uptake rate of MMP-digested CPP-Dox in NCI/ADR was almost doubled compared with the uptake rates of free Dox and ELP-CPP-Dox. These results suggest that MMP-cleaved CPP-Dox can penetrate and kill even Dox-resistant cancer cells, probably with the help of a CPP (Tat peptide). One limitation of this experiment is that 4 μM of a doxorubicin-equivalent dose is the maximum concentration that can be reached from the current cleavage assay protocol; further optimization of the protocol may enable the generation of a higher concentration of each drug and calculation of IC50 to compare the cytotoxicity of each treatment.

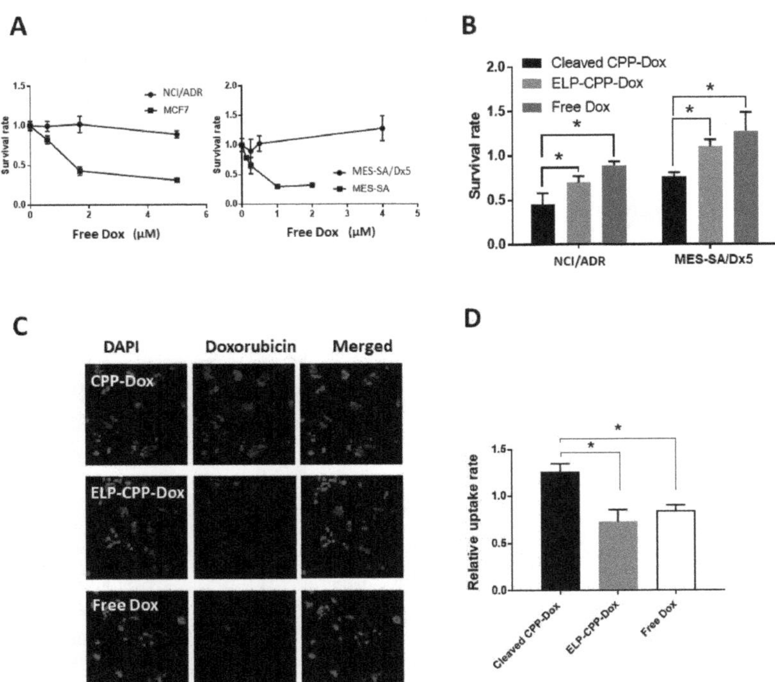

Figure 4. Cytotoxicity of CPP-Dox against Dox-resistant cancer cells. (**A**) Free Dox killed Dox-sensitive cancer cells (MCF7 and MES-SA), while it spared Dox-resistant NCI/ADR and MES-SA/Dx5. (**B**) Cytotoxicities of CPP-Dox and ELP-CPP-Dox in NCI/ADR and MES-SA/5DX at 4 μM Dox equivalence. (**C**) Confocal microscopic images show that CPP-Dox penetrated into NCI/ADR. (**D**) Flow cytometry, 60% increased uptake in CPP-Dox in comparison with ELP-CPP-Dox and free Dox. * $p < 0.05$.

2.5. MMP-Releasing HT-1080 Can Cleave ELP-mmpL-CPP-rho and Take Up Cleaved CPP-rho

Given that an MMP-cleaved CPP-Dox can inhibit proliferation in Dox-resistant cancer cell lines, this ELP-mmpL-CPP-Dox system was further validated using HT-1080, a fibrosarcoma cancer cell producing endogenous MMP-2 and MMP-9. This experiment showed that the ELP-mmpL-CPP construct could also be digested by the endogenous MMP enzyme and release CPP cargo molecules. MMP-releasing HT-1080 cells were incubated with either ELP-mmpL-CPP-rho or ELP-CPP-rho for 4 h, and each group of treated cells was processed either for flow cytometry or fluorescence microscopy. In flow cytometry, cells incubated with the ELP-mmpL-CPP-rho group had twice the rhodamine signal of the ELP-CPP-rho group. However, this increased uptake was reversed by pretreatment with GM6001, an MMP catalytic inhibitor (Figure 5A). This finding was further confirmed by fluorescence microscopy, with the rhodamine particles being found in the nucleus of HT-1080 cells treated with ELP-mmpL-CPP-rho (Figure 5B). Uptake of these particles, as in the flow cytometry experiment, was also abolished by GM6001 pretreatment. GM6001 prevents MMP digestion, and undigested ELP-mmpL-CPP-rho was likely washed off the cells during the rinsing step. These results indicate that ELP-mmpL-CPP-rho was digested by intrinsic MMP released from HT-1080 cells, and that the resultant cleaved CPP-rho penetrated the HT-1080 cells.

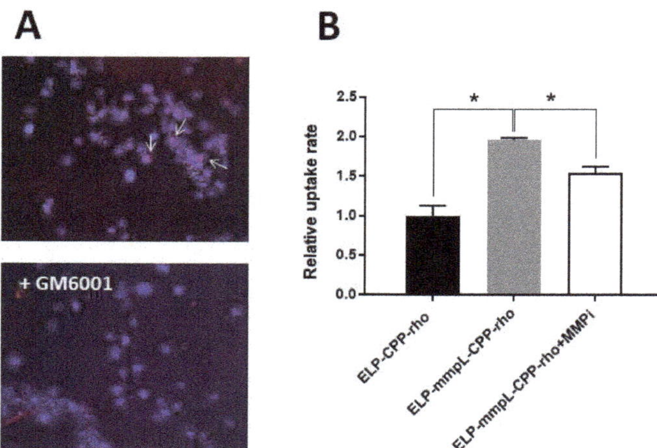

Figure 5. Cellular uptake rate of CPP-rhodamine in MMP-expressing HT-1080 cells. (**A**) Localization of CPP-rho (fluorescence microscopy, 20x) in cultured HT-1080 cells. The arrows indicate the CPP-rhodamine in the cells. (**B**) Flow cytometry showing increased uptake in cleaved CPP-rho in cells. * $p < 0.05$.

3. Discussion

Our tumor-targeted drug delivery system using an ELP delivers anticancer cargo molecules specifically to the tumor site by exploiting the enhanced permeability and retention (EPR) effect along with the active thermal targeting approach [15–17,20]. This thereby increases the relative concentration of the cargo drugs in tumors and improves the therapeutic index of the drugs, alleviating unacceptable toxicity to the patients [12,21]. A striking example of this targeting can be found in previous studies [20], in which fluorescently labeled CPP-ELPs were administered into S2013 tumor-bearing mice. One group of animals received hyperthermic treatment with infrared (IR) lasers on tumors immediately after injection of CPP-ELP so that the temperature in the tumor core reached 42 °C, while the other group was exempt from hyperthermic treatment. This study demonstrated that the IR heating of tumors created 2–3 times greater tumor accumulation of CPP-ELP as well as the penetration of this protein into the tumor tissues. Given the thermoresponsive behavior of the ELP, the aggregation of CPP-ELP in hyperthermic tumors resulted in an increase of the construct's concentration in the tumors. These results strongly suggest that the ELP preferentially accumulated in tumors in response to local hyperthermia.

We further improved the ELP system in this study to release payloads in response to additional external stimuli—that is, matrix metalloproteinases. This novel system is comprised of four components: an ELP, an MMP-2-cleavable linker, CPP (Tat peptide), and doxorubicin as a payload. The linker is a substrate of MMP-2, designed to be cleaved by MMP-2 so that the ELP system can eventually release a complex of the payload (doxorubicin) and a CPP (CPP-Dox) in tumor tissues. The involvement of MMPs, which are zinc-containing endopeptidases, in cancer biology has been extensively discussed in a variety of publications [22–25]. Especially, increased expression and activity of MMP-2 and MMP-9 subtypes in tumors are known to be related to the degradation of basement membranes—an essential step in tumor invasion and in enhancing angiogenesis. For example, Tutton et al. reported that MMP-2 expression was significantly increased in colorectal cancer tissues compared to matched normal colon tissue as measured by ELISA [26]. High levels of MMPs in tumors will facilitate the release of CPP-Dox out of the renovated ELP complex and provide an additional, secondary tumor-targeting opportunity compared to the previous ELP delivery system. This system thus becomes a triple-targeting strategy when used along with the EPR effect and local hyperthermia. Specifically, this cleavable ELP construct still contains the ELP molecule until it is digested by MMPs at tumor tissues.

This late cleavage process will allow CPP-Dox to benefit from ELP by EPR, and from the thermo-targeting. ELP is expected to allow the proposed construct to be initially targeted to the tumor site by the local application of mild heat. Then, ELP-mmpL-CPP-Dox will be fully digested by MMP to release CPP-Dox (Figure 1B), followed by improved cellular uptake by cancer cells and increased cancer-cell death.

This MMP-cleavable system displays a couple of other advantages in delivering chemotherapeutic molecules. First, when the MMP-cleavable ELP-CPP-drug is digested by MMP in tumor tissue, small fragments (CPP-Dox) will be produced. Since the molecular weight of the released CPP-Dox (<3 kDa) is one-twentieth that of the parental ELP construct (60 kDa), it will quickly infiltrate into adjacent tumor cells, as can be seen in other studies [27,28]. This hypothesis was examined by cell-uptake assays in this study. Cells treated with CPP-rho, which is a digested product from ELP-mmpL-CPP-rho, showed more rhodamine uptake than the cells treated with undigested ELP-CPP-rho. It is also demonstrated in Figure 5 that the ELP-mmpL-CPP-rho could be digested by endogenous MMPs and taken up by HT-1080 cells. Reversal of this uptake by GM6001 (an MMP inhibitor) indicates that the cell uptake of rhodamine by HT-1080 relies on the catalytic activity of MMP. This increased uptake was reflected in the enhanced cytotoxicity of MMP-cleaved CPP-Dox in breast cancer cells. After MMP-2 digestion, ELP-mmpL-CPP-Dox killed more cancer cells than did ELP-CPP-Dox (Figure 3C).

A second advantage of this system is that the released CPP-Dox still takes advantage of the abilities of the CPP to facilitate uptake by the cells and to penetrate physiological barriers like the blood–brain barrier. More importantly, there is an increasing number of studies reporting the role of CPPs in overcoming the multidrug resistance (MDR) of cancer cells, which has been one of main hurdles that doxorubicin has faced. The use of doxorubicin, one of the most effective chemotherapy agents since the 1960s, has been compromised by the development of MDR in patients [29,30]. MDR involves increased efflux, decreased uptake, and enzymatic drug metabolism (e.g., glutathione S-transferase) of chemotherapeutic drugs such as doxorubicin [31]. An elevated expression of active drug transporters in cancer cells is known to be a major resistance mechanism [32]. The coupling of chemotherapeutic drugs to peptides such as CPPs have been suggested as the solution for these problems, since this strategy may alter the cellular uptake pathway and circumvent ABC-transporter-mediated drug efflux, allowing drugs to accumulate at high concentrations in drug-resistant cells, leading to an improved therapeutic index and fewer adverse effects [28,33–35]. Specifically, CPP-Dox developed by Liang et al. [35] showed a 59% uptake rate in Dox-resistant MCF7 cells, while 90% of free Dox was lost during cell penetration, leading to a considerable improvement in the IC_{50} of doxorubicin. In line with these reports, our current study also demonstrates that cleaved CPP-Dox showed greater cellular uptake by Dox-resistant NCI/ADR in comparison with free Dox and ELP-CPP-Dox (Figure 4C). Like Dox-sensitive breast cancer cells, this increased uptake also led to an enhanced cytotoxicity of cleaved CPP-Dox against NCI/ADR and MES-SA/Dx5 (Figure 4B).

In summary, the modified ELP-CPP-Dox was cleaved by incubation with an intrinsic or extrinsic MMP enzyme. MMP digestion produced CPP-Dox (or rhodamine), which showed better membrane penetration and induced more cancer cell death *in vitro*. This CPP-complexed Dox released from an ELP killed even Dox-resistant cells more efficiently than both free doxorubicin and non-cleaved ELP-CPP-Dox. This pilot study emphasizes the extra functionalities of the ELP drug delivery system. The novelty of this study is improvement in the drug delivery efficiency of ELP and demonstration of ELPs' potential in multiple-targeting strategies.

4. Materials and Methods

4.1. Design of Construct and Protein Preparation

The ELP used in this study consists of 150 repeats of VPGXG with guest residues (amino acid at position X) of Val, Gly, and Ala in a 5:3:2 ratio. ELP coding sequences were modified by the addition of the "Tat" cell-penetrating peptide sequence (YGRKKRRQRRR), an MMP-cleavable sequence (PLGALG), and three Gly-Gly-Cys residues to the C-terminus of the ELP for the conjugation with Dox (Table 1). For an MMP-uncleavable control, six Gly residues were used instead of the MMP-cleavable sequence. All constructs were expressed in the *Escherichia coli* strain BLR (DE3) using pET 25b as an expression vector, and were purified by repeated inverse transition cycling.

Table 1. Construct sequences.

Construct	Sequence
MMP-cleavable ELP-mmpL-CPP	ELP-(PLGALG)-CPP-(GGC)3
MMP-uncleavable ELP-CPP	ELP-GGGGGG-CPP-(GGC)3

4.2. Conjugation of ELP Constructs with Doxorubicin or Fluorescent Probes

Protein (100 μM) in PBS was reduced with 1 mM of tris-(2-carboxyethyl) phosphine (TCEP, Molecular probes) for 30 min at room temperature. Conjugation with 200 μM of the 6-maleimidocaproyl amide derivative of doxorubicin or tetramethylrhodamine-5-iodoacetamide dihydroiodide (molecular probes) was followed by incubation at 4 °C overnight. Conjugated peptides were purified by inverse transition purification as described previously [9], and the concentration and the labeling efficiency were assessed by UV–visible spectrometry (UV-1600, Shimadzu). Concentrations of labeled ELP polypeptides were determined using the following equations:

for Dox conjugation

$$\text{protein, M} = \frac{\text{Abs 280nm} - (0.71 \times \text{Abs 495nm})}{6890\text{M} - 1\text{cm}^{-1}}$$

for tetra-methyl-rhodamine conjugation

$$\text{protein, M} = \frac{\text{Abs 280nm} - (0.17 \times \text{Abs 541nm})}{5690\text{M} - 1\text{cm}^{-1}}$$

4.3. Cell Culture

MDA-MB-231, MCF7, NCI/ADR, MES-SA, MES-SA/Dx5, and SKBR3 cell lines were obtained from ATCC. HT-1080 was a generous gift from Dr. Michael Herbert of the University of Mississippi Medical Center. All cell lines were grown and maintained at 37 °C, 5% CO_2 in Dulbecco's Modified Eagle's Medium with 10% fetal bovine serum.

4.4. Cleavage Assays

Recombinant human pro-MMP-2 (Enzo life science) was activated with 2.5 mM 4-aminophenylmercuric acetate at 37 °C for 2 h. Then, 1 μg of each ELP construct was incubated with the pre-activated MMP-2 (10 pmol) for 4 h in a reaction buffer (50 mM Tris, 200 mM NaCl, 10 mM $CaCl_2$, and 10 mM $ZnCl_2$, pH 7.5). After the reactions, the samples were loaded and separated on an SDS-PAGE gel, and each peptide's cleavage pattern was confirmed by silver staining and by scanning the fluorescence of the gel with the IVIS Live Animal Imager (Caliper Life Sciences).

4.5. Flow Cytometry Analysis of Cellular Uptake

Cells were incubated with each treatment for 2 h at 37 °C, rinsed with PBS, and collected by trypsinization. Intracellular fluorescence was measured using a Gallios Flow Cytometer (Beckman Coulter) after trypan-blue quenching as described previously [36]. Forward- versus side-scatter gating was used to remove cell debris from the analysis, and the mean cellular fluorescence intensity was recorded. The mean cellular fluorescence was corrected for differences in labeling efficiencies among polypeptides, and the results shown are an average of at least 3 experiments with bars representing the standard error of the mean (SEM).

4.6. Cytotoxicity Test

Cells were plated in a 96-well plate and treated with a range of concentrations of each treatment for 24 h at 37 °C. After further incubation with fresh media for 48 h, cell viability was assessed using the MTT assay (Sigma). Briefly, a 0.5 mg/mL solution of thiazolyl blue tetrazolium bromide dissolved in PBS was added to each well and the plates were incubated for 4 h at 37 °C. Formazan formed by mitochondrial reduction was dissolved in 100 μL of DMSO, and its absorbance was read at 570 nm. The survival rate of each group was calculated in comparison to a vehicle-treated control group.

4.7. Confocal Microscopy

Cells (10^3 cells/chamber) were plated in 2-well Lab-Tek CC2 chamber slides (Nunc). After 24 h incubation at 37 °C, the cells were treated with each treatment for two hours at 37 °C. The cells were washed three times with PBS, fixed with cold 4% paraformaldehyde, and stained with DAPI (molecular probe) for 10 min at room temperature to visualize the nucleus. Distribution of each molecule was investigated by laser scanning confocal microscopy with a 60x oil immersion objective (Leica).

Author Contributions: Conceptualization, J.S.R.; methodology, J.S.R.; investigation, J.S.R.; resources, D.R. and F.K.; writing, J.S.R.; funding acquisition, D.R. All authors have read and agreed to the published version of the manuscript.

Funding: Research reported in this publication was supported by the National Science Foundation (NSF, IIP- 1640519), the National Cancer Institute (1R21CA229943-01A1) and by the National Institute of General Medical Sciences of the National Institutes of Health under Award Number P20GM121334. The content is solely the responsibility of the authors and does not necessarily represent the official views of the National Institutes of Health.

Data Availability Statement: The data that support the findings of this study are available from the corresponding author upon reasonable request.

Acknowledgments: We would like to express our great appreciation to Rebecca B. Mackey of the UMMC for her editorial support. Access to animal imaging equipment was provided by the Animal Imaging Core Facility at the University of Mississippi Medical Center.

Conflicts of Interest: Drazen Raucher is a founder/owner of Thermally Targeted Therapeutics, Inc. Felix Kratz is Vice President of CytRx Corporation.

Abbreviations

ELP	Elastin-like polypeptide
MMP	Matrix metalloproteinase
CPP	Cell-penetrating peptide
Dox	Doxorubicin
MDR	Multidrug resistance
EPR	Enhanced permeability and retention
Rho	Rhodamine
mmpL	MMP-cleavable linker

References

1. Sun, S.; Schiller, J.H.; Spinola, M.; Minna, J.D. New molecularly targeted therapies for lung cancer. *J. Clin. Investig.* **2007**, *117*, 2740–2750. [CrossRef] [PubMed]
2. Li, G.; Gao, Y.; Gong, C.; Han, Z.; Qiang, L.; Tai, Z.; Tian, J.; Gao, S. Dual-Blockade Immune Checkpoint for Breast Cancer Treatment Based on a Tumor-Penetrating Peptide Assembling Nanoparticle. *ACS Appl. Mater. Interfaces* **2019**, *11*, 39513–39524. [CrossRef] [PubMed]
3. Zheng, K.Z.; Loh, K.Y.; Wang, Y.; Chen, Q.S.; Fan, J.Y.; Jung, T.; Nam, S.H.; Suh, Y.D.; Liu, X.G. Recent advances in upconversion nanocrystals: Expanding the kaleidoscopic toolbox for emerging applications. *Nano Today* **2019**, *29*. [CrossRef]
4. Baskar, R.; Lee, K.A.; Yeo, R.; Yeoh, K.W. Cancer and Radiation Therapy: Current Advances and Future Directions. *Int. J. Med. Sci.* **2012**, *9*, 193–199. [CrossRef] [PubMed]

5. Patra, J.K.; Das, G.; Fraceto, L.F.; Campos, E.V.R.; Rodriguez-Torres, M.D.P.; Acosta-Torres, L.S.; Diaz-Torres, L.A.; Grillo, R.; Swamy, M.K.; Sharma, S.; et al. Nano based drug delivery systems: Recent developments and future prospects. *J. Nanobiotechnol.* **2018**, *16*. [CrossRef]
6. Tharkar, P.; Varanasi, R.; Wong, W.S.F.; Jin, C.T.; Chrzanowski, W. Nano-Enhanced Drug Delivery and Therapeutic Ultrasound for Cancer Treatment and Beyond. *Front Bioeng. Biotechnol.* **2019**, *7*, 324. [CrossRef]
7. Floss, D.M.; Schallau, K.; Rose-John, S.; Conrad, U.; Scheller, J. Elastin-like polypeptides revolutionize recombinant protein expression and their biomedical application. *Trends Biotechnol.* **2010**, *28*, 37–45. [CrossRef]
8. Massodi, I.; Bidwell, G.L., 3rd; Davis, A.; Tausend, A.; Credit, K.; Flessner, M.; Raucher, D. Inhibition of ovarian cancer cell metastasis by a fusion polypeptide Tat-ELP. *Clin. Exp. Metastasis* **2009**, *26*, 251–260. [CrossRef]
9. Raucher, D.; Chilkoti, A. Enhanced uptake of a thermally responsive polypeptide by tumor cells in response to its hyperthermia-mediated phase transition. *Cancer Res.* **2001**, *61*, 7163–7170.
10. Chilkoti, A.; Dreher, M.R.; Meyer, D.E.; Raucher, D. Targeted drug delivery by thermally responsive polymers. *Adv. Drug Deliv. Rev.* **2002**, *54*, 613–630. [CrossRef]
11. Dreher, M.R.; Raucher, D.; Balu, N.; Michael Colvin, O.; Ludeman, S.M.; Chilkoti, A. Evaluation of an elastin-like polypeptide-doxorubicin conjugate for cancer therapy. *J. Control Release* **2003**, *91*, 31–43. [CrossRef]
12. Ryu, J.S.; Raucher, D. Elastin-like polypeptide for improved drug delivery for anticancer therapy: Preclinical studies and future applications. *Expert Opin. Drug Deliv.* **2015**, *12*, 653–667. [CrossRef] [PubMed]
13. Raucher, D.; Ryu, J.S. Cell-penetrating peptides: Strategies for anticancer treatment. *Trends Mol. Med.* **2015**, *21*, 560–570. [CrossRef] [PubMed]
14. Bidwell, G.L., 3rd; Raucher, D. Cell penetrating elastin-like polypeptides for therapeutic peptide delivery. *Adv. Drug Deliv. Rev.* **2010**, *62*, 1486–1496. [CrossRef]
15. Bidwell, G.L., 3rd; Perkins, E.; Raucher, D. A thermally targeted c-Myc inhibitory polypeptide inhibits breast tumor growth. *Cancer Lett.* **2012**, *319*, 136–143. [CrossRef]
16. Ryu, J.S.; Raucher, D. Anti-tumor efficacy of a therapeutic peptide based on thermo-responsive elastin-like polypeptide in combination with gemcitabine. *Cancer Lett.* **2014**, *348*, 177–184. [CrossRef]
17. Bidwell, G.L., 3rd; Perkins, E.; Hughes, J.; Khan, M.; James, J.R.; Raucher, D. Thermally targeted delivery of a c-Myc inhibitory polypeptide inhibits tumor progression and extends survival in a rat glioma model. *PLoS ONE* **2013**, *8*, e55104. [CrossRef]
18. Moktan, S.; Perkins, E.; Kratz, F.; Raucher, D. Thermal targeting of an acid-sensitive doxorubicin conjugate of elastin-like polypeptide enhances the therapeutic efficacy compared with the parent compound in vivo. *Mol. Cancer Ther.* **2012**, *11*, 1547–1556. [CrossRef]
19. Kessenbrock, K.; Plaks, V.; Werb, Z. Matrix metalloproteinases: Regulators of the tumor microenvironment. *Cell* **2010**, *141*, 52–67. [CrossRef]
20. Ryu, J.S.; Raucher, D. Elastin-like polypeptides: The influence of its molecular weight on local hyperthermia-induced tumor accumulation. *Eur. J. Pharm. Biopharm.* **2014**, *88*, 382–389. [CrossRef]
21. Ryu, J.S.; Kuna, M.; Raucher, D. Penetrating the cell membrane, thermal targeting and novel anticancer drugs: The development of thermally targeted, elastin-like polypeptide cancer therapeutics. *Ther. Deliv.* **2014**, *5*, 429–445. [CrossRef] [PubMed]
22. Merdad, A.; Karim, S.; Schulten, H.J.; Dallol, A.; Buhmeida, A.; Al-Thubaity, F.; Gari, M.A.; Chaudhary, A.G.; Abuzenadah, A.M.; Al-Qahtani, M.H. Expression of matrix metalloproteinases (MMPs) in primary human breast cancer: MMP-9 as a potential biomarker for cancer invasion and metastasis. *Anticancer Res.* **2014**, *34*, 1355–1366. [PubMed]
23. Roomi, M.W.; Monterrey, J.C.; Kalinovsky, T.; Rath, M.; Niedzwiecki, A. Patterns of MMP-2 and MMP-9 expression in human cancer cell lines. *Oncol. Rep.* **2009**, *21*, 1323–1333.
24. Zucker, S.; Vacirca, J. Role of matrix metalloproteinases (MMPs) in colorectal cancer. *Cancer Metastasis Rev.* **2004**, *23*, 101–117. [CrossRef]
25. Rundhaug, J.E. Matrix metalloproteinases, angiogenesis, and cancer: Commentary re: A. C. Lockhart et al., Reduction of wound angiogenesis in patients treated with BMS-275291, a broad spectrum matrix metalloproteinase inhibitor. *Clin. Cancer Res.* **2003**, *9*, 551–554. [PubMed]
26. Tutton, M.G.; George, M.L.; Eccles, S.A.; Burton, S.; Swift, R.I.; Abulafi, A.M. Use of plasma MMP-2 and MMP-9 levels as a surrogate for tumour expression in colorectal cancer patients. *Int. J. Cancer* **2003**, *107*, 541–550. [CrossRef] [PubMed]
27. Shang, L.; Nienhaus, K.; Nienhaus, G.U. Engineered nanoparticles interacting with cells: Size matters. *J. Nanobiotechnol.* **2014**, *12*, 5. [CrossRef] [PubMed]
28. Aroui, S.; Brahim, S.; Waard, M.D.; Kenani, A. Cytotoxicity, intracellular distribution and uptake of doxorubicin and doxorubicin coupled to cell-penetrating peptides in different cell lines: A comparative study. *Biochem. Biophys. Res. Commun.* **2010**, *391*, 419–425. [CrossRef]
29. Smith, L.; Watson, M.B.; O'Kane, S.L.; Drew, P.J.; Lind, M.J.; Cawkwell, L. The analysis of doxorubicin resistance in human breast cancer cells using antibody microarrays. *Mol. Cancer Ther.* **2006**, *5*, 2115–2120. [CrossRef]
30. Cox, J.; Weinman, S. Mechanisms of doxorubicin resistance in hepatocellular carcinoma. *Hepat. Oncol.* **2016**, *3*, 57–59. [CrossRef]
31. Schultz, M.; Dutta, S.; Tew, K.D. Inhibitors of glutathione S-transferases as therapeutic agents. *Adv. Drug Deliver Rev.* **1997**, *26*, 91–104. [CrossRef]

32. Eckford, P.D.; Sharom, F.J. ABC efflux pump-based resistance to chemotherapy drugs. *Chem. Rev.* **2009**, *109*, 2989–3011. [CrossRef] [PubMed]
33. Bajo, A.M.; Schally, A.V.; Halmos, G.; Nagy, A. Targeted doxorubicin-containing luteinizing hormone-releasing hormone analogue AN-152 inhibits the growth of doxorubicin-resistant MX-1 human breast cancers. *Clin. Cancer Res.* **2003**, *9*, 3742–3748. [PubMed]
34. Bidwell, G.L., 3rd; Davis, A.N.; Fokt, I.; Priebe, W.; Raucher, D. A thermally targeted elastin-like polypeptide-doxorubicin conjugate overcomes drug resistance. *Investig. New Drugs* **2007**, *25*, 313–326. [CrossRef] [PubMed]
35. Liang, J.F.; Yang, V.C. Synthesis of doxorubicin-peptide conjugate with multidrug resistant tumor cell killing activity. *Bioorg. Med. Chem. Lett.* **2005**, *15*, 5071–5075. [CrossRef] [PubMed]
36. Moktan, S.; Raucher, D. Anticancer activity of proapoptotic peptides is highly improved by thermal targeting using elastin-like polypeptides. *Int. J. Pept. Res. Ther.* **2012**, *18*, 227–237. [CrossRef]

International Journal of
Molecular Sciences

Article

Transferrin-Bound Doxorubicin Enhances Apoptosis and DNA Damage through the Generation of Pro-Inflammatory Responses in Human Leukemia Cells

Monika Jedrzejczyk [1], Katarzyna Wisniewska [1], Katarzyna Dominika Kania [2], Agnieszka Marczak [1] and Marzena Szwed [1,*]

1 Department of Medical Biophysics, Institute of Biophysics, Faculty of Biology and Environmental Protection, University of Lodz, Pomorska 141/143 Street, 90-236 Lodz, Poland; mon_ag_wer@wp.pl (M.J.); kasiaawisniewska@interia.pl (K.W.); agnieszka.marczak@biol.uni.lodz.pl (A.M.)

2 Laboratory of Transcriptional Regulation, Institute for Medical Biology, PAS, Lodowa 106 Street, 93-232 Lodz, Poland; kkania@cbm.pan.pl

* Correspondence: marzena.szwed@biol.uni.lodz.pl; Tel.: +48-635-4481 or +48-50661-2428

Received: 15 November 2020; Accepted: 7 December 2020; Published: 10 December 2020

Abstract: Doxorubicin (DOX) is an effective antineoplastic drug against many solid tumors and hematological malignancies. However, the clinical use of DOX is limited, because of its unspecific mode of action. Since leukemia cells overexpress transferrin (Tf) receptors on their surface, we proposed doxorubicin–transferrin (DOX–Tf) conjugate as a new vehicle to increase drug concentration directly in cancer cells. The data obtained after experiments performed on K562 and CCRF-CEM human leukemia cell lines clearly indicate severe cytotoxic and genotoxic properties of the conjugate drug. On the other hand, normal peripheral blood mononuclear cells (PBMCs) were more resistant to DOX–Tf than to DOX. In comparison to free drug, we observed that Tf-bound DOX induced apoptosis in a TRAIL-dependent manner and caused DNA damage typical of programmed cell death. These fatal hallmarks of cell death were confirmed upon morphological observation of cells incubated with DOX or DOX–Tf. Studies of expression of *TNF-α, IL-4, and IL-6*at the mRNA and protein levels revealed that the pro-inflammatory response plays an important role in the toxicity of the conjugate. Altogether, the results demonstrated here describe a mechanism of the antitumor activity of the DOX–Tf conjugate.

Keywords: leukemias; doxorubicin; inflammation

---

1. Introduction

Despite the key achievements obtained with novel biological agents developed during the last decade, leukemia remains an incurable disease [1]. One of the major challenges related to the treatment of leukemia is the huge diversity of these neoplasms [2,3]. The pathogenesis of the most common types of hematological malignancies, acute lymphoblastic leukemia (ALL), and chronic myelogenous leukemia (CML), involves successive genomic alterations in multipotent hematopoietic stem cells, and leads to their abnormal proliferation. Although CML is characterized by the constitutively activated BCR/ABL tyrosine kinase that confers resistance to apoptosis induced by anticancer drugs [4], ALL is typically of unknown etiology. ALL can develop from underlying hematological disorders (e.g., myelodysplastic syndromes) or from exposure to genotoxic agents (e.g., Type II topoisomerases, alkylating agents, or radiation) [5]. Another challenge with chemotherapy for leukemia is the development of treatment resistance [6], and accordingly there is a constant search for new therapeutic targets to improve patient survival rate and quality of life.

Anthracyclines, including doxorubicin (DOX), are commonly used for the treatment of both solid tumors and hematologic malignancies [7]. DOX exerts its cytotoxicity by producing free radicals, intercalating with DNA base pairs, and interacting with several molecular targets such as DNA topoisomerase II. There are several side effects that have been observed in patients treated with DOX, such as skin irritation, nausea, fever, and severe cardiac toxicity, leading to increased risk of heart failure [8]. Several years ago, it was demonstrated that the DOX-induced pro-inflammatory response may play a crucial role in the symptoms associated with anthracycline therapy [9,10]. DOX triggers inflammatory signaling cascades by promoting the release of different cytokines, including interleukin-1 (IL-1) and tumor necrosis factor-α (TNF-α), in combination with the activation of various signaling pathways, such as nuclear factor-κB (NF-κB), p38-MAPK, and autophagy pathways [11]. Moreover, the cytokines and inflammatory markers released after DOX stimulation are associated with the induction of the extrinsic apoptotic pathway [12,13]. For instance, TNF-related apoptosis-inducing ligand (TRAIL) is highly expressed after various anticancer drug treatments. Upon binding to TNF death receptors, TRAIL leads to their aggregation and recruits Fas-associated death domain-containing protein (FADD). The death domain of FADD binds to an analogous domain of caspase-8 to form the death-inducing signaling complex (DISC). This process activates downstream caspases and leads to programmed cell death [14]. The extrinsic apoptotic pathway may remain functional and successfully kill cancer cells, unless DOX treatment results in nonspecific cytotoxic action towards normal bone marrow, renal, and heart tissue.

Transferrin receptors (TfR) are overexpressed on many malignant tumor cells [15,16], whereas normal cells are deficient in this type of receptor [17]. Importantly, erythroid precursors, as well as their malignant clones represent those groups of cells that have the highest expression of TfRs [18].

Therefore, we suggested that transferrin (Tf) may be used as a carrier for DOX. For this reason, we used two different human leukemia cell lines CCRF-CEM (1×10^5 TfR per cell [19]) and K562 (1.5×10^5 TfR per cell, [20]) that represent ALL and chronic myeloid leukemia, respectively.

In parallel, the experiments were carried out on normal, non-cancer peripheral blood mononuclear cells (PBMCs) acquired from healthy donors. By conjugating DOX with transferrin (DOX–Tf conjugate), we proposed that Tf-bound DOX would improve drug cytotoxicity toward human cancer cells and that normal cells would remain intact. This hypothesis was partially proven in our previous study, showing that the conjugate was better able to induce programmed cell death compared to free DOX [21]. Moreover, when we compared the activities of different caspases involved in conjugate-mediated induction of apoptosis, we noticed the activation of casapse-8. This may indicate a contribution of the extrinsic apoptotic pathway to cytotoxicity with the conjugate. Therefore, we chose to explore the involvement of cell death receptor machinery as a component of the DOX–Tf-induced pro-death response in human leukemia cells.

In the current study, we demonstrate that the toxicity of DOX–Tf is much higher than that of free DOX in CCRF-CEM and K562 human leukemia cells in vitro. In addition, the conjugate is less cytotoxic to normal peripheral mononuclear cells than it is to each of the leukemia cell lines. We confirm the selective activity of Tf-bound DOX towards cancer cells by assessing genotoxic properties of the conjugate. Interestingly, our previous studies performed on solid tumor cells as well as hematological malignancies clearly show the predominant properties of DOX–Tf conjugate to induce apoptosis [22,23]. Here we find that pro-inflammatory cytokines participate in the cytotoxic reaction triggered by the conjugate, and this may be associated with the active delivery of a higher DOX dosage to cancer cells via a transferrin-dependent pathway.

2. Results

2.1. *Various Forms of Doxorubicin Exert Differential Cytotoxicity in Human Leukemia Cells*

To assess the cytotoxicity of Tf-bound DOX and free drug in human leukemia cells, we measured the mitochondrial activity of respiratory chain oxidoreductases that are active only in living cells. Reduced cell viability was observed in CCRF-CEM and K562 cell lines following DOX treatment

(Figure 1A), and this was further enhanced in cells treated with the conjugate. In contrast, free DOX was more cytotoxic to PBMCs than the conjugate. This agrees with our previous finding showing that both forms of DOX displayed diverse cytotoxicity in solid tumor cell lines, as determined by the varied panel of viability assays. As shown in Figure 1B, K562 and CCRF-CEM leukemia cell lines were consistently more sensitive to DOX–Tf than to DOX. In contrast, normal PBMCs were 2-fold less sensitive to DOX–Tf conjugate than to free drug, and this difference was significant. This observation is supported by microscopy studies (Figure 1C) showing that DOX–Tf affects the cellular morphology and induces morphological changes such as cell shrinkage and membrane protrusions.

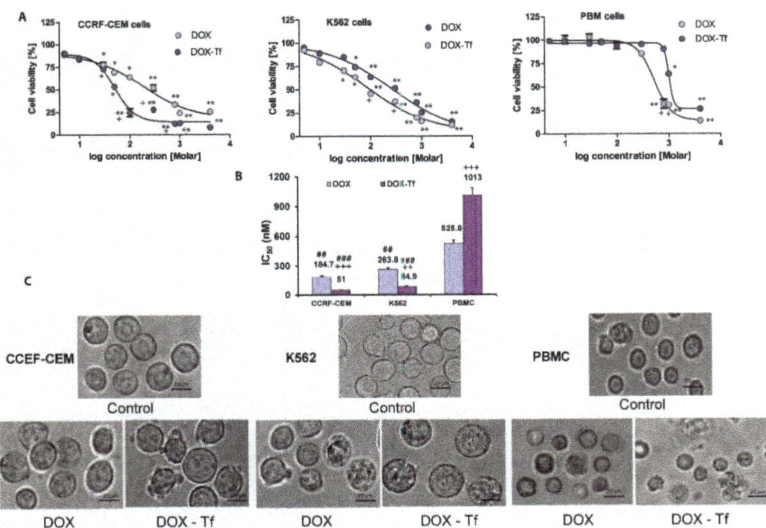

Figure 1. Cytotoxicity of free doxorubicin and doxorubicin–transferrin conjugate in human leukemia cell lines and peripheral blood mononuclear cells. (**A**): Viability of CCRF-CEM, K562, and PBMC cells after 48 h incubation with increasing concentrations of doxorubicin–transferrin (DOX–Tf) conjugate (violet symbols) or free doxorubicin (DOX) (blue symbols). Cell viability was measured by XTT assay. Values represent the means ± SD, ($n = 5$) * $p < 0.05$, ** $p < 0.01$ in comparison to untreated, control cells, (+) statistically significant differences noted between the probes incubated with free DOX compared to the conjugate, + $p < 0.05$. (**B**): Comparison of the cytotoxicity of free DOX and transferrin-bound DOX in CCRF-CEM and K562 cell lines or PBMCs. ## $p < 0.05$, ### $p < 0.01$ in comparison to normal, non-cancer cells, (++, +++) statistically significant differences noted between the probes incubated with free DOX compared to the conjugate, ++ $p < 0.01$, +++ $p < 0.001$. The values are the IC_{50} mean [nM] ± SD of five independent experiments with a 95% confidence interval. (**C**): Morphological changes observed with microscopy. Inverted phase contrast microscopy images were obtained following treatment of CCRF-CEM and K562 cells or PBMCs for 48 h with DOX–Tf or free DOX with the IC_{50} concentrations shown in the photos. Images were captured at 20× magnification, and the scale bars represent 20 µm.

2.2. DOX–Tf Conjugate Generates the Accumulation of γH2AX Phosphorylation

The reduction in cell viability triggered by the conjugate may be related to the various features of DOX–Tf toxicity, such as genotoxicity. Therefore, we measured the phosphorylation of histone H2AX, which is a molecular marker of dsDNA breaks. Our previous findings showed that Tf-bound DOX significantly induced DNA damage in both solid tumor and leukemia cell lines [24], demonstrating that the conjugate caused DNA lesions and the formation of alkali-labile sites. Here, we aimed to determine whether DOX–Tf triggered dsDNA breaks in two malignant cell lines, versus noncancerous PBMCs. As shown in Figure 2A, we found a significant increase in phosphorylation of histone H2A, predominantly in CCRF-CEM cells after 6 and 48 h of drug treatment. Under the same conditions, we observed a predominant role of the conjugate that induced 1.2- and 1.4-fold increases in intracellular γH2AX levels. In contrast, 1.3-fold growth was elicited by free DOX in K562 cells after a 24 h incubation. Furthermore, mostly in the CCRF-CEM cell line, DOX–Tf conjugate treatment led to an increase in histone *H2AX* transcription as the first cellular response to DNA lesions (Figure 2B).

2.3. Conjugate-Dependent DNA Damage/Lesions Are Connected to Apoptotic Cell Death

Intrigued by the increasing level of histone H2AX, we further analyzed whether the DNA damage induced by DOX was the molecular consequence of activated programmed cell death pathways. With transferase dUTP nick end labeling (TUNEL) assay, we measured pro-apoptotic DNA fragmentation to estimate the fraction of cells that exhibited single- and dsDNA fragments with possible label-free 3′-OH ends following treatment with DOX or DOX–Tf conjugate.

As shown in Figure 3A,B, the population of TUNEL-positive cells increased significantly when treated with free or conjugated DOX. The presence of DNA fragments with possible label-free 3′-OH ends was the highest in CCRF-CEM cells, and apoptosis increased from 20% to approximately 75% after 6 and 24 h of incubation with free or conjugated DOX. However, upon analysis of TUNEL-positive cells in the PBMC culture, it was revealed that DOX alone markedly increased the population of cells displaying DNA fragmentation, which is typical of apoptotic cell death.

2.4. The Extrinsic TRAIL-Dependent Apoptotic Pathway Is Triggered by DOX–Tf Conjugate in Human Leukemia Cells

Based on our previous data that the conjugate causes a reduction of mitochondrial membrane potential, provokes cytochrome c leakage, and mediates the activation of caspase-3 [21,23] we next asked the question whether the extrinsic pathway of apoptosis is induced by DOX–Tf treatment in human leukemia cells. In addition to our previous findings, this hypothesis was developed based on caspase-8 activity measurements that revealed the possibility that the conjugate induced a TRIAL-dependent [21] mechanism of cell death. Indeed, an increase in TRAIL activity was observed in both leukemia cell lines after 24 and 48 h incubation with DOX or DOX–Tf. As shown in Figure 4A, we observed 2.6- and 1.4-fold higher TRAIL expression between samples incubated with the conjugate or with free DOX in CCRF-CEM and K562 cells, respectively. In contrast, the increase in TRAIL ligand (TRAIL-L) levels in normal PBMC cultures was induced only after 24 h continuous incubation with free DOX. To determine whether the effect of the conjugate on TRAIL ligand production was attributed to changes at the mRNA level, we decided to measure *TRIAL–L* gene transcript. Figure 4B shows that both forms of DOX stimulated *TRAIL-L* expression in PBMCs, which suggests a substantial influence of free drug and the conjugate to induce of the extrinsic apoptotic pathway in human leukemia cells. However, a statistically significant difference in *TRAIL-L* mRNA level was observed between DOX-treated and conjugate-treated samples only in CCRF-CEM cell lines, confirming a superior effect of the conjugate in acute leukemia cells.

2.5. The Growth of TNF-α Expression Is Initiated by Tf-Bound DOX and Free Drug in Human Leukemia Cells

In multiple myeloma cells, transcriptional regulation of TRAIL is often triggered by a TNF-α dependent pathway [25]. TNF-α is a multifunctional cytokine capable of inducing several biological responses such as apoptosis, inflammation, and the stress response [26]. Having shown that DOX–Tf conjugate leads to increased mRNA and protein expression of TRAIL, we wished to examine whether TNF-α production is induced by transferrin-bound DOX or free drug. Indeed, the strongest effect of both examined forms of drug was observed in the CCRF-CEM cell line (Figure 5A). These cells displayed more than a 1.5-fold increase in TNF-α production after 24 h incubation with DOX or after 48 h incubation with the conjugate. It is well reported that TNF-α expression can also be regulated at the transcriptional level [27]. By using quantitative real-time RT-PCR, we found that mRNA levels of *TNF-α* increased 1.2-, 1.4- and 1.6-fold following treatment with free DOX in CCRF-CEM, K562, and PBMCs, respectively. However, as shown in Figure 5B, the conjugate-dependent increase in *TNF-α* transcript was observed only in normal PBMCs and showed more than a 2-fold increase in comparison to non-treated cells.

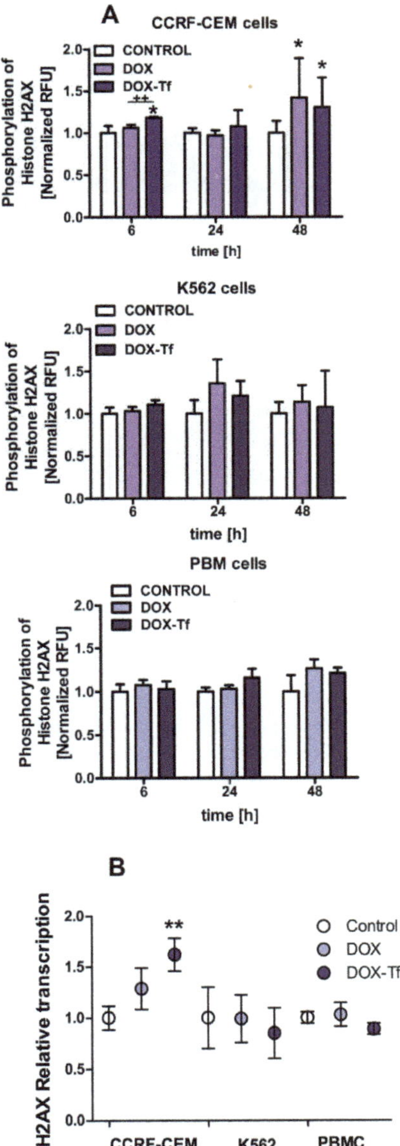

Figure 2. Doxorubicin–transferrin conjugate induced modifications of histone H2AX in human leukemia cells (**A**):The ratio of phosphorylation of histone H2AX (γH2AX) in comparison to total cellular content of this protein after treatment of CCRF-CEM and K562 cells or PBMCs with IC_{50} concentrations of doxorubicin (DOX) alone and doxorubicin–transferrin (DOX–Tf) conjugate for 6, 24, or 48 h. All values were normalized to untreated control cells, taken as 1. Data are expressed as the means ± SD, (n = 3). * $p < 0.05$ indicates statistically significant difference compared with control cells; and ++ $p < 0.01$ shows a difference of γH2AX level between cells treated with DOX or DOX–Tf. (**B**): The level of mRNA transcripts for the histone *H2AX* gene in the examined human leukemia cell lines as well as PBMCs exposed to IC_{50} concentrations of free DOX or DOX–Tf for 24 h. Data are expressed as the means ± SD, (n = 3). Asterisks refer to the level of significant difference (** $p < 0.01$) in mRNA level in the conjugate-treated cells compared to untreated control cells.

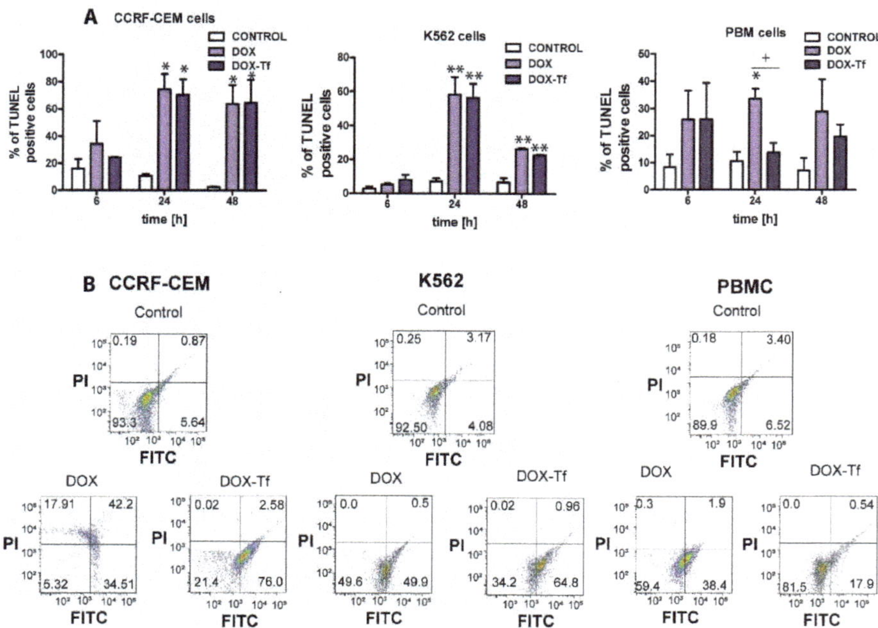

Figure 3. DNA damage is correlated with doxorubicin–transferrin-induced apoptosis in CCRF-CEM and K562 cell lines and in PBMCs (**A**) The influence free doxorubicin (DOX) and transferrin (Tf)-bound DOX (IC_{50} concentrations, respectively) on the induction of apoptosis in human leukemia cells as well as normal PBMCs was determined by transferase dUTP nick end labeling (TUNEL) assay. Quantitative results, calculated as the level of TUNEL-positive cells, are presented as mean ± SD (≥3). * $p < 0.05$, ** $p < 0.01$ relative to the control (untreated cells); + $p < 0.05$ indicates statistically significant difference between cells treated with DOX or DOX–Tf. (**B**) Typical cytometric dot-blots obtained after 24 h of incubation of human cancer and normal cells with IC_{50} concentrations of DOX alone or DOX–Tf conjugate. Lower left down corner, TUNEL-negative cells, propidium iodide (PI)-negative; lower right corner, TUNEL-positive cells, PI-negative; left upper corner, TUNEL-negative, PI-positive cells; upper right corner, TUNEL-positive, PI-positive cells.

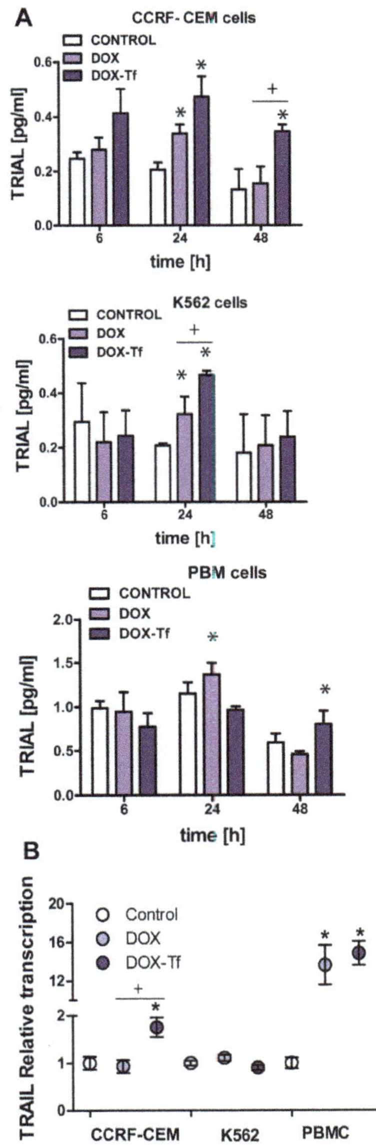

Figure 4. Doxorubicin -transferrin conjugate alters TRAIL ligand expression in human leukemia cells. (**A**) The production of TRAIL-ligand (TRAIL-L) in cancer and normal cells incubated with IC_{50} concentrations of doxorubicin (DOX) and doxorubicin–transferrin (DOX–Tf) conjugate for 6, 24, or 48 h. The results represent mean ± SD of three independent experiments. * $p < 0.05$ in comparison to respective control cells taken as 100%, (+) statistically significant difference observed between cells incubated with DOX in comparison to DOX–Tf, + $p < 0.05$. (**B**) *TRAIL-L* transcript levels (relative to *Hydroxymethylbilane synthase (HPRT1)* and *hydroxymethylbilane synthase (HMBS)* housekeeping genes) in CCRF-CEM and K562 cell lines and PBMCs exposed to either form of DOX for 24 h at IC_{50} concentrations, $n = 3$. Asterisks refer to significant differences (* $p < 0.05$) in the transcription levels in cells treated with free drug or DOX–Tf compared to the untreated cells. + indicates statistically significant changes between samples incubated with DOX alone or with DOX–Tf (+ $p < 0.05$).

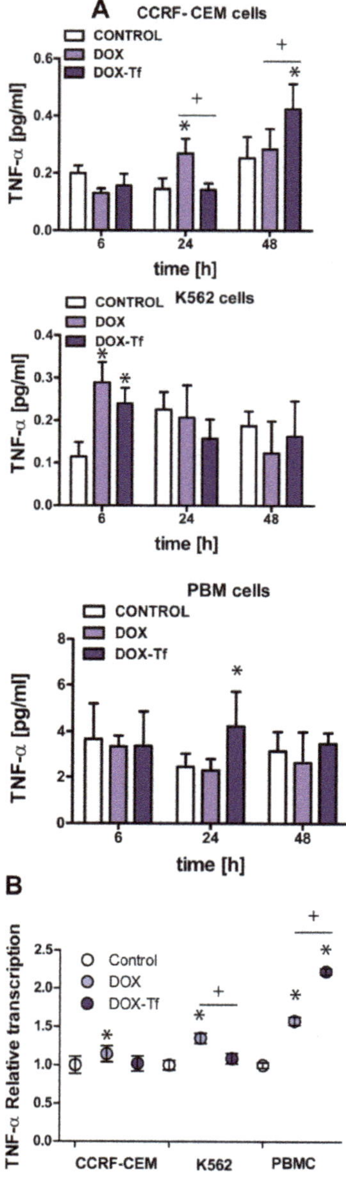

Figure 5. Changes in TNF-α expression in K562 and CCRF-CEM cell lines, and in PBMCs treated with transferrin-bound doxorubicin or free drug (**A**) The effect of IC_{50} concentrations of doxorubicin (DOX) and doxorubicin–transferrin (DOX–Tf) conjugate on TNF-α expression. The data represent mean ± SD of three independent experiments. * $p < 0.05$ in comparison to untreated, control cells, (+) statistically significant differences noted between the probes incubated with free DOX compared to the conjugate, + $p < 0.05$. (**B**) Expression of *TNF-α* gene transcript (relative to *HPRT1* and *HMBS* housekeeping genes) in investigated human leukemia cell lines and PBMCs exposed to IC_{50} concentrations of DOX alone and DOX–Tf conjugate. Asterisks refer to the level of significance (* $p < 0.05$, $n = 3$) whereas symbol "+" indicates a difference between expression in the cells treated with DOX or DOX–Tf, + $p < 0.05$.

2.6. The Involvement of IL-6 and IL-4 Cytokines in DOX–Tf Cytotoxicity

Apart from TNF–α, another major cytokine that mediates in the induction of apoptosis and the pro-inflammatory response is IL-6 [28]. Normal and malignant human cells secrete these cytokines in response to different types of external stress factors such as radiotherapy, toxins, and chemotherapeutic drugs [29]. It was therefore important to elucidate the role of IL-6 in the toxicity and induction of apoptosis caused by DOX–Tf conjugate. As shown in Figure 6A, the IL-6 expression was up-regulated in PBMC cultures treated with DOX for up to 6 h (1.5-fold increase) and in CCRF-CEM cells incubated with drugs for 24 and 48 h (approximately 2-fold increase with DOX and with the conjugate, respectively). Furthermore, the mRNA level of the gene for *IL-6* was also up-regulated (Figure 6C) in CCRF-CEM cells and in PBMCs treated with either form of DOX.

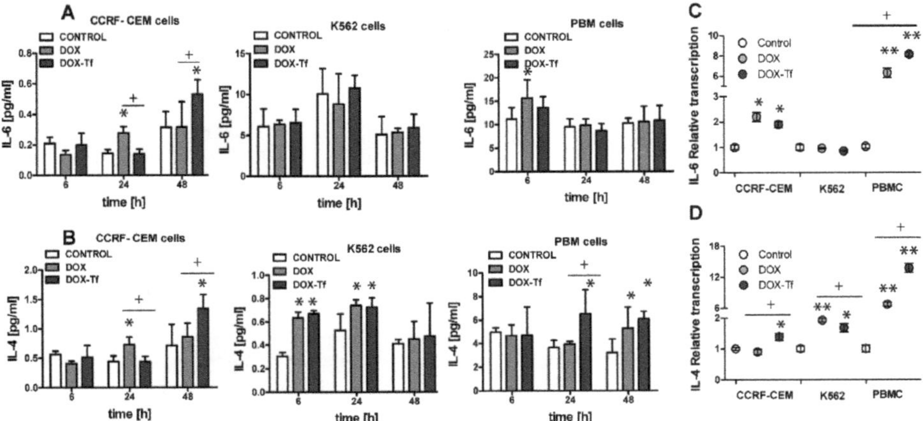

Figure 6. The evaluation of pro-inflammatory cytokines level after leukemia cells treatment with DOX and DOX–Tf conjugate. (**A**,**B**): The concentrations of interleukin 4 (IL-4) and interleukin 6 (IL-6) secreted in the supernatants of CCRF-CEM, K562, and PBMC cultures treated with IC_{50} concentrations of doxorubicin–transferrin (DOX–Tf) conjugate and DOX alone for 6, 24, or 48 h. The results show mean ± SD of three independent experiments. * $p < 0.05$,in comparison to untreated, control cells, (+) statistically significant differences observed between the probes incubated with free DOX in comparison to the conjugate, + $p < 0.05$. (**C**,**D**): Level of mRNA transcripts for *IL-6* and *IL-4* genes in CCRF-CEM, K562, and PBMCs exposed to IC_{50} concentrations of DOX–Tf conjugate and free doxorubicin for 48 h. Data (ΔΔCt values) were transformed into relative copy number values (the number of mRNA copies of the examined genes per housekeeping gene index, calculated as the average Ct value of the *HPRT1* and *HMBS* housekeeping genes) and standardized to the level of mRNA transcripts in untreated cells, taken as 1. Results are presented as the means ± SD, ($n = 6$). Asterisks refer to the level of significant difference (* $p < 0.05$, ** $p < 0.01$) in mRNA level in the drug treated cells compared to untreated cells (control), whereas—symbol "+" (+ $p < 0.05$) refers to statistically significant differences between the cells incubated with DOX–Tf conjugate in comparison to the cells incubated with a reference drug (free DOX).

Having ruled out the possibility that IL-6 is always involved in the conjugate-dependent induction of the pro-inflammatory response, we next explored whether IL-4 may participate in the inflammation process that is triggered by these drugs. IL-4 is a cytokine that is produced by T lymphocytes, basophiles, and mastocytes, as well as different types of cancer cells. It was recently reported that IL-4 is crucial during development of leukemia cell sensitivity to chemotherapeutic agents [30]. Figure 6B shows that the conjugate caused a 1.5-fold increase in IL-4 levels in PBMCs after 24 and 48 h of treatment. In human leukemia cell lines, a 1.8-fold and 1.3-fold increase of IL-4 expression was observed in K562 cells after 6 and 24 h incubation with either drug, respectively. In the most sensitive CCRF-CEM cell line, we noticed a time dependence between DOX and conjugate modes of action. Free drug triggered double expression of IL-4 after 24 h treatment, whereas Tf-bound DOX took an additional 24 h to achieve the same effect. A simplified model of the differential effects of DOX–Tf on autophagy is shown in Figure 7.

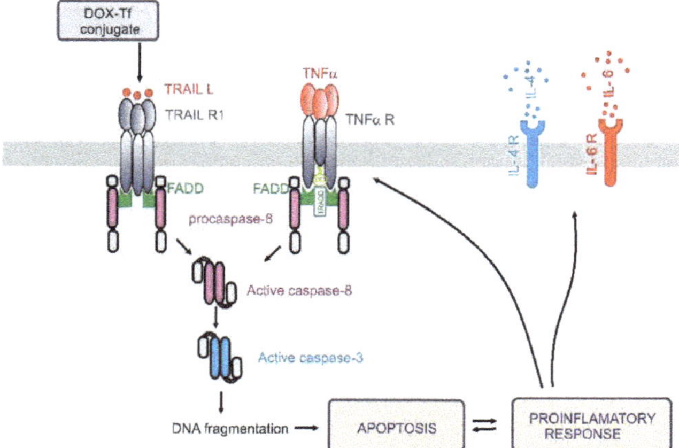

Figure 7. A simplified model of the differential effects of the doxorubicin–transferrin (DOX–Tf) conjugate on the pro-inflammatory response and induction of apoptosis via the extrinsic pathway. DOX–Tf conjugate stimulates the production of TNF-α and TRAIL ligand, the cell signaling proteins. They are also known as two major extrinsic mediators of apoptosis. Acute lymphoblastic cells and chronic myeloid malignancies have different isoreceptors for these cytokines. Binding of TRAIL or TNF-α to their receptors, results in death receptors aggregation (DR4, DR5) and Fas-associated death domain-containing protein (FADD). The death domain of FADD binds to an analogous domain of caspase-8 to form the death-inducing signaling complex (DISC). This process activates downstream caspases and leads to apoptosis. In parallel, in the presence of typical hallmarks of apoptosis, the DOX–Tf conjugate affects cellular genome and mediates the production of pro-inflammatory response. The transcription of IL-4 and IL-6 genes mediates the secretion of these cytokines to the extracellular matrix.

The data obtained from enzyme-linked immunosorbent assays (ELISA) assays agree with our quantitative RT results (Figure 6D). Therefore, there is an increase in *IL-4* mRNA and protein expression following treatment with either drug in all normal and cancer cell types investigated. Interestingly, there was a predominant IL-4-dependent response of K562 cells to both free DOX and DOX–Tf treatment. This further confirmed the function of IL-4 in the sensitivity of leukemia cells to chemotherapeutic agents.

3. Discussion

In the present paper, we address the mechanisms whereby DOX–Tf conjugate kills human leukemia cells in vitro. In particular, we elucidate the involvement of the pro-inflammatory response pathway and its downstream target TRAIL. The main findings and conclusions of the study are summarized in Figure 7. Our results show that the induction of apoptosis via the extrinsic pathway of apoptosis may significantly contribute to the toxicity of the conjugate. Previous studies have demonstrated a potent role of DOX–Tf to cause programmed cell death. Consequently, DOX attached to Tf triggers activation of caspases 3, 8, and 9, decreases mitochondrial membrane potential, produces reactive oxygen species (ROS), and leads to single or dsDNA breaks [21,24].

The fatal consequences of DNA damage are often referred to as oligonucleosomal fragmentation, which can be visualized during DNA agarose gel electrophoresis as a ladder-shape pattern.

We observed this phenomenon in A549 and HepG2 cells treated with the conjugate for up to 48 h [22]. During apoptosis, the DNA fragments retain free 3′-OH ends, which can be detected via the TUNEL method. It has been previously described that drugs belonging to the first and second generations of anthracyclines induce apoptotic DNA degradation [31]. Moreover, the strong dependence between DNA lesions and apoptosis induced by DOX–Tf conjugate was observed by Lubgan et al [32] in human HL-60-promyelocytic leukemia cells. It was reported that in the multidrug resistant cell line, DOX–Tf induced more dsDNA breaks compared to the free drug. The study also showed that DOX–Tf induced dsDNA breaks more efficiently than free DOX. Furthermore, data described by the same group agreed that the apoptotic rates triggered by DOX–Tf were greater than in the DOX-sensitive parental cell line. Here in our study, we observed that Tf-bound DOX significantly augmented the number of TUNEL-positive cells. Interestingly, the CCRF-CEM cell line, which provides a molecular model of ALL, was the most sensitive to genotoxic DOX–Tf properties with respect to histone H2AX phosphorylation. On the other hand, the K562 cell line, which represents CML, was more resistant to DNA damage induced by either form of DOX compared with CCRF-CEM cells. These data agree with the general evaluation of DOX–Tf conjugate cytotoxicity. Here we show the results obtained after XTT assay readouts, but during earlier experiments we used MTT [33], Neutral red [34], or Alamar Blue [35] assays, which refer to mitochondria, lysosomes, and redox homeostasis, respectively. Even though these assays have different molecular targets, the direction of cellular resistance towards DOX–Tf treatment was the same (K562 < CCRF-CEM < PBMCs). The various sensitivity of the examined human leukemia cells to the conjugate treatment might be related with the TfR expression. The number of TfR per cell is tightly regulated by many different factors such as intracellular iron level, cell proliferation or erythropoiesis at levels of receptor recycling, transcriptional, or post-transcriptional control. Additionally, it was proved that phytohemagglutinin stimulated peripheral blood lymphocytes showed a marked increase in TfR expression a few hours before the initiation of replication [36].

The issue differing sensitivity of normal and cancer cells to the conjugate is more complex and is related to apoptosis induction, DNA damage, TfR expression, and oxidative stress, as well as initiation of local inflammation. Cytokine production has been linked to the pro-inflammatory response, apoptosis, and stimulation of cells towards autophagy [37]. Little is known about cytokines that negatively affect leukemia cell growth and survival. For instance, Pena Martines showed a previously unrecognized role of IL-4 as an inhibitor of the growth and survival of primitive acute myelogenous leukemia (AML) cells. The same group revealed that IL-4 induced apoptosis of AML cells in a Stat6-dependent manner [30]. Therefore, it was interesting to find that the DOX–Tf conjugate-mediated production of IL-4 in both of our leukemic cell lines. Indeed, we noticed that mRNA and protein IL-4 expression was enhanced by free drug and by the conjugate. However, in various types of lymphomas, IL-4 may behave differently. IL-4 increased the sensitivity of diffuse large B-cell lymphoma (DLBCL) subtype to doxorubicin-induced apoptosis and complement-dependent rituximab cell death. In contrast, IL-4 protected ABC-like DLBCL from the cytotoxic effects of doxorubicin and rituximab [38]. The same double-edged sword was demonstrated for IL-6, a cytokine that regulates not only immune and inflammatory responses, but also participates in hepatic acute phase protein synthesis, hematopoiesis, and bone metabolism [39]. IL-6 is known to activate both STAT3 and NF-κB in primary chronic lymphocytic leukemia (CLL) cells. The IL-6 production by CLL B-cells is associated with worse clinical outcome for patients with CLL [40]. Regarding the connection between the cytotoxicity of DOX and IL-6 production, it was found that IL-6 was significantly up-regulated in DOX-treated tissues and cells [41]. We propose that the DOX–Tf-mediated induction of the pro-inflammatory response may be part of a programmed cell death-initiating response (Figure 7). This conclusion agrees with our data showing an increase of IL-6 levels in the supernatant of normal and cancer cells incubated with the conjugate. Furthermore, the overproduction of IL-6 can often appear in parallel to TNF-α secretion [42]. Surprisingly, we observed the same molecular sequence; TNF-α and IL-6 were simultaneously secreted in leukemia cells and in PBMCs following treatment with free DOX or the conjugate. However, TNF-α is not only important in the pro-inflammatory process. As a death ligand, TNF-α can be used for anticancer treatment to induce the extrinsic apoptosis pathway, which is dependent upon TNF-α binding to the death receptor FAS/APO-1, also known as CD95 [43]. The application of antibodies against the CD95 receptor may be a promising therapeutic tool for inducing apoptosis in tumor cells, unless acute and lethal hepatic toxicity appears during cancer treatment [44]. Based on its sequence homology to TNF-α, it has been subsequently identified that TRAIL-L [45] did not affect normal cells and had similar apoptotic effects as TNF-α [46]. However, TRAIL does not always act alone. It was shown that TRAIL in combination with DOX or 4-hydroxy-IFO had highly toxic and pro-apoptotic effects in the TNF-α-sensitive rhabdomyosarcoma cell line KYM-1 [47]. Moreover, DOX can trigger overexpression of *TRAIL-L* mRNA and protein in different types of hepatoma cancer cells [48]. We observed a similar pattern with DOX and DOX–Tf treatment in human leukemia cell lines as well as in normal PBMCs. However, this does not necessarily mean that both forms of DOX trigger pro-death pathways in non-cancer cells. Following the explanation provided by Rogalska et al. [49], we propose that DOX alone, or DOX conjugated to Tf, may activate many effector cells such as cytotoxic T lymphocytes, NK cells, neutrophils, monocytes. or macrophages. Consequently, these drugs might initiate apoptosis through the engagement of death receptors.

In conclusion, we showed that the DOX–Tf conjugate is a novel therapeutic that increases the sensitivity of cancer cell lines and does not affect normal cells. As a promising drug delivery system, Tf-bound DOX induces DNA damage and triggers programmed cell death with the engagement of the TRAIL-dependent, extrinsic pathway of apoptosis. The involvement of TNF-α and other cytokines suggests that pro-inflammatory effects of the conjugate are closely related with its cytotoxicity

4. Materials and Methods

4.1. Reagents

DOX was purchased from Sequoia Research Products (Pangbourne, United Kingdom). Transferrin, 2,3-Bis (2-methoxy-4-nitro-5-sulfophenyl)-2*H*-tetrazolium-5-carboxanilide inner salt (XTT) and all reagents for carrying out the conjugation procedure were purchased from Sigma- Aldrich (Darmstadt, Germany). Doxorubicin was bounded to Tf using the modified conjugation procedure developed by Berczi et al. [50], and specified in Patent Claim No WIPO ST 10/C PL 402896. The conjugate was analyzed using spectrofluorometry, mass spectrometry [51], and sodium dodecyl sulfate–polyacrylamide gel electrophoresis (SDS–PAGE), according to Lubgan et al. [32]. TNF-α, TRAIL, IL-6, IL-4, and Human Phospho-Histone H2AX immunoassays were supplied by R&D Systems (Oxford, United Kingdom), whereas the terminal deoxynucleotidyl TUNEL assay was from BioVision (Milpitas, CA, USA). Dulbecco's Modified Eagle's Medium (DMEM) and fetal bovine serum (FBS) were supplied by Cambrex (Basel, Switzerland). All other chemicals and solvents used in this study were of the highest analytical grade.

4.2. Cell Culture

ALL cells (CCRF-CEM cell line, ATCC CCL-119™) were provided by Prof G. Bartosz (Department of Molecular Biophysics, University of Lodz, Lodz, Poland). Myelogenous erythroleukemia cells (K562 cell line, ATCC CCL-243™) were a kind gift from Prof J. Robert at Institute Bergonie, Bordeaux, France. In parallel, the experiments were performed on normal PBMCs. The blood used to isolate these cells was obtained from the Blood Bank in Lodz, Poland. The PBMCs were isolated by centrifugation in a density gradient of Histopaque (300× *g* for 30 min at 22 °C), as described previously [51]. All cells were cultured in RPMI 1640 Cambrex (Basel, Switzerland) containing 2 mM L-glutamine (Invitrogen, Carlsbad, CA, USA), supplemented with 10% FBS (Sigma–Aldrich, Darmstadt, Germany), 100 U/mL penicillin (Invitrogen, Carlsbad, CA, USA), and 100 µg/mL streptomycin (Invitrogen, Carlsbad, CA, USA). The cells were grown in standard conditions (37 °C, 100% humidity, and in an atmosphere containing 5% CO_2 and 95% normal air).

4.3. Cell Cytotoxicity Assay

The viability of CCRF-CEM, K562, and PBMCs was measured using the XTT cytotoxicity test. The principle of this assay is that viable cells reduce the tetrazolium salt XTT 2,3-bis(2-methyloxy-4-nitro-5-sulfophenyl)-2*H*-tetrazolium-5 carboxanilide to a medium-soluble product (Tunney et al., 2004). Briefly, cells were seeded on 96-well plates at a density of 1×10^4 (CCRF-CEM, K562) or 1×10^5 (PBMC) in each well in 0.1 mL of culture medium. Subsequently, various concentrations of 0.05 mL DOX or DOX–Tf were added to appropriate wells, and cells were treated with drugs for 48h. At the end of the incubation, cells were centrifuged (230× *g* at 4 °C, for 10 min), medium was gently removed, and the cell cultures were resuspended in 50 μL XTT at a final concentration of 0.3 mg/mL medium. The microplates were incubated in a CO_2 incubator for 4 h and the reduction of XTT was measured at 492 nm using a microtiter plate reader (Awareness Technology Inc., Palm City, FL, USA). The percentage of viable cells was calculated by comparing the reduction of XTT in drug treated cells to non-treated control cells. In parallel, the IC_{50} was quantified (using GraphPad Prism software, Graphpad Inc, San Diego, CA, USA), and this defines the drug concentration causing a 50% reduction of cell viability relative to the control. In parallel with the cytotoxicity experiments, the cellular morphology changes were observed after treatment with free DOX or DOX–Tf by phase contrast microscopy using an IX73 Olympus microscope (Olympus, Tokyo, Japan) equipped with a 20× objective and a Digital Sight camera (Olympus, Tokyo, Japan).

4.4. H2AX Assay

The level H2AX protein with phosphorylated Ser139 was measured using a Human Phospho-Histone H2AX DuoSet IC assay kit, according to the manufacturer's instructions (R&D Systems, Oxford, UK). A total of 2×10^6 cells were incubated for 6 h, 24 h, or 48 h with DOX–Tf or DOX alone. Following incubation, the cells were permeabilized with buffer supplied by the assay manufacturer and incubated with antibodies. Antibodies included primary rabbit monoclonal antibody specific to H2AX (phospho S139; dilution 1:100) and anti-rabbit secondary antibody solution (dilution 1:1000). The fluorescence measurements were obtained on a Fluoroskan Ascent plate reader (Fluoroskan Ascent FL, Stockholm, Sweden) with filter pairs of 540 nm/600 nm and 360 nm/450 nm. The results are presented as the fluorescence ratio measured at 540 nm/600 nm to that measured at 360 nm/450 nm (phosphorylated form to total histone H2AX concentration).

4.5. TUNEL Assay

To examine DNA damage associated with apoptosis, a TUNEL assay was used [31]. We performed a detection of the early stages of apoptosis by labeling 3′-OH ends of single- and double-stranded DNA (dsDNA) fragments with bromo-deoxyuridine triphosphate nucleotides (Br-dUTP) according to the Apo-BrdU In Situ DNA Fragmentation Assay Kit protocol supplied by the manufacturer (BioVision Milpitas, CA, USA). Human leukemia cells as well as PBMCs were seeded in 6-well plates at a density of 2×10^4 per well, and were subsequently treated with free DOX or DOX–Tf conjugate for 6 h, 24 h, or 48 h in cell culture growth conditions. After the incubation was complete, cells were collected, washed with; 10 mM phosphate, 0.15 M NaCl, pH 7.4 (PBS), fixed with freshly prepared 4% paraformaldehyde, and incubated for 1 h at 37 °C in DNA Labeling mixture containing TdT Reaction Buffer and terminal deoxynucleotidyl transferase (TdT). Subsequently, the cells were stained with Br-dUTP and anti-BrdU-FITC antibody solution in total darkness for 30 min at room temperature, and then the incubation with propidium iodide/RNase A solution was carried out. Finally, the fluorescence intensity was measured using a Becton Dickinson LSR II flow cytometer (BD Biosciences, Franklin Lakes, N.J., USA) equipped with green and red lasers. The number of TUNEL-positive cells was expressed as a percentage of the total number of cells in the sample.

4.6. ELISA

The level of human TRAIL, TNF-α, IL-6, and IL-4 in cell-free culture supernatants was determined in duplicate by ELISAs specific for TRAIL, TNF α, and IL-6 (RnD Systems, Oxford, UK) according to the manufacturer's recommendation. The absorbance at 450 nm and 560 nm were determined in a microtiter plate reader (Awareness Technology Inc., Palm City, FL, USA). The wavelength values of 450 nm were subtracted from those at 560 nm. Antibody signal intensity was normalized to the total protein amount using the Lowry method.

4.7. Quantitative Real-Time RT-PCR

Total RNA was isolated using the TRI Reagent (Sigma–Aldrich, Darmstadt, Germany), according to the manufacturer's instructions [22]. Total RNA (5 μg) was used for cDNA synthesis using the Maxima First Strand cDNA Synthesis Kit for RT-qPCR (Thermo Fisher Scientific, Inc., Carlsbad, CA, USA). The real-time PCR analysis was run on a LightCycler 480 SYBR Green I Master Mix (Roche Diagnostics GmbH, Mannheim, Germany) and a Roche LightCycler 480 Instrument (Roche Diagnostics GmbH). The cycling conditions were 95 °C for 1 min, followed by 40 cycles of 95 °C for 5 s, 55 °C for 5 s, and 72 °C for 5 s. Relative gene expression was normalized to the housekeeping genes hydroxymethylbilane synthase (*HMBS*) and hypoxanthine phosphoribosyl transferase (*HPRT*), and was calculated using the ΔΔCt method. The study of mRNA expression included the following genes: *IL-4, IL-6, TNF-α, H2X, and TRAIL-L*. The primer sequences are shown in Table 1.

Table 1. Primer sequences used for RT-PCR.

Gene	Strand	Sequence 5′to 3′
Hypoxanthine-guanine Phosphoribosyltransferase (HPRT1)	Forward Reverse	TGACACTGGCAAAACAATGCA GGTCCTTTTCACCAGCAAGCT
Hydroxymethylbilane synthase (HMBS)	Forward Reverse	CAAGGACCAGGACATCTTGGAT CCAGACTCCTCCAGTCAGGTACA
H2AX	Forward Reverse	GGCCTCCAGTTCCCAGTG TCAGCGGTGAGGTACTCCAG
TRAIL-L	Forward Reverse	ACCAACGAGCTGAAGCAGAT CAAGTGCAAGTTGCTCAGGA
TNF-α	Forward Reverse	CCCAGGGACCTCTCTCTAATCA GCTACAGGCTTGTCACTCGG
IL-4	Forward Reverse	TTGAACAGCCTCACAGAGCAGA GTTGTGTTCTTGGAGGCAGCA
IL-6	Forward Reverse	TCTCCACAAGCGCCTTCG CTCAGGGCTGAGATGCCG

4.8. Statistical Analysis

Statistics were calculated using STATISTICA.PL software v.12.5 (StatSoft, Poland) [33], and the viability curves were prepared using GraphPad Prism 5.0 software (GraphPad Inc., San Diego, CA, USA). All measurements were performed at least in duplicate with n = 3–6; a p value of 0.05 was considered significant. The data were expressed as the mean ± SD. Statistical significance was evaluated using Student's t-test or one-way ANOVA followed by Tukey's test.

Author Contributions: M.J., K.W.: performed the in vitro experiments, and contributed to drafting the manuscript. K.D.K.: performed RT-PCR experiments and participated in preparation of the manuscript A.M.: participated in preparation of the manuscript. M.S.: planned and supervised the in vitro study and preparation of the manuscript All authors have read and agreed to the published version of the manuscript.

Funding: This work was supported in part by Grant no. B1611000001151.02 of the University of Lodz, Poland.

Acknowledgments: The group in Lodz thanks Marzena Pacholska for technical assistance with in vitro studies.

Conflicts of Interest: The authors declare no conflict of interest. The manuscript has been read and approved by all the authors, the requirements for authorship have been met, and each author believes the manuscript represents honest work.

Abbreviations

ALL	acute lymphoblastic leukemia
CML	chronic myelogenous leukemia
DOX	Doxorubicin
DOX–Tf	Doxorubicin–transferrin conjugate
FBS	Fetal bovine serum
HMBS	Hydroxymethylbilane synthase (HMBS)
HPRT1	Hypoxanthine phosphoribosyltransferase 1
IL-1	interleukin-1
TfR	Transferrin receptors
TNF-α	tumor necrosis factor-α
NF-κB	nuclear factor-κB
TRAIL	TNF-related apoptosis-inducing ligand
FADD	Fas-associated death domain-containing protein
DISC	death-inducing signaling complex

References

1. Dong, Y.; Shi, O.; Zeng, Q.; Lu, X.; Wang, W.; Li, Y.; Wang, Q. Leukemia incidence trends at the global, regional, and national level between 1990 and 2017. *Exp. Hematol. Oncol.* **2020**, *9*, 14. [CrossRef]
2. Muselli, F.; Peyron, J.-F.; Mary, D. Druggable Biochemical Pathways and Potential Therapeutic Alternatives to Target Leukemic Stem Cells and Eliminate the Residual Disease in Chronic Myeloid Leukemia. *Int. J. Mol. Sci.* **2019**, *20*, 5616. [CrossRef] [PubMed]
3. Green, S.D.; Konig, H. Treatment of Acute Myeloid Leukemia in the Era of Genomics—Achievements and Persisting Challenges. *Front. Genet.* **2020**, *11*, 480. [CrossRef] [PubMed]
4. Loscocco, F.; Visani, G.; Galimberti, S.; Curti, A.; Isidori, A. BCR-ABL Independent Mechanisms of Resistance in Chronic Myeloid Leukemia. *Front. Oncol.* **2019**, *9*. [CrossRef] [PubMed]
5. Phelan, K.W.; Advani, A.S. Novel Therapies in Acute Lymphoblastic Leukemia. *Curr. Hematol. Malign Rep.* **2018**, *13*, 289–299. [CrossRef]
6. Imai, K. Acute lymphoblastic leukemia: Pathophysiology and current therapy. *[Rinsho ketsueki] Jpn. J. Clin. Hematol.* **2017**, *58*, 460–470.
7. Sauter, K.A.D.; Wood, L.J.; Wong, J.; Iordanov, M.; Magun, B.E. Doxorubicin and daunorubicin induce processing and release of interleukin-1β through activation of the NLRP3 inflammasome. *Cancer Biol. Ther.* **2011**, *11*, 1008–1016. [CrossRef]
8. Pai, V.B.; Nahata, M.C. Cardiotoxicity of chemotherapeutic agents: Incidence, treatment and prevention. *Drug Saf.* **2000**, *22*, 263–302. [CrossRef]
9. Tien, C.-C.; Peng, Y.-C.; Yang, F.-L.; Subeq, Y.-M.; Lee, R.-P. Slow infusion rate of doxorubicin induces higher pro-inflammatory cytokine production. *Regul. Toxicol. Pharmacol.* **2016**, *81*, 69–76. [CrossRef]
10. Kabel, A.M.; Alzahrani, A.A.; Bawazir, N.M.; Khawtani, R.O.; Arab, H.H. Targeting the proinflammatory cytokines, oxidative stress, apoptosis and TGF-β1/STAT-3 signaling by irbesartan to ameliorate doxorubicin-induced hepatotoxicity. *J. Infect. Chemother.* **2018**, *24*, 623–631. [CrossRef]
11. Sala, V.; Della Sala, A.; Hirsch, E.; Ghigo, A. Signaling Pathways Underlying Anthracycline Cardiotoxicity. *Antioxid. Redox Signal.* **2020**, *32*, 1098–1114. [CrossRef]
12. Singla, D.K.; Johnson, T.A.; Dargani, Z.T. Exosome Treatment Enhances Anti-Inflammatory M2 Macrophages and Reduces Inflammation-Induced Pyroptosis in Doxorubicin-Induced Cardiomyopathy. *Cells* **2019**, *8*, 1224. [CrossRef] [PubMed]
13. Dargani, Z.T.; Singla, D.K. Embryonic stem cell-derived exosomes inhibit doxorubicin-induced TLR4-NLRP3-mediated cell death-pyroptosis. *Am. J. Physiol. Circ. Physiol.* **2019**, *317*, H460–H471. [CrossRef] [PubMed]

14. Xu, L.; Hu, X.; Qu, X.; Hou, K.; Zheng, H.; Liu, Y. Cetuximab enhances TRAIL-induced gastric cancer cell apoptosis by promoting DISC formation in lipid rafts. *Biochem. Biophys. Res. Commun.* **2013**, *439*, 285–290. [CrossRef]
15. Walker, L.; Perkins, E.; Kratz, F.; Raucher, D. Cell penetrating peptides fused to a thermally targeted biopolymer drug carrier improve the delivery and antitumor efficacy of an acid-sensitive doxorubicin derivative. *Int. J. Pharm.* **2012**, *436*, 825–832. [CrossRef] [PubMed]
16. Kratz, F.; Warnecke, A.; Schmid, B.; Chung, D.-E.; Gitzel, M. Prodrugs of Anthracyclines in Cancer Chemotherapy. *Curr. Med. Chem.* **2006**, *13*, 477–523. [CrossRef] [PubMed]
17. Florent, J.C.; Monneret, C. Doxorubicin conjugates for selective delivery to tumors. *Top. Curr. Chem.* **2008**, *283*, 99–140. [PubMed]
18. Kawabata, H. Transferrin and transferrin receptors update. *Free Radic. Biol. Med.* **2019**, *133*, 46–54. [CrossRef] [PubMed]
19. Hedley, D.; Rugg, C.; Musgrove, E.; Taylor, I. Modulation of transferrin receptor expression by inhibitors of nucleic acid synthesis. *J. Cell. Physiol.* **1985**, *124*, 61–66. [CrossRef]
20. Bridges, K.R.; Smith, B.R. Discordance between transferrin receptor expression and susceptibility to lysis by natural killer cells. *J. Clin. Investig.* **1985**, *76*, 913–918. [CrossRef]
21. Szwed, M.; Laroche-Clary, A.; Robert, J.; Jozwiak, Z. Induction of apoptosis by doxorubicin–transferrin conjugate compared to free doxorubicin in the human leukemia cell lines. *Chem. Interact.* **2014**, *220*, 140–148. [CrossRef] [PubMed]
22. Szwed, M.; Kania, K.D.; Jozwiak, Z. Assessment of pro-apoptotic activity of doxorubicin-transferrin conjugate in cells derived from human solid tumors. *Int. J. Biochem. Cell Biol.* **2016**, *70*, 57–67. [CrossRef] [PubMed]
23. Szwed, M.; Laroche-Clary, A.; Robert, J.; Jozwiak, Z. Efficacy of doxorubicin-transferrin conjugate in apoptosis induction in human leukemia cells through reactive oxygen species generation. *Cell. Oncol.* **2016**, *39*, 107–118. [CrossRef] [PubMed]
24. Szwed, M.; Jozwiak, Z. Genotoxic effect of doxorubicin–transferrin conjugate on human leukemia cells. *Mutat. Res. Toxicol. Environ. Mutagen.* **2014**, *771*, 53–63. [CrossRef]
25. Twomey, J.D.; Kim, S.-R.; Zhao, L.; Bozza, W.P.; Zhang, B. Spatial dynamics of TRAIL death receptors in cancer cells. *Drug Resist. Updates* **2015**, *19*, 13–21. [CrossRef] [PubMed]
26. Foghsgaard, L.; Wissing, D.; Mauch, D.; Lademann, U.; Bastholm, L.; Boes, M.; Elling, F.; Leist, M.; Jäättelä, M. Cathepsin B Acts as a Dominant Execution Protease in Tumor Cell Apoptosis Induced by Tumor Necrosis Factor. *J. Cell Biol.* **2001**, *153*, 999–1010. [CrossRef]
27. Falvo, J.V.; Tsytsykova, A.V.; Goldfeld, A.E. Transcriptional Control of the TNF Gene. *Curr. Dir. Autoimmun.* **2010**, *11*, 27–60. [CrossRef]
28. Wojdasiewicz, P.; Poniatowski, Ł.A.; Szukiewicz, D. The Role of Inflammatory and Anti-Inflammatory Cytokines in the Pathogenesis of Osteoarthritis. *Mediat. Inflamm.* **2014**, *2014*. [CrossRef]
29. Jones, V.S.; Huang, R.-Y.; Chen, L.-P.; Chen, Z.-S.; Fu, L.; Huang, R.-P. Cytokines in cancer drug resistance: Cues to new therapeutic strategies. *Biochim. Biophys. Acta (BBA) Rev. Cancer* **2016**, *1865*, 255–265. [CrossRef]
30. Peña-Martínez, P.; Eriksson, M.; Ramakrishnan, R.; Chapellier, M.; Högberg, C.; Orsmark-Pietras, C.; Richter, J.; Hagström-Andersson, A.K.; Fioretos, T.; Järås, M. Interleukin 4 induces apoptosis of acute myeloid leukemia cells in a Stat6-dependent manner. *Leukemia* **2018**, *32*, 588–596. [CrossRef]
31. Rogalska, A.; Szwed, M.; Jozwiak, Z. Aclarubicin-induced apoptosis and necrosis in cells derived from human solid tumours. *Mutat. Res. Toxicol. Environ. Mutagen.* **2010**, *700*, 1–10. [CrossRef]
32. Lubgan, D.; Jóźwiak, Z.; Grabenbauer, G.G.; Distel, L.V.R. Doxorubicin-transferrin conjugate selectively overcomes multidrug resistance in leukaemia cells. *Cell. Mol. Biol. Lett.* **2009**, *14*, 113–127. [CrossRef] [PubMed]
33. Szwed, M.; Wrona, D.; Kania, K.D.; Koceva-Chyla, A.; Marczak, A. Doxorubicin–transferrin conjugate triggers pro-oxidative disorders in solid tumor cells. *Toxicol. Vitr.* **2016**, *31*, 60–71. [CrossRef] [PubMed]
34. Szwed, M.; Kania, K.D.; Jóźwiak, Z. Changes in the activity of antioxidant barrier after treatment of K562 and CCRF-CEM cell lines with doxorubicin–transferrin conjugate. *Biochimie* **2014**, *107*, 358–366. [CrossRef] [PubMed]
35. Szwed, M.; Kania, K.D.; Jóźwiak, Z. Toxicity of doxorubicin-transferrin conjugate is connected to the modulation of Wnt/β-catenin pathway in human leukemia cells. *Leuk. Res.* **2015**, *39*, 1096–1102. [CrossRef] [PubMed]

36. Pelosi, E.; Testa, U.; Louache, F.; Thomopoulos, P.; Salvo, G.; Samoggia, P.; Peschle, C. Expression of transferrin receptors in phytohemagglutinin-stimulated human T-lymphocytes. Evidence for a three-step model. *J. Biol. Chem.* **1986**, *261*, 3036–3042.
37. Jiang, G.; Tan, Y.; Wang, H.; Peng, L.; Chen, H.-T.; Meng, X.-J.; Li, L.-L.; Liu, Y.; Li, W.-F.; Shan, H. The relationship between autophagy and the immune system and its applications for tumor immunotherapy. *Mol. Cancer* **2019**, *18*, 1–22. [CrossRef]
38. Sarosiek, K.A.; Nechushtan, H.; Lu, X.; Rosenblatt, J.D.; Lossos, I.S. Interleukin-4 distinctively modifies responses of germinal centre-like and activated B-cell-like diffuse large B-cell lymphomas to immuno-chemotherapy. *Br. J. Haematol.* **2009**, *147*, 308–318. [CrossRef]
39. Wei, L.-H.; Kuo, M.-L.; Chen, C.-A.; Chou, C.-H.; Cheng, W.-F.; Chang, M.-C.; Hung, M.-C.; Hsieh, C.-Y. The anti-apoptotic role of interleukin-6 in human cervical cancer is mediated by up-regulation of Mcl-1 through a PI 3-K/Akt pathway. *Oncogene* **2001**, *20*, 5799–5809. [CrossRef]
40. Wang, H.; Jia, L.; Li, Y.; Farren, T.; Agrawal, S.G.; Liu, F. Increased autocrine interleukin-6 production is significantly associated with worse clinical outcome in patients with chronic lymphocytic leukemia. *J. Cell. Physiol.* **2019**, *234*, 13994–14006. [CrossRef]
41. Zhang, W.; Yu, J.; Dong, Q.; Zhao, H.; Li, F.; Li, H. A mutually beneficial relationship between hepatocytes and cardiomyocytes mitigates doxorubicin-induced toxicity. *Toxicol. Lett.* **2014**, *227*, 157–163. [CrossRef] [PubMed]
42. Ramírez-Expósito, M.J.; Martínez-Martos, J.M. Anti-Inflammatory and Antitumor Effects of Hydroxytyrosol but Not Oleuropein on Experimental Glioma In Vivo. A Putative Role for the Renin-Angiotensin System. *Biomedicines* **2018**, *6*, 11. [CrossRef] [PubMed]
43. Dai, X.; Zhang, J.; Arfuso, F.; Chinnathambi, A.; Zayed, M.E.; Alharbi, S.A.; Kumar, A.P.; Ahn, K.S.; Sethi, G. Targeting TNF-related apoptosis-inducing ligand (TRAIL) receptor by natural products as a potential therapeutic approach for cancer therapy. *Exp. Biol. Med.* **2015**, *240*, 760–773. [CrossRef] [PubMed]
44. Itoh, N.; Yonehara, S.; Ishii, A.; Yonehara, M.; Mizushima, S.-I.; Sameshima, M.; Hase, A.; Seto, Y.; Nagata, S. The polypeptide encoded by the cDNA for human cell surface antigen Fas can mediate apoptosis. *Cell* **1991**, *66*, 233–243. [CrossRef]
45. Wiley, S.R.; Schooley, K.; Smolak, P.J.; Din, W.S.; Huang, C.-P.; Nicholl, J.K.; Sutherland, G.R.; Smith, T.D.; Rauch, C.; Smith, C.A.; et al. Identification and characterization of a new member of the TNF family that induces apoptosis. *Immunity* **1995**, *3*, 673–682. [CrossRef]
46. Walczak, H.; Miller, R.E.; Ariail, K.; Gliniak, B.; Griffith, T.S.; Kubin, M.; Chin, W.; Jones, J.; Woodward, A.; Le, T.; et al. Tumoricidal activity of tumor necrosis factor–related apoptosis–inducing ligand in vivo. *Nat. Med.* **1999**, *5*, 157–163. [CrossRef]
47. Komdeur, R.; Meijer, C.; Van Zweeden, M.; De Jong, S.; Wesseling, J.; Hoekstra, H.; Van Der Graaf, W. Doxorubicin potentiates TRAIL cytotoxicity and apoptosis and can overcome TRAIL-resistance in rhabdomyosarcoma cells. *Int. J. Oncol.* **2004**, *25*, 677–684. [CrossRef]
48. Shiraki, K.; Yamanaka, T.; Inoue, H.; Kawakita, T.; Enokimura, N.; Okano, H.; Sugimoto, K.; Murata, K.; Nakano, T. Expression of TNF-related apoptosis-inducing ligand in human hepatocellular carcinoma. *Int. J. Oncol.* **2005**, *26*, 1273–1281. [CrossRef]
49. Rogalska, A.; Gajek, A.; Marczak, A. Epothilone B induces extrinsic pathway of apoptosis in human SKOV-3 ovarian cancer cells. *Toxicol. Vitr.* **2014**, *28*, 675–683. [CrossRef]
50. Bérczi, A.; Ruthner, M.; Szüts, V.; Fritzer, M.; Schweinzer, E.; Goldenberg, H. Influence of conjugation of doxorubicin to transferrin on the iron uptake by K562 cells via receptor-mediated endocytosis. *JBIC J. Biol. Inorg. Chem.* **1993**, *213*, 427–436. [CrossRef]
51. Szwed, M.; Matusiak, A.; Laroche-Clary, A.; Robert, J.; Marszalek, I.; Jozwiak, Z. Transferrin as a drug carrier: Cytotoxicity, cellular uptake and transport kinetics of doxorubicin transferrin conjugate in the human leukemia cells. *Toxicol. Vitr.* **2014**, *28*, 187–197. [CrossRef] [PubMed]

International Journal of
Molecular Sciences

Article

Ligand-Based Pharmacophore Modeling, Molecular Docking, and Molecular Dynamic Studies of Dual Tyrosine Kinase Inhibitor of EGFR and VEGFR2

Frangky Sangande [1,2], Elin Julianti [1] and Daryono Hadi Tjahjono [1,*]

1 School of Pharmacy, Bandung Institute of Technology, Jalan Ganesha 10, Bandung 40132, Indonesia; frangky.sangande@gmail.com (F.S.); elin_julianti@fa.itb.ac.id (E.J.)
2 Study Program of Pharmacy, Faculty of Science and Technology, Universitas Prisma, Jalan Pomorow 113, Manado 95126, Indonesia
* Correspondence: daryonohadi@fa.itb.ac.id; Tel.: +62-812-2240-0120

Received: 22 September 2020; Accepted: 16 October 2020; Published: 21 October 2020

Abstract: Epidermal growth factor receptor (EGFR) and vascular endothelial growth factor receptor 2 (VEGFR2) play an important role in cancer growth. Both of them have close relationships. Expression of EGFR will induce an angiogenic factor (VEGF) release for binding with VEGFR2. However, the existence of VEGF up-regulation independent of EGFR leads to cancer cell resistance to anti-EGFR. Therefore, a therapeutic approach targeting EGFR and VEGFR2 simultaneously may improve the outcome of cancer treatment. The present study was designed to identify potential compounds as a dual inhibitor of EGFR and VEGFR2 by the computational method. Firstly, the ligand-based pharmacophore model for each target was setup to screen of ZINC database of purchasable compounds. The hit compounds obtained by pharmacophore screening were then further screened by molecular docking studies. Taking erlotinib (EGFR inhibitor) and axitinib (VEGFR2 inhibitor) as reference drugs, six potential compounds (ZINC08398597, ZINC12047553, ZINC16525481, ZINC17418102, ZINC21942954, and ZINC38484632) were selected based on their docking scores and binding interaction. However, molecular dynamics simulations demonstrated that only ZINC16525481 and ZINC38484632 which have good binding free energy and stable hydrogen bonding interactions with EGFR and VEGFR2. The result represents a promising starting point for developing potent dual tyrosine kinases inhibitor of EGFR and VEGFR2.

Keywords: dual inhibitor; EGFR; VEGFR2; ligand-based pharmacophore; molecular docking; molecular dynamics

---

1. Introduction

Tyrosine kinases are enzymes that catalyze the phosphorylation of tyrosine residues in proteins and activate signal transduction pathways that play an important role in cell proliferation, differentiation, migration, metabolism, and apoptosis. The existence of mutation and overexpression in these enzymes will trigger cancer formation [1]. Epidermal growth factor receptor (EGFR) is a member of tyrosine kinases and is commonly overexpressed in some types of cancer, such as non-small-cell lung cancer, breast, esophageal, cervical, and head and neck cancer [2]. The inhibition of EGFR activity is a rational strategy in design of anticancer [3]. There are two main groups of EGFR inhibitors in clinical use, i.e., monoclonal antibodies (mAbs) and small-molecule tyrosine kinase inhibitors (TKIs). mAbs compete with the endogenous ligand for receptor binding in the extracellular domain of EGFR, while TKIs work by competing with ATP in the intracellular catalytic domain, thereby preventing enzyme activity [4].

In pathologic conditions, EGFR activation will increase the expression of vascular endothelial growth factor (VEGF). When VEGF binds to vascular endothelial growth factor receptor (VEGFR), especially VEGFR2, the process of angiogenesis will begin [5]. VEGFR is also a member of the tyrosine kinase group that demonstrates a role in cancer growth through the angiogenesis mechanism, therefore it is an ideal drug target in cancer therapy. However, there are several mechanisms of expression of VEGF independent of EGFR. Overexpression of VEGF by this pathway is associated with cancer cell resistance to anti-EGFR [6]. Therefore, therapeutic approaches using drugs that inhibit both EGFR and VEGFR2 can increase the efficiency of cancer therapy and overcome the resistance problem [7,8]. In 2011, the US Food and Drug Administration (US FDA) approved the use of vandetanib, targeting EGFR and VEGFR2, for the treatment of medullary thyroid cancer [9]. The inhibition activity of vandetanib is based on the 4-anilino-quinazoline scaffold. Several studies have tried to design and develop new compounds as dual tyrosine kinases inhibitor of EGFR and VEGFR2 by generating the 4-anilino-quinazoline derivatives such as 2-chloro-4-anilino-quinazoline [10], 4-anilino-quinazoline-urea [11]. In other studies, benzimidazole [12] and phthalazine derivatives [13] have been tested for their activity against EGFR and VEGFR2.

In drug discovery, computational methods have been applied widely for identifying new drug candidates against the individual target. In particular, screening of large virtual libraries by molecular docking, pharmacophore modeling, quantitative structure-activity relationship (QSAR), and combination methods have been used for lead discovery of EGFR [14,15] and VEGFR2 inhibitor [16,17] as an individual target. However, very few computational methods were employed for identifying chemical compounds that can inhibit EGFR and VEGFR2 simultaneously. Therefore, in the present study, we designed computational screening methods to identify new potential compounds as dual tyrosine kinase inhibitor of EGFR and VEGFR2. The strategy includes three main stages. Firstly, we generated a validated pharmacophore model for EGFR and VEGFR2, and used it for initial screening of the ZINC database. Secondly, the hit compounds from the first stage were then screened again by molecular docking simulations against both targets using two docking tools. Finally, molecular dynamics simulations were performed for further validation of our screening result.

2. Results and Discussion

2.1. Ligand-Based Pharmacophore Screening

In the present study, we built two ligand-based pharmacophore models, one for EGFR and the other for VEGR2 using LigandScout 4.3 on a set of known EGFR and VEGFR2 inhibitors. A total of 10 pharmacophore models were obtained for each target and the first model was selected as the best model based on its AUC value of receiver operating characteristic (ROC). As shown in Figure 1, both pharmacophore models had AUC100% of ROC ≥ 0.7, indicating that they could distinguish the active molecules from decoys and were categorized as valid.

The selected pharmacophore model of EGFR consists of one hydrophobic group, three aromatic groups, two hydrogen bond acceptors, and one hydrogen bond donor, while the selected pharmacophore model of VEGFR2 consists of one hydrophobic group, one aromatic group, one hydrogen bond acceptor, and one hydrogen bond donor (Figure 2).

The two pharmacophore models were then used to screen the ZINC database. Initially, the ZINC database was screened with pharmacophore models of EGFR, and 18,806 compounds were identified that could be mapped with the pharmacophore features. These compounds were then screened again with the pharmacophore model of VEGFR2, and it was obtained 6896 compounds that satisfied to our requirements for both models.

2.2. Molecular Docking Screening

In order to acquire dependable results, virtual screening was performed by two docking tools, DOCK6 and iGemdock. These docking tools were validated by redocking the native ligand to the active site of the target. The validation results showed that redocking of erlotinib into EGFR using DOCK6 and iGemdock gave a root-mean-square deviation (RMSD) value of 1.44 and 1.75 Å, respectively, while redocking of axitinib to VEGFR2 using DOCK6 and iGemdock gave RMSD value of 0.18 and 0.51 Å. A docking prediction is considered successful if the RMSD value is <2.0 Å for the best-scored conformation [18]. Therefore, DOCK6 and iGemdock have a good accuracy to put back erlotinib and axitinib into the binding pocket of EGFR and VEGFR2. The best-scored conformation of erlotinib after redocked in DOCK6 and iGemdock protocols were −64.26 and −107.61, respectively, whereas for axitinib in DOCK6 and iGemdock were −83.01 and −149.07, respectively.

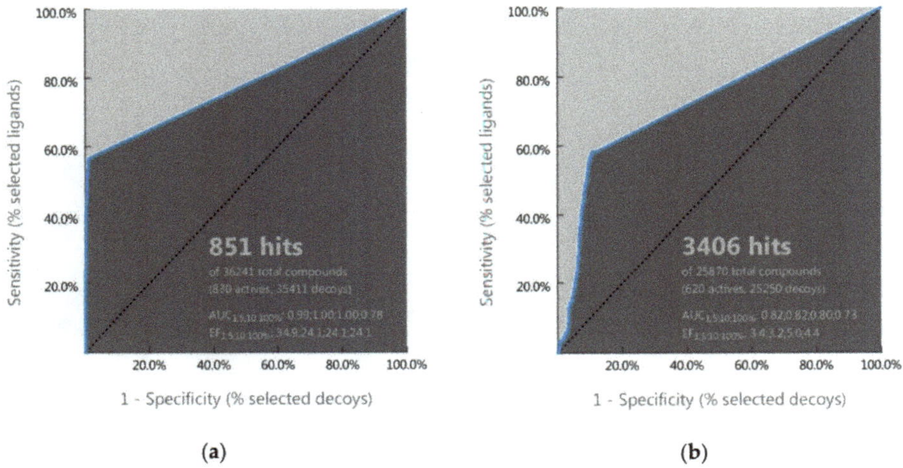

Figure 1. Receiver operating characteristic (ROC) curve validation of the selected pharmacophore model of (**a**) EGFR and (**b**) VEGFR2 using a set of active and decoy molecules.

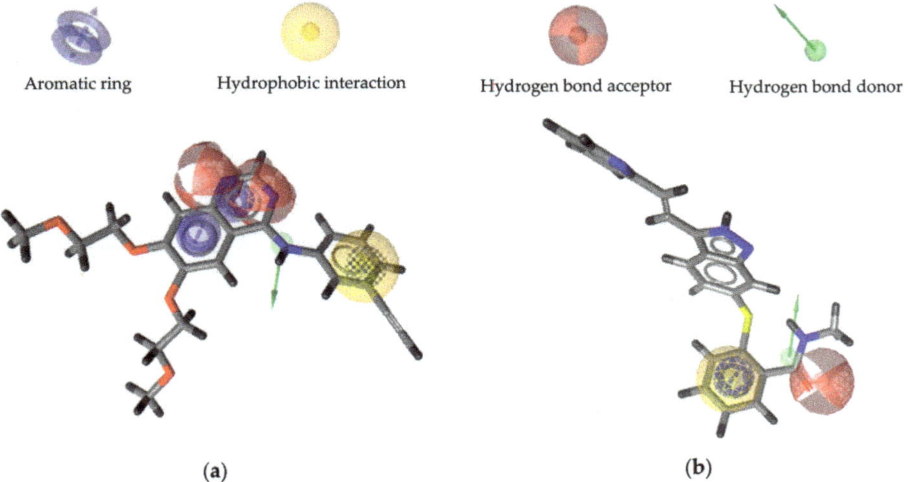

Figure 2. Overlay of erlotinib (**a**) and axitinib (**b**) on the selected pharmacophore model of EGFR and VEGFR2 inhibitor, respectively.

Moreover, analysis of the ROC curve was also performed to evaluate the ability of both docking protocols to discriminate between active and inactive compounds. As shown in Figure 3, DOCK6 and iGemdock have a steeper ROC curve than the random selection curve (diagonal curve). This indicates that these docking protocols are able to discriminate between active and inactive compounds. For verification, the AUC of the ROC curve for both docking methods were calculated. AUC of DOCK6 and iGemdock with EGFR as the target were 0.838 and 0.720, respectively, while toward VEGFR2, DOCK6 and iGemdock gave AUC of 0.809 and 0.874, respectively. AUC value of ≥0.7 could be considered that a method has acceptable discrimination [19]. Therefore, both of them could be applied for virtual screening.

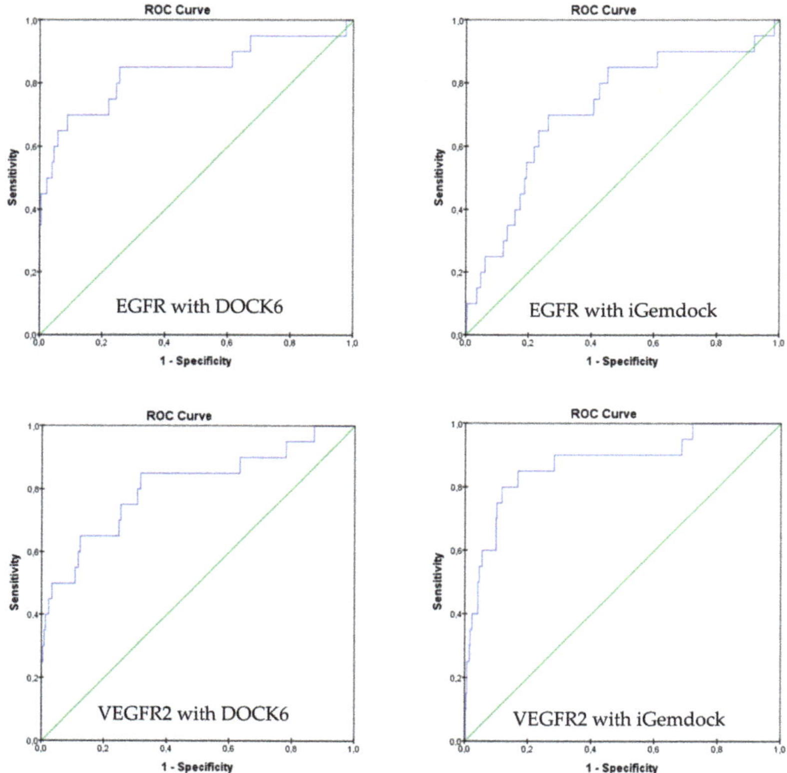

Figure 3. Receiver operating characteristic (ROC) curve validation of docking protocols using DOCK6 and iGemdock.

According to the best-scored conformation of erlotinib and axitinib, the 6896 hit compounds that satisfied the features of EGFR and VEGFR2 pharmacophores were then screened further by molecular docking. Firstly, the hit compounds were docked to EGFR and then continued to VEGFR2 using DOCK6. There were 67 hit compounds with more negative docking scores than those of erlotinib and axitinib. These compounds were then docked again to EGFR and then to VEGFR2 using the second docking tool, iGemdock, and there were 6 hits as potential dual inhibitor compounds.

2.3. Molecular Docking Analysis

We identified 6 compounds, i.e., ZINC08398597, ZINC12047553, ZINC16525481, ZINC17418102, ZINC21942954, and ZINC38484632 that meet all the filtering criteria. Interestingly, based on their chemical structure (Figure 4), some of them contain quinazoline- and phthalazine-based scaffold.

Quinazoline ring has been commonly used in drug design studies for a dual inhibitor of EGFR and VEGFR2 [10,11], while the phthalazine ring is the isostere of the quinazoline ring [13]. These heterocyclic rings could serve as a scaffold that occupies the binding site for adenine of ATP in the hinge region [20].

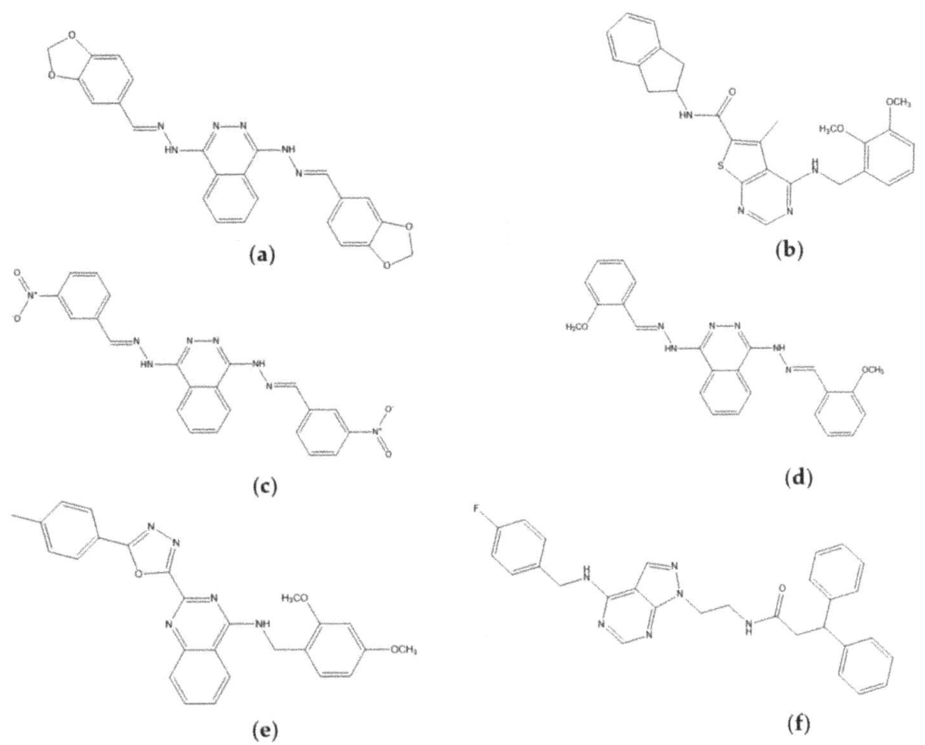

(a) (b) (c) (d) (e) (f)

Figure 4. Chemical structure of (**a**) ZINC08398597, (**b**) ZINC12047553, (**c**) ZINC16525481, (**d**) ZINC17418102, (**e**) ZINC21942954, and (**f**) ZINC3848463.

In order to know the interaction profiles of our potential compounds with EGFR and VEGFR2, their best-scored conformation from docking simulations using DOCK6 were analyzed. We chose DOCK6 due to its accuracy better than iGemdock based on the RMSD value. According to docking studies, generally, our hit compounds were docked successfully into the binding site of EGFR and VEGFR2. As shown in Table 1, there were 4 compounds, i.e., ZINC08398597, ZINC12047553, ZINC21942954, and ZINC38484632 that formed a hydrogen bond with Met769 of EGFR. It must be noted that Met769 is recognized as a hinge region key residue in the binding of EGFR inhibitors with 4-anilinoquinazoline scaffolds, such as erlotinib [21].

In addition, some compounds also formed a hydrogen bond with Lys692, Lys721, Pro770, Cys773, Asp831. The existence of these residues in hydrogen bonds match with previously reported results by Yang et al. [22]. Especially on the Asp831, this residue is a part of the DFG sequence that plays a critical role in ATP binding. A study by Peng et al. demonstrated that the hydrogen bond between their ligands and Asp831 is important for enhancing the inhibitory activity [23]. Among the six compounds, only ZINC17418102 did not form hydrogen bonds with EGFR. However, it had 8 similar residues out of 10 EGFR residues in the erlotinib complex. Thereby, its affinity was predicted tends to be influenced by hydrophobic interaction. Furthermore, the molecular docking studies with VEGFR2 as the target (Table 2) showed that all compounds formed a hydrogen bond with Asp1046. This residue is equivalent to Asp831 at EGFR which is a part of the DFG motif. Four compounds, ZINC08398597, ZINC16525481, ZINC17418102, and ZINC38484632 also formed a hydrogen bond with Glu885 locating in the back pocket. Glu885 and Asp1046 are involved in the interaction of axitinib with VEGFR2, in addition to Glu917 and Cys919 which locate in the hinge region [24].

Table 1. EGFR-ligand interactions recorded during docking.

Ligand	DOCK6 Score	iGemdock Score	H-Bond	Hydrophobic interaction
Erlotinib	−64.26	−107.61	Met769	Leu694, Ala719, Lys721, Leu764, Thr766, Leu768, Gly772, Leu820, Asp831
ZINC08398597	−68.76	−129.93	Met769	Leu694, Ala719, Val702, Leu768, Pro770, Gly772, Asp776, Tyr777, Leu820
ZINC12047553	−72.76	−125.93	Met769	Leu694, Ala719, Val702, Lys721, Met742, Leu764, Pro770, Gly772, Asp776, Leu820
ZINC16525481	−73.03	−128.81	Lys692, Lys721, Asp831	Leu694, Ala719, Lys721, Met742, Pro770, Gly772, Cys773, Leu820
ZINC17418102	−65.89	−122.16	-	Lys692, Leu694, Ala719, Val702, Lys704, Lys721, Leu764, Thr766, Pro770, Gly772, Leu820, Asp831
ZINC21942954	−72.93	−117.69	Met769, Cys773, Asp831	Leu694, Gly695, Lys721, Leu768, Pro770, Gly772, Leu820
ZINC38484632	−72.97	−111.43	Met769, Pro770	Lys692, Leu694, Ala719, Val702, Lys721, Met742, Thr766, Leu768, Gly772, Leu820

In contrast to other compounds, ZINC21942954 and ZINC38484632 also formed an additional hydrogen bond with Cys1045. In the 4AG8 structure, VEGFR2 is in an inactive form, which adopts DFG-out conformation [25]. This conformation provides an additional hydrophobic pocket for ligand interaction [24]. The six potential compounds have a moiety placed at this hydrophobic pocket. Therefore, we assumed that they will stabilize the DFG-out conformation and maintain VEGFR2 in its inactive form, and consequently inhibit the phosphorylation process.

2.4. *Molecular Dynamics Studies*

2.4.1. Stability Analysis

In order to further evaluate the effect of solvent and flexibility of protein toward the binding mode of our candidate compounds, molecular dynamics (MD) simulations were carried out during 50 ns. Based on these simulations, the stability of the ligand-protein complexes was determined by the RMSD value. According to Figure 5, the EGFR backbone of each complex equilibrated at about 0.5 nm, except the complex of ZINC12047553 equilibrated at the higher value. The complex of erlotinib had a lower RMSD value at the start of the simulations. However, it still seemed to fluctuate at the end of the simulation. Furthermore, each complex reached the equilibration at different times. The complex of ZINC16525481 and ZINC12047553 equilibrated faster than others. Nevertheless, the ZINC12047553 complex had the highest RMSD (0.6 nm). These results suggested that the interaction of each compound had different effects on EGFR stability.

Table 2. VEGFR2-ligand interactions recorded during docking.

Ligand	DOCK6 Score	iGemdock Score	H-Bond	Hydrophobic interaction
Axitinib	−83.00	−149.07	Glu885, Glu917, Cys919, Asp1046	Leu840, Ala866, Lys868, Val916, Phe918, Gly992, Leu1035, Phe1047
ZINC08398597	−89.65	−156.25	Glu885, Asp1046	Leu840, Val848, Ala866, Val867, Lys868, Leu889, Val914, Val916, Gly992, Leu1019, His1026, Leu1035, Phe1047
ZINC12047553	−85.20	−168.46	Asp1046	Leu840, Val848, Ala866, Lys868, Glu885, Val899, Val914, Val916, Phe918, Cys919, Gly992, His1026, Leu1035, Ile1044, Cys1045, Phe1047
ZINC16525481	−94.84	−162.89	Glu885, Asp1046	Leu840, Val848, Ala866, Val867, Lys868, Leu889, Val914, Val916, Phe918, His1026, Leu1035, Cys1045, Phe1047
ZINC17418102	−89.39	−155.33	Glu885, Asp1046	Leu840, Val848, Ala866, Val867, Lys868, Leu889, Val914, Val916, Glu917, Phe918, His1026, Leu1035, Ile1044, Phe1047
ZINC21942954	−84.25	−152.80	Cys1045, Asp1046	Leu840, Lys868, Leu889, Val899, Val914, Val916, Phe918, Cys919, Leu1035, Ile1044, Phe1047
ZINC38484632	−85.05	−151.26	Glu885, Cys1045, Asp1046	Leu840, Val848, Ala866, Lys868, Leu889, Ile892, Val916, Cys1024, Leu1019, Ile1025, His1026, Leu1035, Phe1047

For their complex with VEGFR2 (Figure 6), axitinib, and all potential compounds influenced the stability of VEGFR2 at a similar level. However, RMSD of the VEGFR2 backbone in the axitinib-complex seemed higher than the others. Overall, no significant fluctuations were observed in the backbone of VEGFR2, which indicates that all protein structures were stable during 50 ns simulation.

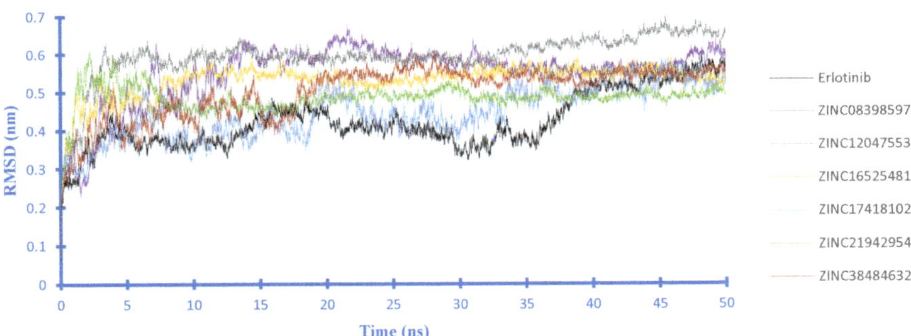

Figure 5. Backbone RMSD of the EGFR complexes.

Furthermore, according to the ligand RMSD in the EGFR complex (Figure 7), there were three compounds (i.e., ZINC21942954, ZINC16525481and ZINC38484632) whose RMSD curves appeared to overlap at the lower value, indicating that their docking pose was more stable than others. Among them, ZINC16525481 had the average RMSD value close to erlotinib during the last 10 ns of MD simulations and quickly reached the equilibrium. In the VEGFR2 complex (Figure 8), five compounds had an average value of RMSD < 0.25 nm. The most stable ligand's RMSD was found in ZINC08398597, with an average RMSD of 0.18 nm. Different from those of other compounds, ZINC17418102 showed a sharp increase of RMSD at around 25 ns. However, according to the visualization, this compound stays inside of the binding pocket, and it was observed that the high RMSD value at that time is related to the conformation change of ligand.

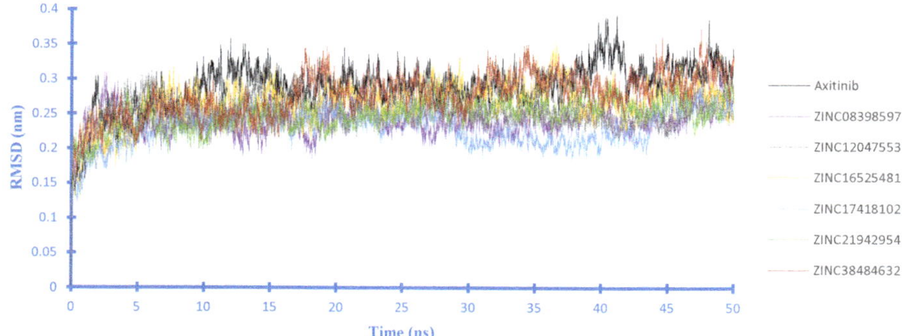

Figure 6. Backbone RMSD of the VEGFR2 complexes.

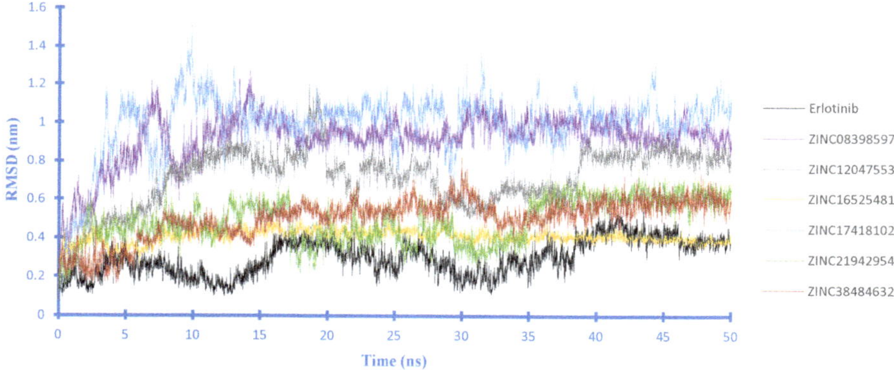

Figure 7. Ligand RMSD in the EGFR complexes.

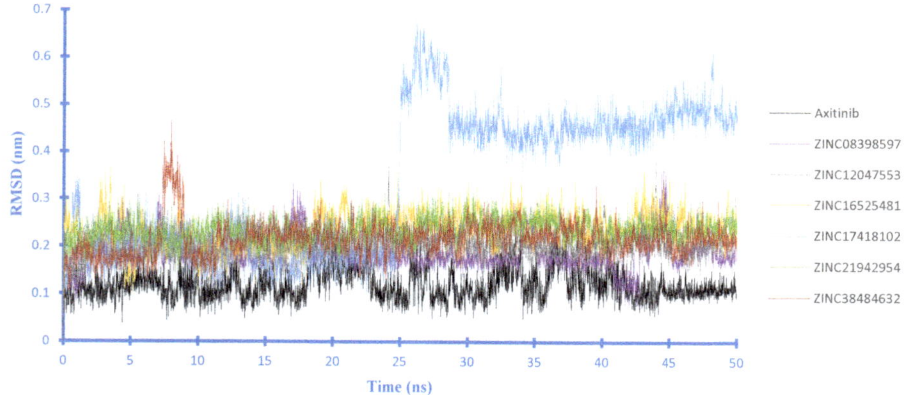

Figure 8. Ligand RMSD in the VEGFR2 complexes.

2.4.2. Hydrogen Bond Analysis

In the present study, we have calculated the occupancy percentages of hydrogen bonds that more than 10% for all complexes during MD simulations. Based on the occupancy percentages, a hydrogen bond is considered to be stable when the occupancy is more than 50% [26]. The results showed that only two compounds, i.e., ZINC16525481, and ZINC38484632, which formed stable hydrogen bonds in their complex with EGFR and VEGFR2 as presented in Table 3. Interestingly, the stable hydrogen bonds were formed with the key residues of both targets. In the case of EGFR complexes, similar to erlotinib, ZINC16525481 showed nicely hydrogen bonds with Met769 through its N1 and N2 atoms of phthalazine ring as acceptors with occupancy 88% and 73.7%, respectively. Additionally, it was also found to form hydrogen with occupancy 74.3% at Gln767, whereas ZINC38484632 bound to Met769 with occupancy 92.1%; however, this interaction did not involve the atom of the heterocyclic ring of ZINC38484632.

Table 3. Occupancy of hydrogen bond during 50 ns simulations.

Ligand	Target			
	EGFR		VEGFR2	
	Donor-Acceptor	Occupancy (%)	Donor-Acceptor	Occupancy (%)
Erlotinib	Met769 (H)—(N2)	28.1	-	-
Axitinib	-	-	(H12)—Glu917 (O) (H1)—Glu885 (OE2) (H1)—Glu885 (OE1) Asp1046 (H)—(O81) Cys919 (H)—(N14)	76.5 30.3 27.7 69.5 88.8
ZINC08398597	Cys773 (H)—(N1) Cys773 (H)—(N2) Met769 (H)—(O4)	21.1 13.4 10.2	(H12)—Glu885 (OE2) (H12)—Glu885 (OE1) Asp1046 (H)—(N6) Asp1046 (H)—(N1)	41.5 34.5 32.5 45.6
ZINC12047553	Met769 (H)—(O1)	25.1	(H15)—Asp1046 (O)	11.5
ZINC16525481	(H5)—Gln767 (O) Phe832 (H)—(O) Met769 (H)—(N1) Met769 (H)—(N2)	74.3 45.4 88 73.7	(H11)—Glu885 (OE2) (H11)—Glu885 (OE1) Asp1046 (H)—(N1) Asp1046 (H)—(N2) Cys919 (H)—(O)	17.6 17.4 70.2 47.7 70.5
ZINC17418102	-	-	Asp1046(H)—(N6)	41.3
ZINC21942954	Cys773 (H)—(O3)	48.7	(H12)—Glu885 (OE2) (H12)—Glu885 (OE1) Asp1046 (H)—(N4) Asp1046 (H)—(N5)	31.5 27.4 37.4 24.3
ZINC38484632	(H16)—Pro770 (O) Cys773 (H)—(N4) Met769 (H)—(O1)	10.5 16.8 92.1	(H9)—Glu885 (OE2) (H9)—Glu885 (OE1) Asp1046 (H)—(O1)	16.8 19.8 65.3

On the other hand, for their complex with VEGFR2, ZINC16525481 had two stable hydrogen bonds with Asp1046 (70.2%) and Cys919 (70.5%), while ZINC38484632 had a hydrogen bond at Asp1046 with occupancy 65%. In general, the results demonstrated that these potential compounds tend to show better binding interactions with VEGFR2 than EGFR. Moreover, Figure 9 showed that they also appear more stable to persist in the VEGFR2 binding pocket. This might be due to their number and occupancy percentages of hydrogen bonds in the VEGFR2 complex were higher than in the EGFR complex. In other words, the low number and percentage occupancy of hydrogen bonds are related to poor interaction stability. As shown in Figure 10, ZINC17418102, which has no hydrogen bond with occupancy of more than 10%, moved far away from its initial position in the binding pocket of EGFR.

Similarly, the change in position of ZINC12047553 in the binding pocket was also observed. However, this compound still had one hydrogen bond with occupancy 25%, therefore its ligand stability appeared to be slightly better than ZINC17418102.

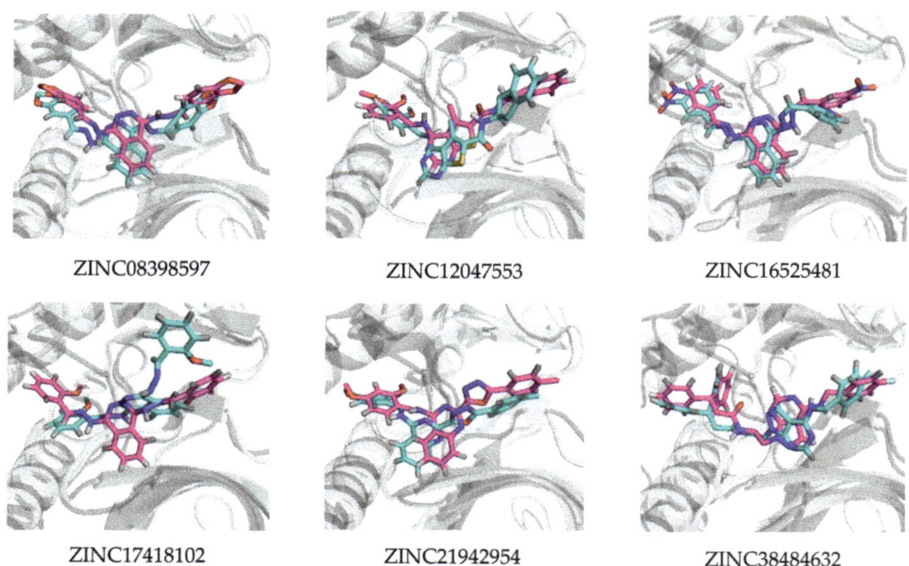

Figure 9. Superimposed ligand binding poses of the potential compounds at 0 ns (magenta) and 50 ns (cyan) in their complex with VEGFR2.

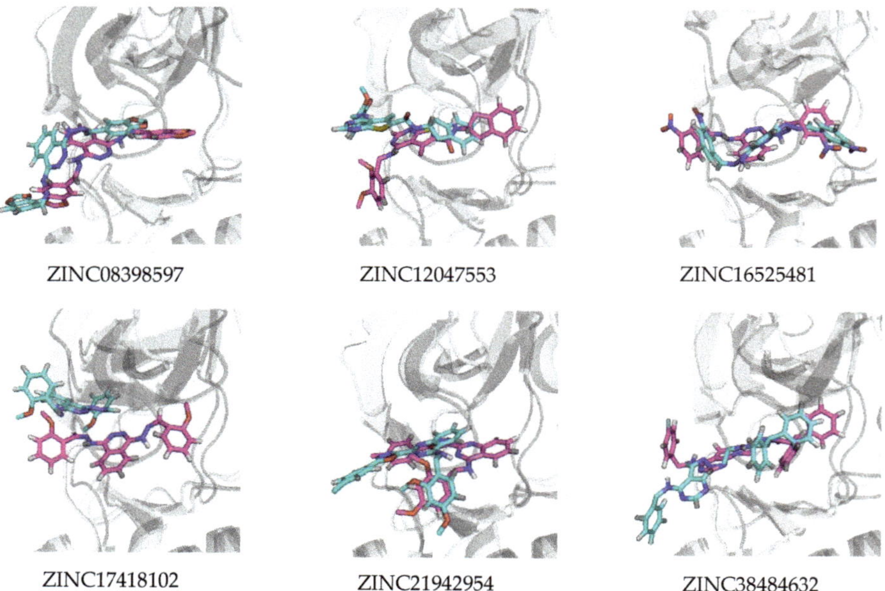

Figure 10. Superimposed ligand binding poses of the potential compounds at 0 ns (magenta) and 50 ns (cyan) in their complex with EGFR.

2.4.3. MMPBSA Analysis

Based on the MD simulations, the binding free energy of the six potential compounds was calculated using The Molecular Mechanics Poisson–Boltzmann Surface Area (MMPBSA) method during their steady-state. As demonstrated in Tables 4 and 5, energy decomposition to individual components showed that van der Waals, electrostatic, and SASA energy were favorable for ligand binding. In their interaction with EGFR, there were three compounds (ZINC08398597, ZINC16525481, ZINC38484632) with binding free energy < −100 kJ/mol like erlotinib. Among them, ZINC08398597 possessed the best binding energy, even better than erlotinib. Unfortunately, this compound exhibited a high ligand RMSD value of approximately 1 nm (10 Å) and it had been confirmed by visualization that it occupied an area slightly away from ATP binding pocket. Furthermore, ZINC38484632 and ZINC1652548 ranked at second and third place, respectively. However, among these three compounds, ZINC1652548 showed the van der Waals and electrostatic energy close to erlotinib, in correspondence with its interaction and RMSD profile. Meanwhile, the lowest free energy was observed in the ZINC17418102-EGFR complex. Based on its interaction profile, this compound did not form hydrogen bonds, either at the docking stage or during MD simulations. For their complex with VEGFR2, all six potential compounds got a better free binding energy than axitinib. This result indicated that they could bind to VEGFR2 with a good affinity. Compared to the other systems, the binding energy of ZINC08398597 was the best, while ZINC21942954 showed the lowest binding energy. Furthermore, based on the individual energy component, van der Waals energy of all the potential compounds was greater than axitinib, showing that the hydrophobic interaction is a dominant contributor for the binding of these potential compounds to VEGFR2. Thus, ZINC16525481 and ZINC38484632 were selected as a potential dual inhibitor for EGFR and VEGFR2 according to several parameters, such as their binding energy, RMSD profiles, hydrogen bond stability, and binding mode to the ATP binding pocket (Figure 11).

Table 4. The predicted binding free energy and the individual energy components (kJ/mol) ligand-EGFR complexes.

Ligands	Van der Waals Energy (ΔE_{vdW})	Electrostatic Energy (ΔE_{elec})	Polar Solvation Energy (ΔG_{polar})	SASA Energy ($\Delta G_{nonpolar}$)	Binding Energy (ΔG_{bind})
Erlotinib	−232.64 ± 13.43	−46.89 ± 10.03	176.32 ± 17.18	−24.18 ± 0.92	−127.38 ± 17.58
ZINC08398597	−220.76 ± 10.87	−24.78 ± 7.60	114.24 ± 14.70	−23.37 ± 1.72	−154.67 ± 13.85
ZINC12047553	−183.44 ± 16.81	−23.33 ± 7.92	131.99 ± 29.16	−19.71 ± 1.58	−94.49 ± 19.15
ZINC16525481	−223.17 ± 12.92	−43.01 ± 12.74	182.64 ± 14.30	−22.30 ± 0.99	−105.84 ± 17.08
ZINC17418102	−131.71 ± 12.28	−16.65 ± 10.30	85.25 ± 20.67	−15.09 ± 1.43	−78.21 ± 11.57
ZINC21942954	−200.15 ± 14.24	−16.94 ± 12.14	143.44 ± 29.17	−19.55 ± 1.2	−93.20 ± 18.56
ZINC38484632	−179.60 ± 11.35	−33.06 ± 8.88	118.60 ± 20.60	−21.25 ± 1.24	−115.31 ± 19.27

Table 5. The predicted binding free energy and the individual energy components (kJ/mol) ligand-VEGFR2 complexes.

Ligands	Van der Waals Energy (ΔE_{vdW})	Electrostatic Energy (ΔE_{elec})	Polar Solvation Energy (ΔG_{polar})	SASA Energy ($\Delta G_{nonpolar}$)	Binding Energy (ΔG_{bind})
Axitinib	−202.38 ± 11.43	−75.49 ± 10.19	163.49 ± 10.45	−20.69 ± 0.63	−135.08 ± 7.94
ZINC08398597	−267.39 ± 13.12	−32.26 ± 8.51	154.24 ± 12.60	−25.56 ± 0.76	−170.96 ± 12.17
ZINC12047553	−258.36 ± 10.04	−24.89 ± 8.73	163.99 ± 21.33	−26.22 ± 1.16	−145.48 ± 23.74
ZINC16525481	−272.50 ± 11.94	−29.68 ± 8.44	184.00 ± 15.54	−26.25 ± 1.03	−144.43 ± 12.32
ZINC17418102	−252.86 ± 10.25	−28.79 ± 6.46	166.57 ± 12.21	−24.63 ± 1.13	−139.71 ± 14.81
ZINC21942954	−268.09 ± 9.85	−42.85 ± 6.59	199.97 ± 25.24	−26.72 ± 0.73	−137.69 ± 25.23
ZINC38484632	−260.64 ± 9.06	−25.80 ± 7.40	167.18 ± 16.01	−27.72 ± 1.15	−146.97 ± 12.43

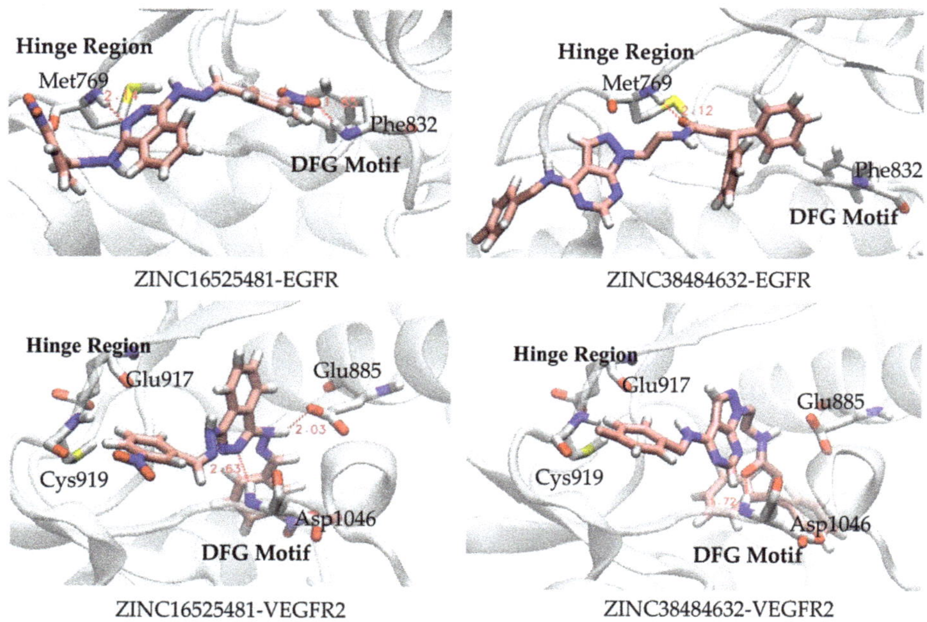

Figure 11. Binding mode of ZINC16525481 and ZINC38484632 to the ATP binding pocket of EGFR and VEGFR2 at 50 ns.

3. Materials and Methods

3.1. Ligand-Based Pharmacophore Modeling

For each target, EGFR and VEGFR2, there were 26 known inhibitors as training set obtained from the literature. In order to generate pharmacophore models of EGFR, we used the same training set contain 26 active compounds that were compiled by Gupta et al. [14], while for VEGFR2, 26 active compounds with $IC_{50} < 100$ nM were selected from published literature by Zhang et al. [27]. Their structure was built and optimized using Hyperchem 8.0 [28] on the semiempirical method-Parametric Method 3 (PM3). These files were saved as .mol2 and loaded to LigandScout 4.3 [29], then generated their conformation using the default settings of iCon best option. Finally, LigandScout converted it into corresponding pharmacophore features.

A total of 10 pharmacophore models were generated and the first model (model 1) was selected and validated for its performance to distinguish active compounds from decoys by screening a set of 830 known actives and 35,411 decoy compounds for target EGFR, and 620 known actives and 25,250 decoy compounds for target VEGFR2 obtained from DUD-E decoys database [30]. The database from DUD-E was converted in the .ldb format before screening by the "create screening database" menu of LigandScout.

Initial screening was carried out using ZINCPharmer [31] for target EGFR. The previously obtained pharmacophore model of EGFR was submitted to ZINCPharmer, to search hits from the ZINC database of "purchasable compounds" consist of 22,723,923 compounds [32]. A maximum of 0.5 Å RMSD from sphere centers, 10 rotatable bonds cut-off, and the molecular weight no more than 500 Daltons were used as input parameters for ZINCPharmer. The database of hit compounds from the ZINCPharmer was downloaded and submitted to LigandScout in the .ldb format, and were then screened again using the pharmacophore model of VEGFR2.

3.2. Molecular Docking

The hit compounds obtained from the pharmacophore modeling were used as input for further screening by docking simulations toward EGFR and VEGFR2 using DOCK6 [33] and iGemdock 2.1 [34]. The crystal structure of EGFR (PDB code: 1M17) and VEGFR2 (PDB code: 4AG8) were downloaded from the RSCB protein data bank (PDB). In the 1M17 structure, EGFR is in complex with a native ligand, erlotinib, while in the 4AG8, VEGFR2 is in complex with axitinib. Swiss–Model was used for adding missing residues of the proteins [35].

For docking using DOCK6, the proteins without water molecules were separated from their native ligands using Chimera [36]. The proteins and native ligands were prepared by the dock prep tool of Chimera. This stage includes the addition of hydrogen atoms and standard charges. Specifically for the proteins, the molecular surface was generated with the DMS tool, as provided by DOCK6 software, with a default probe radius of 1.4 Å. The DMS output with default settings of Sphegen was applied to generate sphere, while the sphere selector tool and an 8 Å radius about the native ligands was used to define the active site. Moreover, 5 Å extra margin in all six directions was applied to support the box around the active site. At the last step, flexible ligand docking was used for virtual screening with default input parameters according to the DOCK6 tutorial.

For docking with iGemdock, only water molecules were removed from protein-ligand complexes. The complexes were loaded to iGemdock and their binding sites were identified at a distance 5 Å of the bounded ligands. The stable docking module was used for virtual screening with the default parameters such as population size = 300, generations = 80, number of solutions = 10.

The accuracy of both docking protocols was validated by redocking native ligand into the binding pocket of the protein and its RMSD value was calculated. These docking protocols were also validated for their performance in distinguishing between active and decoy compounds by analyzing AUC of ROC after screening the DUD-E database of EGFR and VEGFR2.

3.3. Molecular Dynamics Simulations

The topologies of native ligands and candidate compounds were generated using Automated Topology Builder (ATB) 3.0 [37]. All MD simulations were performed using the GROMACS 2016.3 package [38] under simple point charge (SPC) water mode and GROMOS96 54A7 force field. The simulation of system was run under the periodic boundary condition with a dodecahedron periodic box, which was also solvated by SPC water molecules. The surface of the enzyme was covered with a water shell of 1.0 nm. The balance of system charge was achieved by adding sodium and chloride ions. The relaxation of complex system was conducted by energy minimization under 1000 kcal/mol/nm by using the steepest descent method. A 250 ps NVT (constant number of particles, volume, and temperature) ensemble was then performed to stabilize the system at 310 K and subsequently, 250 ps NPT (constant number of particles, pressure, and temperature) equilibration stabilized the system pressure by using coupling reference pressure of 1 bar. The last step, MD simulation, was conducted for 50 ns with a time step of 2 fs, and the corresponding coordinates were stored every 2 ps. Particle Mesh Ewald (PME) was applied to calculate the long-range electrostatics during the simulation. In the present work, to calculate the binding energy of each complex from the MD trajectories, we used the g_mmpbsa tool and its protocols as reported by Kumar et al. [39]. The calculation of binding energies was based on the extracted 20 snapshots from the last 5 ns trajectory of 50 ns simulation results.

4. Conclusions

In the present study, we developed a computational method combining ligand-based pharmacophore filtering and molecular docking for identifying the potential compounds that can inhibit tyrosine kinase of EGFR and VEGFR2, simultaneously. The results showed that there were six compounds mapped well onto the features of our selected pharmacophore models and had a better docking score than erlotinib and axitinib as the reference drugs. Additionally, the binding mode analysis revealed that they also had a similar interaction with the reference drugs. However, after careful analysis of their binding mode stability during molecular dynamics simulations, it was found that there were two compounds, ZINC16525481 and ZINC38484632 retained the stable hydrogen bonds (occupancy > 50%) with the essential residues of both targets, with a good binding free energy, indicating that these compounds were able to steadily anchor to the binding pocket of EGFR and VEGFR2 to exert an inhibitory effect. Thus, the results suggested that ZINC16525481 and ZINC38484632 could be considered as a potential dual inhibitor for EGFR and VEGFR2 for further study. The study of biological activities of two compounds is currently underway and will be reported elsewhere.

Author Contributions: D.H.T. conceived and designed the experiments (methodology), analyzed the data, and corrected the manuscript; F.S. contributed with methodology, performed the experiments, collected the data, and drafted the manuscript; and E.J. analyzed the data and corrected the manuscript. D.H.T. also contributed with resources, funding acquisition and supervision. All authors have read and agreed to the published version of the manuscript.

Funding: This research was supported in part by P3MI (Research, Community Service and Innovation Program) 2019, Bandung Institute of Technology, Indonesia.

Conflicts of Interest: The authors declare no conflict of interest.

References

1. Broekman, F. Tyrosine kinase inhibitors: Multi-targeted or single-targeted? *World J. Clin. Oncol.* **2011**, *2*, 80–93. [CrossRef] [PubMed]
2. Harari, P.M. Epidermal growth factor receptor inhibition strategies in oncology. *Endocr. Relat. Cancer* **2004**, *11*, 689–708. [CrossRef] [PubMed]
3. Seshacharyulu, P.; Ponnusamy, M.P.; Haridas, D.; Jain, M.; Ganti, A.K.; Batra, S.K. Targeting the EGFR signaling pathway in cancer therapy. *Expert Opin. Ther. Targets* **2012**, *16*, 15–31. [CrossRef] [PubMed]
4. Yewale, C.; Baradia, D.; Vhora, I.; Patil, S.; Misra, A. Epidermal growth factor receptor targeting in cancer: A review of trends and strategies. *Biomaterials* **2013**, *34*, 8690–8707. [CrossRef]
5. Pennell, N.A.; Lynch, T.J. Combined Inhibition of the VEGFR and EGFR Signaling Pathways in the Treatment of NSCLC. *Oncologist* **2009**, *14*, 399–411. [CrossRef] [PubMed]
6. Huang, L.; Fu, L. Mechanisms of resistance to EGFR tyrosine kinase inhibitors. *Acta Pharm. Sin. B* **2015**, *5*, 390–401. [CrossRef]
7. Tabernero, J. The role of VEGF and EGFR inhibition: Implications for combining Anti-VEGF and Anti-EGFR Agents. *Mol. Cancer Res.* **2007**, *5*, 203–220. [CrossRef]
8. Sarkar, S.; Mazumdar, A.; Dash, R.; Sarkar, D.; Fisher, P.B.; Mandal, M. ZD6474, a dual tyrosine kinase inhibitor of EGFR and VEGFR-2, inhibits MAPK/ERK and AKT/PI3-K and induces apoptosis in breast cancer cells. *Cancer Biol. Ther.* **2010**, *9*, 592–603. [CrossRef]
9. Grande, E.; Kreissl, M.C.; Filetti, S.; Newbold, K.; Reinisch, W.; Robert, C.; Schlumberger, M.; Tolstrup, L.K.; Zamorano, J.L.; Capdevila, J. Vandetanib in advanced medullary thyroid cancer: Review of adverse event management strategies. *Adv. Ther.* **2013**, *30*, 945–966. [CrossRef]
10. Barbosa, M.L.; Lima, L.M.; Tesch, R.; Sant'Anna, C.M.; Totzke, F.; Kubbutat, M.H.; Schächtele, C.; Laufer, S.A.; Barreiro, E.J. Novel 2-chloro-4-anilino-quinazoline derivatives as EGFR and VEGFR-2 dual inhibitors. *Eur. J. Med. Chem.* **2014**, *71*, 1–14. [CrossRef]
11. Zhang, H.Q.; Gong, F.H.; Li, C.G.; Zhang, C.; Wang, Y.J.; Xu, Y.G.; Sun, L.P. Design and discovery of 4-anilinoquinazoline-acylamino derivatives as EGFR and VEGFR-2 dual TK inhibitors. *Eur. J. Med. Chem.* **2016**, *109*, 371–379. [CrossRef]
12. Li, Y.; Tan, C.; Gao, C.; Zhang, C.; Luan, X.; Chen, X.; Liu, H.; Chen, Y.; Jiang, Y. Discovery of benzimidazole derivatives as novel multi-target EGFR, VEGFR-2 and PDGFR kinase inhibitors. *Bioorg. Med. Chem.* **2011**, *19*, 4529–4535. [CrossRef]
13. Amin, K.M.; Barsoum, F.F.; Awadallah, F.M.; Mohamed, N.E. Identification of new potent phthalazine derivatives with VEGFR-2 and EGFR kinase inhibitory activity. *Eur. J. Med. Chem.* **2016**, *123*, 191–201. [CrossRef] [PubMed]

14. Gupta, A.K.; Bhunia, S.S.; Balaramnavar, V.M.; Saxena, A.K. Pharmacophore modelling, molecular docking and virtual screening for EGFR (HER 1) tyrosine kinase inhibitors. *SAR QSAR Environ. Res.* **2011**, *22*, 239–263. [CrossRef] [PubMed]
15. Li, S.; Sun, X.; Zhao, H.; Tang, Y.; Lan, M. Discovery of novel EGFR tyrosine kinase inhibitors by structure-based virtual screening. *Bioorg. Med. Chem. Lett.* **2012**, *22*, 4004–4009. [CrossRef] [PubMed]
16. Lee, K.; Jeong, K.W.; Lee, Y.; Song, J.Y.; Kim, M.S.; Lee, G.S.; Kim, Y. Pharmacophore modeling and virtual screening studies for new VEGFR-2 kinase inhibitors. *Eur. J. Med. Chem.* **2010**, *45*, 5420–5427. [CrossRef] [PubMed]
17. Li, J.; Zhou, N.; Luo, K.; Zhang, W.; Li, X.; Wu, C.; Bao, J. In silico discovery of potential VEGFR-2 inhibitors from natural derivatives for anti-angiogenesis therapy. *Int. J. Mol. Sci.* **2014**, *15*, 15994–16011. [CrossRef]
18. Hevener, K.E.; Zhao, W.; Ball, D.M.; Babaoglu, K.; Qi, J.; White, S.W.; Lee, R.E. Validation of molecular docking programs for virtual screening against dihydropteroate synthase. *J. Chem. Inf. Model.* **2009**, *49*, 444–460. [CrossRef]
19. Hosmer, D.W.; Lemeshow, S.; Sturdivant, R.X. *Applied Logistic Regression*, 3rd ed.; John Wiley & Sons, Inc.: Hoboken, NJ, USA, 2013; p. 177.
20. Zhang, J.; Yang, P.L.; Gray, N.S. Targeting cancer with small molecule kinase inhibitors. *Nat. Rev. Cancer* **2009**, *9*, 28–39. [CrossRef]
21. Nasab, R.R.; Mansourian, M.; Hassanzadeh, F.; Shahlaei, M. Exploring the interaction between epidermal growth factor receptor tyrosine kinase and some of the synthesized inhibitors using combination of in-silico and in-vitro cytotoxicity methods. *Res. Pharm. Sci.* **2018**, *13*, 509–522.
22. Yang, Y.A.; Tang, W.J.; Zhang, X.; Yuan, J.W.; Liu, X.H.; Zhu, H.L. Synthesis, molecular docking and biological evaluation of Glycyrrhizin analogs as anticancer agents targeting EGFR. *Molecules* **2014**, *19*, 6368–6381. [CrossRef]
23. Peng, Y.H.; Shiao, H.Y.; Tu, C.H.; Liu, P.M.; Hsu, J.T.; Amancha, P.K.; Wu, J.S.; Coumar, M.S.; Chen, C.H.; Wang, S.Y.; et al. Protein kinase inhibitor design by targeting the Asp-Phe-Gly (DFG) motif: The role of the DFG motif in the design of epidermal growth factor receptor inhibitors. *J. Med. Chem.* **2013**, *56*, 3889–3903. [CrossRef] [PubMed]
24. Sanphanya, K.; Wattanapitayakul, S.K.; Phowichit, S.; Fokin, V.V.; Vajragupta, O. Novel VEGFR-2 kinase inhibitors identified by the back-to-front approach. *Bioorg. Med. Chem. Lett.* **2013**, *23*, 2962–2967. [CrossRef] [PubMed]
25. Structures, C.; Indices, P.; Studies, E.; Parade, C. Selecting protein structure/s for docking-based virtual screening: A case study on type II inhibitors of VEGFR-2 Kinase. *Int. J. Pharm. Sci. Res.* **2019**, *10*, 2998–3011.
26. Desheng, L.; Jian, G.; Yuanhua, C.; Wei, C.; Huai, Z.; Mingjuan, J. Molecular dynamics simulations and MM/GBSA methods to investigate binding mechanisms of aminomethylpyrimidine inhibitors with DPP-IV. *Bioorg. Med. Chem. Lett.* **2011**, *21*, 6630–6635. [CrossRef] [PubMed]
27. Zhang, Y.; Yang, S.; Jiao, Y.; Liu, H.; Yuan, H.; Lu, S.; Ran, T.; Yao, S.; Ke, Z.; Xu, J.; et al. An integrated virtual screening approach for VEGFR-2 inhibitors. *J. Chem. Inf. Model.* **2013**, *53*, 3163–3177. [CrossRef] [PubMed]
28. Froimowitz, M. HyperChem: A software package for computational chemistry and molecular modeling. *Biotechniques* **1993**, *14*, 1010–1013.
29. Wolber, G.; Langer, T. LigandScout: 3-D pharmacophores derived from protein-bound ligands and their use as virtual screening filters. *J. Chem. Inf. Model.* **2005**, *45*, 160–169. [CrossRef] [PubMed]
30. Mysinger, M.M.; Carchia, M.; Irwin, J.J.; Shoichet, B.K. Directory of useful decoys, enhanced (DUD-E): Better ligands and decoys for better benchmarking. *J. Med. Chem.* **2012**, *55*, 6582–6594. [CrossRef]
31. Koes, D.R.; Camacho, C.J. ZINCPharmer: Pharmacophore search of the ZINC database. *Nucleic Acids Res.* **2012**, *40*, 409–414. [CrossRef]
32. Irwin, J.J.; Shoichet, B.K. ZINC-A free database of commercially available compounds for virtual screening. *J. Chem. Inf. Model.* **2005**, *45*, 177–182. [CrossRef] [PubMed]
33. Allen, W.J.; Balius, T.E.; Mukherjee, S.; Brozell, S.R.; Moustakas, D.T.; Lang, P.T.; Case, D.A.; Kuntz, I.D.; Rizzo, R.C. DOCK 6: Impact of new features and current docking performance. *J. Comput. Chem.* **2015**, *36*, 1132–1156. [CrossRef]
34. Yang, J.M.; Chen, C.C. GEMDOCK: A Generic Evolutionary Method for Molecular Docking. *Proteins* **2004**, *55*, 288–304. [CrossRef] [PubMed]

35. Waterhouse, A.; Bertoni, M.; Bienert, S.; Studer, G.; Tauriello, G.; Gumienny, R.; Heer, F.T.; de Beer, T.; Rempfer, C.; Bordoli, L.; et al. SWISS-MODEL: Homology modeling of protein structures and complexes. *Nucleic Acids Res.* **2018**, *46*, 296–303. [CrossRef] [PubMed]
36. Pettersen, E.F.; Goddard, T.D.; Huang, C.C.; Couch, G.S.; Greenblatt, D.M.; Meng, E.C.; Ferrin, T.E. UCSF Chimera-A visualization system for exploratory research and analysis. *J. Comput. Chem.* **2004**, *25*, 1605–1612. [CrossRef]
37. Stroet, M.; Caron, B.; Visscher, K.M.; Geerke, D.P.; Malde, A.K.; Mark, A.E. Automated Topology Builder Version 3.0: Prediction of Solvation Free Enthalpies in Water and Hexane. *J. Chem. Theory Comput.* **2018**, *14*, 5834–5845. [CrossRef]
38. Hess, B.; Kutzner, C.; Van Der Spoel, D.; Lindahl, E. GROMACS 4: Algorithms for highly efficient, load-balanced, and scalable molecular simulation. *J. Chem. Theory Comput.* **2008**, *4*, 435–447. [CrossRef]
39. Kumari, R.; Kumar, R.; Lynn, A. g-mmpbsa-A GROMACS tool for high-throughput MM-PBSA calculations. *J. Chem. Inf. Model.* **2014**, *54*, 1951–1962. [CrossRef]

MDPI
St. Alban-Anlage 66
4052 Basel
Switzerland
Tel. +41 61 683 77 34
Fax +41 61 302 89 18
www.mdpi.com

International Journal of Molecular Sciences Editorial Office
E-mail: ijms@mdpi.com
www.mdpi.com/journal/ijms

www.ingramcontent.com/pod-product-compliance
Lightning Source LLC
LaVergne TN
LVHW071615170726
843515LV00009B/2400

* 9 7 8 3 0 3 6 5 5 9 2 8 5 *